ISBN 978-0-428-54423-2
PIBN 11302687

1 MONTH OF
FREE
READING

at
www.ForgottenBooks.com

By purchasing this book you are eligible for one month membership to ForgottenBooks.com, giving you unlimited access to our entire collection of over 1,000,000 titles via our web site and mobile apps.

To claim your free month visit:
www.forgottenbooks.com/free1302687

English
Français
Deutsche
Italiano
Español
Português

www.forgottenbooks.com

Mythology Photography **Fiction**
Fishing Christianity **Art** Cooking
Essays Buddhism Freemasonry
Medicine **Biology** Music **Ancient
Egypt** Evolution Carpentry Physics
Dance Geology **Mathematics** Fitness
Shakespeare **Folklore** Yoga Marketing
Confidence Immortality Biographies
Poetry **Psychology** Witchcraft
Electronics Chemistry History **Law**
Accounting **Philosophy** Anthropology
Alchemy Drama Quantum Mechanics
Atheism Sexual Health **Ancient History**
Entrepreneurship Languages Sport
Paleontology Needlework Islam
Metaphysics Investment Archaeology
Parenting Statistics Criminology
Motivational

NIST Special Publication 305
Supplement 21

Publications of the National Institute of Standards and Technology 1989 Catalog

United States Department of Commerce
National Institute of Standards and Technology

NIST Special Publication 305
Supplement 21

Publications of the National Institute of Standards and Technology 1989 Catalog

Rebecca J. Pardee and
Ernestine T. Gladden, Editors

Information Resources and Services Division
National Institute of Standards and Technology
Gaithersburg, MD 20899

Issued July 1990

U.S. Department of Commerce
Robert A. Mosbacher, Secretary

National Institute of Standards and Technology
John W. Lyons, Director

National Institute of Standards and Technology Special Publication 305 Supplement 21
to Accompany National Bureau of Standards Special Publication 305 and Its Supplements 1 through 20
Natl. Inst. Stand. Technol. Spec. Publ. 305 Suppl. 21, 413 pages (July 1990)

CODEN: NSPUE2

U.S. GOVERNMENT PRINTING OFFICE
WASHINGTON: 1990

For sale by the Superintendent of Documents, U.S. Government Printing Office, Washington, DC 20402.

CONTENTS

CATALOG STRUCTURE AND USE

Full bibliographic citations including keywords and abstracts for National Institute of Standards and Technology (NIST) (formerly National Bureau of Standards (NBS)) papers published and entered into the National Technical Information Service (NTIS) collection are cited in the "NIST Publications Announcements" section of this catalog. (Also included are papers published prior to 1989 but not reported in previous supplements of this annual catalog.) Entries are arranged by NTIS subject classifications which consist of 38 broad subject categories (see back cover) and over 350 subcategories. Within a subcategory, entries are listed alphanumerically by NTIS order number.

Four indexes are included to allow the user to identify papers by personal author, keywords, title, and NTIS order/report number. Each entry lists the appropriate title, the NTIS order number, and the abstract number.

Papers may also be identified by searching the NTIS database either online via commercially available systems such as DIALOG, or in the issues of NTIS's *Government Reports Announcements and Index* and its *Government Reports Annual Index*.

AVAILABILITY AND ORDERING INFORMATION

The highest quality and least expensive copies of NIST publications published as Government documents are available from the Superintendent of Documents, U.S. Government Printing Office, Washington, DC 20402. Publications cited with stock numbers (SN) should be ordered by these numbers. GPO will accept payment by check, money order, VISA, MasterCard, or deposit account. For availability and price, write to the GPO or telephone (202) 783-3238. Should an NIST publication be out of print at the GPO, its continued availability is assured at NTIS which sells publications in microfiche or paper copy reproduced from microfiche.

If an entry has a price code, such as PC A04/MF A01, the publication may be ordered from NTIS in paper copy (PC) or microfiche (MF) or both if both codes are given. Order from the National Technical Information Service, 5285 Port Royal Road, Springfield, VA 22161. A copy of the latest price code schedule is available from NTIS. NTIS will accept payment by check, money order, VISA, American Express, MasterCard, or deposit account. NTIS is the sole source of Federal Information Processing Standards (FIPS), Interagency Reports (IRs), and Grant/Contract Reports (GCRs). For more information call (703) 487-4650.

Papers noted "Not Available NTIS" may be obtained directly from the author or from the external publisher

cited. Such papers are not for sale by either the GPO or NTIS.

Two other sources for NIST publications are depository libraries (libraries designated to receive Government publications) and Department of Commerce District Offices. The depository libraries listed in Appendix A receive selected NIST publications (see inside back cover for a description of the various NIST publication series). While not every Government publication is sent to all depository libraries, certain depositories designated as Regional Depositories receive and retain one copy of all Government publications made available. Contact the depository library in your area to obtain information on what is available and where.

Department of Commerce District Offices listed in Appendix B provide ready access at the local level to publications, statistical data and summaries, and surveys. Each District Office serves as an official sales agency of the Superintendent of Documents, U.S. Government Printing Office. A wide range of Government publications can be purchased from these offices. In addition, the reference library of each District Office contains review copies of many Government publications.

NIST PUBLICATIONS ANNOUNCEMENTS

SAMPLE ENTRY

ADMINISTRATION & MANAGEMENT

Public Administration & Government

900,001
PB89-161905   PC A04/MF A01
National Inst. of Standards and Technology (NEL),
Gaithersburg, MD. Center for Computing and Applied
Mathematics.
Internal Revenue Service Post-of-Duty Location Modeling System: Programmer's Manual for PASCAL Solver.
P. D. Domich, R. H. F. Jackson, M. A. McClain, and D. M. Tate. Feb 89, 66p NISTIR-86/3472-1
See also PB87-165171 and PB89-161913. Sponsored by Internal Revenue Service, Washington, DC.

Keywords: *Facilities management, *Programming manuals, Mathematical models, Computer programs, Heuristic methods, Lagrangian functions, *Internal Revenue Service, *Site selection, User manuals(Computer programs), PASCAL subroutines.

The report is a programmer's manual for a microcomputer system designed at the National Institute of Standards and Technology for selecting optimal locations of IRS posts-of-duty. The mathematical model is the uncapacitated, fixed charge, facility location model which minimizes travel and facility costs. The package consists of two sections of code, one in FORTRAN and the other in PASCAL. The FORTRAN driver handles graphics displays and controls input and output for the solution procedure. The report discusses the mathematical techniques used to solve the mathematical model developed and includes a Greedy procedure, an interchange procedure, and a Lagrangian approach to the related linear program. A description of these PASCAL routines and definitions of key data structures and variables are provided.

900,002
PB89-161913   PC A04/MF A01
National Inst. of Standards and Technology (NEL),
Gaithersburg, MD. Center for Computing and Applied
Mathematics.

Internal Revenue Service Post-of-Duty Location Modeling System: Programmer's Manual for FORTRAN Driver Version 5.0.
P. D. Domich, R. H. F. Jackson, and M. A. McClain.
Feb 89, 66p NISTIR-86/3473-1
See also PB89-161905. Sponsored by Internal Revenue Service, Washington, DC.

Keywords: *Facilities management, Regional planning, Cost engineering, Computer graphics, *Internal Revenue Service, *Site selection, Fortran subroutines, Input output processing.

The report is a programmer's manual for a microcomputer package which was designed by the National Institute of Standards and Technology to assist the Internal Revenue Service in choosing locations for its posts-of-duty which will minimize costs to the IRS and to the taxpayer. The package was written in two sections of code, one in FORTRAN and the other in PASCAL. The manual describes the FORTRAN driver which handles graphics displays and controls input and output for the solution procedure.

1

AERONAUTICS & AERODYNAMICS

Research Program Administration & Technology Transfer

Research Program Administration & Technology Transfer

900,003
PB89-166094   PC A06/MF A01
National Inst. of Standards and Technology (IMSE), Gaithersburg, MD. Polymers Div.
Institute for Materials Science and Engineering, Polymers: Technical Activities 1988.
Annual rept. 1 Oct 87-30 Sep 88.
L. E. Smith, and B. M. Fanconi. Nov 88, 120p
NISTIR-88/3842
See also PB87-136693.

Keywords: *Research program administration, *Polymers, Chemical properties, Mechanical properties, Standards, Composite materials, Processing, Durability, Blends, National Institute of Standards and Technology, Technical activities.

The Technical Activities of the Polymers Division for FY 88 are reviewed in the report. Included are descriptions of the 6 Tasks of the Division, project reports, publications, and other technical activities.

900,004
PB89-189294   PC A03/MF A01
National Inst. of Standards and Technology, Gaithersburg, MD.
National Engineering Laboratory's 1989 Report to the National Research Council's Board on Assessment of NIST (National Institute of Standards and Technology) Programs.
Rept. for Apr 88-Apr 89.
G. Ehrlich. Mar 89, 45p NISTIR-89/4060

Keywords: *Research management, *Technology innovation, Manufacturing, Buildings, Chemical engineering, Computers, Measurements, Application of mathematics, Energy methods, Law(Jurisdiction), Standards, Inventions, *National Institute of Standards and Technology.

The 1989 report to the National Research Council's (NRC's) Board on Assessment of the National Institute of Standards and Technology (NIST) programs provides an overview of the National Engineering Laboratory (NEL). It describes the climate that influences NEL's work, program and budget trends, and the external interactions with industry, academia, and trade and professional organizations. Descriptions of NEL's program activities with accompanying lists of recent accomplishments, trends, and significant budget changes are also included. The programs described are Electronic and Electrical Measurements, Manufacturing Research and Standards, Building Research, Fire Research, Chemical Engineering Metrology, Mathematical Sciences, Computing Support, Energy Related Inventions, and Law Enforcement Standards. The impact of the recently enacted Omnibus Trade and Competitiveness Act of 1988 is discussed.

900,005
PB89-189310   PC A03/MF A01
National Inst. of Standards and Technology, Gaithersburg, MD.
NIST (National Institute of Standards and Technology) Research Reports, March 1989.
Special pub.
1989, 36p NIST/SP-761
See also PB89-133565.

Keywords: *Research projects, *Technology innovation, Electrooptics, Telecommunication, Manufacturing, Commerce, Government policies, Quality control, Adhesive bonding, Temperature measurement, Lasers, Clocks, Neutrons, Diamonds, Dental materials, *National Institute of Standards and Technology, Photonics, Small businesses.

The report contains a number of articles which discuss the following subjects: Research update; Light: The wave of the future; New centers to aid industry; Commercialization of technology: Whose job; NIST 1990 budget proposed; Quest for quality; Too hot to handle, but not to measure; The beauty of time; Cold neutron facility dedicated; Diamond films: new gems in advanced materials; Fracture test on thick steel plate sets U.S. record; New dental bonding system licensed; New publications; and Conference calendar.

900,006
PB89-218382   PC A15/MF A01
National Inst. of Standards and Technology, Gaithersburg, MD. Information Resources and Services Div.

Publications of the National Institute of Standards and Technology, 1988 Catalog.
Rept. for Jan-Dec 88.
R. J. Pardee. Jun 89, 348p NIST/SP-305-SUPPL-20
Also available from Supt. of Docs. as SN003-003-02940-8. See also PB86-240007. Library of Congress catalog card no. 48-47112.

Keywords: *Catalogs(Publications), *Bibliographies, Science, Technology, Research management, *National Institute of Standards and Technology.

The 20th Supplement to Special Publication 305 contains full bibliographic citations including keywords and abstracts for National Institute of Standards and Technology (NIST) (formerly National Bureau of Standards (NBS)) 1988 papers published and entered into the National Technical Information Service (NTIS) collection. Also included are NBS/NIST papers published prior to 1988 but not reported in previous supplements of this annual catalog. Four indexes are included to allow the user to identify NBS/NIST papers by author, keywords, title, and NTIS order/report number.

900,007
PB89-235113   PC A03/MF A01
National Inst. of Standards and Technology, Gaithersburg, MD.
NIST (National Institute of Standards and Technology) Research Reports, June 1989.
Special pub.
Jun 89, 36p NIST/SP-765
Also available from Supt. of Docs. as SN003-003-02956-4. See also PB89-189310.

Keywords: *Research projects, *Reviews, Fatigue(Materials), Aircraft, Hot pressing, Isostatic pressing, Geology, Composite materials, Time measurement, Sterilization, Smoke abatement, *National Institute of Standards and Technology, Technology transfer, Computer security, Materials science, Computer aided manufacturing.

Contents: Computer security: protection is the name of the game; 'Standard crack' helps detect metal fatigue in aircraft; Building quality into advanced materials during processing; 'HiPing': from metal powders to reliable materials; Exploring earth's formation; NIST: helping industry to compete; Tracking time; Movies reveal secrets of materials; Technique sterilizes clinical instruments in seconds; and Can smoke control systems save lives and property.

General

900,008
PB89-221147   PC A19/MF A01
National Inst. of Standards and Technology, Gaithersburg, MD. Office of Standards Code and Information.
Directory of International and Regional Organizations Conducting Standards-Related Activities.
Final rept.
M. Breitenberg. May 89, 443p NIST/SP-767
Also available from Supt. of Docs. as SN003-003-02937-8. Supersedes PB84-203439. Library of Congress catalog card no. 89-600735.

Keywords: *Directories, *Standardization, *Organizations, Standards, Trade associations, Technical societies, *International organizations.

The directory contains information on 338 international and regional organizations which conduct standardization, certification, laboratory accreditation, or other standards-related activities. The volume describes their work in these areas, the scope of each organization, national affiliations of members, U.S. participants, restrictions on membership, as well as the availability of any standards in English. The volume is designed to serve the needs of federal agencies and standards writers for information on international and regional organizations involved in standardization and related activities. It may also be useful to manufacturers, engineers, purchasing agents, and others.

AERONAUTICS & AERODYNAMICS

Aeronautics

900,009
PB89-172886   PC A04/MF A01
National Bureau of Standards (NEL), Gaithersburg, MD. Center for Fire Research.
Ignition and Flame Spread Measurements of Aircraft Lining Materials.
M. Harkleroad. May 88, 64p NBSIR-88/3773
Sponsored by Federal Aviation Administration Technical Center, Atlantic City, NJ.

Keywords: *Ignition, *Linings, *Flammability, *Aircraft, Measurement, Test methods, Composite materials, Heat transfer, Panels, Epoxy resins, Fiberglass reinforced plastics, Aircraft interiors.

Experimental tests were conducted to study the lateral and upward flame spread behavior of eight aircraft lining materials. The results are tabulated in terms of parameters useful in predicting ignition and flame spread in the presence of an ignition source under exposure to an external radiant source. Experimental and derived results are graphically compared. Derived material properties related to and indicative of the propensity to support flame spread are presented.

Test Facilities & Equipment

900,010
PB89-175293
     (Order as PB89-175194, PC A06)
European Molecular Biology Lab., Heidelberg (Germany, F.R.).
Computational Analysis of Protein Structures: Sources, Methods, Systems and Results.
Bi-monthly rept.
A. M. Lesk, and A. Tramontano. 1989, 9p
Prepared in cooperation with Medical Research Council, Cambridge (England). Lab. of Molecular Biology. Sponsored by Istituto Internazionale di Genetica e Biofisica, Naples (Italy).
Included in Jnl. of Research of the National Institute of Standards and Technology, v94 n1 p85-93 Jan-Feb 89.

Keywords: *Proteins, *Molecular structure, Information retrieval, Information systems, Models, *Proteins conformation, Molecular biology, Data banks, Computer graphics.

Computational molecular biology is a relatively new field that has arisen in response to the very large amount and quality of data currently being produced, including gene and protein sequences and nucleic acid and protein structures. Many important biological investigations can be carried out only through effective computational access to the entire corpus of data. This has stimulated the development of data banks and information retrieval systems. The article describes the kinds of inferences that are possible if such a relationship is found.

AGRICULTURE & FOOD

Food Technology

900,011
PB89-186399   Not available NTIS

2

National Bureau of Standards (NML), Gaithersburg, MD. Ionizing Radiation Physics Div.
Comprehensive Dosimetry for Food Irradiation.
Final rept.
W. L. McLaughlin. 1988, 16p
Pub. in Health Impact, Identification, and Dosimetry of Irradiated Foods, p384-399 Jun 88.

Keywords: *Food processing, *Ionizing radiation, *Food irradiation, Quality control, Radiation protection, Dosimetry, *Radiopreservation, *Gamma dosimetry, Gamma radiation, Reprints.

When ionizing radiation is used to process food of many kinds, comprehensive dosimetry is the most expeditious method of assuring that the process has been accomplished within specifications and without excess energy deposition. The term 'comprehensive dosimetry' then implies that, before, during, and after the radiation process, a number of modifying factors with some complexity must be considered and applied. This is particularly the case with the radiation food processing, where achieving success and maintaining wholesomeness and health and safety of the consumer are paramount. The main steps and considerations for practical and relevant food irradiation dosimetry and documentation are reviewed, which will satisfy regulatory authorities and encourage successful trade and safe marketing of food commodities. The process properly controlled with the aid of good dosimetry gives the opportunity to circumvent unsafe chemical additives and treatments to enable means of meeting legal requirement of quarantines and to improve public health through improved diets.

900,012
PB90-107046   PC A03/MF A01
National Inst. of Standards and Technology, Gaithersburg, MD.
Glass Bottles for Carbonated Soft Drinks: Voluntary Product Standard PS73-89.
Product std.
B. M. Meigs. Jul 89, 18p
Also available from Supt. of Docs. as SN003-003-02958-1. Sponsored by Glass Packaging Inst., Inc., Washington, DC.

Keywords: *Food packaging, *Packaging materials, *Bottles, Design standards, Silica glass, Manufacturing, Carbonation, Defects, Impact strength, Tolerances(Mechanics).

The Voluntary Product Standard covers conventional refillable and nonrefillable glass bottles that are manufactured from soda-lime-silica glass that have a nominal capacity of up to and including 36 fluid ounces, and that are intended for use in the packaging of soft drinks carbonated to a maximum of five volumes. Manufacturing requirements for bottles are provided for temper number, thermal shock resistance, internal pressure, strength, impact resistance, abrasion resistance, detection of visual defects, wall thickness, dimensional tolerances for height and maximum outside diameter, tolerances for capacity and mass (weight), perpendicularity, bottom characteristics, and bottle identification. A model statement is included for use on manufacturing orders and invoices specifying the maximum carbonation volumes intended for the bottles. Terms are defined or described that include trade terms and methods for identifying bottles that conform to the Standard.

ASTRONOMY & ASTROPHYSICS

Astronomy & Celestial Mechanics

900,013
PB89-171268   Not available NTIS
National Bureau of Standards (NML), Boulder, CO.
Quantum Physics Div.

Microarcsecond Optical Astrometry: An Instrument and Its Astrophysical Applications.
Final rept.
R. D. Reasenberg, R. W. Babcock, J. F. Chandler, M. V. Gorenstein, J. P. Huchra, M. R. Pearlman, I. I. Shapiro, R. S. Taylor, R. Bender, A. Buffington, B. Carney, J. A. Hughes, K. J. Johnston, B. F. Jones, and L. E. Matson. 1988, 15p
Grants NSF-PHY84-09671, NSF-AST85-19763
Sponsored by National Science Foundation, Washington, DC., and National Aeronautics and Space Administration, Washington, DC.
Pub. in Astronomical Jnl. 96, n5 p1731-1745 Nov 88.

Keywords: *Astrometry, *Optical interferometers, Gravitation, Reprints, Laser metrology, POINTS interferometer.

POINTS, an optical astrometric interferometer to be operated in space, would be a means of performing a wide variety of astrophysical studies, including a vastly improved deflection test of general relativity, a precise and direct calibration of the Cepheid distance scale, and the determination of stellar masses. The nominal 5 microarcsecond uncertainty in the measurement of the angular separation of two stars in the sky and the estimated measurement rate of 60 star pairs per day would support a rich mixture of scientific projects during the nominal mission life of ten years. Useful results would be available after less than a year. The key to the instrument's success is the control of systematic error, which is addressed by instrumentation and postanalysis of the astrometric data.

Astrophysics

900,014
N89-16535/1   PC A19/MF A01
European Space Agency, Paris (France).
Proceedings of the Celebratory Symposium on a Decade of UV (Ultraviolet) Astronomy with the IUE Satellite, Volume 2.
E. J. Rolfe. cJun 88, 427p ESA-SP-281-V-2
Symposium Held in Greenbelt, MD, 12-15 Apr. 1988; Sponsored by NASA, Esa, the United Kingdom Science and Engineering Research Council, and the American Astronomical Society.

Keywords: *Conferences, *IUE, *Spaceborne astronomy, *Ultraviolet astronomy, *Ultraviolet spectra, Active galactic nuclei, Data bases, Planetary nebulae, Pre-main sequence stars, Quasars, Star formation, Stellar winds, *Foreign technology, *Meetings.

No abstract available.

900,015
PB89-149199   Not available NTIS
National Bureau of Standards (NML), Boulder, CO.
Quantum Physics Div.
Doppler Imaging of AR Lacertae at Three Epochs.
Final rept.
F. M. Walter, J. E. Neff, J. L. Linsky, and M. Rodono. 1988, 3p
Pub. in A Decade of UV Astronomy with the IUE Satellite, v1 p295-297 Jun 88.

Keywords: Stellar atmospheres, Doppler effect, Ultraviolet spectra, Magnetism, Reprints, *AR Lacertae star, *Stellar chromospheres, Image analysis, Late stars, IUE.

Doppler imaging analysis allows use of the information contained in a time sequence of spectral line profiles to deduce the size, location, and surface flux of regions of contrasting brightness on rotating stars. The authors have used IUE observations to study the structure of the lower chromosphere of AR Lacertae in the light of Mg II k. They have obtained sequences of LWR/P-HI images distributed around the binary period at three epochs. Discrete plage-like regions of enhanced Mg II surface flux in this system were identified. Even with the limited S/N attainable with the IUE, one can map the gross structures of active stellar atmospheres. With such information, one can begin to study the true 3-D structure of the atmospheres of late-type stars.

900,016
PB89-149207   Not available NTIS
National Bureau of Standards (NML), Boulder, CO.
Quantum Physics Div.

Late Stages of Close Binary Systems-Clues to Common Envelope Evolution.
Final rept.
R. F. Webbink. 1985, 44p
Pub. in Proceedings of Beijing Colloquium on Models for Close Binary Systems, Beijing, China, November 7-13, 1985, p397-440.

Keywords: *Binary stars, Mass flow, Planetary nebulae, *Stellar envelopes, Cataclysmic variables, Symbiotic stars, Barium stars.

Those circumstances are outlined which theoretically lead to engulfment of one star by its companion, creating a common envelope binary. This evolutionary course is believed to lead to the formation of cataclysmic binaries, and other very compact systems containing degenerate components. The implications of these and other evolved binaries which, directly or indirectly, provide insight into the occurrence and nature of common envelope evolution, are examined.

900,017
PB89-157663   Not available NTIS
National Bureau of Standards (NEL), Boulder, CO. Scientific Computing Div.
Proper Motion vs. Redshift Relation for Superluminal Radio Sources.
Final rept.
B. W. Rust, S. G. Nash, and B. J. Geldzahler. 1989, 30p
Pub. in Astrophysics Space Science 152, p141-170 1989.

Keywords: *Radio sources(Astronomy), *Red shift, *Quasars, Reprints, *Proper motion, Superluminal motion, Astronomical models.

Two models for superluminal radio sources predict sharp lower bounds for the apparent velocities of separation. The light echo model predicts a minimum velocity $v(min) = 2c$, and the dipole field model predicts $v(min) = 4.446c$. Yahil (1979) has suggested that, if either of these models is correct, then $v(min)$ provides a 'standard velocity' which can be used to determine the cosmological parameters H and q sub 0. This is accomplished by estimating a lower envelope for the proper motion vs redshift relation. Yahil also argued that the procedure could easily be generalized to include a nonzero cosmical constant Lambda. The authors derive the formulas relating the proper motion theta dot to the redshift z in a Friedmann universe with a nonzero Lambda. They show that the determination of a lower envelope for a given sample of measured points (z sub i, theta dot sub i) yields an estimate of the angle of inclination phi sub i for each source in the sample. The authors formulate the estimation of the lower envelope as a constrained maximum likelihood problem with the constraints specified by the expected value of the largest order statistic for the estimated phi sub i. The authors solve this problem numerically using an off-the-shelf nonlinearly constrained nonlinear optimization program from the NAg library.

900,018
PB89-171573   Not available NTIS
National Bureau of Standards (NML), Boulder, CO.
Quantum Physics Div.
Computer Program for Calculating Non-LTE (Local Thermodynamic Equilibrium) Model Stellar Atmospheres.
Final rept.
I. Hubeny. 1988, 30p
Pub. in Computer Physics Communications 52, p103-132 1988.

Keywords: *Stellar atmospheres, *Thermodynamic equilibrium, Astrophysics, Spectral lines, Reprints, Radiative transfer.

The program calculates model stellar atmospheres, assuming plane-parallel, horizontally homogeneous atmosphere in radiative and hydrostatic equilibrium and allowing for departures from local thermodynamic equilibrium (LTE) for a set of occupation numbers of selected atomic and ionic energy levels. The program is very flexible as to the choice of chemical.

900,019
PB89-171615   Not available NTIS
National Bureau of Standards (NML), Boulder, CO.
Quantum Physics Div.

ASTRONOMY & ASTROPHYSICS

Astrophysics

Rotational Modulation and Flares on RS CVn and BY Dra Stars IX. IUE (International Ultraviolet Explorer) Spectroscopy and Photometry of II Peg and V711 Tau during February 1983.
Final rept.
A. D. Andrews, C. J. Butler, M. Rodono, S. Catalano, J. L. Linsky, A. Brown, F. Scaltriti, M. Busso, I. S. Nha, J. Y. Oh, M. C. D. Henry, J. L. Hopkins, H. J. Landis, and S. Engelbrektso. 1988, 16p
Contract NASA-NAG5-82
See also PB88-189055. Sponsored by National Aeronautics and Space Administration, Washington, DC.
Pub. in Astronomy and Astrophysics 204, p177-192 1988.

Keywords: *Binary stars, Stellar atmospheres, Ultraviolet spectra, Reprints, Stellar flares, Starspots, Faculae, IUE.

Evidence is presented for spots, plages and flares on the non-eclipsing RS CVn system II Peg and V711 Tau, based on sixty spectra obtained with the IUE satellite between 2-7 February 1983 and on supporting ground-based photometry. The large spot originally found on II Peg in 1981.8 could still be identified in 1983. Two spectroscopic flares of II Peg were detected. On V711 Tau at least two flares were observed. For both stellar systems the fluxes from the higher temperature emission lines showed the greatest variations. The ratio of the Mg II k and h fluxes was the same for the active component in each stellar system and was closely similar to the solar chromospheric value.

900,020
PB89-202592   Not available NTIS
National Bureau of Standards (NML), Boulder, CO. Quantum Physics Div.
Photospheres of Hot Stars. 3. Luminosity Effects at Spectral Type O9.5.
Final rept.
S. A. Voels, B. Bohannan, D. C. Abbott, and D. G. Hummer. 1989, 18p
Grants NSF-AST85-05919, NAGW-766
See also PB87-153680. Sponsored by National Science Foundation, Washington, DC., and National Aeronautics and Space Administration, Washington, DC.
Pub. in Astrophysical Jnl. 340, p1073-1090, 15 May 89.

Keywords: *Stellar atmospheres, Line spectra, Hydrogen, Helium, Luminosity, Reprints, *Hot stars, Early stars, Stellar winds, O stars.

The authors have observed hydrogen and helium line profiles with high signal-to-noise ratios for four stars of spectral type O9/5 (alpha Cam, xc' Ori A, delta Ori A, AE Aur) that form a sequence in luminosity: Ia, Ib, II, V. The basic stellar parameters of these stars are determined by fitting the observed line profiles of weak photospheric absorption lines with profiles from models which. include the effect of radiation scattered back onto the photosphere from their stellar winds, an effect referred to as wind blanketing. The stellar parameters derived for these four O9.5-type stars vary in a monotonic way with luminosity class. The authors argue that helium enrichment, and by implication CNO processed material as well, is probably a general characteristic of stars with Of and O Ia spectral classifications.

900,021
PB89-202616   Not available NTIS
National Bureau of Standards (NML), Boulder, CO. Quantum Physics Div.
Rotational Modulation and Flares on RS Canum Venaticorum and BY Draconis Stars X: The 1981 October 3 Flare on V711 Tauri (=HR 1099).
Final rept.
J. L. Linsky, J. E. Neff, A. Brown, B. D. Gross, T. Simon, A. D. Andrews, M. Rodono, and P. A. Feldman. 1989, 14p
See also PB89-171615.
Pub. in Astronomy and Astrophysics 211, p173-184 1989.

Keywords: *Stellar spectra, Ultraviolet spectra, Microwave spectra, Binary stars, Reprints, *Stellar flares, Stellar chromospheres, Starspots, IUE.

A unique set of high resolution spectra of V711 Tauri = HR 1099 (G5V + K1IV) is presented, obtained with both the SWP and LWR cameras of IUE, together with simultaneous 6.4 GHz microwave emission and optical photometry, during a bright flare on 3 October 1981.

900,022
PB89-202626   Not available NTIS
National Bureau of Standards (NML), Boulder, CO. Quantum Physics Div.

Stellar Winds of 203 Galactic O Stars: A Quantitative Ultraviolet Survey.
Final rept.
I. D. Howarth, and R. K. Prinja. 1989, 66p
Pub. in Astrophysical Jnl. Supplement Series 69, p527-592 Mar 89.

Keywords: Ultraviolet spectra, Reprints, *Stellar winds, *O stars, Early stars, IUE.

The paper presents a homogeneous set of column densities and maximum observed velocities derived from the C IV, N V, and Si IV resonance doublets of the 203 O stars observed at high resolution with IUE prior to 1987 January 1. In addition, fundamental parameters (T(eff), L(*), M(*)) are estimated for 201 of these stars. The relationships between observed velocities and physically relevant velocities in the wind and between column densities and mass-loss rates are discussed.

900,023
PB89-212054   Not available NTIS
National Bureau of Standards (NML), Boulder, CO. Quantum Physics Div.
Interpretation of Emission Wings of Balmer Lines in Luminous Blue Variables.
Final rept.
I. Hubeny, and C. Leitherer. 1989, 4p
Grants NSF-AST85-20278, NSF-AST88-02937
Sponsored by National Science Foundation, Washington, DC.
Pub. in Publications of the Astronomical Society of the Pacific 101, n635 p114-117 Jan 89.

Keywords: *Variable stars, Line spectra, Stellar atmospheres, Reprints, *Balmer lines, Blue stars, Stellar winds.

The authors discuss H(alpha) line profiles calculated with plane-parallel, hydrostatic NLTE model atmospheres. in their lowest log g models the profiles show extended emission wings. Qualitatively, these wings are similar to the extended wings generated by electron scattering of line photons in the stellar wind. It is proposed that the line wings observed in luminous blue variables may be due to a combination on the NLTE effect discussed. here and the traditional scattering mechanism.

900,024
PB89-228373   Not available NTIS
National Bureau of Standards (NML), Boulder, CO. Quantum Physics Div.
IUE Observation of the Interstellar Medium Toward Beta Geminorum.
Final rept.
J. Murthy, J. B. Woffard, R. C. Henry, H. W. Moos, A. Vidal-Madjar, J. L. Linsky, and C. Gry. 1989, 5p
Pub. in Astrophysical Jnl. 336, p949-953, 15 Jan 89.

Keywords: *Interstellar matter, Ultraviolet spectra, Reprints, Galactic center, IUE.

The authors present a high-dispersion (Delta lambda = 0.1A) IUE spectrum of the hydrogen Ly(alpha) emission line of the nearby late-type star beta Gem from which they have derived the density, velocity dispersion and bulk velocity of the interstellar H1 in that direction. While interstellar deuterium Ly(alpha) is clearly seen in absorption in the line profile, the authors do not obtain a useful limit on the ratio. The present result is important in confirming the generality of the 'emptiness' of the interstellar medium away from the Galactic center.

900,025
PB89-228506   Not available NTIS
National Inst. of Standards and Technology (NML), Gaithersburg, MD. Molecular Spectroscopy Div.
Laboratory Measurement of the 1(sub 01)-0(sub 00) Transition and Electric Dipole Moment of SiC2.
Final rept.
R. D. Suenram, F. J. Lovas, and K. Matsumura. 1989, 3p
Pub. in Astrophysical Jnl. 342, pL103-L105, 15 Jul 89.

Keywords: *Interstellar matter, *Silicon carbides, Microwave spectroscopy, Reprints, *Electric dipole moment.

The 1(sub01)-0(sub00) transitions of (28)SiC2, (29)SiC2, and (30)SiC2 have been measured in the laboratory using a laser-ablation source coupled to a pulsed nozzle Fabry-Perot Fourier transform microwave spectrometer. The measured frequencies are

23600.242(4) MHz, 23257.511(8) MHz and 22937.583(8) MHz, respectively. The electric dipole moment for (28)SiC2 has been measured to be mu = mu sub alpha = 2.393(6) debye. Using this value for the dipole moment, the column density of SiC2 in IRC + 10216 has been recalculated to be 2.7x10 to the 14th power/sq cm(-2). which is approximately a factor of 2 larger than the previous estimate for which a theoretical value of the dipole moment was used.

900,026
PB89-234298   Not available NTIS
National Inst. of Standards and Technology (NML), Boulder, CO. Quantum Physics Div.
Rotational Modulation and Flares on RS Canum Venaticorum and BY Draconis Stars. XI. Ultraviolet Spectral Images of AR Lacertae in September 1985.
Final rept.
J. E. Neff, F. M. Walter, M. Rodono, and J. L. Linsky. 1989, 13p
See also PB89-202618.
Pub. in Astronomy and Astrophysics 215, p79-91 1989.

Keywords: *Binary stars, Ultraviolet spectra, Red shift, Reprints, *AR Lacertae stars, Stellar chromospheres, Stellar flares, Faculae.

Using a series of high-resolution ultraviolet spectra, the authors have derived a series of images of the chromosphere of AR Lacertae. In September 1985, neither star in this system was uniformly bright. The trailing hemisphere of the K0IV star was globally brighter than the leading hemisphere. The position, size, and surface flux of three distinct plage regions on the K star were measured with the spectral imaging procedure. The factor of 3 variability in total emission from the G2IV star was interpreted as due to a large chromospherically inactive region on its surface. The authors were able to constrain the position, size, and surface flux of a flaring region on the G star and to measure a significant redshift and broadening of the line emission from the flaring region. They used the rotational modulation of the integrated low-resolution line fluxes to determine the far-ultraviolet spectra of the global K and G stars and of the plage and flare regions alone.

900,027
PB90-118118   Not available NTIS
National Inst. of Standards and Technology (NML), Boulder, CO. Quantum Physics Div.
Solar and Stellar Magnetic Fields and Structures: Observations.
Final rept.
J. L. Linsky. 1989, 10p
Grant NGL-06-003-057, Contract NAG5-82
Sponsored by National Aeronautics and Space Administration, Washington, DC.
Pub. in Solar Physics 121, p187-196 1989.

Keywords: *Solar magnetic fields, *Stellar magnetic fields, Zeeman effect, Circular.polarization, Reviews, Reprints, Linear polarization.

The review of stellar magnetic field measurements is both a critique of recent spectral diagnostic techniques and a summary of important trends now appearing in the data. Both the Zeeman broading techniques that have evolved from Robinson's original approach, and techniques based on circular and linear polarization data are discussed. The review concludes with an ambitious agenda for developing self-consistent models of the magnetic atmosphere of active stars.

900,028
PB90-118142   Not available NTIS
National Inst. of Standards and Technology (NML), Boulder, CO. Quantum Physics Div.
Helium Resonance Lines in the Flare of 15 June 1973.
Final rept.
J. G. Porter, K. B. Gebbie, and L. J. November. 1989, 93p
Contract AFOSR-ISSA-79-0002
Sponsored by Air Force Office of Scientific Research, Bolling AFB, DC.
Pub. in Solar Physics 120, p909-341 1989.

Keywords: *Solar flares, *Helium, Solar spectrum, Line spectra, Far ultraviolet radiation, Solar ultraviolet radiation, Reprints.

Time sequences of HeI and HeII resonance line intensities at several sites within the flare of 15 June 1973 are derived from observations obtained with the Naval Research Laboratory's Slitless Spectroheliograph on Skylab. The data are compared with predictions in six model flare atmospheres based on two values for the heating rate and three for the flux of photoionizing coronal X-rays and EUV. A peak ionizing flux more than 1000 times that in the quiet Sun is indicated. Implications for the common practice of deriving stellar coronal fluxes from HeII 1640 A fluxes are indicated, assuming dominance of the recombination mechanism.

900,029
PB90-123787 Not available NTIS
National Inst. of Standards and Technology (NML), Gaithersburg, MD. Molecular Spectroscopy Div.
Millimeter- and Submillimeter-Wave Surveys of Orion A Emission Lines in the Ranges 200.7-202.3, 203.7-205.3, and 330-360 GHz.
Final rept.
P. R. Jewell, J. M. Hollis, F. J. Lovas, and L. E. Snyder. 1989, 32p
Pub. in the Astrophysical Jnl. Supplement Series 70, n4 p833-864 Aug 89.

Keywords: Radio astronomy, Millimeter waves, Submillimeter waves, Emission spectra, Microwave spectra, Molecular spectra, Spectral lines, Surveys, Reprints, *Orion A, *Interstellar radiation.

The authors have conducted a continuous spectral line survey of the Orion A position from 330.5 to 360.1 GHz. This survey covers nearly the entire 870 micrometer atmospheric window accessible from ground-based observations. Approximately 160 distinct spectral features composed of about 180 lines were detected, 29 of which could not be readily identified. In addition, they also surveyed Orion A from 200.7 to 202.3 GHz and from 203.7 to 205.3 GHz and detected 42 distinct, new spectral lines, including four that are unidentified at present. These data sets are the first thorough survey results in these spectral regions. The new interstellar lines in the survey bands are tabulated and displayed graphically. Moreover, the data are being made available to the Astronomical Data Center at the Goddard Space Flight Center for distribution by request to the astronomical community.

ATMOSPHERIC SCIENCES

Meteorological Instruments & Instrument Platforms

900,030
PB90-163862
(Order as PB90-163874, PC A04)
National Inst. of Standards and Technology, Gaithersburg, MD.
Reduction of Uncertainties for Absolute Piston Gage Pressure Measurements in the Atmospheric Pressure Range.
B. W. Welch, R. E. Edsinger, V. E. Bean, and C. D. Ehrlich. 5 Sep 89, 4p
Included in Jnl. of Research of the National Institute of Standards and Technology, v94 n6 p343-346 1989.

Keywords: *Manometers, *Atmospheric pressure, *Nitrogen, Measurement, Calibrating, Metrology, Gas thermometry.

NIST pressure calibration services with nitrogen are now based on two transfer standard piston gages for which the effective areas have been determined by calibration with the manometer developed at NIST for gas thermometry. Root-sum-squared three sigma uncertainties for the areas for the two gages are 3.05 ppm and 4.18 ppm.

Physical Meteorology

900,031
PB90-118035 Not available NTIS
National Inst. of Standards and Technology (NEL), Gaithersburg, MD. Electrosystems Div.
Interactions between Two Dividers Used in Simultaneous Comparison Measurements.
Final rept.
Y. X. Zhang, R. H. McKnight, and R. E. Hebner. 1989, 9p
Sponsored by Department of Energy, Washington, DC.
Pub. in IEEE (Institute of Electrical and Electronics Engineers) Transactions on Power Delivery 4, n3 p1586-1594 Jul 89.

Keywords: *Standards, *Lightning, *Measurement, Electric current, Regulations, Atmospheric electricity, Thunderstorms, Specifications, Reprints.

A revised international standard for the measurement of lightning and front-chopped lightning impulses is presently under consideration. The standard states that the accuracy of these measuring systems is to be determined by comparison to reference systems maintained by appropriate national laboratories. Investigations have been made of the interactions between two systems configured for simultaneous measurements and of methods for minimizing these interactions. Step responses were measured for different configurations and a model developed to predict divider response. Simultaneous measurements were made of full and chopped lightning impulses using different divider systems to determine the effects of divider interactions on measurements.

900,032
PB90-123951 Not available NTIS
National Inst. of Standards and Technology (NML), Gaithersburg, MD. Gas and Particulate Science Div.
High-Accuracy Gas Analysis via Isotope Dilution Mass Spectrometry: Carbon Dioxide in Air.
Final rept.
R. M. Verkouteren, and W. D. Dorko. 1989, 7p
Pub. in Analytical Chemistry 61, n21 p2416-2422, 1 Nov 89.

Keywords: *Gas analysis, *Carbon dioxide, *Mass spectroscopy, Error analysis, Temperature control, Performance evaluation, Concentration(Composition), Reprints, *Atmospheric chemistry, *Isotope dilution.

An absolute method, based on isotope dilution mass spectrometry, is described for the determination of atmospheric concentrations of carbon dioxide (CO_2) in dry air. In the study, the relative amounts of sample and spike gases are measured manometrically and temperature control before blending. In the study, the major contributors to uncertainty and imprecision are the predetermination of the gas volume ratio and the measurement of the isotopic composition of the blended CO_2, respectively.

BIOMEDICAL TECHNOLOGY & HUMAN FACTORS ENGINEERING

Biomedical Instrumentation & Bioengineering

900,033
PATENT-4 832 745 Not available NTIS
Department of Health and Human Services, Washington, DC.
Non-Aqueous Dental Cements Based on Dimer and Trimer Acids.
Patent.
J. M. Antonucci. Filed 16 Oct 86, patented 23 May 89, 11p PB89-219281, PAT-APPL-6-922 811
See also PB85-203628. Prepared in cooperation with National Inst. of Standards and Technology, Gaithersburg, MD.

This Government-owned invention available for U.S. licensing and, possibly, for foreign licensing. Copy of patent available Commissioner of Patents, Washington, DC 20231 $1.50.

Keywords: *Patents, *Dental materials, *Acid bonded reaction cements, Polymerization, Crosslinking(Chemistry), Cations, Carboxylic acids, PAT-CL-106-35.

Non-aqueous polycarboxylic acids such as dimer and trimer acids are reacted with a variety of polyvalent metal bases to yield a new, versatile class of cements. Many of these cements have unique energy-absorbing properties and excellent dimensional stability yielding mechanically tough and ductile materials. They also do not inhibit the polymerization of resin-based dental materials and thus can be formulated to yield hybrid resin-composite-cement materials. The bulky, hydrophobic nature of these acids with their relatively low carboxylic content results in cements that are low shrinking, hydrolytically resistant and biocompatible.

900,034
PB89-146716 Not available NTIS
National Bureau of Standards (IMSE), Gaithersburg, MD. Polymers Div.
Effects of Purified Ferric Oxalate/Nitric Acid Solutions as a Pretreatment for the NTG-GMA and PMDM Bonding System.
Final rept.
R. L. Blosser, and R. L. Bowen. 1988, 7p
Sponsored by American Dental Association Health Foundation, Chicago, IL.
Pub. in Dental Materials 4, p225-231 1988.

Keywords: *Dental materials, *Acid bonded reaction cements, *Nitric acid, Dentin, Adhesives, Oxalates, Iron, Bonding strength, Electron microscopy, Reprints.

Nitric acid, found to be a contaminant left over from the synthesis of ferric oxalate, has been shown by this in vitro study to be responsible for the cleansing effect previously associated with ferric oxalate pretreatments. When 2.5% (w/w) nitric acid solution was substituted for ferric oxalate solution and then used in conjunction with the experimental dental bonding system of NTG-GMA (the adduct of N(p-tolyl)glycine and glycidyl methacrylate) and PMDM (the adduct of pyromellitic acid dianhydride and 2-hydroxyethyl methacrylate), the result was strong adhesion of both dentin and enamel surfaces to Adaptic (registered trademark) Dental Restorative. When a solution containing 3.4% (w/w) purified ferric oxalate was used with this system, the average bond strengths on both dentin and enamel decreased (p is less than 0.001) as compared with 3.4% (w/w) ferric oxalate solution containing 2.5% (w/w) nitric acid. The effects of pretreatment solutions containing 2.5% (w/w) nitric acid and varying concentrations of purified ferric oxalate on dentin and enamel were demonstrated by adhesion testing and scanning electron microscopy.

900,035
PB89-146732 Not available NTIS
National Bureau of Standards (IMSE), Gaithersburg, MD. Polymers Div.
Bonding Agents and Adhesives: Reactor Response.
Final rept.
R. L. Bowen. 1986, 3p
Sponsored by American Dental Association Health Foundation, Chicago, IL.
Pub. in Advances in Dental Research 2, n1 p155-157 Aug 88.

Keywords: *Dental materials, *Acid bonded reaction cements, Dentin, Adhesives, Composite materials, Dimensional stability, Polymers, Reprints.

Adhesive materials must form multiple bonds with sound tooth substrates for maximum adhesion. Adhesive resins can be applied in incremental layers to bond composite materials to enamel and dentin. Hardening shrinkage and stress concentrations are factors that have detrimental effects on adhesive bonding with resins and composites. Improvements in dimensional stability of composites can therefore allow for better bonding and sealing of preventive and restorative materials.

900,036
PB89-146757 Not available NTIS
National Bureau of Standards (NML), Gaithersburg, MD. Organic Analytical Research Div.

5

BIOMEDICAL TECHNOLOGY & HUMAN FACTORS ENGINEERING

Biomedical Instrumentation & Bioengineering

Liposome-Enhanced Flow Injection Immunoanalysis.
Final rept.
R. A. Durst, L. Locascio-Brown, A. L. Plant, and M. V. Brizgys. 1988, 2p
See also PB88-217914.
Pub. in Clinical Chemistry 34, n9 p1700-1701 1988.

Keywords: Antibodies, Fluorescein, Bioassay, Feedback control, Antigens, Reprints, *Immunoassay, *Biomedical engineering, Liposomes, Immobilized cells.

The development of a repetitive immunoassay is important for monitoring and feedback control in bioprocessing. The assay should be fast and the immunoreactor, which uses immobilized antibodies, regenerable. The automated system employs flow injection analysis for solution manipulation and contains an immunoreactor column with covalently bound Fab' fragments. Recognition of the antigen is through competitive binding on this column of sample antigen and antigen contained in the membrane of liposomes. The concentration of sample antigen is proportional to the amount of liposomes competitively excluded from the column. Liposomes are spherical structures composed of a phospholipid bilayer, and are detectable through the marker species which is entrapped inside their aqueous compartment. Detection is through a fluorescent marker, carboxyfluorescein. Liposomes excluded from the column in the competitive assay flow downstream where they are chemically disrupted, and the contents measured. The signal enhancement provided by the liposome marker is approximately 100,000 per analyte binding event making this technique competitive in sensitivity with radioimmunoassays.

900,037
PB89-157127 Not available NTIS
National Bureau of Standards (IMSE), Gaithersburg, MD. Polymers Div.
Interaction of Cupric Ions with Calcium Hydroxylapatite.
Final rept.
D. N. Misra. 1988, 5p
Sponsored by American Dental Association Health Foundation, Chicago, IL.
Pub. in Materials Research Bulletin 23, n11 p1545-1549 1988.

Keywords: *Copper, *Calcium, *Dental materials, Ions, Phosphates, Hydroxides, X-ray analysis, Chemical composition, Bonding, Reprints.

The interaction of aqueous cupric ions with calcium hydroxylapatite produces cupric orthophosphate, Cu3(PO4)2 times 3H2O, and libethenite, Cu2(OH)PO4. The latter product was identified and characterized by chemical analysis and powder X-ray diffraction. The orthophosphate itself changes to libethenite under various experimental conditions. Contrary to a published report, it was not cupric hydroxylapatite that was previously identified. These findings are important to understand the chemistry of and composite bonding to bone and teeth.

900,038
PB89-157150 Not available NTIS
National Bureau of Standards (IMSE), Gaithersburg, MD. Polymers Div.
Biological Evaluations of Zinc Hexyl Vanillate Cement Using Two In vivo Test Methods.
Final rept.
J. C. Keller, B. D. Hammond, K. K. Kowalyk, and G. M. Brauer. 1988, 10p
Contract PHS-DE-06675
Sponsored by National Institutes of Health, Bethesda, MD.
Pub. in Dental Materials 4, p341-350 1988.

Keywords: *Dental materials, *Adhesives, *Histology, *Pathology, Tissue extracts, Cells(Biology), Implantation, Peritoneum, Connective tissue, Inflammation, Zinc oxides, Phosphates, Vanillin, In vivo analysis, Reprints, Eugenol.

The cellular and tissue responses to 3 dental cements were studied by 2 methodologies, the connective tissue implantation technique (CTI), recommended by the ADA, and the peritoneal cavity implantation technique (PCI), which has emerged as a method to quantitatively study the cellular response to implanted materials. While similar histopathological results were obtained for the implantation of cements using both methodologies, the PCI technique offers a more thorough investigation of cellular and tissue responses to

implanted materials. In addition to histopathological evaluation, the PCI technique allows quantitative investigation of the specific cells responding to the implants, and provides a mechanism, using chemical analysis techniques, to quantify the concentration of specific degradative products within the retrieved cells and host tissue. Finally, the results from these 2 methodologies demonstrated the acceptable biological performance of zinc hexyl vanillate cement compared with the clinically acceptable zinc phosphate and zinc oxide-eugenol cement formulations.

900,039
PB89-157168 Not available NTIS
National Bureau of Standards (IMSE), Gaithersburg, MD. Polymers Div.
Adhesion to Dentin by Means of Gluma Resin.
Final rept.
E. Asmussen, J. M. Antonucci, and R. L. Bowen. 1988, 6p
Sponsored by American Dental Association Health Foundation, Chicago, IL.
Pub. in Scandinavian Jnl. of Dental Research 96, p584-589 1988.

Keywords: *Dental materials, *Resins, *Bonding strength, *Dentin, *Adhesives, Composite materials, Pyruvates, Enamels, Amino acids, Camphor, Quinones, Methacrylic acid, Glycine, Glutarates, Reprints.

In its present version, the Gluma system for bonding restorative resin to dentin involves the application of an enamel bonding agent prior to the composite resin. Conceivably, pretreating the dentin with solutions of amino acids, and incorporating camphorquinone and selected methacrylic monomers into the Gluma adhesive would nullify the need for the enamel bonding agent. A bond strength to dentin of 13.4 MPa was obtained in the control experiment. Using a solution of pyruvic acid and glycine as pretreatment, and an optimized adhesive mixture containing glutaraldehyde, HEMA, BIS-GMA, camphorquinone, and water, bond strengths to dentin of 14.5 MPa and to enamel of 23.3 MPa were obtained. Thus, the new Gluma bonding system gave acceptable bond strengths without the prior application of enamel bonding agents.

900,040
PB89-176077 Not available NTIS
National Bureau of Standards (IMSE), Gaithersburg, MD. Polymers Div.
Mesh Monitor Casting of Ni-Cr Alloys: Element Effects.
Final rept.
J. A. Tesk, O. Okuno, and R. Penn. 1986, 1p
Pub. in Jnl. of Dental Research 65, p301 1986.

Keywords: *Dental materials, *Castings, *Nickel, *Chromium, Metal alloys, Computerized simulation, Temperature, Silicon, Beryllium, Solidus, Design, Molybdenum, Niobium, Boron, Aluminum, Reprints.

A mesh monitor is used for quantitatively evaluating the casting of dental alloys. A castability value, C(sub v), is defined as the fraction of completely cast grid segments. For statistical analysis, a transformed castability value, C(sub vt), is used and one equation for C(sub vt) was found to fit all of the Ni-Cr alloy data at the 95% confidence level: $C(sub\ vt) = a + bT(sub\ sub\ 1/2) T(sub\ M)(sub\ 2)$ where, C(sub vt) = ln (2/3 + the square root of C(sub v))/(2/3 + the square root of 1 - C(sub v)) -- a and b are characteristic constants for each alloy. T(sub A) = T(sub c) - T(sub s), with T(sub c) the casting temperature, T(sub s) the solidus temperature and T(sub M) the mold temperature. A series of alloys were selected to determine effects of critical elements on casting. Compositions were chosen to assure the avoidance of correlated effects. Assuming a linear dependence, the following equation was found to describe C(sub vt): C(sub vt) = K(sub 0) Ni/Cr + K(sub 1)(Mo) + K(sub 2)(Si) + K(sub 3)(Nb) + K(sub 4)4(B) + K(sub 5)(Al) + K(sub 6)(Be) + K(sub 7)(Be x Si) where () are elemental concentrations in weight percent, and K(sub i) is a coefficient for the ith term. Because composition is constant for each alloy, K(sub i) = f(sub i) (T(sub A), T(sub M)) = g(sub i) (T(sub c), T(sub m)). The temperature dependent coefficients were determined for seven elements and the (Ni)/(Cr) ratio. It was also found that Si and Be produce a synergistic effect. The results can be used in computer aided design of Ni-Cr alloys.

900,041
PB89-179220. Not available NTIS
National Bureau of Standards (IMSE), Gaithersburg, MD. Polymers Div.

Adsorption of 4-Methacryloxyethyl Trimellitate Anhydride (4-META) on Hydroxyapatite and Its Role in Composite Bonding.
Final rept.
D. N. Misra. 1989, 6p
Sponsored by American Dental Association Health Foundation, Chicago, IL.
Pub. in Jnl. of Dental Research 68, n1 p42-47 Jan 89.

Keywords: Adsorption, Dental materials, Reprints, *Hydroxyapatite, *Composite bonding, *Methacryloxyethyl trimellitate anhydride, Absorption isotherm.

The adsorption of 4-methacryloxyethyl trimellitate anhydride (4-META) was studied from ethanol and dichloromethane onto synthetic hydroxyapatite (containing about 1.5 monolayers of physisorbed water) in order to study its role in restorative composite bonding to teeth. The adsorption isotherm of 4-META was S-shaped and reversible from ethanol and followed the Langmuir plot at lower concentrations. The isotherm was irreversible from dichloromethane and a constant amount of adsorbate was removed from the solutions above a certain concentration. The irreversibly adsorbed compound was completely removed by washing with ethanol. Therefore, the bonding between teeth and the restorative resin containing 4-META as a coupling agent is micromechanical and not chemical in nature. An analysis of isotherms showed that the benzene rings of the adsorbate molecules lie flat on the surface for both solvents.

900,042
PB89-179253 Not available NTIS
National Bureau of Standards (IMSE), Gaithersburg, MD. Polymers Div.
Oligomers with Pendant Isocyanate Groups as Adhesives for Dentin and Other Tissues.
Final rept.
C. H. Lee, and G. M. Brauer. 1989, 5p
Pub. in Jnl. of Dental Research 68, n3 p484-488 Mar 89.

Keywords: *Dentin, *Adhesives, Dental materials, Bonding strength, Isocyanates, Reprints, *Oligomers, isocyanic acid/dimethylbenzyl-isopropenyl, Methacrylate/isocyanatoethyl.

Oligomers containing pendant isocyanate groups were synthesized from various vinyl monomers, m-isopropenyldimethylbenzyl isocyanate (TMI), and 2-isocyanatoethyl methacrylate (IEM). The liquids were characterized by their refractive indices, infrared spectra, and percentage of isocyanate groups in the molecule. Adhesive properties of these compounds were compared with those of oligomers prepared from methacrylate esters, IEM, and/or TMI, which had been synthesized previously. These adhesive compositions, especially formulations synthesized from vinyl monomers, adhered at least as well to dentin as did other dentin bonding agents. Oligomers synthesized with methacrylate esters bonded more strongly to bone than did other hard-tissue adhesives. These oligomeric compositions are also excellent soft-tissue adhesives. Provided that their biological properties prove satisfactory, these compositions could find many applications as hard-and soft-tissue adhesives in clinical dentistry.

900,043
PB89-228068 Not available NTIS
National Bureau of Standards (NEL), Boulder, CO. Thermophysics Div.
Thermophysical Properties for Bioprocess Engineering.
Final rept.
N. A. Olien. 1987, 4p
Pub. in Chemical Engineering Progress, p45-48 Oct 87.

Keywords: *Thermophysical properties, Solutions, Biological products, Reprints, *Biotechnology, Data bases.

The commercialization of biotechnology requires the development of processes for routine, scientifically-based scaleup of bioprocesses. Many of the engineering problems associated with scaleup are in downstream processing, which is comprised primarily of separation and purification. In traditional process industries, accurate data and predictive models for the thermophysical properties of fluid mixtures play significant roles in the design and operation of separation processes. In a workshop it was concluded that several steps must be taken: develop experimental capability for obtaining thermophysical property data on bio-

6

logical solutions; begin theoretical studies which can lead to the development of predictive models; identify candidate biomolecules for experimental measurements and begin acquisition of a database for their solution properties.

900,044
PB89-229256 Not available NTIS
National Inst. of Standards and Technology (IMSE), Gaithersburg, MD. Polymers Div.
Simplified Shielding of a Metallic Restoration during Radiation Therapy.
Final rept.
F. C. Eichmiller, and R. A. Schrack. 1989, 1p
Sponsored by American Dental Association Health Foundation, Chicago, IL.
Pub. in Jnl. of Prosthetic Dentistry 61, n5 p640 May 89.

Keywords: *Radiation protection, *Protective equipment, Radiation shielding, Radiotherapy, Dental equipment, Reprints, Dental restorations.

Radiation-induced lesions directly adjacent to large metallic restorations have been associated with scattered or low energy electrons and positrons generated within the restorative material by high energy incident beams. These normally self-resolving lesions very often result in the interruption of the normal course and timing of the radiation treatment. Demonstrated is a simple absorptive intra-oral shield utilizing materials and techniques adaptable to any dental office or radiotherapy treatment facility. The use of the shielding technique can attenuate the low energy scatter to a level where damage to surface tissues can be avoided. The simple intraoral shielding technique utilizes readily available commercial impression materials and can be easily adapted to any radiotherapy treatment setting or dental office. The elimination of the painful and debilitating side effect of radiotherapy will allow for adherence to the most effective therapy course and improvement in the patient's quality of life.

900,045
PB89-229272 Not available NTIS
National Inst. of Standards and Technology (IMSE), Gaithersburg, MD. Polymers Div.
Ferric Oxalate with Nitric Acid as a Conditioner in an Adhesive Bonding System.
Final rept.
E. N. Cobb, R. L. Blosser, R. L. Bowen, and A. D. Johnston. 1989, 9p
Pub. in Jnl. of Adhesion 28, p41-49 1989.

Keywords: *Dental materials, *Acid bonded reaction cements, *Nitric acid, *Oxalates, Dentin, Iron, Enamels, Composite materials, Tensile strength, Adhesive strength, Surface finishing, Resin cements, Reprints.

Strong adhesive bonding of composite resins to dentin and enamel is obtained by conditioning the surface and applying adhesion-promoting compounds. The study examines tensile adhesive bond strengths and effects of the conditioners having various concentrations of ferric oxalate (FO) and nitric acid. In the first part of the study, the average tensile bond strengths increased with concentrations of commercial FO as received up to about 6.8% and averaged no higher with higher concentrations. After the first part of the testing had been completed, it was discovered that the FO as received contained a small amount of nitric acid. Use of solutions having from 6.8% to 20% ferric oxalate as received yielded bonds with strengths that averaged about 13 MPa psi) to dentin and 16 MPa (2,400 psi) to enamel. In the second part of the study, the FO was stripped of the fortuitous nitric acid and, based on results from the first part of the study, solutions were made up to contain a fixed concentration of purified FO (6.8% Fe2(C2O4)3) and various known concentrations of nitric acid. The highest bond strengths to dentin and enamel were obtained with the purified FO solution which contained approximately 2.5% nitric acid.

900,046
PB89-229284 Not available NTIS
National Inst. of Standards and Technology (IMSE), Gaithersburg, MD. Polymers Div.
Transient and Residual Stresses in Dental Porcelains as Affected by Cooling Rates.
Final rept.
K. Asaoka, and J. A. Tesk. 1989, 17p
Pub. in Dental Materials Jnl. 8, n1 p9-25 Jun 89.

Keywords: *Dental materials, Cooling rate, Residual stress, Computerized simulation, Reprints, *Dental porcelain.

The development of either transient or residual stress in a slab of dental porcelain during cooling was simulated by use of a super-computer. The temperature dependences of the elastic modulus, the thermal expansion coefficient, the shear viscosity, and the cooling rate dependence of the glass transition temperature were considered in the calculation. Internal stress and viscoelastic creep were computed for several cooling rates. Calculated results display stress profiles which agree reasonably well with reported measured profiles in quenched, tempered glasses. The method by which residual stress develops is also discussed. This discussion suggests a method for strengthening of the porcelain by the development of high-compressive residual stress on the surface.

900,047
PB89-231278 Not available NTIS
National Inst. of Standards and Technology (IMSE), Gaithersburg, MD. Polymers Div.
Development of a Microwave Sustained Gas Plasma for the Sterilization of Dental Instruments.
Final rept.
W. G. de Rijk, and L. L. Forsythe. 1988, 3p
Pub. in Proceedings of Southern Biomedical Engineering Conference (7th), Greenville, SC., October 27-28, 1988, p189-191.

Keywords: *Microwaves, *Plasmas(Physics), *Sterilization, Spores.

The development of a low temperature gas plasma sterilizing process was sought using electromagnetic radiation from a conventional microwave oven providing energy for both igniting and sustaining the plasma. It was found that a gas plasma readily ignites, but causes rapid heating of the vacuum chamber contents and its walls. Bacterial spore strips showed that sterilization can be achieved with the gas plasma.

900,048
PB90-117375 Not available NTIS
National Inst. of Standards and Technology (IMSE), Gaithersburg, MD. Polymers Div.
Use of N-Phenylglycine in a Dental Adhesive System.
Final rept.
R. S. Chen, and R. L. Bowen. 1989, 6p
Sponsored by American Dental Association Health Foundation, Chicago, IL.
Pub. in Jnl. Adhesion Sci. Technol. 3, n1 p49-54 1989.

Keywords: *Dental materials, *Adhesive bonding, Bonding strength, Performance evaluation, Substitutes, Reprints, *Glycine/N-phenyl.

A dentin and enamel bonding procedure that used an acidic solution (aqueous ferric oxalate), a surface-active co-monomer (NTG-GMA, the reaction product of N(p-tolyl)glycine and glycidyl methacrylate), and a dimethacrylate (PMDM, the reaction product of hydroxyethyl methacrylate and pyromellitic dianhydride) was described earlier. The present paper reveals the discovery that a much more simple and available compound, N-phenylglycine, can be substituted for NTG-GMA without a sacrifice in bond strength.

900,049
PB90-117516 Not available NTIS
National Inst. of Standards and Technology (IMSE), Gaithersburg, MD. Polymers Div.
In vitro Investigation of the Effects of Glass Inserts on the Effective Composite Resin Polymerization Shrinkage.
Final rept.
K. J. Donly, T. W. Wild, R. L. Bowen, and M. E. Jensen. 1989, 4p
Sponsored by American Dental Association Health Foundation, Chicago, IL.
Pub. in Jnl. of Dental Research 68, n8 p1234-1237 Aug 89.

Keywords: *Dental materials, *Glass, *Composite materials, *Resins, *Polymerization, *Shrinkage, In $vitro$ analysis, Strain gages, Reprints.

An MOD preparation was placed in each of 12 permanent molars, then each tooth was restored with a posterior composite resin by means of six different application techniques: I-polymerization as one complete unit; II-polymerization as one complete unit with glass inserts; III-polymerization in gingivo-occlusal increments; IV-polymerization in gingivo-occlusal increments with glass inserts; V-polymerization in bucco-lingual increments; and VI-polymerization in a gingival increment with glass inserts, then bucco-lingual incre-

ments. After each increment was polymerized, the strain appearing on the strain gauge indicator was recorded. Results demonstrated the average microstrain units to be 127-I, 102-II, 105-III, 86-IV, 72-V, and 66-VI. A randomized block design was the format used for data evaluation. Scheffe's Test indicated that composite resin placement and polymerization in bucco-lingual increments (V) created significantly less cuspal deflection than polymerization as one complete unit, with or without glass inserts (I and II), $p < 0.001$, and gingivo-occlusal increments (III), $p < 0.05$).

900,050
PB90-123696 Not available NTIS
National Inst. of Standards and Technology (IMSE), Gaithersburg, MD. Polymers Div.
Adhesive Bonding of Composites.
Final rept.
R. L. Bowen, F. C. Eichmiller, W. A. Marjenhoff, and N. W. Rupp. 1989, 4p
Sponsored by American Dental Association Health Foundation, Chicago, IL.
Pub. in Jnl. of the American College of Dentists 56, n2 p10-13 1989.

Keywords: *Adhesive bonding, *Dental materials, *Composite materials, Adhesive strength, Enamels, Dentin, Teeth, Dentistry, Reprints.

Although the strength and durability of adhesive bonds to enamel have been adequate for numerous clinical procedures for many years, commercial products that demonstrate a range of effectiveness of adhesion to dentin are relatively new. The paper discusses several of the more popular commercial dentin and enamel bonding systems, noting differences in the chemical components and instructions for use. Since simple technique errors can lead to failure, it is important that practitioners understand not just the sequence of protocol steps, but the function of each system component. Dentists are encouraged to try dentin and enamel adhesives supported by research literature in referred journals to determine which works best for them, and to provide feedback to dental manufacturers and researchers so that the materials and clinical techniques can be further improved.

900,051
PB90-123795 Not available NTIS
National Inst. of Standards and Technology (IMSE), Gaithersburg, MD. Polymers Div.
Substitutes for N-Phenylglycine in Adhesive Bonding to Dentin.
Final rept.
A. D. Johnston, E. Asmussen, and R. L. Bowen. 1989, 8p
Sponsored by American Dental Association Health Foundation, Chicago, IL.
Pub. in Jnl. of Dental Research 68, n9 p1337-1344 Sep 89.

Keywords: *Adhesive bonding, *Dentin, *Surfactants, Surface reactions, Glycine, Reprints.

A number of related compounds were investigated using bond strength measurements in order to elucidate the role of the surface-active ingredient N-phenylglycine (NPG) in experimental two-step and three-step bonding protocols resulting in adhesive bonding to dentin. All active compounds identified for the two-step or the three-step protocol were N-aryl-alpha-amino acids, and the results delineate some of the key features of the NPG molecule for bonding. For the three-step protocol, there was a requirement for a secondary or tertiary aromatic amino group, a carboxylic acid group, and a single (secondary or tertiary) methylene unit between those two functional groups of the amino acid. For the two-step protocol, additional substitutions at the para position of the phenyl ring on the amine improved the bond strength. In both protocols, para-methyl- and para-chloro-substituted NPG analogues ranked higher than NPG. A 'catalytic' effect of the aromatic tertiary amino group on the polymerization of the adhering resin in both procedures could not be ruled out.

900,052
PB90-128711 Not available NTIS
National Inst. of Standards and Technology (IMSE), Gaithersburg, MD. Polymers Div.

BIOMEDICAL TECHNOLOGY & HUMAN FACTORS ENGINEERING

Biomedical Instrumentation & Bioengineering

Dental Materials and Technology Research at the National Bureau of Standards: A Model for Government-Private Sector Cooperation.
Final rept.
J. A. Tesk. 1989, 8p
Pub. in Materials Research Society Symposia Proceedings, v110 p177-184 1989.

Keywords: *Dental materials, *Technical societies, *Research, Composite materials, Acid bonded reaction cements, X ray apparatus, Resins, Prosthetic devices, Weibull density functions.

In 1919 the United States Army commissioned the National Bureau of Standards (NBS) to develop a federal specification for dental amalgam. In 1926, the American Dental Association (ADA) joined forces with the NBS for dental research and in 1964 the National Institute of Dental Research (NIDR) initiated support. Today, the program involves personnel from the NBS, ADA, NIDR, two dental companies and numerous researchers from around the world. Major advances in dentistry have emanated from the collaborative effort; among them are the panoramic x-ray unit, the high speed, turbine, contra-angle handpiece, numerous cements, and the basic formulation of modern composite restoratives. The latter alone has been estimated at saving the American public an annual amount that exceeds the current, combined, appropriated budgets of the NIDR, NBS and the ADA. Current efforts are focusing on tissue adhesives, biocompatible cements, atherosclerotic plaque, new resins for composites, characterization of materials via Weibull Statistics and reliability analysis of dental prosthetic systems.

Prosthetics & Mechanical Organs

900,053
PB89-150890   Not available NTIS
National Bureau of Standards (IMSE), Gaithersburg, MD. Metallurgy Div.
Corrosion of Metallic Implants and Prosthetic Devices.
Final rept.
A. C. Fraker. 1987, 12p
Pub. in Metals Handbook (9th Edition), v13 p1324-1335 1987.

Keywords: *Corrosion resistance, *Metals, *Implantation, *Prosthetic devices, Stainless steels, Titanium, Cobalt, Surgery, Alloys, Reprints.

The paper deals with the corrosion of metallic surgical implants. The history of the development of the use of metals as implants and the associated research are given. The metals and alloys are described in detail with the intent of emphasizing effects of composition, microstructure and other metallurgical factors on the corrosion resistance. Basic principles of corrosion processes, corrosion problems occurring with surgical implants and corrosion test procedures are discussed.

900,054
PB89-157143   Not available NTIS
National Bureau of Standards (IMSE), Gaithersburg, MD. Polymers Div.
Casting Metals: Reactor Response.
Final rept.
J. A. Tesk. 1988, 3p
Pub. in Advances in Dental Research 2, n1 p44-46 Aug 88.

Keywords: *Dental materials, *Prosthetic devices, *Castings, *Metals, Palladium alloys, Titanium, Waxes, Sintering, Reprints.

A commentary on the paper given by Dr. Asgar is presented. Agreement with that paper's contents on the trend toward palladium-based and other lower-cost alloys is rendered. However, with a view toward future competitiveness and quality of prosthetic restorations, the commentary looks for the application of new and emerging technologies. To implement these developments, the National Institute of Dental Research is urged to find avenues to support applications for dentistry in selected instances, despite adverse reviews from an entrenched establishment.

900,055
PB89-202212   Not available NTIS
National Bureau of Standards (IMSE), Gaithersburg, MD. Polymers Div.

Oligomers with Pendant Isocyanate Groups as Tissue Adhesives. 1. Synthesis and Characterization.
Final rept.
G. M. Brauer, and C. H. Lee. 1989, 15p
See also PB89-179253.
Pub. in Jnl. of Biomedical Materials Research 23, p295-309 1989.

Keywords: Synthesis(Chemistry), Methacrylates, Reprints, *Tissue adhesives, *Biocompatible materials, Oligomers.

A series of methacrylate oligomers containing pendant isocyanate groups were synthesized by reacting 2-isocyanatoethyl methacrylate (IEM) and/or m-isopropenyl-alpha, alpha-dimethylbenzyl isocyanate (TMI) in ethoxyethyl acetate with metacrylates ranging from methyl to stearyl methacrylate or allyl-, cyclohexyl-, glycidyl-, i-bornyl-, or dicyclopentenyloxyethyl methacrylate. The oligomers which are stable at room temperature were characterized by IR for NCO, ester, and C=C groups and by their refractive indices. HPLC showed no residual monomer. GPC and intrinsic viscosity of selected oligomers indicated a molecular weight range from 1400 to 2600. Isocyanate groups were determined titrimetrically and ranged from 15.9% to 5.1%. Concurrent studies have demonstrated that these oligomers bond strongly to hard and soft tissues. Thus, subject to their biocompatibility, they could find many applications as tissue adhesives.

900,056
PB89-231245   Not available NTIS
National Inst. of Standards and Technology (IMSE), Gaithersburg, MD. Polymers Div.
Oligomers with Pendant Isocyanate Groups as Tissue Adhesives. 2. Adhesion to Bone and Other Tissues.
Final rept.
G. M. Brauer, and C. H. Lee. 1989, 11p
See also Part 1, PB89-202212.
Pub. in Jnl. of Biomedical Materials Research 23, p753-763 1989.

Keywords: *Methacrylates, *Isocyanates, *Bones, Tissues(Biology), Materials tests, Tensile strength, Reprints, *Tissue adhesives, Biocompatible materials, Dental cements.

The adhesive properties of a series of oligomers prepared from 2-isocyanatoethyl methacrylates (IEM) and/or m-isopropenyl-alpha, alpha-dimethylbenzyl isocyanate (TMI) and various acrylates or methacrylates were studied. The bond strength of bone, dentin, or soft tissue specimens joined with these oligomers respectively to bone, dental composite restorative, or denture base resin were determined by tensile adhesion or shear tests. These oligomers are more effective in forming stronger bonds to bone than are other tissue adhesives. Fracture occurs cohesively, usually within the bone. Thermocycling in water for 1 week between 5 C and 55 C did not decrease adhesion indicating that exposure to water or thermal shock produced no deterioration of the bond. Tensile adhesion of bovine or human dentin joined to composite restorative resin by means of the oligomers is similar to that of the best dental bonding agents such as Gluma (glutaraldehyde and 2-hydroxy-ethyl methacrylate) or ferric oxalate + N-phenylglycine + dimethylacryloxyethyl-pyromellitate. These oligomers also strongly bond soft tissues and calfskin and to acrylic resins and composites.

BUILDING INDUSTRY TECHNOLOGY

Architectural Design & Environmental Engineering

900,057
PB89-150763   Not available NTIS
National Bureau of Standards (NEL), Gaithersburg, MD. Building Physics Div.

Experimental Validation of a Mathematical Model for Predicting Moisture Transfer in Attics.
Final rept.
D. M. Burch. 1985, 10p
Sponsored by Department of Energy, Washington, DC.
Pub. in Proceedings of International Symposium on Moisture and Humidity, Washington, DC., April 15-18, 1985, p287-296.

Keywords: *Mathematical models, *Moisture content, *Residential buildings, *Wood, *Roofs, Proving, Predictions, Heating, Test chambers, Climate, Ventilation, Water vapor, Adsorption, Condensing, Static tests, Dynamic tests, Dew point, Environment simulation, *Attics.

A small test house having a pitched roof/ventilated attic was installed in a high-bay environmental chamber. The test house and its attic were extensively instrumented for measuring heat and moisture transfer. The test house was subsequently exposed to a series of steady diurnal outdoor climatic conditions. Representative conditions of a residence were simulated within the test house. A mathematical model was developed that included the adsorption of water vapor at wood surfaces in the attic. This model closely predicted the attic dewpoint temperatures for both the steady and dynamic outdoor cycle tests. The model showed that wood surfaces of the attic at a moisture content of 12.5% (by weight) adsorbed water vapor and maintained the wood surface dewpoint temperature below the roof sheathing temperature, thereby preventing condensation.

900,058
PB89-151765   PC A03/MF A01
National Inst. of Standards and Technology (IMSE), Gaithersburg, MD. Center for Computing and Applied Mathematics.
ZIP: The ZIP-Code Insulation Program (Version 1.0) Economic Insulation Levels for New and Existing Houses by Three-Digit ZIP Code. Users Guide and Reference Manual.
S. R. Petersen. Jan 89, 42p NISTIR-88/3801
Contract DE-AC05-84OR21400
Also pub. as Oak Ridge National Lab., TN. rept. no. ORNL/TM-11009. Prepared in cooperation with Oak Ridge National Lab., TN. Sponsored by Department of Energy, Washington, DC. Office of Buildings and Community Systems.

Keywords: *Computer systems programs, *Operating systems(Computers), *Residential buildings, *Thermal insulation, Cost engineering, Zip codes, Microcomputer.

ZIP (the ZIP Code Insulation Program) is a computer program developed to support the DoE 'Insulation Fact Sheet' by providing users with customized estimates of economic levels of residential insulation for any location in the United States, keyed to the first three digits of its ZIP Code. The program and supporting files are contained on a single 5-1/4 in. diskette for use with microcomputers having an MS-DOS operating system capability. The ZIP program currently calculates economic levels of insulation for attic floors, exterior wood-frame and masonry walls, floors over unheated areas, slab floors, and basement and crawl-space walls. The economic analysis can be conducted for either new or existing houses. Climate parameters are contained in a file on the ZIP diskette and automatically retrieved when the program is run. Regional energy and insulation price data are also retrieved from the ZIP diskette, but these can be overridden to more closely correspond to local prices. ZIP can be run for a single ZIP Code and specified heating and cooling system. It can also be run in a 'batch' mode for any number of consecutive ZIP Codes in order to provide a table of economic insulation levels for use at the state or national level.

900,059
PB89-157267   Not available NTIS
National Bureau of Standards (NEL), Gaithersburg, MD. Fire Science and Engineering Div.
Computer Model of Smoke Movement by Air Conditioning Systems (SMACS).
Final rept.
J. H. Klote. 1988, 13p
See also PB88-159462.
Pub. in Fire Technology 24, n4 p299-311 Nov 88.

Keywords: *Air conditioning equipment, *Smoke, *Computerized simulation, *Air circulation, Mass flow,

8

Mass transfer, Fans, Ducts, Cooling systems, Mathematical models, Reprints.

A computer model for simulation of smoke movement through air conditioning systems is described. A brief overview of air conditioning systems is presented. The methods of calculation of mass flow, smoke transport, fan flow, and duct and fitting resistances are presented along with a general description of the program logic.

900,060
PB89-159446 CP. D01
National Inst. of Standards and Technology (NCTL), Gaithersburg, MD.
ZIP: **ZIP-Code Insulation Program (for Microcomputers).**
Software.
S. R. Petersen. Jan 89, 1 diskette NBS/SW/DK-89/003
The software is contained on 5 1/4-inch diskettes, double density (360K), compatible with the COMPAQ Portable II microcomputer. The diskettes are in the ASCII format. Price includes documentation, PB89-151765.

Keywords: *Software, *Thermal insulation, *Residential buildings, *Economics, Engineering costs, Climate, Prices, Diskettes, Energy conservation, Energy accounting, Costs, Zip codes, L=BASIC, H=COMPAQ Portable II.

ZIP (the ZIP code Insulation Program) is a computer program developed to support the DoE 'Insulation Fact Sheet' by providing users with customized estimates of economic levels of residential insulation for any location in the United States, keyed to the first three digits of its ZIP Code. The ZIP program currently calculates economic levels of insulation for attic floors, exterior wood-frame and masonry walls, floors over unheated areas, slab floors, and basement and crawlspace walls. The economic analysis can be conducted for either new or existing houses. Climate parameters are contained in a file on the ZIP diskette and automatically retrieved when the program is run. Regional energy and insulation price data are also retrieved from the ZIP diskette, but these can be overridden to more closely correspond to local prices. ZIP can be run for a single ZIP Code and specified heating and cooling system. It can also be run in a 'batch' mode for any number of consecutive ZIP Codes in order to provide a table of economic insulation levels for use at the state or national level. Software Description: The software is written in the Basic programming language for implementation on the COMPAQ Portable II or compatible machines using MS DOS operating system.

900,061
PB89-172340 Not available NTIS
National Bureau of Standards (NEL), Gaithersburg, MD. Building Environment Div.
Control Strategies and Building Energy Consumption.
Final rept.
J. Y. Kao. 1985, 8p
Pub. in ASHRAE (American Society of Heating, Refrigerating and Air-Conditioning Engineers) Transactions 91, pt2 p810-817 1985.

Keywords: *Buildings, *Energy consumption, *Control equipment, *Environmental engineering, Temperature control, Cooling systems, Heating, Ventilation, Management methods, Economic analysis, Air conditioning, Reprints, HVAC systems.

A summary report of building energy studies on basic control strategies applied to air-handling systems of four different buildings in six climatic regions is presented. The building energy program BLAST is used to simulate commonly used air-handling systems for two office buildings, a school, and a retail store. The results of the cooling and the heating energy consumption of these buildings are presented and compared. The energy effects of various economy cycles and temperature resetting strategies applied to reheat, variable air volume, dual-duct, and other systems are discussed.

900,062
PB89-173926 Not available NTIS
National Bureau of Standards (NEL), Gaithersburg, MD. Building Environment Div.
Developments in the Heat Balance Method for Simulating Room Thermal Response.
Final rept.
G. N. Walton. 1985, 18p
Pub. in Proceedings of Workshop on HVAC (Heating, Ventilation, and Air Conditioning) Controls Modeling

and Simulation, Atlanta, GA., February 2-3, 1984, 18p 1985.

Keywords: *Ventilation, *Heat balance, *Buildings, *Computerized simulation, Air flow, Conduction, Heat transfer.

The paper reviews recent developments in the heat balance method for the thermal simulation of the nonmechanical components of buildings. These developments include: improved methods for computing response factors, simplified radiant interchange analysis, and detailed calculations for interroom airflows. Further developments are anticipated in the use of response factors for short time step simulation and multidimensional conduction and more detailed simulation of the room air through the concept of ventilation effectiveness.

900,063
PB89-174114 Not available NTIS
National Bureau of Standards (NEL), Gaithersburg, MD. Center for Building Technology.
Design Quality through the Use of Computers.
Final rept.
R. N. Wright. 1988, 7p
Sponsored by American Society of Civil Engineers, New York.
Pub. in Manual of Professional Practice for Quality in the Constructed Project, Chapter 11, v1, p66-72 1988.

Keywords: *Civil engineering, *Construction, Quality, Design, Reprints, *Computer aided design.

The American Society of Civil Engineers is developing an authoritative and comprehensive guide for quality in construction called the Manual of Professional Practice for Quality in the Constructed Project. The goals for the Manual are to clarify greatly the roles in the construction process and to outline proper procedures and responsibilities for each member of the construction team. Chapter 11 'Design Quality through Use of Computers' describes typical activities and flows of information in design, provides general considerations for quality in the use of computers in design and construction, and gives guidance for the use of computers in each stage of the design process. The designer must use the computer effectively to be competitive both economically and in the quality of his work. The designer must use the computer responsibly to maintain full professional control of his decisions.

900,064
PB89-175905 Not available NTIS
National Bureau of Standards (NEL), Gaithersburg, MD. Office of Energy-Related Inventions.
Field Measurement of Thermal and Solar/Optical Properties of Insulating Glass Windows.
Final rept.
M. E. McCabe, and D. Hill. 1987, 16p
Sponsored by Department of Energy, Washington, DC. Solar Buildings Technology Div.
Pub. in ASHRAE (American Society of Heating, Refrigerating and Air-Conditioning Engineers) Transactions 93, pt1 p1409-1424 1987.

Keywords: *Calorimeters, *Heating load, *Optical properties, *Window glass, Insulation, Transmittance, Coatings, Leakage, Measurement, Emittance, Reprints.

The thermal performance of windows with alternative glazing systems were compared by field testing in a side-by-side arrangement using portable calorimeters. Existing double-hung windows installed in an NBS test building were replaced with new sash and insulating glazings provided by the original window manufacturer. Two of the new glazing units had insulating glass with a low-emittance coating applied to one glass surface. The other two glazing units were identical except there was no coating present. Portable calorimeters were placed on the interior of the two northfacing windows, which included a low-emittance and an uncoated control window. Comparative measurements of air leakage, heat loss and solar transmittance were made for these windows. Nighttime heat loss for the uncoated window was approximately 23% greater than that of the low-emittance window, although the U-values for both windows were greater than indicated by the manufacturer's laboratory test results. Possible causes for the discrepancy between laboratory and field test data are evaluated and the potential use of portable calorimeters for both comparative and absolute measurements of windows is discussed.

900,065
PB89-176127 Not available NTIS

National Bureau of Standards (NEL), Gaithersburg, MD. Building Physics Div.
Indoor Air Quality.
Final rept.
P. E. McNall. 1986, 7p
Pub. in ASHRAE (American Society of Heating, Refrigerating and Air-Conditioning Engineers) Jnl. 28, n6 p39-42, 44, 46, 48 Jun 86.

Keywords: *Buildings, *Ventilation, Design, Air conditioning, Environmental engineering, Maintenance, Air filters, Efficiency, Quality control, Heat recovery, Reprints, *HVAC systems, *Air Quality, *Indoor air pollution.

The paper outlines in a general way, the methods which are available to the HVAC engineer in the practical design, operation and maintenance of systems which impact the air quality in buildings. The general control methods are: exclude the pollutant source; prevent the source from emanating pollutants into the air; dilute the air with purer air (ventilation); remove pollutants from the air (air cleaners); provide other ventilation strategies (ventilation effectiveness, local exhaust, etc.); discuss occupant actions which reduce their exposure. Strategies which can be used now are: ventilation quantities (ASHRAE Standard 62-1981 and 62-1981P); commissioning and testing and balancing; exhaust air reentry; heat recovery; system design for less contamination; particulate filtering; gaseous filtering; local exhaust. Strategies which are under development or need further application knowledge are: use of ventilation effectiveness; better gaseous removal standards; air washers; air ionization; differential pressure control.

900,066
PB89-176499 Not available NTIS
National Bureau of Standards (NEL), Gaithersburg, MD. Building Equipment Div.
Interzonal Natural Convection for Various Aperture Configurations.
Final rept.
B. M. Mahajan, and D. D. Hill. 1986, 7p
Sponsored by Department of Energy, Washington, DC. Passive and Hybrid Solar Energy Div.
Pub. in Proceedings of ASME (American Society of Mechanical Engineers) Winter Annual Meeting, Anaheim, CA., December 7-12, 1986, 7p.

Keywords: *Air flow, *Buildings, Doors, Heating, Heat transmission, Walls, Mathematical models, Mass flow, *Natural convection, *Aperture shape.

Experiments were conducted to study the interzonal natural convection for different aperture configurations for a two-zone set-up. The following four aperture configuration were studied: a center door; a side door; a window; and (4) a split window, i.e., two small windows situated symmetrically about the horizontal bisector of the common wall. One of the two zones was heated with baseboard electric heaters placed adjacent to the floor and the wall opposite to the common wall. For each aperture configuration, tests were conducted with various heat inputs to the warmer of the two zones. The data indicate that the discharge coefficient used in a simple one-dimensional model for interzonal airflow varies with the aperture configuration and status of heat input to the warmer zone. The variations in the discharge coefficient are apparently due to the different flow fields and temperature distributions for each aperture configuration.

900,067
PB89-176614 Not available NTIS
National Bureau of Standards (NEL), Gaithersburg, MD. Building Physics Div.
Ventilation Effectiveness Measurements in an Office Building.
Final rept.
A. K. Persily. 1986, 11p
Sponsored by Department of Energy, Washington, DC.
Pub. in Proceedings of ASHRAE (American Society of Heating, Refrigerating and Air-Conditioning Engineers) Conference on IAQ (Indoor Air Quality): Managing Indoor Air for Health and Energy Conservation, Atlanta, GA., April 20-23, 1986, p548-558.

Keywords: *Ventilation, *Commercial buildings, Air flow, Distribution systems, Environmental engineering, Performance evaluation, Draft(Gas flow), Efficiency, Quality control, Reprints, *Air quality, *HVAC systems, Tracer studies.

To evaluate the impact of ventilation on air quality within mechanical ventilated office buildings, one must examine both the outside air intake or ventilation rate and the performance of the air distribution system in delivering this outside air to the building occupants. Inadequate air distribution may lead to air quality problems, even if a building's total ventilation rate is adequate. Tracer gas measurement procedures have been developed to evaluate these air distribution characteristics of mechanical ventilation systems or their ventilation effectiveness. The paper examines the use of one such technique, tracer gas measurement employing age distribution analysis in modern, mechanically ventilated office buildings. The measurement and analysis techniques are described, along with those features of the building type which impact on their application. The results of the measurements indicate that there is good mixing in a whole building scale, but also provide evidence of local, stagnant zones in the occupied space. These apparently contradictory results are suspected to be due to the fact that the layout of an actual building is much more complex than the measurement theory assumes.

900,068
PB89-177141 Not available NTIS
National Bureau of Standards (NEL), Gaithersburg, MD. Building Environment Div.
Application of Direct Digital Control to an Existing Building Air Handler.
Final rept.
G. E. Kelly, W. B. May, and C. Park. 1985, 26p
Sponsored by Department of Energy, Washington, DC. Office of Buildings and Community Systems, and Naval Civil Engineering Lab., Port Hueneme, CA.
Pub. in Performance of HVAC (Heating, Ventilating and Air Conditioning) Systems and Controls in Buildings, Proceedings of CIB International Symposium at the Building Research Establishment, Garston, England, June 18-19, 1984, p3-28 1985.

Keywords: *Buildings, *Automatic control equipment, *Digital systems, Dynamic response, Systems engineering, Management systems, Environmental engineering, *HVAC systems.

In order to gather reliable information on the dynamic performance of heating, ventilating and air conditioning systems and their controls in buildings, the Center for Building Technology (CBT) at the National Bureau of Standards (NBS) has developed a Building Management and Controls Laboratory. The Laboratory consists of a distributed computerized Energy Management and Control System (EMCS) developed at NBS and used to control a number of different experiments. These experiments currently include the control of an eleven-story office building on the NBS site, a thoroughly instrumented air handler in a laboratory environment, and a large air handler used to condition part of the perimeter of the building housing the Center for Building Technology.

900,069
PB89-177158 Not available NTIS
National Bureau of Standards (NEL), Gaithersburg, MD. Building Environment Div.
Flow Coefficients for Interzonal Natural Convection for Various Apertures.
Final rept.
B. M. Mahajan, and D. D. Hill. 1987, 7p
Sponsored by Department of Energy, Washington, DC. Office of Solar Heat Technologies.
Pub. in Proceedings of ASME-JSME-JSES (American Society of Mechanical Engineers-Japan Society of Mechanical Engineers-Japanese Solar Energy Society) Solar Energy Conference, Honolulu, HI., March 22-27, 1987, p300-306.

Keywords: *Convection, *Air flow, *Heat transfer coefficient, Temperature distribution, Heat measurement, Environmental engineering, Thermal expansion, Tests, Temperature control, Energy conservation.

Experiments to determine the flow coefficients for interzonal natural convection were carried out at the National Bureau of Standards' Passive Solar Test Facility. Interzonal natural convection was studied for ten different apertures. One of the two zones was heated with baseboard electric heaters placed adjacent to the floor along the wall opposite to the common wall. Experiments were conducted with various heat inputs to the warmer of the two zones. The flow coefficients used in simple one-dimensional models for interzonal airflow varies with the aperture configurations, aperture to wall area ratio, and the level of heat input to the warmer zone. Variations in the flow coefficients are apparently due to the different flow fields and temperature distributions for each aperture configuration.

900,070
PB89-177166 Not available NTIS
National Bureau of Standards (NEL), Gaithersburg, MD. Building Environment Div.
HVACSIM+, a Dynamic Building/HVAC/Control Systems Simulation Program.
Final rept.
G. E. Kelly, C. Park, D. R. Clark, and W. B. May. 1985, 19p
Sponsored by Department of Energy, Washington, DC. Office of Buildings and Community Systems, and Naval Civil Engineering Lab., Port Hueneme, CA.
Pub. in Proceedings of Workshop on HVAC (Heating, Ventilating and Air Conditioning) Controls Modeling and Simulation, Atlanta, GA., February 2-3, 1984, p1-19 1985.

Keywords: *Buildings, *Control equipment, *Computerized simulation, Dynamic response, Design, Architecture, Systems engineering, Environmental engineering, Performance evaluation, *HVAC systems.

The dynamic performance of buildings and the service systems within them is becoming more and more important as the use of computerized building management systems become less expensive and increasingly popular. In an effort to understand the dynamic interactions between the building shell, the HVAC system, and building controls, the National Bureau of Standards (NBS) has begun work on a non-proprietary building system simulation program. Called HVACSIM+, which stands for HVAC SIMulation PLUS other systems, this program employs advanced equation solving techniques and a hierarchical, modular approach to simulate the dynamic performance of entire building/HVAC/control systems. The paper discusses the architecture of HVACSIM+, the HVAC component models and building shell model being developed, and other important features of the program. A brief status report on the development of HVACSIM+ and a timetable for completing various portions of the program are also presented.

900,071
PB89-177174 Not available NTIS
National Bureau of Standards (NEL), Gaithersburg, MD. Building Environment Div.
Simulation of a Large Office Building System Using the HVACSIM+ Program.
Final rept.
C. Park, S. T. Bushby, and G. E. Kelly. 1989, 10p
Sponsored by Department of Energy, Washington, DC.
Pub. in ASHRAE (American Society of Heating, Refrigerating and Air-Conditioning Engineers) Transactions, v95 pt1 10p 1989.

Keywords: *Buildings, *Computerized simulation, Systems analysis, Design, Dynamic response, Systems engineering, Environmental engineering, Control equipment, *HVAC systems.

A large office building system located in Gaithersburg, MD was simulated using the HVACSIM+ computer program. A typical floor of this 11-story building was selected and divided into four zones, with one air-handling unit serving each zone. Dynamic interactions between the building zones and the HVAC and control system were studied for several different control strategies during the cooling season. Simulations were performed using the building shell and zone models along with the air handler and control system models. Simulation results are presented and compared with experimental measurements. The effects of three different control schemes on energy consumption are compared with each other. These schemes are: the start/stop control without purging cycle, the start/stop control with purging, and continuous operation.

900,072
PB89-179667 Not available NTIS
National Bureau of Standards (NEL), Gaithersburg, MD. Building Physics Div.
Performance Measurements of Infrared Imaging Systems Used to Assess Thermal Anomalies.
Final rept.
Y. M. Chang, and R. A. Grot. 1986, 14p
Pub. in Proceedings of SPIE (Society of Photo-Optical Instrumentation Engineers), Thermal Imaging, v636 p17-30 1986.

Keywords: *Infrared thermal detectors, *Temperature measuring instruments, *Temperature distribution, Performance evaluation, Thermal stability, Environmental engineering, Buildings, Spatial distribution, Imaging techniques, Energy conservation.

An evaluation of various infrared imaging systems was performed to determine their abilities to identify thermal anomalies in buildings. The systems were tested under environmental temperatures from -20 C to 25 C for their minimum resolvable temperature differences (MRTD) at spatial frequencies between 0.03 to 0.25 cy/mrad. The temperature dependence of MRTD was analyzed and compared with the predicted values in ASHRAE Standard 101-83 for thermal imaging systems. The temperature dependence of infrared systems object temperature calibrations was investigated. The signal transfer functions (SiTF) of infrared sensors are generated to verify and calibrate the dynamic range of each sensor. Also discussed are the results of measurements of modulation transfer function (MTF) of infrared imaging systems, which are based on Fourier Transforms of the line spread function (LSF). It is shown that the results of the MTF calculations can be correlated with their MRTD measurements.

900,073
PB89-189153 PC A07/MF A01
National Inst. of Standards and Technology, Gaithersburg, MD.
Evaluating Office Lighting Environments: Second Level Analysis.
B. L. Collins, W. S. Fisher, G. L. Gillette, and R. W. Marans. Apr 89, 143p NISTIR-89/4069
See also PB88-164512. Prepared in cooperation with Lighting Research Inst., New York, and Michigan Univ., Ann Arbor. Sponsored by International Council for Educational Development, New York, and New York State Energy Research and Development Authority, Albany.

Keywords: *Office buildings, *Lighting equipment, Lamps, Energy consumption, Energy conservation, Evaluation, Illuminance, Luminance.

Data from a post-occupancy evaluation (POE) of 912 work stations with lighting power density (LPD), photometric, and occupant response measures were examined in a detailed, second-level analysis. Seven types of lighting systems were identified with different combinations of direct and indirect ambient lighting, and task lighting and daylight. The mean illuminances at the primary task location were within the IES target values for office task with a range of mean illuminances from 32 to 75 fc, depending on the lighting system. The median LPD was about 2.36 watts/sq ft, with about one-third the work stations having LPD's at or below 2.0 watts/sq ft. Although a majority of the occupants (69%) were satisfied about their lighting, the highest percentage of those expressing dissatisfaction (37%) with lighting had an indirect fluorescent furniture mounted (IFFM) system. The negative reaction of so many people to the IFFM system suggests that the combination of task lighting with an indirect ambient system had an important influence on lighting satisfaction, even though task illuminances tended to be higher with the IFFM system. Concepts of lighting quality, visual health, and control were explored, as well as average luminance to explain the negative reactions to the combination of indirect lighting with furniture mounted lighting.

900,074
PB89-189237 PC A04/MF A01
National Inst. of Standards and Technology (NEL), Gaithersburg, MD. Center for Building Technology.
Illumination Conditions and Task Visibility in Daylit Spaces.
S. J. Treado. Mar 89, 55p NISTIR-88/4014

Keywords: *Buildings, *Daylighting, *Architecture, *Windows, Design, Illuminating, Computerized simulation, Light(Visible radiation), Brightness, Environmental engineering, Electric lighting, Comparisons, Graphs(Charts), Energy conservation.

Illumination conditions are evaluated in typical building spaces based on detailed computer simulations, in order to characterize and quantify the effects of daylighting on task visibility. Examined are the effects of fenestration location and type on task contrast under daylit, electric-lit and combined conditions. The implications of the illumination conditions with daylighting on lighting and daylighting system design are discussed.

900,075
PB89-193247 PC A03/MF A01
National Inst. of Standards and Technology, Gaithersburg, MD.
Rating Procedure for Mixed Air-Source Unitary Air Conditioners and Heat Pumps Operating in the Cooling Mode. Revision 1.
P. A. Domanski. May 89, 25p NISTIR-89/4071
See also PB86-166279. Sponsored by Department of Energy, Washington, DC. Office of Buildings and Community Systems.

Keywords: *Heat pumps, *Air conditioners, *Residential buildings, Cooling systems, Cooling rate, Standards, Evaporators, Condensing, Efficiency.

A procedure is presented for rating split, residential air conditioners and heat pumps operating in the cooling mode which are made up of an evaporator unit combined with a condensing unit which has been rated under current procedures in conjunction with a different evaporator unit. The procedure allows calculation of capacity at the 95 F rating point and seasonal energy efficiency ratio without performing laboratory tests of the complete system. The procedure has been prepared for the Department of Energy for consideration in the rule making process. It is a revised version of the original version of the procedure published in 1986.

900,076
PB89-193254 PC A05/MF A01
National Inst. of Standards and Technology (NEL), Gaithersburg, MD. Building Environment Div.
AIRNET: A Computer Program for Building Airflow Network Modeling.
G. N. Walton. Apr 89, 85p NISTIR-89/4072
Contract DE-AI01-36CE2101-3
Sponsored by Department of Energy, Washington, DC.

Keywords: *Air flow, *Buildings, *Models, Ventilation, Ventilation fans, Ducts, Air conditioning, Infiltration, Computer applications, User manuals(Computer programs).

In spite of its importance, the analysis of airflows has significantly lagged the modeling of other building features because of limited data, computational difficulties, and incompatible methods for analyzing different flows. Methods have been developed to analyze airflows in HVAC ducts and to estimate infiltration, but the interaction between building HVAC systems and infiltration airflows has seldom been studied. The report describes a computer program for modeling networks of airflow elements such as openings, ducts, and fans. It emphasizes the numerical aspects of an airflow network method which would provide a unified approach to building airflow calculations. It also discusses the limitations of the method and poorly understood factors that could profit from further research.

900,077
PB89-206833 PC A03/MF A01
National Inst. of Standards and Technology (NEL), Gaithersburg, MD. Building Environment Div.
Integral Mass Balances and Pulse Injection Tracer Techniques.
J. Axley, and A. Persily. Oct 88, 40p NISTIR-86/3855
Presented at the Air Infiltration and Ventilation Centre Conference, Effective Ventilation held at Novotel Gent, Belgium, September 12-15, 1988. Sponsored by Department of Energy, Washington, DC.

Keywords: *Air flow, *Ventilation, *Buildings, Ducts, Environmental engineering, Air circulation, Measuring instruments, Air quality, *Tracer techniques, *HVAC systems, Energy conservation.

Tracer gas techniques for measuring airflow rates in building systems are considered. These techniques are classified in terms of tracer gas injection strategies employed and mass balance relationships used to analyze measured tracer concentration data. The discussion focuses on one class of tracer techniques, the pulse injection techniques, based upon pulse injection strategies and integral mass balance relationships. These pulse injection techniques have not been commonly used in the past yet they provide practically useful means for the determination of airflow rates in building systems. Pulse injection techniques are presented for measuring airflows in ducts, and for studying single-zone and multi-zone building airflow systems. Experimental procedures for these three cases are discussed, and preliminary results from field applications of these techniques are presented. The possi-

bility of flow variation is accounted for in all cases, and the sensitivity of the single-zone pulse injection technique to these flow variations is compared to that of the single-zone constant injection technique. This comparison leads to integral formulations of the constant injection technique for duct, single-zone, and multi-zone situations that may provide means to improve the accuracy of the commonly used constant injection tracer technique.

900,078
PB89-211858 Not available NTIS
National Bureau of Standards (NEL), Gaithersburg, MD. Fire Safety Technology Div.
Outline of a Practical Method of Assessing Smoke Hazard.
Final rept.
A. J. Fowell. 1986, 7p
Pub. in Proceedings of Joint Meeting on Progress in Fire Safety - Society of the Plastics Industry and Fire Retardant Chemical Association, Washington, DC., March 19-21, 1986, p139-145.

Keywords: *Buildings, *Fires, *Smoke, Safety engineering, Computer systems programs, Building codes, Fire prevention, Assessments.

The document outlines the form, content, and capabilities of a practical method for assessing smoke hazards planned for delivery by the National Bureau of Standards, Center for Fire Research before the end of 1986. The method will contain a step-by-step procedure, description of generic fires and buildings, data for using with computer programs, and worked examples.

900,079
PB89-228977 PC A07/MF A01
National Inst. of Standards and Technology (NEL), Gaithersburg, MD. Center for Building Technology.
Air Quality Investigation in the NIH (National Institutes of Health) Radiation Oncology Branch.
A. Persily, W. S. Dols, S. J. Nabinger, and D. A. VanBronkhorst. Aug 89, 132p NISTIR-89/4145
Sponsored by National Institutes of Health, Bethesda, MD.

Keywords: *Ventilation, *Air flow, *Radioactive contaminants, Human factors engineering, Design standards, Safety engineering, Gas flow, Graphs(Charts), Recommendations, *National Institutes of Health Radiation Oncology Branch, *Air quality, *HVAC systems, *Indoor air pollution, Building technology, Office buildings.

The Radiation Oncology Branch (ROB) is located in the Clinical Center of the National Institutes of Health (NIH). The occupants of the ROB facility have expressed dissatisfaction with the air quality within the facility for several years. To identify the sources of the air quality problems in the ROB facility and to obtain recommendations for their solution, the Center for Building Technology at the National Institute of Standards and Technology (NIST, formerly The National Bureau of Standards) conducted an indoor quality investigation of the ROB facility. Results revealed several deficiencies in the design and current condition of the ROB ventilation system, such as significant differences between the design airflow rates and those recommended in current standards and guidelines. The airflow measurements showed many instances in which measured airflow rates were different from their design values and revealed the existence of airflows leading to the potential for pollutant transport within the building. The contaminant measurements fell generally well below the maximum values in the ASHRAE air quality standard. Thermal comfort measurements revealed instances when the temperature and relative humidity were outside of ASHRAE comfort limits. Recommendations are made to remedy the deficiencies noted and to control the conditions contributing to the building's air quality problems.

900,080
PB89-229157 Not available NTIS
National Bureau of Standards (NEL), Gaithersburg, MD. Fire Safety Technology Div.
Capabilities of Smoke Control: Fundamentals and Zone Smoke Control.
Final rept.
J. H. Klote, and E. K. Budnick. 1989, 10p
Pub. in Jnl. of Fire Prot. Engr. 1, n1 p1-10 1989.

Keywords: *Smoke abatement, *Air flow, Ventilation, Fire protection, Safety engineering, Fire alarm systems, Buildings, Fire safety, Reprints, *HVAC systems, Smoke detectors.

The paper discusses the principles of smoke control and the practical application of these principles to zoned smoke control systems. Zoned smoke control can use dedicated fans or the fans of a building's heating, ventilating and air conditioning systems. The paper discusses concerns with systems that only purge in an attempt to control smoke movement. Considerations of system activation and acceptance testing are presented.

900,081
PB89-230361 Not available NTIS
National Inst. of Standards and Technology (NEL), Gaithersburg, MD. Building Environment Div.
Investigation of a Washington, DC Office Building.
A. K. Persily, W. A. Turner, H. A. Burge, and R. A. Grot. 1989, 16p
Sponsored by Department of Energy, Washington, DC.
Pub. in Design and Protocol for Monitoring Indoor Air Quality, p35-50 1989.

Keywords: *Commercial buildings, Human factors engineering, Temperature control, Air flow, Ventilation, Monitors, Measuring instruments, Reprints, *Indoor air pollution, *Air quality, *Office buildings, *Sick building syndrome, *HVAC systems, Building technology.

The paper describes the techniques used to study a Washington, D.C. office building with a long history of indoor air quality and thermal comfort complaints. More than twenty investigations, mostly relatively short term, have been conducted since 1978 to determine the causes of the building's problems and to recommend corrective actions. More recently a long term, intensive study of the building has been undertaken to study the building more thoroughly and to investigate the application of several techniques for studying office building air quality. These techniques include tracer gas measurements of air exchange rates, ventilation system performance and ventilation effectiveness, and measurements of the levels of various indoor pollutants including bioaerosols. The paper reviews the previous investigations of the building and describes the procedures used in the current study. Some preliminary results of the current effort are presented.

900,082
PB89-230379 Not available NTIS
National Inst. of Standards and Technology (NEL), Gaithersburg, MD. Building Environment Div.
Airflow Network Models for Element-Based Building Airflow Modeling.
Final rept.
G. N. Walton. 1989, 10p
Sponsored by Department of Energy, Washington, DC. Office of Buildings and Community Systems.
Pub. in ASHRAE (American Society of Heating, Refrigerating and Air Conditioning Engineers) Transactions, v95 pt2 10p Jul 89.

Keywords: *Buildings, *Airflow, Ventilation, Air circulation, Ducts, Temperature control, Reprints, *HVAC systems, *Air quality, Building technology.

In spite of its importance, the analysis of airflows has significantly lagged behind the modeling of other building features because of limited data, computational difficulties, and incompatible methods for analyzing different flows. Methods have been developed to analyze airflows in HVAC ducts and to estimate infiltration, but the interaction between building HVAC systems and infiltration airflows has seldom been studied. The paper emphasizes the numerical aspects of an airflow network method that would provide a unified approach to building airflow calculations. It also discusses the limitations of the method and poorly understood factors that could profit from further research.

900,083
PB89-231005 Not available NTIS
National Inst. of Standards and Technology (NEL), Gaithersburg, MD. Office of Energy-Related Inventions.
Origins of ASHRAE (American Society of Heating, Refrigerating and Air-Conditioning Engineers) Window U-Value Data and Revisions for the 1989 Handbook of Fundamentals.
Final rept.
M. E. McCabe. 1989, 5p
Pub. in Proceedings of the Annual Conference of the Solar Energy Society of Canada (15th), Penticton, B. C., June 19-21, 1989, p273-277.

11

BUILDING INDUSTRY TECHNOLOGY

Architectural Design & Environmental Engineering

Keywords: *Windows, Architecture, Solar energy, Heat transfer, Buildings, Research projects, *U values, Energy conservation, Building technology.

The ASHRAE Handbook of Fundamentals and its predecessor, the ASHVE Guide, have been the authoritative source of technical information on window U-values since the 1920s. The paper discusses the historical origins of window U-values beginning with research conducted in the last century. The technical basis for the modern form of the ASHRAE U-value table, which appeared at about 1950, is described. Revisions to the U-value table are traced during the ensuing years, concluding with a discussion of the current concerns over the data and the changes that will appear in the 1989 ASHRAE Handbook.

900,084
PB89-231047   Not available NTIS
National Inst. of Standards and Technology (NEL), Gaithersburg, MD. Building Environment Div.
Considerations for Advanced Building Thermal Simulation Programs.
Final rept.
G. N. Walton. 1989, 6p
Sponsored by Department of Energy, Washington, DC.
Pub. in Proceedings of Building Simulation '89 Conference, Vancouver, B.C., Canada, June 23-24, 1989, p155-160.

Keywords: *Buildings, *Thermal analysis, Heat loss, Air flow, Ventilation, Environment simulation, Matrix methods, Thermal measurement, *Energy audits, Energy conservation.

In order to assess the applicability of a more modular approach to the development of building thermal analysis programs, the paper begins with a review of some of the basic numerical methods used in simulation. These are discussed with some observations from other fields of study besides building simulation. Two major examples of advanced simulation methods are presented: the use of sparse matrix methods for heat transfer simulation and a modular calculation of building airflows. Their implications on the development of the next generation of building thermal analysis programs are discussed.

900,085
PB89-231161   Not available NTIS
National Bureau of Standards (NEL), Gaithersburg, MD. Fire Safety Technology Div.
Analysis and Prediction of Air Leakage through Door Assemblies.
Final rept.
D. Gross, and W. L. Haberman. 1989, 10p
Pub. in Proceedings of International Symposium on Fire Safety Science (2nd), Tokyo, Japan, June 13-17, 1988, p169-178 1989.

Keywords: *Air flow, *Doors, *Heat loss, Flow rate, Flow measurement, Leakage, Energy dissipation, Comparisons, *Energy conservation.

A generalized relationship is presented for determining air flow rates through narrow gaps around door edges. The relationship provides values of leakage rates for steady laminar flow through gaps over a wide range of pressure difference and eliminates approximations associated with the often inappropriate use of discharge coefficients and exponents in the flow equation Q = C A(delta p)n. The analysis covers straight-through, single bend and double bend gaps of constant thickness, as well as connected gaps of constant thicknesses. Comparison of measured flow rates for installed stairwell door assemblies with those predicted by use of the relationship shows agreement within 20%. The volumetric flow of heated air through simple door gaps has been calculated by use of the relationship. The results show that the flow rate may increase or decrease with temperature depending on gap size and flow region.

900,086
PB89-235881   PC A07/MF A01
National Inst. of Standards and Technology, Gaithersburg, MD.
EVSIM: An Evaporator Simulation Model Accounting for Refrigerant and One Dimensional Air Distribution.
P. A. Domanski. Aug 89, 142p NISTIR-89/4133
Sponsored by Department of Energy, Washington, DC. Office of Buildings and Community Systems.

Keywords: *Computerized simulation, *Air conditioning equipment, *Evaporators, Air conditioners, Heat

exchanges, Refrigerants, Air circulation, Thermodynamics, Models, Computer programs.

The report describes a computer model, EVSIM, of a refrigerant-to-air heat exchanger of the type used in residential air conditioning as an evaporator. The model provides performance predictions of a one-slab or two-slab evaporator for a given refrigerant enthalpy at the coil inlet, saturation temperature and superheat at the coil outlet, and at imposed one dimensional air mass flow distribution over the coil face. The model accounts for air distribution and for complex refrigerant circuitry designs by simulating refrigerant distribution. Performance of the coil is calculated employing a tube-by-tube scheme. The report includes a User's Guide and a listing written in FORTRAN 77.

900,087
PB90-112368   PC A06/MF A01
National Inst. of Standards and Technology (NEL), Gaithersburg, MD. Center for Building Technology.
Proposed Methodology for Rating Air-Source Heat Pumps That Heat, Cool, and Provide Domestic Water Heating.
B. P. Dougherty. Aug 89, 108p NISTIR-89/4154, EPRI-RP-2033-26
Sponsored by Electric Power Research Inst., Palo Alto, CA., and Department of Energy, Washington, DC.

Keywords: *Residential buildings, *Hot water heating, Ratings, Performance standards, Heat exchangers, Air conditioning, Heating, Seasonal variations, Mathematical models, *Air source heat pumps.

The work at NIST has centered upon developing a proposed rating methodology for integrated appliances that heat water in a water heating only mode or while simultaneously air conditioning or space heating. Despite the emphasis, the proposed methodology provides a framework for rating other types of integrated heat pump/water heating appliances. The laboratory testing, the calculation procedure, and the method for reporting performance are described. The testing is an adaptation of the laboratory tests conducted when rating conventional heat pumps and water heaters. Seasonal estimates of energy consumption rates are calculated using a bin type approach. Combined performance factors and operating costs are used for reporting performance.

900,088
PB90-112384   PC A08/MF A01
National Inst. of Standards and Technology (NEL), Gaithersburg, MD. Center for Building Technology.
Post-Occupancy Evaluation of Several U.S. Government Buildings.
B. L. Collins, G. L. Gillette, M. S. Dahir, and P. J. Goodin. Sep 89, 159p NISTIR-89/4175
Sponsored by Army Communications-Electronics Support Facility, Vint Hill Farms Station, VA.

Keywords: *Public buildings, *Office buildings, *Human factors engineering, *Environmental engineering, *Comfort, Assessments, Noise(Sound), Performance evaluation, Questionnaires, Lighting equipment, Air quality, Indoor air pollution, HVAC systems.

A post-occupancy evaluation was performed on five small, low-rise U.S. government buildings at a site south of Washington, D.C. to evaluate environmental conditions including lighting, space, noise, and indoor air quality, and provide recommendations for change. In addition, a comparison was made of environmental conditions before and after renovation of one of the buildings. The study employed a questionnaire about the environmental conditions, physical measures of the space (lighting, space, noise, temperature, etc.) and interviews with personnel at the site. A total of 306 people participated (including measures before and after the renovation) and physical measures were taken at 92 work stations. Analysis of the physical measurement data indicated problems with limited space, lack of adjustable task lighting, and perceptions of poor indoor air quality in two of the buildings. The renovation was perceived to have improved the appearance of one building substantially, however. Suggestions for improvements to the buildings at the site were also made.

900,089
PB90-118043   Not available NTIS
National Inst. of Standards and Technology (NEL), Gaithersburg, MD. Mathematical Analysis Div.

Advanced Heat Pumps for the 1990's Economic Perspectives for Consumers and Electric Utilities.
Final rept.
S. R. Petersen. 1989, 6p
Sponsored by Electric Power Research Inst., Palo Alto, CA.
Pub. in ASHRAE (American Society of Heating, Refrigerating and Air-Conditioning Engineers) Jnl., p36, 38, 40, 42, 44, and 46, Sep 89.

Keywords: *Heat pumps, *Electric power demand, Hot water heating, Electric utilities, Economic analysis, Air conditioning equipment, Heating, Cooling systems, Reprints.

Advanced heat pumps promise improved energy efficiency and reduced peak-power demand for space heating, space cooling and water heating in houses. Economic analysis shows that they can be cost effective in much of the United States, both for consumers and electric utilities. Consumers will benefit from greatly reduced electric bills; utilities will benefit from improved load management and the increased competitiveness of electric heating. A new program, HPEAK (Heat Pump/Economic Analysis of Kilowatt-hours), developed at the National Institute of Standards and Technology, evaluates the hourly performance of conventional and advanced heat pump systems, both from consumer and utility perspectives. HPEAK analyses demonstrate the economic benefits of an advanced air-source heat pump with adjustable-speed control and integrated water-heating capability in five locations in the United States.

900,090
PB90-128158   Not available NTIS
National Inst. of Standards and Technology (NEL), Gaithersburg, MD. Building Environment Div.
Measured Air Infiltration and Ventilation Rates in Eight Large Office Buildings.
Final rept.
R. A. Grot, and A. K. Persily. 1986, 33p
Pub. in Measured Air Leakage of Buildings, ASTM STP 904, p151-183 1986.

Keywords: *Office buildings, *Ventilation, *Air flow, *Flow measurement, Safety engineering, Human factors engineering, Measuring instruments, Design standards, Building codes, Environmental engineering, Reprints.

Air infiltration and ventilation rate measurements were made during all seasons of the year in eight federal office buildings using an automatic air infiltration system designed at the National Bureau of Standards. The eight federal office buildings were located in Anchorage, Alaska; Ann Arbor, Michigan; Columbia, South Carolina; Fayetteville, Arkansas; Huron, South Dakota; Norfolk, Virginia; Pittsfield, Massachusetts; and Springfield, Massachusetts. These buildings ranged in size from 1730 sq m (18,600 sq ft) for the building in Pittsfield to 45,500 sq m (490,000 sq ft) for the Anchorage federal building. All were constructed within the last 10 years. Air infiltration rates were found to vary from 0.2 to 0.7 air changes per hour and constituted from 23% to 61% of the building design load. Minimum ventilation rates in the tighter buildings were found to be less than what would be recommended for occupied offices.

Building Equipment, Furnishings, & Maintenance

900,091
PB89-157010   Not available NTIS
National Bureau of Standards (NEL), Gaithersburg, MD. Building Environment Div.
Control System Simulation in North America.
Final rept.
G. E. Kelly. 1988, 10p
Pub. in Energy and Buildings 10, p193-202 1988.

Keywords: *Control equipment, *Buildings, *Simulation, Systems engineering, Industrial buildings, Commercial buildings, Management systems, Research projects, Maintenance management, Reviewing, Evaluation, Reprints.

The paper presents a historical review of control system simulation in North America from around 1970 until the present. The subject is divided into the topics of Regulation, Supervisory Control, and Optimized Building Controls. Different research efforts and simulation programs that have made a significant contribution to advancing the state of the art are reviewed. The current emphasis and what the author believes will be some future trends in this important field are also discussed.

900,092
PB89-189146 PC A04/MF A01
National Inst. of Standards and Technology (NEL), Gaithersburg, MD. Center for Building Technology.
Assessment of Robotics for Improved Building Operations and Maintenance.
B. M. Mahajan, J. M. Evans, and J. E. Hill. Nov 88, 56p NISTIR-88/4006
Portions of this document are not fully legible. Prepared in cooperation with Transitions Research Corp., Hartford, CT. Sponsored by General Services Administration, Washington, DC.

Keywords: *Buildings, *Maintenance, *Operations, Automation, Robots, Cleaning, Delivery, Security, Services, Barriers, *Robotics.

The report provides a state-of-the-art survey of robotic technology useful for building operation and maintenance. Floor cleaning, mail delivery, security, and storage facility operations represent current and near term opportunities for robotic application. Likely future applications may include: bathroom, office and window cleaning; lawn mowing; trash handling; wall painting; and miscellaneous material handling. Potential barriers to the use of robotics within buildings are identified. Amendments to the GSA Handbook on 'Quality Standards for Design and Construction' to accommodate the use of service robots in Federal buildings are suggested. The suggested amendments are not exhaustive and as the knowledge base expands, should be refined and augmented.

900,093
PB89-193288 PC A11/MF A01
National Inst. of Standards and Technology (NEL), Gaithersburg, MD. Center for Fire Research.
False Alarm Study of Smoke Detectors in Department of Veterans Affairs Medical Centers (VAMCS).
P. M. Dubivsky, and R. W. Bukowski. May 89, 235p NISTIR-89/4077
Sponsored by Department of Veterans Affairs, Washington, DC., Department of the Air Force, Washington, DC., and Underwriters' Labs., Inc,, Northbrook, IL.

Keywords: *Fire alarm systems, *Smoke, *False alarms, *Hospitals, *Military facilities, Warning systems, Medical centers, Veterans Administration.

A study of 133 VA Medical Centers (VAMC) out of a total of 172 throughout the U.S. coupled with visits to 20 facilities was conducted to gather data on false alarms of smoke detectors. Data collected included name of the detector manufacturer and model number, control unit manufacturer and model number, number and type of detectors installed, where installed, number of false and real alarms for preceding year, date of installation, and policies on smoking, testing, cleaning, and maintenance. VAMC personnel involved with the installations were requested to indicate the maximum level of false alarms that could be tolerated and to provide any recommendations to reduce their occurrence. The study included a total of approximately 37,000 system type smoke detectors of which 69% were of ionization (ion) type and 31% photoelectric, 3000 duct detectors (90% ion and 10% photo), and 1100 smoke detector modules (80% ion and 20% photo) integral with door holder closers. Also included are approximately 100 single station smoke alarms. Analysis of data collected from operating facilities through forms, site visits, and staff interviews resulted in a series of recommendations which could result in a substantial reduction in observed false alarms.

900,094
PB89-229009 PC A03/MF A01
National Inst. of Standards and Technology (NEL), Gaithersburg, MD. Center for Fire Research.
Estimating the Environment and the Response of Sprinkler Links in Compartment Fires with Draft Curtains and Fusible Line-Actuated Ceiling Vents. Part 2. User Guide for the Computer Code Lavent.
W. D. Davis, and L. Y. Cooper. Aug 89, 43p NISTIR-89/4122
See also PB86-215462. Sponsored by American Architectural Mfrs. Association, Des Plaines, IL.

Keywords: *Fires, *Buildings, *Sprinkler systems, Ventilation, Ducts, Computer systems programs, Computerized simulation, Safety engineering, Environment simulation, Fire safety, Graphs(Charts).

Presented is a User Guide for the computer code LAVENT (Link-Actuated VENTs) and an associated graphics code GRAPH. LAVENT has been developed to simulate the environment and the response of sprinkler links in compartment fires with draft curtains and fusible-link-actuated ceiling vents. The use of LAVENT is presented by a series of exercises in which the reader reviews and modifies a default input data file which describes vent and sprinkler actuation during fire growth in an array of wood pallets located in a curtained warehouse-type of configuration. Results of the default simulation are discussed. LAVENT is written in FORTRAN 77. The executable code operates on PC-compatible computers and requires a minimum of 252 kilobytes of memory.

900,095
PB89-231187 Not available NTIS
National Inst. of Standards and Technology (NEL), Gaithersburg, MD. Fire Safety Technology Div.
Test Results and Predictions for the Response of Near-Ceiling Sprinkler Links in a Full-Scale Compartment Fire.
Final rept.
L. Y. Cooper, and D. W. Stroup. 1989, 10p
See also PB88-113741. Sponsored by Fire Administration, Emmitsburg, MD.
Pub. in Proceedings of International Symposium Fire Safety Science (2nd), Tokyo, Japan, June 13-17, 1988, p623-632, 1989.

Keywords: *Sprinkler systems, *Fire tests, *Safety engineering, Fire prevention, Smoke, Heat transfer, Buildings, Fire detection systems, Comparisons, Defectors.

Data acquired during tests involving full-scale, sprinklered compartment fires are presented and analyzed. Attention is focused on key features of the typical sprinkler link deployment/response problem. It is found that the elevated-temperature smoke layer which develops inevitably in compartment fires can have a major impact on the thermal response of sprinkler links. It is shown that traditionally accepted methods of predicting sprinkler link response which do not account for this upper layer can be totally inadequate. Link response predictions used here involve a method of calculation which does take account of the smoke layer. Favorable comparisons between predictions and experiment are obtained and further validation of the method is recommended. Finally, it is found that sprinkler link-to-ceiling spacing can have a significant effect on the thermal response of links and it is recommended that a method which accounts for this effect be developed and validated.

900,096
PB89-235873 PC A04/MF A01
National Inst. of Standards and Technology (NEL), Gaithersburg, MD. Center for Fire Research.
Development of a Multiple Layer Test Procedure for Inclusion in NFPA (National Fire Protection Association) 701: Initial Experiments.
S. Davis, and K. M. Villa. Aug 89, 63p NISTIR-89/4138

Keywords: *Curtains, *Flammability testing, Fire tests, Fabrics, Textiles, Tables(Data), Graphs(Charts).

The research program investigated the flammability behavior of multiple layer fabric assemblies used for draperies and developed a laboratory-scale test protocol for predicting full-scale fire behavior. The need for such a study arose from recent findings that showed multiple layers of fabrics, comprised of individual fabrics which meet the requirements of National Fire Protection Association (NFPA) 701, may present a serious fire hazard. Eight combinations of four drapery fabrics and two lining fabrics were examined using variants of two established test procedures for single layers: the ASTM D3659 Semi-Restraint Test Method and the NFPA 701 Large-Scale Test Method. The study concludes that neither of the methods, as currently written, adequately predict the full-scale fire behavior of multiple layer fabric assemblies. Based on the results of this study, it is too early to recommend any test protocol for inclusion in NFPA 701.

900,097
PB90-117813 Not available NTIS

National Inst. of Standards and Technology (NEL), Gaithersburg, MD. Fire Science and Engineering Div.
Experimental Fire Tower Studies of Elevator Pressurization Systems for Smoke Control.
Final rept.
G. T. Tamura, and J. H. Klote. 1989, 10p
Pub. in Elevator World 37, n6 p80-89 Jun 89.

Keywords: *Elevators(Lifts), *Fire safety, *Smoke abatement, *Pressure control, Pneumatic valves, Relief valves, Ventilation, Fire tests, Air flow, Safety engineering, Buildings, Reprints.

Tests were conducted in the experimental fire tower at the National Research Council of Canada to study smoke movement caused a large fire and to determine the effectiveness of mechanical pressurization in keeping the elevator shaft and lobbies tenable for evacuation of the handicapped and fire-fighting. The tests indicated that pressure control is required to cope with loss of pressurization due to open doors. Equations were developed to assist in designing pressure control systems involving either a variable supply air rate with feedback control or relief dampers in the walls of the elevator shaft or lobbies. Tests conducted in the tower indicated that for both methods of pressure control, comparison of measured and calculated values of supply air rates and pressure differences are in good agreement.

Building Standards & Codes

900,098
PB89-149132 Not available NTIS
National Bureau of Standards (NEL), Gaithersburg, MD. Structures Div.
Earthquake Hazard Mitigation through Improved Seismic Design.
Final rept.
C. G. Culver. 1984, 6p
Sponsored by Geological Survey, Reston, VA.
Pub. in Proceedings of Conference of a Workshop on Geologic Hazards in Puerto Rico (24th), San Juan, Puerto Rico, April 4-6, 1984, p125-130.

Keywords: *Earthquake resistant structures, *Structural design, Design standards, Safety engineering, Structural engineering, Concrete construction, Steel construction, Collapse, Failure.

Buildings and other structures represent a substantial portion of a nation's wealth. For example, the total construction value of buildings and other structures in the United States was estimated at $2.3 Trillion in 1980. These facilities support a variety of activities ranging from providing basic shelter to facilities housing commercial and industrial functions. Safety and economy are two important factors that must be considered in the design and construction of buildings. Most deaths and injuries during earthquakes result from the failure of man-made structures. Building collapse, falling debris within and around buildings, and the loss of life support systems represent hazards. Immediate and long-term economic losses are a direct consequence.

900,099
PB89-149140 Not available NTIS
National Bureau of Standards (NEL), Gaithersburg, MD. Structures Div.
Earthquake Resistant Design Criteria.
Final rept.
C. G. Culver. 1984, 5p
Sponsored by Geological Survey, Reston, VA.
Pub. in Proceedings of Workshop on Earthquake Hazards in the Virgin Islands Region (25th), St. Thomas, U.S. Virgin Islands, April 9-10, 1984, p103-107.

Keywords: *Earthquake resistant structures, *Design criteria, Building codes, Specifications, Regulations, Concrete construction, Steel construction.

History shows that properly designed and constructed facilities can withstand earthquakes. The design and construction of ordinary buildings are governed by building codes, which are legal documents that specify minimum standards of construction and are adopted by government agencies. A summary of design requirements for earthquake resistant construction included in the building regulations of various countries throughout the world is available. It is important that

13

BUILDING INDUSTRY TECHNOLOGY

Building Standards & Codes

such design requirements be reviewed periodically and updated to incorporate the results of research and knowledge gained from the performance of buildings in earthquakes.

900,100
PB89-186894 Not available NTIS
National Bureau of Standards (NEL), Gaithersburg, MD. Structures Div.
Probabilistic Models for Ground Snow Accumulation.
Final rept.
B. Ellingwood. 1984, 10p
Pub. in Proceedings of Eastern Snow Conference, v41 p49-58 1984.

Keywords: *Roofs, *Design standards, *Snow, *Loads(Forces), *Meteorological data, Buildings, Structural design, Probability theory, Statistical distribution, Climatology.

Snow loads specified in modern structural design standards, e.g., American National Standard A58 and the National Building Code of Canada, are calculated as the product of a ground snow load and a ground-to-roof conversion factor. The design-basis ground snow loads are sensitive to the choice of probability distribution, since they must be obtained by extrapolating into the upper tail of the distribution beyond the range covered by the historical data. The paper considers the selection of probability distributions for modeling annual extreme ground loads, sampling errors caused by limitations in the data, and the sensitivity of nominal design-basis snow loads to these factors. Extensive water-equivalent data from weather stations in the northern United States and Canada are analyzed. The analysis strongly suggests that the log-normal probability distribution is preferable for describing the annual ground snow loads at sites in the north-central United States, while the Type I distribution appears preferable in the northeast United States and in Canada.

900,101
PB89-231062 Not available NTIS
National Inst. of Standards and Technology (NEL), Gaithersburg, MD. Center for Building Technology.
Research as the Technical Basis for Standards Used in Building Codes.
Final rept.
J. G. Gross. 1989, 6p
Pub. in Proceedings of Pacific Rim Conference of Building Officials, Honolulu, HI, April 9-13, 1989, p51-56.

Keywords: *Building codes, *Design standards, *Government policies, Regulations, Construction industry, Research, *Center for Building Technology.

Most of the technical requirements in building codes find their basis in national standards which may be adopted by reference or incorporated directly into the body of building codes. National Consensus Standards, which are most widely used and accepted, have their technology based on experience and research. The primary activity of the Center for Building Technology (CBT) at the National Institute of Standards and Technology is research. The research results are provided to standards committees which use the findings for the preparation of National Consensus Standards. Such standards are widely used by the building community for both construction specifications and regulatory requirements. CBT does not directly produce standards, nor does it have any regulatory authority. This is left to other agencies of federal, state, and local government.

Construction Management & Techniques

900,102
PB89-172399 Not available NTIS
National Bureau of Standards (NEL), Gaithersburg, MD. Mathematical Analysis Div.
Building Economics in the United States.
Final rept.
H. E. Marshall. 1987, 10p
Sponsored by Foras Forbartha Teoranta, Dublin (Ireland).
Pub. in Construction Management and Economics 5, pS43-S52 1987.

Keywords: *Investments, *Construction industry, Benefit cost analysis, Cost engineering, Management methods, Project planning, Thermal efficiency, Cost effectiveness, Return on investment, Reprints.

Building economics is described in the narrow context methods of capital investment analysis applied to building investments in the United States. The common characteristic of all the methods described is that they consider benefits (savings) and costs over the project's life cycle or study period. Eight steps involved in making an economic evaluation are presented. The process is illustrated with a problem in choosing the economically efficient thermal resistance level of attic insulation. Appropriate applications are described for the following methods: life-cycle cost; net benefits; benefit-to-cost and savings-to-investment ratios; and internal rate of return. Federal and state agencies that use these methods are identified. The role of standards societies and professional organizations in encouraging the use of these methods is described. Three projects in which these methods have been used are examined briefly in terms of methodology to illustrate the use of building economics methods. A description of difficulties in applying the methods concludes the paper.

900,103
PB89-173819 Not available NTIS
National Bureau of Standards (NEL), Gaithersburg, MD. Mathematical Analysis Div.
Survey of Selected Methods of Economic Evaluation for Building Decisions.
Final rept.
H. E. Marshall. 1987, 35p
Pub. in Proceedings of CIB International Symposium on Building Economics (4th), Copenhagen, Denmark, September 14-17, 1987, p23-57.

Keywords: *Construction industry, Technology assessment, Economic analysis, Return on investment, Decision making, Benefit cost analysis, *Building technology, Life cycle costs.

The building community needs technically correct, but practical, methods and guidelines for evaluating the economic performance of alternative building technologies. Some of the methods described in the literature and used in practice do not provide the technically correct economic measure. Improved methods help the building community achieve building performance objectives at affordable costs. The paper provides formulas for computing and guidelines on the appropriate use of life-cycle costing, net benefits, benefit-to-cost ratio, internal-rate-of-return, and payback methods. These methods are evaluated for making building decisions on accepting or rejecting a given building investment, the cost-effective design or size of a building or component, and the economically efficient combination of projects competing for a limited budget. Techniques available for measuring uncertainty and risk when applying the methods are surveyed. Some experiences of the United States in applying these methods are described.

900,104
PB89-174106 Not available NTIS
National Bureau of Standards (NEL), Gaithersburg, MD. Center for Building Technology.
Trends for Building Technology in North America.
Final rept.
R. N. Wright. 1988, 4p
Pub. in Proceedings of Canadian Building and Construction Congress (5th), Montreal, Canada, November 27-29, 1988, p303-306.

Keywords: *North America, *Construction industry, *Trends, Automation, Forecasting, Competition, Design criteria, Management methods.

Trends in building technology for North America will be dominated by advances in building process technologies: advanced computation and automation. These will facilitate effective responses to demands for increasing the international competitiveness of North American commerce and industry, supporting new industries, commerce and life styles, improving safety and health, and conserving energy and the environment. Advance of information technologies will affect organization of the building process to allow better attention in design to issues such as constructability, maintainability, and productivity of constructed facilities. Automation will advance in design, construction and operation (intelligent buildings) of constructed facilities.

900,105
PB89-191670 PC A03/MF A01
National Bureau of Standards (NEL), Gaithersburg, MD. Center for Building Technology.
Potential Applications of a Sequential Construction Analyzer.
L. W. Masters. May 87, 24p NBSIR-87/3599
Sponsored by Construction Engineering Research Lab. (Army), Champaign, IL.

Keywords: *Sequential analysis, *Construction industry, *Quality assurance, *Construction management, Management methods, Computer systems programs, Maintenance management, Tables(Data).

The need exists in construction applications for improved methods by which (1)quality can be assured throughout the construction process, (2)the degree of construction progress can be assessed and documented and (3)the performance of systems and materials can be assessed over time to aid in maintenance decision-making. Although these aspects of construction processes have traditionally been addressed empirically, recent advances in computer technology have provided new opportunities for improving upon the traditional methods. The Construction Engineering Research Laboratory (CERL) of the U.S. Army's Corps of Engineers, for example, is exploring the use of a sequential construction analyzer to aid in quality assurance, tracking construction progress, and obtaining data for maintenance decision-making. The study was carried out to identify potential applications of the sequential construction analyzer in three areas of construction, buildings, construction sites and paving.

900,106
PB89-191985 PC A05/MF A01
National Bureau of Standards (NEL), Gaithersburg, MD. Center for Building Technology.
Use of Artificial Intelligence Programming Techniques for Communication between Incompatible Building Information Systems.
W. F. Danner. Apr 87, 100p NBSIR-87/3529

Keywords: *Construction industry, *Information systems, *Communicating, *Computer programming, Compatibility, Artificial intelligence, Design, Construction, Data base management systems, Data bases, Protocols, Knowledge representation, *Data transfer protocols.

A communication capability between incompatible information systems is presented. The research develops an interface based on a format for the exchange of knowledge needed by each system to understand the other and a format for the exchange of information in the context of that knowledge. Particular emphasis has been placed on developing protocols supporting the transfer of analytical data. These data are seen as comprising not only facts but also the semantics associated with those facts. Two artificial intelligence programming techniques have been employed: frame-based knowledge representation and object-oriented programming capabilities as an integral part of the frame-based representation. These techniques make self-descriptive formats possible that provide for a virtual extension of an information management system. Such an extension provides access to information without requiring a detailed understanding of specific system operations.

900,107
PB90-112376 PC A03/MF A01
National Inst. of Standards and Technology (NEL), Gaithersburg, MD. Center for Building Technology.
Report of Roof Inspection: Characterization of Newly-Fabricated Adhesive-Bonded Seams at an Army Facility.
W. J. Rossiter, J. F. Seiler, and P. E. Stutzman. Oct 89, 31p NISTIR-89/4155
See also PB89-131916. Sponsored by Corps of Engineers, Baltimore, MD. Baltimore District.

Keywords: *Roofs, *Adhesive bonding, Strength, Seams(Joints), Elastomers, Membranes, Tests, Evaluation, Military facilities, Inspection, Maryland, Aberdeen Proving Ground.

The investigation was a limited study of seams in an EPDM rubber membrane of the roof of the new 'Wheeled Vehicle Facility' located at Aberdeen Proving Ground, Maryland. The study was initiated at the request of the Corps of Engineers (CoE) to provide data that could contribute to a data base on the characterization of newly-prepared field seams. The investigation was beneficial to the National Institute of Standards and Technology (NIST) because it complemented laboratory research on test methods for evaluating seams of vulcanized rubber roof membranes.

Construction Materials, Components, & Equipment

900,108
PB89-148126   **PC A03/MF A01**
National Bureau of Standards (NEL), Gaithersburg, MD. Center for Fire Research.
Calculating Flows through Vertical Vents in Zone Fire Models under Conditions of Arbitrary Cross-Vent Pressure Difference.
L. Y. Cooper. May 88, 17p NBSIR-88/3732

Keywords: *Mathematical models, *Air flow, Algorithms, Pressure gradients, Pressure vessels, Fires, Buildings, Fire safety, *Fire studies.

In typical compartment fire scenarios, ratios of cross-vent absolute pressures are close to 1. When such is the case, algorithms are available to predict the resulting cross-vent room-to-room flows. There are, however, important situations where this pressure condition does not prevail, for example, in fire scenarios involving relatively small penetrations in otherwise hermetically-sealed compartments of fire origin. It is important for a versatile compartment fire model to have a capability of predicting vent flows for the entire range of possible cross-vent pressure conditions. The paper develops a unified analytic description for flows through vertical vents between pairs of two-layer room fire environments under conditions of arbitrary cross-vent pressure difference. The analysis, which takes advantage of generally useful modeling approximations, leads to a concise result which is not significantly more complicated than the result for simple, low-pressure-difference cases.

900,109
PB89-148514   **PC A08/MF A01**
National Inst. of Standards and Technology (NEL), Gaithersburg, MD. Building Environment Div.
Preliminary Performance Criteria for Building Materials, Equipment and Systems Used in Detention and Correctional Facilities.
R. D. Dikkers, R. J. Husmann, J. H. Webster, J. P. Sorg, and R. A. Holmes. Jan 89, 157p NISTIR-89/4027
Prepared in cooperation with Omni Signal, Inc., Capitola, CA., Webster (James H.), Arlington, VA., Sorg (John P.), Annandale, VA., and Sure-Lock Homes, Inc., Albany, NY.

Keywords: *Construction materials, *Materials specifications, Performance standards, Criteria, Security, Internal security, Design standards, Facilities management, Planning, Site surveys, Warning systems, Communications equipment, *Correctional institutions.

In a National Institute of Corrections (NIC) sponsored study, many important criteria and standards which need to be developed for improving the selection of materials, equipment and systems for use in detention and correctional facilities were identified. The preliminary performance criteria for materials, equipment, and systems contained in the report have the following objectives: establish performance levels which are consistent with the security and custody levels used in detention and correctional facilities; and establish standard performance measures with regard to security, safety and durability. Part I contains general criteria pertaining to the overall facility. Part II contains requirements and criteria relating to the perimeter security of the facility. Part III includes requirements and criteria pertaining to structural systems, doors, windows, glazing, locks, control center, alarms and communication systems.

900,110
PB89-150734   **Not available NTIS**
National Bureau of Standards (NEL), Gaithersburg, MD. Building Materials Div.

Knowledge Based System for Durable Reinforced Concrete.
Final rept.
J. R. Clifton. 1986, 6p
Pub. in Proceedings of Corrosion/86 Symposium on Computers in Corrosion Control, Houston, TX., p110-115 1986.

Keywords: *Reinforced concrete, *Concrete durability, Durability, Corrosion, Reinforcing steel, Cement aggregate reactions, Sulfate resisting cements, Recommendations, Industrial engineering, Systems engineering, *Expert systems.

DURCON is a prototype expert system being developed to give recommendations on the selection of constituents for durable concrete. Four major concrete deterioration problems are being covered by DURCON: corrosion of reinforcing steel, freeze-thaw, sulfate attack, and cement-aggregate reactions. In this report, the portion of DURCON dealing with corrosion of reinforcing steel is discussed. The factual knowledge based for DURCON is based on the American Concrete Institute Guide to Durable Concrete. Heuristic knowledge is being obtained from experts on the durability of concrete. The approach taken in developing DURCON is discussed. Then a model expert system for the corrosion of reinforcing steel is described.

900,111
PB89-157275   **Not available NTIS**
National Bureau of Standards (NEL), Gaithersburg, MD. Fire Science and Engineering Div.
Analytical Methods for Firesafety Design.
Final rept.
J. G. Quintiere. 1986, 20p
See also PB88-153333.
Pub. in Fire Technology 24, n4 p333-352 Nov-88.

Keywords: *Fire safety, *Building codes, Fire prevention, Fires, Safety engineering, Mathematical models, Design criteria, Building, Standards, Reprints.

The ability to predict aspects of fire and its impact on a building's structure, contents, and people is discussed in terms of its application to safety design. It is presented from the perspective of how research has addressed the prediction of fire phenomena. A review of the state of the art on the capability for predicting the fire, its impact and response, is given. Examples are cited to illustrate the scope and accuracy of predictive methods and how they are being incorporated into some codes and standards.

900,112
PB89-158000   **Not available NTIS**
National Bureau of Standards (NEL), Gaithersburg, MD. Building Materials Div.
Prediction of Service Life of Building Materials and Components.
Final rept.
L. W. Masters. 1986, 6p
Pub. in Mater. Struct. 19, n114 p417-422 Nov/Dec 86.

Keywords: *Construction materials, *Service life, Predictions, Estimates, Maintenance, Selection, Durability, Research projects, Degradation, Reprints.

Data on service life are essential to the effective selection, use and maintenance of materials and components used in buildings. Changing technologies and requirements in recent years have re-emphasized the need to advance the state of knowledge of service life prediction, resulting in a number of new internationally sponsored activities including a joint Technical Committee of CIB and RILEM. In the paper, recent activities of the CIB/RILEM committee are described by summarizing the technical barriers to service life prediction and outlining courses of direction and types of research which can help overcome the barriers.

900,113
PB89-162580   **PC A06/MF A01**
National Inst. of Standards and Technology (NEL), Gaithersburg, MD. Center for Building Technology.
Corrosion of Metallic Fasteners in Low-Sloped Roofs: A Review of Available Information and Identification of Research Needs.
W. J. Rossiter, M. A. Streicher, and W. E. Roberts. Feb 89, 105p NISTIR-88/4008
Prepared in cooperation with Webster Farm, Wilmington, DE. Sponsored by Department of Energy, Washington, DC.

Keywords: *Fasteners, *Corrosion, *Roofs, Joining, Durability, Deterioration, Roofing, Research projects.

BUILDING INDUSTRY TECHNOLOGY
Construction Management & Techniques

The paper presents the results of a study conducted to summarize available information on the corrosion issue, and to identify research needed to correct problems. In particular, the incidence of loss of fastener securement due to corrosion could not be established because of the inaccessibility of installed fasteners within roofs. In reviewing factors affecting fastener corrosion, water was the only one that stood out on the basis of the information obtained. Uniform corrosion (rust on some or all of the surface) was the predominant type that inspectors have observed in service. Nevertheless, some evidence of localized corrosion processes (e.g., crevice corrosion) has also been observed. Both types of corrosion may lead to loss of fastener securement in service. The results of the study indicated that there are three major gaps in the knowledge base: (1) evaluation test procedures for the corrosion resistance of fasteners are limited and need to be improved; (2) a data base on field performance of fasteners is lacking, and; (3) non-destructive diagnostic procedures for assessing the condition of inplace fasteners are not available.

900,114
PB89-168025   **PC A04/MF A01**
National Inst. of Standards and Technology (NEL), Gaithersburg, MD. Building Materials Div.
Interim Criteria for Polymer-Modified Bituminous Roofing Membrane Materials.
Final rept.
W. J. Rossiter, and J. F. Seiler. Feb 89, 52p NIST/BSS-167
Also available from Supt. of Docs. as SN003-003-02922-0. Library of Congress catalog card no. 89-600702. Sponsored by Tri-Service Building Materials Committee, Washington, DC.

Keywords: *Roofing, *Bituminous coatings, *Polymers, Bitumens, Stability, Fire resistance, Viscosity, Mechanical properties, Absorption, Moisture content, Performance standards, Military facilities.

The report presents the results of a study to develop interim criteria for the selection of polymer-modified bituminous roofing membrane materials. The criteria are based on a review of existing standard specifications and related documents. They are intended for use by the construction agencies of the Department of Defense in specifying polymer-modified bituminous roofing membrane materials until voluntary consensus standards are developed in the United States. The suggested interim criteria are generally presented using a performance criteria format. The membrane characteristics for which performance criteria are suggested are: dimensional stability, fire, flow resistance, hail impact, moisture content and absorption, pliability, strain energy, uplift resistance, and weathering resistance (heat exposure). Prescriptive criteria for five membrane characteristics are used to complement the suggested performance criteria. The approach of using complementary prescriptive criteria is taken to incorporate in the performance criteria test methods which can be relatively rapidly performed for characterization or identification of the membrane material.

900,115
PB89-171748   **Not available NTIS**
National Bureau of Standards (NEL), Gaithersburg, MD. Structures Div.
Damage Accumulation in Wood Structural Members Under Stochastic Live Loads.
Final rept.
J. Murphy, B. Ellingwood, and E. Hendrickson. 1987, 11p
Pub. in Wood and Fiber Science 19, n4 p453-463 1987.

Keywords: *Structural forms, *Construction materials, *Loads(Forces), *Wood products, *Damage, *Creep rupture tests, Stochastic processes, Stress concentration, Specifications, Reprints.

Damage accumulation in wood structural members is assessed using realistic stochastic models of live load. It is found that practically all damage occurs at times when the live load intensity is equal, or nearly equal to the nominal live load, L sub n, required by codes for design. The time spent at or above the nominal life load, L sub n, is about 40 days during a reference period of 50 years, and not the presently assumed time of 10 years. Using 10 years as the basis for setting allowable stresses for wood should be reexamined.

Construction Materials, Components, & Equipment

900,116
PB89-172514   Not available NTIS
National Bureau of Standards (NEL), Boulder, CO.
Chemical Engineering Science Div.
Specific Heat of Insulations.
Final rept.
J. G. Hust, J. E. Callanan, and S. A. Sullivan. 1985,
18p
Sponsored by Oak Ridge National Lab., TN.
Pub. in Thermal Conductivity 19, p533-550 1985.

Keywords: *Thermal insulation, *Specific heat, Thermophysical properties, Latent heat, Heat of vaporization, Moisture content, Temperature, Thermal conductivity, Measurement, Thermal diffusivity, Reprints, *Standard Reference Materials.

The specific heats of insulation Standard Reference Materials(SRMs) have been measured for temperatures from 250 to 400 K for both dry and moist specimens. Measurements were performed on SRM 1450b, SRM 1451; a high temperature SRM candidate, and on specimens of the phenolic binder used in SRM 1451. The measured specific heats of moist specimens are significantly larger than the specific heats of dry specimens due to the effect of latent heat of vaporization and desorption of the moisture. These results are analyzed in terms of the differences that may be observed between steady-state and transient thermal conductance measurements.

900,117
PB89-172548   Not available NTIS
National Bureau of Standards (NEL), Boulder, CO.
Chemical Engineering Science Div.
Insulation Standard Reference Materials of Thermal Resistance.
Final rept.
J. G. Hust. 1985, 9p
Pub. in Thermal Conductivity 19, p261-269 1985.

Keywords: *Thermal insulation, *Thermal resistance, Heat transmission, Thermophysical properties, Thermal conductivity, Fiberboards, Low temperature tests, Measurement, Thermal diffusivity, Reprints, *Standard Reference Materials.

The National Bureau of Standards recently established two insulation Standard Reference Materials(SRMs) of thermal resistance. These SRMs are available from the Office of Standard Reference Materials. The paper provides a brief description of these SRMs and provides comparisons to similar materials for temperatures from 100 to 330 K. A brief description of present and planned research for the establishment of further SRMs is also provided.

900,118
PB89-173454   Not available NTIS
National Bureau of Standards (NEL), Gaithersburg, MD. Electrosystems Div.
Coupling, Propagation, and Side Effects of Surges in an Industrial Building Wiring System.
Final rept.
F. D. Martzloff. 1988, 10p
Pub. in Conference Record of the IEEE (Institute of Electrical and Electronics Engineers)-IAS (Industry Applications Society) Annual Meeting, Pittsburgh, PA., October 3-6, 1988, p1467-1476.

Keywords: *Surges, *Industrial buildings, *Power lines, Measurement, Electric wire, Electromagnetic interference, Electromagnetic radiation, Coupling circuits, Wave propagation.

Measurements were made in an industrial building to determine the propagation characteristics of surges in the AC power wiring of the facility. The surges, of the unidirectional type or the ring-wave type described in ANSI/IEEE C62.41-1980, were injected at one point of the system and the resulting surges arriving at other points were measured. The results show how unidirectional surges couple through transformers and produce a ring wave component in the response of the system. An unexpected side effect of these surges, applied to the power lines only, was apparent damage suffered by the data line input components of some computer-driven printers.

900,119
PB89-174916   Not available NTIS
National Bureau of Standards (NEL), Gaithersburg, MD. Building Environment Div.

Thermal Resistance Measurements and Calculations of an Insulated Concrete Block Wall.
Final rept.
D. M. Burch, B. A. Licitra, D. F. Ebberts, and R. R. Zarr. 1989, 7p
Sponsored by Department of Energy, Washington, DC.
Pub. in ASHRAE (American Society of Heating, Refrigerating and Air-Conditioning Engineers) Transactions 95, pt1 7p 1989.

Keywords: *Concrete blocks, *Thermal resistance, *Thermal insulation, *Thermal measurements, *Walls, Temperature gradients, Polystyrene, Temperature, Reprints.

Thermal resistance measurements of an insulated concrete block wall were conducted using a calibrated hot box at four different mean temperatures. The hollow concrete block wall was insulated by installing partial-size inserts composed of expanded polystyrene insulation with reflective air spaces into the cores of the blocks. The thermal resistance measurements were compared with the ASHRAE isothermal plane and parallel path methods. The isothermal plane method was subsequently used to calculate the thermal resistance of uninsulated concrete block, concrete block with full-size insulation inserts, and concrete block with partial-size insulation inserts. Both ways of insulating the concrete block increased the thermal resistance of the uninsulated block by more than a factor of three.

900,120
PB89-175848   Not available NTIS
National Bureau of Standards (NEL), Gaithersburg, MD. Building Materials Div.
Prediction of Service Life of Construction and Other Materials.
Final rept.
G. Frohnsdorff, and L. W. Masters. 1986, 4p
Pub. in Communications on the Materials Science and Engineering Study, p61-64 1986.

Keywords: *Construction materials, *Service life, Durability, Design criteria, Reliability, Forecasting, Quality control, Maintenance, Standards, Cost engineering, Reprints.

A major need in construction technology is to be able to predict the service lives of materials. The need is shared with other technologies, and prediction of the service life of construction materials can use techniques of reliability engineering used in other industries. In the last decade, several important steps have been taken toward development and standardization of a methodology for service life prediction of construction materials and components. These activities have influenced the way service lives of construction materials are evaluated, but more needs to be done. It is recommended that the materials science and engineering community encourage the development of improved methodologies for service life prediction and provide guidance on how the methodologies should be applied.

900,121
PB89-175889   Not available NTIS
National Bureau of Standards (NEL), Gaithersburg, MD. Office of Energy-Related Inventions.
U-Value Measurements for Windows and Movable Insulations from Hot Box Tests in Two Commercial Laboratories.
Final rept.
M. E. McCabe, W. Ducas, R. W. Cholvibul, and P. Wormser. 1986, 21p
Sponsored by Department of Energy, Washington, DC.
Office of Solar Heat Technologies.
Pub. in ASHRAE (American Society of Heating, Refrigerating and Air-Conditioning Engineers) Transactions 92, pt1A p453-473 1986.

Keywords: *Windows, *Thermal insulation, Solar energy, Design, Architecture, Tests, Wind velocity, Performance evaluation, Simulation, Comparisons, Reprints, *Energy conservation.

Different laboratory test procedures (ASTM 236, ASTM C976 and AAMA 1503.1) are discussed in the context of measuring the U-Value of direct gain fenestration (DGF) components. Four representative DGF components, including a multiple-glazed window and a single-glazed window with movable insulation systems, were purchased and prepared for testing. U-Values were measured for each test article at two commercial testing laboratories, using the ASTM 236 and AAMA 1502.6 test methods for a range of simulat-

ed outdoor conditions. For the three movable insulation test articles, manufacturer's claims for energy savings were also overstated, suggesting the need for a standard method for estimating energy performance. Test results between individual laboratories are compared and significant differences are noted where wind speeds of 6.7 m/s (15 mph) were simulated. The differences are attributed to the different directions (parallel and perpendicular) used by each laboratory for simulating wind. The suitability of a single test condition for measuring U-Value is discussed in relation to the use of that test data for estimating seasonal thermal performance.

900,122
PB89-176168   Not available NTIS
National Bureau of Standards (NEL), Gaithersburg, MD. Fire Safety Technology Div.
Fundamentals of Enclosure Fire 'Zone' Models.
Final rept.
J. G. Quintiere. 1988, 47p
Pub. in Proceedings of National Fire Protection Association Annual Meeting for Society of Fire Protection Engineers, Cincinnati, OH., May 16, 1988, 47p.

Keywords: *Fires, *Enclosures, Fire protection, Safety engineering, Model tests, Open channel flow, Fire control.

The conservation laws are presented in control volume form and applied to the behavior of fire in enclosures. The behavior of enclosure fires are discussed and the assumptions for justifying the use of the control volume or 'zone' modeling approach are presented. The governing equations are derived and special solutions are given. Flow through wall vents, room filling, and growing fires are analyzed.

900,123
PB89-176309   Not available NTIS
National Bureau of Standards (NEL), Gaithersburg, MD. Building Materials Div.
Thermographic Imaging and Computer Image Processing of Defects in Building Materials.
Final rept.
J. W. Martin, M. E. McKnight, and D. P. Bentz. 1986, 4p
Pub. in Proceedings of SPIE (Society of Photo-Optical Instrumentation Engineers) International Conference on Thermal Infrared Sensing for Diagnostics and Control, Cambridge, MA., September 17-20, 1985, p152-155 1986.

Keywords: *Construction materials, *Defects, *Thermography, Degradation, Nondestructive tests, *Image processing, Image enhancement, Image analysis.

An image processing system has been coupled to either a thermographic or a video camera for quantifying defects in images of building materials. Several applications to building materials are presented including the detection of delaminations in single-ply roofing membrane seams, the characterization of the extent of corrosion under pigmented organic coatings on metallic substrates, the determination of the fractal dimensions of a sandblasted metallic substrate, and the determination of the percent porosity in hydrated cement. It is concluded that infrared thermography and image processing are useful analysis tools in detecting and quantifying defects in building materials.

900,124
PB89-180004   Not available NTIS
National Bureau of Standards (NEL), Gaithersburg, MD. Fire Science and Engineering Div.
Recent Activities of the American Society for Testing and Materials Committee on Fire Standards.
Final rept.
D. Gross. 1985, 2p
Pub. in Fire and Materials 9, n2 p109-110 1985.

Keywords: *Fires, *Fire tests, Fire protection, Fire prevention, Fire safety, Standards, Flammability tests, Reprints.

A brief summary of selected actions and activities at the recent meeting of ASTM Committee E5 on Fire Standards is presented.

900,125
PB89-184527   PC A08/MF A01
California Univ., Los Angeles. Dept. of Mechanical, Aerospace and Nuclear Engineering.

Fire Risk Analysis Methodology: Initiating Events.
Final rept.
M. D. Brandyberry, and G. E. Apostolakis. Mar 89,
165p NIST/GCR-89/562
Grant NANB-6-D0649
Sponsored by National Inst. of Standards and Technology (NEL), Gaithersburg, MD. Center for Fire Research.

Keywords: *Fires, *Ignition, *Space heaters, *Upholstery, Heat transfer, Drawings, Tables(Data), Models, Probability distribution functions, *Risk assessment.

The report outlines a method for assessing the frequency of ignition of a consumer product in a building and shows how the method would be used in an example scenario utilizing upholstered furniture as the product and radiant auxiliary heating devices (electric heaters, wood stoves) as the ignition source. Deterministic thermal models of the heat transport processes are coupled with parameter uncertainty analysis of the models and with a probabilistic analysis of the events involved in a typical scenario. This leads to a distribution for the frequency of ignition for the product.

900,126
PB69-188635 PC A04/MF A01
National Inst. of Standards and Technology (NEL), Gaithersburg, MD. Center for Fire Research.
Fire Properties Database for Textile Wall Coverings.
M. F. Harkleroad. Apr 89, 54p NISTIR-89/4065
Sponsored by American Textile Manufacturers Inst., Washington, DC.

Keywords: *Data bases, *Fire safety, *Textiles, *Walls, Flammability, Materials, Physical properties, Ignition, Flame propagation, Polyester fibers, Nylon fibers, Rayon, Polypropylene fibers, Charring, Wall coverings.

A technical basis for linking small scale fire property test data to realistic performance has been initiated by the establishment of a small scale fire property database for some textile wall covering materials. The properties are obtained from experimental small-scale tests of materials in a vertical orientation. They include ignition and flame spread properties based on measurements from the Lateral Ignition and Flame spread Test (LIFT) apparatus and energy release rate measurements from the Cone Calorimeter. The database includes fire properties for woven, knit and needle punched polyesters, woven cotton/rayon-and wool/ nylon blends, nylon and polypropylene wall covering materials.

900,127
PB69-189167 PC A09/MF A01
Michigan State Univ., East Lansing. Dept. of Mechanical Engineering.
Effect of Water on Piloted Ignition of Cellulosic Materials.
Doctoral thesis.
M. Abu-Zaid, and A. Atreya. Feb 89, 187p NIST/ GCR-89/561
Grant NANB-5DO578
See also PB87-127732. Sponsored by National Inst. of Standards and Technology (NEL), Gaithersburg, MD: Center for Fire Research.

Keywords: *Water, *Ignition, *Fire protection, Pyrolysis, Fire extinguishing agents, Wood, Flame propagation, Drops(Liquids), Graphs(Charts), Flammability testing, Heat transfer, Porous materials, *Cellulosic materials.

The experimental study is an attempt to quantify the effect of water on extinguishment; thermal decomposition and piloted ignition of wood. In the extinguishment part, cooling of hot porous and non-porous ceramic solids by water droplets was studied. These solids were used to simulate low thermal diffusivity porous and non-porous combustible building materials and were instrumented by several surface and in-depth thermocouples. Temperature measurements in the solid were used to quantify the heat transfer during droplet evaporation. Thermal decomposition of wood in air was also studied as a function of sample moisture content and externally applied radiation prior to the ignition experiments. Simultaneous measurements of weight loss rate; surface, bottom and in-depth temperatures; O2 depletion and production of CO2, CO, total hydrocarbons and water were made. It was found that the presence of moisture delayed the decomposition process and diluted the decomposition products. Piloted ignition experiments were conducted on Doug-

las fir for four different moisture contents and at different levels of externally applied radiation. It was found that the presence of moisture increases the ignition time, surface temperature and the evolved mass flux at ignition. A single equation was derived to correlate all the ignition data. This correlation accounts for the moisture dependent thermal properties and the heat loss from the sample surface.

900,128
PB69-189252 PC A03/MF A01
National Inst. of Standards and Technology (NEL), Gaithersburg, MD. Center for Fire Research.
Calculation of the Flow Through a Horizontal Ceiling/Floor Vent.
L. Y. Cooper. Mar 89, 31p NISTIR-89/4052
Sponsored by Naval Research Lab., Washington, DC.

Keywords: *Air flow, *Vents, *Fire safety, *Buildings, Algorithms, Mathematical models, Pressure gradients, Fires, Temperature.

Calculation of the flow through a horizontal vent located in a ceiling or floor of a multi-room compartment is considered. It is assumed that the environments of the two, vent-connected spaces near the elevation of the vent are of arbitrary relative buoyancy and cross-vent pressure difference, delta p. An anomaly of the standard vent flow model, which uses delta p to predict stable uni-directional flow according to Bernoulli's equation is discussed. The problem occurs in practical vent configurations of unstable hydrostatic equilibrium, where, for example, one gas overlays a relatively lessdense gas, and where delta p is relatively small. In such configurations the cross-vent flow is not uni-directional. Also, it is not zero at delta p = 0. Previously published experimental data on a variety of related flow configurations are used to develop a completely general flow model which does not suffer from the standard model anomaly. The model developed leads to a uniformly valid algorithm, called VENTCL, for horizontal vent flow calculations suitable for general use in zone-type compartment fire models.

900,129
PB69-189260 PC A03/MF A01
National Inst. of Standards and Technology (NEL), Gaithersburg, MD. Center for Fire Research.
Fire Induced Flows in Corridors: A Review of Efforts to Model Key Features.
K. D. Steckler. Feb 89, 26p NISTIR-89/4050
Sponsored by General Services Administration, Washington, DC.

Keywords: *Buildings, *Fire tests, Gas flow, Directional measurement, Fire safety, Mathematical models, *Corridors.

A literature review was undertaken to identify engineering formulas or models which can be used to predict key features of the corridor-filling process. The results of that review are presented and assessed. The filling process is viewed as a series of three events: a forward gravity current moving away from the fire source, a reflected or return gravity current moving toward the source, followed by uniform filling of the entire corridor. Recommendations for estimating the filling during each of these stages are presented.

900,130
PB69-189328 PC A04/MF A01
National Inst. of Standards and Technology (NEL), Gaithersburg, MD. Center for Building Technology.
Friability of Spray-Applied Fireproofing and Thermal Insulations: Field Evaluation of Prototype Test Devices.
W. J. Rossiter, W. E. Roberts, and R. G. Mathey.
Mar 89, 64p NISTIR-88/4012
See also PB69-131924. Sponsored by General Services Administration, Washington, DC.

Keywords: *Thermal insulation, *Fire resistant materials, *Friability, Mechanical tests, Brittleness, Toughness, Shear, Tests, Abrasion, Compressive strength.

The report describes results of the third and final phase of a study conducted for the General Services Administration (GSA) to develop a field test method to measure the friability of spray-applied fireproofing and thermal insulation materials. Field tests were conducted on 17 fibrous and 2 cementitious spray-applied materials to assess surface and bulk compression/shear, indentation, abrasion, and impact properties. The tests were performed using prototype devices developed in an earlier phase of the study. As expected, the field specimens displayed varying response to dislodgment

or indentation in the tests. The field tests confirmed that the goal of the study had been achieved.

900,131
PB69-193213 PC A05/MF A01
National Inst. of Standards and Technology (NEL), Gaithersburg, MD. Center for Building Technology.
Building Technology Project Summaries 1989.
N. J. Raufaste. Apr 89, 86p NISTIR-89/4068
See also PB88-215512.

Keywords: *Construction industry, *Construction materials, *Project management, Research management, Cost engineering, Contract administration, Performance analysis, *Building technology.

The Center for Building Technology (CBT) of the National Institute of Standards and Technology (NIST) is the national building research laboratory. CBT works cooperatively with other organizations, private and public, to improve building practices. It conducts laboratory, field, and analytical research to predict, measure, and test the performance of building materials, components, systems, and practices. CBT's technologies are widely used in the building industry and adopted by governmental and private organizations that have standards and codes responsibilities. The report summarizes the research underway in the Center during 1989.

900,132
PB69-193304 PC A03/MF A01
National Inst. of Standards and Technology (NEL), Gaithersburg, MD. Center for Fire Research.
Fire Research Publications, 1988.
N. H. Jason. May 89, 42p NISTIR-89/4081
See also PB86-199641.

Keywords: *Fire safety, *Bibliographies, Safety engineering, Burning rate, Smoke, Ventilation, Sprinkler systems, Flame propagation, Fire resistant materials, Heat measurement.

The document is a supplement to previous editions. Only publications prepared by members of the Center for Fire Research (CFR), by other National Institute of Standards and Technology (NIST) (formerly National Bureau of Standards (NBS)) personnel for CFR, or by external laboratories under contract or grant from the CFR are cited.

900,133
PB69-195671 PC A05/MF A01
National Inst. of Standards and Technology (NEL), Gaithersburg, MD. Center for Fire Research.
Considerations of Stack Effect in Building Fires.
J. H. Klote. May 89, 83p NISTIR-89/4035
Sponsored by Fire Administration, Emmitsburg, MD.

Keywords: *Fire safety, *Fires, *Buildings, Smoke, Air flow, Convection, Mathematical models, Elevators, Combustion, Vents.

The following driving forces of smoke movement in buildings are discussed: stack effect, buoyancy of combustion gases, expansion of combustion gases, wind effect, and elevator piston effect. Based on an analysis of elevator piston effect, it is concluded that the likelihood of smoke being pulled into an elevator shaft due to elevator car motion is greater for single car shafts than for multiple car shafts. Methods of evaluating the location of the neutral plane are presented. It is shown that the neutral plane between a vented shaft and the outside is located between the neutral plane height for an unvented shaft and the vent elevation.

900,134
PB69-200091 PC A04/MF A01
National Inst. of Standards and Technology (NEL), Gaithersburg, MD. Center for Fire Research.
Executive Summary for the Workshop on Developing a Predictive Capability for CO Formation in Fires.
W. M. Pitts. Apr 89, 72p NISTIR-89/4093

Keywords: *Fires, *Carbon monoxide, *Meetings, Combustion, Gases, Fire hazards, Toxicity.

The proceedings and recommendations of a workshop entitled 'Workshop on Developing a Predictive Capability for CO Formation in Fires' are summarized. The meeting took place on December 3-4, 1988 in Clearwater, Florida. Several brief technical presentations are critiqued. Short summaries for each talk are included in an appendix. Findings of two working groups constituted to address the fundamental and engineering aspects of the workshop topic are discussed. The most important areas of research required to fulfill the workshop topic are provided as final recommendations. Six specific areas are listed. Many workshop details are included in appendices.

900,135
PB89-201149   Not available NTIS
National Bureau of Standards (NEL), Gaithersburg, MD. Statistical Engineering Div.
Statistical Analysis of Experiments to Measure Ignition of Cigarettes.
Final rept.
K. R. Eberhardt. 1986, 10p
Pub. in Jnl. of the Washington Academy of Sciences 78, n4 p323-332 Dec 88.

Keywords: *Ignition, *Statistical analysis, *Flammability testing, Combustion, Experimental data, Chi square test, Probability theory, Reprints, *Cigarettes.

Under the Cigarette Safety Act of 1984, NIST was given the task of studying several types of commercial and experimental cigarettes to determine their relative propensities to ignite soft furnishings. The analysis of the data came under close scrutiny by the Technical Study Group appointed to oversee the research. In one experiment where the usual chi-squared test could not be readily justified, an extension of Fisher's Exact Test to 2 x 12 contingency tables was adopted. In another experiment, a modification of the angular transformation for count data was used along with normal probability plots of the effects to analyze a 2(sup 5) factorial experiment.

900,136
PB89-209316   PC A03/MF A01
National Inst. of Standards and Technology (NEL), Gaithersburg, MD. Center for Building Technology.
Results of a Survey of the Performance of EPDM (Ethylene Propylene Diene Terpolymer) Roofing at Army Facilities.
W. J. Rossiter, and J. F. Seiler, Jun 89, 28p NISTIR-89/4085
Sponsored by Construction Engineering Research Lab. (Army), Champaign, IL.

Keywords: *Roofs, *Ethylene copolymers, *Performance evaluation, Maintenance, Military facilities, Mechanical properties, Weathering, Seams(Joints), Surveys.

The report presents a summary of a survey to obtain information on the performance of EPDM roofing at Army facilities. Emphasis in the survey was on the performance of seams fabricated with unaged rubber and also patches made on existing, aged rubber. The results are intended to help provide guidelines for the maintenance of EPDM roofs at Army facilities, as well as to define research needs to overcome problems identified. Based on the results of the survey, it is recommended that studies be carried out to provide the technical basis for preparing the surfaces of aged rubber membranes before making seams or patches. The effect of aging on the surface characteristics of EPDM rubber has received little attention in the roofing literature.

900,137
PB89-212005   Not available NTIS
National Bureau of Standards (NEL), Gaithersburg, MD. Fire Science and Engineering Div.
Refinement and Experimental Verification of a Model for Fire Growth and Smoke Transport.
Final rept.
W. W. Jones, and R. D. Peacock. 1989, 10p
Pub. in Proceedings of International Symposium on Fire Safety Science (2nd), Tokyo, Japan, June 13-17, 1988, p897-906 1989.

Keywords: *Fires, Ignition, Fire safety, Fire tests, Experimental data, Algorithms, Numerical analysis, Verifying, *Fire models.

There is considerable interest in modeling the growth of fires and the spread of toxic gases in multicompartment structures. Much of the attention is focused on the development of numerical models which are fast and robust, but able to make reasonably accurate predictions from the onset of ignition. The authors have constructed such a model (FAST) and performed a series of validation experiments to test it. The paper discusses some of the improvements which have been made to physical algorithms and the underlying numer-

ical basis of the model, a description of some of the experiments used to verify the refined model, and of some additions which the authors intend to incorporate.

900,138
PB89-212120   Not available NTIS
National Bureau of Standards (NEL), Gaithersburg, MD. Building Materials Div.
Tests of Adhesive-Bonded Seams of Single-Ply Rubber Membranes.
Final rept.
W. J. Rossiter. 1987, 10p
Pub. in Proceedings of ASTM (American Society for Testing and Materials) Symposium on Roofing Research and Standards Developm t, New Orleans, LA., December 3, 1986, p53-62 1987.

Keywords: *Roofing, *Ethylene copolymers, *Membranes, *Mechanical tests, Elongation, Loads(Forces), High temperature tests, Shear properties, Stress relaxation, Seams(Joints), Failure.

Commercially-available ethylene propylene diene terpolymer (EPDM), neoprene, and chlorosulphonated polyethylene (CSME) roofing membrane specimens with adhesive-bonded seams were tested in tension in a lap-shear configuration. Some T-peel tests of an EPDM material were also conducted. The lap-shear tests were conducted at temperatures ranging from -20 to 75 C (-4 to 167 F), and at rates of loading from 0.05 to 50 cm/min (0/02 to 20 in/min). In most cases, the EPDM and neoprene specimens failed by seam delamination; otherwise, they failed by membrane rupture with partial delamination of the seam. The CSME specimens always failed by membrane rupture without seam delamination. In general, the results indicated that, as the temperature of the test increased, the ultimate load and elongation at failure decreased. Also, at a given temperature, the ultimate load and elongation general decreased, for most specimens, as the rate of loading decreased. For stress relaxation experiments, specimens strained to 15% remained intact for over 15 months, whereas other specimens strained to 30% failed in about five weeks. The results of the tests are discussed with regard to the development of tests for seams in single-ply membranes.

900,139
PB89-212203   Not available NTIS
National Bureau of Standards (NEL), Gaithersburg, MD. Building Materials Div.
Strain Energy of Bituminous Built-Up Membranes: A New Concept in Load-Elongation Testing.
Final rept.
W. J. Rossiter, and D. P. Bentz. 1987, 10p
See also PB87-136376.
Pub. in Proceedings of Conference on Roofing Technology (8th) 'Applied Technology for Improving Roof Performance', Gaithersburg, MD., April 16-17, 1987, p40-49.

Keywords: *Roofing, *Bitumens, *Strain energy methods, *Tensile strength, Elongation, Mechanical properties, Loads(Forces), Polyester fibers, Performance standards, Mechanical tests.

The study was conducted to revise the performance criterion for tensile strength of bituminous built-up membranes. Bituminous membrane samples, fabricated from polyester fabric, polyester-glass composite fabric, and single plies of APP- and SBS-modified bitumen, were tested in tension to determine their load-elongation properties and to measure their strain energy. The results of the tensile tests of the new bituminous membranes indicated wide variability of load and elongation among the different types of materials. As an alternative to the criterion that a bituminous built-up membrane have a tensile strength of 200 lbf/in (35 kN/m), it was recommended that the strain energy should be a minimum of 3 lbf/in/in (14 N/m/m), when tested at 0 F (-18°C) in the weaker direction.

900,140
PB89-212260   Not available NTIS
National Bureau of Standards (NEL), Gaithersburg, MD. Building Materials Div.
ASTM (American Society for Testing and Materials) Committee Completes Work on EPDM Specification.
Final rept.
W. J. Rossiter. 1987, 4p
Pub. in Handbook of Commercial Roofing Systems, p26-29 1987.

Keywords: *Roofing, *Materials specifications, *Membranes, Ethylene copolymers, Bitumens, Design

standards, Waterproofing, Construction materials, Reprints.

The use of single-ply membrane materials as the waterproofing component of low-sloped roofing systems has become commonplace in the United States. The growth in their use has not been without concern. One of the key issues confronting the roofing industry since the single-ply materials first appeared has been a lack of standard specifications to aid material selection as well as roofing systems design. ASTM Committee D08 on Roofing, Waterproofing, and Bituminous Materials is the lead organization to which the U.S. roofing industry turns for guidance in developing voluntary standards. The paper presents an update of the latest developments at ASTM regarding single-ply standards, and in particular, details concerning the expected appearance of the long-awaited specification for EPDM.

900,141
PB89-214787   PC A04/MF A01
Pennsylvania State Univ., University Park. Dept. of Mechanical Engineering.
Upward Flame Spread on Vertical Walls.
Final rept.
A. K. Kulkarni. Jun 89, 70p NIST/GCR-89/565
Grant NANB-4-D0037
Sponsored by National Inst. of Standards and Technology (NEL), Gaithersburg, MD. Center for Fire Research.

Keywords: *Walls, *Pyrolysis, *Flame propagation, Burning rate, Particle boards, Fire safety, Fires, Mathematical models.

Reported is a study of upward flame spread on vertical walls. First, a detailed review of literature on upward flame spread is presented. A 'complete procedure' for predicting upward flame spread on practical materials, which can be used in a global fire hazard assessment model, is then described. Experimental results on upward flame spread on various materials are obtained and the validity of the model is established.

900,142
PB89-215354   PC A05/MF A01
Dayton Univ., OH. Research Inst.
Validated Furniture Fire Model with FAST (HEM-FAST).
Technical rept. Jul 87-Sep 88.
M. A. Dietenberger. Dec 88, 90p UDR-TR-88-136, NIST/GCR-89/564
Grant NANB-S-D0556
Sponsored by National Inst. of Standards and Technology (NEL), Gaithersburg, MD. Center for Fire Research.

Keywords: *Furniture, *Fire safety, *Models, Burning rate, Fires, Computer program applications, Upholstery, *Validation.

The technical document reports on the validation of the furniture fire model with the furniture calorimeter data and on the restructure of the program 'HEMFAST'. Significant restructuring of the model and its code resolve various problems associated with the first version of HEMFAST. Comprehensive descriptions of the current model and its code structure benefit the HEMFAST users. The descriptions include: data processing of the bench scale fire tests database, effective time integrations of surface temperatures, flame spreads, and burn time, effective couple solutions of pyrolysis rates, burnrates, soot production, and thermal radiation, and the effective interfacing between the furniture fire model and FAST. The model is validated with fire tests for a 4-cushion mockup fire with three different fabric/foam cushion types. The comparisons include: fire area fractions of each cushion as a function of time, burnrate of the mockup as a function of time with fire test data from the furniture calorimeter, mass loss rate of the furniture as a function of time, and the overall levels of soot production.

900,143
PB89-215404   CP D05
National Inst. of Standards and Technology (NEL), Gaithersburg, MD. Center for Fire Research.
HAZARD I Fire Hazard Assessment Method.
Software.
R. W. Bukowski, R. Peacock, and W. Jones. May 89, 11 diskettes NBS/SW/DK-89/005
The software is contained on 3 1/2 and 5 1/4-inch diskettes, double density (360K and 720K), compatible

with the IBM PC XT/AT microcomputer. The diskettes are in the ASCII format.

Keywords: *Software, *Fire hazards, *Fire losses, Fires, Fire damage, Fire safety, Models, Evacuating(Transportation), Human behavior, Diskettes, L=Fortran;Basic;Assembly;Clipper, H=IBM PC/XT/AT;IBM PS2.

The Center for Fire Research has developed HAZARD 1, a method for predicting the hazards to building occupants from a fire. Within prescribed limits, HAZARD 1 allows the user to predict the outcome of a fire in a building populated by a representative set of occupants in terms of which persons successfully escape and which are killed, including the time, location, and likely cause of death for each. Specifically, the microcomputer program involves four procedures that combine expert judgment and calculations to estimate the consequences of a specified fire. Software description: The software is written in Fortran, Basic, Assembly and Clipper programming languages for implementation on an IBM-PC (XT, AT, PS/2) microcomputer under MS-DOS 3.0 (or higher). Memory requirement is 640K. A math co-processor (8087, 80287, or 80387) and a 2Mb hard disk drive are required to operate the system.

900,144
PB89-218325   **PC A03/MF A01**
National Inst. of Standards and Technology (NEL), Boulder, CO. Chemical Engineering Science Div.
Interlaboratory Comparison of Two Types of Line-Source Thermal-Conductivity Apparatus Measuring Five Insulating Materials.
J. G. Hust, and D. R. Smith. Jan 89, 25p NISTIR-89/3908
Sponsored by Oak Ridge National Lab., TN.

Keywords: *Thermal conductivity, *Thermal insulation, *Thermal measurements, Heat transmission, Thermophysical properties, Thermal resistance, Test facilities, Comparison, Standard deviation.

Measurements of apparent thermal conductivity performed by five different laboratories are compared. Subcommittee C-16.30 (Thermal Measurements) of the American Society for Testing and Materials (ASTM) sponsored the interlaboratory comparison. Two different types of line-source apparatus were used: the needle and the hot wire. The five laboratories measured thermal conductivity of Ottawa silica sand, perlite wax, and three insulating materials (fibrous glass, expanded polystyrene, and extruded polystyrene). Comparison of the test results illustrates the interlaboratory reproducibility. The standard deviation of the thermal conductivity results for the needle apparatus is 26%, whereas the standard deviation of the results for the hot-wire apparatus is 17%.

900,145
PB89-229215   Not available NTIS
National Inst. of Standards and Technology (NEL), Gaithersburg, MD. Office of Energy-Related Inventions.
Window U-Values: Revisions for the 1989 ASHRAE (American Society of Heating, Refrigerating and Air-Conditioning Engineers) Handbook - Fundamentals.
Final rept.
M. E. McCabe. 1989, 7p
Pub. in ASHRAE (American Society of Heating, Refrigerating and Air-Conditioning Engineers) Jnl. 31, n6 p56-62 Jun 89.

Keywords: *Window glazing, *Thermal insulation, *Heat transfer, *Storm windows, Solar energy, Temperature control, Solar radiation, Performance evaluation, Reprints, *U values, Energy conservation.

Recently, high-performance insulated glass has been introduced into the window market. However, with reduced heat flow in the central glazed portion of a window, heat conduction in the edge spacer and in the frame and sash members has become more important in determining the overall U-Value of the system. Characterization of frame heat transfer coefficients is complicated by the variety of frame configurations for operable windows, the different materials, and the different product sizes available. Standard methods for measuring frame heat transfer are not generally available. Chapter 27 in the ASHRAE Handbook of Fundamentals is the authoritative source of technical information on fenestration products such as windows, patio doors, skylights, shading devices, etc. The ASHRAE Handbook is revised every four years based on updated technical information. A number of changes have been made to the window U-Value table appearing in the 1989 Handbook. The article discusses the technical basis for the changes to appear in the 1989 Handbook and compares data with that from the 1985 Handbook.

900,146
PB90-111667   **PC A04/MF A01**
National Inst. of Standards and Technology (NEL), Gaithersburg, MD.
Robot Crane Technology.
Technical note (Final).
N. G. Dagalakis, J. S. Albus, K. R. Goodwin, J. D. Lee, T. M. Tsai, H. Abrishamian, R. Bostelman, and C. Yancey. Jul 89, 65p NIST/TN-1267
Also available from Supt. of Docs. as SN003-00302953-0. Sponsored by Defense Advanced Research Projects Agency, Arlington, VA.

Keywords: *Cranes(Hoists), *Robots, *Automation, Construction equipment, Kinematics, Dynamics, Stiffness, Stability, Constraints, Hoisting, Models, Tests.

The effort to develop kinematically constrained, dynamically stabilized, robot cranes capable of lifting, moving and positioning heavy loads over large volumes, capable of supporting fabrication tools and the inspection of large size and difficult to reach structures, is described in the report. The approach taken was to build on previous work at the NIST Robot Systems Division which has analyzed and measured the stiffness of a small model six-cable suspension system. The system is a modified Stewart platform. Under DARPA sponsorship, the author has: Extended the work to measure and optimize the stiffness of full-size models; Actively damped oscillations in a small scale six-cable suspension platform; Constructed an intermediate sized six-cable suspension platform for an industrial robot.

900,147
PB90-117573   Not available NTIS
National Inst. of Standards and Technology (NEL), Gaithersburg, MD. Fire Science and Engineering Div.
Note on Calculating Flows Through Vertical Vents in Zone Fire Models Under Conditions of Arbitrary Cross-Vent Pressure Difference.
Final rept.
L. Y. Cooper. 1989, 8p
Pub. in Combustion Science and Technology 64, n1-3 p43-50 1989.

Keywords: *Vents, *Pressure gradients, *Airflow, *Fire tests, Ventilation, Algorithms, Pressure, Fluid flow, Reprints.

In typical compartment fire scenarios, ratios of cross-vent absolute pressure are very close to 1. When such is the case, algorithms are available to predict the resulting cross-vent room-to-room flows. There are, however, important situations where this pressure condition does not prevail. For example, in fire scenarios involving relatively small penetrations in otherwise hermetically-sealed compartments of fire origin, cross-penetration pressure differences can be of the order of an atmosphere and pressure ratios, outside-to-insider, can be several tenths less than one. It is important for a versatile compartment fire model to have a capability of predicting vent flows for the entire range of possible cross-vent pressure conditions. The paper presents a unified analytic description of flows through uniform-width vertical vents connecting pairs of two-layer room fire environments under conditions of arbitrary cross-vent pressure difference. The algorithm presented is not significantly more complicated than previously-available algorithms which are restricted to low-pressure-difference cases.

900,148
PB90-118050   Not available NTIS
National Inst. of Standards and Technology (NEL), Gaithersburg, MD. Fire Science and Engineering Div.
Upward Turbulent Flame Spread on Wood under External Radiation.
Final rept.
K. Saito, F. A. Williams, I. S. Wichman, and J. G. Quintiere. 1989, 8p
Grant NSF-INT84-03848
See also PB87-126005. Sponsored by National Science Foundation, Washington, DC.
Pub. in Jnl. of Heat Transfer - Transactions of the ASME (American Society of Mechanical Engineers) 111, p436-445 May 89.

Keywords: *Fire tests, *Douglas fir wood, *Flame propagation, *Turbulence, *Thermal radiation, Burning rate, Temperature, Combustion, Ignition, Experimental data, Thermal measurement, Reprints.

Experiments were performed to obtain histories of surface temperatures and rates of upward flame spread for vertically oriented, thermally thick wood slabs exposed to surface fluxes of thermal radiation up to 2.6 W/sq cm. Above a critical irradiance sustained upward flame spread occurred for Douglas-fir particle board with pilot initiation at the base of the fuel face. Data obtained included temperatures, flame heights, pyrolysis-front heights, combustion duration, and char-layer thickness for various irradiances and preheat times. The measurements were compared with theory.

900,149
PB90-118068   Not available NTIS
National Inst. of Standards and Technology (NEL), Gaithersburg, MD. Fire Science and Engineering Div.
Scaling Applications in Fire Research.
Final rept.
J. G. Quintiere. 1989, 27p
Pub. in Fire Safety Jnl. 15, p3-29 1989.

Keywords: *Fire safety, *Scale(Ratio), Models, Fire tests, Froude number, Pressure, Analogs, Reprints.

The principles for scaling fire phenomena are examined from the dimensionless groups derived from the governing differential equations. A review of the literature shows examples of where correlations have been successfully developed for a wide range of fire phenomena in terms of the significant dimensionless groups. Scaling techniques based on Froude modeling, pressure modeling and analog modeling are described and illustrated. The use of small geometric models ranging from fire plumes to enclosure fires are illustrated.

900,150
PB90-118076   Not available NTIS
National Inst. of Standards and Technology (NEL), Gaithersburg, MD. Fire Science and Engineering Div.
Heat Transfer in Compartment Fires Near Regions of Ceiling-Jet Impingement on a Wall.
Final rept.
L. Y. Cooper. 1989, 6p
Pub. in Jnl. of Heat Transfer 111, p455-460 May 89.

Keywords: *Fire tests, *Fires, *Heat transfer, *Jet flow, Ceilings(Architecture), Walls, Enthalpy, Two-dimensional flow, Flux(Rate), Impingement, Momentum, Reprints.

The problem of heat transfer to walls from fire-plume-driven ceiling jets during compartment fires is introduced. Estimates are obtained for the mass, momentum, and enthalpy flux of the ceiling jet immediately upstream of the ceiling-wall function. An analogy is drawn between the flow dynamics and heat transfer at ceiling-jet/wall impingement and at the line impingement of a wall and a two-dimensional plane free jet. Using the analogy, results from the literature on plane free-jet flows and corresponding wall-stagnation heat transfer rates are recast into a ceiling-jet/wall-impingement-problem formulation. This leads to a readily usable estimate for the heat transfer from the ceiling jet as it turns downward and begins its initial descent as a negatively buoyant flow along the compartment walls. Available data from a reduced-scale experiment provide some limited verification of the heat transfer estimate.

900,151
PB90-128232   Not available NTIS
National Inst. of Standards and Technology (NEL), Gaithersburg, MD, Fire Measurement and Research Div.
Effects of Thermal Stability and Melt Viscosity of Thermoplastics on Piloted Ignition.
Final rept.
T. Kashiwagi, and A. Omori. 1989, 10p
Pub. in Proceedings of International Symposium on Combustion (22nd), Seattle, WA., August 14-19, 1988, p1329-1338 1989.

Keywords: *Ignition, *Flammability, *Thermoplastic resins, Polystyrene, Polymethyl methacrylate, Fire tests, Temperature, Flux(Rate), Radiant flux density.

The effects of material characteristics on piloted ignition were studied by using two different polystyrene, PS, samples and two different poly(methyl methacrylate), PMMA, samples. The difference between the

19

Construction Materials, Components, & Equipment

two PS samples was melt viscosity due to two different initial molecular weights and that between the two PMMA samples was thermal stability and melt viscosity also due to two different initial molecular weights. Ignition delay times and time histories of surface temperature and sample weight changes were measured in the external radiant flux range of 0.9-3.0 W/sq cm. A comparison of results between the two PS samples and between the two PMMA samples was made. The comparison indicates that the transport process of in-depth degradation products through the molten polymer layer to the sample surface has negligible effects on piloted ignition. However, the thermal stability of the material has significant effects on the piloted ignition delay time and the surface temperature at ignition.

900,152
PB90-128570   Not available NTIS
National Inst. of Standards and Technology (NEL), Gaithersburg, MD. Fire Science and Engineering Div.
Fire Growth and Development.
Final rept.
J. G. Quintiere. 1989, 24p
Pub. in Proceedings of International Symposium on Fire Safety and Engineering, Sydney and Melbourne, Australia, April 27, 1989, p1-24.

Keywords: *Fire tests, *Furniture, *Fires Fire resistance, Fire safety, Fire hazards, Flammability, Fla-shower, Combustion.

The phenomena of fire initiation and development is described and discussed. Fire behavior in closed and vented compartments is considered along with the factors important to spread beyond the compartment. The role of furnishings and contents is reviewed and fla-shower is found to be principally responsible for hazardous conditions, both thermal and toxic. Exceptions to this are discussed. The use of current and emerging test methods are described. Emphasis is on the interpretation and use of engineering measurements needed in innovative design hazard assessment, in contrast to the conventional practice of performance rankings of materials by a test method.

900,153
PB90-130311   PC A04/MF A01
National Inst. of Standards and Technology (NEL), Boulder, CO. Chemical Engineering Science Div.
Microporous Fumed-Silica Insulation as a Standard Reference Material of Thermal Resistance at High Temperature.
D. R. Smith. Aug 89, 65p NISTIR-89/3919
See also PB89-148373. Sponsored by Oak Ridge National Lab., TN.

Keywords: *Silicon dioxide, *Insulating boards, *Standards, *Porosity, *Thermal conductivity, Fumes, Measurement, High temperature tests, Air pressure, Density, Atmospheric pressure, Graphs(Charts), Standard reference materials.

Measurements of apparent thermal conductivity of microporous fumed-silica insulation board, already certified as a Standard Reference Material (SRM) of thermal resistance, are reported here to extend the range of certification of the material to higher temperatures and lower pressures. Apparent thermal conductivities of five different pairs of specimens ranging in mean density from 300 to 348 kg/cu.m were measured with a high-temperature,guarded hot plate 25 cm in diameter. The measurements cover a range of mean specimen temperatures from 318 to 733 K (45 to 460C), and of environmental air pressures from 26.7 to 83.5 kPa (200 to 626 Torr). Detailed analyses are given. The microporous fumed silica (at an ambient pressure of 83 kPa and a density of 300 kg/cu m) has an apparent thermal conductivity of 19.8 mW/(m K) at 300 K and is suitable for use as an SRM of very low conductivity from 297 to 735 K (24 to 460C). Adsorbed moisture within the material must be driven off by prolonged heating at 110C before its conductivity is measured.

900,154
PB90-132705   PC A03/MF A01
National Inst. of Standards and Technology (NEL), Gaithersburg, MD. Center for Building Technology.
Gypsum Wallboard Formaldehyde Sorption Model.
S. Silberstein. Nov 89, 22p NISTIR-89/4028

Keywords: *Gypsum, *Wallboard, *Formaldehyde, *Houses, Sorption, Safety engineering, Human factors engineering, Air entrainment, Concentration(Composition), Contaminants, Environmental engineering, *Indoor air pollution.

Gypsum wallboard was shown to absorb formaldehyde in a prototype house and in a measuring chamber, as reported previously by researchers at Oak Ridge National Laboratory (ORNL). Also as reported previously, formaldehyde concentrations attained equilibrium in two phases in response to a change in the air exchange rate or to the removal of the formaldehyde source. A rapid initial phase was followed by a slow phase lasting several days. A formaldehyde sorption model that accounts for the biphasic concentration pattern is presented here. Experiments for testing the predictability of the model are proposed.

900,155
PB89-136805   Not available NTIS
National Inst. of Standards and Technology (NEL), Gaithersburg, MD. Fire Measurement and Research Div.
Flammability of Upholstered Furniture with Flaming Sources.
Final rept.
V. Babrauskas. 1989, 27p
Pub. in Cellular Polymers 8, p198-224 1989.

Keywords: *Furniture, *Upholstery, *Flammability testing, Fire tests, Ignition, Fire safety, Fire resistance, Calorimeters, Reprints.

A number of countries and localities have either recently adopted furniture flammability regulations, or are actively considering them. In addition, a number of furniture flammability test methods have been developed in recent years in the course of research. Some of the methods share certain similarities; but, even so, many different testing philosophies exist. The paper compares the more widely used of the methods and examines their advantages and limitations. The impact of some recent research results on test method design is also considered. The methods are discussed only on their technical features and not on their regulatory aspects. The scope is limited to methods for testing the behavior under flaming fire conditions, and excludes tests for determining the cigarette ignition resistance. Some unresolved areas where further research is desirable are also cited.

Structural Analyses

900,156
PB89-148092   PC A03/MF A01
National Bureau of Standards (NEL), Gaithersburg, MD. Center for Building Technology.
Guidelines and Procedures for Implementation of Executive Order on Seismic Safety.
C. W. C. Yancey, and J. Greenberg. Jan 88, 32p NBSIR-88/3711
Also pub. as Interagency Committee on Seismic Safety in Construction rept. no. ICSSC/RP-2. Prepared in cooperation with Interagency Committee on Seismic Safety in Construction. Sponsored by Federal Emergency Management Agency, Washington, DC.

Keywords: *Earthquakes, *Safety engineering, *Earthquake resistant structure, Design criteria, Specifications, Concrete construction, Construction management, Readiness, Accident prevention, Project management, Guidelines, Earthquake Hazards Reduction Act of 1977.

The 'Earthquake Hazards Reduction Act of 1977,' Public Law 95-125, was passed by Congress to foster the reduction of life and property risks from future earthquakes in the United States through the establishment and maintenance of an effective earthquake hazards reduction program. A proposed Executive Order on Seismic Safety has been drafted that would implement the provisions of the Act by requiring Federal preparedness and mitigation activities to be implemented. The required activities would include the development and promulgation of specifications, building standards, design criteria, and construction practices for new and existing buildings and lifelines. The guidelines and procedures described herein have been prepared to support the implementation of the Executive Order on Seismic Safety. It is recommended that each agency concerned with buildings and lifeline that are Federally owned, leased, assisted, or regulated designate an individual or an operating unit as the Agency Seismic Coordinator. It would be the responsibility of the Agency Seismic Coordinator to coordinate all aspects of the agency seismic safety program.

900,157
PB89-154835   PC A21/MF A01
National Inst. of Standards and Technology (NEL), Gaithersburg, MD. Center for Building Technology.
Wind and Seismic Effects. Proceedings of the Joint Meeting of the U.S.-Japan Cooperative Program in Natural Resources Panel on Wind and Seismic Effects (20th) Held in Gaithersburg, Maryland on May 17-20, 1988.
Final rept.
N. J. Raufaste. Jan 89, 486p NIST/SP-760
Also available from Supt. of Docs. as SN003-003-02917-3. See also PB86-183963. Library of Congress catalog card no. 88-600610.

Keywords: *Meetings, *Bridges(Structures), *Buildings, *Earthquakes, *Wind pressure, *Ocean waves, Tsuamis, Forecasting, Seismic waves, Earth movements, Dynamic structural analysis, Storm surges, Design criteria, Dynamic loads, Soil mechanics, Standards, Structural engineering, *Seismic design, *Earthquake engineering, Ground motion, Risk assessments.

The 20th Joint Meeting of the U.S.-Japan Panel on Wind and Seismic Effects was held at the National Bureau of Standards, Gaithersburg, Maryland from May 17-20, 1988. The proceedings of the Joint Meeting, includes the program, list of members, panel resolutions, task committee reports, and technical papers. The papers covered five themes: Wind engineering, Earthquake engineering, Storm Surge and Tsunamis, Summary of U.S.-Japan Cooperative Research Program, and Two decades of accomplishments and challenges for the future.

900,158
PB89-175715   Not available NTIS
National Bureau of Standards (NEL), Gaithersburg, MD. Structures Div.
Progressive Collapse: U.S. Office Building in Moscow.
Final rept.
F. Y. Yokel, R. N. Wright, and W. C. Stone. 1989, 20p
Pub. in Jnl. of Performance of Constructed Facilities 3, n1 p57-76 Feb 89.

Keywords: *Structural analysis, *Loads(Forces), *Collapse, Dynamic response, Failure, Supports, Structural members, Stress analysis, Structural design, Reprints, *Office buildings.

As part of a structural assessment of the new U.S. Embassy Office Building being constructed in Moscow, United Soviet Socialist Republics, the National Bureau of Standards determined the susceptibility of the building to progressive collapse, which might be triggered by a local failure of a primary load supporting structural member. The building is a precast concrete structure that uses a standardized Soviet building system. The paper discusses criteria for the progressive collapse analysis, mechanisms for alternative load paths, analysis techniques used, and recommended retrofit measures. Although the building system was not designed to provide continuity in structural connections, it is possible to protect the building against progressive collapse with relatively modest retrofit measures.

900,159
PB89-175723   Not available NTIS
National Bureau of Standards (NEL), Gaithersburg, MD. Structures Div.
Pore-Water Pressure Buildup in Clean Sands Because of Cyclic Straining.
Final rept.
R. S. Ladd, R. Dobry, P. Dutko, F. Y. Yokel, and R. M. Chung. 1989, 10p
Pub. in Geotechnical Testing Jnl. 12, n1 p77-86 Mar 89.

Keywords: *Pore pressure, *Water pressure, *Sands, *Liquefaction, Shear strain, Soil properties, Mechanical tests, Dynamic response, Reprints, *Earthquake engineering.

The prediction of pore-water pressure buildup in sands caused by undrained cyclic loading is one of the key items in evaluating the potential for liquefaction of sandy sites during earthquakes. Presented herein are data indicating that, in strain-controlled tests, there is a predictable correlation between cyclic shear strain, number of cycles, and pore-water pressure buildup; this correlation is much less sensitive to factors, such as relative density and fabric than comparable results obtained from stress-controlled tests. The data indicate that, for clean sands, this threshold shear strain, as well as the pore-water pressure buildup for strains slightly above the threshold, are basically independent of relative density, grain size distribution, fabric, and method of testing (triaxial and direct simple shear). However, both threshold shear strain and pore-water pressure buildup do depend on the overconsolidation ratio.

900,160
PB89-187504 Not available NTIS
National Bureau of Standards (NEL), Gaithersburg, MD. Center for Building Technology.
Brick Masonry: U.S. Office Building in Moscow.
Final rept.
J. G. Gross, R. G. Mathey, C. Scribner, and W. C. Stone. 1989, 22p
Pub. in Jnl. of Performance of Constructed Facilities 3, n1 p35-56 Feb 89.

Keywords: *Brick structures, *Structural analysis, *Walls, Lateral stability, Cracks, Defects, Masonry, Mechanical properties, Assessments, Reprints, *Office buildings.

The National Bureau of Standards conducted a structural assessment of the new U.S. Embassy Office Building being built in Moscow. The paper reports the portion of the assessment dealing with the brick masonry walls. It describes the walls and provides summaries of two site visits, laboratory studies, and an analysis of the exterior walls, parapet walls, penthouse walls, and interior brick masonry core walls. Numerous cracks were found in exterior walls, incomplete construction of interior core walls was documented, and inadequate lateral strength of parapet walls was identified. Remedial measures were recommended for correction of deficiencies. Companion papers provide background information about the structure, the investigation, the assessment of the primary structural system, and the potential for progressive collapse.

900,161
PB89-188627 PC A03/MF A01
National Inst. of Standards and Technology (NEL), Gaithersburg, MD. Center for Building Technology.
Guidelines for Identification and Mitigation of Seismically Hazardous Existing Federal Buildings.
H. S. Lew. Mar 89, 17p NISTIR-89/4062
Also pub. as interagency Committee on Seismic Safety in Construction rept. no. ICSSC/RP-3. Prepared in cooperation with interagency Committee on Seismic Safety in Construction. Sponsored by Federal Emergency Management Agency, Washington, DC.

Keywords: *Hazards, *Buildings, *Earthquakes, Identifying, Safety, Instructions, Requirements, Vulnerability, Evaluation, Federal agencies, Earthquake Hazards Reduction Act of 1977.

The report includes Guidelines for identification and Mitigation of Seismically Hazardous Existing Federal Buildings, and was prepared by the Interagency Committee on Seismic Safety in Construction in support of the National Earthquake Hazards Reduction Program, the President's plan to implement the Earthquake Hazards Reduction Act of 1977 (Public Law 95-124). The Guidelines are intended for consideration and use, as appropriate, by Federal agencies in their plans for mitigation of seismic hazards in existing buildings. Some Federal agencies have their mitigation plan in operation.

900,162
PB89-235865 PC A04/MF A01
National Inst. of Standards and Technology (NEL), Gaithersburg, MD. Center for Building Technology.
Sensors and Measurement Techniques for Assessing Structural Performance.
R. D. Marshall. Aug 89, 67p NISTIR/89-4153
Proceedings of an International Workshop held in Gaithersburg, MD. on September 8-9, 1988. Sponsored by National Science Foundation, Washington, DC., United States-Japan Cooperative Program in Natural Resources. Panel on Wind and Seismic Effects, and American Society of Civil Engineers, New York. Performance of Structures Research Council.

Keywords: *Meetings, *Structural analysis, Loads(Forces), Detectors, Measurement, Research, Structural engineering, Earthquake engineering.

The report identifies research and development efforts needed to advance the state-of-the-art in instrumentation and measurement techniques for assessing structural performance. Four topic areas consisting of seismic effects; wind effects; effects due to occupancy, traffic, snow and other loads; and sensor technology were addressed by respective task groups during a two-day meeting of international experts. The forty-eight specific recommendations presented in the report are intended to serve as a research agenda for use by universities, research establishments and funding agencies.

900,163
PB90-117631 Not available NTIS
National Inst. of Standards and Technology (IMSE), Boulder, CO. Fracture and Deformation Div.
Measurement of Applied J-Integral Produced by Residual Stress.
Final rept.
D. T. Read. 1989, 7p
Sponsored by Naval Sea Systems Command, Washington, DC.
Pub. in Engineering Fracture Mechanics 32, n1 p147-153 1989.

Keywords: *Plates(Structural members), *Strain measurement, *Determination of stress, *Welded joints, *Stress analysis, Mechanical properties, Bending, Elasticity, Cracks, Surface defects, Fractures(Materials), Reprints.

An approximate method for measuring the applied J-integral produced by residual stresses was developed and applied to four wide plates. The technique uses multiple strain measurements during the cutting of a notch in the weld. Results for welded and post-weld-heat-treated (PWHT) plates with semi-elliptical surface cracks and tube hole ligament cracks were compared. The PWHT plates had much lower J levels. Comparisons with the Newman-Raju linear elastic calculation for surface cracks in bending indicated that the present results are reasonable. Comparison of the present result for the ratio of stress intensity factor to crack mouth-opening displacement with values calculated for through and surface cracks provides additional confidence in the correctness of the present results.

General

900,164
PB89-173983 Not available NTIS
National Bureau of Standards (NEL), Gaithersburg, MD. Fire Measurement and Research Div.
Hand Calculations for Enclosure Fires.
Final rept.
E. K. Budnick, and D. D. Evans. 1986, 6p
Pub. in Fire Protection Handbook (16th Edition), Section 21, Chapter 3, p21-19-21-24 1986.

Keywords: *Enclosures, *Fire protection, *Mathematical models, Containers, Expansion, Growth, Forecasting, Reprints.

In the chapter, a brief discussion of enclosure fire effects is presented, along with equations that can be evaluated using hand calculators to provide estimates of particular effects. Generally, the equations presented are well documented and are widely used for such estimates. However, the user is cautioned that most of the equations were developed based on data from experiments that were conducted for very specific, and sometimes idealized, conditions. Therefore, some judgment must be exercised when applying these equations to complex conditions occurring in enclosure fires of general interest.

900,165
PB89-173991 Not available NTIS
National Bureau of Standards (NEL), Gaithersburg, MD. Fire Measurement and Research Div.
Computer Fire Models.
Final rept.
E. K. Budnick, and W. D. Walton. 1986, 6p
Pub. in Fire Protection Handbook (16th Edition), Section 21, Chapter 4, p21-25-21-30 1986.

Keywords: *Mathematical models, *Fire protection, *Computerized simulation, Fire safety, Building codes, Fire prevention, Regulations, Design standards, Reprints.

In recent years, increasing attention has been given to the development and use of computer fire models.

They have been used by engineers and architects for building design, by building officials for plan review, by the fire service for prefire planning, by investigators for post fire analysis, by groups writing fire codes, and by materials manufacturers, fire researchers, and educators. While these models are not a replacement for the building and fire codes they can be a valuable tool for fire professionals. The report focuses on a representative selection of models. At the end of the chapter an in-depth review of models is given.

900,166
PB89-174130 Not available NTIS
National Bureau of Standards (NEL), Gaithersburg, MD. Fire Science and Engineering Div.
Creation of a Fire Research Bibliographic Database.
Final rept.
N. H. Jason. 1986, 6p
Pub. in Proceedings of International Meeting of Fire Research and Test Centres, Avila, Spain, October 7-9, 1986, p669-674.

Keywords: *Bibliographies, *Information retrieval, *Information systems, Data retrieval, Indexes(Documentation), Fire protection, Fire prevention, Fire hazards, Fire resistant materials, *Fire research information services, *Data bases.

It is difficult to perform a comprehensive literature survey in several technical fields, and in fire research the problem is compounded by the diverse nature of the field. Fire research cuts across many boundaries, e.g., chemistry, physics, fluid mechanics, mechanical engineering. In an effort to enhance the retrieval rate of information from the Fire Research Information Services (FRIS) collection, the Center for Fire Research, National Bureau of Standards decided to automate its literature collection. Using available supermicro hardware and software, FIREDOC was created as the on-line bibliographic database for the FRIS collection. Analysis of the alternatives which led to the creation of FIREDOC will be discussed, as well as some retrieval methods to locate relevant information in the database.

900,167
PB89-163222 PC A03/MF A01
National Inst. of Standards and Technology (NEL), Gaithersburg, MD. Center for Fire Research.
Engineering View of the Fire of May 4, 1988 in the First Interstate Bank Building, Los Angeles, California.
H. E. Nelson. Mar 89, 40p NISTIR-89/4061

Keywords: *Fires, *Banks(Buildings), Flame propagation, Safety, Fire damage, Burning rate, Diagrams, Investigations, Smoke, Office buildings, Case studies.

The course of the fire is traced in terms of developing fire phenomena. Special emphasis is given to burning rate of building furnishings, smoke layer temperature, layer level, oxygen consumption, combustion efficiency, flashover, exterior fire propagation, detector response, sprinkler operation, smoke movement and some contamination.

900,168
PB89-202584 Not available NTIS
National Bureau of Standards (NEL), Gaithersburg, MD. Center for Building Technology.
Effects of Research on Building Practice.
Final rept.
R. N. Wright. 1989, 8p
Pub. in Construction Specifier 42, n5 p98-105 May 89.

Keywords: *Research management, *Safety engineering, *Building codes, *Economic analysis, International relations, Trends, Competition, Construction industry, Value engineering, Reprints, *Center for Building Technology.

The Center for Building Technology of the National Institute of Standards and Technology (formerly the National Bureau of Standards) is the U.S. national building research laboratory. The Center collaborates with other organizations of the building community to advance building technology to increase the usefulness, safety and economy of buildings and to enhance the international competitiveness of U.S. building products and services. Benefits of research, development and application efforts are described by examples of successful programs with which the Center has been associated. Trends requiring building research and application are noted and corresponding aspects of the Center's current and planned programs are cited.

900,169
PB90-112327   PC A03/MF A01
National Inst. of Standards and Technology, Gaithersburg, MD. National Voluntary Lab. Accreditation Program.
NVLAP (National Voluntary Laboratory Accreditation Program) Program Handbook Construction Testing Services. Requirements for Accreditation.
R. L. Gladhill. Mar 89, 48p NISTIR-89/4039

Keywords: *Construction industry, *Laboratories, *Test facilities, Performance evaluation, Professional personnel, Systems analysis, Scientists, Concretes, *Accreditation, Certification.

The document explains the operation and technical requirements of the Laboratory Accreditation Program for Construction Testing Services. All of the steps leading to accreditation are discussed. Technical requirements are explained including how the National Voluntary Laboratory Accreditation Program criteria are applied. It is intended for use by staff of accredited laboratories, those seeking accreditation, other laboratory accreditation systems, and others needing information on the requirements for NVLAP accreditation.

900,170
PB90-128620   Not available NTIS
National Inst. of Standards and Technology (NEL), Gaithersburg, MD. Fire Science and Engineering Div.
Comparisons of NBS/Harvard VI Simulations and Full-Scale, Multiroom Fire Test Data.
Final rept.
J. A. Rockett, M. Morita, and L. Y. Cooper. 1989, 10p
Pub. in Proceedings of International Symposium on Fire Safety Science (2nd), Tokyo, Japan, June 13-17, 1988, p481-490 1989.

Keywords: *Fire tests, *Computerized simulation, *Buildings, Safety engineering, Fire safety, Model tests, Environmental engineering, Comparisons.

The NBS/Harvard VI multi-room fire model was used to simulate results of previously reported full-scale multi-room fire experiments. The tests and simulations involved: four different compartment configurations of two or three rooms connected by open doorways, four different fire types generated by a methane burner and up to four different doorway openings between the burn room and adjacent space. A total of nineteen different tests were carried out and simulated. Selected comparisons between simulated and measured parameters of the fire-generated environments are reviewed. While the computer code is found to provide generally favorable simulations for the entire range of tests, several areas of modeling detail are identified for further improvement.

BUSINESS & ECONOMICS

Domestic Commerce, Marketing, & Economics

900,171
PB89-120742   PC A05/MF A01
National Governors' Association, Washington, DC.
Promoting Technological Excellence: The Role of State and Federal Extension Activities.
M. K. Clarke, and E. N. Dobson. c1989, 69p ISBN-1-55877-069-0, NIST/GCR-89-567
Grant NANB-8-D0868
Sponsored by National Inst. of Standards and Technology (NEL), Gaithersburg, MD. Center for Fire Research.

Keywords: *Technology transfer, *Businesses, Surveys, State government, National government, Organizations, Improvement, Extension services.

The report presents the findings of a Nationwide survey of state and federal organizations providing business and technology assistance to small and medium-sized businesses. Information was collected on the nature of the services provided by these organizations, the type of firms being assisted, and methods used to reach potential clients. In addition, the survey solicited the views of program managers regarding the needs of small and medium-sized businesses for information on new and existing technologies and ways to improve the transfer of federal technology to potential users in the small business community. The report also contains short descriptions of specialized technology extension services in seven states. Finally, the report recommends actions states should take in expanding their technology assistance efforts and the federal government should take to support these efforts.

International Commerce, Marketing, & Economics

900,172
PB89-166128   PC A03/MF A01
National Inst. of Standards and Technology, Gaithersburg, MD. Office of the Associate Director for Industry and Standards.
Effect of Chinese Standardization on U.S. Export Opportunities.
Y. Lin. Dec 88, 16p NISTIR-88/4000

Keywords: *China, *International trade, *Standards, Requirements, Electric devices, Economic development, Technology transfer, Exports, Information exchange.

The paper describes the standardization system as it exists in the Peoples Republic of China and identifies the role of the China State Bureau of Standards (CSBS) in the standards coordination process. The standards development and approval process is also described. The implementation of the IECQ program in China for producing electronic products in conformance with internationally recognized quality requirements is explained and the organizations responsible for the several aspects of the system are identified. The paper advocates technical information exchange programs between the U.S. and the Peoples Republic of China and more U.S. trade missions to China to encourage the adoption of U.S. standards as well as to increase trade between the two countries.

900,173
PB89-191977   PC A03/MF A01
National Inst. of Standards and Technology, Gaithersburg, MD. Office of the Associate Director for Industry and Standards.
GATT (General Agreement on Tariffs and Trade) Standards Code Activities of the National Institute of Standards and Technology 1988.
Annual Rept.
J. R. Overman. Mar 89, 39p NISTIR-89/4074
See also PB88-201611.

Keywords: *Standards, Notifications, Regulations, Foreign countries, *GATT standards code, Technical assistance.

The report describes the GATT Standards Code activities conducted by the Office of Standards Code and Information, National Institute of Standards and Technology (NIST), for calendar year 1988. NIST responsibilities include operating the U.S. GATT inquiry point for information on standards and certification activities; notifying the GATT Secretariat of proposed U.S. Federal government standards-based rules that might significantly -affect trade; assisting U.S. industry with standards-related trade problems; and responding to inquiries about proposed foreign and U.S. regulations.

CHEMISTRY

Analytical Chemistry

900,174
PB89-146807   Not available NTIS
National Bureau of Standards (IMSE), Gaithersburg, MD. Ceramics Div.
Speciation Measurements of Butyltins: Application to Controlled Release Rate Determination and Production of Reference Standards.
Final rept.
W. R. Blair, G. J. Olson, and F. E. Brinckman. 1986, 5p
Sponsored by Civil Engineering Lab. (Navy), Port Hueneme, CA.
Pub. in Oceans 86--Conference Record, Washington, DC., September 23-25, 1986, p1141-1145.

Keywords: *Chemical analysis, *Gas chromatography, *Wooden piles, Standards, Extraction, Hydrolization, Tin organic compounds, Reprints, *Tin/butyl, *Tin/dibutyl, *Tin/tributyl, *Tin/tetrabutyl, *Flame photometry.

The paper describes methods and results of the determination of release rates for organotin species released from organotin impregnated wood pilings. The analytical method consists of simultaneous extraction/ hydridization of aqueous lechate samples, with organotin speciation by gas chromatography coupled with tin selective flame photometric detection. The sensitivity of the flame photometric detector to the butyltin family of organotins is 0.1 to 0.2 ng, depending on the species. Chromatographic separation of the butyltins provides speciation of mono-through tetrabutyltin within a 15 min chromatogram, with the additional capability of identifying any methylbutyltin compounds that may be present in the sample. Water samples were collected from the piling leaching tanks immediately upon immersion of the pilings and continued to be collected for approximately 1 year. Speciation and release rate data were obtained on both the early, first order stage of release, and the latter, zeroth order phase of controlled release. Instrument calibrations were performed using a specially prepared organotin research material.

900,175
PB89-146905   Not available NTIS
National Bureau of Standards (NML), Gaithersburg, MD. Inorganic Analytical Research Div.
Determining Picogram Quantities of U in Human Urine by Thermal Ionization Mass Spectrometry.
Final rept.
W. R. Kelly, J. D. Fassett, and S. A. Hotes. 1987, 6p
Pub. in Health Physics 52, n3 p331-336 Mar 87.

Keywords: *Uranium 238, *Quantitative analysis, *Urine, ionization, Mass spectroscopy, Isotope separation, Health physics, Reprints.

The U concentration in SRM 2670, Toxic Metals in Freeze-Dried Urine, and the urine of two pre-school age children was determined by measuring the chemically separated U by isotope dilution thermal ionization mass spectrometry using ion counting detection. This procedure can detect about 1% of the atoms in the sample and has a total chemical blank of about 5 pg U. The U concentration in SRM 2670 was found to be 113 plus or minus 2 pg (sup 238) U/ml (1s). At this level a 1 ml sample is sufficient for a determination with a total uncertainty of less than 5%. The U concentration in the two children was 3.1 plus or minus 0.9 and 3.6 plus or minus 0.9 pg (sup 238) U/g. These low values suggest that the U concentration in urine of unexposed persons may be at this level or lower.

900,176
PB89-150858   Not available NTIS
National Bureau of Standards (NML), Gaithersburg, MD. Inorganic Analytical Research Div.
Comparison of Detection Limits in Atomic Spectroscopic Methods of Analysis.
Final rept.
M. S. Epstein. 1988, 17p
Pub. in ACS (American Chemical Society) Symposium Series 361, p109-125 1988.

Keywords: *Quantitative analysis, *Atomic spectroscopy, Detectors, Lasers, Fluorescence, Absorption, Plasma radiation, Accuracy, Reprints.

The comparison of detection limits is a fundamental part of many decision-making processes for the analytical chemist. Despite numerous efforts to standardize methodology for the calculation and reporting of detection limits, there is still a wide divergence in the way they appear in the literature. The paper discusses valid and invalid methods to calculate, report, and compare detection limits using atomic spectroscopic techniques. 'Noises' which limit detection are discussed for analytical methods such as plasma emission spectroscopy, atomic absorption spectroscopy and laser excited atomic fluorescence spectroscopy.

900,177
PB89-151773 PC A09/MF A01
National Inst. of Standards and Technology (NML), Gaithersburg, MD. Center for Analytical Chemistry.
Technical Activities, 1988, Center for Analytical Chemistry.
B. I. Diamondstone, R. A. Durst, and H. S. Hertz.
Nov 88, 178p NISTIR-88/3875
See also report for 1985, PB86-178902.

Keywords: *Chemical analysis, *Research projects, Standards, Inorganic compounds, Organic compounds, Particles, Gases, Standards, Biological extracts, *Standard reference materials.

The report summarizes the technical activities of the Center for Analytical Chemistry at the National institute of Standards and Technology. It emphasizes activities over the Fiscal Year 1988 in the Inorganic Analytical Research Division, the Organic Analytical Research Division, and the Gas and Particulate Science Division. In addition, it describes certain special activities in the Center including quality assurance and voluntary standardization coordination.

900,178
PB89-156889 Not available NTIS
National Bureau of Standards (NML), Gaithersburg, MD. Organic Analytical Research Div.
Standard Reference Materials for the Determination of Polycyclic Aromatic Hydrocarbons.
Final rept.
S. A. Wise, L. R. Hilpert, R. E. Rebbert, L. C. Sander, M. M. Schantz, S. N. Chesler, and W. E. May. 1988, 10p
Pub. in Fresenius' Zeitschrift fuer Analytische Chemie 332, p573-582 1988.

Keywords: *Aromatic polycyclic hydrocarbons, *Chemical analyses, *Spectrum analyses, Standards, Gas chromatography, Concentration(Composition), Reprints, *Standard reference materials, Certification, Air pollution detection.

Since 1980 a number of Standard Reference Materials (SRMs) have been issued by the National Bureau of Standards (NBS) to assist in validating measurements for the determination of polycyclic aromatic hydrocarbons (PAH) and other polycyclic aromatic compounds (PAC). These SRMs are certified for selected PAC and range in analytical difficulty from calibration solutions to complex natural matrix materials, such as air and diesel particulate matter, shale oil, and crude oil. In the past year three new SRMs have been introduced: (1) SRM 1647a 'Priority Pollutant PAH in Acetonitrile'; (2) SRM 1491 'Aromatic Hydrocarbons in Hexane/Toluene'; and (3) SRM 1597 'Complex Mixture of PAH from Coal Tar.' The SRMs available from NBS for use in the determination of PAC are described and the concentrations of PAC determined in the natural matrix SRMs are summarized and compared. The primary analytical techniques used for the measurement of PAC in these SRMs were gas chromatography, liquid chromatography, and gas chromatography/mass spectrometry.

900,179
PB89-156913 Not available NTIS
National Bureau of Standards (NML), Gaithersburg, MD. Inorganic Analytical Research Div.
Radiochemical Procedure for Ultratrace Determination of Chromium in Biological Materials.
Final rept.
R. R. Greenberg, and R. Zeisler. 1988, 16p
Pub. in Jnl. of Radioanalytical and Nuclear Chemistry 124, n1 p5-20 1988.

Keywords: *Chromium, *Biological surveys, *Neutron activation analysis, Chemical analysis, Radiochemistry, Concentration(Composition), Separation, Blood analysis, Solvent extraction, Trace elements, Reprints, Standard reference materials.

Chromium is one of the most difficult elements to accurately determine at the naturally occurring, ultratrace levels normally found in uncontaminated biological samples. In view of the importance of Cr, both as an essential and as a toxic element, efforts have focused on developing a simple, yet reliable, radiochemical procedure for Cr determination using neutron activation analysis. A number of problem areas have been identified in earlier methods, and an improved radio-

chemical separation procedure, based upon the liquid/liquid extraction of Cr(VI) into a solution of tribenzylamine/chloroform, has been developed. The fast neutron interference from Fe has been evaluated for the highly thermal FT-4 facility of the NBS Research Reactor, and Cr concentrations have been determined in samples of whole human blood collected under clean conditions and in two certified reference materials.

900,180
PB89-156921 Not available NTIS
National Bureau of Standards (NML), Gaithersburg, MD. Inorganic Analytical Research Div.
Neutron Activation Analysis of the NIST (National Institute of Standards and Technology) Bovine Serum Standard Reference Material Using Chemical Separations.
Final rept.
R. R. Greenberg, R. Zeisler, H. M. Kingston, and T. M. Sullivan. 1988, 5p
Pub. in Fresenius' Zeitschrift fuer Analytische Chemie 332, p652-656 1988.

Keywords: *Neutron activation analysis, *Chemical analysis, *Separation, *Biological surveys, Standards, Trace elements, Toxicology, Human nutrition, Reprints, *Standard reference materials, Certification.

The U.S. National Institute of Standards and Technology is currently in the process of certifying a Bovine Serum Standard Reference Material. In addition to elements normally considered to be of clinical interest, a number of other elements, which are analytically more difficult to determine yet are of importance from either a nutritional or toxicological viewpoint, are being determined by a variety of analytical techniques. Neutron activation analysis in combination with appropriate pre- or post-irradiation chemical separations has been used to determine many of these difficult elements.

900,181
PB89-156939 Not available NTIS
National Bureau of Standards (NML), Gaithersburg, MD. Inorganic Analytical Research Div.
Long-Term Stability of the Elemental Composition in Biological Materials.
Final rept.
R. Zeisler, R. Greenberg, S. Stone, and T. Sullivan. 1988, 4p
Sponsored by Environmental Protection Agency, Washington, DC.
Pub. in Fresenius' Zeitschrift fuer Analytische Chemie 332, p612-615 1988.

Keywords: *Chemical analysis, *Biological surveys, *Tissue extracts, Standards, Concentration(Composition), Chemical stabilization, Neutron activation analysis, Zinc, Selenium, Arsenic, Trace elements, Liver extracts, Reprints, *Standard reference materials.

Lyophilized and radiation sterilized biological certified reference materials (CRMs) are believed to be stable in their chemical composition. Generally, the certifying agencies consider the certificates of these biological CRMs valid for a 5-year shelf life, i.e., apart from measurable moisture content, the chemical composition should not change during that time. The long-term behavior of fresh frozen materials is not known. In the study the elemental compositions of the Bovine Liver Standard Reference Material (SRM 1577) and human liver tissue samples are evaluated over a time period of more than 7 years. The concentrations of selected elements were determined by neutron activation analysis at various times. The initial evaluation of zinc, selenium and arsenic results gives no indication of changes during 7 years storage of fresh frozen tissues, however, a trend towards lower arsenic concentrations has been observed in SRM 1577 during a 10-year period.

900,182
PB89-156947 Not available NTIS
National Bureau of Standards (NML), Gaithersburg, MD. Inorganic Analytical Research Div.
High-Accuracy Differential-Pulse Anodic Stripping Voltammetry with Indium as an Internal Standard.
Final rept.
K. W. Pratt, and W. F. Koch. 1988, 8p
Pub. in Analytica Chimica Acta 215, p21-28 1988.

Keywords: *Indium, *Standards, *Cadmium, *Copper, *Lead, *Voltmeters, Anodic polarization, Chemical analysis, Reprints.

Indium (III) is used as an internal standard for the determination of cadmium, copper and lead at the 20 ng/

g level by using differential-pulse anodic stripping voltammetry; the supporting electrolyte is 1.0 mol/l ammonium bromide/0.25 mol/l nitric acid. For each solution, each stripping peak of interest is normalized to the corresponding peak height obtained in the same voltammogram for a known, added concentration of indium (III). A calibration curve is prepared for each element by using these normalized peak heights. The technique is demonstrated for NBS SRM 1643b (Trace Elements in Water). The relative standard deviations for six independent determinations of Cd, Cu, and Pb at the 20 ng/g level are 1.9%, 5.4%, and 1.2%, respectively. The imprecision for copper is limited by the sloping baseline at its stripping potential. The detection limit for each element is less than 1 ng/g.

900,183
PB89-156970 Not available NTIS
National Bureau of Standards (NML), Gaithersburg, MD. Inorganic Analytical Research Div.
Activation Analysis Opportunities Using Cold Neutron Beams.
Final rept.
R. M. Lindstrom, R. Zeisler, and M. Rossbach. 1987, 10p
Pub. in Jnl. of Radioanalytical and Nuclear Chemistry 112, n2 p321-330 1987.

Keywords: *Radioactivation analysis, Neutron beams, Reprints, *Activation analysis, *Cold neutrons, Neutron capture, Prompt gamma radiation.

Guided beams of cold neutrons being installed at a number of research reactors may become increasingly available for analytical research. A guided cold beam will provide higher neutron fluence rates and lower background interferences than in present facilities. In an optimized facility, fluence rates of one billion n/sq cm/sec are readily obtainable. Focusing a large area beam onto a small target will further increase the neutron intensity. In addition, the shift to lower energies will increase the effective cross sections. The absence of fast neutrons and gamma rays permit detectors to be placed near the sample without intolerable background, and thus the efficiency for counting prompt gamma rays can be much higher than in present systems.

900,184
PB89-157085 Not available NTIS
National Bureau of Standards (IMSE), Gaithersburg, MD. Polymers Div.
Trace Speciation by HPLC-GF AA (High-Performance Liquid Chromatography-Graphite Furnace Atomic Absorption) for Tin- and Lead-Bearing Organometallic Compounds, with Signal Increases Induced by Transition-Metal Ions.
Final rept.
E. J. Parks, F. E. Brinckman, K. L. Jewett, W. R. Blair, and C. S. Weiss. 1988, 10p
Pub. in Applied Organometallic Chemistry 2, p441-450 1988.

Keywords: *Chromatography, *Quantitative analysis, *Environmental tests, *Organometallic compounds, Atomic spectra, Tin, Lead, Metals, Tungsten, Chromium, Manganese, Volatility, Chlorides, Oxides, Ligands, Metal complexes, Trace elements, Effluents, Liquid phases, Absorption spectra, Reprints.

High-performance liquid chromatography coupled with graphite furnace atomic absorption spectroscopy (HPLC-GF AA) gives element-specific detection of environmental samples containing trace amounts of organotin or organolead species. The analyte and a modifier are co-pipetted into a conventional furnace tube, from either a solution of analyte or an HPLC effluent. Oxides of transition metals (e.g., chromium, manganese, or tungsten) are shown to enhance both tin and lead signals, whereas chlorides do not, suggesting the low-temperature formation of relatively involatile metal oxides or volatile metal chlorides, respectively.

900,185
PB89-157739 Not available NTIS
National Bureau of Standards (NML), Gaithersburg, MD. Gas and Particulate Science Div.
Preparation of Accurate Multicomponent Gas Standards of Volatile Toxic Organic Compounds in the Low-Parts-per-Billion Range.
Final rept.
G. C. Rhoderick, and W. L. Zielinski. 1988, 7p
Pub. in Analytical Chemistry 60, n22 p2454-2460 1988.

23

CHEMISTRY

Analytical Chemistry

Keywords: *Standards, *Chemical analysis, Mixtures, Calibrating, Concentration(Composition), Chemical stabilization, Gravimetric analysis, Microanalysis, Reprints, *Volatile organic compounds, *Standard reference materials, *Air pollution detection, Toxic materials.

Methodology is described for the microgravimetric preparation and analytical evaluation of accurate, stable multicomponent gas standards in compressed gas cylinders containing volatile toxic organic compounds in pure nitrogen at the mid- to low-parts-per-billion (ppb) level. Standard mixtures have been prepared containing up to nine organic compounds at concentrations ranging from 1 to 1000 ppb by mole. Current indications are that the number of organic compounds in a single mixture is more limited by analytical capability than by the preparation methodology. Over 100 standards, of which several will be discussed in the paper, have been prepared and evaluated for long-term stability and internal consistency. Over 25 different volatile organic compounds spanning three concentration decades have been studied. The sum of preparative and analytical error compounds spanning three concentration decades have been studied. The sum of preparative and analytical error components of the uncertainty associated with the concentration of the organic analytes at the 95% confidence level typically ranges from 3 to 10% relative, depending upon the compound and its concentration. Intercomparative analyses of new and previously prepared standards have verified that such mixtures are stable for several years.

900,186
PB89-157747   Not available NTIS
National Bureau of Standards (NML), Gaithersburg, MD. Gas and Particulate Science Div.
Moydite, (Y, REE) (B(OH)4)(CO3), a New Mineral Species from the Evans-Lou Pegmatite, Quebec.
Final rept.
J. D. Grice, J. Van Velthuizen, P. J. Dunn, D. E. Newbury, E. S. Etz, and C. H. Nielsen. 1986, 9p
Pub. in Canadian Mineralogist 24, p665-673 Dec 86.

Keywords: *Carbonate minerals, *Qualitative analysis, *Yttrium, *Rare earth minerals, *Borate minerals, Crystal structure, Raman spectroscopy, X-ray analysis, Reprints, *Moydite.

The complete mineralogy of a newly discovered mineral, recently named moydite, is reported from the application of classical techniques of mineral characterization supported by the results of modern microanalytical methods. The mineral is described as a yttrium/rare earth element tetrahydroxoborate carbonate of ideal empirical formula (Y, REE) (B(OH)4) (CO3). The chemical composition and formula of moydite were derived from elemental analysis and x-ray crystal structure determination. The microanalytical techniques employed include electron and ion probe microanalysis, as well as Raman microprobe spectroscopy to infer molecular and vibrational structure relationships. The data on the new mineral are presented and its properties discussed to illustrate the advantages of applying a range of synergistic analytical techniques to the comprehensive characterization of a new mineral species.

900,187
PB89-157879   Not available NTIS
National Bureau of Standards (NML), Gaithersburg, MD. Inorganic Analytical Research Div.
Role of Neutron Activation Analysis in the Certification of NBS (National Bureau of Standards) Standard Reference Materials.
Final rept.
R. R. Greenberg. 1987, 15p
Pub. in Jnl. of Radioanalytical and Nuclear Chemistry 113, n1 p233-247 1987.

Keywords: *Neutron activation analysis, Standards, Chemical analysis, Spectrum analysis, Performance evaluation, Quality assurance, Reprints, *Standard reference materials, Certification.

Neutron activation analysis (NAA) is extensively used at the National Bureau of Standards as one of the analytical techniques in the certification of Standard Reference Materials (SRMs). Characteristics of NAA which make it valuable in this role are: inherent accuracy, multielemental capability, especially in the instrumental mode; ability to assess homogeneity; high sensitivity for many elements, and essentially blank-free nature. Examples of recent SRM analyses illustrating these characteristics are described.

900,188
PB89-157994   Not available NTIS
National Bureau of Standards (NML), Gaithersburg, MD. Inorganic Analytical Research Div.
Voltammetric and Liquid Chromatographic Identification of Organic Products of Microwave-Assisted Wet Ashing of Biological Samples.
Final rept.
K. W. Pratt, H. M. Kingston, W. A. MacCrehan, and W. F. Koch. 1988, 4p
Pub. in Analytical Chemistry 60, n19 p2024-2027, 1 Oct 88.

Keywords: *Liver extracts, *Biological surveys, *Chemical analysis, *Coulometers, *Spectrophotometry, *Nitric acid, Spectrum analysis, Digestion(Decomposition), Polarographic analysis, Reprints, *Wet methods, *Voltammetry, *Liquid column chromatography, Benzoic acid/nitro, Standard reference materials.

Residual organic species in nitric acid digests of freeze-dried bovine liver (NBS SRM 1577a) have been identified by use of voltammetry, liquid chromatography, spectrophotometry, and classical chemical tests. Data from these techniques show that major products of microwave-assisted dissolution by nitric acid include o-, m-, and p-nitrobenzoic acids (NBA). In addition to these compounds, other organic species present in these digests irreversibly complex copper, but not zinc, and result in low values for copper by polarography. The NBAs and these other organic species are all eliminated by refluxing the nitric acid digest in perchloric acid at atmospheric pressure. Polarographic results obtained for copper following treatment with perchloric acid agree with the certified value. The use of voltammetry in the evaluation of wet ashing procedures is discussed.

900,189
PB89-161590   Not available NTIS
National Bureau of Standards (NML), Gaithersburg, MD. Inorganic Analytical Research Div.
Analytical Applications of Resonance Ionization Mass Spectrometry (RIMS).
Final rept.
J. D. Fassett, and J. C. Travis. 1988, 14p
Pub. in Spectrochimica Acta 43B, n713 p1409-1422 1988.

Keywords: *Chemical analysis, *Inorganic compounds, Performance evaluation, Rare gases, Mass spectroscopy, Sampling, Solids, Reprints, *Resonance ionization mass spectroscopy, State of the art, Isotope dilution.

A perspective on the role of resonance ionization mass spectrometry illustrate these capabilities and define the potential of RIMS in the generalized field of chemical analysis. Three areas of application are reviewed here: (1) noble gas measurements; (2) materials analysis using isotope dilution (IDMS); and, (3) solids analysis using direct sampling. The role of RIMS is discussed relative to the more traditional mass spectrometric methods of analysis in these areas. The applications are meant to illustrate the present state-of-the-art as well as point to the future state-of-the-art of RIMS in chemical analysis.

900,190
PB89-171763   Not available NTIS
National Bureau of Standards (IMSE), Gaithersburg, MD. Ceramics Div.
Standard X-ray Diffraction Powder Patterns from the JCPDS (Joint Committee on Powder Diffraction Standards) Research Associateship.
Final rept.
H. F. McMurdie, M. C. Morris, E. H. Evans, B. Paretzkin, W. Wong-Ng, Y. Zhang, and C. R. Hubbard. 1986, 12p
See also PB87-119756. Sponsored by JCPDS-International Centre for Diffraction Data, Swarthmore, PA.
Pub. in Powder Diffraction 1, n4 p334-345 Dec 86.

Keywords: *X ray diffraction, *Powder(Particles), *Standards, Calibrating, Crystal structure, Reprints.

Standard x-ray powder diffraction patterns are presented for 17 substances. These patterns, useful for identification, were obtained by automated diffractometer methods. The lattice constants from the experimental work were refined by least-squares methods, and reflections were assigned hkl indices consistent with space group extinctions. Relative intensities, calculated densities, literature references, and other relevant data are included.

900,191
PB89-171938   Not available NTIS
National Bureau of Standards (NML), Gaithersburg, MD. Inorganic Analytical Research Div.
Chemical Calibration Standards for Molecular Absorption Spectrometry.
Final rept.
R. Mavrodineanu, and R. W. Burke. 1987, 50p
Pub. in Advances in Standards and Methodology in Spectrophotometry, p125-174 1987.

Keywords: *Chemical analysis, *Spectrophotometry, Calibrating, Molecular spectroscopy, Design criteria, Performance evaluation, Reprints, *Standard reference materials.

The publication describes activities undertaken since 1969 within the Center for Analytical Chemistry of the National Bureau of Standards (NBS) in the field of high-accuracy spectrophotometry. It presents a summary of the Standard Reference Materials that have been developed for checking the proper functioning of ultraviolet and visible spectrophotometers and includes a brief description of the high-accuracy spectrophotometer constructed in the Center for Analytical Chemistry and subsequently used for performing all of the transmittance measurements.

900,192
PB89-172498   Not available NTIS
National Bureau of Standards (NML), Gaithersburg, MD. Gas and Particulate Science Div.
Strategy for Interpretation of Contrast Mechanisms in Scanning Electron Microscopy: A Tutorial.
Final rept.
D. E. Newbury. 1986, 5p
Pub. in Microbeam Analysis - 1986, p1-5 1986.

Keywords: *Laboratory equipment, Performance evaluation, Design criteria, Topography, Images, Reprints, *Scanning light microscopy.

The interpretation of images in the scanning electron microscope is based upon prior knowledge of the characteristics of the contrast mechanism coupled with knowledge of the response of the electron detector. Pertinent contrast and detector properties include the mechanism of contrast encoding by signal carrier, number, trajectory, or energy effects and detector sensitivity to these factors. Examples of the interpretation of images of specimens which have compositional and topographic features are given. Compositional features are best visualized with a number-sensitive and trajectory-insensitive detector, while topographic features are detected best with a trajectory-sensitive detector.

900,193
PB89-173643   Not available NTIS
National Bureau of Standards (NML), Gaithersburg, MD. Gas and Particulate Science Div.
Comparison of a Cryogenic Preconcentration Technique and Direct Injection for the Gas Chromatographic Analysis of Low PPB (Parts-per-Billion) (NMOL/MOL) Gas Standards of Toxic Organic Compounds.
Final rept.
G. C. Rhoderick. 1988, 6p
Sponsored by Environmental Protection Agency, Washington, DC.
Pub. in Proceedings of EPA/APCA (Environmental Protection Agency/Air Pollution Control Association) International Symposium on Measurement of Toxic and Related Air Pollutants, Research Triangle Park, NC., May 2-4, 1988, p259-264.

Keywords: *Gas analyses, *Gas chromatography, Halogen organic compounds, Performance evaluation, Cryogenics, Emission spectroscopy, *Air pollution detection, *Volatile organic compounds, *Toxic substances, *Standard reference materials, Environmental monitoring, Electron capture detectors, Flame ionization.

There is an increasing need for multicomponent gas standards containing volatile toxic organic compounds at the low parts-per-billion level for use in environmental monitoring programs. Standards containing many organic compounds, both halogenated and nonhalogenated species within the same mixture, can be very difficult to analyze at the 1-15 ppb concentration level. Analyses of low level multicomponent mixtures have been done using several different techniques. Gas chromatography has been used to separate compounds in simple and complex mixtures. Original work was done using packed columns with a flame-ionization detector (FID) and large sample volumes, 10 mL and an electron-capture detector (ECD) to analyze for halogenated compounds at low ppb levels. Therefore, to measure all the compounds in a single analysis, a cryogenic preconcentration technique was developed to increase the sensitivity of both types of compounds to the FID. Temperature programming was coupled with this cryogenic preconcentration technique to increase the quality of baseline separations.

900,194
PB89-175236
(Order as PB89-175194, PC A06)
National Inst. of Standards and Technology, Gaithersburg, MD.
Numeric Databases for Chemical Analysis.
Bi-monthly rept.
S. G. Lias. 1989, 11p
Included in Jnl. of Research of the National Institute of Standards and Technology, v94 n1 p25-34 Jan-Feb 89.

Keywords: *Chemical analysis, Nuclear magnetic resonance, Infrared spectroscopy, Mass spectroscopy, Evaluation, Identifying, Chemical compounds, *Numerical data bases, Analytical chemistry, Computer applications.

Databases for use with analytical chemistry instrumental techniques are surveyed, with attention to existing databases and collection efforts now underway, as well as needs for new databases. Collections of spectra for use in Nuclear Magnetic Resonance Spectroscopy, infrared spectroscopy, and mass spectroscopy are described. Using mass spectral databases as an example, a critique is presented of automated quality control procedures used to evaluate individual spectra in large collections; the kinds of problems which have been encountered in using these procedures are discussed. Finally, a brief critical review is presented covering the application of computers to the identification of unknown compounds using spectral databases; again, algorithms used with mass spectrometry are taken as the example.

900,195
PB89-175863　　　　　　Not available NTIS
National Bureau of Standards (NML), Gaithersburg, MD. Organic Analytical Research Div.
Development of Electrophoresis and Electrofocusing Standards.
Final rept.
D. J. Reeder. 1987, 15p
Pub. in ACS (American Chemical Society) Symposium Series, v335 p102-116 1987.

Keywords: *Standards, *Electrophoresis, Separation, Quality assurance, Comparison, Reviews.

The work reviews some of the approaches to standardization in several different areas of electrophoretic separations. While no definitive standards have been established, some practical standards have been reported and are being used by researchers. Standards usage is part of quality assurance programs and is necessary for interlaboratory comparison studies.

900,196
PB89-175962　　　　　　Not available NTIS
National Bureau of Standards (IMSE), Gaithersburg, MD. Ceramics Div.
Characterization of Organolead Polymers in Trace Amounts by Element-Specific Size-Exclusion Chromatography.
Final rept.
E. J. Parks, F. E. Brinckman, and L. B. Kool. 1986, 4p
Pub. in Jnl. of Chromatography 370, n1 p206-209, 26 Nov 86.

Keywords: *Chromatographic analysis, *Size separation, *Exclusion, *Lead organic compounds, *Polymers, Spectroscopy, Detectors, Graphite, Atomic spectra, Absorption spectra, Molecular weight, Methacrylates, Organometallic compounds, Ultraviolet spectroscopy, Reprints.

Size exclusion chromatography (SEC), coupled with lead-specific graphite furnace atomic absorption (GFAA) spectroscopy and ultraviolet spectroscopy (UV) detectors was applied to the characterization of a 5:1 copolymer of 4-vinylphenyl, triphenyllead and octadecylmethacrylate. Less than 1.0 micrograms of the polymer, dissolved and injected in tetrahydrofuran

(THF), provided sufficient data to determine the number- and weight-average molecular weights as well as monomer conversion, with approximately 100% recovery of the injected lead. The method is uniquely capable of characterizing trace quantities of organometallic polymers such as may be obtained in preliminary stages of synthetic research, provided the polymers are soluble in column-compatible solvents.

900,197
PB89-176143　　　　　　Not available NTIS
National Bureau of Standards (NML), Gaithersburg, MD. Gas and Particulate Science Div.
Role of Standards in Electron Microprobe Techniques.
Final rept.
D. E. Newbury. 1986, 26p
Pub. in Jnl. of Trace Microprobe Tech. 4, n3 p103-128 1986.

Keywords: *Quantitative analysis, *X ray spectroscopy, *Standards, *Chemical composition, *Electron microscopy, Reprints, *Electron microprobe analysis.

Standards play a vital role in quantitative analysis techniques based upon electron excitation of x-rays. In the technique of electron probe x-ray microanalysis (EPMA), the description of the interaction of electrons and x-rays is sufficiently well known to permit quantitative analysis with a suite of pure element standards and calculated matrix corrections. Such a scheme provides flexibility in responding to problems in analyzing unknowns of arbitrary composition. In analytical electron microscopy (AEM), constraints on the specimen force reliance on an internal reference through the use of relative sensitivity factors coupled with matrix corrections. AEM standards must consist at least of two components and be in the form of thin foils, and each sensitivity factor requires a separate standard, making the standards suite more difficult to obtain.

900,198
PB89-176267　　　　　　Not available NTIS
National Bureau of Standards (NML), Gaithersburg, MD. Inorganic Analytical Research Div.
High-Accuracy Differential-Pulse Anodic Stripping Voltammetry Using Indium as an Internal Standard.
Final rept.
K. W. Pratt, and W. F. Koch. 1986, 1p
Pub. in Abstracts of Papers of the American Chemical Society 192, p98 Sep 86.

Keywords: *Coulometers, *Indium, Chemical analysis, Performance evaluation, Standards, Cadmium, Copper, Lead(Metal), Reprints, *Anodic stripping, *Standard reference materials.

In(III) is employed as an internal standard for the determination of Cd, Cu, and Pb at the 20 micrograms/g level using differential pulse anodic stripping voltammetry. A multi-point calibration curve is prepared for each element using these normalized peak heights. The technique is demonstrated using NBS Standard Reference Material 1643b.

900,199
PB89-176275　　　　　　Not available NTIS
National Bureau of Standards (NEL), Gaithersburg, MD. Statistical Engineering Div.
Tests of the Recalibration Period of a Drifting Instrument.
Final rept.
W. Liggett. 1986, 6p
Sponsored by Environmental Monitoring Systems Lab., Research Triangle Park, NC.
Pub. in Proceedings of Oceans 86, Conference Record on National Monitoring Strategies, Washington, DC, September 23-25, 1986, p923-928.

Keywords: *Continuous sampling, *Samplers, *Calibrating, *Statistical analysis, *Drift(Instrumentation), Sulfur dioxide, Performance evaluation, Quality assurance, Reliability, *Air pollution detection, *Air pollution sampling.

The use of a drifting instrument requires that an adequately short recalibration period be chosen. After several periods, the calibration data can be used to test the adequacy of the choice. The paper discusses statistical tests of the recalibration period and applies these tests to continuous analyzers for sulfur dioxide. The paper presents two tests, a test of the second differences of the calibration sequence for normality and a test of the upper part of the spectrum for flatness. The paper illustrates these tests with two sequences

each consisting of about fifty recalibrations of a sulfur dioxide analyzer. Also, the power of the tests and some approximations made in their formulation are investigated by Monte Carlo experiments.

900,200
PB89-176887　　　　　　Not available NTIS
National Bureau of Standards (NML), Gaithersburg, MD. Gas and Particulate Science Div.
Comparison of Two Transient Recorders for Use with the Laser Microprobe Mass Analyzer.
Final rept.
R. A. Fletcher, and D. S. Simons. 1985, 3p
Pub. in Microbeam Analysis, p319-321 1985.

Keywords: *Laboratory equipment, *Mass spectroscopy, Performance evaluation, Comparison, Reprints, *Data acquisition systems, *Laser microprobe mass analyzers, Laser spectroscopy.

Two transient recorders have been utilized as data acquisition systems for the Laser Microprobe Mass Analyzer (LAMMA). Past work has shown that one transient recorder degrades from about 8 to 4 bit resolution when subjected to high frequency signals. This contributes to errors quantification of peak intensities. The dynamic precision of the two recorders will be compared, analyzing the bit degradation with rapidly varying waveforms. Spectral quality of the two recorders and the precision and accuracy of isotopic ratios from standard samples will be examined. The advantages and disadvantages of each device will be reported.

900,201
PB89-184105
(Order as PB89-184089, PC A04)
National Inst. of Standards and Technology, Boulder, CO.
Supercritical Fluid Chromatograph for Physicochemical Studies.
Bi-monthly rept.
T. J. Bruno. 1989, 8p
Included in Jnl. of Research of the National Institute of Standards and Technology, v94 n2 p105-112 Mar-Apr 89.

Keywords: *Chromatography, Chromatographic analysis, Chemical analysis, Physicochemical properties, Diffusion coefficient, Molecular weight, Polymers, Solubility, *Supercritical fluid chromatography, *Supercritical fluids.

A supercritical fluid chromatograph has been designed and constructed to make physicochemical measurements, while retaining the capability to perform chemical analysis. The physicochemical measurements include diffusion coefficients, capacity ratios, partition coefficients, partial molar volumes, virial coefficients, solubilities, and molecular weight distributions of polymers. In the paper, the apparatus will be described in detail, with particular attention given to its unique features and capabilities. The instrument has recently been applied to the measurement of diffusion coefficients of toluene in supercritical carbon dioxide at a temperature of 313 K, and pressures from 133 to 304 bar (13.3-30.4 MPa). The data are discussed and compared with previous measurements on similar systems.

900,202
PB89-186357　　　　　　Not available NTIS
National Bureau of Standards (NML), Gaithersburg, MD. Inorganic Analytical Research Div.
Analysis of Ultrapure Reagents from a Large Sub-Boiling Still Made of Teflon PFA.
Final rept.
P. J. Paulsen, E. S. Beary, D. S. Bushee, and J. R. Moody. 1989, 4p
Pub. in Analytical Chemistry 61, n8 p827-830, 15 Apr 89.

Keywords: *Chemical analysis, *Mass spectroscopy, Distillates, Purity, Reprints, *Teflon, *Sub-boiling stills, Polytetra-fluoroethylene.

Inductively coupled plasma mass spectrometry was applied to the analysis of distillates of ultrahigh purity from quartz sub-boiling stills, a sub-boiling still made of Teflon TFE, and a sub-boiling still made of Teflon PFA. Although these studies were originally intended to prove the purity of distillates from the PFA still, comparison of distillates from the various stills has led to an interesting observation about the suitability of TFE for these reagents.

25

CHEMISTRY

Analytical Chemistry

900,203
PB89-196795   Not available NTIS
National Bureau of Standards (NML), Gaithersburg,
MD. Temperature and Pressure Div.
**Preparation of Multistage Zone-Refined Materials
for Thermochemical Standards.**
Final rept.
E. Rubinstein, M. E. Glicksman, B. W. Mangum, Q. T.
Fang, and N. B. Singh. 1988, 10p
Sponsored by National Aeronautics and Space Admin-
istration, Washington, DC.
Pub. in Jnl. of Crystal Growth 89, p101-110 1988.

Keywords: *Calibrating, *Standards, *Purity, *Thermal
analysis, *Nitriles, Measurement, Phase diagrams,
Critical temperature, Hermetic compressors, Crystal
structure, Zone melting, Melting points, Chemical equilibrium, Solids, Liquids, Gases, Reprints, Succinonitrile.

The melting, boiling, and triple points of materials have
long served to define the International Practical Temperature Scale (IPTS), which is the embodiment of the
thermodynamic temperature scale as 13 well-defined
fixed points along with interpolative schemes for
standardized instruments used for temperature measurement. Techniques have evolved over the past
seven years in the laboratory for preparing and characterizing ultra-pure organic compounds by multistage
hermetic zone-refining. Methods have matured to the
point where high-purity succinonitrile (SCN) now provides the triple-point equilibrium defining 58.0796 + or
- 0.0015 C relative to the IPTS of 1968. Most recently,
the National Bureau of Standards, through its office of
Standard Reference Materials, has issued certification
for Standard Reference Material 1970, which is offered
as an evacuated minicell containing approximately 60
g of double-stage hermetically zone-purified SCN. The
steps that lead up to the completed minicells filled with
SCN are discussed in detail. Since the cells are evacuated prior to the purification process, the succinonitrile
is under its own vapor pressure. The solid/liquid equilibrium, as determined by melting and freezing point
measurements, is therefore considered to be equivalent to the triple-point. A compilation of recently acquired performance data which assess the statistical
properties of over 100 SCN triple-point cells is presented and discussed. These data firmly establish the reliability of multistage zone-refining methods for achieving reliable thermometric standards using solid/liquid/
vapor thermochemical equilibria.

900,204
PB89-187520   Not available NTIS
National Bureau of Standards (NML), Gaithersburg,
MD. Organic Analytical Research Div.
**Recent Advances in Bonded Phases for Liquid
Chromatography.**
Final rept.
L. C. Sander, and S. A. Wise. 1987, 117p
Pub. in CRC Critical Reviews in Analytical Chemistry
18, n4 p299-415 1987.

Keywords: Chromatography, Chromatographic analysis, Substrates, Reprints, *Liquid chromatography,
Void volume, Bonded phases, Retention.

Theoretical and practical aspects of bonded phase research are reviewed for work carried out over the last
5-10 years. Included in the review is research concerning liquid chromatographic substrates, bonded phase
syntheses, novel bonded phases, methods used in the
characterization of these materials, and progress in
the development of retention theory. Not included is
work dealing with ion exchange chromatography, immobilized enzymes, size exclusion chromatography, or
purely mechanical aspects of liquid chromatography.
An effort was made to include studies representative
of the more important ongoing research in bonded
phase technology.

900,205
PB89-187538   Not available NTIS
National Bureau of Standards (NML), Gaithersburg,
MD. Organic Analytical Research Div.
**Determination of Hydrocarbon/Water Partition
Coefficients from Chromatographic Data and
Based on Solution Thermodynamics and Theory.**
Final rept.
M. M. Schantz, and D. E. Martire. 1987, 17p
Pub. in Jnl. of Chromatography 391, n1 p35-51 1987.

Keywords: *Chromatographic analysis, *Alcohols,
*Water, Chemical analysis, Aromatic monocyclic hy-
drocarbons, Alkane compounds, Alkenes, Bromine,

Reprints, *Partition coefficients, Octanol, Hexade-
cane, Bromoalkanes, Alkylbenzenes.

The octanol/water and hexadecane/water partition
coefficients for series of alkylbenzenes, alkanes, alkenes, alcohols, and bromoalkanes were determined
by the generator column technique and compared favorably to those calculated from the activity coefficients in each phase. A lattice-model theory suggested
and the data confirmed that the logarithms of the partition coefficients and solute molar volume were the
same for all homologous series studied. Furthermore,
the logarithms of the octanol/water partition coefficients were linearly related (r(sup 2) = 0.993) to the
logarithms of the reversed phase liquid chromatographic adjusted retention volumes determined by extrapolation to 100% water as the mobile phase for a
variety of the solutes.

900,206
PB89-187546   Not available NTIS
National Bureau of Standards (NML), Gaithersburg,
MD. Organic Analytical Research Div.
**Preparation of Glass Columns for Visual Demon-
stration of Reversed Phase Liquid Chromatogra-
phy.**
Final rept.
L. C. Sander. 1988, 2p
Pub. in Jnl. of Chemical Education 65, n4 p373-374
1988.

Keywords: Separation, Dyes, Education,
Colors(Materials), Reprints, *Reverse phase liquid
chromatography, *Column packings, Glass column.

The preparation of a reversed phase glass column is
described for demonstration of chromatographic principles. The column is prepared from large diameter
particles of the type used in solid phase extraction. A
simple demonstration is outlined for the separation of
various food color dyes.

900,207
PB89-187553   Not available NTIS
National Bureau of Standards (NML), Gaithersburg,
MD. Inorganic Analytical Research Div.
**Design Principles for a Large High-Efficiency Sub-
Boiling Still.**
Final rept.
J. R. Moody, C. E. Wissink, and E. S. Beary. 1989,
5p
Pub. in Analytical Chemistry 61, n8 p823-827, 15 Apr
89.

Keywords: *Distillation equipment, *Acid treatment,
*Purification, Condensers(Liquefiers), Tetrafluoroethy-
lene resins, Trace elements, Metals, Stills, Chemical
analysis, Reprints, *Sub-boiling stills.

The sub-boiling method of acid purification for low
trace element blank has now been in use for about 20
years. However, to achieve commercially useful yields,
distillates must be produced at approximately 25-100
L/day. Aspects of still design that include throughput
have been examined, including variables such as con-
denser area, temperatures, distance between the con-
denser and the liquid in the still pot, and coolant temperature. Designs for several new stills are given and
compared to those of prior sub-boiling stills. Perfluor-
oalkoxy resin construction was used both for ease of
fabrication and for the distinct distillate purity.

900,208
PB89-201602   Not available NTIS
National Bureau of Standards (NML), Gaithersburg,
MD. Gas and Particulate Science Div.
**High-Purity Germanium X-ray Detector on a 200 kV
Analytical Electron Microscope.**
Final rept.
E. B. Steel. 1986, 10p
Pub. in Microbeam Analysis - 1986, p439-448 1986.

Keywords: *Electron microscopes, Thin films, Reprints, *High-purity Ge detectors, *X-ray detection, Li-
drifted Si detectors.

A JEOL 200CX AEM with a zero-take-off-angle germanium detector and with a high-take-off-angle Be
window Si(Li) detector was used to collect data for the
paper. The window thicknesses were 13 micrometers
Be on the Si(Li), and 50 micrometers Be on the HPGE.
The detectors were connected to either a TN-2000
with a PDP-11 or a multiplexed Ortec 918 multichannel
analyzer connected to a VAX-11/730. NBS synthetic
glasses, natural minerals, and pure metals were examined at an accelerating potential of 200 kV. The speci-

mens were prepared as particles or sputtered films on
thin carbon films supported by 200 mesh copper grids.

900,209
PB89-201610   Not available NTIS
National Bureau of Standards (NML), Gaithersburg,
MD. Gas and Particulate Science Div.
**Continuum Radiation Produced in Pure-Element
Targets by 10-40 keV Electrons: An Empirical
Model.**
Final rept.
J. A. Small, S. D. Leigh, D. E. Newbury, and R. L
Myklebust. 1986, 3p
Pub. in Microbeam Analysis - 1986, p289-291 1986.

Keywords: *x ray analysis, *Microanalysis, *Bremsstrahlung, Electron beams, Mathematical models, Targets, Reprints, KeV range 10-100.

A new global relation has been developed for predicting electron-excited continuum intensities over a wide
range of accelerating voltages 10-40 keV, atomic numbers 4-92, and x-ray energies 1.5-20 keV. The new relation was determined empirically from the mathematical modeling of extensive data and is designed for calculating continuum intensities in analytical procedures,
such as those requiring peak-to-background measurements, where the direct measurement of the continuum intensities is impracticable. The distribution of
errors between the data and the model is symmetrical,
centered around zero error with 63% of the values falling between plus or minus 10% relative error.

900,210
PB89-201628   Not available NTIS
National Bureau of Standards (NML), Gaithersburg,
MD. Gas and Particulate Science Div.
**Laser Microprobe Mass Spectrometry: Description
and Selected Applications.**
Final rept.
D. S. Simons. 1986, 15p
See also PB86-193232.
Pub. in Appl. Surf. Sci. 31, n1 p103-117 1986.

Keywords: *Mass spectroscopy, Particles, Aerosols,
Surfaces, Reprints, *Laser microprobe mass spectroscopy, Time-of-flight spectrometers.

Laser microprobe mass spectrometry (LMMS) uses a
high power density pulsed laser beam to ablate a mi-
crovolume of material. The fraction of this material that
is ionized can be detected using a time-of-flight mass
spectrometer. Two different instrumental configurations, 'transmission' and 'reflection', that satisfy different analytical requirements are characterized by the
geometry of ion collection from the specimen. The features of LMMS include a spatial resolution of about 1
micrometer, high mass range, isotopic selectivity, ppm
detection limits for many elements from picograms of
material, and molecular structure information from organic materials. The major application areas for this
technique are in the analysis of biological tissues and
cells, organic materials, particles and aerosols, and
surfaces of metals, semiconductors, and dielectric materials.

900,211
PB89-201651   Not available NTIS
National Bureau of Standards (NML), Gaithersburg,
MD. Gas and Particulate Science Div.
Performance Standards for Microanalysis.
Final rept.
E. Steel, A. Hartman, G. Hembree, P. Sheridan, and
J. Small. 1986, 15p
Pub. in Jnl. of Trace Microprobe Tech. 4, n3 p147-161
1986.

Keywords: *Microanalysis, *Standards, *Electron microscopy, *Asbestos, Optical microscopes, Magnification, Reprints, Polystyrene, Spheres, Shape, Size
determination, Reprints, Scanning electron microscopy, Transmission electron microscopy.

Several standards for characterizing the performance
of microanalysis procedures and instruments that are
not directly related to chemical analysis are described.
These standards may help analysts in calibration and
quality assurance procedures in a wide variety of applications. The magnification, particle size and shape, microscope performance, and asbestos analysis standards discussed are standards developed and characterized at NBS that are specifically designed for electron and light microscopy.

26

900,212
PB89-201677 Not available NTIS
National Bureau of Standards (NML), Gaithersburg, MD. Gas and Particulate Science Div.
Uncertainties in Mass Absorption Coefficients in Fundamental Parameter X-ray Fluorescence Analysis.
Final rept.
B. A. R. Vrebos, and P. A. Pella. 1988, 10p
Pub. in X-ray Spectrometry 17, n1 p3-12 1988.

Keywords: Reprints, *X-ray fluorescence analysis, Mass absorption coefficients, Intercomparison, Uncertainty.

Various compilations of mass absorption coefficients are currently used in fundamental parameter computer programs for correction of interelement effects in quantitative x-ray fluorescence analysis. Statistically significant differences in results of analysis were observed for certain sample types when three most commonly used compilations were compared, especially when only pure element standards were used for calibration. When type standards combined with a second degree polynomial fit to the standards are used for calibration, differences between compilations become negligible in the analysis results.

900,213
PB89-202063 Not available NTIS
National Bureau of Standards (NML), Gaithersburg, MD. Inorganic Analytical Research Div.
Quality Assurance in Metals Analysis Using the Inductively Coupled Plasma.
Final rept.
R. L. Watters. 1987, 20p
Pub. in ASTM (American Society for Testing and Materials) Special Technical Publication 944, p108-127 1987.

Keywords: *Quality assurance, *Chemical analysis, *Metals, Accuracy, Precision, Reprints, *Inductively coupled plasma.

The inductively coupled plasma (ICP) technique is a useful approach for multielement analysis of a wide variety of materials. Published reports have described the successful application of the ICP technique for the analysis of trace, minor, and major elements in metal alloys. When assessing the quality of analytical results using the ICP or any other technique, one must consider the contribution of various parts of the measurement process to the total random error. In addition, a careful evaluation of sources of systematic error must be undertaken, so that the appropriate corrections may be applied to the analytical results. The inherent linearity of the ICP technique offers a corrections scheme for dealing with spectral interferences, but the method of measuring correction factors and accounting for their variability warrants close examination. Most descriptions of ICP applications report relative freedom from matrix effects. Although the magnitude of systematic errors may be less than for other spectrometric techniques, such errors can cause analytical bias which can appreciably affect the final results. For example, it has been shown that differences in final acid concentration between pure element standards and the sample can cause systematic error. This type of problem can occur when complex alloys are dissolved for ICP analysis. Examples of these kinds of errors and approaches to correcting for them will be presented.

900,214
PB89-202246 Not available NTIS
National Bureau of Standards (IMSE), Gaithersburg, MD. Ceramics Div.
Standard X-ray Diffraction Powder Patterns from the JCPDS (Joint Committee on Powder Diffraction Standards) Research Association.
Final rept.
H. F. McMurdie, M. C. Morris, E. H. Evans, B. Paretzkin, W. Wong-Ng, and C. R. Hubbard. 1986, 11p
See also PB89-171763. Sponsored by JCPDS-International Centre for Diffraction Data, Swarthmore, PA.
Pub. in Powder Diffract 1, n3 p265-275 Sep 86.

Keywords: *X ray diffraction, *Powder(Particles), *Standards, Qualitative analysis, Crystal structure, Lattice parameters, Reprints.

Standard x-ray powder diffraction patterns are presented for 15 substances. The patterns, useful for identification, were obtained by automated diffractom-

eter methods. The lattice constants from the experimental work were refined by least-squares methods, and reflections were assigned hk(sub l) indices consistent with space group extinctions. Relative intensities, calculated densities, literature references, and other relevant data are included.

900,215
PB89-211940 Not available NTIS
National Bureau of Standards (NML), Gaithersburg, MD. Gas and Particulate Science Div.
Preparation of Standards for Gas Analysis.
Final rept.
G. C. Rhoderick, and E. E. Hughes. 1987, 10p
Sponsored by Institute of Gas Technology, Chicago, IL.
Pub. in Natural Gas Energy Measurement, p45-54 1987.

Keywords: *Gas analysis, *Gravimetric analysis, Concentration(Composition), Chemical analysis, Natural gas, Methane, Reprints, *Standard reference materials.

A primary standard is prepared by an absolute method with the gravimetric method being the preferred technique. The basic method can be modified to prepare samples at very low concentrations (parts per trillion) and containing very specific and complex mixtures such as hydrocarbons. As will be illustrated, it is very important to have standards which are very close and bracket the concentration of the unknown in question. This is especially important when analyzing natural gas. In this case, two methods may be used to determine the concentration of methane in natural gas. However, one method will lead to a much lower uncertainty and thus, a lower uncertainty in the BTU value of the natural gas.

900,216
PB89-229106 Not available NTIS
National Bureau of Standards (NML), Gaithersburg, MD. Inorganic Analytical Research Div.
Expert-Database System for Sample Preparation by Microwave Dissolution. 1. Selection of Analytical Descriptors.
Final rept.
F. A. Settle, B. I. Diamondstone, H. M. Kingston, and M. A. Pleva. 1989, 7p
Grant NSF-CHE85-17147
Sponsored by National Science Foundation, Washington, DC.
Pub. in Jnl. of Chemical Information and Computer Sciences 29, p11-17 1989.

Keywords: *Chemical analysis, Dissolving, Microwaves, Samples, Reprints, *Computer applications, Expert systems, Data bases.

A hybrid expert-database system is being developed to provide advice on the preparation of samples for elemental analysis. The paper describes an expert system component that is designed to assist the analyst in the identification of four analytical descriptors necessary to develop procedures for sample preparation. When completed, the system will be able to furnish information on the dissolution of the sample. Future versions of the system will also provide advice on separations that may be required prior to the analytical measurement. A PC-AT microcomputer and commercially available software were used to develop the system. A compiled version of the system will run on PC-compatible computers.

900,217
PB89-229116 Not available NTIS
National Bureau of Standards (NML), Gaithersburg, MD. Inorganic Analytical Research Div.
Microwave Digestion of Biological Samples: Selenium Analysis by Electrothermal Atomic Absorption Spectrometry.
Final rept.
K. Y. Patterson, C. Veillon, and H. M. Kingston. 1988, 12p
Pub. in Introduction to Microwave Sample Preparation, Chapter 7, p155-166 1988.

Keywords: *Atomic absorption, *Selenium, *Analytical chemistry, *Microwaves, Zeeman effect, Tissues(Biology), Isotopic labelling, Standards, Temperature, Pressure, Reprints.

Closed vessel, microwave-heated digestions are used to rapidly destroy the organic matrix of biological samples with nitric acid at elevated temperatures and pressures. The system allows controlled, uniform applica-

tion of microwave power and monitoring of temperature and pressure, and permits reproducible conditions while not exceeding the pressure or temperature limitations of the container. Samples are digested and analyzed for selenium by electrothermal atomic absorption spectrometry using matrix modification and Zeeman background correction. Analyte recoveries are established by using a radiotracer (75)Se and accuracy is verified with standard reference materials of biological origin.

900,218
PB89-229173 Not available NTIS
National Bureau of Standards (NML), Gaithersburg, MD. Radiation Physics Div.
Improved Low-Energy Diffuse Scattering Electron-Spin Polarization Analyzer.
Final rept.
M. R. Scheinfein, D. T. Pierce, J. Unguris, J. J. McClelland, and R. J. Celotta. 1989, 11p
Sponsored by Office of Naval Research, Arlington, VA.
Pub. in Review of Scientific Instruments 60, n1 p1-11 Jan 89.

Keywords: *Electron spin, *Polarization(Spin alignment), *Measurement instruments, Diffusion, Energy, Gold, Polycrystalline, Asymmetry, Trajectories, Reprints.

An improved low-energy diffuse scattering electron-spin polarization analyzer is described. It is based on the low-energy (150eV) diffuse scattering of polarized electrons from polycrystalline evaporated Au targets. By collecting large solid angles and efficiently energy filtering the scattered electrons, a maximum figure of merit, FOM = S sup 2 I/I sub o = 2.3 x 10(sub -4) is achieved. Maximum measured values of the Sherman function were S = 0.15. Further, the instrumental (false) asymmetry due to changes in the trajectory of the incident electron beam has been minimized by balancing the angular and displacement asymmetries. A total residual scan asymmetry as low as 0.0035/mm has been measured over 4-mm scan fields at the Au target in the detector. This instrumental asymmetry would produce a maximum error in the polarization in a SEMPA experiment of less than 0.3% for a 100-micron full-field scan. Details of the design and performance of the new detector are given.

900,219
PB89-230312 Not available NTIS
National Inst. of Standards and Technology (NML), Gaithersburg, MD. Inorganic Analytical Research Div.
Introduction to Supercritical Fluid Chromatography. Part 2. Applications and Future Trends.
Final rept.
M. D. Palmieri. 1989, 7p
Pub. in Jnl. of Chemical Education 66, n5 pA141-A147 May 89.

Keywords: *Chromatographic analysis, Chemical analysis, Bioassay, Organic compounds, Separation, Trends, Forecasting, Reprints, *Supercritical fluid chromatography.

The article describes applications and future trends of supercritical fluid chromatography in analytical chemistry. Examples are given showing the types of organic and biological compounds which can be separated using supercritical fluid chromatography. Present and future areas of research in supercritical fluid chromatography are described.

900,220
PB89-230338 Not available NTIS
National Inst. of Standards and Technology (NML), Gaithersburg, MD. Inorganic Analytical Research Div.
Isotope Dilution Mass Spectrometry for Accurate Elemental Analysis.
Final rept.
J. D. Fassett, and P. J. Paulsen. 1989, 7p
Pub. in Analytical Chemistry, p643A-649A, 15 May 89.

Keywords: *Chemical analysis, *Mass spectroscopy, *Inorganic compounds, Reviews, Reprints, *Isotope dilution.

The technique of isotope dilution mass spectrometry (IDMS) as it is applied to inorganic analysis is reviewed. The principles of isotopic spiking are discussed as well as the possibilities of high accuracy, high precision measurement. Recent improvements in instrumentation are cited and special mention is made of the use of IDMS with inductively coupled plasma mass spectrometry. The present role of IDMS in ana-

CHEMISTRY

Analytical Chemistry

lytical chemistry is assessed and a future; expanded role promoted.

900.221
PB89-231070   Not available NTIS
National Inst. of Standards and Technology (NML), Gaithersburg, MD. Inorganic Analytical Research Div.
Development of the NBS (National Bureau of Standards) Beryllium Isotopic Standard Reference Material.
Final rept.
J. D. Fassett, K. G. W. Inn, and R. Watters. 1988, 4p
See also DE88002671.
Pub. in RIS 88, Institute of Physics, Conference Series No. 94: Section 11, Gaithersburg, MD, April 10-15, 1988, p379-382.

Keywords: *Beryllium isotopes, *Mass spectroscopy, *Atomic spectroscopy, Spectrum analysis, Chemical analysis, *Standard reference materials, National Institute of Standards and Technology.

A beryllium isotopic standard is being developed for the Accelerator Mass Spectrometry community. A critical aspect in this exercise is the evaluation of the isotopic composition of the (10)Be enriched source material. A procedure is presented here for the indirect determination of mass spectrometric isotopic discrimination, a potential systematic error in this evaluation. The procedure is demonstrated through the combined application of resonant ionization mass spectrometry and inductively coupled plasma atomic spectroscopy.

900.222
PB89-235642
                    (Order as PB89-235634, PC A04)
National Inst. of Standards and Technology, Gaithersburg, MD.
Determination of Trace Level Iodine in Biological and Botanical Reference Materials by Isotope Dilution Mass Spectrometry.
Bi-monthly rept.
J. W. Gramlich, and T. J. Murphy. 1989, 6p
Included in Jnl. of Research of the National Institute of Standards and Technology, v94 n4 p215-220 Jul/Aug 89.

Keywords: *Iodine, *Trace elements, *Mass spectrometry, Chemical analysis, Milk, Plants(Biology), Isotopes, Lanthanum compounds, *Standard reference materials.

A method has been developed for the determination of trace level iodine in biological and botanical materials. The method consists of spiking a sample with (129)I, equilibration of the spike with the natural iodine, wet ashing under carefully controlled conditions, and separation of the iodine by co-precipitation with silver chloride. Measurement of the (129)I/(127)I ratio is accomplished by negative thermal ionization mass spectrometry using LaB6 for ionization enhancement. The application of the method to the certification of trace iodine in two Standard Reference Materials is described.

900.223
PB89-235659
                    (Order as PB89-235634, PC A04)
National Inst. of Standards and Technology, Gaithersburg, MD.
Spectrum of Doubly Ionized Tungsten (W III).
Bi-monthly rept.
L. Iglesias, M. I. Cabeza, F. R. Rico, and O. Garcia-Riquelme. and V. Kaufman. 1989, 38p
Prepared in cooperation with Consejo Superior de Investigaciones Cientificas, Madrid (Spain). Inst. de Optica.
Included in Jnl. of Research of the National Institute of Standards and Technology, v94 n4 p221-258 Jul/Aug 89.

Keywords: *Tungsten, *Spectroscopy, *Ionization, Tables(Data).

The spectrum of doubly ionized tungsten (W III) was produced in a sliding-spark discharge and recorded photographically on the NIST 10.7-m normal-incidence vacuum spectrograph in the 600-2680 A spectral region. The analysis has led to the establishment of 71 levels of interacting 5d(4), 5d(3)6s(2) even configurations and 164 levels of the interacting 5d(3)6p and 5d(2)6s6p odd ones. A total of 2636 lines have been classified as transitions between the 235 experimentally determined levels. Comparison between the observed levels and those computed from matrix diagonalizations with least-squares fitted parameters shows an rms deviation of + or - 87/cm for the even configurations and + or - 450/cm for the odd ones.

900.224
PB90-117300   Not available NTIS
National Inst. of Standards and Technology (NEL), Gaithersburg, MD. Semiconductor Electronics Div.
Interlaboratory Determination of the Calibration Factor for the Measurement of the Interstitial Oxygen Content of Silicon by Infrared Absorption.
Final rept.
A. Baghdadi, W. M. Bullis, M. C. Croarkin, Y. Li, R. I. Scace, R. W. Series, P. Stallhofer, and M. Watanabe. 1989, 10p
Pub. in Jnl. of the Electrochemical Society 136, n7 p2015-2024 Jul 89.

Keywords: *Silicon, *Quantitative analysis, *Calibrating, *Oxygen, *Infrared spectroscopy, *Interstitials, Reproducibility, Accuracy, Reprints.

An international interlaboratory dual experiment was performed to determine the calibration factor used to calculate the interstitial oxygen content of silicon from room-temperature (300 K) infrared (IR) absorption measurements. Round robins were conducted for both the infrared and the absolute measurements on the same or equivalent specimens. The calibration factor for computing the oxygen content of silicon in parts per million atomic (ppma) from a room-temperature measurement of the absorption coefficient at 1107/cm was determined to be 6.28 + or - 0.18 ppma/cm. The IR round robin showed a reproducibility on the order of 3%.

900.225
PB90-117912   Not available NTIS
National Inst. of Standards and Technology (NML), Gaithersburg, MD. Chemical Thermodynamics Div.
Differential Scanning Calorimetric Study of Brain Clathrin.
Final rept.
F. P. Schwarz, C. J. Steer, and W. H. Kirchhoff.
1989, 7p
Pub. in Archives of Biochemistry and Biophysics 273, n2 p433-439 Sep 89.

Keywords: *Brain, *Chemical analysis, Cattle, Enthalpy, Reprints, *Cell membrane coated pits, *Clathrin, *Differential scanning calorimetry.

The thermal denaturation of clathrin-coated vesicles isolated from bovine brain tissue has been studied by differential scanning calorimetry and has been compared to basket structures reformed from isolated triskelion trimers of clathrin and to isolated triskelions. The coated vesicles and reformed vesicles exhibited a single denaturation transition peak at 55.9 + or - 0.1 C, skewed to low temperatures whereas the thermograms for the reformed baskets exhibited a broad transition peak at 53.1 + or - C and a peak at 56.3 + or - 0.1 C. Neither transition was reversible. The specific transition enthalpy was 11.5 + or 1.0 J/g for the coated vesicles and the total transition enthalpy was 9.1 + or - 0.3 J/g for the reformed baskets. In contrast, isolated triskelions showed no thermal transition between 15 and 90 C. Although the coated vesicles and the reformed baskets have similar stability reflecting their similar structures, the coated vesicles appear to be marginally more stable than the reformed baskets. The complexity of the transition profiles and their lack of symmetry suggest the existence of several, somewhat independent, domains unique to the cage-like structure of the coated vesicles and reformed baskets.

900.226
PB90-118175   Not available NTIS
National Inst. of Standards and Technology (NML), Gaithersburg, MD. Inorganic Analytical Research Div.
Preconcentration of Trace Transition Metal and Rare Earth Elements from Highly Saline Solutions.
Final rept.
D. M. Strachan, S. Tymochowicz, P. Schubert, and H. M. Kingston. 1989, 7p
Contract DE-AC06-76RL01830
Sponsored by Department of Energy, Washington, DC.
Pub. in Analytica Chimica Acta 220, p243-249 1989.

Keywords: *Quantitative analysis, *Brines, *Transition metals, *Rare earth elements, Concentration(Composition), Chelation, Reprints, *Trace amounts.

Quantitative recovery and preconcentration of trace amounts of Ce(III), Co(III), Eu(III), Fe(III), Gd(III), Mn(II), Y(III) and Zn(II) ions from nearly saturated brines on the chelating resin Chelex-100 are described. Carrier-

free radioactive isotopes were used. Only manganese was significantly affected by the high ionic strength of the brines. Chromium(III) was retained quantitatively by the resin but not eluted quantitatively. The results indicate that transition metal and rare earth ions can be quantitatively preconcentrated from solutions of low and high ionic strength.

900.227
PB90-118191   Not available NTIS
National Inst. of Standards and Technology (NML), Gaithersburg, MD. Inorganic Analytical Research Div.
Introduction to Microwave Acid Decomposition.
Final rept.
L. B. Jassie, and H. M. Kingston. 1988, 6p
Pub. in Introduction to Microwave Acid Decomposition, Chapter 1, p1-6 1988.

Keywords: *Acid treatment, *Decomposition, *Microwave equipment, Chemical analysis, Reviews, Reprints, *Sample preparation.

A chronological review of the literature on microwave-assisted sample preparations is presented. Temperature measurements and other innovations in the field of sample decomposition using microwave systems are surveyed.

900.228
PB90-123399   Not available NTIS
National Inst. of Standards and Technology (IMSE), Gaithersburg, MD. Polymers Div.
Calcium Hydroxyapatite Precipitated from an Aqueous Solution: An International Multimethod Analysis.
Final rept.
J. Arends, J. Christoffersen, M. R. Christoffersen, H. Eckert, B. O. Fowler, J. C. Heughebaert, G. H. Nancollas, J. P. Yesinowski, and S. J. Zawacki.
1987, 18p
Pub. in Jnl. of Crystal Growth 84, n3 p515-532 1987.

Keywords: *Chemical   analysis, *Precipitation(Chemistry), Synthesis(Chemistry), Infrared spectroscopy, Nuclear magnetic resonance, Differential thermal analysis, Area, Crystal growth, X-ray diffraction, Reprints, *Standard reference materials, *Apatite/(calcium-salt)-hydroxyl, Scanning electron microscopy.

Calcium hydroxylapatite (Ca10(PO4)6(OH)2), prepared for use as a standard reference material from aqueous solutions at 70 C, was analyzed by various techniques in six different institutes to determine its purity and composition. The techniques employed were: chemical analyses, x-ray diffraction, infrared analysis, magic angle spinning nuclear magnetic resonance, differential thermal analysis, scanning electron microscopy, surface area and crystal size distribution measurements, crystal growth and crystal dissolution measurements.

900.229
PB90-123464   Not available NTIS
National Inst. of Standards and Technology (NML), Gaithersburg, MD. Inorganic Analytical Research Div.
Comparison of Microwave Drying and Conventional Drying Techniques for Reference Materials.
Final rept.
E. S. Beary. 1988, 5p
Pub. in Analytical Chemistry 60, n8 p742-746 1988.

Keywords: *Ovens, *Microwave equipment, Substitutes, Comparison, Performance evaluation, Laboratory equipment, Chemical analysis, Reprints, *Sample preparation, *Standard reference materials.

A microwave drying oven was evaluated for potential use as an alternate method of sample pre-treatment in an analytical laboratory. The high precision drying data achieved suggests that this instrument is well suited to industrial applications such as quality control and merits further study. However, differences in conventional drying and microwave drying makes the current implementation of the microwave drying oven unfeasible in a reference laboratory. In addition, a stable sample composition must be confirmed before the method could be recommended as a drying procedure preceeding sample analysis.

900.230
PB90-123472   Not available NTIS
National Inst. of Standards and Technology (NML), Gaithersburg, MD. Inorganic Analytical Research Div.

28

Determination of Selenium and Tellurium in Copper Standard Reference Materials Using Stable Isotope Dilution Spark Source Mass Spectrometry.
Final rept.
E. S. Beary, P. J. Paulsen, and G. M. Lambert. 1988, 4p
Pub. in Analytical Chemistry 60, n7 p733-736 1988.

Keywords: *Selenium, *Tellurium, *Mass spectroscopy, *Chemical analysis, Copp, Reprints, *Isotope dilution, *Standard reference materials, Tracer studies, Selenium 75.

The concentration of Se and Te have been determined by isotope dilution spark source mass spectrometry (ID SSMS). Discrepancies in Se concentrations prompted an extensive evaluation of the copper benchmarks series. Because certain forms of Se are volatile, a (75)Se tracer study aided in the identification of areas where Se losses could occur. In isotope dilution, slight losses of Se after equilibration of the natural and enriched isotopes would not alter the calculated concentrations. They are based on data generated from the measured isotopic ratio which is established during equilibration. Agreement of Se concentrations between copper samples dissolved in covered teflon beakers with those dissolved in carius tubes assured that no Se was volatilized prior to equilibration. The Cu samples analyzed had Se and Te concentration ranging from 0.59-480 micrograms/g and 0.29-196 micrograms/g respectively with about a 2% relative standard deviation. The blank contribution was negligible.

900,231
PB90-128539 Not available NTIS
National Inst. of Standards and Technology (NML), Gaithersburg, MD. Organic Analytical Research Div.
Comparison of Liquid Chromatographic Selectivity for Polycyclic Aromatic Hydrocarbons on Cyclodextrin and C18 Bonded Phases.
Final rept.
M. Olsson, L. C. Sander, and S. A. Wise. 1989, 14p
Pub. in Jnl. of Chromatography 477, p277-290 1989.

Keywords: *Aromatic polycyclic hydrocarbons, *Chemical analysis, *Chromatographic analysis, *Dextrins, Comparison, Separation, Performance evaluation, Molecular structure, Reprints, *Liquid chromatography, Retention, Selection rules.

Selectivity towards polycyclic aromatic hydrocarbons (PAHs) was studied on cyclodextrin bonded phases and compared to selectivity observed on C18 phases. The study included the separation of eleven five-ring PAH isomers on each of three phase types; monomeric C18, polymeric C18 and cyclodextrin. Retention of PAHs ranging in size from three to six condensed rings was also investigated. Retention on the cyclodextrin phase is based on inclusion complexing between the solute and cyclodextrin cavity, resulting in a strong shape dependence. However, the shape selectivity exhibited by the cyclodextrin phase is different from that exhibited by either the monomeric or polymeric C18 phases; retention on the cyclodextrin phase is strongly dependent on the shape and shows very little molecular weight dependence. Calculations of solute molecular widths were performed to predict the isomers' ability to enter the cyclodextrin cavity. The effect of sample solvent and injection volume was also investigated for the cyclodextrin phase. A retention model based on the solute shape is proposed for PAH isomers on Beta-cyclodextrin phase.

900,232
PB90-128653 Not available NTIS
National Inst. of Standards and Technology (NML), Gaithersburg, MD. Gas and Particulate Science Div.
Determination of Experimental and Theoretical k(sub ASI) Factors for a 200-kV Analytical Electron Microscope.
Final rept.
P. J. Sheridan. 1989, 21p
Pub. in Jnl. of Electron Microscopy Technique 11, p41-61 1989.

Keywords: *Quantitative analysis, *Electron microscopy, Glass, X ray analysis, Grinding(Comminution), Standards, Particle size, Reprints.

The relative sensitivity of an analytical electron microscope and energy-dispersive x-ray detector to x-rays of various elements is investigated through an extensive k(sub ASI) factor study. Elemental standards, primarily National Bureau of Standards multielement research glasses, were dry-ground into submicrometer-

sized particles and analyzed at 200 kV accelerating potential. The effect of self-absorption of x-rays by the particle has been corrected for, allowing the experimental k(sub ASI) factors from the study to approximate those that could be obtained from 'infinitely thin' specimens. Whenever possible, elemental k-factors were determined by the analysis of many (up to a maximum of nine) different standard materials. Experimental k(sub ASI) factors were calculated for a wide range of K(sub alpha), L(sub alpha), and M(sub alpha) x-ray lines. For comparison, theoretical k(sub ASI) factors, employing a variety of ionization cross sections, were computed. Good agreement is obtained between several of the theoretical k-factor models and the experimental results. Mass volatilization of Na and K from the small glass particles during analysis is discussed, as are observations that the grinding and/or dispersing of standard materials in a liquid (such as ethanol) may promote leaching of certain elements from the particle matrix.

900,233
PB90-128695 Not available NTIS
National Inst. of Standards and Technology (NML), Gaithersburg, MD. Organic Analytical Research Div.
Synthesis and Characterization of Novel Bonded Phases for Reversed-Phase Liquid Chromatography.
Final rept.
A. M. Stalcup, D. E. Martire, L. C. Sander, and S. A. Wise. 1989, 7p
Pub. in Chromatographia 27, n9/10 p405-411 May 89.

Keywords: *Chromatographic analysis, *Aromatic polycyclic hydrocarbons, *Synthesis(Chemistry), *Chemical analysis, Comparison, Chemical bonds, Separation, Reprints, *Liquid chromatography, Retention, Selection rules.

The results of a chromatographic comparison of newly synthesized reversed-phase bonded stationary phases incorporating various degrees of unsaturation or rigidity are presented. The phases studied included monomeric n-decyl, polymeric n-decyl, adamantyl, norpinyl, decadiynyl, phenylbutyl and Beta-naphthyl bonded phases. The effect of bonded phase structure on retention is discussed for selected aromatic solutes.

900,234
PB90-128786 Not available NTIS
National Inst. of Standards and Technology (NML), Gaithersburg, MD. Ionizing Radiation Physics Div.
Pattern Recognition Approach in X-ray Fluorescence Analysis.
Final rept.
L. I. Yin, J. I. Trombka, and S. M. Seltzer. 1989, 8p
Pub. in Nuclear Instruments and Methods in Physics Research A277, p619-626 1989.

Keywords: *Pattern recognition, *Alloys, Computation, Reprints, *X-ray fluorescence analysis.

In many applications of X-ray fluorescence (XRF) analysis, quantitative information on the chemical components of the sample is not of primary concern. Instead, the XRF spectra are used to monitor changes in the composition among samples, or to select and classify samples with similar compositions. The authors propose in this paper that the use of pattern recognition technique in such applications may be more convenient than traditional quantitative analysis. The pattern recognition technique discussed here involves only one parameter, i.e., the normalized correlation coefficient and can be applied directly to raw data. Its computation is simple and fast, and can be easily carried out on a personal computer. The efficacy of this pattern recognition approach is luuustrated withthe analysis of experimental XRF spectra obtained from geological and alloy samples.

900,235
PB90-135922 Not available NTIS
National Inst. of Standards and Technology (NML), Gaithersburg, MD. Gas and Particulate Science Div.
Low Pressure, Automated, Sample Packing Unit for Diffuse Reflectance Infrared Spectrometry.
Final rept.
A. A. Christy, J. E. Tvedt, T. V. Karstang, and R. A. Velapoldi. 1988, 4p
Pub. in Review of Scientific Instruments 59, n3 p423-426 Mar 88.

Keywords: *Sample preparation, *Infrared spectroscopy, Chemical analysis, Low pressure tests, Solids,

Design criteria, Particle size distribution, Performance evaluation, Automatic control equipment, Reprints, *Fourier transform spectroscopy, *Packing.

The authors have designed and built an automatic, low pressure packing unit to prepare ground, solid samples for diffuse reflectance infrared Fourier transform spectroscopy (DRIFTS). Use of this unit coupled with sample rotation during measurement, and control of time, pressure, particle size and size distribution provides excellent precision in obtaining DRIFTS spectra. Thus representative DRIFTS spectra can be obtained quickly and efficiently, with a single spectrum as opposed to previous efforts requiring the averaging of several spectra.

900,236
PB90-136540 Not available NTIS
National Inst. of Standards and Technology (NML), Gaithersburg, MD. Gas and Particulate Science Div.
Identification of Carbonaceous Aerosols via C-14 Accelerator Mass Spectrometry, and Laser Microprobe Mass Spectrometry.
Final rept.
L. A. Currie, R. A. Fletcher, and G. A. Klouda. 1987, 8p
Pub. in Nuclear Instruments and Methods in Physics Research Section B - Beam Interactions with Materials and Atoms 29, n1-2 p346-354 1987.

Keywords: *Chemical analysis, *Aerosols, *Carbon 12, *Carbon 14, *Mass spectroscopy, Particles, Sampling, Performance evaluation, Reprints, *Laser spectroscopy, Atmospheric chemistry.

Carbon isotopic measurements of ((12)C, (14)C), derived from chemical measurements of total carbon plus AMS measurements of (14)C/(12)C have become an accepted means for estimating fossil and contemporary carbon source contributions to atmospheric carbon. Because of the limited sensitivity of these techniques, however, such measurements are restricted to 'bulk' samples comprising at least 10 - 100 micrograms of carbon. Laser mass spectrometry offers an important complementary opportunity to investigate the chemical nature of individual particles as small as 0.1 micrometers in diameter. Although there is little hope to measure (14)C/(12)C in such small samples, the compositional and structural information available with the laser microprobe is of interest for possible source discrimination. Also, the analysis of individual particles, which may reflect individual sources, yields significant potential increases in spatial, temporal and source resolution, in comparison to bulk sample analysis.

900,237
PB90-136979 Not available NTIS
National Inst. of Standards and Technology (NML), Gaithersburg, MD. Organic Analytical Research Div.
Spectroelectrochemistry of a System Involving Two Consecutive Electron-Transfer Reaction.
Final rept.
W. T. Yap, G. Marbury, E. A. Blubaugh, and R. A. Durst. 1989, 5p
Pub. in Jnl. of Electroanalytical Chemistry 271, p325-329 1989.

Keywords: *Spectrum analysis, *Electrochemistry, *Electron transfer, Nernst effect, Nonlinear systems, Numerical analysis, Chemical reactions, Separation, Chemical analysis, Reprints, Potentials.

Analysis of the spectroelectrochemistry of a multi-step electrochemical process in a thin layer cell is presented. Typical characteristic Nernst plots for various formal potential separations and various extinction coefficients of the species are illustrated. A method for calculating the formal potentials from these non-linear Nernst plot is suggested, and an illustrative application is given.

Basic & Synthetic Chemistry

900,238
PB89-146963 Not available NTIS
National Bureau of Standards (NEL), Gaithersburg, MD. Building Materials Div.

CHEMISTRY

Basic & Synthetic Chemistry

Synthesis and Characterization of Ettringite and Related Phases.
Final rept.
L. Struble. 1986, 7p
Sponsored by Department of Energy, Washington, DC. Passive and Hybrid Solar Energy Div.
Pub. in Proceedings of International Congress on the Chemistry of Cement (8th), Rio de Janeiro, Brazil, September 22-27, 1986, p582-588.

Keywords: *Dehydration, *Phase diagrams, *Energy storage, Synthesis, Enthalpy, Compositions, Assessments, Evaluation, Reprints, *Ettringite, *Energy storage materials.

Ettringite and related phases were studied to assess their performance as energy storage materials, utilizing the dehydration reaction. Ettringite and four isostructural phases were synthesized and characterized, particularly with regard to thermal properties, to study the effect of chemical composition on such thermal parameters as the temperature and the enthalpy change of the dehydration reaction. Synthesis procedures were developed for following phases: (Ca3Al(OH)6)2(SO4)3x26 (ettringite), (Ca3Fe(OH)6)2(SO4)3x26H2O, (Ca3Cr(OH)6)2(SO4)3x26H2O, (Ca3Al(OH)6)2(CO3)3x26H2O, and (Ca3Si(OH)6)2(SO4)2(CO3)2x24H2O (thaumasite). Each synthesized material was examined for bulk chemical composition, phase composition, unit cell parameters, morphology, and thermal properties. Each phase showed a low-temperature dehydration with initial temperature ranging from 6 C to 33 C. The change in enthalpy associated with the dehydration ranged from 60 to 200 cal/g sample. Thus both the dehydration temperature and the enthalpy change varied significantly for the isostructural phases studied.

900,239
PB89-228423   Not available NTIS
National Inst. of Standards and Technology (NML), Gaithersburg, MD. Surface Science Div.
Cr(110) Oxidation Probed by Carbon Monoxide Chemisorption.
Final rept.
N. D. Shinn. 1989, 13p
Sponsored by Department of Energy, Washington, DC.
Pub. in Surface Science 214, p174-186 1989.

Keywords: *Oxidation reduction reactions, *Chromium oxides, *Chemisorption, *Carbon monoxide, Electron energy, Annealing, Oxygen, Vibrational spectra, Surface chemistry, Reprints, Binding sites.

High resolution electron energy spectroscopy of chemisorbed carbon monoxide is used to distinguish between an annealed, oxidized Cr(110) surface, an intermediate, subsurface Cr(110) oxide, and oxygen-dosed Cr(110). On the annealed oxide, weak signals are observed from both the dissociation-precursor (alpha 1-CO) and the terminally-bonded (alpha 2-CO) molecular CO binding states, but only after relatively high CO exposures, reflecting greatly reduced sticking probabilities. On the surface with subsurface oxygen, both binding states are sequentially populated with sticking probabilities comparable to that of clean Cr(110); also, an increase in nuCO from approximately 1975 to 2035/cm is observed at high coverages of terminally-bonded alpha sub 2-CO. Both the annealed-oxide and subsurface-oxide CO chemisorption results are in contrast to the selective poisoning of only alpha sub 1-CO by chemisorbed atomic oxygen on Cr(110). Comparisons among the three oxygen containing surfaces show how vibrational spectroscopy of chemically-inequivalent molecular binding states may be used as a probe of surface oxidation and provide insights into the oxygen-CO surface chemistry.

900,240
PB90-123753   Not available NTIS
National Inst. of Standards and Technology (NML), Gaithersburg, MD. Organic Analytical Research Div.
Facile Synthesis of 1-Nitropyrene-d9 of High Isotopic Purity.
Final rept.
A. J. Fatiadi, and L. R. Hilpert. 1989, 8p
Pub. in Jnl. of Labelled Compounds and Radiopharmaceuticals XXVII, n2 p129-136 1989.

Keywords: *Chemical reactions, *Isotopes, Aromatic polycylic hydrocarbons, Gas chromatography, Mass spectroscopy, Reprints, *Standard reference materials, *Nitropyrenes, Liquid chromatography.

1-Nitropyrene-d9, 1,3-dinitropyrene-d8, 1,6-dinitropyrene-d8, and 1,8-dinitropyrene-d8, used as internal

standards for the determination of nitro-polycyclic aromatic hydrocarbons in simple and complex mixtures and required for GC/MS measurements used in the certification of NBS Standard Reference Material 1596 (a mixture of nitropyrenes), were synthesized in one step from commercially available pyrene-d10 and nitric acid-d1. The electron impact mass spectra and isotopic purity of 1-nitropyrene-d8 and the deuterated dinitropyrenes were determined; the compounds were characterized by gas chromatography-mass spectrometry and by high performance liquid chromatography.

Industrial Chemistry & Chemical Process Engineering

900,241
PB89-156376   PC A08/MF A01
National Inst. of Standards and Technology (NML), Gaithersburg, MD. Center for Chemical Technology.
Center for Chemical Technology: 1988 Technical Activities.
Summary rept. 1 Oct 87-30 Sep 88.
J. Hord. Dec 88, 162p NISTIR-88/3907
See also PB88-164272.

Keywords: *Chemical engineering, *Research projects, Measurement, Thermophysical properties, Standards, Reaction kinetics, Proteins, Calibrating, Fluid mechanics, Separation, Thermodynamics, Biotechnology.

Technical research activities performed by the Center for Chemical Technology during the Fiscal Year 1988 are summarized herein. These activities include work in the general categories of measurements (standards, processes, and equipment design), properties (thermophysical, thermochemical, and kinetic), and biotechnology (protein engineering and separations). They embody: development and improvement of measurement standards, measurement principles, and calibration services for pressure, temperature, volumetric and mass flow rates, liquid volume and density, humidity, and airspeed; generation (via accurate measurements and advanced predictive models) of reliable reference data for thermophysical, thermochemical, and kinetic properties of pure fluids, fluid mixtures, and solids of industrial and environmental importance; provision of fundamental understanding of protein structure-function and advanced technology for commercial scale separation of proteins; and development of improved correlations, models, and measurement techniques for complex flows, heat and mass transport, mixing, and chemically reacting flows of interest in modern unit operations.

900,242
PB89-174908   Not available NTIS
National Bureau of Standards (NEL), Boulder, CO. Chemical Engineering Science Div.
Latent Heats of Supercritical Fluid Mixtures.
Final rept.
M. C. Jones, and P. J. Giarratano. 1988, 4p
Pub. in AIChE (American Institute of Chemical Engineers) Jnl. 34, n12 p2059-2062 Dec 88.

Keywords: *Latent heat, *Decanes, Equations of state, Aliphatic hydrocarbons, Mixtures, Enthalpy, Carbon dioxide, Reprints, *Supercritical fluids, Heat of condensation, Supercritical pressures.

The Peng-Robinson equation of state was used to calculate partial molar enthalpies for coexisting vapor and liquid phases in the mixture N-decane-CO2 above the critical pressures of the pure components. From these, the heats of condensation in the retrograde region could be calculated.

900,243
PB89-176416   Not available NTIS
National Bureau of Standards (NEL), Gaithersburg, MD. Chemical Process Metrology Div.
Laser Excited Fluorescence Studies of Black Liquor.
Final rept.
J. J. Horvath, and H. G. Semerjian. 1986, 7p
Pub. in Proceedings of SPIE (Society of Photo-Optical Instrumentation Engineers)-Optical Techniques for Industrial Inspection, v665 p258-264 1986.

Keywords: *Black liquors, *Pulping, Near ultraviolet radiation, Process control, Papers, Pulps, Inspection, Lignin, *Laser induced fluorescence.

Laser excited fluorescence of black liquor was investigated as a possible monitoring technique for pulping processes. A nitrogen pumped dye laser was used to examine the fluorescence spectrum of black liquor solutions. Various excitation wavelengths were used between 290 and 403 nm. Black liquor fluorescence spectra were found to vary with both excitation wavelength and black liquor concentration. Laser excited fluorescence was found to be a sensitive technique for measurement of black liquor with good detection limits and linear response over a large dynamic range.

900,244
PB89-176481   Not available NTIS
National Bureau of Standards (NEL), Boulder, CO. Chemical Engineering Science Div.
Influence of Reaction Reversibility on Continuous-Flow Extraction by Emulsion Liquid Membranes.
Final rept.
D. L. Reed, A. L. Bunge, and R. D. Noble. 1987, 21p
Grant EPA-R-811247
Sponsored by Environmental Protection Agency, Washington, DC. Grants Administration Div.
Pub. in ACS (American Chemical Society) Symposium Series 347, p62-82 1987.

Keywords: *Solvent extraction, *Mathematical models, *Liquid filters, *Membranes, *Chemical reactivity, *Chemical equilibrium, Performance evaluation, Irreversible processes, Emulsions, Continuum mechanics, Reprints.

The paper examines theoretically the continuous flow extraction by emulsion globules in which the transferring solute reacts with an internal reagent. The reversible reaction model is used to predict performance. These results are compared with advancing front calculations which assume an irreversible reaction. A simple criterion which indicates the importance of reaction reversibility on performance is described. Calculations show that assuming an irreversible reaction can lead to serious underdesign when low solute concentrations are required. For low solute concentrations an exact analytical solution to the reversible reaction problem is possible. For moderate solute concentrations, the authors have developed an easy parameter adjustment of the advancing front model which reasonably approximates expected extraction rates.

900,245
PB89-176739   Not available NTIS
National Bureau of Standards (NEL), Gaithersburg, MD. Thermophysics Div.
Modelling of Impurity Effects in Pure Fluids and Fluid Mixtures.
Final rept.
J. S. Gallagher. 1986, 20p
Pub. in Proceedings of American Institute of Chemical Engineers Spring National Meeting and Petroleum Expo'86, New Orleans, LA., April 6-10, 1986, 20p.

Keywords: *Impurities, Mixtures, Ethylene, Thermodynamic properties, *Fluid modelling, *Impurity effects, Isobutane, Isopentane.

An extended corresponding-states model for the Helmholtz free energy for two-component mixtures, based upon existing accurate representations of the principal component as the reference function, has been used to model the effects of impurities. The results give a clearer and more accurate view of the errors caused in the measurement of thermodynamic properties by small amounts of impurity than was previously obtained with simplifying approximations. The model is also extended to include three-component mixtures to allow the estimation of the effects of impurities in two-component mixtures. The model is applied to two systems of practical importance: methane as an impurity in ethylene, and n-butane as an impurity in isobutane-isopentane mixtures. Both of these systems are of commercial importance and commonly used in their critical regions where the impurities can cause large errors in the thermodynamic properties used and where existing procedures for the estimation of the impurity effects break down.

900,246
PB89-179584   Not available NTIS
National Bureau of Standards (NEL), Gaithersburg, MD. Chemical Process Metrology Div.

Dynamic Light Scattering and Angular Dissymmetry for the In situ Measurement of Silicon Dioxide Particle Synthesis in Flames.
Final rept.
M. R. Zachariah, D. Chin, H. G. Semerjian, and J. L. Katz. 1989, 7p
Pub. in Applied Optics 28, n3 p530-536, 1 Feb 89.

Keywords: *Silicon dioxide, *Particle size, *Light scattering, Diffusion flames, Optical measurement, Process control, Reprints, *Synthesis(Chemistry).

Particle size measurements have been made of silica formation in a counterflow diffusion flame reactor using dynamic light scattering and angular dissymmetry methods. The results suggest that the techniques compare quite favorably in conditions of high signal to noise. However, the dynamic light scattering technique degrades rapidly as the signal strength declines, resulting in erroneously small particle diameters. As a general rule, dynamic light scattering does not seem to possess the versatility and robustness of the classical techniques as a possible on-line diagnostic for process control. The drawbacks and limitations of the two techniques are also discussed.

Photo & Radiation Chemistry

900,247
PB89-146682   Not available NTIS
National Bureau of Standards (IMSE), Gaithersburg, MD. Polymers Div.
Radiation-Induced Crosslinks between Thymine and 2-D-Deoxyerythropentose.
Final rept.
M. Farahani, and W. L. McLaughlin. 1988, 4p
Pub. in Radiation Physics and Chemistry 32, n6 p731-734 1988.

Keywords: *Crosslinking, *Molecular structure, *Radiation chemistry, Free radicals, Gas chromatography, Mass spectroscopy, Deoxyribonucleic acids, Reprints, *Hydroxyl radicals, *Erythropentose/deoxy, *Thymine, Chemical reaction mechanisms.

Hydroxyl radicals generated by ionizing radiation in aqueous solutions of thymine (Thy) and 2-D-deoxyerythropentose (dR) induce crosslinking between Thy and dR. The crosslinked products were identified by gas chromatography--mass spectrometry (GC-MS) and their yields determined by GC. The mechanisms of their formation are discussed and may serve as a model for radiation-induced or free-radical induced intra-DNA crosslinks.

900,248
PB89-147490   Not available NTIS
National Bureau of Standards (NML), Gaithersburg, MD. Ionizing Radiation Physics Div.
Dichromate Dosimetry: The Effect of Acetic Acid on the Radiolytic Reduction Yield.
Final rept.
M. Al-Sheikhly, M. H. Hussmann, and W. L. McLaughlin. 1988, 7p
Pub. in Radiation Physics and Chemistry 32, n3 p545-551 1988.

Keywords: *Reduction(Chemistry), *Potassium chromates, *Gamma irradiation, *Visible spectrum, Acetic acid, Nitrogen, Oxygen, Dosimetry, Silver, Hydrogen peroxide, Reprints.

The radiation chemical yield for the reduction of dichromate, Cr(VI) to Cr(III), in an acidic aqueous perchloric acid solution of potassium dichromate, may be increased from 0.04 to greater than 0.2 micro mol/J sub -1) by adding acetic acid. The increased yield, G, is about the same in N(sub 2-) and O(sub 2-) saturated solutions. The molar linear absorption coefficient at 350 n also is the same in both solutions. Epsilon sub-m = 2800-M(sup -1)cm(sup -1) at pH 0.4. The proposed mechanism to explain the enhanced response in N(sub 2-) saturated solutions involves the efficient reaction of acetic acid with hydroxyl radicals by the abstraction of H from the methyl group; the resulting acid radicals react with relatively high yield to reduce Cr(VI). In O(sub 2-) saturated solution, the acetic acid radical apparently goes through an acetic acid peroxyl radical by a bimolecular reaction to the tetroxide intermediate of acetic acid, which releases H2O2 with relatively high yield by a Bennett-type reaction. This additional H2O2, as a reducing agent, reacts slowly with dichromate and

boosts the value of G. The negative slope of the response continues to increase during the period immediately after irradiation of oxygenated solution, due to slow reaction of radiolytically-produced H2O2 with dichromate. There is also in both O(sub 2-) and N(sub 2-) saturated solution a long-term slow reaction involving oxidation of the organic substrate (in this case, acetic acid). Because of these instabilities, the solutions cannot readily be used for dosimetry without the presence of silver ions, which in the oxidized state, Ag(sup 2+), act to stabilize the solution after irradiation.

900,249
PB89-156954   Not available NTIS
National Bureau of Standards (NML), Gaithersburg, MD. Inorganic Analytical Research Div.
High Accuracy Determination of (235)U in Nondestructive Assay Standards by Gamma Spectrometry.
Final rept.
R. R. Greenberg, and B. S. Carpenter. 1987, 21p
Pub. in Jnl. of Radioanalytical and Nuclear Chemistry 111, n1 p177-197 Apr 87.

Keywords: *Uranium 235, *Chemical analysis, *Assaying, *Gamma ray spectroscopy, *Accountability, Nondestructive tests, Standards, Accuracy, Uranium oxides, Radioactive isotopes, Mineral deposits, Reprints.

High precision gamma spectrometry measurements have been made on five sets of uranium isotope abundance reference materials for nondestructive assay (NDA). These sets are intended for international safeguards use as primary reference materials for the determination of the 235U abundance in homogeneous uranium bulk material by gamma spectrometry. The measurements were made to determine the count rate uniformity of the 235U 185.7 keV gamma-ray as well as the 235U isotope abundance for each sample. Since the samples were packaged such that the U308 is infinitely thick for the 185.7 keV gamma-ray, the measured count rate was not dependent on the material density. In addition, the activity observed by the detector was collimated to simulate calibration conditions used to measure bulk material in the field. The results of the study indicate that accuracy of 235U determination via gamma spectrometry, in the range of few hundredths of a percent (2 sigma), is achievable. The main requirement for achieving this level of accuracy is a set of standards whose 235U isotope abundances are known to within 0.01% (2 sigma).

900,250
PB89-157135   Not available NTIS
National Bureau of Standards (IMSE), Gaithersburg, MD. Polymers Div.
Effect of pH on the Emission Properties of Aqueous tris (2,6-dipicolinato) Terbium (III) Complexes.
Final rept.
T. K. Trout, J. M. Bellama, R. A. Faltynek, E. J. Parks, and F. E. Brinckman. 1989, 3p
Pub. in Inorganica Chimica Acta 155, p13-15 1989.

Keywords: *Metal complexes, *Terbium, *Luminosity, Emissivity, Pyridines, Carboxylic acid esters, pH, Ligands, Dissociation, Reprints.

Emission characteristics of the complexes formed between terbium(III) and pyridine-2,6-dicarboxylic acid have been measured as a function of solution pH and the molar ratio of ligand to metal present. The liability of these complexes was clearly demonstrated via correlation of solution pH to the pKa values of the complexes and the non-coordinated ligand.

900,251
PB89-157556   Not available NTIS
National Bureau of Standards (NML), Gaithersburg, MD. Chemical Kinetics Div.
Dehydrogenation of Ethanol in Dilute Aqueous Solution Photosensitized by Benzophenones.
Final rept.
P. Green, W. A. Green, A. Harriman, M. C. Richoux, and P. Neta. 1988, 19p
Pub. in Jnl. of the Chemical Society, Faraday Transactions I 84, pt6 p2109-2127 1988.

Keywords: *Ethyl alcohol, *Dehydrogenation, *Photochemical reactions, *Benzophenones, Solutions, Chemical properties, Molecular energy levels, Free radicals, Radiolysis, Catalysts, Reaction kinetics, Reprints, *Hydrogen production.

The photochemical properties of a series of water-soluble benzophenones have been evaluated in dilute

aqueous solution. The compounds possess lowest energy singlet and triplet excited states demonstrating considerable n, *character. As such, irradiation of the compounds in aqueous solution containing ethanol (2% v/v) results in pinacol formation via a triplet state hydrogen abstraction process. In the presence of a colloidal Pt catalyst, the intermediate ketyl and 1-methyl-1-hydroxyethyl radicals can be used to reduce water to H2. The rate of H2 formation and its total yield depend upon the nature of the substituent used to solubilize the benzophenone. The rate at which the ketyl radical transfers an electron to the Pt particles can be rationalized in terms of thermodynamic and electrostatic factors.

900,252
PB89-157697   Not available NTIS
National Bureau of Standards (NML), Gaithersburg, MD. Surface Science Div.
Synchrotron Radiation Study of BaO Films on W(001) and Their Interaction with H2O, CO2, and O2.
Final rept.
D. R. Mueller, A. Shih, E. Roman, T. E. Madey, R. Kurtz, and R. Stockbauer. 1988, 5p
Sponsored by Office of Naval Research, Arlington, VA.
Pub. in Jnl. of Vacuum Science and Technology A 6, n3 p1067-1071 May/Jun 88.

Keywords: *Oxygen, *Carbon dioxide, *Water, *Barium oxides, *Adsorption, *Tungsten, *Photoelectric emission, Ultraviolet spectroscopy, Hydroxides, Carbonates, Thin films, Cathodes, Radiation effects, Synchrotrons, Substrates, Reprints.

The interaction of O2, CO2, and H2O with bulk BaO and BaO adlayers adsorbed on W(001) has been examined using ultraviolet photoelectron spectroscopy. H2O reacts with bulk BaO to form Ba(OH)2, while CO2 forms a surface layer of BaCO3. Water and carbon dioxide also-react with a (2 sup(1/2))R 45-BaO monolayer adsorbed on W(001) to produce adsorbed OH and CO3 species bound to the tungsten substrate. The interaction of O2 with W(001) is enhanced by the presence of a BaO monolayer on the substrate. The observations are compared with the results of previous studies.

900,253
PB89-171193   Not available NTIS
National Bureau of Standards (NML), Boulder, CO. Quantum Physics Div.
Photodissociation of Methyl Iodide Clusters.
Final rept.
D. J. Donaldson, S. Sapers, V. Vaida, and R. Naaman. 1987, 11p
Pub. in Large Finite Systems, p253-263 1987.

Keywords: *Dissociation, *Photochemical reactions, *Iodoalkanes, Absorption spectra, Rare gases, Dimerization, Clustering, Ionization, Reprints, Methyl iodide, Rydberg states.

The dissociation of methyl iodide molecules 'solvated' in clusters has been investigated using both direct absorption spectroscopy and multiphoton ionization methods. Clusters of both neat methyl iodide and of methyl iodide with rare gases were studied in a molecular jet. It was found that dimerization slows the predissociation rate from the Rydberg states of CH3I, whereas in large clusters the direct dissociation from the valence state is slowed. A model is presented that explains the effect of CH3I dimer formation on the predissociation dynamics. Evidence is also presented for electron delocalization in higher clusters after excitation into the Rydberg states.

900,254
PB89-171201   Not available NTIS
National Bureau of Standards (NML), Boulder, CO. Quantum Physics Div.
Production of 0.1-3 eV Reactive Molecules by Laser Vaporization of Condensed Molecular Films: A Potential Source for Beam-Surface Interactions.
Final rept.
L. M. Cousins, and S. R. Leone. 1988, 11p
Contract DAAG29-85-K-0033
Sponsored by Army Research Office, Arlington, VA.
Pub. in Jnl. of Materials Research 3, n6 p1158-1168 Nov/Dec 88.

Keywords: *Laser beams, *Vaporizing, *Thin films, Molecular theory, Pulse generators, Radiation effects, Cryogenics, Substrates, Condensing, Chlorine, Nitro-

CHEMISTRY

Photo & Radiation Chemistry

gen oxide(NO), Energy levels, Microelectronics, Reprints.

A versatile, repetitively pulsed source of translationally fast, reactive molecules suitable for materials processing experiments is described. The pulsed beams are generated by excimer laser vaporization of cryogenic molecular films that are continuously condensed on transparent substrates. The generation of fast, energy variable pulsed molecular sources of Cl2 and NO is demonstrated. The most probable translational energies of Cl2 and NO molecules can be reproducibly varied monotonically by adjusting the laser fluence or film thickness. Here, the most probable translational energy is quoted as the energy corresponding to the maximum of the time-of-flight trace. Using laser fluences of 2-25 mJ/sq cm from a 193 nm excimer laser, the most probable translational energies of Cl2 are 0.4-2 eV. Significant fractions of molecules with translational energies greater than 3 eV are observed at the leading edges of the distributions. Very similar results are obtained by vaporizing Cl2 with 248 and 351 nm radiation. Pulses of translationally fast NO molecules are generated in a similar manner; most probable energies from 0.1-0.4eV, with the fastest molecules up to 0.8 eV, are obtained using laser fluences of 1-11 mJ/sq cm at 193 nm. Approximately 10 sup 13 to 10 sup 14 molecules per sq cm of the film are vaporized per laser pulse, depending on film thickness and laser fluence.

900,255
PB89-172423   Not available NTIS
National Bureau of Standards (NML), Gaithersburg, MD. Molecular Spectroscopy Div.
Picosecond Laser Study of the Collisionless Photodissociation of Dimethylnitramine at 266 nm.
Final rept.
J. C. Mialocq, and J. C. Stephenson. 1986. 4p
Pub. in Chemical Physics Letters 123, n5 p390-393, 24 Jan 86.

Keywords: *Nitramines, *Photolysis, Near ultraviolet radiation, Nitrogen dioxide, Reprints, *Photodissociation, Laser induced fluorescence, Picosecond pulses.

A picosecond pump-probe study of the uv photolysis of gaseous dimethylnitramine at 266 nm, efficient monophotonic collision-free photodissociation occurs within 6 ps. NO2 fragments are formed in the ground electronic state and in a fluorescent excited state; the quantum yields for both of these channels are estimated.

900,256
PB89-176754   Not available NTIS
National Bureau of Standards (NML), Gaithersburg, MD. Chemical Kinetics Div.
Chemical Kinetics of Intermediates in the Autoxidation of SO2.
Final rept.
R. E. Huie. 1986, 9p
Pub. in ACS (American Chemical Society) Symposium Series 319, p284-292 1986.

Keywords: *Kinetics, *Oxidation, *Sulfur dioxide, *Free radicals, *Acidity, *Rainfall, Sulfites, Scrubbers, Flue gases, Peroxy esters, Sulfuric acid, Reprints, Acid rain.

The autoxidation of SO2 solutions is known to involve free radicals. Recent work on the reaction of free radicals with sulfite and bisulfite and on the reaction of the sulfite and peroxysulfite radicals is beginning to allow the complex system to be understood better. It is particularly true for the effect of added chemicals on SO2 autoxidation and chemical transformations induced by the system.

900,257
PB89-176945   Not available NTIS
National Bureau of Standards (NML), Gaithersburg, MD. Radiometric Physics Div.
Vibrationally Resolved Photoelectron Studies of the 7(sigma) (-1) Channel In N2O.
Final rept.
T. A. Ferrett, A. C. Parr, S. H. Southworth, J. E. Hardis, and J. L. Dehmer. 1989, 6p
Contract W-31109-eng-38
Sponsored by Department of Energy, Washington, DC.
Pub. in Jnl. of Chemical Physics 90, n3 p1551-1556, 1 Feb 89.

Keywords: *Nitrogen oxide(N2O), *Vibrational spectra, *Photoelectrons, Energy, Fluorescence, Molecular

structure, Synchrotron radiation, Molecular spectra, Reprints.

Vibrationally resolved photoelectron studies of the 000, 100, 200, and 001 modes of the A state (7sigma(-1)) of N2O+ in the 17.4-26 eV photon-energy range were performed. The vibrational branching ratios sigma(100)/sigma(000) and sigma(001)/sigma(000) agree very well with fluorescence measurements by Kelly et al. and qualitatively with recent theoretical predictions of Braunstein and McKoy. The large non-Franck-Condon variations in the sigma(100)/sigma(000) and sigma(200)/sigma(000) branching ratios are associated with a predicted 7 sigma -> Epsilon sigma shape resonance near 20 eV. Overall, the vibrational branching ratios imply lower resonant energies for the stretching modes (100 and 200) and a similar resonant energy for the asymmetric stretch (001), compared with the 000 mode. The vibrational asymmetry parameters (beta) display a strong variation with energy which is qualitatively reproduced by theory; however, the experimental values for beta(100) and beta(001) exhibit additional structure around 20 eV. When combined with theory and recent fluorescence data, these results help to demonstrate a correlation of shape resonance energy with overall molecular length (RN-N) + R(N-O); this important result implies a resonant state which is localized on the entire triatomic molecular frame rather than on the N-N or N-O components.

900,258
PB89-179762   Not available NTIS
National Bureau of Standards (NML), Boulder, CO. Quantum Physics Div.
Photodissociation Dynamics of C2H2 at 193 nm: Vibrational Distributions of the CCH Radical and the Rotational State Distribution of the A(010) State by Time-Resolved Fourier Transform Infrared Emission.
Final rept.
T. R. Fletcher, and S. R. Leone. 1989, 9p
Contract DE-AC02-79ER10396
Sponsored by Department of Energy, Washington, DC.
Pub. in Jnl. of Chemical Physics 90, n2 p871-879, 15 Jan 89.

Keywords: *Acetylene, *Chemical radicals, Carbon, Hydrogen, Reprints, *Photodissociation, Rotational states, Vibrational states, Fourier transform infrared spectroscopy.

Time-resolved Fourier transform infrared (FTIR) emission is used to study the formation of CCH in the photodissociation of C2H2 at 193 nm. Excitation of C2H2 at 193 nm is known to populate the 10 mu sub 3 level of the trans-bent electronically excited state of acetylene, which undergoes decomposition. State-resolved infrared emission is obtained from the CCH radicals that are produced. Only vibronic levels which originate or borrow oscillator strength from the low-lying electronically excited state of CCH, A(doublet Pi), are observed in the study. The relative intensities of these bands are measured and the rotational state distribution for the A(010) state is obtained. A kinematic model which can account for a rotational cooling effect in the A(010) state is described.

900,259
PB89-201222   Not available NTIS
National Bureau of Standards (NML), Gaithersburg, MD. Surface Science Div.
Bond Selective Chemistry with Photon-Stimulated Desorption.
Final rept.
J. A. Yarmoff, and S. A. Joyce. 1989, 6p
Pub. in Materials Research Society Symposia Proceedings, v143 p91-96 1989.

Keywords: *Desorption, *Fluorine, *Ions, *Synchrotrons, *Silicon, Photons, Electron transitions, Surface chemistry, Irradiation.

Photon stimulated desorption of fluorine ions from silicon surfaces was studied via excitation of the Si 2p core level with synchrotron radiation. The results showed that the process is chemically selective in that the removal of a fluorine ion from a silicon species in a given oxidation state can be enhanced by tuning the photon energy to the excitation wavelength corresponding to a transition from the 2p core level of the bonding atom to the conduction band minimum. The process was studied as a possible means for the production of surfaces with selected compositions of species. The results of selective exposures of fluorinated surfaces to monochromatized radiation indicated that

secondary desorption processes and the inherent chemistry of the surface reactions can override the effects of selective desorption. Other possibilities for selective surface reactions via core-level excitations are discussed.

900,260
PB89-202634   Not available NTIS
National Bureau of Standards (NML), Boulder, CO. Quantum Physics Div.
Time-of-Flight Measurements of Hyperthermal Cl(sub 2) Molecules Produced by UV Laser Vaporization of Cryogenic Chlorine Films.
Final rept.
L. M. Cousins, and S. R. Leone. 1989, 6p
Contract DAAG29-85-K-003
Sponsored by Army Research Office, Washington, DC.
Pub. in Chemical Physics Letters 155, n2 p162-167, 24 Feb 89.

Keywords: *Chlorine, *Vaporizing, *Thin films, Ultraviolet lasers, Condensates, Cryogenics, Time, Reprints

Time-of-flight distributions are obtained for the repetitive, unfocused laser vaporization of cryogenic Cl2 films condensed on a transparent substrate held at 25-110K. Translationally fast molecules are observed using 193, 248, 351 and 355 nm, but no vaporization occurs for 532 or 1064 nm. The most probable kinetic energies of the Cl2 molecules, estimated from the maxima of the Cl2 time-of-flight traces, range from 0.4 to 2 eV for 193 nm vaporization at laser powers of 0.2-2 MW/sq cm. The energies increase monotonically with increasing laser fluence and film thickness. Less than 7% of the desorbed flux is due to Cl atoms formed by dissociation in the desorption process. Energies above 5 eV are obtained with focused 355 nm radiation.

900,261
PB89-202642   Not available NTIS
National Bureau of Standards (NML), Boulder, CO. Quantum Physics Div.
Time-Resolved FTIR Emission Studies of Molecular Photofragmentation.
Final rept.
S. R. Leone. 1989, 6p
Grants NSF-PHY86-04504, NSF-CHE84-08403
Sponsored by National Science Foundation, Washington, DC., and Department of Energy, Washington, DC.
Pub. in Accounts of Chemicals Research 22, n4 p139-144 Apr 89.

Keywords: *Infrared spectroscopy, *Polyatomic molecules, *Photochemical reactions, Vibrational spectra, Rotational spectra, Excitation, Chemical bonds, Fourier transformation, Molecular structure, Photolysis, Steric hindrance, Reprints.

A number of dynamical experiments were conducted in which low resolution, time-resolved infrared fluorescence detection was employed. The experiments used 30-60/cm resolution tunable interference filters for wavelength selectivity. The lack of rotational resolution in those experiments limited the dynamical detail. In the present account, new experiments are described which utilize a high resolution, time-resolved FTIR emission method to study photofragmentation dynamics of large polyatomic molecules. The capability to obtain vibrational and rotational state details of the excited dynamics of simple bond-breaking processes. The results include information about the timescales for breaking several bonds in a single molecule and the constraints on rotational motion due to steric hindrance and energy and angular momentum considerations.

900,262
PB89-231336   Not available NTIS
National Inst. of Standards and Technology (NML), Gaithersburg, MD. Surface Science Div.
Synchrotron Photoemission Study of CO Chemisorption on Cr(110).
Final rept.
N. D. Shinn. 1988, 11p
Sponsored by Office of Naval Research, Arlington, VA., and Department of Energy, Washington, DC.
Pub. in Physical Review B 38, n17 p12248-12258, 15 Dec 88.

Keywords: *Carbon monoxide, *Chemisorption, *Photoelectric emission, *Ultraviolet radiation, *Chemical bonds, Dipole moments, Chromium, iron, Molybdenum, Synchrotron radiation, Reprints.

Angle-integrated ultraviolet photoemission studies of carbon monoxide chemisorption on Cr(110) confirm the sequential population of two electronically inequivalent molecular binding states (alpha 1-CO and alpha 2-CO) at 90 K. They are distinguished by differences in the CO 4 sigma binding energies (0.8 eV) and photoemission cross sections. Work-function measurements indicate that the surface dipole moment associated with alpha 1-CO is significantly less than that for alpha 2-CO. CO/Cr interaction data exhibit oxygen-induced alpha 1-CO site blocking and alpha 1-CO alpha 2-CO binding mode conversion. The results support current models for the CO binding geometries on Cr(110) and related binding states on Fe(100) and Mo(100).

900,263
PB90-117425   Not available NTIS
National Inst. of Standards and Technology (NML), Gaithersburg, MD. Molecular Spectroscopy Div.
Dissociation Lifetimes and Level Mixing in Overtone-Excited HN3 (X tilde (sup 1) A').
Final rept.
B. R. Foy, M. P. Casassa, J. C. Stephenson, and D. S. King. 1989, 9p
Sponsored by Air Force Office of Scientific Research, Bolling AFB, DC.
Pub. in Jnl. of Chemical Physics 90, n12 p7037-7045, 15 Jun 89.

Keywords: *Photochemical reactions, *Dissociation, *Nitrides, *Hydrides, Vibrational spectra, Molecular spectroscopy, Electron transitions, Reprints.

Vibrational overtone photodissociation is used to examine the spectroscopy and vibrational predissociation lifetimes of HN3 in its ground electronic state. Direct overtone pumping of the N-H stretching levels 5v(sub NH) and 6v(sub NH) prepares molecules in selected states (v,J,K) near 15,100 and 17,700/cm of vibrational energy; spin-forbidden NH dissociation fragments are detected by laser-induced fluorescence. Photodissociation spectra of beam-cooled HN3 display mixing of individual rotational levels of the nv(sub NH) vibrations with several background states, with derived coupling matrix elements in the range 0.01-0.1/cm. Vibrational predissociation lifetimes of mixed components of 5v(sub NH) are state specific, with variations of a factor of 2 for only 0.1/cm energy differences. Average lifetimes for low J, K are 210 ns for 5v(sub NH) and 0.95 ns for 6v(sub NH). The ratio of decay rates for the two overtone levels, k(6v(sub NH))/k(5v(sub NH)) = 220, is much greater than predicted by statistical theory, which gives a ratio of 4.

900,264
PB90-123498   Not available NTIS
National Inst. of Standards and Technology (NML), Gaithersburg, MD. Chemical Kinetics Div.
Reaction of (Ir(C(3), N bpy)(bpy)2)(2+) with OH Radicals and Radiation Induced Covalent Binding of the Complex to Several Polymers in Aqueous Solutions.
Final rept.
D. Behar, P. Neta, J. Silverman, and J. Rabani. 1987, 8p
Pub. in Radiation Physics and Chemistry 29, n4 p253-260 1987.

Keywords: *Polymers, Polyelectrolytes, Radiolysis, Reprints, *Iridium complexes, Crossings, Dynamics, Electron energy, *Electronic states, *Energy transfer, Graphs, *Lasers, Rare gases, Reactivities, Reprints, *Strontium.

Irradiation of aqueous solutions of (Ir(C sup 3, N bpy)(bpy)2)(2+) (IrP) produces a variety of OH adducts to IrP. The OH adducts decay by two second order processes separated in time. In the presence of both IrP and a soluble polymer, the OH radicals are shared between the IrP and the polymer, simultaneously producing OH adducts and polymer radicals. This is followed by radical-radical reactions. The rate constants of the various reactions between the OH adducts and the polymer radicals have been determined. The products of reactions of the OH adducts are discussed. IrP behaves similarly to Ru(bpy)3(2+) which was studied before. This indicates that the radiation method may have a general use in the preparation of polymers with pendant bpy complexes.

900,265
PB90-123704   Not available NTIS
National Inst. of Standards and Technology (NML), Gaithersburg, MD. Chemical Kinetics Div.

Fluid Flow in Pulsed Laser Irradiated Gases; Modeling and Measurement.
Final rept.
W. Braun, T. J. Wallington, and R. J. Cvetanovic. 1988, 15p
Pub. in Jnl. of Photochemistry and Photobiology A-Chemistry 42, n2-3 p207-221 1988.

Keywords: *Gases, *Fluid flow, Energy transfer, Irradiation, Light pulses, Mixtures, Mercury, Mathematical models, Computerized simulation, Temperature, Reprints, Laser radiation, Tracer techniques, Density.

Fluid flow processes following laser irradiation of gaseous mixtures have been modeled in order to facilitate quantitative interpretation of experimental observations. Results of computer simulations are compared with experimental measurements of the temporal changes of temperature and density obtained using the recently developed Hg tracer technique. The model outlined in the work accurately reproduces the experimentally observed changes in temperature and density in pulsed laser irradiated gas mixtures using a variety of cell and laser irradiation geometries. Comparison of the model result with experimental data shows how Hg tracer technique measurements may be influenced by fluid flow.

900,266
PB90-136680   Not available NTIS
National Inst. of Standards and Technology (NML), Boulder, CO. Quantum Physics Div.
Time-Resolved FTIR Emission Studies of Molecular Photofragmentation Initiated by a High Repetition Rate Excimer Laser.
Final rept.
T. R. Fletcher, and S. R. Leone. 1987, 4p
Pub. in Proceedings of International Laser Science Conference (3rd), Atlantic City, NJ., p595-598 Nov 87.

Keywords: *Photolysis, Acetylene, Emission spectra, *Fourier transform infrared spectroscopy, Eximer lasers.

The availability of high repetition rate (>300 Hz) excimer lasers provides new opportunities for studies of molecular processes by time-resolved FTIR spectroscopy. An overview of the technique is given and state resolved infrared emission results are presented for the triatomic radical, C2H, generated by photolysis of C2H2 at 193 nm. Electronic emission from the low lying (A tilde) doublet Pi state of C2H is observed, along with high vibrational levels of the ground state which gain intensity by coupling with the vibrationless level of (A tilde) doublet Pi.

Physical & Theoretical Chemistry

900,267
AD-A202 820/7   PC A03/MF A01
Colorado Univ. at Boulder. Dept. of Chemistry.
Alignment Effects in Electronic Energy Transfer and Reactive Events.
S. R. Leone. 1988, 20p AFOSR-TR-88-1316
Grant AFOSR-86-0018
Pub. in Selectivity in Chemical Reactions, p245-263 1988.

Keywords: Accuracy, Alignment, *Atomic orbitals, *Calcium, *Pumping(Electronics), Crossings, Dynamics, Electron energy, *Electronic states, *Energy transfer, Graphs, *Lasers, Rare gases, Reactivities, Reprints, *Strontium.

The rates of electronic curve crossing processes depend critically on the alignment of atomic orbitals, which determine the symmetries of the electronic potentials participating in the reaction or energy transfer event. Recent work from our laboratory is presented on the effect of orbital alignment in near resonant energy transfer processes of electronically excited Ca and Sr atoms. Several energy transfer events are carried out on aligned p-states in collisions with rare gases. The simplicity of the rare gas systems in terms of their symmetry and nonreactive nature is advantageous for comparison to accurate theoretical treatment. In the context of understanding chemical phenomena, collisions of these atoms with molecular partners are also investigated. This opens the possibility to study the correlation of alignment dependent effects in competing reactive and energy transfer pathways. Remarkably state-specific alignment effects are also ob-

---

served when two or more independent energy transfer pathways are accessible. Keywords: Pumping (electronics); alignment; Energy transfer; Laser; Reaction dynamics; Calcium; Strontium; Reprints. (JHD)

900,268
PB89-145114   Not available NTIS
American Chemical Society, Washington, DC.
Journal of Physical and Chemical Reference Data, Volume 17, Number 4, 1988.
Quarterly rept.
c1988, 468p
See also PB89-145122 through PB89-145189 and PB88-156435. Prepared in cooperation with American Inst. of Physics, New York. Sponsored by National Bureau of Standards (ICST), Gaithersburg, MD.
Available from American Chemical Society, 1155 Sixteenth St., NW, Washington, DC 20036.

Keywords: *Physical properties, *Chemical properties, Data.

Contents: Evaluated chemical kinetic data for the reactions of atomic oxygen O(3P) with sulfur containing compounds; New international skeleton tables for the thermodynamic properties of ordinary water substance; Benzene thermophysical properties from 279 to 900 K at pressures to 1000 bar; Estimation of the thermodynamic properties of hydrocarbons at 298.15 K; Wavelengths and energy level classifications of scandium spectra for all stages of ionization; Atomic weights of the elements 1987; The 1986 CODATA recommended values of the fundamental physical constants. (Copyright (c) by the U.S. Secretary of Commerce, 1988.)

900,269
PB89-145122   Not available NTIS
National Research Council of Canada, Ottawa (Ontario). Div. of Chemistry.
Evaluated Chemical Kinetic Data for the Reactions of Atomic Oxygen O(3P) with Sulfur Containing Compounds.
Quarterly rept.
D. L. Singleton, and R. J. Cvetanovic. c1988, 60p
Sponsored by National Inst. of Standards and Technology (NML), Gaithersburg, MD. Chemical Kinetics Div.
Included in Jnl. of Physical and Chemical Reference Data, v17 n4 p1377-1437 1988. Available from American Chemical Society, 1155 Sixteenth St., NW, Washington, DC 20036.

Keywords: *Reaction kinetics, *Oxygen, *Sulfur compounds, Chemical reactions, Arrhenuis parameters.

Chemical kinetic data for reactions of O(3P) atoms with sulfur containing compounds are compiled and critically evaluated. Specifically, the reactions considered include the interactions of the ground electronic state of oxygen atoms, O(3P), with S2, SF2, SF5, SOF, S2O, SO2, SO3, SH, H2S, D2S, H2SO4, CS, CS2, COS, CH3SH, C2H5SH, CH7SH, C4H9SH, C5H11SH, CH3SCH3, cy-CH2SCH2, cy-CHCH5OHCH, CH3SSCH3, SCF2, SCCi2, and cy-CF2SCF2S. With one exception, the liquid phase reaction O(3P) + H2SO4 -> products, all the data considered are for gas phase reactions. Where possible, 'recommended' values of the rate parameters have been assessed and conservative uncertainty limits assigned to them.

900,270
PB89-145130   Not available NTIS
Keio Univ., Yokohama (Japan). Dept. of Mechanical Engineering.
New International Skeleton Tables for the Thermodynamic Properties of Ordinary Water Substance.
Quarterly rept.
H. Sato, M. Uematsu, K. Watanabe, A. Saul, and W. Wagner. c1988, 100p
Prepared in cooperation with Ruhr. Univ., Bochum (Germany, F.R.). Inst. fuer Thermo- und Fluiddynamik. Included in Jnl. of Physical and Chemical Reference Data, v17 n4 p1439-1540 1988. Available from American Chemical Society, 1155 Sixteenth St., NW, Washington, DC 20036.

Keywords: Thermodynamic properties, *Water, Enthalpy, Density(Mass/volume), Steam, Vapor pressure.

The current knowledge of thermodynamic properties of ordinary water substance is summarized in a condensed form of a set of skeleton steam tables, where

CHEMISTRY

Physical & Theoretical Chemistry

the most probable values with the reliabilities on specific volume and enthalpy are provided in the range of temperatures from 273 to 1073 K and pressures from 101.325 kPa to 1 GPa and at the saturation state from the triple point to the critical point. About 17,000 experimental thermodynamic data were assessed and classified previously by Working Group 1 of IAPS. About 10,000 experimental data were collected and evaluated in detail and especially about 7000 specific-volume data among them were critically analyzed with respect to their errors using the statistical method originally developed at Keio University by the first three authors. As a result, specific volume and enthalpy values with associated reliabilities were determined at 1455 grid points of 24 isotherms and 61 isobars in the single-fluid -phase state and at 54 temperatures along the saturation curve. The background, analytical procedure, and reliability of IST-85 as well as the assessment of the existing experimental data and equations of state are also discussed in the paper.

900,271
PB89-145148 Not available NTIS
National Inst. of Standards and Technology (NEL), Boulder, CO. Thermophysics Div.
Benzene Thermophysical Properties from 279 to 900 K at Pressures to 1000 Bar.
Quarterly rept.
R. D. Goodwin. c1988, 96p
Included in Jnl. of Physical and Chemical Reference Data, v17 n4 p1541-1636 1988. Available from American Chemical Society, 1155 Sixteenth St., NW, Washington, DC 20036.

Keywords: *Thermodynamic properties, *Benzene, Enthalpy, Density, Entropy, Vapor pressure, Equations of state, Joule-Thompson effect, Heat of vaporization, Specific heat.

The thermodynamic data for benzene have been evaluated and fit to a highly constrained, nonanalytic equation of state. Comparisons of the equation with the selected PVT and derived property data are given. Extensive tables are presented providing tabular values for coexisting liquid and vapor as well as for the single phase along isobars. The equation state and tables cover the range from the triple point (278.68 K) to 900 K, with pressures to 1000 bar.

900,272
PB89-145155 Not available NTIS
National Inst. of Standards and Technology (NML), Gaithersburg, MD. Chemical Thermodynamics Div.
Estimation of the Thermodynamic Properties of Hydrocarbons at 298.15 K.
Quarterly rept.
E. S. Domalski, and E. D. Hearing. c1988, 42p
Included in Jnl. of Physical and Chemical Reference Data, v17 n4 p1637-1678 1988. Available from American Chemical Society, 1155 Sixteenth St., NW, Washington, DC 20036.

Keywords: *Thermodynamic properties, *Hydrocarbons, Liquid phases, Solid phases, Enthalpy, Entropy.

An estimation method developed by S.W. Benson and coworkers, for calculating the thermodynamic properties of organic compounds in the gas phase, has been extended to the liquid and solid phases for hydrocarbon compounds at 298.15 K. The second order approach which includes nearest neighbor interactions has been applied to the condensed phase. A total of 1311 comparisons are made between experimentally determined values and those calculated using additive group values. The good agreement between experimental and calculated values shows that the Benson group additivity approach to the estimation of thermodynamic properties of organic compounds is applicable to the liquid and solid phases as well as the gas phase. Appendices provide example calculations of the thermodynamic properties selected hydrocarbon compounds, total symmetry numbers, and methyl repulsion corrections.

900,273
PB89-145163 Not available NTIS
National Inst. of Standards and Technology (NML), Gaithersburg, MD. Atomic and Plasma Radiation Div.
Wavelengths and Energy Level Classifications of Scandium Spectra for All Stages of Ionization.
Quarterly rept.
V. Kaufman, and J. Sugar. c1988, 111p
Included in Jnl. of Physical and Chemical Reference Data v17 n4 p1679-1790 1988. Available from American Chemical Society, 1155 Sixteenth St., NW, Washington, DC 20036.

Keywords: *Energy levels, *Wavelengths, *Scandium, Ionization, Spectra, Atomic energy levels.

Wavelengths and their classifications are compiled for the spectra of scandium, Sc I through Sc XXI. Selections of data are based on the critical evaluations in compilation of energy levels by Sugar and Corliss. These are updated by a thorough search of the subsequent literature. All classifications are verified with predictions made by differencing the energy levels. Spectra are ordered by ionization stage and listed by wavelength. Two finding lists are included, one containing Sc I to Sc III and the other Sc IV to Sc XXI.

900,274
PB89-145171 Not available NTIS
Curtin Univ. of Technology, Bentley (Australia).
Atomic Weights of the Elements 1987.
Quarterly rept.
J. R. De Laeter. c1988, 3p
Included in Jnl. of Physical and Chemical Reference Data, v17 n4 p1791-1793 1988. Available from American Chemical Society, 1155 Sixteenth St., NW, Washington, DC 20036.

Keywords: *Atomic mass, Transuranium elements, Isotopes, Atomic weights.

The International Union of Pure and Applied Chemistry Commission on Atomic Weights and Isotopic Abundances has reviewed recent literature and confirmed the atomic weight values published in 1985, with one minor change. The current table of standard atomic weights is presented.

900,275
PB89-145189 Not available NTIS
Rockwell International, Thousand Oaks, CA. Science Center.
CODATA (Committee on Data for Science and Technology) Recommended Values of the Fundamental Physical Constants, 1986.
Quarterly rept.
E. R. Cohen, and B. N. Taylor. c1988, 9p
Included in Jnl. of Physical and Chemical Reference Data, v17 n4 p1795-1803 1988. Available from American Chemical Society, 1155 Sixteenth St., NW, Washington, DC 20036. Sponsored by National Inst. of Standards and Technology, Gaithersburg, MD.

Keywords: *Physical properties, Physics, Chemistry, *Physical constants, CODATA, Task Group of Fundamental Constants.

Presented here are the values of the basic constants and conversion factors of physics and chemistry resulting from 1986 least-squares adjustment of the fundamental physical constants as published by the CODATA (Committee on Data for Science and Technology) Task Group on Fundamental Constants and recommended for international use by CODATA. The 1986 CODATA set of values replaces its predecessor published by the Task Group and recommended for international use by CODATA in 1973.

900,276
PB89-145197 Not available NTIS
American Chemical Society, Washington, DC.
Journal of Physical and Chemical Reference Data, Volume 17, 1988, Supplement No. 3. Atomic Transition Probabilities Scandium through Manganese.
Quarterly rept.
G. A. Martin, J. R. Fuhr, and W. L. Wiese. c1988, 531p ISBN-0-88318-585-7
See also PB89-145205 and PB89-135735. Library of Congress catalog Card no. 88-72277. Prepared in cooperation with American Inst. of Physics, New York. Sponsored by National Bureau of Standards, Gaithersburg, MD.
Available from American Chemical Society, 1155 Sixteenth St., NW, Washington, DC 20036.

Keywords: *Atomic spectra, *Scandium, *Titanium, *Vanadium, *Chromium, *Manganese, *Electron transitions, *Transition probabilities, Dipoles, Ionization, Quantum interactions, Accuracy, Tables(Data), Selection rules(Physics).

Atomic transition probabilities for about 8,800 spectral lines of five iron-group elements, Sc(Z = 21) to Mn(Z = 25), are critically compiled, based on all available literature sources. The data are presented in separate tables for each element and stage of ionization and are further subdivided into allowed (i.e., electric dipole-E1) and forbidden (magnetic dipole--M1, electric quadrupole--E2, and magnetic quadrupole--M2) transi-

tions. Within each data table the spectral lines are grouped into multiplets, which which are in turn arranged according to parent configurations, transition arrays, and ascending quantum numbers. For each line the transition probability for spontaneous emission and the line strength are given, along with the spectroscopic designation, the wavelength, the statistical weights, and the energy levels of the upper and lower states. For allowed lines the absorption oscillator strength is listed, while for forbidden transitions the type of transition is identified (M1, E2, etc.). In addition, the estimated accuracy and the source are indicated. In short introductions, which precede the tables for each ion, the main justifications for the choice of the adopted data and for the accuracy rating are discussed.(Copyright (c) 1988 by the U.S. Secretary of Commerce.)

900,277
PB89-145205 Not available NTIS
Rensselaer Polytechnic Inst., Troy, NY. Dept. of Chemistry.
Journal of Physical and Chemical Reference Data, Volume 17, 1988, Supplement No. 2. Thermodynamic and Transport Properties for Molten Salts: Correlation Equations for Critically Evaluated Density, Surface Tension, Electrical Conductance, and Viscosity Data.
Quarterly rept.
G. J. Janz. c1988, 327p ISBN-0-88318-587-3
See also PB89-145197. Library of Congress catalog card no. 88-82581. Prepared in cooperation with American Chemical Society, Washington, DC., and American Inst. of Physics, New York. Sponsored by National Bureau of Standards, Gaithersburg, MD.
Available from American Chemical Society, 1155 Sixteenth St., NW, Washington, DC 20036.

Keywords: *Thermodynamics, *Tables(Data), *Fused salts, Interfacial tension, Density, Viscosity, Electrical resistance.

Critically evaluated results for two thermodynamic properties (density and surface tension) and two transport properties (electrical conductance and viscosity) are reported for one and two component salt systems in the molten state. For each system, the recommended results are reported in the form of equations, together with uncertainty estimates, and flagged comments on value judgments and related matters. Results for a limited number of higher multi-component systems are included. The NSRDS-NBS critically evaluated data series have been upgraded as part of the work, and the collection and evaluations of the available experimental data have been systematically extended to 1985.(Copyright (c) 1988 by the U.S. Secretary of Commerce.)

900,278
PB89-146658 Not available NTIS
National Bureau of Standards (NML), Gaithersburg, MD. Chemical Kinetics Div.
Rate Constants for the Reaction HO2+NO2+N2->HO2NO2+N2: The Temperature Dependence of the Fall-Off Parameters.
Final rept.
M. J. Kurylo, and P. Ouellette. 1987, 4p
Pub. in Jnl. of Physical Chemistry 91, n12 p3365-3368 1987.

Keywords: *Reaction kinetics, *Hydroperoxides, *Nitrogen, *Nitrogen dioxide, Ultraviolet spectroscopy, Low pressure research, Reprints, Temperature dependence.

Rate constants for the title reaction were measured via flash photolysis UV absorption spectroscopy at N2 pressures of 25, 50, and 100 Torr over the temperature range 288 - 358K. The data of the study were at low enough pressure to yield a precise determination of n (which describes the temperature dependence of the low pressure limiting third-order rate constant) but were less sensitive to the determination of m (associated with the limiting high pressure rate constant). For this reason, the final analysis utilizes a composite fit of the temperature dependent data along with similar data at 100 and 700 Torr of N2 obtained by Sander and Peterson.

900,279
PB89-146666 Not available NTIS
National Bureau of Standards (NML), Gaithersburg, MD. Chemical Kinetics Div.

34

Multiphoton Ionization Spectroscopy and Vibrational Analysis of a 3p Rydberg State of the Hydroxymethyl Radical.

Final rept.

C. S. Dulcey, and J. W. Hudgens. 1986, 9p

Pub. in Jnl. of Chemical Physics 84, n10 p5262-5270, 15 May 86.

Keywords: *Energy levels, *Vibrational spectra, *Spectrum analysis, Free radicals, Excitation, Deuterium compounds, Isomerization, Reprints, *Hydroxymethyl radicals, *Rydberg states, *Resonance enhanced multiphoton ionization spectroscopy, Electronic structure.

The resonance enhanced multiphoton ionization (REMPI) spectra of CH2OH, CH2OD, CD2OH, and CD2OD between 420-495 nm are reported. Analysis of the excited state vibrational band progressions shows that the spectrum originates from simultaneous two photon absorption to form a 3p Rydberg state (v(sub 00) = 40,064/cm) of the radical. Absorption of a third photon ionized the radical. A normal mode analysis of the REMPI spectra enabled assignments of six active vibrational modes in the excited state. The rate that methoxy radicals isomerize into hydroxymethyl radicals was estimated to be less than 2.9/sec at 300K.

900,280
PB89-146674   Not available NTIS
National Bureau of Standards (NML), Gaithersburg, MD. Chemical Kinetics Div.
Gas Phase Proton Affinities and Basicities of Molecules: A Comparison between Theory and Experiment.

Final rept.

D. A. Dixon, and S. G. Lias. 1987, 46p

Pub. in Molecular Structure and Energetics, v2 p269-314 1987.

Keywords: *Heat of formation, Vapor phases, Thermochemistry, Alkalinity, Comparison, Gases, Reviews, Reprints, *Proton affinity, Ion molecule interactions.

A review of a recently-published evaluation of the scale of gas phase proton affinities (S. G. Lias, J. F. Liebman, and R. D. Levin, J. Phys. Chem. Ref. Data, 13, 695 (1984)). The discussion includes (1) a description of the rationale used in evaluating different data sets from the literature, and (2) detailed evaluations of the heats of formation of MH(1+) ions used in the assignment of absolute values to the relative thermodynamic ladder which constitutes the gas phase proton affinity scale.

900,281
PB89-146864   Not available NTIS
National Bureau of Standards (NML), Gaithersburg, MD. Surface Science Div.
Adsorption Properties of Pt Films on W(110).

Final rept.

R. A. Demmin, S. M. Shivaprasad, and T. E. Madey. 1988, 5p

Sponsored by Department of Energy, Washington, DC.

Pub. in Langmuir 4, n5 p1104-1106 1988.

Keywords: *Platinum, *Thin films, *Tungsten, *Carbon monoxide, *Chemisorption, Desorption, Reprints.

The surface chemistry of ultrathin Pt films on a W(110) substrate has been investigated by using CO chemisorption. Carbon monoxide temperature-programmed desorption experiments show that molecular CO is more weakly bound to a monolayer Pt film deposited at 90 K than to either bulk Pt or the W substrate, similar to conclusions drawn from experiments on other metal thin films. Carbon monoxide is also weakly adsorbed on films annealed to 1500 K, even for initial Pt coverages much greater than one monolayer. This has been interpreted as strong evidence for substantial thermally induced structural changes in multilayer films that result in a W surface that is covered by a monolayer Pt film with unique CO chemisorption properties. Platinum films of at least one monolayer also prevent the dissociative adsorption of CO normally occurring on the W(110) surface. For submonolayer films annealed to 1500 K, the total amount of dissociative adsorption of CO decreases linearly with increasing Pt coverage, reaching zero at one monolayer of Pt. This implies that the inhibition of CO dissociation by Pt is very localized. Previously proposed explanations for CO adsorption behavior common to a variety of overlayer-substrate systems are discussed.

900,282
PB89-146898   Not available NTIS

National Bureau of Standards (NML), Gaithersburg, MD. Surface Science Div.
Core-Level Binding-Energy Shifts at Surfaces and in Solids.

Final rept.

W. F. Egelhoff. 1986, 163p

Pub. in Surface Science Reports 6, n6-8 p253-415 1986.

Keywords: *X-ray analysis, *Energy levels, *Photoelectrons, Cores, Surfaces, Solids, Spectroscopy, Reprints.

The review presents an overview of the theory and of various successful approaches to the interpretation of core-level binding-energy shifts observed in photoelectron spectroscopy. The review specially concentrates on shifts since most of the chemical and physical insights provided by core levels are derived not from the core-level binding energies themselves but from shifts they exhibit. The theoretical background is presented at a level readily accessible to the general reader. Particular attention is paid to relative merits of the two basically different conceptual frameworks for interpreting core-level binding-energy shifts, the initial-state-final approach and the equivalent-core approach.

900,283
PB89-146922   Not available NTIS
National Bureau of Standards (NEL), Gaithersburg, MD. Semiconductor Electronics Div.
Multiple Scattering in the X-ray-Absorption Near-Edge Structure of Tetrahedral Ge Gases.

Final rept.

C. E. Bouldin, G. Bunker, D. A. McKeown, R. A. Forman, and J. J. Ritter. 1988, 4p

Pub. in Physical Review B 38, n15 p10 816-10 819, 15 Nov 88.

Keywords: *X-ray analysis, *Germanium halides, *Germanium hydrides, Scattering, Gases, Absorption spectrum, Molecular structure, Reprints.

X-ray absorption fine-structure (XAFS) measurements of GeCl4, GeH3Cl, and GeH4 were experimentally isolated by comparison of the spectra of the three compounds. The multiple-scattering (MS) amplitude is comparable to the single-scattering (SS) amplitude only within 15 eV of the edge. Beyond 40 eV the MS to SS amplitude ratio is less than 0.06. Calculations are in qualitative agreement with the experiment. Results suggest that XAFS data in the range 1 less than k less than 3 reciprocal Angstroms can be analyzed in a SS picture in many cases.

900,284
PB89-147011   Not available NTIS
National Bureau of Standards (NML), Gaithersburg, MD. Atomic and Plasma Radiation Div.
Fundamental Configurations in Mo IV Spectrum.

Final rept.

M. T. Fernandez, I. Cabeza, L. Iglesias, O. Garcia-Riquelme, F. Rico, and V. Kaufman. 1987, 8p

Pub. in Physica Scripta 35, n6 p819-826 1987.

Keywords: *Molybdenum, *Energy levels, *Line spectra, Spectrum analysis, Spectral energy distribution, Reprints, Configurations.

The spectrum of Mo 4 was produced in a sliding-spark discharge and photographed with the 10.7m normal-incidence vacuum spectrograph at the NBS in the 600 - 3200 A spectral region. All but one of the 35 levels of the 4d(3) and 4d(2)5S even configurations and all 45 of the levels of the 4d(2)5P odd configuration have been established from the 514 line classifications in the 600 - 3150 A region. Parametric calculations have been made for the even level systems with configuration interaction and for the odd configuration.

900,285
PB89-147060   Not available NTIS
National Bureau of Standards (NEL), Boulder, CO. Chemical Engineering Science Div.
Thermodynamics of Ammonium Scheelites. 6. An Analysis of the Heat Capacity and Ancillary Values for the Metaperiodates KIO4, NH4IO4, and ND4IO4.

Final rept.

R. J. C. Brown, J. E. Callanan, R. D. Weir, and E. F. Westrum. 1987, 10p

See also PB88-238498.

Pub. in Jnl. of Chemical Thermodynamics 19, p1173-1182 1987.

Keywords: *Specific heat, Scheelites, Thermodynamic properties, Anisotropy, Thermal expansion, Computa-

tion, Rotation, Deuterium compound, Reprints, *Ammonium scheelites, Ammonium periodate, Deuterated ammonium compounds.

An analysis of the heat capacity of NH4IO4, ND4IO4, and KIO4 has been carried out in which the effects of the anisotropy of the thermal expansion have been considered, an approach hitherto used successfully for the perrhenates KReO4, NH4ReO4, and ND4ReO4. In the ammonium scheelites, the axial expansivities are very large, but of opposite sign, and as a result the molar volume of the scheelite lattice is nearly independent of temperature. It is shown that the correction from constant stress to constant strain results in a major contribution to the heat capacity of this highly anisotropic lattice. The difference between the experimental and calculated heat capacities, referred to as Delta C P.M, is expressed as the sum of the contributions from the anisotropy and the rotational heat capacity. The results of the analysis show that the rotational contribution is much smaller than had previously been thought. However, the exact contribution of the anisotropy cannot yet be calculated because the elastic constants are not known. In calculating the heat capacity, maximum use has been made of external optical-mode frequencies derived from spectroscopic measurements.

900,286
PB89-147110   Not available NTIS
National Bureau of Standards (NML), Gaithersburg, MD. Molecular Spectroscopy Div.
Unimolecular Dynamics Following Vibrational Overtone Excitation of HN3 v1 = 5 and v1 = 6:HN3(x tilde;v,J,K,) -> HN((X sup 3)(Sigma (1-));v,J,Omega) + N2(x sup 1)(Sigma sub g (1+)).

Final rept.

B. R. Foy, M. P. Casassa, J. C. Stephenson, and D. S. King. 1988, 2p

Sponsored by Air Force Office of Scientific Research, Bolling AFB, DC.

Pub. in Jnl. of Chemical Physics 89, n1 p608-609, 1 Jul 88.

Keywords: *Hydrazoic acid, Reprints, *Hydrogen azides, Laser induced fluorescence, Predissociation, Vibrational states, Unimolecular structures, Lifetime.

Excitation of the NH-stretch overtone transitions of HN3 to v1 = 5 and 6 resulted in predissociation to HN(X) and N2(X) with lifetimes of 80(+60,-30) and less than or = 3 ns, respectively. Following excitation of either overtone, the NH-fragments were formed predominantly in the symmetric E1, F3 spin-rotation states, with less than 4% population in the anti-symmetric F2 levels. Fragment Doppler profiles confirmed that most of the available energy (greater than 96%) went into translational motion.

900,287
PB89-147128   Not available NTIS
National Bureau of Standards (NML), Gaithersburg, MD. Molecular Spectroscopy Div.
Electronic Structure of Diammine (Ascorbato) Platinum(II) and the Trans Influence on the Ligand Dissociation Energy.

Final rept.

H. Basch, M. Krauss, and W. J. Stevens. 1986, 3p

Pub. in Inorganic Chemistry 25, n26 p4777-4779 1986.

Keywords: *Platinum, *Ligands, *Reaction kinetics, *Ascorbic acid, *Ammonia, Complex compounds, Molecular orbitals, Electrons, Dissociation energy, Reprints.

The electronic structure of the cis-Pt(ammonia)sub 2 (ascorbate) molecule has been studied by valence electron self-consistent-field calculations. A simpler molecule is used to model the ascorbate in a calculation of ligand binding energies. A comparison of localized ligand bonding orbital charge centroids and densities supports the validity of the model compound to represent the bonding in the ascorbate. The Pt-C bond energy is calculated to exceed that for Pt-O by about 40 kcal/mole. The dissociation energies for the ammonia ligands exhibit a strong trans-influence with a low dissociation energy for the ammonia trans to the Pt-C bond. These results suggest that this ammonia ligand and the Pt-O bond are suitable for exchange in these molecules.

900,288
PB89-147375   Not available NTIS

CHEMISTRY

Physical & Theoretical Chemistry

National Bureau of Standards (NML), Gaithersburg, MD. Molecular Spectroscopy Div.
Electric-Dipole Moments of H2O-Formamide and CH3OH-Formamide.
Final rept.
G. T. Fraser, R. D. Suenram, and F. J. Lovas. 1988, 8p
Pub. in Jnl. of Molecular Structure 189, p165-172 1988.

Keywords: *Dipole moments, *Water, *Formamides, *Carbinols, *Microwave spectroscopy, Hydrogen bonds, Electric charge, Reprints.

Electric-dipole moments have been determined for water-formamide and methanol-formamide by pulsed-nozzle Fourier-transform microwave spectroscopy. For water-formamide mu sub a = 1.050(1) D and mu sub b = 2.135(3) D while for methanol-formamide mu sub a = 1.0668(1) D and mu sub b = 2.091(2) D. The corresponding complexation-induced moments are estimated as 0.55 D and 0.34 D for water-formamide and 0.58 D and 0.10 D for methanol-formamide. The results are compared with theoretical calculations.

900,289

PB89-147417   Not available NTIS
National Bureau of Standards (NML), Gaithersburg, MD. Molecular Spectroscopy Div.
Vibrational Predissociation of the Nitric Oxide Dimer.
Final rept.
M. P. Casassa, J. C. Stephenson, and D. S. King. 1986, 10p
See also PB87-126294.
Pub. in Faraday Discussions of the Chemical Society 82, p251-260 1986.

Keywords: *Nitrogen oxides, *Vibrational spectra, *Dissociation energy, Reaction kinetics, Lasers, Reprints.

Details of experimental measurements of the total energy distribution and time dependence of the vibrational predissociation of the nitric oxide dimer excited to v1 are presented. Energy-disposal measurements indicated the fragments are described by an average rotational energy (E sub R) = 75 cm(sub -1), full equilibration of the lambda doublet species, approximately equal populations in both spin-orbit states, no significant degree of alignment, an isotropic flux distribution and an average kinetic energy of (E sub K) = 400 cm(sup -1) per fragment. Although about 75% of the available energy went into the fragment translation, the predissociation proceeded at a rate greater than 10(sup 8)s(sup -1).

900,290

PB89-147441   Not available NTIS
National Bureau of Standards (NEL), Gaithersburg, MD. Thermophysics Div.
Molecular Dynamics Study of a Dipolar Fluid between Charged Plates.
Final rept.
S. H. Lee, J. C. Rasaiah, and J. B. Hubbard. 1986, 6p
Pub. in Jnl. of Chemical Physics 85, n9 p5232-5237 1986.

Keywords: Polarization(Charge separation), Electric fields, Thin films, Autocorrelation, Dipoles, Reprints, *Dipolar fluids, Dielectric saturation.

Recent experiments and computer simulations of thin films have observed the segregation of nonpolar molecules into layers or sheets parallel to the confining walls. The authors discuss a molecular dynamics study of a thin film of Stockmayer molecules between Lennard-Jones plates and find that, in the absence of an electric field, the dipoles are mainly oriented parallel to the plates in each layer. The polarization density profile, with an electric field perpendicular to the plates, is also studied, and is found to oscillate from layer to layer, with a magnitude that is in excess of that predicted by the Debye theory of dielectric saturation by a factor nearly equal to the ratio of the local density to the average bulk density.

900,291

PB89-147458   Not available NTIS
National Bureau of Standards (NEL), Gaithersburg, MD. Thermophysics Div.

Microwave Measurements of the Thermal Expansion of a Spherical Cavity.
Final rept.
M. B. Ewing, J. B. Mehl, M. R. Moldover, and J. P. M. Trusler. 1988, 9p
Sponsored by North Atlantic Treaty Organization, Brussels (Belgium).
Pub. in Metrologia 25, p211-219 1988.

Keywords: *Gases, *Microwaves, *Measurement, *Volume, *Spherical shells, Expansion, Heating, Water, Gallium, Resonance, Temperature, Reprints.

Microwave resonances have been used to measure the volumetric thermal expansion of a spherical cavity between the temperature of the triple point of water (T sub t) and the temperature of the triple point of gallium (T sub g). Using the TM 1,1 and TM 1,2 modes, we find 10 sup 6(V(T sub g)/V(T sub t)-1) = 1418.5 plus or minus 1.0 and 1418.1 plus or minus 0.6, respectively. These results are in agreement with the value 1416.6 plus or minus 1.5 obtained by filling the cavity with mercury and using it as a dilatometer. The microwave measurements are sufficiently accurate that they can be used for primary gas or acoustic thermometry and for measuring the changes in volume standards with temperature, pressure, or time. There is evidence that microwave measurements can be used to determine the volume of a spherical cavity with an uncertainty of the order of 30 ppm and further improvements are likely.

900,292

PB89-147474   Not available NTIS
National Bureau of Standards (NML), Boulder, CO. Time and Frequency Div.
Far-Infrared Laser Magnetic Resonance Spectrum of Vibrationally Excited C2H(1).
Final rept.
J. M. Brown, and K. M. Evenson. 1988, 11p
Pub. in Jnl. of Molecular Spectroscopy 131, p161-171 1988.

Keywords: *Laser beams, *Magnetic resonance, *Methane, *Molecular vibration, *Far infrared radiation, Spectrum analysis, Fluorine, Excitation, Electronic spectra, Reprints.

Some previously observed but unassigned lines in the laser magnetic resonance spectrum of the F + CH4 flame at 490.4 micro m have been assigned to the N = 6 to 7, - to + transition of the C2H radical in its (010) vibrational level. The spectrum is an interesting example of a magnetic resonance spectrum of a radical displaying Hund's case (b) coupling and arises because the laser frequency is very close to the zero-field molecular frequency. The measured resonances are combined with recently available millimeter-wave observations of C2H in the same vibrational level to give an improved set of molecular parameters.

900,293

PB89-148407   PC A04/MF A01
National Inst. of Standards and Technology (NEL), Boulder, CO. Thermophysics Div.
Experimental Thermal Conductivity, Thermal Diffusivity, and Specific Heat Values of Argon and Nitrogen.
Final rept.
H. M. Roder, R. A. Perkins, and C. A. Nieto de Castro. Oct 88, 54p NISTIR-88/3902

Keywords: *Thermal conductivity, *Thermal diffusivity, *Specific heat, *Argon, *Nitrogen, Hot wire anemometers, Tables(Data), Temperature, Pressure.

Experimental measurements of thermal conductivity and thermal diffusivity as obtained in a transient hot-wire apparatus for argon and nitrogen are reported. Values of the specific heat are calculated from these measured values and the density associated with each measurement. The measurements were made at temperatures between 80 and 320 K with pressures between 0.1 and 70 MPa. The density range is 0 to 36 mol/L for argon and 0 to 32 mol/L for nitrogen. The total number of points recorded is 1484 for argon and 1423 for nitrogen.

900,294

PB89-149258   Not available NTIS
National Bureau of Standards (NML), Gaithersburg, MD. Organic Analytical Research Div.

Optical Rotation.
Final rept.
B. Coxon. 1987, 8p
Pub. in Recommended Reference Materials for the Realization of Physicochemical Properties, Chapter 14, p419-426 1987.

Keywords: *Optical activity, Optical properties, Polarimetry, Utilization, Sources, Sucrose, Glucose, Quartz, Light sources, Physicochemical properties, Reprints, *Reference materials.

An introduction to polarimetry is given, including the application of manual and automatic polarimeters and laser light sources. Definitions of specific optical rotation and specific rotatory power are given and the units of these quantities are discussed. Reference materials for the measurement of optical rotation are discussed, including the usage, sources of supply, and pertinent polarimetric data for sucrose, anhydrous dextrose, and quartz con_r0) plates.

900,295

PB89-150767   Not available NTIS
National Bureau of Standards (NML), Gaithersburg, MD. Radiation Physics Div.
Autodetaching States of Negative Ions.
Final rept.
C. W. Clark. 1986, 6p
Sponsored by Air Force Office of Scientific Research, Arlington, VA.
Pub. in Proceedings of International Laser Science Conference Advances in Laser Science-1 (1st), Dallas, TX., p379-384 1986.

Keywords: *Anions, *Resonance absorption, Stability, Laser spectroscopy, Rare gases, Alkalinity, Ions, *Electron-atom collisions.

Autodetaching states (or resonances) of negative ions are very sensitive to electron correlation effects, and the general rules which determine their energetics and stability are not well understood. The experimental techniques used to study them depend strongly on electronic species. For instance, laser photodetachment spectroscopy is one of the preferred ways of investigating alkali negative ions, whereas negative ion resonances of noble gases can only be produced by particle impact. This tends to obscure possible correspondences between the resonance spectra of different elements. However, evidence of common properties of certain classes of autodetaching states is accumulating. The paper describes a few simple examples.

900,296

PB89-150817   Not available NTIS
National Bureau of Standards (NML), Gaithersburg, MD. Radiation Physics Div.
Resonance Ionization Mass Spectrometry of Mg: The 3pnd Autoionizing Series.
Final rept.
R. E. Bonanno, C. W. Clark, J. D. Fassett, and T. B. Lucatorto. 1986, 2p
Pub. in Proceedings of International Laser Science Conference (1st), Dallas, TX., November 18-22, 1985, p409-410 1986.

Keywords: *Magnesium, *Ionization, Mass spectroscopy, Photons, Excitation, Rydberg series.

Stepwise multiphoton excitation was utilized to observe autoionizing Rydberg states of magnesium. The 3pnd series which converges to the 3p2P0 limits of the Mg ion was observed. Measurements extended up to n-40. A preliminary quantum defect analysis is presented.

900,297

PB89-150908   Not available NTIS
National Bureau of Standards (NML), Gaithersburg, MD. Chemical Kinetics Div.
Reactions of Phenyl Radicals with Ethene, Ethyne, and Benzene.
Final rept.
A. Fahr, W. G. Mallard, and S. E. Stein. 1986, 7p
Sponsored by Gas Research Inst., Chicago, IL.
Pub. in Proceedings of International Symposium on Combustion (21st), Munich, West Germany, August 3-8, 1986, p825-831.

Keywords: *Reaction kinetics, *Ethylene, *Benzene, Displacement reactions, Deuterium compounds, Chemical reactions, *Phenyl radicals, *Ethyne.

36

In a low pressure flow reactor, rates for the displacement of H atoms from unsaturated molecules by phenyl radicals have been measured relative to phenyl radical recombination. Assuming a rate constant for the latter process of $10(\text{sup } 0.3)/M/s$, the rate constants at 1030 K for displacement reaction from ethyne, ethene, and benzene, respectively, are: $Ph^* + C2H4 = PhC2H3 + H$, where $k = 1.2 \times 10(\text{sup } 8)/M/s$; $Ph^* + C2H2 = PhC2H + H$, where $k = 1.6 \times 10(\text{sup } 8)/M/s$; $Ph^* + PhH = Ph\text{-}Ph + H$, where $k = 3.0 \times 10(\text{sup } 7)/M/s$. The role of the reversibility of the formation of the initial radical-molecule complex is investigated by determining D-displacement rates for deuterated ethyne and benzene. Reversibility is important in the latter case, but not in the former.

900,298
PB89-150916   Not available NTIS
National Bureau of Standards (NML), Gaithersburg, MD. Molecular Spectroscopy Div.
Final-State-Resolved Studies of Molecule-Surface Interactions.
Final rept.
D. S. King, and R. R. Cavanagh. 1986, 59p
Contract DE-AI05-84ER13150
Sponsored by Department of Energy, Washington, DC.
Pub. in Chemistry and Structure at Interfaces: New Laser and Optical Techniques, Chapter 2, p25-83 1986.

Keywords: Desorption, Fluorescence, Molecular beams, Ionization, Scattering, Platinum, Ruthenium, Silver, Reprints, *Surface reactions, Molecule collisions, Nitric oxide, Laser applications.

A critical review is provided of recent experimental research in the area of final state resolved studies of molecule surface interactions.

900,299
PB89-150981   Not available NTIS
National Bureau of Standards (NML), Gaithersburg, MD. Chemical Kinetics Div.
Relative Acidities of Water and Methanol and the Stabilities of the Dimer Anions.
Final rept.
M. Mautner, and L. W. Sieck. 1986, 4p
Pub. in Jnl. of Physical Chemistry 90, n25 p6687-6690 1986.

Keywords: *Water, *Anions, *Acidity, *Dimerization, *Chemical stabilization, pH, Dissociation energy, Temperature, Reprints, *Methyl alcohol.

The difference between delta H(sub acid)(600) of H2O and CH3OH was directly measured to be $9.6 + \text{ or } - .2$ kcal/mol using variable-temperature pulsed high pressure mass spectrometry. This result defines delta H(sub acid)(CH3OH) at 300 K as $381.6 + \text{ or } - .7$ kcal/mol and also confirms published values of EA(CH3OH) and delta H(sub D)(CH3O-H). H2O was also used as a reference to measure delta H(sub acid)(C6H6) as $400.7 + \text{ or } - .8$ kcal/mol. The dissociation energies of the hydrogen bonded dimers OH(sup -1) H2O (28.8 kcal/mol), CH3O(sup -1):H2O (23.9 kcal/mol) and CH3O(sup -1):CH3OH (28.8 kcal/mol) were found to be in very good agreement with published ab initio results of Ikuta.

900,300
PB89-150999   Not available NTIS
National Bureau of Standards (NML), Gaithersburg, MD. Chemical Kinetics Div.
Redox Chemistry of Water-Soluble Vanadyl Porphyrins.
Final rept.
P. Hambright, P. Neta, M. C. Richoux, Z. Abou-Gamra, and A. Harriman. 1987, 11p
Pub. in Jnl. of Photochemistry 36, n3 p255-265 1987.

Keywords: *Oxidation reduction reactions, Porphyrins, Photochemical reactions, Radiolysis, Anions, Cations, Reprints, *Porphyrin/vanadyl, *Vanadium porphyrin, Pulsed radiation, Chemical reaction mechanisms.

In aqueous solution, pulse radiolytic studies have shown that vanadyl porphyrins undergo redox reactions only at the porphyrin ring. The resultant porphyrin radical anions and cations are unstable with respect to disproportionation. Steady-state reduction, both radiolytic and photochemical, of the vanadyl porphyrins results in formation of phlorins, porphodimethenes, and chlorins depending upon pH and the nature of the porphyrin periphery groups. There is no evidence to show formation of products in which the central vanadyl ion has been reduced.

900,301
PB89-151005   Not available NTIS
National Bureau of Standards (NML), Gaithersburg, MD. Chemical Kinetics Div.
Reactions of Magnesium Prophyrin Radical Cations in Water. Disproportionation, Oxygen Production, and Comparison with Other Metalloporphyrins.
Final rept.
A. Harriman, P. Neta, and M. C. Richoux. 1986, 5p
Pub. in Jnl. of Physical Chemistry 90, n15 p3444-3448 1986.

Keywords: *Oxidation reduction reactions, *Cations, *Water, *Oxygen, Porphyrine, Comparison, Radiolysis, pH, Chemical stabilization, Disproportionation, Reaction kinetics, Reprints, *Magnesium porphyrin, Pulsed radiation.

Under pulse radiolytic conditions, B22(sup -1) oxidases water-soluble magnesium porphyrins. The stability of the resultant porphyrin radical cations depends upon the relative electron density residing on the porphyrin ring, which is controlled by the nature of the water-solubilizing groups. Decreasing the electron density on the porphyrin ring by attaching N-methyl-4-pyridyl groups at the meso positions or by replacing the central Mg(II) ion with a cation possessing a higher ionization potential renders the radical cation unstable with respect to disproportionation. The rate and total yield of evolved O2 reach a maximum at pH 12.

900,302
PB89-151013   Not available NTIS
National Bureau of Standards (NML), Gaithersburg, MD. Chemical Kinetics Div.
Rate Constants for One-Electron Oxidation by Methylperoxyl Radicals in Aqueous Solutions.
Final rept.
R. E. Huie, and P. Neta. 1986, 7p
Pub. in International Jnl. of Chemical Kinetics 18, n10 p1185-1191 1986.

Keywords: *Reaction kinetics, *Oxidation, Radiolysis, Substitution reactions, Solutions, Reprints, *Peroxylmethyl radicals, Pulsed radiation.

Rate constants for one-electron oxidation by the methylperoxyl radicals, CH3O2, HOCH2O2, (sup -1)O2CCH2O2, and CCl3O2, in aqueous solutions have been measured by pulse radiolysis and found to be in the range of 5x10 to the 5th power to 6x10 to the 8th power/M/s for compounds with redox potentials between 0.6 and 0.1 V. Substitution on the methylperoxyl radical OH or CO2(sup -1) has only a minor effect on the rate of oxidation but substitution with three chlorines increases the rate constants by two orders of magnitude. The redox potential of the CH3O2 radical is estimated to be 0.6-0.7 V.

900,303
PB89-156723   Not available NTIS
National Bureau of Standards (NML), Gaithersburg, MD. Chemical Kinetics Div.
Qualitative MO Theory of Some Ring and Ladder Polymers.
Final rept.
J. P. Lowe, S. A. Kafafi, and J. P. LaFemina. 1986, 9p
Pub. in Jnl. of Physical Chemistry 90, n25 p6602-6610 1986.

Keywords: *Substitution reactions, *Aromatic polycyclic hydrocarbons, *Aliphatic acyclic hydrocarbons, *Polymers, *Band spectra, Polyphenyl hydrocarbons, Thermodynamics, Reprints, *Molecular orbital theory, *Band theory, Polyphenylene, Polypyrrole, Polyacetylene.

The authors apply qualitative molecular orbital techniques to the electronic pi-band structure of substitutionally related ring and ladder polymers of the types represented by poly(paraphenylene), poly(pyrrole), and poly(acene). They show how the band structure for a polymer can be qualitatively constructed from a knowledge of the MOs and energies of the monomer, and then go on to show how the band structure is altered by chemical substitution. There are numerous bands whose edge energies should be insensitive to chemical substitution in certain positions. Relative band gap changes resulting from chemical substitution can be understood as well, although this sometimes requires consideration of orbital mixing. They show how qualitative theory gives insight into the relative thermodynamic stabilities of isomeric polymers and

also into the structures of polarons. In general, qualitative techniques account well for band structures compared at the level of extended Huckel theory and are relevant to results of ab initio calculations as well.

900,304
PB89-156731   Not available NTIS
National Bureau of Standards (NML), Gaithersburg, MD. Chemical Kinetics Div.
New Photolytic Source of Dioxymethylenes: Criegee Intermediates Without Ozonolysis.
Final rept.
R. I. Martinez. 1987, 2p
Pub. in Jnl. of Physical Chemistry 91, n6 p1345-1346 1987.

Keywords: *Photochemistry, *Oxidation, Alkenes, Reprints, *Chemical reaction mechanisms, *Criegee intermediates, *Dioxymethylenes.

The work of Akimoto and coworkers on the matrix photoxidation of alkenes in O2 matrices at 10K is reinterpreted in terms of a generalized mechanism for a new photolytic source of dioxymethylenes.

900,305
PB89-156749   Not available NTIS
National Bureau of Standards (NML), Gaithersburg, MD. Chemical Kinetics Div.
Kinetics of Electron Transfer from Nitroaromatic Radical Anions in Aqueous Solutions. Effects of Temperature and Steric Configuration.
Final rept.
M. Mautner, and P. Netá. 1986, 3p
Pub. in Jnl. of Physical Chemistry 90, n19 p4648-4650 1986.

Keywords: *Reaction kinetics, *Anions, *Nitrogen organic compounds, *Electron transfer, Solutions, Molecular structure, Temperature, Aromatic compounds, Radiolysis, Spectrophotometry, Reprints, *Molecular configurations.

Rate constants for electron transfer from various nitroaromatic radical anions to other nitroaromatic compounds in aqueous solutions have been determined by kinetic spectrophotometric pulse radiolysis. In general, nitroaromatic radical anions donate electrons much more slowly than other radical anions, in reactions with similar driving forces, due to low self-exchange rates for ArNO2/ArNO2(sup -1). The kinetics show no anomalies in supercooled solutions.

900,306
PB89-156756   Not available NTIS
National Bureau of Standards (NML), Gaithersburg, MD. Chemical Kinetics Div.
Hyperconjugation: Equilibrium Secondary Isotope Effect on the Stability of the t-Butyl Cation. Kinetics of Near-Thermoneutral Hydride Transfer.
Final rept.
M. Mautner. 1987, 4p
Pub. in Jnl. of the American Chemical Society 109, n26 p7947-7950 1987.

Keywords: *Reaction kinetics, *Cations, *Thermochemistry, *Isotope effect, *Deuterium compounds, *Hydrides, Temperature, Mass spectroscopy, Entropy, Enthalpy, Gibbs free energy, Chemical equilibrium, Reprints, *Butyl ions, *Hyperconjugation.

The thermochemistry of the hydride transfer equilibrium (CD3)3C(sup +1) + (CH3)3CH yields (CH3)3C(sup +1) + (CD3)3CH was measured by pulsed high pressure mass spectrometry. The direction of the observed isotope effect is consistent with C-H bond weakening in the ion due to hyperconjugation. The kinetics of the reaction show a slow rate and a large negative temperature coefficient, with $K(300) = 0.36$ and $k(600) = 0.00625 \times 10$ to the -10th power cu cm/sec, i.e., reaction efficiencies of about 0.03 to 0.0005. The observed negative temperature coefficient, $k = \text{AT}(\text{sup } -5.8)$ is larger than those observed for more exothermic hydride transfer reactions. The approach to collision rate with decreasing temperature is abrupt.

900,307
PB89-156764   Not available NTIS
National Bureau of Standards (NML), Gaithersburg, MD. Chemical Kinetics Div.

CHEMISTRY

Physical & Theoretical Chemistry

Detection of Gas Phase Methoxy Radicals by Resonance Enhanced Multiphoton Ionization Spectroscopy.
Final rept.
G. R. Long, R. D. Johnson, and J. W. Hudgens.
1986, 3p
Pub. in Jnl. of Physical Chemistry 90, n21 p4901-4903 1986.

Keywords: *Deuterium compounds, *Free radicals, Mass spectroscopy, Ultraviolet spectroscopy, Reprints, *Methyl alcohol, *Resonance ionization mass spectroscopy, *Hydroxylmethyl radicals.

Resonance enhanced multiphoton ionization spectra of CH3O and CD3O between 317-331 nm are reported. Methoxy radicals were generated by the reaction of F atoms with CH3OH, CH3OD, and CD3OD in a flow reactor. The most prominent band of these spectra resides at 324.3 nm in CH3O and 323.2 nm in CD3O. Mass spectra of both isotopic analogues showed that methoxy molecular ions do not fragment.

900,308
PB89-156772   Not available NTIS
National Bureau of Standards (NML), Gaithersburg, MD. Chemical Kinetics Div.
Decay of High Valent Manganese Porphyrins in Aqueous Solution and Catalyzed Formation of Oxygen.
Final rept.
A. Harriman, P. A. Christensen, G. Porter, K. Morehouse, P. Neta, and M. C. Richoux. 1986, 17p
Pub. in Jnl. of the Chemical Society, Faraday Transactions 1 82, n10 p3215-3231 1986.

Keywords: *Oxidation reduction reactions, *Reaction kinetics, *Oxygen, Porphyrins, Catalysts, Radiolysis, Catalysis, Chemical stabilization, pH, Reprints, *Manganese porphyrin, *Chemical reaction mechanisms, Pulsed radiation.

Manganese (III) porphyrins (Mn(III)Ps) are easily oxidized to the corresponding Mn(IV)Ps in alkaline aqueous solution. At pH<5, the oxidation· product is a Mn(III)P pi-radical· cation. These oxidized metalloporphyrins have limited stability in water and they revert to the original Mn(III)P upon standing in the dark. The rate and mechanism of this inherent reduction process depends upon pH, with lower pH giving the higher rates. The inherent reduction appears to involve disproportionation and rearrangement of the Mn(IV)Ps but it does not lead to formation of molecular O2. Addition of colloidal RuO2.2H2O, a good O2-evolving catalyst, has a pronounced effect upon the reduction process. The oxidized metalloporphyrin is bound to the catalyst particles by electrostatic forces and, at pH<11, the bound material decays more slowly than the free compound. For 8<pH<11, decay of the bound metalloporphyrin involves oxidation of water to O2 but the yield of O2 is much less than the stoichiometric values.

900,309
PB89-157176   Not available NTIS
National Bureau of Standards (NML), Gaithersburg, MD. Chemical Thermodynamics Div.
Analytical Expression for Describing Auger Sputter Depth Profile Shapes of Interfaces.
Final rept.
W. H. Kirchhoff, G. P. Chambers, and J. Fine. 1986, 5p
Pub. in Jnl. of Vacuum Science and Technology A 4, n3 p1666-1670 1986.

Keywords: *Interfaces, Width, Least squares method, Mathematical models, Reprints, Logistic functions, Depth profiles, Auger electron spectroscopy, Computer applications.

The composition versus depth distribution of a solid/solid interface as determined by Auger sputter depth profiling can be described by a logistic function. A least squares fitting program has been written which fits measured Auger spectral intensities to the above equation to within measurement error (approximately 1%). The statistics associated with the least squares fit allow confidence limits to be placed on the measured widths of interface regions and on the asymmetry associated with each such region.

900,310
PB89-157192   Not available NTIS
National Bureau of Standards (NEL), Boulder, CO. Thermophysics Div.

PVT of Toluene at Temperatures to 673 K.
Final rept.
G. C. Straty, M. J. Ball; and T. J. Bruno. 1986, 3p
Pub. in Jnl. of Chemical and Engineering Data 33, n2 p115-117 1988.

Keywords: *Pressure, *Compressing, *Gas laws, *Toluene, *Liquid phases, Volume, Temperature, Density(Mass/volume), Reprints.

Measurements of the PVT behavior of compressed gaseous and liquid toluene are reported. Pressure versus temperature observations were made along paths of very nearly constant density (pseudoisochores) in the temperature range from about 348 to over 673 K and at pressures to about 35 MPa. Twenty-seven pseudoisochores were determined ranging in density from about 1.7 to near 9 mol/dm sup 3.

900,311
PB89-157200   Not available NTIS
National Bureau of Standards (NEL), Boulder, CO. Thermophysics Div.
PVT Measurements on Benzene at Temperatures to 723 K.
Final rept.
G. C. Straty, M. J. Ball, and T. J. Bruno. 1987, 4p
Pub. in Jnl. of Chemical and Engineering Data 32, n2 p163-166 Apr 87.

Keywords: *Pressure, *Gas laws, *Benzene, *Liquid phases, Volume, Temperature, Compressing, Density(Mass/volume), Reprints.

Measurements of the PVT behavior of compressed gaseous and liquid benzene are reported. Pressure vs. temperature observations were made along paths of very nearly constant density (pseudoisochores) in the temperature.range from about 425 K to over 720 K and at pressures to about 35 MPa. Twenty-four pseudoisochores were determined ranging in density from about 1 mol/dm sup 3 to over 9 mol/dm sup 3.

900,312
PB89-157218   Not available NTIS
National Bureau of Standards (NEL), Gaithersburg, MD. Thermophysics Div.
Molecular Dynamics Study of a Dipolar Fluid between Charged Plates. 2.
Final rept.
S. H. Lee, J. C. Rasaiah, and J. B. Hubbard. 1987, 11p
See also PB89-147441.
Pub. in Jnl. of Chemical.Physics 86, n4 p2383-2393 1987.

Keywords: Polarization(Charge· separation), Electric fields, Thin films, Dipoles, Reprints, *Dipolar fluids, Stockmayer fluids.

Further molecular dynamics simulations of thin films of Stockmayer molecules between Lennard-Jones plates are discussed when the distance h between the plates ranges from 2.25 sigma to 9.5 sigma, where sigma is the molecular diameter, and the electric field E ranges between 0 and 10 billion V/m. The solvation force is calculated as a function of the plate separation h when E=0 and E=one billion V/m and as a function of the field E when h=4.0 sigma and 7.5 sigma. While, in the absence of a field, the molecules tend to form loops and chain-like structures with the dipoles parallel to the wall, a strong external field orients the dipoles along the field so that the long-range repulsive interaction appears to induce a transition to an imperfect (two-dimensional) triangular lattice at low temperature. In between these states, at low temperatures and high fields, the molecules are packed in parallel chains with their moments perpendicular to the field and in 'ferroelectric domains' of opposite polarization.

900,313
PB89-157226   Not available NTIS
National Bureau of Standards (NEL), Boulder, CO. Thermophysics Div.
Local Order in a Dense Liquid.
Final rept.
H. J. M. Hanley, T. R. Welberry, D. J. Evans, and G. P. Morriss. 1988, 4p
Sponsored by Department of Energy, Washington, DC.
Pub. in Physical Review A 38, n3 p1628-1631, 1 Aug 88.

Keywords: *Liquids, Hexagonal lattices, Relaxation time, Anisotropy, Disks(Shapes), Reprints, Order parameters, Structure factors, Transients.

Transient local hexagonal order in a dense liquid is observed in a molecular-dynamics simulation of two-di-

mensional soft disks. Instantaneous local structure factors at a particular density are obtained using an optical transform of the particle coordinates for a single configuration. Local hexagonal order is quantified using an order parameter, and its lifetime characterized by the decay of the order-parameter autocorrelation function. This type of transient structure is significant at densities greater than 7/10 of the freezing density, and becomes persistent near the freezing density.

900,314
PB89-157242   Not available NTIS
National Bureau of Standards (NEL), Gaithersburg, MD. Thermophysics Div.
Hydrodynamics of Magnetic and Dielectric Colloidal Dispersions.
Final rept.
J. B. Hubbard, and P. J. Stiles. 1986, 14p
Pub. in Jnl. of Chemical Physics 84, n12 p6955-6968, 15 Jun 86.

Keywords: *Hydrodynamics, Brownian movement, Magnetic fields, Electric fields, Anisotropy, Reprints, *Colloidal dispersions.

The hydrodynamic behavior of colloidal suspensions is considered, the particles of which possess electric or magnetic moments. Particular attention is given to the case where an external electric or magnetic field acts on a system in which the polarization does not relax instantaneously, so that reorientational Brownian motion is coupled to both the field and to hydrodynamic degrees of freedom. Magnetosonic and magnetoviscous effects are derived, with emphasis on anisotropy with respect to the external field.

900,315
PB89-157291   Not available NTIS
National Bureau of Standards (NEL), Gaithersburg, MD. Molecular Spectroscopy Div.
Population Relaxation of CO(v=1) Vibrations in Solution Phase Metal-Carbonyl Complexes.
Final rept.
E. J. Heilweil, R. R. Cavanagh, and J. C. Stephenson. 1987, 8p
Pub. in Chemical Physics Letters 134, n2 p181-188 1987.

Keywords: *Vibrational spectra, *Infrared analysis, *Metal complexes, *Carbonyl compounds, Carbon monoxide, Tungsten, Carbon tetrachloride, Chloroform, Hexanes, Benzene,· Chromium, Rhodium, Phosphines, Iridium, Chloromethanes, Chemical bonds, Molecular structure, Solvents, Chlorides, Reprints.

Picosecond infrared saturation-recovery experiments were performed to obtain measurements of the vibrational energy lifetimes (T1) of CO(v=1) vibrations (v approximately = 1920 - 1985 cm(sup -1)) of carbonyl-containing metal complexes in dilute, room temperature solutions. For relaxation of the F(1u) CO-stretching vibration of W(CO)6 in CCl4, CHCl3, n-hexane and benzene, T1 was found to be T1(ps) = 800 plus or minus 200, 480 plus or minus 50, 140 plus or minus 15 and 60 plus or minus 6, respectively, while the same mode of Cr(CO)6 in these solvents gave T1(ps) = 440 plus or minus 70, 295 plus or minus 30, 145 plus or minus 25 and 59 plus or minus 6. Monocarbonyl complexes with coordinated triphenylphosphine groups (TPP) have shorter CO(v=1) lifetimes: T1(ps) = 71 plus or minus 12, 50 plus or minus 19 and 29 plus or minus 6 for Rh(CO)Cl(TPP)2, Ir(CO)Cl(TPP)2, and Ir(CO)H(TPP)3 in CHCl3; T1 = 37 (+20, -10) ps for Ir(CO)H(TPP)3 and T1 less than or equal to 20 ps for Rh(CO)H(TPP)3 in CH2Cl2. These observations are rationalized in terms of molecular structure, intramolecular bonding, solvent interaction, and energy accepting vibrational modes.

900,316
PB89-157309   Not available NTIS
National Bureau of Standards (NML), Gaithersburg, MD. Molecular Spectroscopy Div.
Picosecond Studies of Vibrational Energy Transfer in Molecules on Surfaces.
Final rept.
E. J. Heilweil, M. P. Casassa, R. R. Cavanagh, and J. C. Stephenson. 1986, 2p
Pub. in Jnl. of the Optical Society of America B-Optical Physics 3, n8 p140-141 1986.

Keywords: *Surface· chemistry, *Chemisorption, *Infrared· spectroscopy, Energy transfer, Infrared lasers,

Silicon dioxide, Colloids, Reprints, Vibrational spectra, Picosecond pulses, Tunable lasers, Zeolites.

Pump-probe experiments using tunable infrared picosecond pulses measured vibrational energy relaxation times, T sub 1, for bonds of chemisorbed molecules. The results have important implications for surface chemistry and spectroscopy. Damping mechanisms determining T sub 1 are discussed.

900,317
PB89-157317   Not available NTIS
National Bureau of Standards (NML), Gaithersburg, MD. Molecular Spectroscopy Div.
Picosecond Study of the Population Lifetime of CO(v=1) Chemisorbed on SiO2-Supported Rhodium Particles.
Final rept.
E. J. Heilweil, R. R. Cavanagh, and J. C. Stephenson. 1988, 2p
Sponsored by Air Force Office of Scientific Research, Arlington, VA.
Pub. in Jnl. of Chemical Physics 89, n8 p5342-5343, 15 Oct 88.

Keywords: *Carbon monoxide, *Electron transitions, *Metals, *Infrared spectroscopy, *Chemisorption, Particle size, Rhodium, Reprints. .

Infrared pump-probe characterization of the excited state lifetime reveals that CO bound to isolated metal sites (T1 = 140 plus or minus 20 ps) persists longer than the signal observed for CO bound to approximately 35 Angstroms diameter metal particles (less than or equal to 18 ps), suggesting participation of electron-hole excitations in the larger metal particles.

900,318
PB89-157325   Not available NTIS
National Bureau of Standards (NML), Gaithersburg, MD. Molecular Spectroscopy Div.
Microwave Spectrum, Structure, and Electric Dipole Moment of the Ar-Formamide van der Waals Complex.
Final rept.
R. D. Suenram, G. T. Fraser, F. J. Lovas, C. W. Gillies, and J. Zozom. 1988, 6p
Pub. in Jnl. of Chemical Physics 89, n10 p6141-6146, 15 Nov 88.

Keywords: *Formamides, Microwave spectra, Dipole moments, Molecular structure, Reprints, *Argon complexes, Electric dipoles, Fourier transform spectroscopy.

The microwave spectrum of the Ar-formamide van der Waals complex has been obtained using a pulsed-nozzle Fourier-transform microwave spectrometer. The rotational constants of the complex are: A = 10725.7524(46) MHz, B = 1771.0738(22) MHz, and C = 1548.9974(16) MHz. The complex is shown to be nonplanar. The Ar atom is located at 3.62 A from the center of mass of the formamide unit at Ar-O, Ar-N, and Ar-C distances of 3.55, 3.79, and 3.93 A, respectively. The shortest Ar-H distance is 3.25 A which is similar to that observed for Ar-vinyl cyanide (3.21 A). Stark effect and hyperfine analysis yield values for the electric dipole moment components and (14)N quadrupole coupling constants for the complex.

900,319
PB89-157333   Not available NTIS
National Bureau of Standards (NML), Gaithersburg, MD. Molecular Spectroscopy Div.
Microwave Spectrum and (14)N Quadrupole Coupling Constants of Carbazole.
Final rept.
R. D. Suenram, F. J. Lovas, G. T. Fraser, and P. S. Marfey. 1988, 7p
Pub. in Jnl. of Molecular Structure 190, p135-141 1988.

Keywords: *Microwave spectroscopy, *Carbazoles, *Molecular structure, *Rotational spectra, *Electron transitions, Spectra, Pyrroles, Indoles, Quadrupole moment, Nitrogen, Molecular beams, Reprints.

The microwave spectrum of carbazole was observed and analyzed in the 8-14 GHz region using a pulsed molecular-beam Fabry Perot microwave spectrometer. Carbazole was vaporized in a heated nozzle source and was entrained in neon carrier gas before expansion into the Fabry Perot cavity. The rotational transitions were fitted using a rigid rotor Hamiltonian without centrifugal distortion parameters. The rotational constants are A = 2253.1985(2) MHz, B = 594.1861(2)

MHz and C = 470.3503(1) MHz. The inertial defect is small (-0.36 micro Angstroms sup 2) and consistent with a planar molecule. The high resolution available with the instrument (approximately 10 kHz allowed the determination of the N14 nuclear quadrupole coupling constants as chi sub aa = 2.0697(40) MHz, chi sub bb = 1.8719(35) MHz and chi sub cc = -3.9416(35) MHz. A comparison of the electronic environment of the nitrogen atom was made for the series pyrrole, indole and carbazole.

900,320
PB89-157341   Not available NTIS
National Bureau of Standards (NML), Gaithersburg, MD. Molecular Spectroscopy Div.
Infrared and Microwave Investigations of Interconversion Tunneling in the Acetylene Dimer.
Final rept.
G. T. Fraser, R. D. Suenram, F. J. Lovas, A. S. Pine, J. T. Hougen, W. J. Lafferty, and J. S. Muentor. 1988, 18p
Grant NSF-CHE87-20139
Sponsored by National Science Foundation, Washington, DC.
Pub. in Jnl. of Chemical Physics 89, n10 p6028-6045 Nov 88.

Keywords: *Acetylene, Infrared spectroscopy, Microwave spectroscopy, Hydrogen bonds, Electron tunneling, Reprints, Dimers.

A sub-Doppler infrared spectrum of (HCCH)2 has been obtained in the region of the acetylene C-H stretching fundamental using an optothermal molecular-beam color-center laser spectrometer. Microwave spectra were obtained for the ground vibrational state using a pulsed-nozzle Fourier transform microwave spectrometer. In the infrared spectrum, both a parallel and perpendicular band are observed with the parallel band being previously assigned to a T-shaped C(2 mu) complex by Prichard, Nandi, and Muenter and the perpendicular band to a C(2h) complex by Bryant, Eggers, and Watts.

900,321
PB89-157374   Not available NTIS
National Bureau of Standards (NEL), Boulder, CO. Chemical Engineering Science Div.
CO2 Separation Using Facilitated Transport Ion Exchange Membranes.
Final rept.
R. D. Noble, J. J. Pellegrino, E. Grosgogeat, D. Sperry, and J. D. Way. 1988, 15p
Contract DE-AI21-86MC23120
Sponsored by Department of Energy, Morgantown, WV.
Pub. in Separation Science and Technology 23, n12-13 p1595-1609 1988.

Keywords: *Carbon dioxide, *Separation, *Ion exchange resins, Performance evaluation, Membranes, Transport properties, Thin films, Reprints, *Gas transport, Sulfonic acid/fluoro, Polybenzimidazoles, Sodium ions.

The use of ion-exchange membranes as supports for facilitated transport of CO2 is demonstrated. Two different ionomer films were evaluated. The ionomers were a perfluorosulfonic acid film and a sulfonated poly-lybenzimidazole film. Sodium (Na(sup +1)) was exchanged into the membrane for diffusion experiments and ethylenediamine (EDA) was exchanged for facilitated transport experiments. The results indicate that thin perfluorosulfonic acid membranes provide the best CO2 flux and can also provide exceptionally high selectivity.

900,322
PB89-157416   Not available NTIS
National Bureau of Standards (NML), Boulder, CO. Time and Frequency Div.
Calibration Tables Covering the 1460- to 1550-cm(-1) Region from Heterodyne Frequency Measurements on the nu(sub 3) Bands of (12)CS2 and (13)CS2.
Final rept.
J. S. Wells, M. Schneider, and A. G. Maki. 1986, 7p
Sponsored by National Aeronautics and Space Administration, Washington, DC.
Pub. in Jnl. of Molecular Spectroscopy 132, p422-428 1988.

Keywords: *Carbon disulfide, *Frequency measurement, Infrared spectra, Band spectra, Carbon 12, Tables(Data), Heterodyning, Reprints, Carbon 13, Calibration.

Heterodyne frequency measurements have been made on the nu sub 3 band of both (12)CS2 and (13)CS2 near 1500/cm. The data were fitted and new molecular constants determined. Values for the constants and newly calculated frequency calibration tables are presented. The calibration tables cover the range from 1460 to 1550/cm.

900,323
PB89-157440   Not available NTIS
National Bureau of Standards (NML), Gaithersburg, MD. Molecular Spectroscopy Div.
Ozonolysis of Ethylene. Microwave Spectrum, Molecular Structure, and Dipole Moment of Ethylene Primary Ozonide (1,2,3-Trioxolane).
Final rept.
J. Z. Gillies, C. W. Gillies, R. D. Suenram, and F. J. Lovas. 1988, 9p
Pub. in Jnl. of the American Chemical Society 110, n24 p7991-7999 1988.

Keywords: *Molecular structure, *Dipole moments, *Spectrum analysis, *Microwave spectroscopy, *Ozonization, *Ethylene, Formaldehyde, Stereochemistry, Deuterium compounds, Reprints, *Trioxolane, Molecular conformation, Chemical reaction mechanisms, Dioxirane.

The gas-phase structure of ethylene primary ozonide (CH2CH2O3) has been determined from millimeter wave spectra of five isotopic species. The electric dipole moment of the normal isotopic species is 3.43 (4) D. The barrier to pseudorotation is estimated to be high (greater than 300 to 400/cm) in agreement with ab initio MO calculations. Ethylene primary ozonide, dioxirane, formaldehyde, and ethylene secondary ozonide (CH2OCH2O2) are observed as products of the ozone-ethylene reaction in the low-temperature microwave cell. A mechanism of the ozonolysis of ethylene is presented which suggests that the reaction occurs primarily in the condensed phase on the surface of the cell. Microwave techniques utilizing cis- and trans- C2H2D2 show that ozone adds stereospecifically to e h ene in the formation of ethylene primary ozonide. yl

900,324
PB89-157499   Not available NTIS
National Bureau of Standards (IMSE), Gaithersburg, MD. Reactor Radiation Div.
Neutron Vibrational Spectroscopy of Disordered Metal Hydrogen Systems.
Final rept.
R. Hemplemann, and J. J. Rush. 1986, 20p
Pub. in Hydrogen in Disordered and Amorphous Solids, p283-302 1986.

Keywords: *Neutron spectroscopy, *Vibrational spectroscopy, *Hydrides, Inelastic scattering, Neutron scattering, Physical properties, Chemical bonds, Hydrogen bonds, Reprints, *Metallic glasses, *Metal hydrides.

A review is presented of some recent applications of neutron vibrational spectroscopy to the study of disordered metal-hydrogen systems. The examples discussed cover a range of systems from 'simple' dilute solutions in bcc or fcc metals to amorphous alloy hydrides. It is shown that neutron inelastic scattering studies of the vibrational density of states provide a powerful and sensitive probe of the local potentials and bonding sites of hydrogen in metals and often reveal critical information on the novel microscopic physical properties and behavior of disordered metals-hydrogen systems, including those influenced by interstitial or substitutional defects.

900,325
PB89-157507   Not available NTIS
National Bureau of Standards (NML), Gaithersburg, MD. Chemical Kinetics Div.
Validation of Absolute Target Thickness Calibrations in a QQQ Instrument by Measuring Absolute Total Cross-Sections of NE(1+) (NE,NE)NE(1+).
Final rept.
R. I. Martinez, and S. Dheandhanoo. 1986, 10p
Sponsored by Georgetown Univ., Washington, DC. Dept. of Chemistry.
Pub. in International Jnl. of Mass Spectrometry and Ion Processes 74, n2-3 p241-250 1986.

Keywords: *Mass spectroscopy, *Targets, Thickness, Total cross sections, Dissociation, Reprints, eV range 01-10, eV range 100-100, Calibration, Neon ions.

CHEMISTRY

Physical & Theoretical Chemistry

A methodology has been developed for the measurement of total reaction cross-sections sigma(E) in the range of collision energies E(coll) approx = 5-60 eV (LAB) used for collision-induced dissociation (CID) in triple-quadrupole (QQQ) tandem mass spectrometers. This methodology has been calibrated by making pseudo-first order kinetics measurements of the symmetric (resonant) charge transfer cross-sections sigma(E) for Ne(1+) (Ne,Ne)Ne(1+). The sigma(E), for E approx = 5-60 eV and P approx = 0.04-1 mtorr, agree with the Rapp-Francis theory to within 15%. The authors measured identical sigma(E) from both the rate of reactant decay and the rate of product formation. Hence, it should be possible to readily determine branching ratios for different reactive channels. The methodology provides a means whereby the absolute target thickness ((actual path length traversed by the ion) x (effective number density of the CID target)) of gas targets can be accurately calibrated in-situ for collision energies in the range of ca. 5-60 eV (LAB).

900,326
PB89-157515   Not available NTIS
National Bureau of Standards (NML), Gaithersburg, MD. Chemical Kinetics Div.
Stopped-Flow Studies of the Mechanisms of Ozone-Alkene Reactions in the Gas Phase: Tetramethylethylene.
Final rept.
R. I. Martinez, and J. T. Herron. 1987. 8p.
Pub. in Jnl. of Physical Chemistry 91, n4 p946-953 1987.

Keywords: *Ozonation, Mass spectroscopy, Ozone, Reprints, *Chemical reaction mechanisms, *Acetone/hydroxy, *Glyoxal/methyl, Ethylene/tetramethyl.

The reaction of ozone with tetramethylethylene has been studied in the gas phase at 294K and 530 Pa (4 torr) using a stopped-flow reactor coupled to a photoionization mass spectrometer. The concentrations of reactants and products were determined as a function of reaction time. The major products were (CH3)2CO, H2CO, CH3C(O)CH2OH (hydroxyacetone) and CH3C(O)C(O)H (methylglyoxal). On the basis of computer modeling calculations, the mechanism was proposed.

900,327
PB89-157523   Not available NTIS
National Bureau of Standards (NML), Gaithersburg, MD. Chemical Kinetics Div.
S2F10 Formation in Computer Simulation Studies of the Breakdown of SF6.
Final rept.
J. T. Herron. 1987, 3p
Pub. in IEEE (Institute of Electrical and Electronics Engineers) Transactions on Electrical Insulation 22, n4 p523-525 1987.

Keywords: *Sulfur fluorides, *Reaction kinetics, *Mathematical models, *Dielectric breakdown, Sulfur hexafluoride, Water vapor, Reprints, *Di(sulfur pentafluoride).

The chemistry subsequent to the dielectric breakdown of SF6 under 'mild' conditions has been modeled on the basis of known or estimated chemical kinetic data for the neutral reactive species postulated to be formed in the breakdown process. The emphasis is on the significance of S2F10 as an end product of dielectric breakdown and on the role of water vapor in S2F10 formation.

900,328
PB89-157531   Not available NTIS
National Bureau of Standards (NML), Gaithersburg, MD. Chemical Kinetics Div.
Ionic Hydrogen Bond and Ion Solvation. 5. OH...(1-)O Bonds. Gas Phase Solvation and Clustering of Alkoxide and Carboxylate Anions.
Final rept.
M. Mautner, and J. W. Sieck. 1986, 5p
See also PB88-238629.
Pub. in Jnl. of American Chemical Society 108, n24 p7525-7529 1986.

Keywords: *Chemical bonds, *Hydrogen bonds, *Anions, *Dissociation energy, *Dimerization, Enthalpy, Temperature, Solvation, Reprints.

Dissociation energies delta H(sub D) of RO(sup -1) HOR, RCOO(sup -1) HOR and RCOO(sup -1) HOOCR range from 14 to 29 kcal/mol. Large values of delta H(sub D) are observed for the symmetric dimers CH3O(sup -1) HOCH3 (28.8 kcal/mol) and

CH3COO(sup -1) HOOCCH3 (29.3 kcal/mol). Delta H(sub D) decreases as the difference between the acidities of the components increase; e.g., for dimers with large delta H(sub acid) such as CH3O(sup -1) - H2O and HCOO(sup -1) - H2O (delta H(sub acid) = 42.2 and 45.5 kcal/mol, respectively), delta H(sub D) = 16.0 kcal/mol. For 13 dimers, a linear correlation of the form delta H(sub D) = 26.4 - 0.29 delta H(sub acid) is obtained.

900,329
PB89-157549   Not available NTIS
National Bureau of Standards (NML), Gaithersburg, MD. Chemical Kinetics Div.
Filling of Solvent Shells About Ions. 1. Thermochemical Criteria and the Effects of Isomeric Clusters.
Final rept.
M. Mautner, and C. V. Speller. 1986, 9p
See also AD-A178 006.
Pub. in Jnl. of Physical Chemistry 90, n25 p6616-6624 1986.

Keywords: *Solvation, Temperature, Enthalpy, Entropy, Clustering, Ammonium, Water, Anions, Cations, Reprints, *Electronic structure, *Ion pairs, Hydroxyl radicals, Peroxy radicals.

Solvent shells can build up by the stepwise attachment of solvent molecules to gas-phase ions. The filling of a solvent shell by the s-th solvent molecule is indicated by a discontinuous drop in the attachment energy at the s+1-th solvent molecule, i.e., a discontinuous drop in the plots of H(sup n-1, n) vs. n (i.e., enthalpy sequences) after n = s. Another indication is a gap in the spacing of consecutive van't Hoff plots after n = s: Thermochemical criteria based on enthalpy sequences and the spacing of consecutive van't Hoff plots are developed quantitatively and applied to data on clustering about metal ions and onium ions. Satisfactory evidence for the distinct filling of shells is found in 11, and tentative evidence is found in another 15 out of the 45 systems where data were examined.

900,330
PB89-157580   Not available NTIS
National Bureau of Standards (IMSE), Gaithersburg, MD. Metallurgy Div.
Phase Equilibrium in Two-Phase Coherent Solids.
Final rept.
W. C. Johnson, and P. W. Voorhees. 1987, 16p
Pub. in Metallurgical Transactions A-Physical Metallurgy and Materials Science 18, n7 p1213-1228 1987.

Keywords: *Phase diagrams, *Metals, *Thermodynamic equilibrium, Metastable state, Coherence, Solid phases, Stresses, Elastic properties, Free energy, Reprints, Bifurcation theory.

Phase equilibrium in a two-phase coherent solid is analyzed using the conditions necessary for thermodynamic and mechanical equilibrium in nonhydrostatically stressed coherent solids. Subject to the constraints of constant temperature and external pressure, a bulk alloy composition is chosen and the corresponding volume fractions and phase compositions that satisfy the equilibrium conditions are obtained. It is demonstrated that, unlike fluids, a number of equilibrium states (volume fractions) may exist that yield energy minima for a given temperature, pressure and alloy composition and that these multiple metastable states may lead to a nonuniqueness in the observed physical state of the system. These results are a consequence of the elastic and system energies being a function of the volume fraction in a field of two-phase coexistence in coherent solids. As it is difficult to display these effects on a coherent phase diagram, the concept of a phase stability diagram is introduced for both displaying and analyzing the equilibrium conditions in coherent solids. The influence of elastic inhomogeneity and the form of the free energy curves as a function of composition in the absence of stress on phase equilibrium is examined.

900,331
PB89-157689   Not available NTIS
National Bureau of Standards (NML), Gaithersburg, MD. Surface Science Div.
Theoretical Study of the Vibrational Lineshape for CO/Pt(111).
Final rept.
T. S. Jones, J. W. Gadzuk, and S. Holloway. 1987, 10p
Pub. in Jnl. of Surface Science 184, p L421-L430 1987.

Keywords: *Vibrational spectra, *Carbon monoxide, *Adsorbates, Platinum, Adsorption, Reprints, Potential energy surfaces, Chaos.

The spectral analysis method of obtaining vibrational spectra for adsorbates using a classical trajectory approach is presented. A potential-energy-surface (PES) is proposed, which includes anharmonic coupling between the C-O stretching mode and the CO-surface hindered translation. The various dynamical consequences of this PES topology are discussed by examining trajectories, surfaces of section, and the resulting vibrational spectra.

900,332
PB89-157705   Not available NTIS
National Bureau of Standards (NML), Gaithersburg, MD. Surface Science Div.
Status of Reference Data, Reference Materials and Reference Procedures in Surface Analysis.
Final rept.
J. T. Grant, P. Williams, J. Fine, and C. J. Powell. 1988, 5p
Pub. in Surface and Interface Analysis 13, p46-50 1988.

Keywords: *Surface properties, Assessments, Experimental data, Experimental design, Reprints, *Standard reference materials.

A brief synopsis is given of recent developments and current needs for reference data, reference materials and reference procedures in surface analysis. This assessment is based largely on three presentations and related discussion at the Second Topical Conference on Quantitative Surface Analysis held at Monterey, California on 30-31 October, 1987. While a reasonable start has been made in recent years in providing needed data, materials, and procedures, many important needs remain.

900,333
PB89-157713   Not available NTIS
National Bureau of Standards (NML), Gaithersburg, MD. Surface Science Div.
Semiclassical Way to Molecular Dynamics at Surfaces.
Final rept.
J. W. Gadzuk. 1988, 30p
Pub. in Annual Review of Physical Chemistry 39, p395-424 1988.

Keywords: *Molecular energy levels, *Surface chemistry, *Photoelectric emission, Dynamics, Spectra, Desorption, Wave equations, Reprints.

Time-dependent molecular dynamics at surfaces is considered within the framework of semiclassical wave packet dynamics of Heller. Aspects of photoemission spectroscopy, stimulated desorption, electron energy loss spectroscopy and molecule/surface collisions are considered within a unified picture.

900,334
PB89-157762   Not available NTIS
National Bureau of Standards (NEL), Gaithersburg, MD. Semiconductor Electronics Div.
Application of Multiscattering Theory to Impurity Bands in Si:As.
Final rept.
J. R. Lowney. 1988, 5p
Pub. in Jnl. of Applied Physics 64, n9 p4544-4548, 1 Nov 88.

Keywords: *Additives, *Silicon, *Arsenic, *Electron transitions, Impurities, Computation, Electron scattering, Reprints.

Impurity bands in arsenic-doped silicon have been calculated for doping densities of 3.3 x 10(sup 17), 1.2 x(10 sup 18), and 8.0 x 10(sup 18) cm(sup -3). A multiscattering approach is used with a model potential which provides both electronic screening and the proper bound-state energy for the isolated center. The results are in good agreement with previous calculations based on electron hopping among hydrogenic centers. An advantage of the multiscattering approach is that it treats the conduction-band states as well and shows the loss of these states to the formation of the impurity band. This is a new result and affects the density for the Mott transition. Calculations are also performed for the states associated with the binding of an extra electron to unionized arsenic centers, the so-called D sup minus band. The overall results are in

40

good agreement with the observed Mott transition in Si:As near 8 x 10 (sup 18) cm(sup -3).

900,335
PB89-157846   Not available NTIS
National Bureau of Standards (NML), Gaithersburg, MD. Office of Standard Reference Data.
Shimanouchi, Takehiko and the Codification of Spectroscopic Information.
Final rept.
D. R. Lide. 1986, 7p
Pub. in Jnl. of Molecular Structure 146, p1-7 Aug 86.

Keywords: *Spectrum analysis, *Molecular structure, *Spectroscopic analysis, Molecular vibration, Tables(Data), Forms(Paper), Information systems, Data acquisition, Reviews, Reprints, Standardized terminology.

The contributions of Takehiko Shimanouchi to the compilation of spectroscopic and structural data and to the adoption of standard terminology and formats for reporting data are reviewed. Included are his work on the Tables of Molecular Vibrational Frequencies, spectral database projects, and international organizations concerned with scientific information.

900,336
PB89-157853   Not available NTIS
National Bureau of Standards (IMSE), Gaithersburg, MD. Office of Nondestructive Evaluation.
Infrared Absorption of SF6 from 32 to 3000 cm(-1) in the Gaseous and Liquid States.
Final rept.
C. Chapados, and G. Birnbaum. 1988, 29p
Pub. in Jnl. of Molecular Spectroscopy 132, p323-351 1988.

Keywords: *Sulfur hexafluoride, *Infrared spectra, Gases, Liquids, Pressure, Vibrational spectra, Reprints.

Infrared spectra from 32 to 3000/,cm of SF sub 6 were recorded at several pressures from below atmospheric pressure up to 20 atm and in the liquid phase at temperatures from 228 to 284 K. The infrared active fundamentals, difference bands, combination and harmonic bands, and the collision-induced band in the far-infrared region were observed. Integrated intensities of 37 bands were measured at several densities. In the gas phase, a weak band containing v2 + v6 was always found at higher frequencies than the much stronger bands containing v3. In the liquid phase, the positions of these bands are lowered in frequency and the intensities tend toward equalization. The pair v3 and v2 + v6 itself is the exception: while the v3 band is displaced and its intensity lowered, the v2 + v6 band is practically not modified. A possible explanation of the modification in the intensity of the bands in passing from the gas to the liquid is the effect of the interaction with the liquid environment on the Fermi resonance connecting v3 and v2 + v6. Within experimental error no collision-induced component was identified in the mid-infrared region in the gas phase. In the liquid phase, the integrated absorption of the bands increases with temperature, surpassing in many cases the gas-phase values.

900,337
PB89-157929   Not available NTIS
National Bureau of Standards (NML), Gaithersburg, MD. Surface Science Div.
Secondary-Electron Effects in Photon-Stimulated Desorption.
Final rept.
D. E. Ramaker, T. E. Madey, R. L. Kurtz, and H. Sambe. 1988, 13p
Pub. in Physical Review B 38, n3 p2099-2111, 15 Jul 88.

Keywords: *Desorption, Chemisorption, X rays, Reprints, *Electron stimulated desorption, *Photon stimulated desorption, *Secondary electrons, Oxygen ions, Hydrogen ions, Nitrogen ions.

The magnitude of secondary-electron contributions to electron- or photon-stimulated desorption (ESD or (PSD) yields is considered. In particular, the authors have reexamined three systems where a dominant x-ray-induced ESD (XESD) effect has been postulated. Recent ESD ion-angular-distribution data on the NH3/Ni system and a detailed determination of the mechanisms involved in H(1+) desorption indicate that all of the features previously attributed to the XESD effect may in fact arise from direct core-level processes. A reexamination of the PSD N(1+) and O(1+) yields

from condensed N2-O2 reveals that the indirect XESD mechanism contributes just one-third of the N(1+) yield, but dominates the O(1+) desorption. This arises because the direct Auger-stimulated desorption (ASD) process following core-hole excitation is inactive for O(1+) desorption, but remains active for N(1+). Finally, a detailed interpretation of H(1+) desorption from OH/Ti and OH/Cr, and comparison with the system OH/TbO-Sm indicates that the direct ASD process is also inactive in the latter case.

900,338
PB89-157937   Not available NTIS
National Bureau of Standards (NML), Gaithersburg, MD. Surface Science Div.
Optically Driven Surface Reactions: Evidence for the Role of Hot Electrons.
Final rept.
S. A. Buntin, L. J. Richter, R. R. Cavanagh, and D. S. King. 1988, 4p
Contract DE-AI05-84ER13150
Sponsored by Department of Energy, Washington, DC.
Pub. in Physical Review Letters 61, n11 p1321-1324, 12 Sep 88.

Keywords: *Desorption, *Nitrogen oxide(NO), Platinum, Reprints, *Laser induced desorption, Hot electrons.

Evidence is presented for the role of excited conduction electrons in the laser-induced desorption of NO from Pt(111). State-specific detection of the desorbed NO establishes that the rotational distributions are non-Boltzmann, that the spin-orbit population is inverted, and that both the translational and vibrational distributions are uncorrelated with the laser-induced surface-temperature jump. The role of optically excited substrate electrons in driving the desorption process is evidenced by a dramatic dependence of the vibrational and translation energy distributions on the desorption-laser wavelength.

900,339
PB89-157945   Not available NTIS
National Bureau of Standards (NML), Gaithersburg, MD. Surface Science Div.
Universality Class of Planar Self-Avoiding Surfaces with Fixed Boundary.
Final rept.
U. Glaus, and T. L. Einstein. 1987, 7p
Pub. in Jnl. of Physics A 20, n2 pL105-L111, 1 Feb 87.

Keywords: *Surfaces, Monte Carlo method, Polymers, Reprints, *Self avoiding surfaces.

Using a modified version of a Monte Carlo algorithm proposed by Sterling and Greensite, the authors obtain the exponents theta = 1.51 plus or minus 0.25 and mu = 0.502 plus or minus 0.024 for planar self-avoiding surfaces with fixed boundary in three dimensions, consistent with the conjectured exact values for branched polymers. It is shown how the modifications are needed to obtain a viable distribution of surfaces.

900,340
PB89-157952   Not available NTIS
National Bureau of Standards (NML), Gaithersburg, MD. Surface Science Div.
Non-Boltzmann Rotational and Inverted Spin-Orbit State Distributions for Laser-Induced Desorption of NO from Pt(111).
Final rept.
L. J. Richter, S. A. Buntin, R. R. Cavanagh, and D. S. King. 1988, 2p
Contract DE-AI05-84ER13150
Sponsored by Department of Energy, Washington, DC.
Pub. in Jnl. of Chemical Physics 89, n8 p5344-5345, 15 Oct 88.

Keywords: *Desorption, *Nitrogen oxide(NO), Platinum, Near infrared radiation, Near ultraviolet radiation, Reprints, Laser induced desorption, Laser induced fluorescence, Visible radiation.

The internal state distributions of NO desorbed from a Pt(111) surface by visible and near-visible laser radiation (355, 532 and 1064 nm) were measured by laser-induced fluorescence. Non-Boltzmann rotational state distributions and inverted spin-orbit populations were observed and were found to be relatively insensitive to the desorption-laser wavelength. This is in contrast with the kinetic energy distributions and the vibrational state populations, both of which exhibit a significant dependence on desorption-laser wavelength.

900,341
PB89-157978   Not available NTIS
National Bureau of Standards (NML), Gaithersburg, MD. Surface Science Div.
Calculations of Electron Inelastic Mean Free Paths for 31 Materials.
Final rept.
S. Tanuma, C. J. Powell, and D. R. Penn. 1988, 13p
Pub. in Surface and Interface Analysis 11, p577-589 1988.

Keywords: *Mean free path, *Elastic properties, *Electrons, *Surface properties, Ceramics, Metals, Scattering, Energy dissipation, Dielectric properties, Algorithms, Gases, Reprints.

New calculations of electron inelastic mean free paths (IMFPs) for 200-2000 eV electrons in 27 elements (C, Mg, Al, Si, Ti, V, Cr, Fe, Ni, Cu, Y, Zr, Nb, Mo, Ru, Rh, Pd, Ag, Hf, Ta, W, Re, Os, Ir, Pt, Au and Bi) and four compounds (LiF, SiO2, ZnS and Al2O3) are presented. These calculations are based on an algorithm due to Penn which makes use of experimental optical data (to represent the dependence of the inelastic scattering probability on energy loss) and the theoretical Lindhard dielectric function (to represent the dependence of the scattering probability on momentum transfer). Our calculated IMFPs were fitted to the Bethe equation for inelastic electron scattering in matter; the two parameters in the Bethe equation were then empirically related to several material constants. The resulting general IMFP formula is believed to be useful for predicting the IMFP dependence on electron energy for a given material and the material-dependence for a given energy. The new formula also appears to be a reasonable but more approximate guide to electron attenuation lengths.

900,342
PB89-158026   Not available NTIS
National Bureau of Standards (IMSE), Gaithersburg, MD. Ceramics Div.
Effects of Pressure on the Vibrational Spectra of Liquid Nitromethane.
Final rept.
P. J. Miller, S. Block, and G. J. Piermarini. 1989, 5p
Sponsored by Army Research Office, Arlington, VA., and Naval Surface Warfare Center, Silver Spring, MD.
Pub. in Jnl. of Physical Chemistry 93, n1 p462-466, 12 Jan 89.

Keywords: *Nitromethane, *Vibrational spectra, *Spectrum analysis, *Deuterium compounds, Pressure, Intermolecular forces, Pyrolysis, Decomposition reactions, Reprints, Chemical shifts.

A complete normal-coordinate calculation for nitromethane is given using the vibrational frequencies measured from the three isotopes CH3NO2, CD3NO2, and CH3(sup 15)NO2. Pressure effects on the vibrational normal modes of the superpressed liquid to 2.0 GPa also were measured. A softening of the frequency of the asymmetric stretching mode of NO2 with increasing pressure indicated a strong intermolecular interaction. By calculating a minimum energy pair configuration and applying two-site exiton theory, the authors calculated the shifts of the normal modes of the NO2 stretching vibrations as a function of density and pressure and found them to agree qualitatively with the experimentally measured shifts. The results support the bimolecular nature of the mechanism for the thermal decomposition of nitromethane under pressure.

900,343
PB89-158083   Not available NTIS
National Bureau of Standards (NML), Gaithersburg, MD. Atomic and Plasma Radiation Div.
Stark Broadening of Spectral Lines of Homologous, Doubly Ionized Inert Gases.
Final rept.
N. Konjevic, and T. L. Pittman. 1987, 8p
Pub. in Jnl. of Quantitative Spectroscopy and Radiative Transfer 37, n3 p311-318 Mar 87.

Keywords: *Spectral lines, *Stark effect, *Line width, *Rare gases, *Line spectra, Spectrum analysis, Argon, Numerical analysis, Ionization, Experimental design, Comparison, Reprints.

In the paper the authors report electron impact widths of 36 lines that belong to the sequence of homologous, doubly ionized atoms of inert gases. Only four Ar III lines in this sequence were previously measured and they are in good agreement with the present re-

CHEMISTRY

Physical & Theoretical Chemistry

sults. Good agreement is also found between these experimental results and the theoretical calculations using a modified semiempirical approach. Average ratio between experiment and theory is 1.06 with maximum deviation not exceeding + or - 20%. They used present results to investigate trends of the Stark widths of analogous lines.

900,344
PB89-158125 Not available NTIS
National Bureau of Standards (NML), Gaithersburg, MD. Atomic and Plasma Radiation Div.
Atomic Transition Probabilities of Argon: A Continuing Challenge to Plasma Spectroscopy.
Final rept.
W. L. Wiese. 1988, 7p
Pub. in Jnl. of Quantitative Spectroscopy and Radiative Transfer 40, n3 p421-427 1988.

Keywords: *Argon, *Spectrum analysis, *Spectral lines, *Emission spectroscopy, Performance evaluation, Atomic properties, Reviews, Reprints, *Atom transport.

Determination of the atomic transition probabilities for prominent spectral lines of Ar I and II are classical cases for testing the capabilities of the emission spectroscopy method. Despite numerous attempts to measure these data accurately, differences of about 30-40% remain in the numerical results, and the available material actually suggests two scales for the transition probabilities, differing by about 90%, as has been pointed out repeatedly. A critical analysis of all emission experiments undertaken in the study is able to remove these differences satisfactorily. A single transition probability value with an error estimated of only + or - 5% is established for a typical Ar I transition; however, discrepancies for the Ar II data are not removed and remain as a challenge to future experiments.

900,345
PB89-158133 Not available NTIS
National Bureau of Standards (NEL), Gaithersburg, MD. Thermophysics Div.
Van der Waals Equation of State Around the Van Laar Point.
Final rept.
P. H. E. Meijer. 1989, 9p
Pub. in Jnl. of Chemical Physics 90, n1 p448-456, 1 Jan 89.

Keywords: *Van der Waals equation, *Equations of state, *Binary systems(Materials), *Critical points, Phase diagrams, Separation, Thermodynamic properties, Reprints, *Van Laar point.

The discovery of van Laar that the van der Waals equation of state for binary mixtures can be solved analytically at the double point on the critical line, provided one introduces the geometric-mean condition on the interaction parameters, is used to obtain the explicit expression for the critical line for this case. The critical line is expressed as a function of the density variables, and the origin is shifted to this double point: the van Laar point. By doing so it is possible to show that this double point is also a tricritical point. This is different from the lattice gas, where the double point is always a tricritical point. Small deviations from this point in parameter space induce very different phase diagrams. The influence of excursions from the van Laar point is expressed as a function of the state variables. Both the k factor (the deviation from the geometric-mean rule) and the 'asymmetry' coefficient e (the deviation from the crossing point) are introduced. The results are given in the form of polynomials in local coordinates in density space. Conditions under which the double point is maintained are given and the differences between the Scott and van Konynenburg classes II, III, IV, and IV* are explained.

900,346
PB89-161574 Not available NTIS
National Bureau of Standards (NML), Gaithersburg, MD. Molecular Spectroscopy Div.
Far-Infrared Spectrum of Methyl Amine. Assignment and Analysis of the First Torsional State.
Final rept.
N. Ohashi, K. Takagi, J. Hougen, W. Olson, and W. Lafferty. 1988, 19p
Pub. in Jnl. of Molecular Spectroscopy 132, p242-260 1988.

Keywords: *Methylamine, *Spectrum analysis, *Infrared spectroscopy, *Molecular rotation, *Stress relax-

ation tests, Electron tunneling, Far infrared radiation, Deformation, Reprints.

The far-infrared spectrum of methyl amine has been studied in the 40- to 350/cm region with a resolution of 0.005/cm or better. The pure rotational spectrum in the first excited torsional state, as well as the fundamental torsional band, has been assigned. The data obtained have been combined with microwave data from the literature, and a global fit has been carried out, based on a group theoretical formalism developed previously. Some aspects of the torsional potential function and inversion potential function in this molecule are briefly discussed.

900,347
PB89-161582 Not available NTIS
National Bureau of Standards (NML), Gaithersburg, MD. Chemical Kinetics Div.
Critical Review of the Chemical Kinetics of Sulfur Tetrafluoride, Sulfur Pentafluoride, and Sulfur Fluoride (S2F10) in the Gas Phase.
Final rept.
J. T. Herron. 1987, 14p
Pub. in International Jnl. of Chemical Kinetics 19, n2 p129-142 1987.

Keywords: *Sulfur fluorides, *Reaction kinetics, *Gases, *Dielectric breakdown, Pyrolysis, Reprints.

The gas phase chemical kinetics of SF4, SF5 and S2F10 are reviewed with particular emphasis on relevance to the general problem of the dielectric breakdown of SF6. Specific reaction systems treated are SF4 + F2, SF5 + SF5 and the pyrolysis of S2F10. Computer modeling calculations were carried out to arrive at best estimates of rate parameters. Based on the results of these calculations, sets of recommended rate parameters are provided. The major discrepancies and problems in establishing the kinetic data base are described. Thermochemical consequences of different model calculations are given.

900,348
PB89-161608 Not available NTIS
National Bureau of Standards (NML), Gaithersburg, MD. Chemical Kinetics Div.
Rate Constants for Hydrogen Abstraction by Resonance Stabilized Radicals in High Temperature Liquids.
Final rept.
M. J. Manka, R. L. Brown, and S. E. Stein. 1987, 15p
Pub. in International Jnl. of Chemical Kinetics 19, n10 p943-957 1987.

Keywords: *Reaction kinetics, *High temperature tests, *Free radicals, Transfer characteristics, Chemical equilibrium, Recombination reactions, Energy levels, Reprints, *Hydrogen transfer.

Benzylic H-atom abstraction rates by diphenylmethyl radicals from a series of donors were determined in nonpolar liquids at elevated temperatures. Relative rates were converted to absolute rates via available equilibrium constant data. Abstraction by diphenylmethyl from 1,2,3,4-tetrahydronaphthalene (tetralin) was studied over the temperature range 489-573K. Similar reactions with the fluorenyl radical were also studied. In this case, relative rates were converted to absolute rates with an equilibrium constant determined from the observed homolysis rate of the dimer and an assumed recombination rate. In addition, forward and reverse rate measurements yielded the equilibrium constant for hydrogen transfer between fluorenyl and diphenylmethyl.

900,349
PB89-161889 PC A07/MF A01
National Inst. of Standards and Technology (NML), Gaithersburg, MD. Surface Science Div.
Technical Activities 1988, Surface Science Division.
C. J. Powell. Jan 89, 142p NISTIR-89/4025
See also rept for 1987, PB88-169453.

Keywords: *Surface chemistry, Standards, Surface properties, Catalysts, Electron spectra, Atomic structure, Adsorption, Surface physics.

The report summarizes technical activities of the NIST Surface Science Division during Fiscal Years 1987 and 1988. These activities include surface-standards work, experimental and theoretical research in surface science, the development of improved measurement methods, and applications to important scientific and national problems. A listing is given of publications,

talks, professional committee participation, and professional interactions by the Division staff.

900,350
PB89-171219 Not available NTIS
National Bureau of Standards (NML), Boulder, CO. Quantum Physics Div.
Infrared Spectrum of D2HF.
Final rept.
C. M. Lovejoy, D. D. Nelson, and D. J. Nesbitt. 1988, 9p
Grants NSF-CHE86-05970, NSF-PHY86-04504
Sponsored by National Science Foundation, Washington, DC.
Pub. in Jnl. of Chemical Physics 89, n12 p7180-7186, 15 Dec 88.

Keywords: *Infrared spectra, *Hydrogen, *Fluorine, Deuterium, Isotopic labeling, Lasers, Vibration, Energy levels, Gases, Reprints.

Ultrasensitive infrared laser absorption spectroscopy in a slit supersonic expansion is used to obtain the spectrum of the HF stretching fundamental of D2HF. Both a Pi to Pi band due to para-D2HF and a Sigma to Sigma band due to ortho-D2HF are observed, in contrast to the H2HF spectrum which consists of the Pi to Pi band alone. Analysis of the spectrum indicates that the D2HF Pi states are more strongly bound than the Sigma states. Doublet splittings in the Pi to Pi band are analyzed to determine barriers to internal rotation of D2 within the complex. The vibrational predissociation rate of D2HF is approximately 25 times faster than that of H2HF, suggesting the opening of a channel which results in vibrational excitation of the D2 fragment.

900,351
PB89-171227 Not available NTIS
National Bureau of Standards (NML), Boulder, CO. Quantum Physics Div.
Infrared Spectrum of NeHF.
Final rept.
D. C. Clary, C. M. Lovejoy, S. V. ONeil, and D. J. Nesbitt. 1988, 4p
Sponsored by National Science Foundation, Washington, DC.
Pub. in Physical Review Letters 61, n14 p1576-1579, 3 Oct 88.

Keywords: *Neon, *Hydrogen, *Fluorine, *Molecular orbitals, Gases, Quantum theory, Vibration, Infrared spectra, Absorption spectra, Computation, Reprints.

The infrared-absorption spectrum of the previously unobserved NeHF molecule has been predicted from an ab initio quantum-mechanical calculation and subsequently determined for the first time by direct measurement. The two procedures yield remarkable agreement in the positions, widths, and intensities of the infrared spectral lines. The calculations predict, and the experiments confirm, highly unusual vibrational dynamics.

900,352
PB89-171243 Not available NTIS
National Bureau of Standards (NML), Boulder, CO. Quantum Physics Div.
Laser Probing of Ion Velocity Distributions in Drift Fields: Parallel and Perpendicular Temperatures and Mobility for Ba(1+) in He.
Final rept.
R. A. Dressler, J. P. M. Beijers, H. Meyer, S. M. Penn, V. M. Bierbaum, and S. R. Leone. 1988, 9p
Grant AFOSR-84-0272
Sponsored by Air Force Office of Scientific Research, Bolling AFB, DC.
Pub. in Jnl. of Chemical Physics 89, n8 p4707-4715, October 15, 1988.

Keywords: *Barium, *Helium, *Electric fields, *Ionic mobility, Ions, Drift, Velocity, Lasers, Fluorescence, Doppler tracking, Transport properties, Probes, Reprints.

Measurements of ion velocity distributions are presented for Ba+ drifted in helium under well characterized conditions using single-frequency laser-induced fluorescence probing. The reduced mobilities and the Doppler profiles parallel and perpendicular to the electric field vector as a function of the ratio of the field strength (E) to the buffer gas density (N) up to 33.5 Td are presented. The reduced mobility decreases monotonically with increasing E/N from the zero-field value of 16.7 plus or minus 0.4 sq cm/V/S at 313 K. The parallel and perpendicular ion temperatures are in very

42

good agreement with both a repulsive Maxwell model and a parametrized version of the three-temperature theory of Lin et al. The parallel temperature is always higher than the perpendicular one. Effects of optical pumping on the Doppler profiles are also presented.

900,353
PB89-171250 Not available NTIS
National Bureau of Standards (NML), Boulder, CO.
Quantum Physics Div.
Laser Probing of Product-State Distributions in Thermal-Energy Ion-Molecule Reactions.
Final rept.
S. R. Leone, and V. M. Bierbaum. 1987, 11p
Grants NSF-PHY86-04504, NSF-CHE86-08043
Sponsored by National Science Foundation, Washington, DC., and Air Force Office of Scientific Research, Bolling AFB, DC.
Pub. in Faraday Discussions of the Chemical Society 84, p253-263 1987.

Keywords: *Molecular rotation, *Molecular vibration, *Electron transfer, Ionization, Nitrogen, Carbon monoxide, Molecular energy levels, Reprints, *Ion-molecule collisions, *Laser induced fluorescence, Franck-Condon principle.

Vibrational and rotational product-state distributions are determined for thermal-energy charge-transfer reactions and Penning ionization processes using laser-induced fluorescence detection in both a flowing afterglow apparatus and a single-collision molecular beam device. The reactions investigated are the charge transfers between N(+)+CO, A(+)+N2, Ar(+)+CO, and the Penning ionization of N2 by Ne(3P2). Vibrational distributions provide direct information on major features of the dynamics, such as whether a Franck-Condon mechanism is dominant, whether collision complex formation is important, or if selective vibrational passageways exist between the electronic potential-energy surfaces. The rotational distributions show a variety of additional discriminating dynamical effects, including corroborating evidence for Franck-Condon channels, pinpointing separate mechanisms for different vibrational product states and detecting microscopic bimodalities within individual vibrational levels, which are indicative of multiple entrance- or exit-channel pathways.

900,354
PB89-171284 Not available NTIS
National Bureau of Standards (NML), Gaithersburg, MD. Temperature and Pressure Div.
Three-State Lattice Gas as Model for Binary Gas-Liquid Systems.
Final rept.
P. H. E. Meijer, and M. Napiorkowski. 1987, 7p
Pub. in Jnl. of Chemical Physics 86, n10 p5771-5777 1987.

Keywords: *Binary systems(Materials), *Phase diagrams, *Critical field, Gases, Liquids, Lattice parameters, Clustering, Reprints.

The paper deals with the three-state lattice model as applied to binary liquid-gas systems. Schouten, ten Seldam and Trappeniers (Physica 73,556 (1974)) used the model to describe the transition between the liquid phases as well as the gas-gas separation when varying the interaction parameters of the model. The analysis is aimed at the behavior of the system in the vicinity of its tricritical point in the generalized field space. Different shapes of the critical lines are calculated by using the molecular field method as well as by using the lowest approximation of the cluster variation method. The equivalence between the two approaches is demonstrated. In the immediate neighborhood of the tricritical point, the shapes of the critical lines are also determined analytically.

900,355
PB89-171292 Not available NTIS
National Bureau of Standards (NML), Gaithersburg, MD. Temperature and Pressure Div.
High Resolution Inverse Raman Spectroscopy of the CO Q Branch.
Final rept.
G. J. Rosasco, L. A. Rahn, W. S. Hurst, R. E. Palmer, J. P. Looney, and J. W. Hahn. 1988, 13p
Sponsored by Department of Energy, Washington, DC., and Army Research Office, Arlington, VA.
Pub. in Pulsed Single-Frequency Lasers: Technology and Applications, v912 p171-183 1988.

Keywords: *Vibrational spectra, *Raman spectroscopy, *Carbon monoxide, *Pressure, Gases, Lasers, Reprints.

Preliminary results of a high resolution spectroscopic study of the pressure dependence of the Raman vibrational Q-branch spectrum of pure CO are reported. Measurements are made at room temperature over the pressure range 0.5 to 6 atm. The technique of quasi-cw inverse Raman spectroscopy utilizing a pulsed single-frequency laser source is employed. This approach gives enhanced sensitivity compared to earlier work which employed cw lasers, allowing extension of that work to higher accuracy, higher J states, and higher pressure. The goal of this work is to test the accuracy of a modified exponential-gap rate law model which is used to predict the pressure dependent spectra.

900,356
PB89-171300 Not available NTIS
National Bureau of Standards (NML), Gaithersburg, MD. Temperature and Pressure Div.
Dynamics of a Spin-One Model with the Pair Correlation.
Final rept.
M. Keskin, and P. H. E. Meijer. 1986, 10p
Pub. in Jnl. of Chemical Physics 85, n12 p7324-7333 1986.

Keywords: *Phase diagrams, *Transition temperature, *Spin spin interactions, Metastable state, Quenching(Cooling), Reprints.

A spin-1 or three state system will undergo a first or second order phase transition depending on the ratio of coupling parameter alpha. Using the pair correlation approximation, the transition temperature is determined in order to obtain the unstable, the metastable, as well as the stable states of this cooperative system. The dynamics of the system is studied by means of the most probable path method and the flow lines and fixed points of the system are given for zero field. The choice of possible initial conditions is discussed. The role of the unstable points in the phase diagram, as separators between the stable and metastable points, is described and they are computed for a number of cases.

900,357
PB89-171532 Not available NTIS
National Bureau of Standards (NML), Gaithersburg, MD. Chemical Kinetics Div.
Absolute Rate Constants for Hydrogen Abstraction from Hydrocarbons by the Trichloromethylperoxyl Radical.
Final rept.
S. Mosseri, Z. B. Alfassi, and P. Neta. 1987, 9p
Pub. in International Jnl. of Chemical Kinetics 19, n4 p309-317 1987.

Keywords: *Reaction kinetics, *Hydrogen, *Spectrophotometry, *Free radicals, Cyclohexane, Cyclohexene, Oxidation, Metal containing organic compounds, Porphyrins/Hydrocarbons, Reprints, *Peroxyl radicals, Benzene/hexamethyl.

Absolute rate constants have been measured for the reactions of trichloromethylperoxyl radicals with cyclohexane, cyclohexene, and hexamethylbenzene. The CCl3O2 radicals contained various amounts of the hydrocarbons. The rate constants were determined by competition with the one-electron oxidation of metalloporphyrins, using the rate of formation of the metalloporphyrin radical cation absorption to monitor the reaction by kinetic spectrophotometry.

900,358
PB89-171920 Not available NTIS
National Bureau of Standards (NEL), Gaithersburg, MD. Scientific Computing Div.
Modeling Chemical Reaction Systems on an IBM PC.
Final rept.
W. Braun, J. Herron, and D. Kahaner. 1988, 4p
Pub. in ACCESS, the Jnl. of Microcomputer Applications 7, n4 p45-48 Jul/Aug 88.

Keywords: *Chemical reactions, *Reaction kinetics, Reprints, Microcomputers, Computer applications.

The paper provides an informal description of a pair of computer programs, AC40 and ACPLOT, for the efficient solution of reaction rate problems in chemical kinetics.

900,359
PB89-171961 Not available NTIS
National Bureau of Standards (NML), Gaithersburg, MD. Inorganic Analytical Research Div.

Microwave Energy for Acid Decomposition at Elevated Temperatures and Pressures Using Biological and Botanical Samples.
Final rept.
H. M. Kingston, and L. B. Jassie. 1986, 8p
Pub. in Analytical Chemistry 58, n12 p2534-2541 1986.

Keywords: *Bioassay, *High temperature tests, *High pressure tests, *Microwave equipment, *Acid treatment, *Digesters, Sampling, Digestion(Decomposition), Nitric acid, Sulfuric acid, Hydrochloric acid, Hydrofluoric acid, Comparison, Reprints, Standard reference materials.

A closed vessel microwave digestion system is described. In situ measurement of elevated temperatures and pressures in closed Teflon PFA vessels during acid decomposition of organic samples is demonstrated. Temperature profiles for the acid decomposition of biological and botanical standard reference materials are modeled by the decomposing acid. Microwave power absorption of nitric, hydrofluoric, sulfuric, and hydrochloric acids is compared. An equation is applied to acid microwave interactions to predict the time needed to reach target temperatures during sample dissolutions. Reaction control techniques and safety precautions are recommended.

900,360
PB89-172365 Not available NTIS
National Bureau of Standards (NML), Gaithersburg, MD. Atomic and Plasma Radiation Div.
Aluminumlike Spectra of Copper through Molybdenum.
Final rept.
J. Sugar, V. Kaufman, and W. L. Rowan. 1988, 7p
Sponsored by Department of Energy, Washington, DC. Office of Magnetic Fusion Energy, and Naval Research Lab., Washington, DC.
Pub. in Jnl. of the Optical Society of America B 5, n10 p2183-2189 Oct 88.

Keywords: *Copper, *Spectrum analysis, Molybdenum, Rubidium, Strontium, Experimental design, Comparison, Numerical analysis, Dirac equation, Hartree-Fock approximation, Reprints, Isoelectronic atoms.

Spectra of copper through molybdenum were generated in a tokamak plasma and were photographed with a 2.2-m grazing-incidence spectrograph. The doublet system of the Al I isoelectronic sequence was derived from these data and compared with Dirac-Fock calculations of the wavelengths. The smooth variation of the differences was used to improve the wavelengths and to predict those of rubidium and strontium, which were not observed.

900,361
PB89-172373 Not available NTIS
National Bureau of Standards (NML), Gaithersburg, MD. Atomic and Plasma Radiation Div.
Spectrum and Energy Levels of Singly Ionized Cesium. 2. Interpretation of Fine and Hyperfine Structures.
Final rept.
C. J. Sansonetti, K. L. Andrew, and J. F. Wyart. 1988, 11p
See also PB86-200979.
Pub. in Jnl. of the Optical Society of America B 5, n10 p2076-2086 Oct 88.

Keywords: *Atomic structure, *Cesium, Molecular orbitals, Hyperfine structure, Electron transitions, Eigenvectors, Reprints.

The theoretical interpretation of Cs II has been extended and now includes the 5p(sup 5)ns (n=6-12), 5p(sup 5)np (h=6-8), 5p(sup 5)nd (n=5-11), 5p(sup 5)nf (n=4-8), 5p(sup 5)ng (n=5-10), and 5p(sup 5)nh (n=6-8) configurations. Most levels are well represented in the single configuration approximation when far configuration interactions are included through effective electrostatic parameters. Explicit interactions of low-lying 5p(sup 5)nd + 5p(sup 5)(n+1)s configurations have been determined. For most configurations, good jK coupling is found. Purities of the levels in jK coupling and the LS composition of the eigenvectors are given. The intermediate-coupling eigenvectors have been used to calculate magnetic-dipole hyperfine-splitting factors, and these are compared with 167 experimentally determined values from earlier work.

900,362
PB89-172415 Not available NTIS

43

CHEMISTRY

Physical & Theoretical Chemistry

National Bureau of Standards (NML), Gaithersburg, MD. Molecular Spectroscopy Div.
Infrared Spectrum of the nu6, nu7, and nu8 Bands of HNO3.
Final rept.
A. G. Maki, and W. B. Olson. 1989, 11p
Pub. in Jnl. of Molecular Spectroscopy 133, p171-181 1989.

Keywords: *Infrared spectroscopy, *Spectrum analysis, *Nitric acid, *Band spectra, Bandwidth, Molecular spectroscopy, Molecular rotation, Numerical analysis, Reprints.

The high-resolution spectrum of nitric acid (HNO3) has been measured from 500 to 800/cm. The nu8 band is a C-type out-of-plane vibration and both nu6 and nu7 are in-plane vibrations giving rise to A-type bands. The band centers are at 580.3035(5)/cm for nu7, 646.6262(5)/cm for nu6, and 763.1543(5)/cm for nu8. Constants are given that allow one to accurately calculate the infrared line positions for all three bands. The intensity of the nu8 band is anomalous with an R branch that is much stronger than the P branch. Effective rotational intensity correction terms are given for both the nu8 and the nu6 band. An estimate is given for the relative transition moments for the nu6 and nu7 bands.

900,363
PB89-172431   Not available NTIS
National Bureau of Standards (IMSE), Gaithersburg, MD. Reactor Radiation Div.
Neutron Powder Diffraction Structure and Electrical Properties of the Defect Pyrochlores Pb1.5M2O6.5 (M = Nb, Ta).
Final rept.
F. W. Beech, W. M. Jordon, C. R. A. Catlow, A. Santoro, and B. C. H. Steele. 1988, 14p
Pub. in Jnl. of Solid State Chemistry 77, p322-335 1988.

Keywords: *Crystal structure, *Neutron diffraction, *Electrical properties, Minerals, Lead inorganic compounds, Reprints, *Lead niobates, *Lead tantalates, *Pyrochlore.

Powder neutron diffraction and Rietveld analysis were used to investigate the crystal structures of the defective pyrochlores Pb1.5Nb2O6.5 and Pb1.5Ta2O6.5. Both materials crystallize with the symmetry of space group Fd3m, with lattice parameters a = 10.5647(2) and a = 10.5558(2) A, respectively. No evidence has been observed of oxygen or lead vacancy ordering in these compounds. This result is interpreted in terms of a model in which all lead present in the structure has sevenfold pyramidal coordination and forms domains separated by regions of lead vacancies with hexagonal or bipyramidal configurations of the oxygen atoms. This model, built on the assumption that the driving force in the formation of this type of defect pyrochlore is the coordination of lead, leads us to conclude that the system Pb(1+2)M2O(6+x)(M = Nb, Ta) may exist over a range of compositions with 0.33 < or = x < or = 0.5, and may also explain results obtained in other studies of related materials. The electric measurements show that both compounds are predominantly electronic conductors and that the ionic contribution to the total conductivity is very small even at the highest temperatures used in the study.

900,364
PB89-172506   Not available NTIS
National Bureau of Standards (NML), Gaithersburg, MD. Gas and Particulate Science Div.
Dependence of Interface Widths on Ion Bombardment Conditions in SIMS (Secondary Ion Mass Spectrometry) Analysis of a Ni/Cr Multilayer Structure.
Final rept.
M. Moens, F. C. Adams, and D. S. Simons. 1987, 12p
Pub. in Analytical Chemistry 59, n11 p1518-1529 1987.

Keywords: *Chromium, *Nickel, *Mass spectroscopy, *Spectrum analysis, Secondary emission, Ion beams, Ion irradiation, Surface chemistry, Interfaces, Reprints, *Secondary ion mass spectroscopy, Scanning electron microscopy.

The effect of varying ion bombardment conditions on interface widths in a Ni/Cr multilayer structure is described. SIMS depth profiles of the structure were obtained using different primary ions, accelerating voltages and ambient oxygen pressures. The depth profiles are displayed together with scanning electron mi-

crographs of the craters formed in an effort to relate the induced surface roughness to the interface widths obtained. Secondary ion yield variations, especially at the interfaces, have been studied.

900,365
PB89-174007   Not available NTIS
National Bureau of Standards (NML), Gaithersburg, MD. Inorganic Analytical Research Div.
Twenty Five Years of Accuracy Assessment of the Atomic Weights.
Final rept.
I. L. Barnes, and H. S. Peiser. 1986, 10p
Pub. in Proceedings of Workshop and Symposium National Conference of Standards Laboratories, Gaithersburg, MD., October 6-9, 1986, p22-1-22-10.

Keywords: *Atomic mass, *Chemical elements, *Minerals, *Accuracy, Isotopes, Abundance.

The atomic weights of the chemical elements in their natural terrestrial occurrences are among the most widely used measurement data affecting science, technology, and commerce. Uncertainties in these values arise not only from the remaining experimental imprecision and bias in determining these values, but also from variations between different sources of a given element. In the tables of the Standard Atomic Weights revised biennially by the International Union of Pure and Applied Chemistry, an attempt is made to publish a single value for each atomic weight with the largest number of significant figures that can be stated with a single digit uncertainty. In this way users can, at a glance, perceive the atomic weight to the highest precision that is safely applicable to any source and corresponding to the best current knowledge as assessed by the Commission on Atomic Weights and Isotopic Abundances. In the paper the steps are traced by which this procedure has been developed.

900,366
PB89-174015   Not available NTIS
National Bureau of Standards (NEL), Boulder, CO. Thermophysics Div.
High Temperature Thermal Conductivity Apparatus for Fluids.
Final rept.
H. M. Roder. 1987, 8p
Sponsored by Department of Energy, Washington, DC. Engineering and Geosciences Div.
Pub. in Proceedings of DOE (Department of Energy) Symposium on Energy Engineering Science (5th), Boulder, CO., p61-68 1987.

Keywords: *High temperature tests, *Thermal measuring instruments, *Thermal conductivity, *Thermal diffusion, *Fluids, Performance evaluation, Design criteria, Specific heat.

A new apparatus for measuring both thermal conductivity and thermal diffusivity of fluids at high temperatures is described. The technique employed is that of the transient hot wire. Measurements are made with a 12.7 micrometer diameter platinum wire at times of up to 1 second. The data acquisition system is controlled by a microcomputer and includes several programmable digital voltmeters. The hot wire and a shorter compensating hot wire are arranged in different arms of a Wheatstone bridge. The cell containing the core of the apparatus is designed to accommodate pressures from near zero to 70 MPa and temperatures form 0 to 500 C. For thermal conductivity, the precision of the new system is expected to be around 0.3% and the accuracy 1.0%. For thermal diffusivity the accuracy is estimated to be around 5%. From the two variables measured, the author can obtain values of the specific heat, Cp, of the fluid, provided that the density is either measured, or available through an equation of state.

900,367
PB89-174098   Not available NTIS
National Bureau of Standards (NEL), Boulder, CO. Time and Frequency Div.
Study of Long-Term Stability of Atomic Clocks.
Final rept.
D. W. Allan. 1987, 5p
Pub. in Proceedings of Annual Precise Time and Time Interval (PTTI) Applications and Planning Meeting (19th), Redondo Beach, CA., December 1-3, 1987, p375-379.

Keywords: *Frequency standards, *Atomic clocks, Time measurement, Masers, Hydrogen, Cesium, Mercury(Metal), Beams(Radiation), Tracking(Position), Frequency stability, Pulsation, Spacecraft tracking, Pulsars.

The importance of long-term frequency stability has increased significantly in recent years because of the discovery of the millisecond pulsar, PSR 1937 + 21. In addition, long-term stability is extremely useful not only in evaluating primary frequency standards and the performance of national timing centers, but also in addressing questions regarding autonomy for the Global Positioning System (GPS) and syntonization for Jet Propulsion Laboratory's (JPL) Deep Space Tracking Network. Over the last year, NBS has carried out several studies addressing questions regarding the long-term frequency stability of different kinds of atomic clocks as well as of principal timing centers. These analyses cover commercial and primary cesium-beam frequency standards, active hydrogen master frequency standards and the new commercial mercury-ion frequency standards at United States Naval Observatory (USNO). Fractional frequency stabilities of parts in 10 sup 14 down to parts in 10 sup 15 were observed for various of these standards. It is believed that frequency stabilities on the order of a part in 10 sup 15 will be necessary to measure the effects of gravity waves on frequency stability between millisecond pulsars and atomic clocks on or near the earth.

900,368
PB89-174932   Not available NTIS
National Bureau of Standards (NEL), Boulder, CO. Thermophysics Div.
Vapor-Liquid Equilibrium of Nitrogen-Oxygen Mixtures and Air at High Pressure.
Final rept.
J. C. Rainwater, and R. T. Jacobsen. 1988, 10p
Pub. in Cryogenics 28, p22-31 Jan 88.

Keywords: *High pressure tests, *Nitrogen, *Oxygen, *Binary systems(Materials), *Mathematical models, *Thermodynamic equilibrium, *Phase diagrams, Phase transformations, Vapor phases, Liquid phases, Critical point, Reprints.

The vapor-liquid equilibrium surface of the binary mixture nitrogen-oxygen is correlated over an extended critical region with the Leung-Griffiths model as modified by Rainwater and Moldover. No single comprehensive experimental measurement of the coexistence surface is available. However, several different experiments along isopleths, isotherms and isobars collectively provide enough data to make possible the development of a reasonable correlation. The model is optimized to modern data and is shown to be consistent with pioneering measurements done before 1930 to within experimental and temperature-scale uncertainties. It is shown that air in the critical region can be accurately modelled as a nitrogen-oxygen binary mixture by including the small argon component with oxygen. Ancillary equations for the saturation properties of a as functions of temperature are also constructed.ir

900,369
PB89-174940   Not available NTIS
National Bureau of Standards (NEL), Gaithersburg, MD. Thermophysics Div.
Non-Equilibrium Theories of Electrolyte Solutions.
Final rept.
J. B. Hubbard. 1987, 34p
Pub. in Physics and Chemistry of Aqueous Ionic Solutions, p95-128 1987.

Keywords: *Aqueous electrolytes, *Nonequilibrium flow, Mathematical models, Light scattering, Electrical resistance, Friction, Ions, Reprints, Space-time model, Solvent properties, Ion-ion collisions, Debye Falkenhagen Onsager method, Laser spectroscopy.

Dynamic aspects of electrolyte solution theory are explored through a generalized Langevin approach as well as through a van Hove/Smoluchowski description. The classical theory of ion-ion dynamical interactions is presented at the Debye-Falkenhagen-Onsager level, while the non-equilibrium ion-solvent interaction is analyzed from both a microscopic and a continuum viewpoint. Emphasis is placed on understanding the physics of simple models on which explicit calculations can be performed. These include electrical conductance, ionic friction coefficients, space-time correlation functions, and laser light scattering spectra for simple electrolytes and charged macromolecules.

900,370
PB89-174957   Not available NTIS
National Bureau of Standards (NEL), Gaithersburg, MD. Thermophysics Div.

Molecular Dynamics Investigation of Expanded Water at Elevated Temperatures.
Final rept.
R. D. Mountain. 1989, 5p
Pub. in Jnl. of Chemical Physics 90, n3 p1866-1870, 1 Feb 89.

Keywords: *High temperature tests, *Water, *Dynamic structural analysis, Molecular structure, Hydrogen bonds, Density(Mass/volume), Thermodynamic properties, Reprints, Expansion.

The structure of expanded states of TIP4P water has been examined over a range of densities running from 1000 to 100 kg/cu m and for a range of temperatures running from the coexistence temperature up to supercritical temperatures. The main result is that hydrogen bonding, as evidenced by the maximum in g(sub OH) (R) at 0.18 nm, persists to supercritical temperatures over the entire density range examined. For most liquid densities, the number of hydrogen bonds per molecule scales as a single function of the temperature but does not scale for dense vapor densities.

900,371
PB89-175228

(Order as PB89-175194, PC A06)
National Inst. of Standards and Technology (NML), Gaithersburg, MD. Chemical Thermodynamics Div.
Numeric Databases in Chemical Thermodynamics at the National Institute of Standards and Technology.
Bi-monthly rept.
M. W. Chase. 1989, 4p
Included in Jnl. of Research of the National Institute of Standards and Technology, v94 n1 p21-24 Jan-Feb 89.

Keywords: *Thermodynamics, *Thermochemical properties, Chemical elements, Specific heat, Enthalpy, Vapor pressure, Phase transformations, Bibliographies, Oxides, *Numerical data bases.

During the past year the activities of the Chemical Thermodynamics Data Center and the JANAF Thermochemical Tables project have been combined to obtain an extensive collection of thermodynamic information for many chemical species, including the elements. Currently available are extensive bibliographic collections and data files of heat capacity, enthalpy, vapor pressure, phase transitions, etc. Future plans related to materials science are to improve the metallic oxide temperature dependent tabulations, upgrade the recommended values periodically, and maintain the bibliographic citations and the thermochemical data current. The recommended thermochemical information is maintained on-line and tied to the calculational routines within the data center. Recent thermodynamic evaluations on the elements and oxides will be discussed, as well as studies in related activities at the National Institute of Standards and Technology.

900,372
PB89-175418
PC A07/MF A01
National Inst. of Standards and Technology (NML), Gaithersburg, MD. Center for Atomic, Molecular and Optical Physics.
Technical Activities 1986-1988, Molecular Spectroscopy Division.
A. Weber. Mar 89, 136p NISTIR-89/4051
See also PB87-140224.

Keywords: *Molecular spectroscopy, *Research programs, Experimental design, Numerical analysis, Van der Waals equation, Environmental surveys, Mathematical models, Reaction kinetics, Surface chemistry, Standards, Molecular structure, Hydrogen bonds, Thermodynamics, Chemical reactions, *National Institute of Standards and Technology, State of the art, Molecular conformation.

The report summarizes the technical activities of the NIST Molecular Spectroscopy Division for the Fiscal Year 1987 and 1988. The activities span experimental and theoretical research in high resolution molecular spectroscopy, quantum chemistry, and laser photochemistry, and include the development of frequency standards, critically evaluated spectral data, applications of spectroscopy to important scientific and technological problems, and the advancement of spectroscopy measurement methods and techniques. A listing is given of publications and talks by the Division staff.

900,373
PB89-175681
Not available NTIS
National Bureau of Standards (NEL), Boulder, CO. Thermophysics Div.

Decorated Lattice Gas Model for Supercritical Solubility.
Final rept.
G. C. Nielson, and J. M. H. Levelt Sengers. 1987, 10p
Pub. in Jnl. of Physical Chemistry 91, n15 p4078-4087 1987.

Keywords: *Solubility, *Thermodynamic equilibrium, *Supercritical flow, Phase diagrams, Solutes, Solvents, Temperature, Experimental design, Phase transformations, Solid phases, Liquid phases, Reprints.

The authors describe a self-consistent nonclassical model for supercritical solubility enhancement near the solvent's critical point. A Widom-Wheel-Mermin decorated lattice gas transformation is used to obtain properties of a dilute supercritical solution from known properties of the pure solvent. Phase equilibria between the solution and the additional solid or liquid phase are described. A semiquantitative representation of solubility data for three different solute-solvent pairs at several temperatures was obtained. Infinite-dilution partial molal volumes based on parameter sets obtained from fits to solubility data did not agree very well with experiment.

900,374
PB89-175707
Not available NTIS
National Bureau of Standards (NML), Gaithersburg, MD. Temperature and Pressure Div.
Low Range Flowmeters for Use with Vacuum and Leak Standards.
Final rept.
K. E. McCulloh, C. D. Ehrlich, F. G. Long, and C. R. Tilford. 1987, 6p
Pub. in Jnl. of Vacuum Science and Technology A-Vacuum Surfaces and Films 5, n3 p376-381 1987.

Keywords: *Flowmeters, *Standards, *Pressure sensors, *Gas flow, Calibrating, Comparison, Performance evaluation, Leakage, Vacuum apparatus, Reprints.

Vacuum pressure standards of the orifice-flow type require known gas flows of .000001 mol/s (.01 std cc/s) and less. Known gas flows can also be used to calibrate 'standard' leaks by comparing the pressures generated when flows from the leak and the flowmeter are alternately passed through a constant conductance. Described here are two constant-pressure, piston displacement flowmeters developed at NBS that can generate flows between .000001 and 10 to the -10th power mol/s with an estimated uncertainty of 0.6 to 2%. Comparisons of the flowmeters with alternate calibration techniques, and repeated low range leak and vacuum gage calibrations, have been used to confirm the estimated uncertainty and random errors of the flowmeter.

900,375
PB89-175996
Not available NTIS
National Bureau of Standards (NML), Gaithersburg, MD. Surface Science Div.
Dynamics of Molecular Collisions with Surfaces: Excitation, Dissociation, and Diffraction.
Final rept.
S. Holloway, M. Karikorpi, and J. W. Gadzuk. 1987, 18p
Pub. in Nuclear Instruments and Methods in Physics Research B27, n1 p37-54 1987.

Keywords: Excitation, Dissociation, Diffraction, Scattering, Reprints, *Molecular collisions, *Surface reactions, Charge transfer, EV range 01-10, Vibrational states.

Aspects of molecular collisions with surfaces are discussed which are important in the chemically relevant energy range of 1-10 eV. In particular, the role of charge transfer, potential energy surface topology and intra-molecular ground and excited-state potential curves are investigated as they pertain to internal vibrational excitation, dissociative absorption or scattering, and diffractive scattering. The modeling and analysis is based on classical trajectories and semi-classical wavepacket dynamics, both for intra-molecular and translational motion.

900,376
PB89-176093
Not available NTIS
National Bureau of Standards (NML), Gaithersburg, MD. Chemical Kinetics Div.

One-Electron Transfer Reactions of the Couple SO2/SO2(1-) in Aqueous Solutions. Pulse Radiolytic and Cyclic Voltammetric Studies.
Final rept.
P. Neta, A. Harriman, and R. E. Huie. 1987, 6p
Pub. in Jnl. of Physical Chemistry 91, n6 p1606-1611 1987.

Keywords: *Reaction kinetics, *Sulfur dioxide, *Free radicals, *Radiolysis, *Reduction(Chemistry), pH, Porphyrins, Reprints, *Pulse techniques, *Voltammetry.

Rate constants for one-electron reduction of SO2 by several radicals and for reduction of several compounds by SO2(sup -1) radicals were determined by pulse radiolysis at pH 1. The reduction potentials for SO2 and for porphyrins were determined by cyclic voltammetry under identical conditions. These reduction potentials were used along with the rate constants and previously reported self-exchange rates to estimate the self-exchange rate for the couple SO2/SO2(sup -1) in acidic solutions. The calculated values were found to vary over many orders of magnitude, similar to the situation reported before for the O2/O2(sup -1) couple.

900,377
PB89-176101
Not available NTIS
National Bureau of Standards (NML), Gaithersburg, MD. Chemical Kinetics Div.
Ion Kinetics and Energetics.
Final rept.
S. G. Lias. 1987, 18p
Pub. in Encyclopedia of Physical Science and Technology, v7 p1-18 1987.

Keywords: *Ions, *Reaction kinetics, *Ionization, Vapor phases, Reviews, Chemical reactions, Thermodynamics, Reprints.

The article, written for a non-specialist audience (undergraduate and graduate students, scientists and engineers from other specialties), presents a broad review of the thermodynamics of ions in the gas phase, as well as the unimolecular and bimolecular kinetics of the chemical reactions of ions in the gas phase.

900,378
PB89-176150
Not available NTIS
National Bureau of Standards (NML), Gaithersburg, MD. Gas and Particulate Science Div.
Defocus Modeling for Compositional Mapping with Wavelength-Dispersive X-ray Spectrometry.
Final rept.
R. L. Myklebust, D. E. Newbury, R. B. Marinenko, and D. S. Bright. 1986, 3p
Pub. in Microbeam Analysis - 1986, p495-497 1986.

Keywords: *x ray spectroscopy, *Molecular structure, *Chemical composition, *Quantitative analysis, Spectrochemical analysis, Numerical analysis, Reprints, *Electron microprobe analysis.

Quantitative compositional mapping by electron probe microanalysis requires the use of corrections for spectrometer defocusing when wavelength-dispersive is employed. A correction procedure has been developed which is based on calculating a synthetic map of the standard from an angular scan across the peak. Equations have been developed which relate the distance in a matrix scan from the line-of-best-focus of the spectrometer to the equivalent angular position in a peak scan. The transmission of the spectrometer can be directly determined from the peak scan and appropriate corrections can be applied to the intensity map of an unknown. The method is effective down to magnifications as low as 150x, where statistical considerations become limiting.

900,379
PB89-176242
Not available NTIS
National Bureau of Standards (NML), Gaithersburg, MD. Chemical Kinetics Div.
Rate Constants for Reactions of Nitrogen Oxide (NO3) Radicals in Aqueous Solutions.
Final rept.
P. Neta, and R. E. Huie. 1986, 5p
Pub. in Jnl. of Physical Chemistry 90, n19 p4644-4648 1986.

Keywords: *Chemical radicals, *Nitrogen oxides, Chemical reactions, Radiolysis, Reprints, *Rate constants, *Nitrate radicals, *Aqueous solutions, Hydroxyl radicals, Sulfate radicals.

CHEMISTRY

Physical & Theoretical Chemistry

Rate constants for reactions of NO3 radicals in neutral and acidic aqueous solutions were determined by pulse radiolysis. The values for hydrogen abstraction from several organic compounds were in the range of 10(5) - 10(7)/Ms and were dependent on the C-H bond strength. The rate constants for addition to double bonds ranged up to 5x10(7)/Ms. Rate constants for the one-electron oxidation of several organic compounds and inorganic ions varied over a wide range. From these, it was possible to estimate the redox potential for the NO3/NO3 couple to be 2.3-2.6U. Several rate constants for SO4(2-) were measured as well and the behavior of NO3, SO4(2-), and OH radicals are compared.

900,380
PB89-176408   Not available NTIS
National Bureau of Standards (NML), Gaithersburg, MD. Molecular Spectroscopy Div.
Picosecond Coherent Anti-Stokes Raman Scattering (CARS) Study of Vibrational Dephasing of Carbon Disulfide and Benzene in Solution.
Final rept.
J. W. Perry, A. M. Woodward, and J. C. Stephenson. 1986, 8p
Pub. in Proceedings of SPIE (Society of Photo-Optical Instrumentation Engineers), Laser Applications in Chemistry and Biophysics, v620 p7-14 1986.

Keywords: *Raman spectra, *Coherent scattering, *Vibrational spectra, *Carbon disulfide, *Benzene, *Lasers, Phases, Mixtures, Carbon tetrachloride, Nitrobenzenes, Ethanols, Dilution, Polarization, Solvents, Mathematical models.

The vibrational dephasing of the 656 cm(-1) mode (nu1,a1g) of carbon disulfide and the 991 cm(-1) mode (nu2, a1g) of benzene have been studied as a function of concentration in mixtures with a number of solvents using a picosecond time-resolved coherent anti-Stokes Raman scattering (CARS) technique. The technique employs two tunable synchronously-pumped model-locked dye lasers in a stimulated Raman pump, coherent anti-Stokes Raman probe time-resolved experiment. Results are obtained for CS2 in carbon tetrachloride, benzene, nitrobenzene and ethanol, and for benzene nu2 in CS2. The dephasing rates of CS2 nu1 increase on dilution with the polar solvents and decrease or remain constant on dilution with the nonpolar solvents. The CS2/benzene solutions show a contrasting behavior with the CS2 nu1 dephasing rate being nearly independent of concentrations whereas the benzene nu2 dephasing rate decreases on dilution. These results are compared to theoretical models for vibrational dephasing of polyatomic molecules in solution.

900,381
PB89-176473   Not available NTIS
National Bureau of Standards (NML), Gaithersburg, MD. Surface Science Div.
Influence of Electronic and Geometric Structure on Desorption Kinetics of Isoelectronic Polar Molecules: NH3 and H2O.
Final rept.
T. E. Madey, C. Benndorf, and S. Semancik. 1987, 7p
Pub. in Springer Ser. Surf. Sci. 8, p175-181 1987.

Keywords: *Water, *Ammonia, *Reaction kinetics, *Desorption, Chemisorption, Surface chemistry, Chemical bonds, Adsorption, Polarity, Dipoles, Reprints, *Electronic structure, isoelectronic atoms.

The thermal desorption kinetics of isoelectronic NH3 and H2O from surfaces provide an interesting contrast, due to geometrical and electronic structural effects in the adsorbed layers. Both NH3 and H2O are polar molecules bonded to the surface via lone pair orbitals on N and O, respectively. Hydrogen-bonding attractive interactions between neighboring H2O molecules lead to formation of 2-d and 3-d clusters; thermal desorption kinetics of H2O are characterized by sharp desorption peaks over narrow temperature ranges (delta T less than 10 K in some cases). In distinction, lateral interactions between neighboring NH3 molecules are largely repulsive (dipole-dipole interactions) and the thermal desorption spectra are considerably broader in temperature than for H2O (delta T approximates 80 K to 180 K, depending on substrate).

900,382
PB89-176523   Not available NTIS
National Bureau of Standards (NML), Gaithersburg, MD. Center for Basic Standards.

Near-Threshold X-ray Fluorescence Spectroscopy of Molecules.
Final rept.
D. W. Lindle, P. L. Cowan, R. E. LaVilla, T. Jach, R. D. Deslattes, R. C. C. Perera, and B. Karlin. 1988, 5p
Sponsored by Department of Energy, Washington, DC.
Pub. in Proceedings of SPIE (Society of Photo-Optical Instrumentation Engineers), X-ray and VUV Interaction Data Bases, Calculations, and Measurements, v911 p54-58 1988.

Keywords: *Molecular energy levels, *x ray spectroscopy, *Spectrum analysis, *Emission spectroscopy, *X ray fluorescence, X ray absorption, Excitations, Ionization, *X-ray fluorescence analysis.

The coupling of high-energy-resolution x-ray absorption and emission spectroscopies with a high-intensity and high-resolution synchrotron-radiation beamline has opened up several new avenues of research in inner-shell molecular physics. Of particular importance is the ability to tune the incident photon energy throughout the near-threshold region for the atomic core levels of the molecules. All of these phenomena can be used in a complementary way to uncover some of the underlying molecular dynamics which occur upon inner-shell excitation and ionization. Some of the latest results from the NBS X-Ray Spectroscopy Beamline at NSLS will be presented to highlight these effects.

900,383
PB89-176598   Not available NTIS
National Bureau of Standards (NML), Gaithersburg, MD. Gas and Particulate Science Div.
Luminescence Standards for Macro- and Microspectrofluorometry.
Final rept.
R. A. Velapoldi, and M. S. Epstein. 1989, 29p
Pub. in ACS (American Chemical Society) Symposium Series 383, Chapter 7, p96-126 1989.

Keywords: *Fluorimeters, *Spectrum analysis, *Luminescence, Standards, Calibrating, Assessments, Performance evaluation, Quantum efficiency, *Standard reference materials.

Requirements for standards used in macro- and microspectrofluorimetry differ, depending on whether they are used for instrument calibration, standardization, or assessment of method accuracy. Specific examples are given of standards for quantum yield, number of quanta, and decay time, and for calibration of instrument parameters, including wavelength, spectral responsivity (determining correction factors for luminescence spectra), stability, and linearity. Differences in requirements for macro- and micro-standards are considered, and specific materials used for each are compared. Pure compounds and matrix-matched standards are listed for standardization and assessment of method accuracy and existing Standard Reference Materials are discussed.

900,384
PB89-176788   Not available NTIS
National Bureau of Standards (NML), Gaithersburg, MD. Radiometric Physics Div.
Feasibility of Detector Self-Calibration in the Near Infrared.
Final rept.
J. Geist, M. J. Nofziger, and G. H. Olsen. 1986, 2p
Pub. in CPEM 86 Digest, Proceedings of Conference on Precision Electromagnetic Measurements, Gaithersburg, MD., June 23-27, 1986, p138-139.

Keywords: *Photodiodes, *Semiconductor devices, *Indium phosphides, Feasibility, Performance evaluation, Near infrared radiation, Indium arsenides, Gallium arsenides, Calibrating, Gallium indium arsenides.

A recent study of the self-calibration of InP/InGaAs heterodiodes in the 1000 to 1600 nm spectral region is reported.

900,385
PB89-176952   Not available NTIS
National Bureau of Standards (NML), Gaithersburg, MD. Radiometric Physics Div.
Vibrationally Resolved Photoelectron Angular Distributions for H2 in the Range 17 eV < or = h(nu) < or = 39 eV.
Final rept.
A. C. Parr, J. E. Hardis, S. H. Southworth, C. S. Feigerle, T. A. Ferrett, D. M. P. Holland, F. M. Quinn, B. R. Dobson, J. B. West, G. V. Marr, and J. L. Dehmer. 1988, 7p
Sponsored by Department of Energy, Washington, DC.

Pub. in Physical Review A 37, n2 p437-443, 15 Jan 88.

Keywords: *Hydrogen, Far ultraviolet radiation, Reprints, *Photoelectron spectroscopy, *Photoionization, Autoionization, eV range 10-100, Extreme ultraviolet radiation, Angular distribution.

Vibrationally resolved photoelectron angular distributions have been measured for photoionization of H2 over the range 17 eV = or < h(mu) = or < 39 eV using independent instrumentation at two synchrotron radiation facilities. The present data greatly extend and add vibrational resolution to earlier variable-wavelength measurements. The average magnitude of the asymmetry parameter continues to lie lower than the best independent-electron calculations. Broad structure is observed for the first time, possibly indicating the effects of channel interaction with dissociative, doubly excited states of H2. Neither the average magnitude nor the gross wavelength-dependent structure vary strongly with the final vibrational channel.

900,386
PB89-176960   Not available NTIS
National Bureau of Standards (NML), Gaithersburg, MD. Radiometric Physics Div.
Autoionization Dynamics in the Valence-Shell Photoionization Spectrum of CO.
Final rept.
J. E. Hardis, T. A. Ferrett, S. H. Southworth, A. C. Parr, P. Roy, J. L. Dehmer, P. M. Dehmer, and W. A. Chupka. 1986, 8p
Grant NSF-CHE83-18419, Contract W-31109-eng-38
Sponsored by Department of Energy, Washington, DC., and National Science Foundation, Washington, DC.
Pub. in Jnl. of Chemical Physics 89, n2 p812-819, 15 Jul 88.

Keywords: *Carbon monoxide, Far ultraviolet radiation, Reprints, *Photoionization, *Autoionization, Photoelectron spectroscopy, Rydberg states, Ionization cross sections, Angular distribution, eV range 10-100.

Autoionizing Rydberg series in the valence-shell spectrum of CO have been studied by determining the high resolution relative photoionization cross section of cooled CO in the energy region 14.0-20.0 eV and by determining the vibrational branching ratios and the photoelectron angular distributions for production of CO(1+) X doublet Sigma (+), vu+ = 0-2 in the energy region 16.75-18.75 eV. Of particular interest are three prominent spectral features between 17.0 and 17.5 eV that result from interactions involving Rydberg series converging to the excited A doublet Pi and B doublet Sigma (+) states of the ion. The results are discussed in the context of recent two-step multichannel quantum defect theory calculations by Leyh and Raseev.

900,387
PB89-179105   Not available NTIS
National Bureau of Standards (NML), Gaithersburg, MD. Atomic and Plasma Radiation Div.
Spectra and Energy Levels of the Galliumlike Ions Rb VII-Mo XII.
Final rept.
U. Litzen, and J. Reader. 1989, 8p
Sponsored by Department of Energy, Washington, DC.
Pub. in Physica Scripta 39, p73-80 1989.

Keywords: *Gallium, *Ultraviolet spectra, *Plasmas(Physics), *Ions, *Electron transitions, Rubidium, Strontium, Yttrium, Zirconium, Niobium, Molybdenum, Reprints.

Spectra of the galliumlike ions Rb VII, Sr VIII, Y IX, Zr X, Nb XI, and Mo XII emitted from sparks and laser-produced plasmas have been recorded in the region 235-665 Angstroms. All levels of 4s(sup2)4p, 4s(sup2)4d, 4s4p(sup2), 4p3 and, in Rb VII, Sr VIII and Y IX, 4s(sup2)4f have been established. The level structure has been studied by means of ab initio and parametric calculations.

900,388
PB89-179113   Not available NTIS
National Bureau of Standards (NML), Gaithersburg, MD. Molecular Spectroscopy Div.

Vibrational Exchange upon Interconversion Tunneling in (HF)2 and (HCCH)2.
Final rept.
G. T. Fraser. 1989, 12p
Pub. in Jnl. of Chemical Physics 90, n4 p2097-2108, 15 Feb 89.

Keywords: *Acetylene, *Hydrogen fluoride, Hydrogen bonds, Infrared spectra, Reprints, Vibrational states, Selection rules, Dimers.

Model calculations are presented to interpret the large H-F and C-H stretching vibrational dependencies of the interconversion tunneling splittings and the corresponding infrared vibrational-tunneling state selection rules in (HF)2 and (HCCH)2. The model consists of two potential curves in the tunneling coordinate, coupled by an interaction term that allows the vibrational excitation to be exchanged between the two monomer units, permitting tunneling to occur. The interaction term is approximated by resonant infrared transition-dipole coupling. The magnitudes of the calculated vibrational dependencies, their isotopic shifts, and the predicted selection rules are in agreement with previous experimental observations.

900,389
PB89-179121 Not available NTIS
National Bureau of Standards (NML), Gaithersburg, MD. Molecular Spectroscopy Div.
Infrared and Microwave Spectra of OCO-HF and SCO-HF.
Final rept.
G. T. Fraser, A. S. Pine, R. D. Suenram, D. C. Dayton, and R. E. Miller. 1989, 7p
Pub. in Jnl. of Chemical Physics 90, n3 p1330-1336, 1 Feb 89.

Keywords: *Complex compounds, *Infrared spectra, *Microwave spectra, Hydrogen bonds, Hydrogen fluoride, Carbon dioxide, Bolometers, Molecular beams, Reprints, Vibrational states, Fourier transform microwave spectroscopy.

The H-F stretching bands of the OCO-HF and SCO-HF complexes have been studied by optothermal (bolometer-detected) molecular-beam spectroscopy. Both species exhibit spectra of a quasilinear molecule red shifted from free HF by 52.1 and 57.57cm, respectively. The principal band in both molecules is accompanied by a slightly red-shifted doublet-type subsidiary band that can be interpreted as a hot band of a low frequency bending vibration or a K = 1 subband of a bent molecule. Accurate doublet splittings in the ground H-F vibrational state have been measured by pulsed-nozzle Fourier-transform microwave spectroscopy.

900,390
PB89-179154 Not available NTIS
National Bureau of Standards (NML), Gaithersburg, MD. Gas and Particulate Science Div.
Wheatleyite, Na2Cu(C2O4)2 . 2H2O, a Natural Sodium Copper Salt of Oxalic Acid.
Final rept.
R. C. Rouse, D. R. Peacor, P. J. Dunn, W. B. Simmons, and D. Newbury. 1986, 3p
Pub. in American Mineralogist 71, n9-10 p1240-1242 1986.

Keywords: *Sodium, *Copper, *Oxalates, *Crystal structure, *X ray diffraction, Mohs hardness, Synthesis(Chemistry), Triclinic lattices, Galena, Sphalerite, Powder(Particles), Minerals, Reprints, Wheatleyite.

Wheatleyite, Na2Cu(C2O4)2 . 2H2O, occurs as aggregates of blue acicular crystals associated with galena and sphalerite at the Wheatley mine, near Phoenixville, Pennsylvania. It is triclinic P1, with a = 7.559(3), b = 9.665(4), c = 3.589(1) angstroms, alpha = 76.65(2) degrees, beta = 103.67(2) degrees, gamma = 109.10(2) degrees, and Z = 1. The strongest powder X-ray diffraction lines are (dobs, Iobs, hkl) 7.04(8)(010), 6.539(10)(110), 3.655(5)(210), 3.169(9)(121), 2.799(4)(221), 2.538(3)(021), 2.497(3)(231), and 2.344(3)(300). Measured and calculated densities are 2.27(4) and 2.250 g/cu m, respectively. The Mohs hardness is 1 to 2, and there is a perfect (100) cleavage. Optically, wheatleyite is biaxial positive with indices alpha = 1:400(4), beta = 1.499(2), gamma = 1.667(2), and 2V = 63(5) degrees. Dispersion is r < v; pleochroism is X equal colorless, Y = pale blue, Z equal dark blue; absorption is Z > Y > X. A crystal-structure determination shows wheatleyite to be identical to synthetic Na2Cu(C2O4)2 . 2H2O.

900,391
PB89-179196 Not available NTIS
National Bureau of Standards (NML), Gaithersburg, MD. Temperature and Pressure Div.
Effects of Velocity and State Changing Collisions on Raman Q-Branch Spectra.
Final rept.
G. J. Rosasco, and W. S. Hurst. 1987, 33p .
Sponsored by Army Research Office, Research Triangle Park, NC. .
Pub. in Spectral Line Shapes, p535-567 1987.

Keywords: *Raman spectroscopy, *Molecular vibration, *Line width, *Molecular energy levels, *Velocity, *Molecular rotation, Temperature, Pressure, Performance evaluation, Spectral lines, Reprints, *High resolution.

Recent progress in both the experimental characterization and theoretical understanding of Raman vibrational Q-branch spectra is reviewed. Nonlinear Raman spectroscopy, in particular stimulated Raman gain/loss spectroscopy, has produced high resolution (1-150 MHz) spectra for many systems over quite large ranges of temperature and pressure. The high resolution capability allows testing theoretical predictions to a previously unattainable level of accuracy. In turn, the use of nonlinear Raman spectroscopies for optical measurements of temperature, pressure, and species concentration in hostile environments requires very accurate prediction of spectra as functions of these variables. The level of current understanding and ability to predict these diagnostic spectra is reviewed. Emphasis is placed on the role of velocity and state changing collisions, since these primarily determine the shapes and widths of Raman Q-branch spectra.

900,392
PB89-179568 Not available NTIS
National Bureau of Standards (NEL), Gaithersburg, MD. Chemical Process Metrology Div.
Remote Sensing Technique for Combustion Gas Temperature Measurement in Black Liquor Recovery Boilers.
Final rept.
S. R. Charagundla, and H. G. Semerjian. 1986, 8p
Contract DE-AI01-76PR06010
Sponsored by Department of Energy, Washington, DC. Office of Industrial Programs.
Pub. in Proceedings of SPIE (Society of Photo-Optical Instrumentation Engineers), Optical Techniques for Industrial Inspection, v665 p298-305 1986.

Keywords: *Remote sensing, *Emission spectroscopy, *Black liquors, *Combustion products, *Gas analysis, *Temperature measurement, Potassium, Boilers, Materials recover, Spectral lines.

A remote sensing technique, based on the principles of emission spectroscopy, is being developed for application to temperature measurement in black liquor recovery boilers. Thus far, several tests have been carried out, both in the laboratory and at a number of recovery boilers, to characterize the emission spectra in the wavelength range of 300 micrometers to 800 micrometers. These tests pointed out the potential for the line intensity ratio technique based on a pair of emission lines at 404.4 micrometers and 766.5 micrometers observed in the recovery boiler combustion zone; these emission lines are due to potassium, a common constituent found in all the black liquors. Accordingly, a fiber optics based four-color system has been developed. The in-situ, nonintrusive technique together with some of the results obtained are described in the paper.

900,393
PB89-179576 Not available NTIS
National Bureau of Standards (NEL), Gaithersburg, MD. Chemical Process Metrology Div.
Surface Properties of Clean and Gas-Dosed SnO2 (110).
Final rept.
D. F. Cox, T. B. Fryberger, J. W. Erickson, and S. Semancik. 1987, 2p
Pub. in Jnl. of Vacuum Science and Technology A 5, n4 pt2 p1170-1171 Jul/Aug 87. .

Keywords: *Band spectra, *Tin oxides, *Photoelectric emission, Hydrogen, Oxygen, Water, Gases, Adsorption, Ultraviolet detection, Electron diffraction, Surface properties, Desorption, Reprints.

Surface-sensitive techniques have been used to investigate the mechanisms responsible for gas-induced

changes in the electronic behavior of tin oxide (SnO2). Analytical measurements were made before and after dosing a crystalline specimen to controlled amounts of H2, O2, and H2O. Data are presented and discussed which illustrate band bending and band gap emission effects produced by adsorption.

900,394
PB89-179600 Not available NTIS
National Bureau of Standards. (NEL), Boulder, CO. Chemical Engineering Science Div.
Thermal Conductivity of Liquid Argon for Temperatures between 110 and 140 K with Pressures to 70 MPa.
Final rept.
H. M. Roder, C. A. Nieto de Castro, and U. V. Mardolcar. 1987, 20p
See also PB87-203725.
Pub. in International Jnl. of Thermophysics 8, n5 p521-540 1987.

Keywords: *Argon, *Thermal conductivity, *High pressure tests, *Cryogenics, Gases, Liquids, Density(Mass/volume), Standards, Reprints.

The paper presents new experimental measurements of the thermal conductivity of liquid argon for four temperatures between 110 and 140 K with pressures to 70 MPa and densities between 29 and 36 mol/L. The measurements were made with a transient hot wire apparatus. A curve fit of each isotherm allows comparison of the present results to those of others and to correlations. The results are sufficiently detailed to illustrate several features of the liquid thermal conductivity surface. If these details are taken into account, the comparisons show the accuracy of the present results to be 1%. The present results along with several other sets of data are recommended for selection as standard thermal conductivity data along the saturated liquid line of argon, extending the standards into the cryogenic temperature range. The results cover a fairly wide range of densities. A hard sphere model cannot represent the data within the estimated experimental accuracy.

900,395
PB89-179618 Not available NTIS
National Bureau of Standards (NEL), Boulder, CO. Chemical Engineering Science Div.
Facilitated Transport of CO2 through Highly Swollen Ion-Exchange Membranes: The Effect of Hot Glycerine Pretreatment.
Final rept.
J. J. Pellegrino, R. Nassimbene, and R. D. Noble. 1988, 5p
See also PB87-233532.
Pub. in Gas Separation and Purification 2, p126-130 Sep 88.

Keywords: *Carbon dioxide, *Gases, *Swelling, *Ion exchange membrane electrolytes, *Separation, Glycerol, Hydration, Ethylenediamine, Transport properties, Hydrogen sulfide, Carriers, Diffusion, Permeability, Reprints.

A pretreatment for perfluorosulphonic acid ion-exchange films, using glycerine and heating, causes the membranes to swell and imbibe greater (up to three-fold) amounts of water than obtained using standard membrane hydration techniques. The permeation of CO, CO2 and H2S has been tested for three membranes in the diffusive transport mode and with carrier-mediated (facilitated) transport, using ethylenediamine (EDA). When compared to 'normally' hydrated membranes the results indicate that these membranes are up to 50% thicker, have four to six times higher flux in both diffusive and facilitated transport, maintain a high degree of facilitation for CO2 and H2S; and maintain similar selectivity versus non-carrier reactive gases as previously observed. Experimental results show the effects of pretreatment temperature on the CO2 flux.

900,396
PB89-179659 Not available NTIS
National Bureau of Standards (IMSE), Gaithersburg, MD. Reactor Radiation Div.
Chemical Physics with Emphasis on Low Energy Excitations.
Final rept.
J. J. Rush, and J. M. Rowe. 1986, 14p
Pub. in Physica B + C 137, n1-3 p169-182 1986.

Keywords: *Neutron spectroscopy, Neutron scattering, Cyanides, Reprints, Chemical physics, Low

CHEMISTRY

Physical & Theoretical Chemistry

energy, High resolution, Molecular dynamics, Ionic conductivity, Clathrates, Milli EV range.

Some examples of recent applications of low energy and high resolution neutron spectroscopy in chemical physics are discussed. These include brief descriptions of research on translation-rotation coupling in solids, focused on alkali cyanides and alkali cyanide mixed salts, a discussion of the problem of non-bonded potentials in molecular solids by a study of rotational modes and tunnel splittings, and the use of low energy neutrons to probe the mechanism of the translational diffusion and reorientation in molecular complexes, ionic conductors and intercalation compounds. The illustrative examples described demonstrate the application of modern neutron scattering methods to low energy spectroscopy over five orders of magnitude in time and energy (0.1-10,000 (micro)eV). The importance of future research using neutron methods to study molecular dynamics in dilute and low-dimensional systems is stressed.

900,397
PB89-179691 Not available NTIS
National Bureau of Standards (NEL), Boulder, CO. Thermophysics Div.
Second Viscosity and Thermal-Conductivity Virial Coefficients of Gases: Extension to Low Reduced Temperature.
Final rept.
J. C. Rainwater, and D. G. Friend. 1987, 5p
Pub. in Physical Review A 36, n8 p4062-4066, 15 Oct 87.

Keywords: *Gases, Thermal conductivity, Viscosity, Benzene, Reprints, Virial coefficients, Methanol, Density dependence.

A recent theory of the initial density dependences of both viscosity and thermal conductivity has been extended to include lower reduced temperatures. New data on the second viscosity virial coefficients of some organic vapors are found to be in substantial agreement with the theory even at the lowest temperatures. The authors present in tabular form the numerical values for both transport virial coefficients in the reduced temperature range 0.5 = or less than T* = or less than 100 and include values for the constituent two-monomer, three-monomer, and monomer-dimer contributions. A brief discussion of the theoretical approach and calculational methods is also given.

900,398
PB89-179758 Not available NTIS
National Bureau of Standards (NML), Gaithersburg, MD. Chemical Kinetics Div.
Mechanism and Rate of Hydrogen Atom Attack on Toluene at High Temperatures.
Final rept.
D. Robaugh, and W. Tsang. 1986, 5p
Pub. in Jnl. of Physical Chemistry 90, n17 p4159-4163 1986.

Keywords: *Reaction kinetics, *High temperature tests, *Decomposition reactions, Toluene, Methane, Ethane, Butenes, Reprints, *Chemical reaction mechanisms, *Ethane/hexamethyl, *Hydrogen atoms.

Hexamethylethane has been decomposed in the presence of large excesses of toluene and methane/toluene mixtures in single pulse shock tube experiments in the temperature range of 950-1100K and at 2-5 atm. Methane, ethane, isobutene and benzene are the main light hydrocarbon reaction products. In the presence of sufficiently large excess of methane, benzene yields are reduced. This is due to the competitive process.

900,399
PB89-179766 Not available NTIS
National Bureau of Standards (NML), Boulder, CO. Quantum Physics Div.
Quenching and Energy Transfer Processes of Single Rotational Levels of Br2 B triplet Pi(O(sub u)(+)) v' =24 with Ar under Single Collision Conditions.
Final rept.
K. Yamasaki, and S. R. Leone. 1989, 13p
Grants NSF-CHE84-04803, NSF-CHE86-04504
Sponsored by National Science Foundation, Washington, DC.
Pub. in Jnl. of Chemical Physics 90, n2 p964-976, 15 Jan 89.

Keywords: *Bromine, *Molecular energy levels, Argon, Energy transfer, Fluorescence, Reprints, Atom-molecule collisions, Rotational states, Vibrational states, Dye lasers, Quenching.

State-specific total quenching rate constants are measured for selected rotational levels of Br2 under single collision conditions with argon at 296 K. A strict criterion is used to obtain single collision conditions in a cell experiment. A 0.04/cm bandwidth, etalon-narrowed pulsed dye laser excites single rovibronic transitions of the B triplet Pi((O sub u)(+)) state. Fluorescence decay traces with and without the argon collision partner are analyzed at early times to extract total quenching rate constants.

900,400
PB89-179790 Not available NTIS
National Bureau of Standards (NML), Boulder, CO. Quantum Physics Div.
Alignment Effects in Ca-He(5(sup 1)P(sub 1) - 5(sup 3)P(sub J)) Energy Transfer Collisions by Far Wing Laser Scattering.
Final rept.
K. C. Lin, P. D. Kleiber, J. x. Wang, W. C. Stwalley, and S. R. Leone. 1988, 6p
Grant NSF-CHE86-15118
Sponsored by National Science Foundation, Washington, DC.
Pub. in Jnl. of Chemical Physics 89, n8 p4771-4776, 15 Oct 88.

Keywords: *Calcium, Absorption spectra, Alignment, Energy transfer, Helium, Reprints, *Atom-atom collisions.

The far wing absorption profiles for excitation on the Ca(4s sup 2 singlet S sub 0-4s5p singlet p (sup 0, sub 1)) atomic transition, broadened in collisions with He are measured. Strong absorption was observed in both wings and a blue wing satellite was observed near the delta approx. 125/cm. This satellite was tentatively identified as due to a maximum in the CaHe(4s sup 2 singlet Sigma(+) - 4s5p singlet Sigma(+)) difference potential. These line-broadening techniques are used to study electronic energy transfer in the spin-changing collisions of Ca with He: Ca(5p singlet p(sup 0, sub 1)) + He yields Ca(5p triplet p(sup 0, sub J)) + He + delta E.

900,401
PB89-185755 Not available NTIS
National Bureau of Standards (NEL), Gaithersburg, MD. Thermophysics Div.
Van der Waals Fund, Van der Waals Laboratory and Dutch High-Pressure Science.
Final rept.
J. M. H. Levelt Sengers, and J. V. Sengers. 1989, 14p
Pub. in Physica A 156, p1-14 1989.

Keywords: Fluids, Light scattering, Molecular theory, Nuclear magnetic resonance, Phase transformations, Transport properties, Netherlands, Reprints, *Foreign technology, *Van der Waals Fund, *High pressure, Calibration, Diamond anvils.

The history of the van der Waals Fund is traced from events in the late 19th century leading to its establishment in 1898 to the present time. The impact of the fund on the development of high-pressure science and industry in the Netherlands is discussed. The course of high-pressure research at the University of Amsterdam is sketched in the period of van der Waals and Kohnstamm, 1898 to 1920; that of Michels, 1920 to 1961; and that of Trappeniers, 1961 to 1987. The themes are: test of molecular theories at high pressures; quest for accuracy; high-pressure design; impact on Dutch industry; expanding the ranges of pressure and temperature; expanding the capabilities of property measurement; international recognition; and funding, including industrial support.

900,402
PB89-185888 Not available NTIS
National Bureau of Standards (NML), Boulder, CO. Quantum Physics Div.
One Is Not Enough: Intra-Cavity Spectroscopy with Multi-Mode Lasers.
Final rept.
P. E. Toschek, and V. M. Baev. 1987, 89p
Pub. in Laser Spectroscopy and New Ideas, A Tribute to Arthur L. Shawlow, 89p Sep 87.

Keywords: *Spectrum analyses, Multiplexing, Design criteria, Performance evaluation, Quantum chemistry, Quantum interactions, Reprints, *Laser spectroscopy, *Intercavity spectroscopy, Mode selection.

The scope of the presentation is necessarily limited: the authors have not aspired to offer a complete ac-

count of work on or with intra-cavity spectroscopy (ICS). In particular, attempting to cover all the pertinent work that has been done in molecular spectroscopy and chemical kinetics would far exceed the limits of the spectroscopy and chemical kinetics would far exceed the limits of the article. Instead, the authors have tried to point out two facts that are, it seems, not fully appreciated by the broader spectroscopists' community. First, the considerable measure of understanding they now have of the fundamentals that underlie ICS, and second, the level of sophistication and maturity the field has acquired recently. In adaptation of, but also challenging Arthur Schawlow's popular definition of a diatomic molecule, the authors claim that there indeed exist niches in physics where ONE is not enough. Eventually, what is lost in precious simplicity of the system under scrutiny is, perhaps, more than regained with inherent features that allow the spectroscopists to sometimes extend spectacularly the limits previously put on their techniques.

900,403
PB89-185696 Not available NTIS
National Bureau of Standards (NML), Boulder, CO. Quantum Physics Div.
Structure and Dynamics of Molecular Clusters via High Resolution IR Absorption Spectroscopy.
Final rept.
D. J. Nesbitt. 1987, 5p
Grant NSF-PHY86-04504
Sponsored by National Science Foundation, Washington, DC.
Pub. in Proceedings of SPIE (Society of Photo-Optical Instrumentation Engineers), Laser Applications to Chemical Dynamics, Los Angeles, CA., January 13-14, 1987, v742 p16-20.

Keywords: *Infrared spectroscopy, *Absorption spectra, *Molecular structure, *Complex compounds, *Chemical dynamics, *Spectrum analysis, Molecular vibration, Molecular rotation, Hydrogen bonds, Van der Waals equation, Clustering, *High resolution, *Laser spectroscopy.

The combination of high resolution (.001/cm) tunable difference frequency generation (2.2-4.2 micrometers) with high sensitivity .000001/((Hz)1/2) long path length absorption techniques in pulsed slit supersonic jets has permitted spectroscopic investigation of many weakly bound molecular complexes. Discussion focuses on three complementary areas of experimentation. Cluster formation in the molecular beam is probed via sub-Doppler, velocity resolved absorption profiles of monometer species. Spatially dependent beam clustering is strongly manifested through loss of simple van der Waals molecules such as ArHF are obtained in the nu1 HF stretching region. Information on all modes in the complex is extracted. IR spectra of hydrogen bonded complexes such as HFCO2 are observed which exhibit large changes in average molecular geometry as a function of vibrational state. Surprisingly low intermolecular bending frequencies are evidenced in the spectra via hot bands and provide dynamical information on coupled vibrational-rotational motion in floppy molecular systems.

900,404
PB89-185938 Not available NTIS
National Bureau of Standards (NML), Boulder, CO. Quantum Physics Div.
High Resolution Optical Multiplex Spectroscopy.
Final rept.
K. P. Dinse, M. P. Winters, and J. L. Hall. 1987, 2p
Pub. in Proceedings of International Conference on Laser Spectroscopy (8th), Are, Sweden, June 22-26, 1987, p388-389.

Keywords: *Spectrum analysis, *Doppler effect, *Optical spectra, Calibrating, Excitation, Performance evaluation, Multiplexing, *High resolution, *Laser spectroscopy.

Cross correlation techniques using stochastic sources have been applied in many fields to determine the impulse response functions of a system. The work applies such noise-based techniques for the first time in the optical domain to obtain high resolution sub-Doppler spectra. Interesting advantages of the method include a spectral multiplex advantage which can be a few powers of 10 relative to sequentially-scanned spectroscopy as well as absolute calibration of the frequency axis of the spectrum.

48

900,405
PB89-185946   Not available NTIS
National Bureau of Standards (NML), Boulder, CO.
Quantum Physics Div.
Laser Spectroscopy of Inelastic Collisions.
Final rept.
A. Gallagher. 1987, 11p
Pub. in Proceedings of Symposium on Laser Spectros-
copy (1st), Pecs, Hungary, August 28-30, 1986, p19-29
1987.

Keywords: *Inelastic cross sections, *Doppler effect,
Experimental design, Inelastic scattering, *Atom-atom
collisions, *Electron-atom collisions, *Laser spectroscopy.

The potential for wing recoil Doppler spectroscopy to
measure differential inelastic collision cross sections is
described. Experiments in the author's laboratory, investigating atom-atom and electron-atom collisions by
that technique, are discussed.

900,406
PB89-185961   Not available NTIS
National Bureau of Standards (NML), Boulder, CO.
Quantum Physics Div.
Surface Reactions In Silane Discharges.
Final rept.
A. Gallagher. 1986, 9p
Sponsored by Scientific Energy Research inst.,
Golden, CO.
Pub. in Proceedings of International Symposium on the
Physics of Ionized Gases, Sibernik, Yugoslavia, Sep-
tember 1-5, 1986, p229-237.

Keywords: *Surface chemistry, *Silane, *Silicon hydrides, Experimental design, Sputtering, Thin films,
Ionization, Discharges.

A qualitative model for the gas and surface reactions in
silane discharges leading to a-Si:H films and gas constituents is presented. Several experiments on which
this is based are then described.

900,407
PB89-186415   Not available NTIS
National Bureau of Standards (NML), Gaithersburg,
MD. Atomic and Plasma Radiation Div.
**Scheme for a 60-nm Laser Based on Photopump-
ing of a High Level of Mo(6+) by a Spectral Line of
Mo(11+).**
Final rept.
U. Feldman, and J. Reader. 1989, 3p
Sponsored by Department of Energy, Washington, DC.
Pub. in Jnl. of the Optical Society of America B 6, n2
p264-266 Feb 89.

Keywords: *Spectrum analysis, *Atomic energy levels,
*Molecular energy levels, *Molybdenum, *Ultraviolet
spectroscopy, Ions, Plasmas(Physics), Excitation, Reprints, *Laser spectroscopy.

A near coincidence between lines of two molybdenum
ions at 13.650 nm creates conditions that satisfy requirements for producing a short-wavelength photo-
pumped laser in a plasma containing ions of Mo(6+).
The scheme is based on the use of radiation from the
4s(doublet)4p(doublet)P(3/2)-
4s(doublet)5s(doublet)S(1/2) transition of galliumlike
MO(11+) to pump the 4s(doublet)4p(hexlet)(1)S(0)-
4s(doublet)4p(heplet)6s 1/2(1/2)1 transition of kryp-
tonlike Mo(6+). The Mo(6+) ion would then lase
through several possible 4s(doublet)4p(heplet)5p-
4s(doublet)4p(heplet)6s transitions near 60 nm.

900,408
PB89-186449   Not available NTIS
American Chemical Society, Washington, DC.
**Journal of Physical and Chemical Reference Data,
Volume 17, Number 1, 1988.**
Quarterly rept.
D. R. Lide. c1988, 269p
See also PB89-186456 through PB89-186480 and
PB89-145114. Errata sheet inserted. Prepared in co-
operation with American Inst. of Physics, New York.
Sponsored by National Bureau of Standards (ICST),
Gaithersburg, MD.
Available from American Chemical Society, 1155 16th
St., NW, Washington, DC 20036.

Keywords: *Physical properties, *Chemical properties,
*Experimental data, Vapor pressure, Density(Mass/
volume), Equations of state, Steam, Numerical analysis, Cross section, Ionization, Photochemical reactions, Absorption, Molecular energy levels, Molybde-

num, Thermodynamic properties, Aromatic polycyclic
hydrocarbons, Isomerization, Atomic energy levels,
Benzene.

Contents: Pressure and density series equations of
state for steam as derived from the Haar-Gallagher-
Kell formulation; Absolute cross sections for molecular
photoabsorption, partial photoionization, and ionic
photofragmentation processes; Energy levels of molybdenum, Mo I through Mo XLII; Standard chemical
thermodynamic properties of polycyclic aromatic hydrocarbons and their isomer groups I. Benzene series.
(Copyright (c) by the U.S. Secretary of Commerce,
1988.)

900,409
PB89-186456   Not available NTIS
Brown Univ., Providence, RI.
**Pressure and Density Series Equations of State for
Steam as Derived from the Haar-Gallagher-Kell
Formulation.**
Quarterly rept.
R. A. Dobbins, K. Mohammed, and D. A. Sullivan.
c1988, 8p
Included in Jnl. of Physical and Chemical Reference
Data, v17 n1 p1-8 1988. Available from American
Chemical Society, 1155 16th St., NW, Washington, DC
20036.

Keywords: *Vapor pressure, *Density(Mass/volume),
*Steam, *Experimental data, Physical properties,
Equations of state, Numerical analysis, Thermodynamic properties, Metastable state, Comparison, Virial coefficients.

Two equations of state for the properties of steam,
which are in the form of power series in pressure and
density, are developed from the HGK84 formulation.
These equations are of high accuracy in the equilibrium region where extensive measurements exist. They
also accurately represent the extrapolated data in the
metastable region between the vapor saturation and
spinodal lines. The accuracy of the representations as
a function of the number of terms of the series is presented. Their greatest utility is their use for high accuracy calculations that involve small to moderate departures from ideal-gas behavior. Conversion relationships for the second through the tenth coefficients of
the pressure and density series, which apply to the
corresponding virial coefficients, are presented. The
pressure and density expansions are advantageous
for efficient numerical calculations of water vapor
properties in the equilibrium and metastable regions.

900,410
PB89-186464   Not available NTIS
Joint Inst. for Lab. Astrophysics, Boulder, CO.
**Absolute Cross Sections for Molecular Photoab-
sorption, Partial Photoionization, and Ionic Photo-
fragmentation Process.**
Quarterly rept.
J. W. Gallagher, C. E. Brion, J. A. R. Samson, and P.
W. Langhoff. c1988, 145p
Prepared in cooperation with British Columbia Univ.,
Vancouver. Dept. of Chemistry, Nebraska Univ.-Lin-
coln. Dept. of Physics and Astronomy, and Florida
State Univ., Tallahassee. Sponsored by National
Bureau of Standards (ICST), Gaithersburg, MD.
Included in Jnl. of Physical and Chemical Reference
Data, v17 n1 p9-149 1988. Available from American
Chemical Society, 1155 16th St., NW, Washington, DC
20036.

Keywords: *Experimental data, *Cross sections, *Ionization, *Photochemical reactions, *Absorption, *Fragmentation, Physical properties, dipole moments, Molecular energy levels, Numerical analysis, Electronic
spectra, Graphs(Charts).

A compilation is provided of absolute total photoabsorption and partial-channel photoionization cross
sections for the valence shells of selected molecules,
including diatomics (H2, N2, O2, CO, NO) and triato-
mics (CO2, N2O), simple hydrides (H2O, NH3, CH4),
hydrogen halides (HF, HCl, HBr, HI), sulfur compounds
(H2S, CS2, OCS, SO2, SF6), and chlorine compounds
(Cl2, CCl4). The partial-channel cross sections presented refer to production of the individual electronic
states of molecular ions and also to production of
parent and specific fragment ions, as functions of incident photon energy, typically from approximately 20 to
100 eV. Photoelectron anisotropy factors, which together with electronic partial cross sections provide
cross sections differential in photon energy and in
ejection angle, are also reported. There is generally
good agreement between cross sections measured by

the physically distinct optical and dipole electron-
impact methods. The cross sections and anisotropy
factors also compare favorably with selection ab initio
and model potential (X-alpha) calculations which provide a basis for interpretation of the measurements.

900,411
PB89-186472   Not available NTIS
National Bureau of Standards (NML), Gaithersburg,
MD. Center for Radiation Research.
Energy Levels of Molybdenum, Mo 1 through 42.
J. Sugar, and A. Musgrove. c1988, 85p
Sponsored by National Bureau of Standards (ICST),
Gaithersburg, MD.
Included in Jnl. of Physical and Chemical Reference
Data, v17 n1 p155-239 1988. Available from American
Chemical Society, 1115 16th St., NW, Washington, DC
20036.

Keywords: *Atomic energy levels, *Molybdenum, *Ex-
perimental data, *Molecular energy levels, Ionization,
Tables(Data), isoelectronic atoms, Magnetic dipoles,
Physical properties, Excitation.

The energy levels of the molybdenum atom, in all
stages of ionization for which experimental data are
available, have been compiled. Ionization energies,
either experimental or theoretical, and experimental g-
factors are given. Leading components of calculated
eigenvectors are listed.

900,412
PB89-186480   Not available NTIS
Massachusetts Inst. of Tech., Cambridge. Dept. of
Chemistry.
**Standard Chemical Thermodynamic Properties of
Polycyclic Aromatic Hydrocarbons and Their
Isomer Groups 1. Benzene Series.**
Quarterly rept.
R. A. Alberty, and A. K. Reif. c1988, 13p
Sponsored by National Bureau of Standards (ICST),
Gaithersburg, MD.
Included in Jnl. of Physical and Chemical Reference
Data, v17 n1 p241-253 1988. Available from American
Chemical Society, 1155 16th St., NW, Washington, DC
20036.

Keywords: *Experimental data, *Standards, *Thermodynamic properties, *Aromatic polycyclic hydrocarbons, *Benzene, *Isomerization, Chemical properties,
Numerical analysis, Enthalpy, Specific heat, Entropy,
Gibbs free energy, Tables(Data), Molecular structure.

The polycyclic aromatic hydrocarbons may be organized into an infinite number of series in each of which
successive isomer groups differ by C4H2. The first
series starts with benzene, and chemical thermodynamic tables are presented here for C6H6, C10H8,
C14H10, C18H12, C22H14, and C26H16 in the ideal
gas phase. Since chemical thermodynamic properties
are known for only several polycyclic aromatic hydrocarbons, the properties of individual species have
been estimated using Benson group values of Stein
and Fahr for temperatures from 298.15 to 3000 K.
Values of Cp, S, delta H, and delta G have been calculated in joules for a standard state pressure of 1 bar.
The chemical thermodynamic properties of the isomer
groups have also been calculated. This provides a
basis for extrapolating to higher carbon numbers
where it is not feasible to consider individual molecular
species.

900,413
PB89-186746   Not available NTIS
National Bureau of Standards (NML), Gaithersburg,
MD. Chemical Thermodynamics Div.
**Water Structure in Crystalline Solids: Ices to Pro-
teins.**
Final rept.
H. Savage. 1986, 82p
Sponsored by National Inst. of Arthritis, Diabetes, and
Digestive and Kidney Diseases, Bethesda, MD.
Pub. in Water Science Reviews 2, p67-148 1986.

Keywords: *Molecular structure, *Water, Hydrogen
bonds, Crystal structure, Neutron diffraction, Reprints,
*Crystalline hydrates, Non-bonded interactions.

The water structure in hydrate crystals ranging from
the ice polymorphs to complicated macromolecular
protein systems is reviewed with respect to analyzing
the solvent using neutron and X-ray diffraction methods and understanding the resulting experimental
structures. Detailed characteristics of water structure
(ordered and disordered) are discussed with respect to

CHEMISTRY

Physical & Theoretical Chemistry

the more repulsive interactions that overall appear to control the actual geometries present within each system. The repulsive regularities may be used as a set of restraints in analyzing the possible water structure(s) around larger protein systems and also in computer simulations of water and aqueous systems. .

900,414
PB89-186753 · Not available NTIS
National Bureau of Standards (NML), · Gaithersburg,
MD. Chemical Thermodynamics Div.
**Repulsive Regularities of Water Structure in Ices
and Crystalline Hydrates.**
Final rept.
H. F. J. Savage, and J. L. Finney. 1986, 4p
Pub. in Nature 322, n6081 p717-720 1986.

Keywords: *Water, *Liquids, *Neutron diffraction, *Hydrogen bonds, *Crystal structure, Mathematical models, Solvents, Stereochemistry, Water of 'hydration, Reprints.

Hydrogen bonded structures display wide ranges of the various angles (e.g., O-H. . .O) and distances (e.g., H. . .O) used to describe their intermolecular geometry. For example, O. . .O distances are found to vary between -2.5 and 3.2Å, while the hydrogen bond angles normally occur within the 120-180 range. Moreover, although many potential functions exist for describing water-water interactions, none of them are totally satisfactory in reproducing experimental results even for pure water. The initial results of an analysis of water structures in high resolution neutron crystal structures are presented and are dramatically more successful in rationalizing the stereochemistry of water interactions. Rather than considering the structures in terms of weak orientation-dependent attractions, a concentration on the repulsive interactions leads to a set of very much stronger stereochemical constraints which not only rationalize the structures but appear largely to control the orientational correlations in aqueous systems. Looked at in this way, water network structures in crystal hydrates and probably in the liquid itself become for the first time comprehensible. The approach provides a much firmer base from which to build realistic potential functions to model and simulate solvent structure(s).

900,415
PB89-186779 Not available NTIS
National Bureau of Standards (NML), Gaithersburg,
MD. Chemical Thermodynamics Div.
**Biological Standard Reference Materials for the
Calibration of Differential Scanning Calorimeters:
Di-alkylphosphatidylcholine in Water Suspensions.**
Final rept.
F. Schwarz. 1986, 13p
Pub. in Thermochimica Acta 107, p37-49 1986.

Keywords: *Calibrating, *Heat measurement, Enthalpy, Thermodynamic properties, Reprints, *Phosphatidylcholine/dialkyl, *Standard reference materials, Differential scanning calorimeters.

The temperatures and enthalpies of the phase transitions of suspensions of di-alkylphosphatidylcholines in water solutions, prepared and stored under a variety of experimental conditions, were determined in a differential scanning calorimeter (DSC) to evaluate their potential as standard reference materials for the calibration of DSCs. The di-alkylphosphatidylcholine suspensions were 10 mass % 1,2-ditetradecanoyl-sn-glycero-3-phosphocholine (DTPC), and 1,2-dihexadecanoyl-sn-glycero-3-phosphocholine in aqueous buffered solutions at pH 7.0. A subtransition at 8.5 deg C with an enthalpy of 15.5 kJ/mole was observed in the DTPC suspensions after storage of the sample at -5.5 deg C for 2 days. The appearance and the temperature and enthalpy values of the subtransition, the pretransition, and the ice peaks of the suspensions depended on the preparation and storage conditions of the samples. An impreecision of 10% was observed for the values of the main transition and an imprecision of 0.03 deg C for the main transition temperature. Transition temperatures and enthalpies were also determined for suspensions of other di-alkylphosphatidylcholines in buffered water solutions.

900,416
PB89-186910 Not available NTIS
National Bureau of Standards (NML), Boulder, CO.
Quantum Physics Div.

Towards the Ultimate Laser Resolution.
Final rept.
J. L. Hall, D. Hils, O. Salomon, and J. M. Chartier.
1987, 5p
Pub. in Proceedings of International Conference on
Laser Spectroscopy (8th); Are, Sweden, June 22-26,
1987, p376-380.

Keywords: *Optical spectra, Metastable state, Quantum interactions, Performance evaluation, Calibrating, Iodine, Standards, *Laser spectroscopy, *High resolution.

It is indeed possible to lock to a reference cavity with a reproducibility 10 to the -15th power fringe widths and a stability a decade or so better. producing a laser linewidth of 50 milliHertz. Heterodyne tests against an I2-stabilized reference show a frequency drift rate of +6100 Hz/hr stable to .001 per day and 1% for a week. Feed-forward compensation for the smooth drift then offers a source for spectroscopy with a frequency uncertainty 100 milliHertz for about one minute; as will be appropriate to scan over the 'quantum telegraph' optical resonances of single trapped ions with long metastable lifetimes.

900,417
PB89-187561 Not available NTIS
National Bureau of Standards (NML), Gaithersburg,
MD. Gas and Particulate Science Div.
**High Resolution Spectrum of the nu(sub 1) +
nu(sub 2) Band of NO2. A Spin Induced Perturbation in the Ground State.**
Final rept.
R. L. Sams, and W. J. Lafferty. 1987, 16p
Pub. in Jnl. of Molecular Spectroscopy 125, n1 p99-
114 1987.

Keywords: *Nitrogen dioxide, *Rotational spectra, *Electron energy, Ground state, Infrared spectra, Resolution, Vibration, Reprints, Spin orbit interactions.

The spectrum of the nu(sub 1) + nu(sub 2) band of NO2 has been studied with a resolution of 0.025 cm(sup -1). Spin-rotation constants and rotational constants are reported. An interesting perturbation has been found in the ground state of the molecule which occurs when the K(sub a)=0 and K(sub a)=2 levels become accidentally nearly degenerate around N=42. An explanation of this interaction is presented.

900,418
PB89-189351 PC A05/MF A01
National Inst. of Standards and Technology (NML),
Gaithersburg, MD. Chemical Thermodynamics Div.
Logistic Function Data Analysis Program: LOGIT.
W. H. Kirchhoff. Mar 89, 95p NISTIR-88/3603

Keywords: *Hyperbolic functions, *Statistical analysis,
*Data Processing, Computer programs,
Graphs(Charts), Tests, Data, Comparison, Experimental data, Chromium, Nickel, Interfaces, *Logistic functions, *Autocatalysis, *Computer applications, Computer programs, Fortran programming language.

A FORTRAN program has been written for the statistical analysis of experimental data in terms of an extended logistic function that includes non-horizontal asymptotes and asymmetry in the pre- and post-transition portions of growth and decay curves. The program is robust in that situations in which few or no data fall within the transition interval can be analyzed by the program. Individual weighting of the data is allowed for situations where the errors in experimental data are not uniform. The primary parameters describing the transition region include a location parameter, a width parameter, and an asymmetry parameter. Six more parameters describe the two separate asymptotic regions. Sufficient information is given to allow the development of companion subroutines for the graphing of the function and its parameters. Error bars are provided for displaying the original data with error bars. The program provides a means for systematically parameterizing sigmoidal profiles for the comparison of measurements made with different instruments on different systems and for the comparison of measurements with simulation models. The program is extensively documented.

900,419
PB89-189615 Not available NTIS
National Bureau of Standards (IMSE), Gaithersburg,
MD. Ceramics Div.

**Phase Equilibria and Crystal Chemistry in the Ternary System BaO-TiO2-Nb2O5. Part 2. New Barium
Polytitanates with <5 mole % Nb2O5.**
Final rept.
R. S. Roth, L. D. Ettinger, and H. S. Parker. 1987,
10p
See also Part 1, PB89-171797.
Pub. in Jnl. of Solid State Chemistry 68, n2 p330-339
1987.

Keywords: *Barium titanates, *Crystal structure, Titanium oxides, Barium oxides, Niobates, Niobium oxides, Orthorhombic lattices, Monoclinic lattices, Phase diagrams, Ceramics, Reprints, *Phase equilibria.

Four new compounds were found in the BaO-TiO2-Nb2O5 system, each containing <5 mole percent Nb2O5. Ba6Ti14Nb2O39 is an 8-layer orthorhombic phase, Cmcm, with a = 17.138(.011), b = 9.868(.011), c = 18.759(.010)angstroms. The other three phases have similar a and b parameters (a(mon) approximately = b(orth) approximately = 9.9 angstroms, b(mon) approximately = a(orth) approximately = 17 angstroms). Ba14Ti40Nb2O99 is a 20-layer orthorhombic phase, Cmc*, with c approximately = 46.86 angstroms. Ba10Ti28Nb2O72 is a 7-layer monoclinic phase, C2/m, c approximately = 16.72 angstroms, beta approximately = 101.2 deg. Ba18Ti54Nb2O132 is a 13-layer monoclinic phase, C2/m, c approximately = 30.85 angstroms, beta approximately = 96 deg. The compositions were derived by analogy to the layers in Ba4Ti13O30 and Ba6Ti17O40 and are consistent with limited phase equilibria data.

900,420
PB89-189823 Not available NTIS
National Bureau of Standards (NML), Boulder, CO.
Quantum Physics Div.
**Laser-Induced Fluorescence Study of Product Rotational State Distributions in the Charge Transfer
Reaction: Ar(1+)(sup 2 P)(sub 3/2)) + N2 -> Ar +
N2(1+)(X) at 0.28 and 0.40 eV.**
Final rept.
D. M. Sonnenfroh, and S. R. Leone. 1989, 9p
Grants NSF-PHY86-04504, NSF-CHE83-08403
Sponsored by National Science Foundation, Washington, DC., and Air Force Office of Scientific Research,
Bolling AFB, DC.
Pub. in Jnl. of Chemical Physics 90, n3 p1677-1685, 1
Feb 89.

Keywords: *Rotational spectra, *Nitrogen, Ground
state, Fluorescence, Boltzmann equation, Charged
particles, Reprints, *Argon ion, *Charge transfer.

The nascent rotational state distributions of N2(+) produced in the charge transfer reaction of Ar+(doublet P(sub 3/2)) with N2 at 0.28 and 0.40 are remeasured by laser-induced fluorescence. A supersonic expansion is used to reduce the initial rotational angular momentum of the N2. The N2(+) product rotational distributions, in both V=0 and V=1, have low and high energy components. For ease of reference, each distribution is described as a summation of two Boltzmann distributions . At a relative collision energy of 0.28 eV, the Boltzmann temperatures are 100 + or 20 K and 745 + or - 120 K for N2(+) (V=0) and 80 + or - 10 K and 680 + or - 30 K for N2(+) (V = 1). Adiabatic potential energy .curves for the lowest vibronic states are calculated and a simple curve hopping model is presented. Applying the model to the production of N2(+) (V = 1), for example, those reactants that charge transfer on the outgoing leg of a reactive trajectory interact with a deep potential well in the entrance channel for collinear geometry. It is postulated that rotationally excited products result. In comparison, reactants that charge transfer on the ingoing leg (or in perpendicular geometry) do not sample the collinear potential well and the resulting products are less rotationally excited.

900,421
PB89-201081 Not available NTIS
National Bureau of Standards (NML), Gaithersburg,
MD. Surface Science Div.
Interaction of Water with Solid Surfaces: Fundamental Aspects.
Final rept.
P. A. Thiel, and T. E. Madey. 1987, 175p
Pub. in Surface Science Reports 7, n6-8 p211-385
1987.

Keywords: *Surface chemistry, *Water, *Chemisorption, Adsorption, Surface properties, Reprints.

The review compares and discusses recent experimental and theoretical results in the field of H2O-solid interactions. The authors emphasize studies performed on well-characterized, single crystal surfaces of metals, semiconductors and oxides. The authors discuss the factors which influence dissociative vs. associative adsorption pathways. When H2O adsorbs molecularly, it tends to form three-dimensional hydrogen-bonded clusters, even at fractional monolayer coverages, because the strength of the attractive interaction between two molecules is comparable to that of the substrate-H2O bond. The template effect of the substrate is important in determining both the local and long-range order of H2O molecules in these clusters. The influence of surface additive atoms (e.g., O, Br, Na, K) and also surface imperfections (e.g., steps and defects) on the surface structure and chemistry of H2O is examined in detail. Results on single crystal substrates are compared with earlier measurements of H2O adsorption on high-area materials, where available.

900,422
PB89-201115   Not available NTIS
National Bureau of Standards (NEL), Gaithersburg, MD. Thermophysics Div.
Simple Apparatus for Vapor-Liquid Equilibrium Measurements with Data for the Binary Systems of Carbon Dioxide with n-Butane and Isobutane.
Final rept.
L. A. Weber. 1989, 5p
Pub. in Jnl. of Chemical and Engineering Data 34, n2 p171-175 Apr 89.

Keywords: *Test equipment, *Vapor phases, *Liquids, *Equilibrium, *Gases, Measurement, High pressure tests, Carbon dioxide, Butanes, Henrys Law, Reprints.

The design, construction, and testing of a simple vapor-liquid equilibrium apparatus designed for measurements in the range 300-500 K at pressures to 150 bar are described. Data are given for measurements of P, T, x, and y for binary systems of carbon dioxide with n-butane and isobutane in the range 310-394 K.

900,423
PB89-201123   Not available NTIS
National Bureau of Standards (NEL), Gaithersburg, MD. Chemical Process Metrology Div.
NO/NH3 Coadsorption on Pt(111): Kinetic and Dynamical Effects in Rotational Accommodation.
Final rept.
D. Burgess, R. R. Cavanagh, and D. S. King. 1989, 19p
Contract DE-AI05-84ER13150
Sponsored by Department of Energy, Washington, DC.
Pub. in Surface Science 214, p358-376 1989.

Keywords: *Adsorbates, *Ammonia, *Nitrogen oxide, *Platinum, Mass spectroscopy, Lasers, Fluorescence, Desorption, Dissociation, Reaction kinetics, Reprints.

The strongly interacting coadsorbate system of ammonia/nitric oxide on Pt(111) has been examined using both quadrupole mass spectrometer (QMS) and laser-induced fluorescence (LIF) detected temperature programmed desorption (TPD) methods. The QMS/TPD experiments show simultaneous desorption of NO and NH3 as reaction-limited products from the dissociation of an adsorbed NO-NH3 complex. Although NO TPD kinetics are altered by the formation of the complex, LIF/TPD results show that the rotational accommodation, spin-orbit temperature, lambda doublet populations, and molecular alignment of desorbed NO are insensitive to the attractive interaction. It suggests that, following decomposition of the NO-NH3 species, NO resides on the surface sufficiently long that it has no memory of being complexed.

900,424
PB89-201735   Not available NTIS
National Bureau of Standards (NML), Gaithersburg, MD. Molecular Spectroscopy Div.
Microwave Spectrum and Molecular Structure of the Ethylene-Ozone van der Waals Complex.
Final rept.
J. Z. Gillies, C. W. Gillies, R. D. Suenram, F. J. Lovas, and W. Stahl. 1989, 2p
Pub. in Jnl. of the American Chemical Society 111, p3073-3074 1989.

Keywords: *Molecular structure, *Ethylene, *Ozone, Rotational spectra, Ground state, Excitation, Reprints, *Microwave spectra, Reaction kinetics, Dipole moments, Van der Waals complex.

The rotational spectrum of CH2=CH2 ... 03 complex was observed using a pulsed beam Fabry-Perot cavity Fourier transform microwave spectrometer. Internal motions in the complex produced two components for each transition. The two sets of lines were independently fit to a Watson Hamiltonian to give the rotational constants (in MHZ): Ground State - A 8246.841 (2), B 2518.972 (4), C 2044.248 (5); Excited State - A 8241.897 (4), B 2518.941 (9), C 2044.872 (11). Dipole moment measurements determine mu sub c = mu sub total = 0.461 (2) D. The complex structure is described by two parallel planes containing ethylene and ozone in which the two centers of mass are connected by a 3.30 A line perpendicular to the planes. 1,3-dipolar cycloaddition theory and orbital symmetry rules are used in conjunction with ab initio calculations to argue that the CH2=CH2 ... 03 complex lies in a small minimum on the reaction coordinate prior to the transition state which produces ethylene primary ozonide.

900,425
PB89-201743   Not available NTIS
National Bureau of Standards (NML), Gaithersburg, MD. Molecular Spectroscopy Div.
Heterodyne Measurements on OCS Near 1372 cm(-1).
Final rept.
M. Schneider, A. G. Maki, M. D. Vanek, and J. S. Wells. 1989, 5p
Sponsored by National Aeronautics and Space Administration, Washington, DC.
Pub. in Jnl. of Molecular Spectroscopy 134, p349-353 1989.

Keywords: *Demodulation, *Molecular structure, Carbonyl compounds, Reprints, *Carbonyl sulfide, Frequency standards.

Heterodyne frequency measurements are given for the 01 (sup 1)1 - 00 (sup) 0 and 02 (sup 0)1 - 01 (sup 1)0 bands of OCS between 1363 and 1398/cm. These measurements were combined with heterodyne measurements on the 01 (sup 1)1 - 01 (sup 1)0 and 02 (sup 0)1 - 00 (sup 0)0 bands to obtain frequencies for the 01 (sup 1)0 - 00 (sup 0)0 transitions by two independent paths. A table of wavenumbers is given for the nu(sub 2) band of OCS from 488 to 557/cm.

900,426
PB89-201750   Not available NTIS
National Bureau of Standards (NEL), Gaithersburg, MD. Semiconductor Electronics Div.
Infrared Absorption Cross Section of Arsenic in Silicon in the Impurity Band Region of Concentration.
Final rept.
J. Geist. 1989, 7p
Pub. in Applied Optics 28, n6 p1193-1199, 15 Mar 89.

Keywords: *Arsenic, *Silicon, *Infrared spectra, Absorption, Spectral lines, Reprints.

The spectral dependence of the infrared absorption cross section of As in Si near O K has been determined from infrared transmission measurements for three As concentrations (5.3, 8.4, and 15.9 X 10 to the 17th power cm sup -3) in the impurity band regime. The results demonstrate some features of physical interest. With increasing As concentration, the lines associated with the intra-atomic transitions broaden asymmetrically, while the integral of the total absorption cross section over photon energy is conserved as required by the oscillator strength sum rule. It thus appears that the cross section of the intra-atomic transitions is conserved as the lines hybridize with the continuum. Comparison of results with photoionization cross-sectional data suggests that the lines contribute to the cross section for photoionization through field and thermally assisted transitions when they are near the threshold for photoionization.

900,427
PB89-201800   Not available NTIS
National Bureau of Standards (IMSE), Gaithersburg, MD. Office of Nondestructive Evaluation.
Analysis of Roto-Translational Absorption Spectra Induced in Low Density Gases of Non-Polar Molecules: The Methane Case.
Final rept.
P. Dore, M. Moraldi, J. D. Poll, and G. Birnbaum. 1989, 19p
Pub. in Molecular Physics 66, n2 p355-373 1989.

Keywords: *Methane, *Absorption spectra, Far infrared radiation, Reprints, *Nonpolar gases, Temperature dependence.

A new analysis is given of the roto-translational absorption spectrum of gaseous methane for which experimental data are available at different temperatures. The authors consider components of the induced dipole associated at long range with induction by the octopole and hexadecapole multipole moments together with components associated with the gradient of both the octopolar and hexadecapolar fields. A satisfactory description of experimental data is obtained only if short range anisotropic overlap is included with the octopolar and hexadecapolar induction. A better description of the experimental absorption bands is obtained if, in addition, the intensity of the double transitions is slightly increased with increasing temperature. This increase is attributed to anisotropic overlap effects on the double transition spectrum, which have not been explicitly taken into account.

900,428
PB89-202022   Not available NTIS
National Bureau of Standards (NML), Gaithersburg, MD. Radiometric Physics Div.
Exploratory Research in Reflectance and Fluorescence Standards at the National Bureau of Standards.
Final rept.
V. R. Weidner, J. J. Hsia, and K. L. Eckerle. 1986, 3p
Pub. in Optics News 12, n11 p18-20 Nov 86.

Keywords: *Standards, *Spectrophotometry, Carbon black, Fluorescence, Reflectance, Reprints, Polytetrafluoroethylene.

The paper describes current research relating to the development of spectrophotometric standards of NBS. The results of some experimental work on the preparation and analysis of a gray scale for diffuse reflectance, and materials for possible use as fluorescence standards are reviewed. The gray scale for diffuse reflectance was prepared by sintering mixtures of carbon black and polytetrafluoroethylene (PTFE) resin. These are durable solid disks that vary in reflectance from about 3% to over 90%, depending on the concentration of carbon black. The fluorescent samples are prepared by sintering mixtures of inorganic phosphors in the same PTFE resin. These phosphors provide highly stable specimens with blue, green, yellow, and orange emission spectra.

900,429
PB89-202055   Not available NTIS
National Bureau of Standards (NML), Boulder, CO. Time and Frequency Div.
Pure Rotational Far Infrared Transitions of (16)O2 in Its Electronic and Vibrational Ground State.
Final rept.
L. R. Zink, and M. Mizushima. 1987, 5p
Pub. in Jnl. of Molecular Spectroscopy 125, n1 p154-158 1987.

Keywords: *Oxygen, *Vibrational spectra, Rotational spectra, Molecular spectroscopy, Spectral lines, Infrared spectra, Ground state, Adsorption, Reprints, Electron states.

Five lines in the far infrared region, due to N(+2) <- N (Delta J = 0) transitions of the (16)02 molecule in its (X (sup 3)Sigma (sub g))-state are measured at 773.839691, 1466.807133, 1812.405539, 2157.577773, and 2502.323923 GHz, using our new tunable FIR spectrometer. The spectral line shape of the 2.50 THz line is analyzed and the pressure self-broadening parameter of 18.2(32) kHz/Pa (2.43(43) MHz/torr) is obtained.

900,430
PB89-202071   Not available NTIS
National Bureau of Standards (NML), Gaithersburg, MD. Inorganic Analytical Research Div.
Three-Dimensional Atomic Spectra in Flames Using Stepwise Excitation Laser-Enhanced Ionization Spectroscopy.
Final rept.
G. C. Turk, F. C. Ruegg, J. C. Travis, and J. R. DeVoe. 1986, 7p
Pub. in Jnl. of Applied Spectroscopy 40, n8 p1146-1152 1986.

Keywords: *Atomic spectroscopy, Spectroscopic analysis, Atomic spectra, Excitation, Reprints, Laser enhanced ionization.

Stepwise excitation laser-enhanced ionization spectroscopy utilizes two independently tunable dye lasers to populate high lying excited states of atoms in

51

CHEMISTRY

Physical & Theoretical Chemistry

flames. Two atomic resonances are required, with the upper level of the first step transition coinciding with the lower level of the second step transition. Efficient population of a high lying atomic level is achieved from which a high rate of collisional ionization can take place. The double resonance aspect of such excitation adds an extra dimension of spectroscopic selectivity to the measurement. A computer controlled dual wavelength LEI spectrometer, including a Dizeau wavemeter for wavelength verification, is used to record three-dimensional spectra-ionization signal as a function of both first and second step wavelengths. Examples illustrate the accuracy advantage accorded by the three-dimensional survey.

900,431
PB89-202121   Not available NTIS
National Bureau of Standards (NML), Gaithersburg, MD. Atomic and Plasma Radiation Div.
Accurate Energies of nS, nP, nD, nF and nG Levels of Neutral Cesium.
Final rept.
K. H. Weber, and C. J. Sansonetti. 1987, 11p
Pub. in Physical Review A-General Physics 35, n11 p4650-4660 1987.

Keywords: *Cesium, *Ionization potentials, Interferometers, Lasers, Electron transitions, Energy levels, Reprints, Fabry-Perot spectrometers.

Extensive measurements have been performed to determine the absolute energies of the n doublet S(1/2) (n=8-31), n doublet P(1/2) (n=6,9-80), n doublet D(5/2) (n=5,7-36), n doublet F(5/2) (n=6-65) and n doublet G(7/2) (n=6-54) levels of Cs by using non-resonant and resonantly enhanced Doppler-free two-photon spectroscopy. The excitation mechanisms employed include resonantly enhanced dipole-quadrupole and quadrupole-quadrupole transitions. All energies were measured directly with respect to the n=6 doublet S(1/2) ground state. The laser wavelengths were measured by high-precision Fabry-Perot interferometry yielding an uncertainty of 0.0002/cm for most Cs levels. Ionization energies derived by fitting the modified Ritz formula to each of the five series observed coincide within 0.00005/cm. Taking account of possible systematic errors, the Cs ionization energy is 31406.46766(15)/cm.

900,432
PB89-202139   Not available NTIS
National Bureau of Standards (NML), Gaithersburg, MD. Chemical Kinetics Div.
Pi-Electron Properties of Large Condensed Polyaromatic Hydrocarbons.
Final rept.
S. E. Stein, and R. L. Brown. 1987, 9p
Pub. in Jnl. of the American Chemical Society 109, n12 p3721-3729 1987.

Keywords: *Aromatic polycyclic hydrocarbons, *Electron density(Concentration), Electron distribution, Resonance, Energy levels, Heat of formation, Reprints, Bond orders, Hueckel theory.

Hueckel molecular orbital(HMO) theory has been used to calculate energy level density bond orders, electron distributions, free valence, resonance energies, and heats of formation for several homologous series of large, hexagonally symmetric benzenoid polyaromatic molecules with well-defined edge structures containing up to 2300 carbon atoms. When extrapolated to the infinite limit, values for all properties converge to reasonable values. This is in contrast to several other pi-electron theories which do not yield correct graphite limits. Carbon atoms at the edges of such large molecules are predicted to behave like those in small polynuclear aromatic molecules, with properties strongly dependent on local structure. Regardless of edge structure, interior carbons several bond lengths from an edge have properties similar to those in an infinite graphite sheet. Edge structure has a larger influence on heats of formation than that predicted by group additivity methods. Only a weak correlation was found between the position of the highest occupied molecular orbital and the reactivity of the most reactive position.

900,433
PB89-202162   Not available NTIS
National Bureau of Standards (NML), Gaithersburg, MD. Molecular Spectroscopy Div.

Nonadiabatic Theory of Fine-Structure Branching Cross-Sections for Sodium-Helium, Sodium-Neon, and Sodium-Argon Optical Collisions.
Final rept.
L. L. Vahala, P. S. Julienne, and M. D. Havey. 1986, 13p
Pub. in Physical Review A: General Physics 34, n3 p1856-1868 1986.

Keywords: *Collision cross sections, *Helium, *Neon, *Sodium, *Argon, Rare gases, Atomic energy levels, Quantum interactions, Electron transitions, Photons, Optical spectra, Reprints, Branching ratio, Fine structure.

The nonadiabatic close-coupled theory of atomic collisions in a radiation field is generalized to include electron spin and is used to consider the weak-field Na-rare gas (RG) optical collision. Na(doublet S(1/2)) + RG + nh(nu) yields Na(doublet P(j) + RG + (n-1)h(nu). The effects of detuning and incident energy on the branching into the atomic Na 3p doublet P(3/2) and 3p doublet(1/2) states are examined. The cross sections sigma(j) are found to have a strong asymmetry between red and blue detuning as well as a complex threshold and resonance structure dependence on energy. A partial cross-section analysis of sigma(j) shows a significant difference between contributions from states of e and f molecular parity. The theoretically calculated detuning dependence of the branching ratio into each fine structure state is in good agreement with available experimental data for NaAr, NaNe, and NaHe, as well as the total absorption coefficient for the production of Na 3p atoms. The fine structure branching ratio for thermal energy collisions shows considerable variation with rare gas collision partner, due to the different interaction potentials. For sufficiently high collision energy, the branching approaches a recoil limit which is independent of collision partner.

900,434
PB89-202485   Not available NTIS
National Bureau of Standards (NEL), Gaithersburg, MD. Thermophysics Div.
Gas Solubility and Henry's Law Near the Solvent's Critical Point.
Final rept.
M. L. Japas, and J. M. H. Levelt Sengers. 1989, 9p
Pub. in AIChE Jnl. 35, n5 p705-713 May 89.

Keywords: *Gases, *Henrys law, *Critical point, *Solvents, *Solubility, High temperature tests, Thermodynamic properties, Water, Benzene, Mathematical models, Fugacity, Reprints.

It has been experimentally observed, for water and nonaqueous solvents alike, that Henry's constant passes through a maximum and then declines as the temperature is raised from the triple point to the critical point. Exact relations for the value of Henry's constant and its temperature dependence at the solvent's critical point are developed from classical and nonclassical models showing that the decline of the constant is a universal phenomenon. The limiting temperature dependence of Henry's constant can be predicted from the thermodynamic properties of the pure solvent and the initial slope of the critical line. The validity of the prediction is tested by comparing it with experimental solubility data for several gases in high-temperature water and benzene. The predictive model appears valid over a temperature range of at least 15% in temperature below the critical point of the solvent.

900,435
PB89-202493   Not available NTIS
National Bureau of Standards (NEL), Boulder, CO. Thermophysics Div.
Isochoric (p,v,T) Measurements on CO2 and (0.98 CO2 + 0.02 CH4) from 225 to 400 K and Pressures to 35 MPa.
Final rept.
J. W. Magee, and J. F. Ely. 1986, 11p
Pub. in International Jnl. of Thermophysics 9, n4 p547-557 1988.

Keywords: *Carbon dioxide, *Methane, *Density(Mass/volume), *High pressure tests, *Temperature, Measurement, Purity, Gas laws, Mixtures, Volume, Reprints.

Comprehensive isochoric (p, v, T) measurements have been obtained for (0.98 OC2 + 0.02 CH4) at densities from 1 to 26 mol x dm(sup -3). Supplemental isochoric (p, v, T) measurements have been obtained for high-purity CO2 at densities from 12 to 24 mol x dm(sup -3).

Measurements of p(T) cover a broad range of temperature, 225 to 400 K, at pressures to 35 MPa. Comparisons have been made with independent sources and with a predictive method based on corresponding states.

900,436
PB89-202519   Not available NTIS
National Bureau of Standards (NEL), Gaithersburg, MD. Thermophysics Div.
NaCl-H2O Coexistence Curve Near the Critical Temperature of H2O.
Final rept.
A. H. Harvey, and J. M. H. Levelt Sengers. 1989, 3p
Pub. in Chemical Physics Letters 156, n4 p415-417, 7 Apr 89.

Keywords: *Sodium chloride, *Water, *Solubility, Phase transformations, Concentration(Composition), Critical temperature, Reprints, Temperature dependence, Pressure dependences.

Recent high-temperature data on coexisting vapor-liquid compositions as a function of pressure in the CaC1-H2O system have been interpreted as implying a higher-than-expected value of the critical exponent beta. It is suggested that the observed behavior is a consequence of the fact that, as the solvent's critical point is approached, the coexistence curve in pressure-composition coordinates ceases to behave in the same manner as the pressure-density coexistence curve, which is always characterized by the exponent beta. At the critical temperature of the solvent, the pressure-composition coexistence curve is characterized by the much larger exponent beta + 1.

900,437
PB89-202527   Not available NTIS
National Bureau of Standards (NML), Gaithersburg, MD. Chemical Kinetics Div.
Thermochemistry of Solvation of SF6(1-) by Simple Polar Organic Molecules in the Vapor Phase.
Final rept.
L. W. Sieck. 1986, 4p
Pub. in Jnl. of Physical Chemistry 90, n25 p6684-6687 1986.

Keywords: *Solvation, *Sulfur halides, *Thermochemistry, *Dissociation, Electron beams, Mass spectroscopy, Ligands, Complex ions, Kinetics, Reprints, Temperature dependence, Binding energy, Sulfur fluorides.

The stabilities of SF6(-).HR association ions, where HR is a simple aliphatic alcohol, H2O, or Me2SO, have been investigated by the technique of pulsed electron beam high pressure mass spectrometry. Equilibrium constants were determined as a function of temperature in order to define delta H and delta S values for solvation. The binding energies are quite low, ranging from 10.5 kcal mol for SF6(-).H2O to 14.7 kcal/mol for SF6(-).Me2SO. For ligands in which the nature of the bonding is expected to be similar, the binding energies increase nonlinearly with increasing acid strength of the ligand. On the basis of additional measurements involving I(-).HR complexes, as well as existing literature values, the binding energies in SF6(-).HR ions are found to be slightly higher than those for I(-).HR. The SF6(-).HR complexes are less stable, however, due to their more positive dissociation entropies. Some comments are also included concerning the SF6(-).SF6 dimer ion.

900,438
PB89-202543   Not available NTIS
National Bureau of Standards (NEL), Gaithersburg, MD. Electrosystems Div.
Drift Tubes for Characterizing Atmospheric Ion Mobility Spectra Using AC, AC-Pulse, and Pulse Time-of-Flight Measurement Techniques.
Final rept.
M. Misakian, W. E. Anderson, and O. Laug. 1989, 10p
Pub. in Review of Scientific Instruments 60, n4 p720-729 Apr 89.

Keywords: *Ionic mobility, *Atmospheric pressure, *Measurement, *Test equipment, Spectra, Time, Alternating current, Tubes, Drift(Instrumentation), Reprints.

Two drift tubes constructed of insulating cylinders with conductive guard rings on the inside walls are examined to determine their suitability for measuring ion mobility spectra at atmospheric pressure. One drift tube is of the pulse time-of-flight (TOF) type with adjustable drift distance, and the other is an ac-TOF drift tube similar in principle to devices reported by Tyndall and Powell. The latter drift tube is evaluated using sinusoidal and alternating-polarity pulse-voltage waveforms for gating the shutters. Methods for determining the drift velocity of an ion from theoretical fits of the TOF spectrum are described for drift tubes of fixed length exhibiting 'end effects.' Mobility values with uncertainties less than + or - 1% can be obtained with the pulse-TOF drift tube. Comparable results are obtained with the ac drift tube if an alternating-polarity pulse-voltage waveform is used for gating the shutter.

900,439
PB89-202568 Not available NTIS
National Bureau of Standards (NEL), Gaithersburg, MD. Fire Measurement and Research Div.
Component Spectrum Reconstruction from Partially Characterized Mixtures.
Final rept.
M. R. Nyden, and K. Chittur. 1989, 6p
Pub. in Applied Spectroscopy 43, n1 p123-128 1989.

Keywords: *Proteins, Electromagnetic radiation, Infrared radiation, Mathematical models, Reprints, *Infrared absorption.

A mathematical analysis of some existing approaches to component spectrum reconstruction is presented. The analysis leads to the derivation of a generalization of the cross-correlation technique. The effectiveness of these methods is assessed from the quality of the reconstructions obtained with the use of synthetic mixture spectra. Reconstructions of the spectra of the components of aqueous mixtures of immunoglobulin G and albumin are compared to the corresponding spectral reconstructions of the pure proteins in buffer.

900,440
PB89-202956 Not available NTIS
National Bureau of Standards (NEL), Gaithersburg, MD. Chemical Process Metrology Div.
Coadsorption of Water and Lithium on the Ru(001) Surface.
Final rept.
S. Semancik, D. L. Doering, and T. E. Madey. 1986, 18p
Sponsored by Department of Energy, Washington, DC.
Pub. in Surface Science 176, p165-182 1986.

Keywords: *Surface chemistry, *Adsorption, *Water, *Lithium, *Ruthenium, Single crystals, Desorption, Spectroscopy, Auger electrons, Reprints.

The interactions between water and lithium have been studied on the surface of a Ru(001) crystal using thermal desorption spectroscopy electron stimulated desorption ion angular distributions (ESDIAD), Auger spectroscopy and LEED. The presence of Li was found to influence strongly the H+ ESD yield and the ESDIAD patterns from adsorbed water even at Li coverages of 0.02 monolayer; changes in the thermal desorption states for water were also observed at low Li coverages. For coadsorbed Li coverages above 0.05, ESDIAD measurements provided clear evidence of water decomposition, even at surface temperatures near 80 K; evidence for dissociation was also obtained from thermal desorption and Auger measurements. The present results are compared and contrasted to those reported previously for the H2O/Na/Ru(001) system.

900,441
PB89-202980 Not available NTIS
National Bureau of Standards (NML), Gaithersburg, MD. Surface Science Div.
Oxygen Chemisorption on Cr(110): 1. Dissociative Adsorption.
Final rept.
N. D. Shinn, and T. E. Madey. 1986, 16p
See also PB89-202998.
Pub. in Surface Science 173, n2-3 p379-394 1986.

Keywords: *Chemisorption, *Chromium, *Oxygen, Electron diffraction, Auger electrons, Metals, Vibrational spectra, Reprints.

The initial stages of oxygen chemisorption on Cr(110) at 300K have been investigated using high resolution electron energy loss spectroscopy (HREELS), electron stimulated desorption ion angular distributions (ESDIAD), low energy electron diffraction (LEED), and Auger electron spectroscopy (AES). Dissociative chemisorption occurs with near unit sticking probability, leading to an ordered p(4x2)O overlayer at theta(sub

0) approximately equals 1/8. A model with atomic oxygen located in the two-fold symmetric hollow sites is proposed. The previously reported c(3x1)O overlayer was not observed in the study. A disordered oxygen o0layer is found as theta(sub 0) increases. ESDIAD and HREELS data suggest that chemisorbed oxygen atoms occupy inequivalent local binding sites for theta(sub 0)>0.25; subsurface oxygen is not found under these mild oxidation conditions.

900,442
PB89-202998 Not available NTIS
National Bureau of Standards (NML), Gaithersburg, MD. Surface Science Div.
Oxygen Chemisorption on Cr(110): 2. Evidence for Molecular O2(ads).
Final rept.
N. D. Shinn, and T. E. Madey. 1986, 18p
See also PB89-202980.
Pub. in Surface Science 176, n3 p635-652 1986.

Keywords: *Chemisorption, *Chromium, *Oxygen, Electron diffraction, Auger electrons, Metals, Vibrational spectra, Reprints.

Oxygen chemisorption and dissociation on Cr(110) at 120K have been studied using high resolution electron energy loss spectroscopy (HREELS), electron stimulated desorption ion angular distribution (ESDIAD), low energy electron diffraction (LEED) and Auger electron spectroscopy (AES). Dissociative adsorption dominates although vibrational and stimulated desorption data provide evidence for a coexisting minority molecular binding state. An O2(ads) vibrational frequency of 1020/cm and a six beam ESDIAD pattern are suggestive of super-oxo O2(ads) bonding at six local sites each with the 0-0 molecular axis tilted away from the surface normal. These results are compared with data for chemisorbed oxygen on other transition metal surfaces.

900,443
PB89-203004 Not available NTIS
National Bureau of Standards (NML), Gaithersburg, MD. Surface Science Div.
Stimulated Desorption from CO Chemisorbed on Cr(110).
Final rept.
N. D. Shinn, and T. E. Madey. 1987, 18p
Pub. in Surface Science 180, n2-3 p615-632 1987.

Keywords: *Chemisorption, *Carbon monoxide, *Chromium, *Desorption, Oxygen, Ions, Potassium, Ruthenium, Iron, Reprints.

Electron stimulated desorption (ESD) experiments using a time-of-flight pulse counting method are reported for molecular CO chemisorbed on the Cr(110) surface at 80K. Consistent with previous qualitative observations, negligible CO+ and O+ desorption signals were measured from the alpha(1)CO overlayer which saturates at 1/4 monolayer. For theta(CO) 0.25, a terminally-bonded (alpha(2)CO) binding mode is populated in addition to the existing alpha(1)CO binding mode and the ion yield sharply increases. For alpha(2)CO, both O and CO+ ions are observed; the CO+ ions desorb with characteristically lower kinetic energies than O+ ions. Near saturation coverages of CO(ads), an observed decrease in the O+ yield is attributed to adsorbate-adsorbate interactions which reduce the ion desorption probability, as seen in ESD studies of terminally-bonded CO on other metals. These results are considered in the context of two possible models proposed for the alpha(1)CO binding state and related ESD observations for CO chemisorbed on potassium-promoted Ru(001) and Fe(001).

900,444
PB89-203012 Not available NTIS
National Bureau of Standards (NML), Gaithersburg, MD. Surface Science Div.
Time Resolved Studies of Vibrational Relaxation Dynamics of CO(v=1) on Metal Particle Surfaces.
Final rept.
J. D. Beckerle, M. P. Casassa, R. R. Cavanagh, E. J. Heilweil, and J. C. Stephenson. 1989, 2p
Pub. in Jnl. of Chemical Physics 90, n8 p4619-4620, 15 Apr 89.

Keywords: *Vibrational spectra, *Molecular relaxation, *Carbon monoxide, *Chemisorption, *Catalysts, *Metal carbonyls, Infrared radiation, Bleaching, Temperature, Polarization, Silicon dioxide, Surface chemistry, Platinum, Rhenium, Hole mobility, Supports, Reprints.

The vibrational relaxation dynamics of CO chemisorbed on small Pt and Rh particles supported on SiO2 has been investigated by picosecond time-resolved infrared transient bleaching experiments. A vibrational T1 lifetime of 7 + or - 1 ps has been observed for several different samples, independent of polarization, pump intensity, and sample temperature from 100-400K. A 1:3 isotopic dilution has no effect upon T1. The T1 lifetime is a factor of 10-50 times shorter than T1 decay; redistribution of the energy throughout the broad CO vibrational band, and relaxation directly to electron-hole pairs in the metal particles.

900,445
PB89-212013 Not available NTIS
National Bureau of Standards (NML), Gaithersburg, MD. Surface Science Div.
Chemisorption of HF (Hydrofluoric Acid) on Silicon Surfaces.
Final rept.
S. A. Joyce, J. A. Yarmoff, A. L. Johnson, and T. E. Madey. 1989, 6p
Sponsored by Department of Energy, Washington, DC.
Pub. in Proceedings of Materials Research Society Symposium on Chemical Perspectives of Microelectronic Materials, Boston, MA., December 1988, v131 p185-190 1989.

Keywords: *Chemisorption, *Silicon, *Surface properties, *Hydrogen fluoride, *Gases, *Chemical bonds, Single crystals, X ray analysis, Spectroscopy, Dissociation, Photoelectric emission, Microelectronics, Valence.

The interaction of gaseous hydrofluoric acid (HF) with single crystal silicon surfaces was investigated using soft x-ray photoemission spectroscopy and Electron Stimulated Desorption Ion Angular Distributions (ESDIAD). Examination of the Si(2p) core level for surfaces saturated with HF shows the formation of silicon-fluoride bonds indicating the dissociative chemisorption of HF on both Si(111) and Si(100) surfaces. Inspection of the F(2s) and F(2p) valence levels at saturation coverage indicate that only one-half monolayer of fluorine bonds to the silicon. The present data is corroborated by electron bombardment of these surfaces is F+ with only a minor contribution from H+ ESDIAD images from a saturation coverage of HF on stepped Si(100) surfaces reveal F+ desorption primarily along the direction of the terrace dimers. The ESDIAD patterns from HF adsorbed on Si(111) are characterized by strong normal F+ emission with a weak background component of off-normal emission. The results are consistent with the dissociative chemisorption of HF where the ion emission direction is determined by the Si-F bond directions.

900,446
PB89-212153 Not available NTIS
National Bureau of Standards (NML), Gaithersburg, MD. Office of Standard Reference Data.
Activities of the International Association for the Properties of Steam between 1979 and 1984.
Final rept.
H. J. White. 1986, 9p
Pub. in Proceedings of International Conference on the Properties of Steam (10th), Moscow, USSR, September 3-7, 1984, 9p 1986.

Keywords: *Steam, *Thermodynamic properties, Steam tables, Enthalpy, Entropy, Thermodynamics, Transport properties, *International Association for the Properties of Steam.

The paper reviews the activities of the International Association for the Properties of Steam (IAPS) in the interval between the 9th International Conference on the Properties of Steam held in Munich in 1979 and the 10th Conference held in Moscow, 1984. A brief review of the purpose and organization of IAPS is given to place its latest activities in a broader and perhaps more meaningful context. Some of the activities during the period 1979 to 1984 represent the culmination of a program which began with the 8th Conference, which was held in Giens, France, in 1974.

900,447
PB89-212229 Not available NTIS
National Bureau of Standards (NEL), Gaithersburg, MD. Electrosystems Div.

CHEMISTRY

Physical & Theoretical Chemistry

Measurement of Electrical Breakdown in Liquids.
Final rept.
R. E. Hebner. 1988, 19p
Pub. in NATO Advanced Study Institute on the Liquid State and Its Electrical Properties, Sintra, Portugal, July 5-17, 1987, p519-537 1988.

Keywords: *Liquids, *Electrical properties, *Electrical faults, Measuring instruments, Additives, Particle density(Concentration), Pressure, Viscosity, Voltage gain, Cathodes, Anodes, Reprints.

The continuing development of 'light sources, high speed cameras, and high speed electronic measuring systems have made it possible to study the breakdown process in increasing detail. The four measurements described are high speed photography of the breakdown process, measurement of the voltage and current, optical spectroscopy, and the measurement of acoustic emission. Having developed a battery of measurement techniques, understanding of the breakdown process is gained by changing the system in known ways and determining the effect of these changes on the measured results. Parameters which have been investigated include types of liquids, chemical additives, particle density, pressure, viscosity, and the rate of rise of the applied voltage. These investigations have led to the identification of four modes of growth when the streamer initiates at a cathode and three modes when it initiates at an anode.

900,448
PB89-222525 Not available NTIS
American Chemical Society, Washington, DC.
Journal of Physical and Chemical Reference Data, Volume 18, Number 2, 1989.
Quarterly rept.
D. R. Lide. c1989, 548p
See also PB89-222533 through PB89-222582 and PB89-135685. Prepared in cooperation with American Inst. of Physics, New York. Sponsored by National Inst. of Standards and Technology, Gaithersburg, MD.
Available from American Chemical Society, 1155 16th St., NW, Washington, DC 20036.

Keywords: *Thermodynamic properties, Graphs(Charts), Reaction kinetics, Photochemical reactions, Thermal conductivity, Nitrogen, Carbon monoxide, Density(Mass/volume), Methane, Thermophysical properties, Hexoses, Pentoses, Argon, Pressure, Phase transformations, Transport properties, Standards, Tables(Data), Carbohydrates, Organic phosphates, Triplet point, Difluoride/dioxy, Fluoride/dioxy, Atmospheric chemistry.

Contents: The Thermal Conductivity of Nitrogen and Carbon Monoxide in the Limit of Zero Density; Thermophysical Properties of Methane; Thermodynamic Properties of Argon from the Triple Point to 1200 K with Pressures to 1000 MPa; Thermophysical Properties of Dioxygen Difluoride (O2F2) and Dioxygen Fluoride (O2F); Thermodynamic and Transport Properties of Carbohydrates and their Monophosphates: The Pentoses and Hexoses; Evaluated Kinetic and Photochemical Data for Atmospheric Chemistry: Supplement III. (Copyright (c) by the U.S. Secretary of Commerce, 1989.)

900,449
PB89-222533 (Not available NTIS)
International Union of Pure and Applied Chemistry, London (England). Projects Centre.
Thermal Conductivity of Nitrogen and Carbon Monoxide in the Limit of Zero Density.
Quarterly rept.
J. Millat, and W. A. Wakeham. c1989· 17p
Prepared in cooperation with American Chemical Society, Washington, DC., and American Inst. of Physics, New York. Sponsored by National Inst. of Standards and Technology, Gaithersburg, MD.
Included in Jnl. of Physical and Chemical Reference Data, v18 n2 p565-581 1989.

Keywords: *Thermal conductivity, *Nitrogen, *Carbon monoxide, Density(Mass/Volume), Tables(Data), Reaction kinetics, Numerical analysis, Experimental design, Cross sections, Specific heat, Thermodynamic properties.

The paper presents accurate representations for the thermal conductivity of the diatomic gases nitrogen and carbon monoxide in the limit of zero density. These gases were studied because they have nearly the same molecular mass and viscosities. In contrast, the new analysis confirms that the thermal conductivities of the two gases differ remarkably, especially at low temperatures. The theoretically-based correlations provided are valid for the temperature range 220-2100 K.and have associated uncertainties of + or - 1% between 300 and about 500 K, rising to + or - 2.5% at the low- and high-temperature extremes. A comparison with some empirical and semiempirical correlations is given. (Copyright (c) by the U.S. Secretary of Commerce, 1989.)

900,450
PB89-222541 (Not Available NTIS)
National Inst. of Standards and Technology (NEL), Boulder, CO. Thermophysics Div.
Thermophysical Properties of Methane.
Quarterly rept.
D. G. Friend, J. F. Ely, and H. Ingham. c1989, 56p
Prepared in cooperation with American Chemical Society, Washington, DC., and American Inst. of Physics, New York.
Included in Jnl. of Physical and Chemical Reference Data, v18, n2 p583-638 1989.

Keywords: *Thermophysical properties, *Methane, Numerical analysis, Experimental design, Liquid phases, Equations of state, Specific heat, Graphs(Charts), Transport properties, Pressure, Thermal conductivity, Viscosity, Comparison, Virial coefficients.

New correlations for the thermophysical properties of fluid methane are presented. The correlations are based on a critical evaluation of the available experimental data and have been developed to represent these data over a broad range of the state variables. Estimates for the accuracy of the equations and comparisons with measured properties are given. The reasons for the new study of methane include significant new and more accurate data, and improvements in the correlation functions which allow increased accuracy of the correlations especially in the extended critical region. For the thermodynamic properties, a classical equation for the molar Helmholtz energy, which contains terms multiplied by the exponential of the quadratic and quartic powers of the system density, is used. Tables of coefficients and equations are presented to allow the calculation of these and other thermodynamic quantities. (Copyright (c) by the U.S. Secretary of Commerce, 1989.)

900,451
PB89-222558 (Not Available NTIS)
Idaho Univ., Moscow. Center for Applied Thermodynamic Studies.
Thermodynamic Properties of Argon from the Triple Point to 1200 K with Pressures to 1000 MPa.
Quarterly rept.
R. B. Stewart, and R. T. Jacobsen. c1989, 160p
Prepared in cooperation with American Chemical Society, Washington, DC., and American Inst. of Physics, New York. Sponsored by National Inst. of Standards and Technology, Gaithersburg, MD.
Included in Jnl. of Physical and Chemical Reference Data, v18 n2 p639-798 1989.

Keywords: *Thermodynamic properties, *Argon, Helmholtz free energy, Pressure, Equations of state, Enthalpy, Entropy, Specific heat, Tables(Data), Experimental design, Numerical analysis, Comparison, Liquid phases, Vapor phases, *Triple point.

A new thermodynamic property formulation for argon is presented. The formulation includes a fundamental equation explicit in Helmholtz energy, a vapor pressure equation, and estimating functions for the densities of saturated liquid and vapor states. The coefficients of the fundamental equation and ancillary functions were determined by a weighted least-squares fit of selected experimental data using a statistical procedure to select the terms for the equation most appropriate for the representation of the data. The fundamental equation is valid for liquid and vapor phases except near the critical point. Comparisons between the data used to determine the fundamental equation and values calculated from the formulation are given to verify the accuracy of the fundamental equation. Tables of thermodyanmic properties of argon calculated with the formulation presented here are given for fluid states within the range of validity of the correlation. (Copyright (c) by the U.S. Secretary of Commerce, 1989.)

900,452
PB89-222566 (Not Available NTIS)
Los Alamos National Lab., NM.

Thermodynamic Properties of Dioxygen Difluoride (O2F2) and Dioxygen Fluoride (O2F).
Quarterly rept.
J. L. Lyman. c1989, 9p
Prepared in cooperation with American Chemical Society, Washington, DC., and American Inst. of Physics, New York. Sponsored by National Inst. of Standards and Technology, Gaithersburg, MD.
Included in Jnl. of Physical and Chemical Reference Data, v18 n2 p799-807 1989.

Keywords: *Thermodynamic properties, Gibbs free energy, *Specific heat, Entropy, Enthalpy, Tables(Data), Fluorination, Experimental data, Numerical analysis, *Difluoride/dioxy, *Fluoride/dioxy, Oxygluorides.

Recent spectroscopic and chemical kinetic studies have provided sufficient data for construction of reliable thermodynamic tables for both dioxygen defluoride (O2F2; Chemical Abstracts Registry Number, 7783-44-0) and dioxygen fluoride (O2F; Chemical Abstracts Registry Number, 15499-23-7). The paper contains those tables for these species in both SI units (0.1 MPa standard state) and cal K mol units (1.0 atm standard state). The experimental basis includes three recent assignments of the fundamental vibrational frequencies for O2F2, a new set of rotational constants for O2F, an enthalpy change for dissociation of O2F2, and an updated standard enthalpy of formation for O2F2. (Copyright (c) by the U.S. Secretary of Commerce, 1989.)

900,453
PB89-222574 (Not available NTIS)
National Inst. of Standards and Technology (NML), Gaithersburg, MD. Chemical Thermodynamics Div.
Thermodynamic and Transport Properties of Carbohydrates and Their Monophosphates: The Pentoses and Hexoses.
Quarterly rept.
R. N. Goldberg, and Y. B. Tewari. c1989, 72p
Prepared in cooperation with American Chemical Society, Washington, DC., and American Inst. of Physics, New York.
Included in Jnl. of Physical and Chemical Reference Data, v18 n2 p809-880 1989.

Keywords: *Thermodynamic properties, *Transport properties, *Pentoses, *Hexoses, *Carbohydrates, *Organic phosphates, Gibbs free energy, Specific heat, Viscosity, Phase diagrams, Enthalpy, Entropy, Tables(Data), Reviews, Transition temperature.

The review contains recommended values of the thermodynamic and transport properties of the five and six membered ring carbohydrates and their phosphates in both the condensed and aqueous phases. Equilibrium data, enthalpies, heat capacities, and entropies have been collected from the literature. The accuracy of these-data have been assessed, adjusted to 298.15 K and to common standard state, and entered into a catalog of thermochemical reactions. The solution of the reaction catalog yields a set of recommended values for the formation properties of these substances. The volumetric data have also been critically evaluated. Recommended values are presented for standard state molar volumes and the temperature and pressure derivatives of the molar volume, i.e., the expansivity and the compressibility. The excess property data of aqueous solutions of these substances have been correlated to yield recommended values of the parameters of the virial expansion model used to represent the data. The transport data considered here includes both viscosity and diffusion data of aqueous solutions of the carbohydrates. The available phase diagram data and transition temperatures are summarized. (Copyright (c) by the U.S. Secretary of Commerce, 1989.)

900,454
PB89-222582 (Not Available NTIS)
California Univ., Riverside. Statewide Air Pollution Research Center.
Evaluated Kinetic and Photochemical Data for Atmospheric Chemistry. Supplement 3.
Quarterly rept.
R. Atkinson, D. L. Baulch, R. A. Cox, R. F. Hampson, J. A. Kerr, and J. Troe. c1989, 217p
Prepared in cooperation with American Chemical Society, Washington, DC., American Inst. of Physics, New York, and Leeds Univ. (England). Sponsored by National Inst. of Standards and Technology, Gaithersburg, MD.

54

Included in Jnl. of Physical and Chemical Reference Data, v18 n2 p881-1097 1989.

Keywords: *Reaction kinetics, *Photochemical reactions, Tables(Data), Enthalpy, Thermodynamic properties, Dissociation, *Atmospheric chemistry.

The paper updates and extends previous critical evaluations of the kinetics and photochemistry of gas phase chemical reactions of neutral species involved in atmosphere chemistry (J. Phys. Chem. Ref. Data 9, 295 (1980); 11, 327 (1962); 13, 1259 (1984)). The work has been carried out by the authors under the auspecies of the IUPAC Subcommittee on Gas Phase Kinetic Data Evaluation for Atmospheric Chemistry. Data sheets have been prepared for 360 thermal and photochemical reactions, containing summaries of the available experimental data with notes giving details of the experimental procedures. For each reaction, a preferred value of the rate coefficient at 298 K is given together with a temperature dependence where possible. The selection of the preferred value is discussed; and estimates of the accuracies of the rate coefficients and temperature coefficients have been made for each reaction. The data sheets are intended to provide the basic physical chemical data needed as input for calculations which model atmospheric chemistry. A table summarizing the preferred rate data is provided, together with an appendix listing the available data on enthalpies of formation of the reactant and product species. (Copyright (c) by the U.S. Secretary of Commerce, 1989.)

900,455
PB89-226559 . Not available NTIS
American Chemical Society, Washington, DC.
Journal of Physical and Chemical Reference Data, Volume 18, Number 1, 1989.
Quarterly rept.
D. R. Lide. c1989, 567p
See also PB89-226567 through PB-226609, PB89-227797, and PB89-222525. Prepared in cooperation with American Inst. of Physics, New York. Sponsored by National Inst. of Standards and Technology, Gaithersburg, MD.
Available from American Chemical Society, 1155 16th St., NW, Washington, DC 20036.

Keywords: *Thermodynamic properties, Water, Electrode potentials, Temperature coefficient, Cross sections, Oxygen, Thermal conductivity, Molecular energy levels, Excitation, Reaction kinetics, X rays, Refrigerants, Pressure, Aromatic polycyclic hydrocarbons, Isomers, Pyrenes, Tables(Data), Metal complexes, Helium ions, Hydrogen ions, Coronenes, Electron-molecule collisions, Photon-molecule collisions, Ion-atom collisions, K shell.

Contents: Standard Electrode Potentials and Temperature Coefficients in Water at 298.15 K; Cross Sections for Collisions of Electrons and Photon with Oxygen Molecules; Thermal Conductivity of Refrigerants in a Wide Range of Temperature and Pressure; Standard Chemical Thermodynamic Properties of Polycyclic Aromatic Hydrocarbons and Their Isomer Groups. II. Pyrene Series, Naphthopyrene Series, and Coronene Series; Cross Sections for K-Shell X-Ray Production by Hydrogen and Helium Ions in Elements from Beryllium to Uranium; Rate Constants for the Quenching of Excited States of Metal Complexes in Fluid Solution. (Copyright (c) by the U.S. Secretary of Commerce, 1989.)

900,456
PB89-226567 (Not available NTIS)
Southwest Texas State Univ., San Marcos. Dept. of Chemistry.
Standard Electrode Potentials and Temperature Coefficients in Water at 298.15 K.
Quarterly rept.
S. G. Bratsch. c1989, 21p
Prepared in cooperation with American Chemical Society, Washington, DC., and American Inst. of Physics, New York. Sponsored by National Inst. of Standards and Technology, Gaithersburg, MD.
Included in Jnl. of Physical and Chemical Reference Data, v18 n1 p1-21 1989. Available from American Chemical Society, 1155 16th St., NW, Washington, DC 20036.

Keywords: *Electrode potentials, *Temperature coefficient, *Water, Solvents, Electrochemical cells, Tables(Data), Thermodynamic properties, pH, Enthalpy, Entropy, Gibbs free energy.

A great deal of solution chemistry can be summarized in a table of standard electrode potentials of the ele-

ments in the solvent of interest. In the work, standard electrode potentials and temperature coefficients in water at 298.15 K, based primarily on the 'NBS Tables of Chemical Thermodynamic Properties,' are given for nearly 1700 half-reactions at pH = 0.000 and pH = 13.996. The data allow the calculation of the thermodynamic changes and equilibrium constants associated with approximately 1.4 million complete cell reactions over the normal temperature range of liquid water. Estimated values are clearly distinguished from experimental values, and half-reactions involving doubtful chemical species are duly noted. General and specific methods of estimation of thermodynamic quantities are summarized. (Copyright (c) by the U.S. Secretary of Commerce, 1989.)

900,457
PB89-226575 (Not available NTIS)
Institute of Space and Astronautical Science, Tokyo (Japan).
Cross Sections for Collisions of Electrons and Photons with Oxygen Molecules.
Quarterly rept.
Y. Itikawa, A. ichimura, K. Onda, K. Sakimoto, K. Takayanagi, Y. Hatano, M. Hayashi, H. Nishimura, and S. T. Tsurubuchi. c1989, 20p
Prepared in cooperation with Tokyo Inst. of Tech. (Japan), Nagoya Inst. of Tech. (Japan), and American Chemical Society, Washington, DC. Sponsored by National Inst. of Standards and Technology, Gaithersburg, MD.
Included in Jnl. of Physical and Chemical Reference Data, v18 n1 p23-42 1989. Available from American Chemical Society, 1155 16th St., NW, Washington, DC 20036.

Keywords: *Cross sections, *Oxygen, Tables(Data), Graphs(Charts), Photochemical reactions, Ionization, Dissociation, Spectrum analysis, Thermodynamic properties, Molecular rotation, Molecular vibration, *Electron-molecule collisions, *Photon-molecule collisions.

Data have been compiled on the cross sections for collisions of electrons and photons with oxygen molecules (O2). For electron collisions, the processes included are: total scattering, elastic scattering, momentum transfer, excitations of rotational, vibrational, and electronic states, dissociation, ionization, and attachment. Ionization and dissociation processes are considered for photon impact. Cross-section data selected are presented graphically. Spectroscopic and other properties of the oxygen molecule are summarized for understanding of the collision processes. The literature was surveyed through August 1987, but some more recent data are included when available to the authors. (Copyright (c) by the U.S. Secretary of Commerce, 1989.)

900,458
PB89-226583 (Not available NTIS)
Stuttgart Univ. (Germany, F.R.).
Thermal Conductivity of Refrigerants in a Wide Range of Temperature and Pressure.
Quarterly rept.
R. Krauss, and K. Stephan. c1989, 32p
Prepared in cooperation with American Chemical Society, Washington, DC., and American Inst. of Physics, New York. Sponsored by National Inst. of Standards and Technology, Gaithersburg, MD.
Included in Jnl. of Physical and Chemical Reference Data, v18 n1 p43-76 1989. Available from American Chemical Society, 1155 16th St., NW, Washington, DC 20036.

Keywords: *Thermal conductivity, *Refrigerants, *Temperature, *Pressure, Thermodynamic properties, Tables(Data), Graph(Charts), Fluorohydrocarbons, Dichlorodifluoromethane, Gas dynamics, Comparison, Freons, Ethane/dichloro-tetrafluoro, Cyclobutane/fluoro, Ethane/trichloro-trifluoro.

Thermal conductivities of refrigerant 12 (dichlorodifluoromethane), refrigerant 114 (1,1,2-trichloro-1,2,2-trifluoroethane), refrigerant 114 (1,2-dichloro-1,1,2,2-tetrafluoroethane), and refrigerant C318 (Perfluorocyclobutane) were critically evaluated and correlated on the basis of a comprehensive literature survey. Recommended values were established for a wide range of temperatures and pressures, extending up to three times the critical density and excluding the critical region. Using the residual concept, a dilute-gas function and an excess function of simple form were developed for each refrigerant. The average accuracy obtained is approximately 6%. (Copyright (c) by the U.S. Secretary of Commerce, 1989.)

900,459
PB89-226591 (Not available NTIS)
Massachusetts Inst. of Tech., Cambridge. Dept. of Chemistry.
Standard Chemical Thermodynamic Properties of Polycyclic Aromatic Hydrocarbons and Their Isomer Groups. 2. Pyrene Series, Naphthopyrene Series, and Coronene Series.
Quarterly rept.
R. A. Alberty, M. B. Chung, and A. K. Reif. c1989, 31p
Prepared in cooperation with American Chemical Society, Washington, DC., and American Inst. of Physics, New York. Sponsored by National Inst. of Standards and Technology, Gaithersburg, MD.
Included in Jnl. of Physical Chemical Reference Data, v18 n1 p77-110 1989. Available from American Chemical Society, 1155 16th St., NW, Washington, D 20036.

Keywords: *Thermodynamic properties, *Aromatic polycyclic hydrocarbons, *Isomers, *Pyrenes, Standards, Enthalpy, Specific heat, Gibbs free energy, Entropy, Tables(Data), *Coronenes.

The tables in our first paper on polycyclic aromatic hydrocarbons (J. Phys. Chem. Ref. Data 17, 241 (1988)) have been extended by calculating thermodynamic properties for the first four isomer groups in the pyrene series, the first three isomer groups in the naphthopyrene series, and the first three isomer groups in the coronene series. Successive isomer groups in each series differ by C4H2. Since chemical thermodynamic properties are known for only a limited number of polycyclic aromatic hydrocarbons, the properties of individual species have been estimated using Benson group values of Stein and Fahr for temperatures from 298.15 to 3000 K. Values of C(sub p), S, delta(sub f), H, and delta(sub f)G have been calculated in joules for a standard state pressure of 1 bar. The chemical thermodynamic properties of the individual species have also been calculated. The isomer group values provide a basis for extrapolating to higher carbon numbers where it is not feasible to consider individual molecular species. (Copyright (c) by the U.S. Secretary of Commerce, 1989.)

900,460
PB89-226609 (Not available NTIS)
East Carolina Univ., Greenville, NC.
Cross Sections for K-Shell X-ray Production by Hydrogen and Helium Ions in Elements from Beryllium to Uranium.
Quarterly rept.
G. Lapicki. c1989, 108p
Prepared in cooperation with American Chemical Society, Washington, DC., and American Inst. of Physics, New York. Sponsored by National Inst. of Standards and Technology, Gaithersburg, MD.
Included in Jnl. of Physical Chemical Reference Data, v18 n1 p111-218 1989. Available from American Chemical Society, 1155 16th St., NW, Washington, DC 20036.

Keywords: *x rays, *Scattering cross sections, Thermodynamic properties, Tables(Data), Fluorescence, Ionization, Numerical analysis, Data processing, Experimental design, Alkaline earth metals, Actinide series, *K shell, *Ion-atom collisions, *Hydrogen ions, *Helium ions.

Experimental cross sections for K-shell x-ray production by hydrogen and helium ions (Z1 = 1,2) in target atoms from beryllium to uranium (Z2 = 4-92) are tabulated as compiled (7418 cross sections) from the literature (161 references were found) with the search for the data terminated in January 1986. These cross sections are compared with predictions of the first Born approximation and ECPSSR theory for inner-shell ionization. The ECPSSR accounts for the energy loss (E) and Coulomb deflection (C) of the projectile ion as well as for the perturbed stationary state (PSS) and relativistic (R) nature of the target's inner-shell electron. While the first Born approximation generally overestimates the data by orders of magnitude, the ECPSSR theory is confirmed to be, on the average, in agreement with the experiment to within 10%-20%. (Copyright (c) by the U.S. Secretary of Commerce, 1989.)

900,461
PB89-227797 . (Not available NTIS)
Boston Univ., MA. Dept. of Chemistry.

CHEMISTRY

Physical & Theoretical Chemistry

Rate Constants for the Quenching of Excited States of Metal Complexes in Fluid Solution.
Quarterly rept.
M. Z. Hoffman, F. Bolletta, L. Moggi, and G. L. Hug.
c1989, 324p
Prepared in cooperation with Bologna Univ. (Italy), Notre Dame Univ., IN. Radiation Lab., and American Chemical Society, Washington, DC. Sponsored by National Inst. of Standards and Technology, Gaithersburg, MD.
Included in Jnl. of Physical Chemical Reference Data, v18 n1 p219-543 1989. Available from American Chemical Society, 1155 16th St., NW, Washington, DC 20036.

Keywords: *Reaction kinetics, *Metal complexes, Molecular energy levels, Thermodynamic properties, Excitation, Solutions, Quenching media, Tables(Data).

The rate constants for the quenching of the excited states of metal ions and complexes in homogeneous fluid solution are reported in this compilation. Values of K(sub q) for dynamic, collisional processes between excited species and quenchers have been critically evaluated, and are presented with the following information, among others, from the original publications, when available: description of the solution medium, temperature at which K(sub q) was determined, experimental method, range of quencher concentration used, lifetime of the exited state in the absence of quencher, activation parameters, quenching mechanism. Data collection is complete through the end of 1986, and covers the coordination compounds of 26 metals, including the ions and complexes of the inner- and outer-transition metals, and porphyrin complexes of nontransition metals. The introduction to the work contains a discussion of the conceptual background to quenching, including a general treatment of the kinetics, an explanation of the tables, and a list of recent review articles. Uncommon kinetics mechanisms and equations, used to obtain the reported values of K(sub q) are discussed in detail as part of the notes to the tables. Indexes of excited states, quenchers, and authors are appended. (Copyright (c) by the U.S. Secretary of Commerce, 1989.)

900,462

PB89-227896 Not available NTIS
National Inst. of Standards and Technology (NML), Gaithersburg, MD. Chemical Thermodynamics Div.
Second Virial Coefficients of Aqueous Alcohols at Elevated Temperatures: A Calorimetric Study.
Final rept.
D. G. Archer. 1989, 8p
Pub. in Jnl. of Physical Chemistry 93, n1 p5272-5279, 29 Jun 89.

Keywords: *Gibbs free energy, *Enthalpy, *Dilution, *Cyclohexanols, *Sugar alcohols, *Water, Heat measurement, Butanols, Propanols, Specific heat, Antifreezes, High temperature tests, Gas laws, Reprints.

Enthalpies of dilution of cyclohexanol (aq) to 448 K and of myo-inositol (aq) and of cyclohexanol + myo-inositol (aq) to 398 K are reported. The results, along with enthalpies of dilution for 1-butanol (aq) to 448 K and for 2-methyl-2-propanol (aq) to 423 K, were combined with freezing point depression measurements and ambient-temperature enthalpy of dilution and heat capacity measurements in order to provide the excess Gibbs energy for the aqueous solutes from 273 to 523 K. The excess Gibbs energies were then used to provide parameters for an additivity scheme that permits an approximation of the excess Gibbs energy, and thus the solute and solvent activity coefficients, for dilute aqueous alcohols for which high-temperature data do not exist. The excess Gibbs energies were also used to estimate the aqueous-solution second virial coefficients of the alcohols and of cyclohexane (aq), 1-butane (aq), and 2-methylpropane (aq) in the McMillan-Mayer convention. These virial coefficients for the aqueous hydrocarbons, when compared to gas-phase virial coefficients for the hydrocarbons, suggest that the effect of low-temperature (298 K) water is to lessen the pairwise attraction of aqueous hydrocarbon over that found in the gas phase at the same temperature and that the lessening of the pairwise attraction diminishes at higher temperatures.

900,463

PB89-227912 Not available NTIS
National Inst. of Standards and Technology (NML), Boulder, CO. Quantum Physics Div.
Absolute Infrared Transition Moments for Open Shell Diatomics from J Dependence of Transition Intensities: Application to OH.
Final rept.
D. D. Nelson, A. Schiffman, D. J. Nesbitt, and D. J. Yaron. 1989, 12p
Grant AFOSR-84-0272
Sponsored by Air Force Office of Scientific Research, Bolling AFB, DC.
Pub. in Jnl. of Chemical Physics 90, n10 p5443-5454, 15 May 89.

Keywords: *Dipole moments, Vibrational spectra, Diatomic molecules, Reprints, *Hydroxyl radicals, Infrared absorption, Herman Wallis effect.

A general approach to the determination of the dipole moment function and of the absolute vibrational transition moments for diatomic molecules is presented. The method utilizes the variation of intensity with J within a vibrational transition, together with permanent dipole moment information, to extract the absolute transition moments. An essential feature of the model is its use of algebraic expressions for calculating vibration-rotation line intensities. These expressions can be rapidly evaluated in a least squares fit which determines the dipole moment function. This approach is general in that it is not limited to (sup 1)Sigma state molecules, nor to the simplest of Hund's case couplings of spin, orbital and mechanical angular momentum. It is also not limited to molecules with essentially linear dipole moment functions. The model is successfully applied to the OH molecule which violates each of these restrictions. In the accompanying work the authors report experimental measurements of relative infrared absorption intensity measurements for OHmu = 1 <- 0 transitions and the extraction of an experimental mu(r) using the approach presented here.

900,464

PB89-227920 Not available NTIS
National Inst. of Standards and Technology (NML), Boulder, CO. Quantum Physics Div.
Dipole Moment Function and Vibrational Transition Intensities of OH.
Final rept.
D. D. Nelson, A. Schiffman, and D. J. Nesbitt. 1989, 11p
Grant AFOSR-84-0272
Sponsored by Air Force Office of Scientific Research, Bolling AFB, DC.
Pub. in Jnl. of Chemical Physics 90, n10 p5455-5465, 15 May 89.

Keywords: *Dipole moments, Vibrational spectra, Electron transitions, Infrared spectroscopy, Reprints, *Hydroxyl radicals, Infrared absorption, Flash kinetic spectroscopy.

The relative intensities of nine pairs of rovibrational transitions of OH in the nu = 1 <- 0 fundamental have been measured by flash kinetic infrared absorption spectroscopy. Each pair of transitions originates from a common rotational and spin-orbit state, so that relative intensities are independent of the OH number density and quantum state distribution. The relative intensities are strongly J dependent and this dependence provides detailed information about the shape of the OH dipole moment function, mu(r), and hence the absolute infrared transition strengths. In an accompanying paper, the authors present the theoretical basis for extracting mu(r) for an open shell diatomic like OH from relative infrared intensities and permanent dipole moment measurements and determine the OH dipole moment function.

900,465

PB89-227953 Not available NTIS
National Inst. of Standards and Technology (NML), Boulder, CO. Quantum Physics Div.
Rydberg-Klein-Rees Inversion of High Resolution van der Waals Infrared Spectra: An Intermolecular Potential Energy Surface for Ar+HF (v=1).
Final rept.
D. J. Nesbitt, M. S. Child, and D. C. Clary. 1989, 10p
Grants NSF-CHE86-05970, NSF-PHY86-04504
Sponsored by National Science Foundation, Washington, DC.
Pub. in Jnl. of Chemical Physics 90, n9 p4855-4864, 1 May 89.

Keywords: *Infrared spectroscopy, *Potential energy, *Molecular rotation, *Hydrogen fluoride, *Argon, Molecular energy levels, Intermolecular forces, Molecular vibration, Experimental design, Numerical analysis, Reprints, *Van der Waals forces, *Rydberg-Klein-Rees method, *Laser spectroscopy.

A method is described for extraction of two-dimensional (angular and radial) potential energy surfaces for triatomic rare gas-hydrogen halide van der Waals complexes. The approach relies on extensive J rotational term values obtained by high resolution infrared laser jet spectroscopy for a family of bending vibrational states to deduce the radial and angular dependence of the intermolecular potential. First, effective 1D radial potentials for series of bend states are obtained by rotational RKR analysis of experimentally observed rotational progressions. These 1D potentials, which represent vibrational averages over different bending wave functions, are then inverted to determine the radially dependent coefficients of a Legendre expansion to the full surface. This relies on adiabatic angular motion with respect to radial degrees of freedom, the validity of which is discussed. This approach is tested with experimental data. The accuracy of the resulting surface is verified by exact quantum bound state calculations which quantitatively reproduce the rovibrational input data, as well as predict the spectroscopic properties of five other vibrational states observed in the Ar+HF(v = 1) system but not used in the fitting procedure.

900,466

PB89-227961 Not available NTIS
National Inst. of Standards and Technology (NML), Boulder, CO. Quantum Physics Div.
Three Dimensional Quantum Reactive Scattering Study of the I + HI Reaction and of the IHI(1-) Photodetachment Spectrum.
Final rept.
G. C. Schatz. 1989, 8p
Grant NSF-CHE87-15581
Sponsored by National Science Foundation, Washington, DC.
Pub. in Jnl. of Chemical Physics 90, n9 p4847-4854, 1 May 89.

Keywords: *Hydrogen iodide, *Iodine, Quantum interactions, Resonance, Reprints, *Reactive scattering, *Photodetachment, Coupled channel theory, Hyperspherical coordinates.

The author presents results of coupled channel hyperspherical reactive scattering calculations on the reaction I + HI -> IH + I using a semiempirical potential surface. Only the J = O partial wave is considered. Franck-Condon factors associated with photodetachment of IHI(1-) have also been calculated, and these show mainly direct scattering threshold behavior at low energies (E<0.30 eV), with the (100) and (200) resonances contributing only slightly. Resonant behavior is dominant at higher energies (0.3-0.4 eV) where the (002) resonance especially contributes.

900,467

PB89-227979 Not available NTIS
National Bureau of Standards (NML), Boulder, CO. Quantum Physics Div.
Spectroscopic Signatures of Floppiness in Molecular Complexes.
Final rept.
D. J. Nesbitt. 1988, 10p
Grants NSF-CHE83-05970, NSF-PHY86-04504
Sponsored by National Science Foundation, Washington, DC.
Pub. in Structure of Small Molecules and Ions, p49-58 1988.

Keywords: *Molecular structure, Molecular spectroscopy, Molecular rotation, Molecular vibration, Rigidity, Reprints, *Molecular complexes, Van der Waal forces, Supersonic jet flow.

The challenge of correctly inferring even the qualitative features of the potential energy hypersurface from spectroscopic measurements is heightened dramatically in studies of weakly bound molecular complexes where large amplitude motion is present. This is especially true for data obtained from low temperature, supersonic expansions where Boltzmann distributions limit the range of internally excited states that can be investigated. To stress this point, the author presents simulated spectra for two model triatomic systems, a 'pinwheel' and a 'hinge,' with nearly flat potentials that support extremely large amplitude internal rotation and bending, respectively. The results indicate that simple eigenvalue analysis of jet cooled molecular spectra in the absence of hyperfine resolution may not be sufficiently sensitive to large amplitude angular motion, and that data from a variety of techniques may prove necessary to assess the degree of molecular rigidity.

900,468
PB89-228001   Not available NTIS
National Bureau of Standards (NEL), Gaithersburg,
MD. Thermophysics Div.
Low-Q Neutron Diffraction from Supercooled D-Glycerol.
Final rept.
H. J. M. Hanley, G. C. Straty, C. J. Glinka, and J. B.
Hayter. 1987, 10p
Sponsored by Department of Energy, Washington, DC.
Div. of Materials Sciences, and Martin Marietta Energy
Systems, Inc., Oak Ridge, TN.
Pub. in Molecular Physics 62, n5 p1165-1174 1987.

Keywords: *Glycerol, *Neutron diffraction, Supercooling, Scattering cross sections, Hydrogen bonds, Reprints, *Structure factors, Vanadium fluorides.

Neutron diffraction measurements of the structure
factor of supercooled liquid D-glycerol are reported for
the range 0.03 approx = or < Q = or < 4.0/A at
temperatures between 300 and 175 K. It was found
that the structure is essentially unchanged over this
entire range. The authors expected some evidence of
increased hydrogen bonding at the lower temperatures, but no direct evidence for any strong oriented
density correlation at low-Q due to a hydrogen bond is
observed. To provide a contrast, the structure factor of
vanadium pentafluoride is discussed briefly because
vanadium pentafluoride is known to form fluorine-fluo-
rine bridges in the dense state. The authors equate
loosely this bridging with a hydrogen bond. In that
compound there is evidence for F-F bonding at low-Q.
Small angle Placzek corrections to the D-glycerol scattering cross section data are estimated.

900,469
PB89-228043   Not available NTIS
National Bureau of Standards (NEL), Boulder, CO.
Thermophysics Div.
**Shear-Induced Angular Dependence of the Liquid
Pair Correlation Function.**
Final rept.
H. J. M. Hanley, J. C. Rainwater, and S. Hess. 1987,
8p
Sponsored by Department of Energy, Washington, DC.
Pub. in Physical Review A 36, n4 p1795-1802 Aug 87.

Keywords: *Liquids, Couette flow, Relaxation time,
Spherical harmonics, Tensors, Simulation, Reprints,
Correlation functions, Molecular dynamics.

A formal expansion in spherical harmonics or Cartesian tensors of the pair correlation function of a liquid
subjected to a shear rate is discussed. Expressions for
the coefficients to tensor rank 4 are evaluated via a
nonequilibrium molecular-dynamics simulation of an
inverse twelve soft-sphere liquid undergoing Couette
flow. It is shown that the expansion converges slowly if
the product tau gamma 0.05, where tau is the Maxwell
relaxation time and gamma is the shear rate. Further,
the fourth-rank coefficient that represents cubic symmetry is significant for the model system. The microstructure of a shear liquid is demonstrated by intensity
plots of particles around a given central particle. Expressions were derived for the expansion coefficients
using a relaxation-time model, and the comparison between them and simulations is generally very good.

900,470
PB89-228092   Not available NTIS
National Bureau of Standards (NML), Boulder, CO.
Quantum Physics Div.
**Quantum Mechanical Calculations on the Ar(1+)
+ N2 Charge Transfer Reaction.**
Final rept.
D. C. Clary, and D. M. Sonnenfroh. 1989, 8p
Pub. in Jnl. of Chemical Physics 90, n3 p1686-1693
Feb 89.

Keywords: *Nitrogens, Reprints, *Change-exchange
reactions, *Argon ions, Quantum mechanics, Rotational states, Born approximations.

Calculations of cross sections for the charge transfer
reaction Ar(1+)(doublet P(3/2)) + N2(v=O,j) -> Ar
+ N2+(1+)(v' = 1,j) are reported for thermal collision energies. A three-dimensional quantum-mechanical method is used in which separate rotational close-
coupling calculations are performed for the Ar(1+) +
N2(v=0) -> Ar + N2(1+)(v' = 0) and Ar(2+) +
N2(v=1) -> Ar + N2(1+)(v'=1) channels, and the
cross sections for the v=0 -> v'=1 channel are computed using a coupled channel-distorted wave Born
approximation. Potential energy surfaces and cou-

plings are taken from ab initio data. The predicted rotational product distributions for N2(1+)(v'=1,j') agree
fairly well with those measured in a molecular beam
laser-induced fluorescence experiment.

900,471
PB89-228399   Not available NTIS
National Inst. of Standards and Technology (NML),
Boulder, CO. Quantum Physics Div.
Infrared Spectra of Nitrous Oxide-HF Isomers.
Final rept.
C. M. Lovejoy, and D. J. Nesbitt. 1989, 10p
Grants NSF-PHY86-04504, NSF-CHE86-05970
Sponsored by National Science Foundation, Washing-
ton, DC.
Pub. in Jnl. of Chemical Physics 90, n9 p4671-4680, 1
May 89.

Keywords: *Infrared spectroscopy, *Nitrogen
oxide(N2O), *Hydrogen fluoride, Dissociation, Complex compounds, Hydrogen binds, Molecular isomer-
ism, Molecular vibration, Reprints, *Laser spectroscopy.

Two spectroscopically distinct isomers of a hydrogen
bonded complex between nitrous oxide and hydrogen
fluoride are observed by direct infrared laser absorption detection in a slit supersonic expansion. The linear
isomer FH-NNO contains a relatively rigid hydrogen
bond to the nitrogen end of NNO. The bent isomer
NNO-HF has a stronger hydrogen bond to the oxygen
end of NNO, but this bond is characterized by a softer
bending potential and thus the complex exhibits evidence of large amplitude bending motion. Rapid vibrational predissociation, as determined from the homogeneous broadening of the rovibrational absorption
structure, is evidenced in both isomers. The linear
isomer exhibits predissociation lifetimes which show
structure as a function of the upper J' rotational level,
including narrow resonances which suggest excitation
of NNO fragment vibrational modes.

900,472
PB89-228407   Not available NTIS
National Bureau of Standards (NML), Boulder, CO.
Quantum Physics Div.
**Dynamical Simulation of Liquid- and Solid-Metal
Self-Sputtering.**
Final rept.
W. L. Morgan. 1989, 5p
Pub. in Jnl. of Applied Physics 65, n3 p1265-1269, 1
Feb 89.

Keywords: *Liquid metals, Surfaces, Simulation, Reprints, *Self sputtering, Molecular dynamics.

Molecular dynamics simulations of self-sputtering are
performed using the recent picture of a stratified liquid-
metal surface as a model. These results are compared
to those obtained from a liquid model having uniformly
distributed atoms and a crystalline solid model. The
stratified liquid-metal model shows an enhanced low-
energy sputter yield, which falls below those of the
other models for ion-impact energies above several
hundred electron volts. These results are discussed in
light of various published measurements of sputter
yields of metals in their liquid and solid phases.

900,473
PB89-228415   Not available NTIS
National Bureau of Standards (NML), Boulder, CO.
Quantum Physics Div.
**Calculation of Vibration-Rotation Spectra for Rare
Gas-HCl Complexes.**
Final rept.
D. C. Clary, and D. J. Nesbitt. 1989, 14p
Grants NSF-PHY86-04504, NSF-CHE86-05970
Sponsored by National Science Foundation, Washing-
ton, DC.
Pub. in Jnl. of Chemical Physics 90, n12 p7000-7013,
15 Jun 89.

Keywords: *Complex compounds, *Rotational spectra, *Vibrational spectra, *Hydrogen chloride, Infrared
spectra, Schrodinger equation, Computation, Reprints,
*Chemical complexes, *Argon complexes, *Krypton
complexes, *Neon complexes, *Xenon complexes,
Van der Waals forces.

Calculations are described of spectra for the excitation
of the bending and stretching vibration-rotational
energy levels in the van der Waals complexes of HCl
with the rare gases Ne, Ar, Kr, and Xe. The calculations are performed using a basis set method, with distributed Gaussian functions being employed for the coordinate associated with the stretching of the rare gas

atom. Intensities of combination and fundamental transitions for each of the low frequency modes are calculated for total angular momentum up to J = 25. Surprisingly large intensities are predicted for transitions
to states with multiple vibrations excited in the bending
mode. Promising comparisons are obtained with infrared spectra measured recently for the complexes of
HCl with Ne and Ar at low temperatures.

900,474
PB89-228480   Not available NTIS
National Inst. of Standards and Technology (NEL),
Gaithersburg, MD. Semiconductor Electronics Div.
**Multiple Scattering in the X-ray Absorption Near
Edge Structure of Tetrahedral Germanium Gases.**
Final rept.
C. E. Bouldin, D. A. McKeown, R. A. Forman, J. J.
Ritter, and G. Bunker. 1989, 3p
See also PB89-146922.
Pub. in Physica B 158, p362-364 1989.

Keywords: *Germanium halides, *Germanium hydrides, *X-ray absorption, *Molecular structure,
Gases, Electron scattering, Reprints.

X-ray absorption fine structure (XAFS) measurements
of GeCl4, GeH3Cl, and GeH4 are reported. Since
wide-angle multiple scattering involving H atoms is
negligible, the single and multiple scattering (MS) are
isolated. Components terms in the XAFS of GeCl4 by
comparison of the spectra of the three compounds. It
is found that multiple scattering is nowhere dominant
over single scattering (SS), although within 15 eV of
the edge the two are comparable in size. However, the
multiple scattering damps out very quickly with increasing energy above the absorption edge. Beyond
40 eV past the edge the MS/SS ratio is less than 0.06.
The calculations are found to be in qualitative agreement with the experiment, but overestimate the size
and energy range of the MS. The results suggest that
SFS data in the range 1 < K < 3 A sup -1 can be
analyzed in an SS picture in many cases, as long as
good standard compounds are used, and calculations
are used to estimate possible errors due to neglect of
MS. The first evidence of single scattering observed
from H atoms is also reported.

900,475
PB89-228514   Not available NTIS
National Inst. of Standards and Technology (NEL),
Gaithersburg, MD. Molecular Spectroscopy Div.
**Infrared Spectrum of the 1205-cm(-1) Band of
HNO3.**
Final rept.
A. G. Maki. 1989, 4p
Sponsored by National Aeronautics and Space Admin-
istration, Washington, DC.
Pub. in Jnl. of Molecular Spectroscopy 136, p105-108
1989.

Keywords: *Infrared spectroscopy, *Nitric acid, Molecular rotation, Molecular vibration, Reprints.

The high-resolution spectrum of the 1205/cm band of
HNO3 has been measured. It is shown that this is an
A-type band and must be due to the nu8 + nu9 combination. The band is relatively unperturbed and effective rovibrational constants are given that can be used
to calculate the spectrum between 1183 and 1225/
cm.

900,476
PB89-228555   Not available NTIS
National Inst. of Standards and Technology (NEL),
Gaithersburg, MD. Thermophysics Div.
**Capillary Waves of a Vapor-Liquid Interface Near
the Critical Temperature.**
Final rept.
J. V. Sengers, and J. M. J. van Leeuwen. 1989, 10p
Pub. in Physical Review A 39, n12 p6346-6355, 15 Jun
89.

Keywords: *Critical temperature, Liquid phases,
Vapors, Interfaces, Interfacial tension, Reflectivity, Reprints, *Capillary waves, Surface tension.

An attempt is made to develop the picture of an interface near the critical temperature as an intrinsic Fisk-
Widom interface broadened by capillary waves. The
authors propose a method for determining the free parameters appearing in the capillary-wave theory by requiring that the capillary waves smoothly renormalize
the surface tension from the bare to the experimental
value. They evaluate the effect of the capillary waves
on the width of the interface and make a comparison

with experimental reflectivity measurements obtained by Wu and Webb for SF6 and also with the renormalization results of Jasnow and Rudnick.

900,477
PB89-228563   Not available NTIS
National Bureau of Standards (NEL), Boulder, CO.
Thermophysics Div.
Improved Conformal Solution Theory for Mixtures with Large Size Ratios.
Final rept.
M. L. Huber, and J. F. Ely. 1987, 17p
Sponsored by Department of Energy, Washington, DC.
Pub. in Fluid Phase Equilibria 37, p105-121 1987.

Keywords: *Mixtures, *Perturbation theory, Statistics, Molecules, Potential theory, Reprints, *Lennard-Jones gas, Gas viscosity, Gas density.

Previous studies on model Lennard-Jones systems have shown that conformal solution theories tend to fail when mixtures contain molecules with large size differences. This failure can be attributed to deficiencies in the underlying mean density approximation (MDA) for the mixture's radial distribution functions which was originally proposed by Leland and coworkers. An improved mean density approximation is presented which is obtained from a straightforward application of statistical mechanical perturbation theory. Comparisons of the new theory with simulation PVT data for systems with large size ratios show it to be far superior to van der Waals one-fluid and MDA n-fluid theories and an improvement upon hard sphere expansion and perturbation theories as well. Comparisons are also made with chemical potential simulation data, and for these data the new theory is competitive with perturbation theory.

900,478
PB89-229181   Not available NTIS
National Inst. of Standards and Technology (NEL), Boulder, CO. Thermophysics Div.
PVT Relationships in a Carbon Dioxide-Rich Mixture with Ethane.
Final rept.
G. J. Sherman, J. W. Magee, and J. F. Ely. 1989, 13p
Pub. in International Jnl. of Thermophysics 10, n1 p47-59 Jan 89.

Keywords: *Thermochemistry, *Carbon dioxide, *Ethane, Mixtures, Temperature, Isotherms, Pressure, Volume, Reprints.

Comprehensive isochoric PVT measurements have been obtained for the system (0.99 CO2 + 0.01 C2H6). The range of state points studied includes those with densities from 2 to 24 mol times dm(sup -3), temperatures from 245 to 400 K, and pressures to 35 MPa. Extensive comparisons have been made with two predictive conformal solution models, one which uses the 32-term BWR-type equation of Stewart and Jacobsen as a reference and the other using the newer Schmidt-Wagner functional form. Results obtained with the Schmidt-Wagner equation are better in the near-critical region owing to the flatter critical isotherm associated with this functional form.

900,479
PB89-230288   Not available NTIS
National Inst. of Standards and Technology (NML), Gaithersburg, MD. Molecular Spectroscopy Div.
Structure of the CO2-CO2-H2O van der Waals Complex Determined by Microwave Spectroscopy.
Final rept.
K. I. Peterson, R. D. Suenram, and F. J. Lovas. 1989, 7p
Pub. in Jnl. of Chemical Physics 90, n11 p5964-5970, 1 Jun 89.

Keywords: *Molecular structure, *Microwave spectroscopy, *Van der Waals equation, *Carbon monoxide, *Water, *Deuterium compounds, Molecular rotation, Dipole moments, Chemical bonds, Reprints, *Fourier transform spectroscopy.

Rotational spectra of CO2-CO2-H2O, CO2-CO2-D2O, (13)CO2-(13)CO2-H2O and CO2-CO2-H2(18)O have been measured using a pulsed-molecular-beam Fabry-Perot Fourier-transform microwave spectrometer. An asymmetric top spectrum is observed with rotational constants. The dipole moment was obtained. The orientation of the CO2 subunits in CO2-CO2-H2O is very similar to that observed in CO2-CO2 although the C-C bond length is 0.19 A shorter in the trimer. The

hydrogens of the H2O subunit are directed away from the CO2-CO2 plane although their angular orientation around the b axis is not well determined.

900,480
PB89-230296   Not available NTIS
National Inst. of Standards and Technology (NML), Gaithersburg, MD. Molecular Spectroscopy Div.
Infrared Spectrum of Sodium Hydride.
Final rept.
A. G. Maki, and W. B. Olson. 1989, 6p
Pub. in Jnl. of Chemical Physics 90, n12 p6887-6892 Jun 89.

Keywords: *Infrared spectroscopy, *Sodium hydrides, Molecular rotation, Molecular vibration, Molecular spectroscopy, Reprints.

The infrared spectrum of gaseous NaH from 886 to 1245/cm has been measured with a resolution of 0.015/cm at temperatures between 670 to 720 C. The v = 1 <- 0, 2 <- 1, and 3 <- 2 transitions have been observed and combined with rotational transitions measured by others to obtain appropriate rovibrational constants and Dunham potential constants. The Herman-Wallis intensity effect has been measured in order to estimate a transition moment of 0.31 + or - 0.05 D for the v = 1 <- 0 transition.

900,481
PB89-230320   Not available NTIS
National Inst. of Standards and Technology (NML), Gaithersburg, MD. Inorganic Analytical Research Div.
Determination of the Absolute Specific Conductance of Primary Standard KCl Solutions.
Final rept.
Y. C. Wu, K. W. Pratt, and W. F. Koch. 1989, 14p
Pub. in Jnl. of Solution Chemistry 18, n6 p515-528 1989.

Keywords: *Potassium chloride, *Solutions, *Resistance, *Primary standards, Electrolytes, Performance evaluation, Concentration(Composition), Reprints, Specificity.

A determination of the absolute specific conductance of KCl solutions is demonstrated. The measurement is based on the conductance cell with a well defined geometry, having a difference in the removable center tube of accurately measured dimensions. The specific conductance of the solution is obtained from the measured resistances of the cell with and without the center tube and the measured l/A ratio of the center tube. Specific conductances obtained using the cell agree with the previously accepted standards for 0.1 demal and 0.01 demal solutions within 0.02%. Results are also presented for solutions based on molality. The temperature control, bridge, and detector technology used to obtain results of this accuracy are described.

900,482
PB89-231054   Not available NTIS
National Inst. of Standards and Technology (NEL), Gaithersburg, MD. Building Environment Div.
Thermophysical-Property Needs for the Environmentally Acceptable Halocarbon Refrigerants.
Final rept.
M. O. McLinden, and D. A. Didion. 1989, 14p
Pub. in International Jnl. of Thermophysics 10, n3, p563-576, May 89.

Keywords: *Refrigerants, *Thermodynamic properties, *Halohydrocarbons, Transport properties, Mixtures, Environmental impacts, Reprints, *Working fluids.

The need for and uses of thermodynamic and transport properties in the selection of working fluids for the vapor compression cycle and in equipment design are reviewed. A list of hydrogen-containing halocarbons, as well as their mixtures, is presented as alternatives to the environmentally harmful, fully halogenated chlorofluorocarbons. These fluids range from well-characterized, widely available refrigerants to materials available only by custom synthesis about which very little is known. Data priorities for these fluids are presented; most essential are critical point, vapor pressure, liquid density, ideal-gas heat capacity, and vapor p-V-T data. A critical need exists for these data on a number of candidate working fluids in order not to lose the opportunity to select the best set of future refrigerants.

900,483
PB89-231138   Not available NTIS
National Inst. of Standards and Technology (NML), Gaithersburg, MD. Temperature and Pressure Div.

Pressure Fixed Points Based on the Carbon Dioxide Vapor Pressure at 273.16 K and the H2O(I) - H2O(III) - H2O(L) Triple-Point.
Final rept.
N. Bignell, and V. E. Bean. 1988, 10p
Pub. in Proceedings of Seminar on High Pressure Standards, Paris, France, May 24-25, 1988, p175-184.

Keywords: *Pressure measurement, *Carbon dioxide, *Vapor pressure, *Water, *Ice, Metrology, *Triple point, Fixed points.

The vapor pressure of carbon dioxide in equilibrium with the liquid at 273.16 K has been measured and found to be 3.48608 + or - .00017 MPa. Results were found to depend upon the purity of the carbon dioxide. Samples prepared by heating analytical reagent quality sodium bicarbonate were of sufficient purity to be suitable for use as a pressure fixed point. The pressure of the H2O(U)-H2O(III)-H2)(L) triple-point has also been measured and found to be 208.829 + or - 0.025 MPa.

900,484
PB89-231252   Not available NTIS
National Inst. of Standards and Technology (NML), Gaithersburg, MD. Molecular Spectroscopy Div.
Stabilization and Spectroscopy of Free Radicals and Reactive Molecules in Inert Matrices.
Final rept.
M. E. Jacox. 1989, 31p
Pub. in Chemistry and Physics of Matrix-Isolated Species, p75-105 1989.

Keywords: *Free radicals, Infrared spectra, Ultraviolet spectra, Stabilization, Photolysis, Spectroscopy, Reprints, *Matrix isolation, Atom-molecule collisions.

Characteristics of the infrared and ultraviolet spectra of free radicals isolated in inert matrices are surveyed. Technique by which free radicals have been stabilized in matrices in concentration sufficient for optical spectroscopic detection are reviewed and are illustrated by recent studies. Emphasis is placed on the characteristics of atomic diffusion in rare gas matrices and on the consequences of the occurrence of cage recombination in the photolytic generation of free radicals in matrices. The stabilization of free radicals by matrix isolation sampling of discharge systems is discussed. The role of atom-molecule reactions in free radical stabilization is illustrated by specific consideration of reactions involving H, various metal atoms, C, Si, O, S, and F.

900,485
PB89-231294   Not available NTIS
National Inst. of Standards and Technology (NML), Gaithersburg, MD. Surface Science Div.
Electron Transmission Through NiSi2-Si Interfaces.
Final rept.
M. Stiles, and D. R. Hamann. 1989, 4p
Pub. in Physical Review B 40, n2 p1349-1352, 15 Jul 89.

Keywords: *Electron mobility, *Silicon, *Nickel, *Interfaces, Computation, Epitaxy, Orientation, Surface chemistry, Reprints, *Nickel silicides.

Calculations of electron transmission through epitaxial NiSi2/Si(111) interfaces illustrate the versatility of a newly developed first-principles technique. The transmission is poor and very dependent on the interface structure; of the electrons of primary importance for transport, more than 50% are reflected by the type-A orientation interface and more than 80% by the type B.

900,486
PB89-231310   Not available NTIS
National Inst. of Standards and Technology (NML), Gaithersburg, MD. Surface Science Div.
Methodology for Electron Stimulated Desorption Ion Angular Distributions of Negative Ions.
Final rept.
S. A. Joyce, A. L. Johnson, and T. E. Madey. 1989, 6p
Sponsored by Department of Energy, Washington, DC.
Pub. in Jnl. of Vacuum Science and Technology A7, n3 p2221-2226 May/Jun 89.

Keywords: *Desorption, Nitrogen fluorides, Acetone, Ruthenium, Surfaces, Reprints, *Negative ions, Electron stimulated desorption, Phosphorus fluorides, Time-of-flight method, Angular distribution, Hexafluoroacetone.

The electron stimulated desorption angular distributions technique has been extended to measure the angular distributions of negative ions desorbing from surfaces. The apparatus is a modification of an existing display-type detector. Time-of-flight methods have been employed to separate the negative ions from the large background of electrons and to permit mass-resolved measurements of both positive and negative ions. General trends in negative ion desorption from a series of fluorinated molecules (PF3, NF3, and hexafluoroacetone) adsorbed on Ru(0001) are presented. In these systems, negative ions arise primarily from molecularly intact adsorbates while positive ions can arise from both intact and dissociated species with similar probability.

900,487
PB89-234199 Not available NTIS
National Inst. of Standards and Technology (NML), Gaithersburg, MD. Molecular Spectroscopy Div.
Vibrational Spectra of Molecular Ions Isolated in Solid Neon. I. CO(sub 2, sup +) and CO(sub 2, sup -).
Final rept.
M. E. Jacox, and W. E. Thompson. 1989, 7p
Sponsored by Army Research Office, Research Triangle Park, NC.
Pub. in Jnl. of Chemical Physics 91, n3 p1410-1416, 1 Aug 89.

Keywords: *Vibrational spectra, Infrared spectra, Absorption spectra, Solidified gases, Neon, Reprints, *Carbon dioxide ions, Neon atoms, Photoionization, Photodetachment, Matrix isolation, Molecular ions.

When a Ne:CO2 sample was codeposited at approximately 5 K with a beam of neon atoms that had been excited in a microwave discharge, an absorption appeared at 1421.7/cm very near the gas-phase band center for the antisymmetric stretching fundamental (nu sub 3) of CO2(1+). Detailed isotopic substitution studies support the assignment of this absorption to that fundamental of CO2(1+), as well as of an absorption at 1658.3/cm to nu sub 3 of CO2(1-). In earlier studies of the charge transfer interaction of an alkali metal with CO2, this vibration of CO2(1-) had been strongly perturbed by coordination with the alkali metal cation. In the present experiments, the threshold for electron photodetachment from CO2(1-) was observed in the visible spectral region. Evidence was also obtained for the stabilization of the O2C...OCO(1-) cluster anion.

900,488
PB89-234207 Not available NTIS
National Inst. of Standards and Technology (NML), Gaithersburg, MD. Molecular Spectroscopy Div.
Vibrational Predissociation in the H-F Stretching Mode of HF-DF.
Final rept.
G. T. Fraser, and A. S. Pine. 1989, 4p
Pub. in Jnl. of Chemical Physics 91, n2 p633-636, 15 Jul 89.

Keywords: *Hydrogen fluoride, Deuterium compounds, Infrared spectra, Hydrogen bonds, Molecular beams, Vibrational spectra, Reprints, *Complexes, Color center lasers, Laser spectroscopy, Predissociation, Dimers.

The high-resolution infrared spectrum of the K=1-0 subband of the H-F stretching vibrational band of the hydrogen-bonded HF-DF complex has been recorded using a molecular-beam electric resonance optothermal color-center-laser spectrometer. The spectrum exhibits minor perturbations and vibrational predissociation linewidths of 23 + or - 2 MHz full width at half-maximum for comparison to the 11 + or - 1 MHz widths found for the corresponding mode of the homonuclear HF-HF dimer.

900,489
PB89-234215 Not available NTIS
National Inst. of Standards and Technology (NML), Gaithersburg, MD. Molecular Spectroscopy Div.
Microwave and Infrared Electric-Resonance Optothermal Spectroscopy of HF-HCI and HCI-HF.
Final rept.
G. T. Fraser, and A. S. Pine. 1989, 9p
Pub. in Jnl. of Chemical Physics 91, n2 p637-645, 15 Jul 89.

Keywords: *Hydrogen chloride, *Hydrogen fluoride, Infrared spectroscopy, Microwave spectroscopy, Hydrogen bonds, Reprints, *Complexes, Van der Waals forces, Isomers.

Microwave and infrared spectra of HF-HCI and HCI-HF have been obtained using a molecular-beam electric-resonance optothermal spectrometer, which operates by quadrupole-field focusing of polar molecules onto a bolometer detector. The HF-HCI microwave measurements extend to K(alpha)=1, the previous K(alpha)=0 results of Janda, Steed, Novick, and Klemperer, allowing the determination of the K(alpha) dependence and asymmetry of the CI quadrupole coupling constant. For the metastable HCI-HF isomer no previous spectroscopic measurements have been reported. Here, microwave spectra are observed for the K(alpha)=0 and K(alpha)=1 states and interpreted in terms of an L-shaped hydrogen-bonded structure for the complex, with a 3.235 center-of-mass separation between the HF and HCI subunits.

900,490
PB89-234256 Not available NTIS
National Inst. of Standards and Technology (NML), Boulder, CO. Quantum Physics Div.
High-Resolution, Slit Jet Infrared Spectroscopy of Hydrocarbons: Quantum State Specific Mode Mixing in CH Stretch-Excited Propyne.
Final rept.
A. Mcilroy, and D. J. Nesbitt. 1989, 10p
Grants NSF-PHY86-04504, NSF-CHE86-05970
Sponsored by National Science Foundation, Washington, DC.
Pub. in Jnl. of Chemical Physics 91, n1 p104-113, 1 Jul 89.

Keywords: *Hydrocarbons, *Infrared spectroscopy, Reprints, *Propyne, Laser spectroscopy, High resolution, Vibrational states, Supersonic jet flow.

A direct absorption, difference frequency, infrared laser spectrometer with 0.0001/cm resolution combined with slit supersonic jet optical pathlengths is presented as a tool for the study of mode-mode vibrational coupling in laser-excited hydrocarbons. These weak mode-mode couplings are evidenced in frequency domain studies by virtue of transitions to isolated upper J states that are split into multiplets under sub-Doppler resolution. Instrument performance is demonstrated by investigating vibrational coupling in the 3000-3300/cm C-H stretch fundamental region of (12)C3 propyne, as well as the (12)C2(13)C propynes observed in natural isotopic abundance. The implications of anharmonic coupling matrix elements of the magnitude found in the study, in overtone vibrational dynamics, are discussed.

900,491
PB89-234264 Not available NTIS
National Inst. of Standards and Technology (NML), Boulder, CO. Quantum Physics Div.
Observation of Translationally Hot, Rotationally Cold NO Molecules Produced by 193-nm Laser Vaporization of Multilayer NO Films.
Final rept.
L. M. Cousins, R. J. Levis, and S. R. Leone. 1989, 4p
Sponsored by Army Research Office, Research Triangle Park, NC.
Pub. in Jnl. of Physical Chemistry 93, n14 p5325-5328 1989.

Keywords: *Nitrogen oxide(NO), Spin orbit interactions, Far ultraviolet radiation, Surfaces, Films, Cryogenics, Reprints, Rotational states, Laser heating, Vaporization, Multilayers.

Results are presented for the rotational and spin-orbit state excitation of NO molecules which are ejected at hyperthermal velocities by 193-nm laser vaporization of cryogenic multilayer NO films condensed on MgF2. For molecules with translational energies of 0.14 and 0.56 eV, the average rotational energies are 0.014 and 0.017 eV, corresponding to temperatures of 160 and 180 K, respectively. The spin-orbit population ratio for the 0.14-eV molecules is 0.35 (T approx = 170 K); however, the population ratio for the 0.56-eV molecules is higher, 0.7 (T approx = 500 K). The disequilibrium between translation and rotation may be due in part to a collisional cooling mechanism (adiabatic expansion) which occurs immediately following vaporization. Other dynamical ejection mechanisms are also discussed.

900,492
PB89-234280 Not available NTIS
National Inst. of Standards and Technology (NML), Boulder, CO. Quantum Physics Div.

Reduced Dimensionality Quantum Reactive Scattering Study of the Insertion Reaction O(1D) + H2 -> OH + H.
Final rept.
J. K. Badenhoop, H. Koizumi, and G. C. Schatz. 1989, 8p
Pub. in Jnl. of Chemical Physics 91, n1 p142-149, 1 Jul 89.

Keywords: Hamiltonian functions, Oxygen, Hydrogen, Reprints, *Insertion reactions, Quantum reactive scattering, Hyperspherical coordinates, Hydroxyl radicals, Two degrees of freedom.

The paper presents a two degree of freedom model for describing the quantum dynamics of the insertion reaction O(singlet D) + H2 in which bend motions are treated with a sudden approximation. Comparison of product state vibrational distributions from a classical version of the model with three dimensional trajectory results indicates that the model is realistic. Quantum/classical comparisons for the model Hamiltonian indicate that recrossing is more important in the quantum dynamics, and as a result, the quantum reaction probability from ground state reagents is lower by as much as 40%. In addition, the quantum vibrational state distribution shows higher excitation than its classical counterpart. This difference in excitation is due to trajectories that produce vibrationally cold products, and it is found that these trajectories always cross the deepest part of the H2O well.

900,493
PB89-235204 PC A19/MF A01
National Inst. of Standards and Technology (NEL), Gaithersburg, MD. Thermophysics Div.
Properties of Lennard-Jones Mixtures at Various Temperatures and Energy Ratios with a Size Ratio of Two.
Technical note.
M. L. Huber, and J. F. Ely. May 89, 429p NIST/TN-1331
Also available from Supt. of Docs. as SN003-003-02952-1.

Keywords: *Mixtures, Computerized simulation, Distribution functions, Tables(Data), *Lennard-Jones fluids, Lennard-Jones potential, Molecular dynamics, Mean density approximation, Correlation functions.

Results are presented of molecular dynamics computer simulations for binary Lennard-Jones mixtures with a size ratio of two. A total of 84 state points were investigated at seven compositions ranging from 5% to 95% of the small molecule energy ratios from 0.5 to 3.0 and three reduced temperatures. In addition to the thermodynamic results, extensive tabular data are given for the radial distribution functions, local compositions and the direct correlation functions. The authors also discuss several theories of dense fluid mixtures and give a new approximation for the radial distribution function of a pair in a mixture, which they call the modified mean density approximation. The predictions of various thermodynamic quantities from the different liquid mixture models are compared with simulation results. For mixtures with large size differences, the modified mean density approximation was found to be competitive with first order, Lee-Levesque type statistical mechanical perturbation theory.

900,494
PB90-117316 Not available NTIS
National Inst. of Standards and Technology (NEL), Gaithersburg, MD. Thermophysics Div.
Application of the Gibbs Ensemble to the Study of Fluid-Fluid Phase Equilibrium in a Binary Mixture of Symmetric Non-Additive Hard Spheres.
Final rept.
J. G. Amar. 1989, 7p
Pub. in Molecular Physics 67, n4 p739-745 1989.

Keywords: *Fluids, Monte Carlo method, Spheres, Separation, Reprints, *Phase studies, Binary mixtures, Gibbs ensemble.

The 'Gibbs' Monte Carlo method introduced by Panagiotopoulos to study coexisting phases is applied to the study of fluid-fluid phase equilibrium in a binary system of symmetric, non-additive hard spheres. It was found that the method is generally applicable, although requiring a significantly larger number of insertions per Monte Carlo step at higher densities than in the previously studied cases of gas-liquid binary mixtures. Also studied was the effects of different initial

59

CHEMISTRY

Physical & Theoretical Chemistry

conditions in the simulations and the effects of varying system size (number of particles) on the results.

900,495
PB90-117342   Not available NTIS
National Inst. of Standards and Technology (NML), Boulder, CO. Time and Frequency Div.
Detection of the Free Radicals FeH, CoH, and NiH by Far Infrared Laser Magnetic Resonance.
Final rept.
S. P. Beaton, K. M. Evenson, T. Nelis, and J. M. Brown. 1988, 3p
Pub. in Jnl. of Chemical Physics 89, n7 p4446-4448, 1 Oct 88.

Keywords: *Interstellar matter, Rotational spectra, Free radicals, Far infrared radiation, Stellar spectra, Infrared detection, Reprints, *Cobalt hydrides, *Iron hydrides, *Nickel hydrides, Laser magnetic resonance, Infrared astronomy.

The authors report the detection of rotational transitions of the free radicals FeH, CoH, and NiH. The measurements were made in the far infrared (FIR) between 50-90/cm using laser magnetic resonance. Signal-to-noise ratios of over 1000 with a 1 s time constant were obtained on the strongest lines of all three species. These metal hydrides are expected to be abundant in the interstellar medium, and optical emission spectra of FeH have been detected in many stellar spectra. Precise ground state rotational transition frequencies are needed for radio and FIR astronomical searches and will be provided from an analysis of the authors' spectra.

900,496
PB90-117359   Not available NTIS
National Inst. of Standards and Technology (NML), Boulder, CO. Time and Frequency Div.
Far-Infrared Laser Magnetic Resonance Spectrum of the CD Radical and Determination of Ground State Parameters.
Final rept.
J. M. Brown, and K. M. Evenson. 1989, 18p
Contracts NASW-15, NASW-047
Sponsored by National Aeronautics and Space Administration, Washington, DC.
Pub. in Jnl. of Molecular Spectroscopy 136, p68-85 1989.

Keywords: Far infrared radiation, Deuterium compounds, Infrared spectra, Free radicals, Ground state, Reprints, *Carbon hydrides, Laser magnetic resonance.

The far-infrared laser magnetic resonance spectrum of the CD radical in the nu=0 level of the X doublet pi state has been studied in detail. Twelve transitions which are accessible with currently available laser lines have been recorded. The measurements have been analyzed and subjected to a single least-squares fit using an effective Hamiltonian. The data provide primary information on the rotational and fine-structure intervals between the lowest rotational intervals. They also yield values for the Lambda-type doubling and deuteron hyperfine splittings in the same levels. Combination of the measurements with the corresponding data for CH allows the two parameters, gamma and A sub D, to be determined separately.

900,497
PB90-117433   Not available NTIS
National Inst. of Standards and Technology (NML), Gaithersburg, MD. Molecular Spectroscopy Div.
Electric-Resonance Optothermal Spectrum of (H2O)2: Microwave Spectrum of the K=1-0 Subband for the E((+ or -)2) States.
Final rept.
G. T. Fraser, R. D. Suenram, L. H. Coudert, and R. S. Frye. 1989, 4p
Pub. in Jnl. of Molecular Spectroscopy 137, p244-247 1989.

Keywords: *Microwave spectroscopy, *Molecular spectroscopy, *Water, Hydrogen bonds, Dimerization, Electron transitions, Reprints, *Electron tunneling.

Microwave spectra of (H2O)2 have been obtained using an electric-resonance optothermal spectrometer (EROS). The microwave measurements extend previous results on the K=0-0 and K=1-0 bands for the A2(+/-) and B2(+/-) rotational-tunneling states and provide the first observations of the weak c-type K=1-0 band for the E2(+/-) states. The results allow a near direct determination of the h2v tunneling matrix element that is associated with the tunneling motion

which interchanges the two hydrogens on the acceptor H2O molecule and the two hydrogens on the donor H2O molecule. It was found that h2v = 743 MHz.

900,498
PB90-117441   Not available NTIS
National Inst. of Standards and Technology (NML), Gaithersburg, MD. Molecular Spectroscopy Div.
Rotational Energy Levels and Line Intensities for (2S+1)Lambda-(2S+1)   Lambda   and (2S+1)(Lambda + or -)-(2S+1)Lambda Transitions in a Diatomic Molecule van der Waals Bonded to a Closed Shell Partner.
Final rept.
W. M. Fawzy, and J. T. Hougen. 1989, 12p
Pub. in Jnl. of Molecular Spectroscopy 137, p154-165 1989.

Keywords: Spin orbit interactions, Diatomic molecules, Hamiltonian functions, Spectral lines, Rare gases, Reprints, *Complexes, *Rotational states, Van der Waals forces, Matrices, Renner-Teller splitting.

Hamiltonian matrix elements needed for calculating rotational energy levels are derived for a planar complex consisting of an open-shell diatomic molecule and a closed-shell partner. These matrix elements take account of spin-orbit interaction and a Renner-Teller-like splitting term, but not of the effects of large-amplitude internal rotation of the diatomic fragment within the complex. Transition-moment matrix elements needed for calculating intensities in spin and orbitally allowed transitions in the open-shell diatomic sup (2S+1)Lambda - sup (2S+1) and sup (2S+1)(Lambda plus or minus 1) sup (2S+1)Lambda, where Lambda = Sigma, Pi, Delta, Phi, eta. A brief discussion of how to use the Hamiltonian and transition-moment matrix elements in a computer program is given.

900,499
PB90-117557   Not available NTIS
National Inst. of Standards and Technology (IMSE), Boulder, CO. Fracture and Deformation Div.
Ultrasonic Separation of Stress and Texture Effects in Polycrystalline Aggregates.
Final rept.
P. P. Delsanto, R. B. Mignogna, and A. V. Clark. 1987, 8p
Sponsored by Office of Naval Research, Arlington, VA.
Pub. in Review of Progress in Quantitative Nondestructive Evaluation, v6B p1533-1540 1987.

Keywords: *Aggregates, *Polycrystalline, *Determination of stress, *Texture, Plates(Structural members), Rayleigh waves, Metal plates, Error analysis, Nondestructive tests, Alloys, Reprints, *Ultrasonic testing.

The general perturbation formalism for the propagation of Rayleigh waves on the surface of initially deformed anisotropic material plates is applied to the problem of separation of material texture and stress. The preferential alignment of crystallographic axes in a polycrystalline material is described in terms of their orientation distribution function. The authors considered explicitly the case of an orthotropic distribution of cubic crystallites, which occurs, for example, in aluminum and steel alloys. The measured values of the Rayleigh wave phase velocity at different angles on the material plate can be used for the determination, in a reference plate with similar texture, of the three coefficients W(400), W(420) and W(440), which characterize the orientation distribution. An extension of the formalism separates the acoustoelastic effects of the initial stresses. An error analysis and a comparison with similar techniques using grazing SH-waves complete the discussion.

900,500
PB90-117656   Not available NTIS
National Inst. of Standards and Technology (NML), Gaithersburg, MD. Molecular Spectroscopy Div.
Photoacoustic Measurement of Differential Broadening of the Lambda Doublets in NO(X (2)PI 1/2, v=2-0) by Ar.
Final rept.
A. S. Pine. 1989, 8p
Sponsored by National Aeronautics and Space Administration, Washington, DC.
Pub. in Jnl. of Chemical Physics 91, n4 p2002-2009, 15 Aug 89.

Keywords: *Nitrogen oxide(NO), Ground state, Argon, Reprints, *Nitric oxide, Photoacoustic spectroscopy, Pressure broadening, Color center lasers, Tunable lasers.

Keywords: *Chemical stabilization, *Infrared spectroscopy, *Neon, *Ionization, *Chemical analysis, Molecular structures, Molecular energy levels, Reviews, Carbon dioxide, Oxygen, Carbonyl compounds, Photochemical reactions, Reprints, *Matrix isolation technique, *Fourier transform spectroscopy.

The paper reviews recent studies of the stabilization and infrared spectra of small molecular ions and cluster ions in a neon matrix. The products of the interaction between the molecule of interest and a beam of excited neon atoms are trapped in an excess of solid neon, and their Fourier transform infrared spectrum is obtained. Detailed isotopic substitution studies aid in product identification, as does the study of infrared spectra obtained after exposure of the sample to filtered mercury-arc radiation. Infrared spectra obtained for CO2+, CO2-, O4+, O4-, N4+, (CO)2+, and (CO)2- demonstrate the usefulness of the technique and yield new information on the structures and molecular energy levels of these species.

900,504
PB90-117763   Not available NTIS
National Inst. of Standards and Technology (IMSE), Gaithersburg, MD. Metallurgy Div.
Formation of the Al-Mn Icosahedral Phase by Electrodeposition.
Final rept.
B. Grushko, and G. R. Stafford. 1989, 6p
Grant N00014-88-F-0091
Sponsored by Office of Naval Research, Arlington, VA.
Pub. in Scripta Metallurgica 23, n7 p1043-1048 1989.

Keywords: *Electrodeposition, *Crystal structures, *Aluminum manganese alloys, Temperature, Fused salts, Diffusion, Metastable state, Reprints, *Icosahedrons, Amorphous state, Quasi-steady states.

Amorphous, quasicrystalline and metastable crystalline structures have been observed in binary aluminum-manganese alloys electrodeposited from chloroaluminate electrolytes. During electrodeposition, the extent of the deviation from equilibrium as well as the degree of ordering is defined by the concurrent processes of new layer formation and surface diffusion. The metastable amorphous phase can be formed quite easily by low temperature electrodeposition. The direct formation of quasicrystals, having a level of free energy between that of an amorphous and a stable crystalline phase, can be achieved by an increase in deposition temperature in a manner somewhat analogous to that which has been reported for sputter deposition. An increase in the temperature causes a gradual increase in the size of icosahedral regions.

900,505
PB90-117771   Not available NTIS
National Inst. of Standards and Technology (NML), Gaithersburg, MD. Radiation Physics Div.
Resonance Enhanced Electron Stimulated Desorption.
Final rept.
J. W. Gadzuk, and C. W. Clark. 1989, 8p
Pub. in Jnl. of Chemical Physics 91, n5 p3174-3181, 1 Sep 89.

Keywords: Resonance scattering, Palladium, Reprints, *Electron stimulated desorption, Oxygen atoms.

A theory is presented which accounts for giant enhancements in electron stimulated desorption (ESD) yields from adsorbate-covered surfaces if the incident electrons become trapped in a shape or Feshbach resonance associated with the adsorbate. The resulting temporary negative ion is displaced inwards towards the surface as a result of the force provided by the image screening charge. Upon reneutralization, the desorbate can be returned high on the dissociative repulsive wall of the neutral-surface potential curve. The process has been modeled within the context of semiclassical Gaussian wave packet dynamics. Recent observations of such giant enhancements in the ESD yields for the system O(alpha)/Pd(111) are explained in terms of the model, and an atomic physics basis for the resonance in atomic oxygen is proposed.

900,506
PB90-117797   Not available NTIS
National Inst. of Standards and Technology (NML), Boulder, CO. Time and Frequency Div.

Heterodyne Measurements on N2O Near 1635 cm(-1).
Final rept.
M. D. Vanek, M. Schneider, J. S. Wells, and A. G. Maki. 1989, 5p
Sponsored by National Aeronautics and Space Administration, Washington, DC.
Pub. in Jnl. of Molecular Spectroscopy 134, p154-158 1989.

Keywords: *Nitrogen oxide(N2O), *Infrared spectroscopy, *Molecular spectroscopy, Calibrating, Reprints, Heterodyning.

Heterodyne frequency measurements have been made on eight lines of the 10(sup 0)0-01(sup 1)0 band N2O between 1591 and 1673/cm. These measurements were combined with other heterodyne frequency measurements to obtain improved frequency values for the 01(sup 1)0-00(sup 0)0 transitions from 520 to 660/cm.

900,507
PB90-117805   Not available NTIS
National Inst. of Standards and Technology (NML), Boulder, CO. Time and Frequency Div.
Heterodyne Frequency and Fourier Transform Spectroscopy Measurements on OCS Near 1700 cm(-1).
Final rept.
J. S. Wells, M. D. Vanek, and A. G. Maki. 1989, 5p
Sponsored by National Aeronautics and Space Administration, Washington, DC.
Pub. in Jnl. of Molecular Spectroscopy 135, p84-88 1989.

Keywords: *Infrared spectroscopy, *Fourier transformation, Demodulation, Calibrating, Molecular spectroscopy, Spectrum analysis, Microwave spectroscopy, Reprints, *Carbon oxysulfide, *Heterodyning.

Heterodyne frequency measurements are given for carbonyl sulfide in the 1700/cm region. The measurements were combined with Fourier transform spectroscopy measurements of the same bands and with other measurements in the literature (microwave, submillimeter wave, and other infrared measurements) in a least-squares fit. The combined data and fit result in improved frequency calibration values for the 1700/cm region and also allow the determination of calibration values for the 00(sup 0)3-00(sup 0)0 band near 2550/cm from application of the Ritz principle.

900,508
PB90-117821   Not available NTIS
National Inst. of Standards and Technology (NML), Gaithersburg, MD. Atomic and Plasma Radiation Div.
Analysis of Magnesiumlike Spectra from Cu XVIII to Mo XXXI.
Final rept.
J. Sugar, V. Kaufman, P. Indelicato, and W. L. Rowan. 1989, 7p
Pub. in Jnl. of the Optical Society of America B 6, n8 p1437-1443 Aug 89.

Keywords: *Spectrum analysis, *Atomic energy levels, *Magnesium, Comparison, Dirac equation, Hartree-Fock approximation, Arsenic, Strontium, Selenium, Rubidium, Niobium, Molybdenum, Bromine, Copper, Gallium, Germanium, Krypton, Yttrium, Zinc, Zirconium, Reprints, *Laser spectroscopy, *Tokamak devices.

The magnesium like spectra of Cu to Mo have been observed with laser- and tokamak-generated plasmas in the range of 1000-300 A. The authors give wavelengths accurate to + or -0.005 A and classifications of transitions among the 3s doublet, 3p doublet, 3d doublet, 3s3p, 3s3d, and 3p3d configurations. Comparisons with Dirac-Fock calculations of the wavelengths are presented.

900,509
PB90-117839   Not available NTIS
National Inst. of Standards and Technology (NML), Gaithersburg, MD. Molecular Spectroscopy Div.
Microwave Spectrum of Methyl Amine: Assignment and Analysis of the First Torsional State.
Final rept.
N. Ohashi, S. Tsunekawa, K. Takagi, and J. T. Hougen. 1989, 14p
Pub. in Jnl. of Molecular Spectroscopy 137, p33-46 1989.

Keywords: *Microwave spectroscopy, *Spectrum analysis, *Torsional strength, Molecular rotation, Electron tunneling, Stark effect, Reprints, *Amine/methyl.

The microwave absorption spectrum of methyl amine has been reinvestigated in the range from 7 to 90 GHz, with the aim of analyzing the first torsional state in more detail. By combining the newly obtained microwave data with the far-infrared and microwave data already available, it was possible to make an analysis of the tunneling-rotational levels of the first torsional state in which three types of delta K = + or - 1 elements were introduced into the Hamiltonian matrix described in the group-theoretical formalism developed previously. Stark effect data, remeasured during the present study, are also examined in connection with the delta K = + or - 1 interaction.

900,510
PB90-117847   Not available NTIS
National Inst. of Standards and Technology (NML), Gaithersburg, MD. Molecular Spectroscopy Div.
Microwave Spectrum, Structure, and Electric Dipole Moment of Ar-CH3OH.
Final rept.
R. D. Suenram, F. J. Lovas, G. T. Fraser, J. Z. Gillies, C. W. Gillies, and M. Onda. 1989, 11p
Pub. in Jnl. of Molecular Spectroscopy 137, p127-137 1989.

Keywords: Microwave spectra, Deuterium compounds, Dipole moments, Stark effect, Molecular structure, Reprints, *Argon complexes, Fourier transform spectroscopy, Van der Waals forces, Electric dipoles, Methanol.

Microwave spectra of Ar-CH3OH, Ar-CD3OH, and Ar-(13)CH3OH have been measured between 7 and 25 GHz using a pulsed-nozzle Fourier transform microwave spectrometer. Two tunneling states are observed which correlate to the A and E internal-rotor states of free methanol. The electric-dipole-moment components are determined. The structure of the complex is found to be T-shaped with an Ar to CH3OH center-of-mass separation of 3.684(14) A. A number of transitions are observed which do not appear to fit an asymmetrical-top Hamiltonian. These are assigned to an E tunneling state of the complex.

900,511
PB90-117862   Not available NTIS
National Inst. of Standards and Technology (NEL), Gaithersburg, MD. Electrosystems Div.
Collisional Electron Detachment and Decomposition Rates of SF6(1-), SF5(1-), and F(1-) in SF6: Implications for Ion Transport and Electrical Discharges.
Final rept.
J. K. Olthoff, R. J. Van Brunt, Y. Wang, R. L. Champion, and L. D. Doverspike. 1989, 8p
Sponsored by Department of Energy, Washington, DC.
Pub. in Jnl. of Chemical Physics 91, n4 p2261-2268, 15 Aug 89.

Keywords: *Sulfur hexafluoride, *Sulfur fluorides, Electric discharges, Decomposition, Reaction kinetics, Reprints, *Fluorine ions, *Electron detachment, Charge transfer, Negative ions, Ion mobility.

Measured cross sections for prompt collisional detachment and decomposition of SF6(1-), SF5(1-), and F(1-) in SF6 are used to calculate detachment coefficients and ion-conversion reaction coefficients as functions of electric field-to-gas density ratio (E/N) for ion drift in SF6. The calculated detachment and reaction coefficients are used in a model which invokes detachment from long-lived energetically unstable states of collisionally excited SF6(1-) to explain the pressure dependence of previously measured detachment coefficients and the high detachment thresholds implied by analysis of electrical-breakdown probability data for SF6.

900,512
PB90-117870   Not available NTIS
National Inst. of Standards and Technology (NEL), Gaithersburg, MD. Electrosystems Div.
Electron-Energy Dependence of the S2F10 Mass Spectrum.
Final rept.
J. K. Olthoff, R. J. Van Brunt, and I. Sauers. 1989, 3p
Sponsored by Department of Energy, Washington, DC.
Pub. in Jnl. of Physics D: Applied Physics 22, p1399-1401 1989.

Keywords: *Sulfur fluorides, *Mass spectra, Sulfur hexafluoride, Gas ionization, Detection, Reprints, Electron impact, Positive ions, eV range 10-100.

CHEMISTRY

Physical & Theoretical Chemistry

The positive-ion mass spectrum of S2F10 has been measured as a function of electron-impact energy in the range 20-70 eV using quadrupole mass spectrometer. Contrary to recent results from mass spectrometric analysis of arc-decomposed SF6 there was no evidence of S2F9(1+) or S2F10(1+) ion formation from S2F10 at any energy. The largest ion observed at all electron energies is SF5(1+). It was found, however, that the appearance potentials for SF5(1+) and SF3(1+), the two most prominent ions from S2F10, are significantly lower than the appearance potentials of the same ions from SF6. The differences between the mass spectra of S2F10 and SF6 are delineated and the implications for detection of S2F10 in the presence of SF6 are discussed.

900,513
PB90-117920   Not available NTIS
National Inst. of Standards and Technology (NML), Gaithersburg, MD. Chemical Thermodynamics Div.
Biological Thermodynamic Data for the Calibration of Differential Scanning Calorimeters: Heat Capacity Data on the Unfolding Transition of Lysozyme in Solution.
Final rept.
F. P. Schwarz. 1989. 21p
Pub. in Thermochimica Acta 147, p71-91 1989.

Keywords: *Thermodynamics, Enthalpy, Chemical analysis, Reprints, *Differential scanning calorimetry, *Muramidase.

Differential scanning calorimetry measurements of the unfolding transition of lysozyme in HCl-glycine buffer solutions were performed over a temperature range from 326 K at pH 2.3 to 349 K at pH 3.9. Van't Hoff transition enthalpies were calculated from the fit of a two-state transition model to the heat capacity measurements (Delta Hvf), from the van't Hoff plot of ln(1/K) vs. 1/T where K is the transition equilibrium constant, and from the ratio of the transition peak height to the area under the transition peak. The best linear fit of the van't Hoff enthalpies to the transition temperatures Tm was obtained with Delta Hvf and was Delta Hvf (kJ/mol) = 432.7 + or - 1.7 + or - 5.81 + or - 0.24 (Tm-337.2). Calculated transition enthalpies were determined from the transition peak area using an extrapolated sigmoidal baseline (Delta Hs) and an extrapolated straight linear baseline. The best linear dependence of the calorimetric enthalpy on Tm was obtained with Delta Hs (kJ/mol)=434.7 + or - 4.1 + 6.39 + or - 0.60 (Tm-337.2). Linear least-squares fits of Delta Hvf and Delta Hs to Tm were independent of the DSC scan rate, the source of the lysozyme, the buffer concentration from 0.1 to 0.2 M and the concentration of lysozyme from 0.26 to 2.8 mM. The transition temperature exhibited a linear dependence on pH and concentration. Cooperativities of the transition ranged from 0.988 + or - 0.0007 at 326 K to 1.012 + or - 0.007 at 34°K.

900,514
PB90-117987   Not available NTIS
National Inst. of Standards and Technology (NML), Gaithersburg, MD. Thermophysics Div.
Vapor Pressures and Gas-Phase PVT Data for 1,1,1,2-Tetrafluoroethane.
Final rept.
L. A. Weber. 1989, 11p
Sponsored by Department of Energy, Washington, DC., and Environmental Protection Agency, Washington, DC.
Pub. in International Jnl. of Thermophysics 10, n3 p617-627 May 89.

Keywords: *Refrigerants, *Vapor pressure, Thermodynamic properties, Pressure, Volume, Reprints, Virial equations, Temperature dependence, Density.

New data is presented for the vapor pressure and PVT surface of 1,1,1,2-tetrafluoroethane (Refrigerant 134a) in the temperature range 40C (313 K) to 150C (423 K). The PVT data are for the gas phase at densities up to one-half critical. Densities of the saturated vapor are derived at five temperatures from the intersections of the experimental isochores with the vapor pressure curve. The data are represented analytically in order to demonstrate experimental precision and to facilitate calculation of thermodynamic properties.

900,515
PB90-117995   Not available NTIS
National Inst. of Standards and Technology (IMSE), Gaithersburg, MD. Reactor Radiation Div.

Structure of V9Mo6O40 Determined by Powder Neutron Diffraction.
Final rept.
R. C. T. Slade, A. Ramanan, B. C. West, and E. Prince. 1989, 5p
Pub. in Jnl. of Solid State Chemistry 82, p65-69 1989.

Keywords: *Crystal structure, *Neutron diffraction, *Vanadium oxides, *Molybdenum oxides, Chemical bonds, Crystallography, Powder(Particles), Reprints, Bond lengths.

The crystal structure of V9Mo6O40 has been determined by Rietveld profile analysis of neutron powder diffraction data. The structure is a monoclinically distorted variant of Nb3O7F (space group C2) consisting of ReO3-type slabs three octahedra thick connected by edge sharing of component octahedra. The octahedra are considerably distorted due to off-center displacement of the metal atoms. Metal-oxygen bond lengths conform with lengths of similar compounds.

900,516
PB90-118001   Not available NTIS
National Inst. of Standards and Technology (IMSE), Gaithersburg, MD. Reactor Radiation Div.
Neutron Diffraction Determination of Full Structures of Anhydrous Li-X and Li-Y Zeolites.
Final rept.
C. Forano, R. C. T. Slade, E. K. Andersen, I. G. K. Andersen, and E. Prince. 1989, 8p
Pub. in Jnl. of Solid State Chemistry 82, p95-102 1989.

Keywords: *Crystal structure, Neutron diffraction, Ion exchange resins, Lithium inorganic compounds, Sodium inorganic compounds, Aluminum inorganic compounds, Silicates, Reprints, *Zeolites.

Virtually monoionic Li-X and Li-Y zeolites have been prepared by LiOH titration of parent NH4 zeolites. Structural studies have been performed at room temperature on the anhydrous zeolites, Li(80.7)H(4.9)Na(0.4)Al(86)Si(106)O(384) and Li(46.0)H(5.8)Na(5.1)K(0.1)Al(57)Si(135)O(384) by powder neutron diffraction and Fourier difference to locate Li(+1) cations. The cell parameters are 24.6716(10) and 24.4498(12) A for Li-X and Li-Y, respectively. Three positions have been found for Li(+1) sites I' and II in the six-ring windows of the sodalite unit and site III' in the supercage for the additional Li(+1) of Li-X.

900,517
PB90-118027   Not available NTIS
National Inst. of Standards and Technology (NML), Gaithersburg, MD. Surface Science Div.
Photon-Stimulated Desorption of Fluorine from Silicon via Substrate Core Excitations.
Final rept.
J. A. Yarmoff, and S. A. Joyce. 1989, 8p
Pub. in Physical Review B 40, n5 p3143-3150, 15 Aug 89.

Keywords: *Silicon, *Desorption, Resonance, Reprints, Photon stimulated desorption, *Fluorine ions, Surface reactions.

Photon-stimulated desorption (PSD) of F(1+) was performed for silicon (111) surfaces terminated with fluorine atoms. The surfaces were prepared by exposure of clean silicon to XeF2. The onset for PSD at the Si 2p edge correlated with the transition from the 2p level of the bonding silicon atom to the conduction-band minimum, and was thus a function of the oxidation state of the bonding atom. The ions originating from a SIF species desorbed along the surface normal while the ions from a SIF3 group desorbed in off-normal directions. Localized 3s and 3p Rydberg-like resonances were observed in the quasimolecular SIF3 moieties. The ion kinetic-energy distributions were measured as an aid to elucidating the desorption mechanism. Measurements of the PSD of F(1+) at the Si 2s edge were used to confirm the 3s and 3p character of the measured resonances.

900,518
PB90-118092   Not available NTIS
National Inst. of Standards and Technology (NEL), Gaithersburg, MD. Thermophysics Div.
Measures of Effective Ergodic Convergence in Liquids.
Final rept.
R. D. Mountain, and D. Thirumalai. 1989, 5p
Grant NSF-CHE86-57356
Sponsored by National Science-Foundation, Washington, DC.

Pub. in Jnl. of Physical Chemistry 93, n19 p6975-6979 1989.

Keywords: *Liquids, Ergodic processes, Dielectric properties, Computerized simulation, Water, Convergence, Diffusion, Reprints, Binary mixtures, Molecular dynamics.

The recently introduced measure for ergodic convergence is used to illustrate the time scales needed for effective ergodicity to be obtained in various liquids. The cases considered are binary mixtures of soft spheres, two-component Lennard-Jones systems, and liquid water. It is shown that various measures obey a dynamical scaling law which is characterized by a single parameter, namely, a novel diffusion constant. The time scales for ergodic behavior are found to be dependent on the particular observable being considered. For example, in water, the diffusion constants for the translational and rotational kinetic energies and for the laboratory frame dipole moments are very different. The implications of these results for the calculation of the dielectric constant of polar liquids by computer simulations are discussed.

900,519
PB90-118100   Not available NTIS
National Inst. of Standards and Technology (NML), Boulder, CO. Quantum Physics Div.
Weakly Bound NeHF.
Final rept.
S. V. O'Neil, D. J. Nesbitt, P. Rosmus, H. J. Werner, and D. C. Clary. 1989, 11p
Grants NSF-PHY86-04504, NSF-CHE86-05970
Sponsored by National Science Foundation, Washington, DC.
Pub. in Jnl. of Chemical Physics 91, n2 p711-721, 15 Jul 89.

Keywords: *Hydrogen fluoride, Far infrared radiation, Infrared spectra, Deuterium compounds, Reprints, *Neon complexes, Van der Waals forces, Potential energy surfaces, Predissociation.

The authors have used ab initio methods to characterize the Ne-HF van der Waals complex. The interaction energy was determined using size consistent, correlated CEPA wave functions expanded in a Gaussian basis chosen to represent both intraatomic effects and the low order multipole moments and polarizabilities of Ne and HF. Converged variational and close-coupling calculations using the ab initio potential surface reveal three bound levels of the Ne-HF stretch mode, and several metastable levels correlating asymptotically with rotationally excited HF(j=1). From the calculated line positions, widths, and intensities, the authors have synthesized far infrared and infrared spectra for Ne-HF and Ne-DF.

900,520
PB90-118126   Not available NTIS
National Inst. of Standards and Technology (NML), Boulder, CO. Quantum Physics Div.
Slit Jet Infrared Spectroscopy of NeHF Complexes: Internal Rotor and J-Dependent Predissociation Dynamics.
Final rept.
D. J. Nesbitt, C. M. Lovejoy, T. G. Lindemann, S. V. ONeil, and D. C. Clary. 1989, 10p
Grants NSF-CHE86-05970, NSF-PHY86-04504
Sponsored by National Science Foundation, Washington, DC.
Pub. in Jnl. of Chemical Physics 91, n2 p722-731, 15 Jul 89.

Keywords: *Hydrogen fluoride, Infrared spectroscopy, Reprints, *Neon complexes, Van der Waals forces, Potential energy surfaces, Supersonic jet flow, Predissociation.

Direct absorption tunable difference frequency IR spectroscopy in a slit jet supersonic expansion has been used to observe complexes of Ne with HF for the first time. Spectra of both the weak HF stretch fundamental and the 10-20 fold more intense bend and stretch combination band transitions are observed, and illustrate several interesting dynamical features. The large ratio of combination band to fundamental intensity is evidence for a highly isotropic potential with respect to HF rotation. The HF bend vibration is thus better thought of as nearly free internal rotor motion with a nearly good space fixed quantum number.

900,521
PB90-118209   Not available NTIS

National Inst. of Standards and Technology (NML), Boulder, CO. Quantum Physics Div.
Intramolecular Dynamics of van der Waals Molecules: An Extended Infrared Study of ArHF.
Final rept.
C. M. Lovejoy, and D. J. Nesbitt. 1989, 18p
Grants NSF-CHE86-05970, NSF-PHY86-04504
Sponsored by National Science Foundation, Washington, DC.
Pub. in Jnl. of Chemical Physics 91, n5 p2790-2807, 1 Sep 89.

Keywords: *Hydrogen fluoride, Near infrared radiation, Infrared spectra, Reprints, *Argon complexes, Van der Waals forces, Molecular dynamics, Supersonic expansion, Laser spectroscopy, Potential energy surfaces.

The near-infrared spectrum of ArHF prepared in a slit supersonic expansion is recorded with a difference frequency infrared laser spectrometer. By virtue of the high sensitivity of the technique, and the lack of appreciable spectral congestion at the 10 K jet temperature, the authors observe 9 of the 11 vibrational states with energies below the $Ar+HF(nu=1,j=0)$ dissociation limit. The spectroscopic information is quite sensitive to the $Ar+HF$ potential energy surface away from the equilibrium configuration, and thus provides a rigorous test of trial potential energy surfaces. Excellent agreement is obtained between experiment and the predictions of a recently reported $Ar+HF(nu=1)$ potential.

900,522
PB90-123555 Not available NTIS
National Inst. of Standards and Technology (NML), Gaithersburg, MD. Surface Science Div.
Adsorption of Water on Clean and Oxygen-Predosed Nickel(110).
Final rept.
C. Benndorf, and T. E. Madey. 1988, 29p
Pub. in Surface Science 194, n1-2 p63-91 1988.

Keywords: *Adsorption, *Water, *Surface properties, *Nickel, Hydrogen bonds, Dissociation, Isotopic labeling, Desorption, Chemisorption, Oxygen, Dimerization, Reprints, Electron stimulated desorption ion angular distribution, Thermal desorption spectroscopy, Low energy electron diffraction.

The adsorption of H2O on both clean and modified Ni(110) surface has been studied using a variety of methods: electron stimulated desorption ion angular distribution (ESDIAD), thermal desorption spectroscopy (TDS), and low energy electron diffraction (LEED). Fractional monolayers, of H2O on clean Ni(110) are associated with a four-spot ESDIAD pattern suggesting that the HO ligands are in specific registry with the substrate. The authors postulate the formation of H2O dimers bound via oxygen lone pair orbitals to Ni substrate atoms and oriented with the O...H-O axis in 001 azimuthal directions. Upon heating to greater than or equal to 200K, a fraction of the H2O dissociates, forming OH(ad). TDS of H2O from clean Ni(110) reveals four binding states. They are related to multilayer desorption (155K), desorption from larger bilayer clusters (210K), desorption from H2O dimer clusters which might be stabilized by OH (245K), and recombination of OH to yield H2O (g) (370K). Isotopic exchange of H2(16)O with (18)O(ad) is observed ever for binding states in which dissociation is believed not to occur and is related to a proton exchange involving H2O(ad) hydrogen bonded to O(ad).

900,523
PB90-123563 Not available NTIS
National Inst. of Standards and Technology (NML), Gaithersburg, MD. Surface Science Div.
Ammonia Adsorption and Dissociation on a Stepped Iron(s) (100) Surface.
Final rept.
C. Benndorf, T. E. Madey, and A. L. Johnson. 1987, 11p
Pub. in Surface Science 187, n2-3 p434-444 1987.

Keywords: *Ammonia, *Adsorption, *Dissociation, *Iron, *Surface properties, Single crystals, Desorption, Crystal structure, Chemisorption, Reprints, Electron stimulated desorption ion angular distribution.

The purpose of the present letter is to provide new information about the molecular structure and configuration of NH3 on Fe single crystal surfaces and to provide insight into the influence of steps of NH3 orientation and dissociation. Information about the NH3 orientation on flat Fe(100) terraces and changes induced by steps were gained in the present study from ESDIAD (Electron Stimulated Desorption Ion Angular Distribu-

tion). The results provide evidence that at low NH3 coverages adsorption takes place primarily at step sites. The adsorbed species stabilized at step sites are believed to include oriented NHx fragments as well as 'inclined' NH3. The molecular NH3 dipoles are believed to be tilted due to the electrostatic field at steps. For higher coverages NH3 adsorption sites on the Fe(100) terraces are populated; from ESDIAD data the authors conclude that NH3 is bonded mainly to terrace sites via the N atom and the H-atoms are pointed away from the surface, with no azimuthal ordering.

900,524
PB90-123597 Not available NTIS
National Inst. of Standards and Technology (NEL), Gaithersburg, MD. Thermophysics Div.
Quantitative Characterization of the Viscosity of a Microemulsion.
Final rept.
R. F. Berg, M. R. Moldover, and J. S. Huang. 1987, 5p
Contract NASA-C-86129-D
Sponsored by National Aeronautics and Space Administration, Cleveland, OH. Lewis Research Center.
Pub. in Jnl. of Chemical Physics 87, n6 p3687-3691 1987.

Keywords: *Quantitative analysis, *Viscosity, Temperature, Volume, Drops(Liquids), Critical point, Water, Decanes, Electrical resistivity, Reprints, *Microemulsions, *Ternary systems, AOT.

The authors measured the viscosity of the 3-component microemulsion water/decane/AOT as a function of temperature and droplet volume fraction. At temperatures well below the critical temperature the viscosity is described by treating the droplets as hard spheres suspended in decane. Upon approaching the 2-phase region from low temperature, there is a large (as much a factor of 4) smooth increase of the viscosity which may be related to the percolation-like transition observed in the electrical conductivity. This increase in viscosity is not completely consistent with either a naive electroviscous model or a simple clustering model. The divergence of the viscosity near the critical point (39 C) is superimposed upon the smooth increase. The magnitude and temperature dependence of the critical divergence are similar to that seen near the critical points of binary liquid mixtures.

900,525
PB90-123852 Not available NTIS
National Inst. of Standards and Technology (NML), Boulder, CO. Quantum Physics Div.
Sub-Doppler Infrared Spectroscopy in Slit Supersonic Jets: A Study of all Three van der Waals Modes in v1-Excited ArHCl.
Final rept.
D. J. Nesbitt, and C. M. Lovejoy. 1988, 8p
Grants NSF-PHY86-04504, NSF-CHE86-05970
Sponsored by National Science Foundation, Washington, DC.
Pub. in Faraday Discussions of the Chemical Society 86, p13-20 1988.

Keywords: *Hydrogen chloride, Infrared spectroscopy, Far infrared radiation, Reprints, *Argon complexes, Van der Waals forces, Laser spectroscopy, Vibrational states, Supersonic jet flow, High resolution.

Direct absorption tunable difference frequency infrared radiation laser spectroscopy has been used to study the vibrational dynamics of ArHCl complexes cooled in a slit supersonic jet expansion. As a result of large-amplitude vibrational motion in these weakly bound complexes, transitions to each of the three van der Waals modes are observed as combination bands built on the fundamental HCl stretch, thus permitting detailed study of low-frequency, intermolecular modes with a near-infrared laser source. The molecular constants are in excellent agreement with both far-infrared radiation experimental results and semiempirical predictions for the $Ar+HCl(v=0)$ surface, indicating only a small change in the intermolecular potential upon vibrational excitation of the HCl.

900,526
PB90-123860 Not available NTIS
National Inst. of Standards and Technology (NML), Boulder, CO. Quantum Physics Div.

Apparent Spectroscopic Rigidity of Floppy Molecular Systems.
Final rept.
D. J. Nesbitt, and R. Naaman. 1989, 9p
Grants NSF-CHE86-05970, NSF-PHY86-04504
Sponsored by National Science Foundation, Washington, DC.
Pub. in Jnl. of Chemical Physics 91, n7 p3801-3809, 1 Oct 89.

Keywords: Hydrogen bonds, Quantum theory, Hamiltonian functions, Infrared spectroscopy, Reprints, *Complexes, Van der Waals forces, High resolution, Supersonic jet flow.

There has been a wealth of recent infrared experimental data on van der Waals and hydrogen bonded complexes obtained under cooled, supersonic jet conditions where only a small fraction of the total bound quantum states can be elucidated. This partial set of data can often be well fit to a traditional Watson Hamiltonian derived from a rigid rotor perspective with low order centrifugal distortion effects included. In the paper, the authors show that even in extremely floppy molecular systems with wide amplitude, vibrational, motion, the quantum term values are very well fit by a rigid or semirigid rotor Hamiltonian over the limited range of energy states accessible in a cooled beam. The authors provide explicit examples of this behavior by full quantum solutions in two extremes of floppy motion: a symmetric triatomic with a square well bending potential ('hinge') and a nearly free internal rotor ('pinwheel'). These results show that potentials with fundamentally different topologies can be consistent with same data, and indicate that even the limits of nearly rigid and floppy internal motion may be difficult to distinguish from a limited set of rovibrational eigenvalues.

900,527
PB90-126236 Not available NTIS
American Chemical Society, Washington, DC.
Journal of Physical and Chemical Reference Data, Volume 18, Number 3; 1989.
Quarterly rept.
D. R. Lide. c1989, 426p
See also PB90-126244 through PB90-126269 and PB89-222525. Prepared in cooperation with American Inst. of Physics, New York. Sponsored by National Inst. of Standards and Technology, Gaithersburg, MD.
Available from American Chemical Society, 1155 16th St., NW, Washington, DC 20036.

Keywords: *Physical properties, *Chemical properties, Water, Organic compounds, Solubility, High temperature tests, Heavy water, Gases, Hydrocarbons, Microwave spectroscopy, Tables(Data), Henrys law, Molecular rotation, Octanols, Distribution(Property).

Contents: Octanol water partition coefficients of simple organic compounds; Evaluation of data on solubility of simple apolar gases in light and heavy water at high temperature; Microwave spectral tables. III. Hydrocarbons, CH to C10H10; Cumulative listing of reprints and supplements. (Copyright (c) by the U.S. Secretary of Commerce, 1989.)

900,528
PB90-126244 Not available NTIS
Sangster Research Labs., Montreal (Quebec).
Octanol-Water Partition Coefficients of Simple Organic Compounds.
Quarterly rept.
J. Sangster. c1989, 118p
Prepared in cooperation with American Chemical Society, Washington, DC., and American Inst. of Physics, New York. Sponsored by National Inst. of Standards and Technology, Gaithersburg, MD.
Included in Jnl. of Physical and Chemical Reference Data, v18 n3 p1111-1228 1989. Available from American Chemical Society, 1155 16th St., NW, Washington, DC 20036.

Keywords: *Organic compounds, Tables(Data), Physical properties, Data processing, Environmental surveys, Thermodynamics, Experimental design, *Octanols, *Water, *Distribution(Property), Biological effects.

Octanol-water partition coefficients (log P) for '611' simple organic compounds representing all principal classes have been retrieved from the literature. Available experimental details of measurement are documented from original articles. Pertinent thermodynamic relations are presented, with a discussion of direct and indirect methods of measurement. Reported log P

CHEMISTRY
Physical & Theoretical Chemistry

data for each compound have been evaluated according to stated criteria, and recommended values (with uncertainty) are given.

900,529

PB90-126251   Not available NTIS
Comision Nacional de Energia Atomica, Buenos Aires (Argentina). Dept. de Quimica de Reactores.
Evaluation of Data on Solubility of Simple Apolar Gases in Light and Heavy Water at High Temperature.
Quarterly rept.
R. F. Prini, and R. Crovetto. c1989, 13p
Prepared in cooperation with Delaware Univ., Newark. Dept. of Chemistry, American Chemical Society, Washington, DC., and American Inst. of Physics, New York. Sponsored by National Inst. of Standards and Technology, Gaithersburg, MD.
Included in Jnl. of Physical and Chemical Reference Data, v18 n3 p1231-1243 1989. Available from American Chemical Society, 1155 16th St., NW, Washington, DC 20036.

Keywords: *Solubility, *Heavy water, *Water, *Gases, Deuterium, Chemical properties, Thermodynamic properties, Henrys law, Data processing, Ethane, Nitrogen, Oxygen, Hydrogen, Tables(Data), High temperature tests, Rare gases, Methane.

The solubility data of apolar gases in light and heavy water over the temperature range covered experimentally have been evaluated, laying particular emphasis to the region above the normal boiling points of the solvents. The systems that have been included in the work are the inert gases and CH4 in light water and heavy water, H2, O2, N2 and C2H6 in light water and D2 in heavy water. Data in the original sources have been brought to the same footing by calculating from the raw experimental data P, T, and x when they were not reported by the author. The step is considered necessary to assess critically the available sets of data. The temperature dependence of Henry's constants for all the binary systems have been expressed in terms of two different polynomial equations. The formulations presented are discussed and the limits of application given.

900,530

PB90-126269   Not available NTIS
National Inst. of Standards and Technology (NML), Gaithersburg, MD.
Microwave Spectral Tables, 3. Hydrocarbons, CH to C10H10.
Quarterly rept.
F. J. Lovas, and R. D. Suenram. c1989, 27p
Prepared in cooperation with American Chemical Society, Washington, DC., and American Inst. of Physics, New York.
Included in Jnl. of Physical and Chemical Reference Data, v18 n3 p1245-1524 1989. Available from American Chemical Society, 1155 16th St., NW, Washington, DC 20036.

Keywords: *Microwave spectroscopy, *Hydrocarbons, *Molecular rotation, Physical properties, Tables(Data), Data processing, Dipole moments, Hyperfine structure, Spectral lines, Isotopes.

All of the rotational spectral lines observed and reported in the open literature for 91 hydrocarbon molecules have been tabulated. The isotopic molecular species, assigned quantum numbers, observed frequency, estimated measurement uncertainty and reference are given for each transition reported. In addition to correcting a number of misprints and errors in the literature cited, the spectral lines for many normal isotopic species have been refit to produce a comprehensive and consistent analysis of all the data extracted from various literature sources. The derived molecular properties, such as rotational and centrifugal distortion constants, hyperfine structure constants, electric dipole moments, and rotational g-factors are listed.

900,531

PB90-126141   Not available NTIS
National Inst. of Standards and Technology (NML), Gaithersburg, MD. Molecular Spectroscopy Div.
Microwave Electric-Resonance Optothermal Spectroscopy of (H2O)2.
Final rept.
G. T. Fraser, R. D. Suenram, and L. H. Coudert. 1989, 9p
Pub. in Jnl. of Chemical Physics 90, n11 p6077-6085, 1 Jun 89.

Keywords: *Microwave spectroscopy, Hydrogen bonds, Molecular beams, Reprints, *Water dimers, *Dimers, *Complexes.

The microwave spectrum of (H2O)2 has been measured between 14 and 110 GHz using a newly developed electric-resonance optothermal spectrometer (EROS) described here. The reported measurements extend previous results on the a-type K sub a = 0-0 and 1-1 bands for the (A sub 2, sup (+-)), B sup 2, sup (+-)), and E sup (+-) rotational-tunneling states and include the first observations of the c-type K sub a = 1-0 band for the (A sub 2, sup (+-)) and (B sup 2, sup +-)) states and the a-type K sub a = 0-0 band for the (A sub 1, sup (+-1)) states. For the (A sub 1, sup (+-1)) states an interconversion tunneling splitting of 22.6 GHz is obtained, compared to the 19.5 GHz value found previously for the K sub a = O (A sub 2, sup (+-)) and (B sup 2, sup (+-)) states.

900,532

PB90-128547   Not available NTIS
National Inst. of Standards and Technology (NML), Boulder, CO. Quantum Physics Div.
Interaction of in Atom Spin-Orbit States with Si(100) Surfaces.
Final rept.
D. J. Oostra, R. V. Smilgys, and S. R. Leone. 1989, 6p
Sponsored by Weapons Lab., Kirtland AFB, NM.
Pub. in Materials Research Society Symposia Proceedings Chemical Perspectives of Microelectronic Materials, v131 p239-244 1989.

Keywords: *Indium, *Silicon, *Desorption, Spin orbit interactions, Scattering, Reprints, Laser induced fluorescence, Auger electron spectroscopy, Surface reactions, Binding energy, Semiconductors.

Scattering and desorption of In from Si (100) is investigated. Laser induced fluorescence is used to probe the desorbing and or scattered species. Auger Electron Spectroscopy is used to study the composition on the surface. The results show that at surface temperatures below 820 K atwo dimensional layer of In desorbs by a half order mechanism. This is explained by assuming two dimensional In islands on the surface. Above 820 K, desorption takes place by a first order mechanism. The desorption parameters appear to be spin-orbit state specific. The desorption energy for In doublet P(3/2) is 2.8 plus or minus 0.4 eV and for In doublet P(1/2) 2.5 plus or minus 0.2 eV. The difference is equal to the difference in the spin-orbit energy. So far no specular scattering of In is observed, suggesting that the sticking coefficients are unity.

900,533

PB90-128729   Not available NTIS
National Inst. of Standards and Technology (NML), Gaithersburg, MD. Molecular Spectroscopy Div.
Vibrational Spectra of Molecular Ions Isolated in Solid Neon. 2. O4(1+) and O4(1-).
Final rept.
W. E. Thompson, and M. E. Jacox. 1989, 12p
Sponsored by Army Research Office, Research Triangle Park, NC.
Pub. in Jnl. of Chemical Physics 91, n7 p3826-3837, 1 Oct.89.

Keywords: *Vibrational spectra, *Infrared spectra, Neon, Molecular structure, Photochemical reactions, Ionization, Absorption, Solids, Reprints, *Matrix isolation technique, *Oxygen ions, Chemical reaction mechanisms.

When a relatively concentrated Ne:O2 sample is composited at approximately 5 K with a beam of excited neon atoms, prominent infrared absorptions appear which are assigned to O4(+) and O4(-). Absorptions of O3 and O3(-) are also present, and their product distributions in isotopic substitution experiments indicate that O-atom production and reaction is a minor channel in this experimental system. Detailed isotopic substitution experiments require that both O4(+) and O4(-) possess two equivalent O2 units. Analysis of the isotopic shifts strongly favors a planar trans configuration (C(2h)) for both molecules. Several combination bands of O4(-) are observed, and give evidence regarding the position of Nu(sub 1)(a(sub g)), which is infrared inactive, and regarding perturbations by combinations of low-frequency fundamentals. The mechanism of photodestruction of the ions in this system is also considered.

900,534

PB90-136375   Not available NTIS

National Inst. of Standards and Technology (NML), Gaithersburg, MD. Chemical Kinetics Div.
Temperature Dependence of the Rate Constant for the Hydroperoxy + Methylperoxy Gas-Phase Reaction.
Final rept.
P. Dagaut, T. J. Wallington, and M. J. Kurylo. 1988, 4p
Pub. in Jnl. of Physical Chemistry 92, n13 p3833-3836 1988.

Keywords: *Reaction kinetics, *Photolysis, *Absorption spectra, Ultraviolet spectroscopy, Flash point, Temperature, Spectrum analysis, Vapor phases, Reprints, *Hydroperoxy radicals, *Methylperoxy radicals, Arrhenius equation, Chemical reaction mechanisms, Peroxy radicals.

The temperature dependence of the reaction between hydroperoxy and methylperoxy radicals was measured in a flash photolysis ultraviolet absorption apparatus over the temperature range 228 - 380 K. (HO2 + CH3O2 -> products.) The data represented by the Arrhenius expression are compared to earlier results and discussed in terms of the reaction mechanism. Due to overlapping absorptions of the two radicals and deviations of the complex reaction system from both pseudo-first and -second order behavior, the rate constants were determined from a detailed modeling of the radical decay curves. A sensitivity analysis of the rate constant determination procedure to the assumed radical absorption cross-sections and correlated changes in the rate constants for the HO2 and CH3O2 self-reactions was performed and the results are reported.

900,535

PB90-136474   Not available NTIS
National Inst. of Standards and Technology (NML), Gaithersburg, MD. Electricity Div.
Fundamental Physical Constants - 1986 Adjustments.
Final rept.
E. R. Cohen, and B. N. Taylor. 1987, 4p
Sponsored by Rockwell International, Thousand Oaks, CA. Science Center.
Pub. in Europhysics News 18, n5 p65-68 May 87.

Keywords: *Fundamental constants, Reviews, Physical properties, Least square method, Revisions, Tables(Data), Comparison, Reprints.

The 1986 adjustment of the fundamental constants, just completed under the aegis of the CODATA Task Group on Fundamental Constants, is reviewed and compared with the previous recommendations published in 1973. The precision of the recommended values for the physical constants has improved by an order of magnitude. Summary tables of these values and associated data are presented.

900,536

PB90-136565   Not available NTIS
National Inst. of Standards and Technology (NML), Gaithersburg, MD. Chemical Kinetics Div.
Flash Photolysis Kinetic Absorption Spectroscopy Study of the Gas Phase Reaction HO2 + C2H5O2 Over the Temperature Range 228-380 K.
Final rept.
P. Dagaut, T. J. Wallington, and M. J. Kurylo. 1988, 4p
Pub. in Jnl. of Physical Chemistry 92, n13 p3836-3839 1988.

Keywords: *Reaction kinetics, *Photolysis, *Absorption spectra, Flash point, Temperature, Spectrum analysis, Vapor phases, Reprints, *Hydroperoxy radicals, *Ethylperoxy radicals, Peroxy radicals, Arrhenius equation.

Flash photolysis kinetic absorption spectroscopy was used to investigate the gas phase reaction between hydroperoxy and ethylperoxy radicals between 228 and 380K in the pressure range 25-400 Torr. (HO2 + C2H5O2 -> products). Due to the large difference between the self reactivities of the two radicals, first or second order kinetic conditions could not be maintained for either species. Thus, the rate constant was determined from computer modeled fits of the radical absorption decay curves recorded at wavelengths between 230 and 280 nm. A reanalysis of earlier measurements of the C2H5O2 self-reaction (C2H5O2 + C2H5O2 -> products) using this expression for k1 resulting in the revised Arrhenius equation produced k2.

900,537
PB90-136573   Not available NTIS
National Inst. of Standards and Technology (NML),
Gaithersburg, MD. Molecular Spectroscopy Div.
**Picosecond Vibrational Energy Transfer Studies of
Surface Adsorbates.**
Final rept.
E. J. Heilweil, M. P. Casassa, R. R. Cavanagh, and J.
C. Stephenson. 1989, 29p
Sponsored by Air Force Office of Scientific Research,
Bolling AFB, DC.
Pub. in Annual Review of Physical Chemistry 40, p143-
171 1989.

Keywords: *Adsorbates, *Molecular relaxation, *Surface chemistry, Infrared spectroscopy, Metals, Dielectrics, Reprints, *Picosecond pulses, Laser applications.

Recent measurements of vibrational relaxation for adsorbates at dielectric and metal particle surfaces are
reviewed. The vibrational lifetime (T (sub 1)) for various
surface and model liquid and solid systems are also
tabulated. Transient infrared picosecond laser methods are described as are relevant theories pertinent to
adsorbate vibrational (v=1) relaxation mechanisms at
surfaces.

900,538
PB90-136599   Not available NTIS
National Bureau of Standards (NML), Gaithersburg,
MD. Chemical Thermodynamics Div.
**Use of an Imaging Proportional Counter in Macro-
molecular Crystallography.**
Final rept.
A. J. Howard, G. L. Gilliland, B. C. Finzel, T. L.
Poulos, D. H. Ohlendorf, and F. R. Salemme. 1987,
5p
Pub. in Jnl. of Applied Crystallography 20, pt5 p383-
387, 1 Oct 87.

Keywords: *Proportional counters, *Crystallography,
*x ray diffraction, *Molecule structure, Data acquisi-
tion, Performance evaluation, Reprints, Fourier analysis.

A multiwire proportional chamber known as an Imaging
Proportional Counter has been used to collect X-ray
intensity data for the determination of several structures by molecular replacement or difference Fourier
analysis and has provided data for numerous other
macromolecular crystallographic projects. Results obtained with an Imaging Proportional Counter mounted
on a rotating anode X-ray generator indicate that the
detector produces accurate intensity information and
that its reliability is high.

900,539
PB90-136755   Not available NTIS
National Inst. of Standards and Technology (IMSE),
Gaithersburg, MD. Office of Nondestructive Evalua-
tion.
**Triplet Dipoles in the Absorption Spectra by Dense
Rare Gas Mixtures. 1. Short Range Interactions.**
Final rept.
B. Guillot, R. D. Mountain, and G. Birnbaum. 1989,
13p
Pub. in Jnl. of Chemical Physics 90, n2 p650-662, 15
Jan 89.

Keywords: *Rare gases, *Dipole moments, Far infra-
red radiation, Absorption spectra, Liquefied gases,
Mixtures, Argon, Krypton, Helium, Hydrogen, Reprints,
Molecular dynamics.

A theory is proposed to evaluate the induced-dipole
moment occurring when three dissimilar atoms are
mutually interacting. Based on the one-effective electron model, the theory predicts that, at short and intermediate distances, the triatomic dipole moment originates from three different processes, namely, overlap,
quadrupole-induced, and dipole-induced. These three-
body dipoles have then been implemented into a molecular dynamics simulation in order to generate the
collision-induced absorption spectra by rare-gas mixtures. For Ar-Kr liquid mixtures, the irreducible three-
body contributions to the spectral density are found so
important that they exceed two-body contributions.
This is due to a profound cancellation between two-
body dipoles which does not occur between irreducible
three-body dipoles. However, the comparison with experimental data is poor because of a shortcoming of
the model calculation which does not take into account long-range dispersion interactions. On the contrary, a quantitative agreement is obtained for the fun-

damental vibration band of compressed H2-He mixtures, indicating that exchange overlap, three-body dipoles play the leading role for such light systems.

900,540
PB90-136904   Not available NTIS
National Bureau of Standards (IMSE), Gaithersburg,
MD. Metallurgy Div.
**Laser Induced Vaporization Time Resolved Mass
Spectrometry of Refractories.**
Final rept.
D. W. Bonnell, P. K. Schenck, and J. W. Hastie.
1988, 13p
Pub. in Proceedings of the Electrochemical Society,
v88-10 p82-94 1988.

Keywords: *Vaporizing, *Mass spectroscopy, *Spectrum analysis, *Refractory materials, Chemical analysis, Boron nitrides, Graphite, High temperature tests,
Thermodynamic equilibrium, Surface temperature,
*Laser-radiation heating.

An experimental approach is described which can
yield information about refractory surfaces by examining the time history of the gas-dynamic process occurring during pulsed Nd/YAG laser induced vaporization.
Specific examples consider BN and graphite vaporization. Time resolved mass spectrometric measurements of evolved species permit direct determination
of gas species identities and concentration, independent of mass spectral cracking problems. Of particular
interest is the observation of local thermodynamic
equilibrium for the observed laser vaporized gas species in both systems from surface temperatures of
2900 K (BN) and 4000 K (graphite). Indirect methods
of determining surface temperature as alternatives to
direct measurement of radiance temperature are presented. Also, a preliminary analysis of the convolution
problem to eliminate amplifier RC response delays and
to extract true species velocity distributions is discussed.

900,541
PB90-136946   Not available NTIS
National Inst. of Standards and Technology (NML),
Boulder, CO. Time and Frequency Div.
**Frequency Measurements of High-J Rotational
Transitions of OCS and N2O.**
Final rept.
M. D. Vanek, D. A. Jennings, and J. S. Wells. 1989,
5p
Pub. in Jnl. of Molecular Spectroscopy 138, p79-83
1989.

Keywords: *Nitrogen oxide(N2O), *Infrared spectroscopy, *Molecular vibration, *Thiocarbamates, *Frequency measurement, Molecular energy levels, Molecular rotation, Far infrared radiation, Performance evaluation, Reprints.

A metal-insulator-metal diode has been used to generate far-infrared radiation as the difference between
two CO2 lasers. With this technique rotational transitions of all the vibrational states below 2000/cm have
been measured for the normal isotopic species of OCS
with an accuracy of 200 kHz or better.

900,542
PB90-163874   PC A04
National Inst. of Standards and Technology, Gaithers-
burg, MD.
**Journal of Research of the National Institute of
Standards and Technology. November-December
1989, Volume 94, Number 6.**
1989, 69p
See also PB90-163882 through PB90-163932 and
Volume 94, Number 4. Also available from Supt. of
Docs. SN703-027-00031-8.

Keywords: *Atmospheric pressure, *Metrology, *Radioactive isotopes, *Metals, Radioactivity, Nickel isotopes, Standards, Measurement, Waveforms, Ionizing
radiation, Atomic mass, Manometers, Mass spectrometers.

Contents: The reduction of uncertainties for absolute
piston gage pressure measurements in the atmospheric pressure range; Absolute isotopic abundance ratios
and atomic weight of a reference sample of nickel; The
absolute isotopic composition and atomic weight of
terrestrial nickel; Report on the 1989 meeting of the
radionuclide measurements section of the consultative
committee on standards for the measurement of ionizing radiations; On measuring the root-mean-square
value of a finite record length periodic waveform; A
search for optical molasses in a vapor cell; General
analysis and experimental attempt.

900,543
PB90-163890
                    (Order as PB90-163874, PC A04)
National Inst. of Standards and Technology, Gaithers-
burg, MD.
**Absolute Isotopic Abundance Ratios and Atomic
Weight of a Reference Sample of Nickel.**
J. W. Gramlich, L. A. Machlan, I. L. Barnes, and P. J.
Paulsen. 1989, 10p
Prepared in cooperation with Curtin Univ. of Technolo-
gy, Bentley (Australia).
Included in Jnl. of Research of the National Institute of
Standards and Technology, v94 n6 p347-456 1989.

Keywords: *Nickel isotopes, *Atomic weight, Calibrating, Mass spectrometers, Assay.

Absolute values have been obtained for the isotopic
abundance ratios of a reference sample of nickel
(Standard Reference Material 986), using thermal ionization mass spectrometry. Samples of known isotopic
composition, prepared from nearly isotopically pure
separated nickel isotopes, were used to calibrate the
mass spectrometers.

900,544
PB90-163908
                    (Order as PB90-163874, PC A04)
National Inst. of Standards and Technology, Gaithers-
burg, MD.
**Absolute Isotopic Composition and Atomic Weight
of Terrestrial Nickel.**
J. W. Gramlich, E. S. Beary, L. A. Machlan, and I. L.
Barnes. 1989, 6p
Included in Jnl. of Research of the National Institute of
Standards and Technology, v94 n6 p357-362 1989.

Keywords: *Nickel isotopes, *Atomic weight,
Concentration(Composition), Mass spectrometers,
Standards, Accuracy.

Twenty-nine samples of high-purity nickel metals, reagent sales and minerals, collected from worldwide
sources, have been examined by high-precision isotope ratio mass spectrometry for their nickel isotopic
composition. These materials were compared directly
with SRM 986, certified isotopic standard for nickel,
using identical measurement techniques and the same
instrumentation.

900,545
PB90-163916
                    (Order as PB90-163874, PC A04)
National Inst. of Standards and Technology, Gaithers-
burg, MD.
**Report on the 1989 Meeting of the Radionuclide
Measurements Section of the Consultative Com-
mittee on Standards for the Measurement of Ioniz-
ing Radiations: Special Report on Standards for
Radioactivity.**
D. D. Hoppes. 1989, 4p
Included in Jnl. of Research of the National Institute of
Standards of Technology, v94 n6 p363-366 1989.

Keywords: *Radioactivity, *Radioactive isotopes, *Ionizing radiation, Standards, Measurement.

The report describes the activities discussed at the
10th meeting of Section II of the Consultative Committee on Standards for the Measurement of Ionizing Radiations held in May 1989 at Sevres (France). Topics
included present and future international comparisons
of activity measurements, the status and possible extension of the international reference system for activity measurements of gamma-ray emitting nuclides, reports from other working groups, accomplishments at
the International Bureau of Weights and Measures.

Polymer Chemistry

900,546
PB89-146724   Not available NTIS
National Bureau of Standards (IMSE), Gaithersburg,
MD. Polymers Div.
**Effect of Crosslinks on the Phase Separation Be-
havior of a Miscible Polymer Blend.**
Final rept.
R. M. Briber, and B. J. Bauer. 1988, 8p
Pub. in Macromolecules 21, n11 p3296-3303 Nov 88.

CHEMISTRY

Polymer Chemistry

Keywords: *Irradiation, *Crosslinking, *Polystyrene, *Vinyl ether resins, Phase diagrams, Neutron scattering, Solubility, Blending, Polymers, Reprints.

The effect of radiation cross-linking on the phase diagram and scattering function for a compatible polymer blend of deuterated polystyrene and poly(vinyl methyl ether) has been examined by small angle neutron scattering. The scattering curves for the cross-linked blends exhibit a maximum at a nonzero q vector, the position of which depends linearly on the square root of the radiation dose or inversely on the square root of the number of repeat units between cross-links. This dependence was predicted by de Gennes, but the measured position of the maximum is smaller than predicted. The spinodal temperature can be determined by plotting the inverse of the scattered intensity at the maximum, $S(q\ sub\ *)$ sup minus one, versus inverse temperature and extrapolating to the point where $S(q\ sub\ *)$ sup minus one = 0. The inverse of the measured spinodal temperature depends linearly with radiation dose or with N sup minus one as predicted. This increases the size of the single phase region of the phase diagram with the extrapolated spinodal temperature increasing from 149 deg C for the uncross-linked blend to 430 deg C for the blend with a radiation dose of 125 Mrad. The theory by de Gennes predicts that the scattered intensity at q = 0 equals 0 for the cross-linked blends which is not observed experimentally.

900,547
PB89-147029   Not available NTIS
National Bureau of Standards (NML),, Gaithersburg, MD. Ionizing Radiation Physics Div.
Hydroxyl Radical Induced Cross-Linking between Phenylalanine and 2-Deoxyribose.
Final rept.
M. Farahani, and M. G. Simic. 1988, 4p
Pub. in Biochemistry 27, n13 p4695-4698, 28 Jun 88.

Keywords: *Chemical reactions, *Crosslinking, *Phenylalanine, Chemical radicals, Oxygen, Free radicals, Separation, Gas chromatography, Mass spectroscopy, Models, Deoxyribonucleic acids, Proteins, Amino acids, Radiation chemistry, Reprints, *Deoxyribose, Oxygen radicals.

Hydroxy radicals induce cross-linking between phenylalanine (Phe) and 2-deoxyribose (dR) via formation of corresponding free radical intermediates. The cross-linked products were separated and identified by capillary gas chromatography-mass spectrometry. When phenylalanine and 2-deoxyribose radicals were generated in a 1:1 ratio, the predominant interaction was between Phe and dR radicals while the Phe-Phe and dR-dR cross-links were less abundant. The newly discovered cross-link between 2-deoxyribose and phenylalanine may serve as a model for radiation or free radical induced cross-linking between DNA and proteins and in general between sugar moieties and amino acids.

900,548
PB89-157093   Not available NTIS
National Bureau of Standards (IMSE), Gaithersburg, MD. Polymers Div.
Resonant Raman Scattering of Controlled Molecular Weight Polyacetylene.
Final rept.
M. A. Schen, J. C. W. Chien, E. Perrin, S. Lefrant, and E. Mulazzi. 1988, 6p
Pub. in Jnl. of Chemical Physics 89, n12 p7615-7620, 15 Dec 88.

Keywords: *Acetylene, *Molecular weight, *Catalysts, *Polymers, Titanium, Synthesis, Organometallic compounds, Metal complexes, Aluminum, Raman spectra, Oxidation, Isomerization, Electrical resistivity, Reprints.

Polyacetylene, (CH) sub x, films of 500, 5300, 10,500, and 100,000 Daltons number average molecular weights (M sub n) were synthesized using the titanium tetra-n-butoxide/triethyl aluminum-catalyst/cocatalyst system and examined using resonant Raman scattering techniques. Before isomerization, trans segments are found to exist mainly as short, isolated sequences independent of M sub n. After thermal isomerization, theoretical analysis of the RRS spectra using the Brivio, Mulazzi model indicate the ratio of long trans conjugated segments (N is greater than or equal to 30) to short trans conjugated segments (N is less than or equal to 30) is significantly larger for 100,000 Dalton polymer in comparison to polymer of 10,500 M sub n and below. For samples below 10,500 Daltons, no clear relationship between actual polymer molecular

weight and G is observed. Optimization of the isomerization conditions for 100,000 Dalton polymer results in trans-(CH) sub x with a G = 0.80. These results suggest that not until very long molecular chains are obtained can samples composed principally of long conjugated segments be obtained. It is proposed that defects which arise during and after the polymerization limit the content of long segments. Ambient, short term oxidation of 100,000 M sub n polymer shows a decrease in G from 0.80 to 0.70. Low-level chain oxidation or doping is shown to preferentially occur within long conjugated segments.

900,549
PB89-157465   Not available NTIS
National Bureau of Standards (NEL), Gaithersburg, MD. Building Materials Div.
Thermal Degradation of Poly (methyl methacrylate) at 50 C to 125 C.
Final rept.
J. W. Martin, D. P. Bentz, W. E. Byrd, B. Dickens, E. Embree, W. E. Roberts, and D. Waksman. 1987, 17p
Sponsored by Department of Energy, Washington, DC.
Pub. in Jnl. of Applied Polymer Science 34, n1 p377-393 1987.

Keywords: *Chemical bonds, *Cleavage, *Polymers, *Polymethyl methacrylate, Thin films, Temperature, Free radicals, Oxidation, Reprints.

Small but significant numbers of chain scissions occur in a commercial poly (methyl methacrylate) film exposed to temperatures between 50 and 125 C. The scission rate is initially rapid and then slows down to a constant rate. The initial rate of chain scissions is temperature dependent, while the long-term rate of chain scissions appears to be temperature independent. Four modes of failure are proposed to explain the results: (1) the presence of unreacted initiators of polymerization; (2) free radicals generated from additives in the commercial film; (3) weak links in the polymer chain; and (4) free radicals generated in the thermal decomposition of an oxidation product of the monomer methyl methacrylate. The mode of failure which is most consistent with the experimental results is the one involving free radicals generated from an oxidation product of monomer.

900,550
PB89-157473   Not available NTIS
National Bureau of Standards (IMSE), Gaithersburg, MD. Polymers Div.
Temperature, Composition and Molecular-Weight Dependence of the Binary Interaction Parameter of Polystyrene/Poly(vinylmethylether) Blends.
Final rept.
C. Han, B. Bauer, J. Clark, Y. Muroga, Y. Matsushita, M. Okada, Q. Tran-Cong, and T. Chang. 1988, 13p
Pub. in Polymer 29, p2002-2014 Nov 88.

Keywords: *Polystyrene, *Polymers, *Neutron scattering, *Free energy, *Vinyl ether resins, Isotopic labelling, Deuterium, Molecular weight, Temperature, Chemical composition, Mixtures, Reprints.

The binary interaction parameter has been obtained for deuterated polystyrene/poly(vinyl methyl ether) blends as a function of temperature, composition and molecular weight from small-angle neutron scattering experiments. The consistency of the correlation length epsilon, the zero-wavenumber scattering intensity S(0) and the chi sub eff parameter with the mean-field prediction has been demonstrated by the q dependence of the static structure factor S(q) and the 1/T dependence of epsilon sup -2, S(0) sup -1 and chi sub eff. The effective interaction parameter chi sub eff can be related to the Flory-Huggins interaction parameter chi sub F. The free-energy function as well as the spinodal curve and cloud-point curve have been constructed.

900,551
PB89-161616   Not available NTIS
National Bureau of Standards (IMSE), Gaithersburg, MD. Polymers Div.
Synthesis   and   Characterization   of Poly(vinylmethyl ether).
Final rept.
B. J. Bauer, B. Hanley, and Y. Muroga. 1989, 3p
Pub. in Polymer Communications 30, p19-21 Jan 89.

Keywords: *Polymerization, *Molecular weight, Reprints, *Poly(ether/vinylmethyl).

High molecular weight poly(vinylmethyl ether) has been produced by cationic polymerization with boron trifluoride ethyl ether complex as an initiator. The poly-

mer has a broad molecular weight distribution, but multiple fractionations with toluene as solvent and heptane as non-solvent produced samples with M(sub w)/M(sub n) as low as 1.2.

900,552
PB89-172449   Not available NTIS
National Bureau of Standards (IMSE), Gaithersburg, MD. Polymers Div.
Concentration Dependence of the Compression Modulus of Isotactic Polystyrene/Cis-Decalin Gels.
Final rept.
J. M. Guenet, and G. B. McKenna. 1986, 10p
Pub. in Jnl. of Polymer Science 24, n11 p2499-2508 Nov 86.

Keywords: *Polystyrene, *Decalin, *Gels, *Compressing, *Quenching(Cooling), *Micelles, Polymers, Modulus of elasticity, Phase diagrams, Reprints.

New results for isotactic polystyrene/cis-decalin gels are presented which suggest that a simple fringed micellar model of the gel may not be appropriate. In measuring the room temperature compressive modulus of gels formed at -20C, it was found that, not only are the gels time dependent but also that the isochronal modulus-concentration diagrams exhibit sharp features reminiscent of a phase diagram rather than the smooth curves (straight lines) typical of swollen rubber and fringed micellar gels.

900,553
PB89-172456   Not available NTIS
National Bureau of Standards (IMSE), Gaithersburg, MD. Polymers Div.
Effects of Solvent Type on the Concentration Dependence of the Compression Modulus of Thermoreversible Isotactic Polystyrene Gels.
Final rept.
G. B. McKenna, and J. M. Guenet. 1986, 10p
Pub. in Jnl. of Polymer Science B, Polymer Physics 26, n2 p267-276 Feb 88.

Keywords: *Polymers, *Polystyrene, *Decalin, *Gels, *Compression, Quenching(Cooling), Chlorohydrocarbons, Modulus of elasticity, Relaxation(Mechanics), Stresses, Reprints.

Solutions of isotactic polystyrene in either trans-decalin or 1-chlorodecane were transformed into gels by quenching from a high temperature (approximately 180 C) to -20 C. The relaxation modulus in compression of these gels was measured over a range of concentrations of from 0.04 g/g to 0.40 g/g. At 22 C the gels show a double logarithmic stress relaxation rate, m, which is higher than for polyvinylchloride and gelatin gel systems. 120s isochronal modulus concentration diagrams exhibit non-power law behavior, i.e., not only is the general trend such that the double logarithmic slope decreases with increasing concentration, but there are also regions in which abrupt changes in modulus occur over narrow ranges in concentrations. These features in the concentration dependence of the modulus are less pronounced than those found previously in isotactic polystyrene/cis-decalin gels. The behavior is interpreted to be inconsistent with a fringed micelle picture of the gel structure.

900,554
PB89-172472   Not available NTIS
National Bureau of Standards (IMSE), Gaithersburg, MD. Polymers Div.
Measurement of the Torque and Normal Force in Torsion in the Study of the Thermoviscoelastic Properties of Polymer Glasses.
Final rept.
G. B. McKenna. 1985, 15p
Pub. in Relaxations in Complex Systems, p129-143 1985.

Keywords: *Torsion tests, *Polymers, *Glass, *Aging tests(Materials), *Volume, *Polymethyl methacrylate, Torque, Viscoelasticity, Thermal properties, Normal strain, Strain measurement, Reprints.

The simultaneous measurement of the torque and normal force responses in torsion experiments on viscoelastic materials provides multidimensional (multiaxial) data from a single simple test geometry. Data are presented which support the important theoretical prediction that in two step torsional deformations where the magnitude of the second step is one-half the magnitude of the first step, the normal stress response is predicted to be independent of the duration of the first step and equal to the single step response to a defor-mation of the same magnitude as the second step. The second program involves a study of physical aging in a freshly quenched polymer glass of polymethyl methacrylate.

900,555
PB89-172480　　　　　　　　Not available NTIS
National Bureau of Standards (IMSE), Gaithersburg, MD. Polymers Div.
Viscosity of Blends of Linear and Cyclic Molecules of Similar Molecular Mass.
Final rept.
G. B. McKenna, and D. J. Plazek. 1986, 3p
Pub. in Polymer Communications 27, n10 p304-306 Oct 86.

Keywords: *Polymers, *Molecular weight, *Polysty-rene, *Melt viscosity, Contamination, Chains, Rings, Linearity, Cyclization, Reprints, Reptation.

Narrow fractions of linear polystyrene were added to cyclic polystyrenes of similar molecular weights. For weight concentrations, theta sub c, of the cyclic poly-mers from 0.85 to 1.0, it is found that the zero shear viscosity varies as theta sub c sup (-5.6). These results suggest that discrepancies between reports in the liter-ature for the molecular weight dependence of the melt viscosity of cyclic polymers can only be partially ac-counted for by contamination of the rings with linear chains.

900,556
PB89-173801　　　　　　　　Not available NTIS
National Bureau of Standards (NEL), Gaithersburg, MD. Building Materials Div.
Preliminary Stochastic Model for Service Life Pre-diction of a Photolytically and Thermally Degraded Polymeric Cover Plate Material.
Final rept.
J. Martin, D. Waksman, D. Bentz, J. A. Lechner, and B. Dickens. 1984, 20p
Pub. in Proceedings of International Conference on the Durability of Building Materials and Components (3rd), Espoo, Finland, August 12-15, 1984, v3 p549-568.

Keywords: *Plates(Structural members), *Durability, *Service life, *Thin films, *Polymethyl methacrylate, Stochastic processes, Mathematical model, Forecast-ing, Thermal degradation, Photodegradation, Failure, Polymers, Molecular weight, Oxidation.

A preliminary stochastic model has been developed and partially validated for predicting the service life of a polymeric film, such as poly(methyl methacrylate) (PMMA), which is subjected to both thermal and pho-tolytic degradation. The exposure conditions under which degradation is induced simulate those expected for a polymeric cover plate material in an active solar collector. Service life for a population of films is de-fined as that time beyond which an unacceptable por-tion of the population fails. The criterion used for fail-ure of a film is that its number average molecular weight has fallen below a specified threshold value.

900,557
PB89-176028　　　　　　　　Not available NTIS
National Bureau of Standards (IMSE), Gaithersburg, MD. Polymers Div.
Theory of Microphase Separation in Graft and Star Copolymers.
Final rept.
M. O. de la Cruz, and I. C. Sanchez. 1986, 8p
Pub. in Macromolecules 19, n10 p2501-2508 1986.

Keywords: *Polymers, *Graft polymerization, *Phases, *Separation, Radius of gyration, Critical point, Re-prints.

Phase stability criteria and static structure factors have been calculated for simple AB graft copolymers, for star copolymers with equal number of A and B arms, and for n arm star diblock copolymers. The A-B inter-actions are characterized by the usual chi parameter. The fraction of A monomer in the graft copolymer is denoted as f and the fractional position along the A chain backbone at which the B graft is chemically linked is denoted as tan. When tan equals 0 or 1 the graft copolymer degenerates to a simple diblock co-polymer. Leibler previously calculated that the critical value, (chi N)c, at which a AB diblock copolymer con-taining N monomer units undergoes microphase sepa-ration is 10.5. The critical value occurs at f equals 0.5 and is the only composition for which the transition is second order. According to the present theory, a graft

copolymer (0 < tau < 1) does not have a critical point for any f; i.e. all transitions are first order.

900,558
PB89-176036　　　　　　　　Not available NTIS
National Bureau of Standards (IMSE), Gaithersburg, MD. Polymers Div.
Solid State (13)C NMR Investigation in Polyoxe-tanes. Effect of Chain Conformation.
Final rept.
E. Perez, and D. L. VanderHart. 1987, 6p
Pub. in Polymer 28, n5 p733-738 1987.

Keywords: *Polymers, *Nuclear magnetic resonance, *Carbon, Solids, Amorphous materials, Crystal struc-ture, Reprints, Carbon 13, Polyoxetanes, Conforma-tional changes.

Carbon-13 nuclear magnetic resonance (NMR) in the solid state has been applied to the study of poly(oxetane) and poly(3,3-dimethyloxetane). Two dif-ferent crystalline modifications, with T2G2 and T3G conformations, have been prepared for each of the polymers. The corresponding (13)C chemical shifts for each sample, both amorphous and crystalline, have been determined. Single resonances for the methyl-ene carbons alpha to the oxygens, are observed for the crystal having T2G2 conformation. Two reson-ances are observed for these carbons in crystals having T3G conformation. The equivalence of these carbons in the T2G2 conformation and their inequiva-lence, within the same monomer unit, in the T3G con-formation is a property of the isolated chain.

900,559
PB89-176044　　　　　　　　Not available NTIS
National Bureau of Standards (IMSE), Gaithersburg, MD. Polymers Div.
Polymer Localization by Random Fixed Impurities: Gaussian Chains.
Final rept.
J. Douglas. 1988, 5p
Pub. in Macromolecules 21, n12 p3515-3519 Dec 88.

Keywords: *Polymers, *Monte Carlo method, *Chains, *Impurities, *Molecular structure, Scalars, Surface chemistry, Electron distribution, Interactions, Random processes, Position(Location), Reprints.

Simple dimensional analysis is employed to discuss the relevance of impurity interactions on the molecular dimensions of flexible polymers in the limits of high-and low-impurity densities. Scaling arguments account for the universal behavior of static properties observed by Baumgartner and Muthukumar in their recent Monte Carlo simulations. An approximate model of the random impurity interaction is introduced by consider-ing the random impurities as being analogous to an ef-fective surface with which the polymer interacts. Quali-tatively the same conclusions are obtained as in the scaling arguments except that the effective surface analogy provides closed form scaling functions de-scribing the variation of the molecular dimensions as a function of the dimensionless disorder interaction. The transition to a collapsed state is found to be character-ized by a critical impurity density which is a function of the chain length.

900,560
PB89-176051　　　　　　　　Not available NTIS
National Bureau of Standards (IMSE), Gaithersburg, MD. Polymers Div.
Morphological Partitioning of Ethyl Branches in Polyethylene by (13)C NMR.
Final rept.
E. Perez, B. Crist, P. R. Howard, and D. L. Vanderhart. 1987, 10p
Pub. in Macromolecules 20, n1 p76-87 1987.

Keywords: *Proton magnetic resonance, *Polybuta-diene, *Polyethylene, Carbon 13, Hydrogen, Alkanes, Crystallinity, Morphology, Polarization, Diffusion, Sepa-ration, Reprints.

A combination of (13)C and proton magnetic reso-nance experiments has been performed on a model ethylene copolymer (hydrogenated polybutadiene) of about 100,000 molecular weight and 17 ethyl branches per 1000 total carbons. The ratio of ethyl branches found in the crystal in this 41%-crystalline sample was 0.06 + or - 0.02, and the ratio of concentrations be-tween the crystalline and non-crystalline regions was correspondingly about 1:10. For reasons of best inte-grability, the methyl resonance of the ethyl branches was used to deduce the concentrations in each mor-phological phase. The same resonance is rather ill-be-

haved in cross-polarization experiments so that sever-al auxiliary experiments were undertaken to deduce the true concentrations attributed to each phase. The experimental technique utilizes cross-polarization as a probe of proton polarization levels; moreover, the suc-cess of the method relies on local proton spin diffu-sion.

900,561
PB89-176069　　　　　　　　Not available NTIS
National Bureau of Standards (IMSE), Gaithersburg, MD. Polymers Div.
Microphase Separation in Blockcopolymer/Homo-polymer.
Final rept.
M. O. de la Cruz, and I. C. Sanchez. 1987, 4p
Pub. in Macromolecules 20, n2 p440-443 1987.

Keywords: *Phases, *Separation, *Polymerization, Scattering, Mixtures, Chemical analysis, Reprints.

Microphase separation in an A-B blockcopolymer with B homopolymer blend is analyzed. Let theta be the concentration of homopolymer, f the fraction compo-nent A along the blockcopolymer, N the degree of po-lymerization of the blockcopolymer and r the ratio of the blockcopolymer to homopolymer degree of polym-erization. The variations of the spinodal temperature are obtained. The variations of the spinodal temperature is obtained. The scattering function for such a system is obtained. The variations of the spinodal temperature (chi N)s and the wave vector at which the scattering function diverges, q*, are obtained as a function of f, theta, and r in the vicinity of the critical point of a pure blockcopolymer melt (q-o, f = 0.5). It is found that (chi N)s and q* can increase or decrease with respect to the values at theta = zero.

900,562
PB89-176085　　　　　　　　Not available NTIS
National Bureau of Standards (IMSE), Gaithersburg, MD. Polymers Div.
Crazes and Fracture in Polymers.
Final rept.
E. Passaglia. 1987, 26p
Pub. in Jnl. of Physics and Chemistry of Solids 48, n11 p1075-1100 1987.

Keywords: *Crazing, *Microstructure, *Thermoplastic resins, *Fractures(Materials), *Crack propagation, Polymers, Surveys, Stress analysis, Bursting, Reprints, Fibrils.

A review of crazes in glassy thermoplastic polymers is presented with particular emphasis on those aspects of craze properties that influence and control fracture behavior. Both crazes as they normally occur and crazes at the tip of cracks are covered. The occur-rence of crazes, their microstructure, the stress distri-bution within them and the nature of craze fibrils are discussed. Theoretical treatments of the effect of crazes on polymer fracture are reviewed.

900,563
PB89-176424　　　　　　　　Not available NTIS
National Bureau of Standards (NEL), Gaithersburg, MD. Chemical Process Metrology Div.
Polymeric Humidity Sensors for Industrial Process Applications.
Final rept.
P. H. Huang. 1988, 11p
Pub. in Proceedings of Annual Sensors Expo Interna-tional (3rd), Chicago, IL., September 13-15, 1988, p106B-1-106B-16.

Keywords: *Detectors, *Humidity, *Polymers, Meas-urement, Moisture meters, Thermoplastic resins, Ther-moset resins, Industrial engineering, Process control.

Humidity/moisture sensing and measurement tech-niques are summarized as an introduction. Various commercially available sensors for the on-line determi-nation of humidity/moisture in gases, liquids and solids are described. The advantages and disadvantages of the individual sensor types are listed. Polymeric humid-ity sensors are now finding increasing use in industrial process applications. This is made possible by the recent advances in polymer science and technology. The paper deals with the current status of research in polymeric humidity sensors which are based on the use of thermoplastics and thermosets.

900,564
PB89-179246　　　　　　　　Not available NTIS
National Bureau of Standards (IMSE), Gaithersburg, MD. Polymers Div.

CHEMISTRY

Polymer Chemistry

Influence of Molecular Weight on the Resonant Raman Scattering of Polyacetylene.
Final rept.
M. A. Schen, S. Lefrant, E. Perrin, J. C. W. Chien, and E. Mulazzi. 1989, 8p
Pub. in Synthetic Metals 28, pD287-D294 1989.

Keywords: *Polymers, *Acetylene, *Molecular weight, *Oxidation, *Raman spectra, Catalysis, Titanium, Aluminum, Mathematical models, Resonance, Reprints.

For the first time, polyacetylene, (CH)x, films of known number average molecular weights, Mn, have been examined using Resonance Raman Scattering techniques. (CH)x samples of 500, 5300, 10500, and 100,000 Daltons were synthesized using the classical titanium tetra-n-butoxide/triethyl aluminum catalyst/cocatalyst system. After thermal isomerization, modeling of the RRS spectra using the Brivio, Mulazzi model indicate that the 100,000 Dalton polymer is composed principally of long transconjugated segments. In contrast, polymers of 10500 Mn and below are seen to contain significantly larger fractions of short trans conjugated segments. For samples below 10500, no clear relationship between actual polymer molecular weight and G, the ratio of long to short segments, is observed. These results suggest that not until very long chains are obtained can samples containing a large fraction of long conjugated segments be obtained. Ambient, short term oxidation of 100,000 Mn polymer shows an increase in satellite band intensities at omega (c-c) and omega (c=c), where omega is frequency, which corresponds to a decrease in G. Low level chain oxidation or doping is shown to preferentially occur within long conjugated segments.

900,565
PB89-186365 Not available NTIS
National Bureau of Standards (IMSE), Gaithersburg, MD. Polymers Div.
Deuterium Magnetic Resonance Study of Orientation and Poling in Poly(Vinylidene Fluoride) and Poly(Vinylidene Fluoride-Co-Tetrafluoroethylene).
Final rept.
M. A. Doverspike, M. S. Conradi, A. S. DeReggi, and R. E. Cais. 1989, 7p
Pub. in Jnl. of Applied Physics 65, n2 p541-547, 15 Jan 89.

Keywords: *Tetrafluoroethylene resins, *Nuclear magnetic resonance, Vinyl copolymers, Orientation, Deuterium, Molecular structure, Reprints, *Vinylidene fluoride polymers, Poling.

Deuteron nuclear magnetic resonance line shapes are reported for oriented samples of poly(vinylidene fluoride) and an (80-20) copolymer of vinylidene fluoride and tetrafluoroethylene. For stretched samples the orientation of the chain axes with respect to the stretch direction is given by a Gaussian distribution of width about 20 degrees. The width of the distribution was slightly smaller in the more highly stretched copolymer. Surprisingly, as the magnetic field is rotated in the plane perpendicular to the stretch direction, the deuterium spectra of the poled copolymer sample do not change. The occurrence of electrical polarization in the absence of orientation dependence of the deuterium line shape indicates molecular reorientation through 180 degrees in the copolymer.

900,566
PB89-188601 PC A06/MF A01
National Bureau of Standards (IMSE), Gaithersburg, MD. Polymers Div.
Institute for Materials Science and Engineering, Polymers: Technical Activities 1987.
Annual rept. 1 Oct 86-30 Sep 87.
L. E. Smith, and B. M. Fanconi. Nov 87, 104p
NBSIR-87/3614
See also PB89-166094.

Keywords: *Polymers, *Research program administration, Chemical properties, Mechanical properties, Standards, Composite materials, Blends, Durability, Processing, Technical activities, National Institute of Standards and Technology.

Technical Activities of the Polymers Division for FY 87 are reviewed. Included are descriptions of the 6 Tasks of the Division, project reports, publications, and other technical activities.

900,567
PB89-200430 Not available NTIS
National Bureau of Standards (IMSE), Gaithersburg, MD. Polymers Div.

Dielectric Measurements for Cure Monitoring.
Final rept.
F. I. Mopsik, S. S. Chang, and D. L. Hunston. 1989, 7p
Pub. in Materials Evaluation 47, n4 p448-453, 465 Apr 89.

Keywords: *Measurement, *Curing, *Epoxy resins, *Monitors, *Dielectric properties, Spectroscopy, Resistance, Viscosity, Ultrasonics, Reprints.

The use of dielectric measurements to monitor the cure of epoxy resins is investigated. Time-domain spectroscopy and an automated wide-dynamic-range AC-conductance measuring system show that dielectric loss and conductance measurements follow the entire cure cycle with good sensitivity and resolution. Comparisons with viscosity, differential scanning calorimetry (DSC) exotherm, and shear-mode ultrasonic attenuation measurements made simultaneously on the same sample show that dielectric methods compare very favorably with the others. In addition, a limit to the validity of a direct, simple relation between viscosity and dielectric loss is discussed. The conditions necessary for meaningful measurements are considered along with possible implementations of the method.

900,568
PB89-201487 Not available NTIS
National Bureau of Standards (IMSE), Gaithersburg, MD. Polymers Div.
Thermal Analysis of VAMAS (Versailles Project on Advanced Materials and Standards) Polycarbonate-Polyethylene Blends.
Final rept.
S. S. Chang. 1989, 13p
Pub. in Thermochimica Acta 139, p313-325 1989.

Keywords: *Thermogravimetry, *Polycarbonate resins, *Polyethylene, *Thermodynamic properties, Differential thermal analysis, Solubility, Melting points, Glass transition temperature, Specific heat, Heat of fusion, Compositions, Blends, Reprints.

Differential scanning calorimetry and thermogravimetric analysis were performed on a series of polycarbonate-polyethylene (PC-PE) blends, which were provided by the Technical Working Party of Polymer Blends, Versailles Project on Advanced Materials and Standards (VAMAS). Detailed comparison of results from cooperating laboratories were summarized. The PC-PE blends were found to be immiscible. Intensive properties such as melting points and glass transition temperatures were found to be independent of the composition. Extensive properties such as specific heat, heat of fusion and delta C(sub p); of glass transition were found to be nearly proportional to the composition. Onset and residues of degradation at different stages in thermogravimetric analysis were also found to be a function of the composition. Multiple melting peaks of low density polyethylene and relaxation peaks of annealed PC glass transition were also studied.

900,569
PB89-202451 Not available NTIS
National Bureau of Standards (IMSE), Gaithersburg, MD. Polymers Div.
13C NMR Method for Determining the Partitioning of End Groups and Side Branches between the Crystalline and Non-Crystalline Regions in Polyethylene.
Final rept.
D. L. VanderHart, and E. Perez. 1986, 8p
Pub. in Macromolecules 19, n7 p1902-1909 1986.

Keywords: *Carbon 13, *Nuclear magnetic resonance, *Crystal defects, *Polyethylene, Polymers, Butenes, Ethylene, Morphology, Separation, Spin lattice relaxation, Molecular structure, Reprints.

A butene-ethylene linear copolymer (BELC) is shown to give rise to a solid-state 13C spectrum, which contains resolvable resonances from vinyl and methyl end groups as well as ethyl side-branches. In contrast to the backbone methylene resonance which shows shifted, separable signals arising from the crystalline and non-crystalline regions, these weak 'defect' resonances show corresponding shifts too small to be definitive. Therefore, the problem of determining the partitioning of defects between the crystalline and non-crystalline regions of polyethylene must be approached indirectly. The method for identifying the morphological origin of defect signals is based on the concept that 13C CP signals are proportional to the local spin polarization levels, which, in turn, are kept

quite uniform over distances of nearest neighbors because of proton spin diffusion. Thus, defect signal intensities are argued to be proportional to backbone signal intensities within a given morphological region. By isolating the backbone resonances corresponding to the pure crystalline and non-crystalline components, one thereby isolates the defect resonances.

900,570
PB89-202923 Not available NTIS
National Bureau of Standards (IMSE), Gaithersburg, MD. Polymers Div.
Polymer Phase Separation.
Final rept.
I. C. Sanchez. 1987, 18p
Pub. in Encyclopedia of Physical Science and Technology, v11 p1-18 1987.

Keywords: *Polymers, *Liquid phases, Thermodynamics, Separation, Agglomeration, Entropy, Reaction kinetics, Reprints.

A homogeneous polymer solution or mixture under certain thermodynamic conditions will separate into two or more liquid phases that differ in composition. Polymer solutions as a rule are very viscous and the time scales required to effect phase separation are much longer than for analogous mixtures involving only small molecules. In addition to kinetic differences, phase separation in polymer solutions differs qualitatively in other important respects. In particular, polymer solutions are more susceptible, especially at elevated temperatures, to phase separation. The propensity for phase separation is related to the small entropy of mixing intrinsic to polymer solutions and a concomitant instability with respect to volume fluctuations. In a solution very dilute in polymer, phase separation can proceed by the development of a polymer rich phase formed by intermolecular aggregation of the polymer chains. However, intramolecular aggregation will precede the intermolecular aggregation process and cause a partial collapse of a coiling polymer chain. The latter process represents phase separation of isolated polymer chains.

900,571
PB89-202949 Not available NTIS
National Bureau of Standards (IMSE), Gaithersburg, MD. Polymers Div.
Surface-Interacting Polymers: An Integral Equation and Fractional Calculus Approach.
Final rept.
J. F. Douglas. 1989, 12p
Pub. in Macromolecules 22, n4 p1786-1797 Apr 89.

Keywords: *Polymers, *Surface properties, Integral calculus, Friction, Eigenvalues, Reprints.

A method due to Feynman and Kac is used to convert the path integral formulation of surface (variable surface dimension) interacting polymers into an equivalent integral equation approach. The integral equation for the surface-interacting chain partition function is determined to be the Volterra analogue of the Fredholm integral equation describing the friction coefficient in the preaveraged Kirkwood-Riseman theory. The approach to surface interacting polymers thus clarifies the close interconnection between the surface interaction and Kirkwood-Riseman theories noted in previous renormalization group calculations. An exact solution of the surface-interacting chain partition function is obtained by using the Riemann-Liouville fractional calculus. Finally, the integral equation and fractional calculus methods are combined to explain some of the most conspicuous features of the renormalization group theory--the mathematical significance of the 'crossover exponent,' 'infrared fixed point,' nontrivial 'critical exponents,' and the pole structure found in the interaction perturbation theory. The integral equation-fractional calculus formalism is also used to examine a point of 'critical instability' defining the adsorption threshold and to examine the failure of the renormalization group and eigenfunction expansion methods to describe scaling functions for values of the interaction near the instability point. The critical instability is readily understood on the basis of the Fredholm alternative.

900,572
PB89-212021 Not available NTIS
National Bureau of Standards (NEL), Gaithersburg, MD. Fire Measurement and Research Div.

Effects of Material Characteristics on Flame Spreading.
Final rept.
T. Kashiwagi, A. Omori, and J. Brown. 1989, 11p
Pub. in Proceedings of International Symposium on Fire Safety Science (2nd), Tokyo, Japan, June 13-17, 1989, p107-117.

Keywords: *Polystyrene, *Polymethyl methacrylate, *Flame propagation, Molecular weight, Thermal stability, Melt viscosity, Spreading, Polymers.

Effects of initial molecular weight and thermal stability of polymer samples on horizontal flame spreading behavior and spread rate were studied by comparing results between two polystyrene (PS) samples with different initial molecular weights and between two poly(methylmethacrylate) (PMMA) samples with different thermal stability and initial molecular weights. The flame spread rate of the higher molecular weight PS sample was about 25% larger than that for the low molecular weight PS sample and the flame spread rate of the higher molecular weight PMMA sample was about four times larger than that for the low molecular weight sample. The sample with high initial molecular weight does not form molten polymer near the flame front and the flame spreads steadily. However, the sample with low initial molecular weight forms molten polymer and the opposed slow fluid motion of molten polymer along the inclined vaporizing surface against the traveling flame significantly affects flame spreading behavior and its rate.

900,573
PB89-226589 Not available NTIS
National Inst. of Standards and Technology (IMSE), Gaithersburg, MD. Ceramics Div.
Reevaluation of Forces Measured Across Thin Polymer Films: Nonequilibrium and Pinning Effects.
Final rept.
R. G. Horn, S. J. Hirz, G. Hadziioannou, C. W. Frank, and J. M. Catala. 1989, 8p
Sponsored by Office of Naval Research, Arlington, VA. Pub. in Jnl. of Chemical Physics 90, n11 p6767-6774, 1 Jun 89.

Keywords: *Polymers, *Fluid friction, *Thin films, *Siloxanes, *Interfacial tension, Measurement, Surface properties, Drag, Viscosity, Molecular structure, Liquid phases, Reprints, *Pinning.

Forces between molecularly smooth solid surfaces separated by thin films of molten polydimethylsiloxane have been measured. A long-range repulsion reported in earlier work is not an equilibrium force, but can be attributed to viscous drag effects. Consistent with previous results, the viscosity of the film can be modeled by assuming that a layer of polymer molecules is immobilized or pinned at each surface for a time longer than the time scale of the measurements. The pinning is a result of entanglement-like effects in the vicinity of a wall.

900,574
PB89-229264 Not available NTIS
National Inst. of Standards and Technology (IMSE), Gaithersburg, MD. Polymers Div.
Shear Effects on the Phase Separation Behaviour of a Polymer Blend in Solution by Small Angle Neutron Scattering.
Final rept.
A. I. Nakatani, H. Kim, Y. Takahashi, and C. C. Han. 1989, 4p
Pub. in Polymer Communications 30, p143-146 May 89.

Keywords: *Polymers, *Blends, *Boundary layer separation, *Neutron scattering, Polystyrene, Poly butadiene, Plastics, Elastomers, Phthalates, Shear rate, Phase, Settling, Reprints.

The phase separation behavior of polystyrene and polybutadiene in dioctyl phthalate has been examined as a function of shear rate by small angle neutrons scattering. A couette geometry shear cell was used to apply a shear field to the solution at various temperatures. Differences in the correlation lengths parallel and perpendicular to the flow direction were observed by assuming mean field behavior of the solution. Dramatic decreases in the spinodal temperature were observed as a function of shear rate in the direction of flow. The results are consistent with the notion of shear stabilization and decoupling of the flow field in orthogonal directions.

900,575
PB89-231286 Not available NTIS
National Inst. of Standards and Technology (IMSE), Gaithersburg, MD. Polymers Div.
Polymerization of a Novel Liquid Crystalline Diacetylene Monomer.
Final rept.
M. A. Schen. 1988, 8p
Pub. in Proceedings of SPIE (Society of Photo-Optical Instrumentation Engineers) - Nonlinear Optical Properties of Organic Materials, San Diego, CA., August 17-19, 1988, v971 p178-185.

Keywords: *Liquid crystals, *Polymers, *Acetylene, *Reaction kinetics, *Activation energy, *Order disorder transformation, Microscopy, Temperature, Heat measurement.

The polymerization of a liquid crystalline diacetylene monomer, 1,6 bis-(N-4-oxybenzylidene 4'-n-octylaniline) 2,4-hexadiyne, 1-OBOA, is reported both within the crystal and liquid crystal phases. The monomer exhibits smectic phase polymorphism at temperatures below the clearing temperature as show by small angle C-ray scattering and optical microscopy. Polymerization kinetic results show first-order disappearance of monomer as seen by differential scanning calorimetry with a thermal activation energy of 31 kcal/mol. in the liquid crystal phase and approximately 71 kcal/mol. in the crystal phase. Poly(1-OBOA) exhibits a 'soft' lamellar-like layer structure reminiscent of the monomer when polymerized within a low temperature liquid crystal phase. Consequently, the topochemical polymerization of diacetylene monomers within a liquid crystal phase to give ordered polymer structures is reported for the first time.

900,576
PB90-117524 Not available NTIS
National Inst. of Standards and Technology (IMSE), Gaithersburg, MD. Polymers Div.
Off-Lattice Simulation of Polymer Chain Dynamics.
Final rept.
D. Eichinger, D. Kranbuehl, and P. Verdier. 1989, 2p
Pub. in Polymer Preprints 30, p45-46 1989.

Keywords: *Polymers, *Computerized simulation, *Lattice parameters, Mathematical models, Monte Carlo method, Coils, Molecular structure, Reprints.

The use of lattice model chains and Monte Carlo techniques has been the basis of numerous studies on the dynamics of random coil polymer chains. The majority of papers have focused on the effect of excluded volume and its associated effects of chain connectivity-entanglement on the dynamic properties of polymer chains as a function of chain length and segment density. At the same time, the increasing use of lattice models and the disagreement generated by the predictions have heightened concern about the possibility of anomalous effects due to the lattice constraints and the choices of bead movement rules. In order to study the effects of both the lattice constraints and the bead movement rules on chain dynamics, simulations in which no lattice constraints are present have been carried out. Overall the off lattice results, as did the recent SC, BCC and FCC simulations show remarkable consistency that excluded volume can cause an additional lengthening of a single chain's longest relaxation time well beyond the value observed for equilibrium properties.

900,577
PB90-123456 Not available NTIS
National Inst. of Standards and Technology (IMSE), Gaithersburg, MD. Polymers Div.
Small Angle Neutron Scattering Studies of Single Phase Interpenetrating Polymer Networks.
Final rept.
B. J. Bauer, R. M. Briber, and C. C. Han. 1987, 2p
Pub. in Polymer Preprints 28, n2 p169-170 1987.

Keywords: *Polymers, *Networks, *Neutron scattering, Vinyl ether resins, Polystyrene, Crosslinking, Deuteration, Phase transformations, Deformation, Stability, Compressing, Solubility, Reprints, Synthesis(Chemistry).

Compatible semi-interpenetrating polymer networks (IPN) have been synthesized from linear poly(vinylmethylether) and deuterated poly(styrene) crosslinked with divinyl benzene. The phase separation behavior of the IPNs has been studied by small angle neutron scattering. At the temperature of synthesis increasing the PSD crosslink density decreases

the miscibility of the system. However, once a compatible IPN has been formed, the presence of the crosslinks increases the single phase stability as the temperature is raised. The influence of deformation on phase separation behavior of the samples was also studied. A sample was deformed to 2.5 times the original length by compressing the sample and constraining it to elongate in one direction only. The scattering data indicate that the miscibility is decreased significantly along the deformation direction when compared to the undeformed sample.

900,578
PB90-136813 Not available NTIS
National Bureau of Standards (IMSE), Gaithersburg, MD. Polymers Div.
Charging Behavior of Polyethylene and Ionomers.
Final rept.
M. G. Broadhurst, A. S. DeReggi, G. T. Davis, and F. I. Mopsik. 1987, 6p
Pub. in Proceedings of Conference on Electrical Insulation and Dielectric Phenomena, Gaithersburg, MD,. October 18-22, 1987, p313-318.

Keywords: *Charging, *Polyethylene, *Ions, Electric charge, Polymeric films, Plastics, Zinc, Methacrylic acid, Copolymers, Dielectric properties, Electrical insulation, Temperature.

The charging behaviors of films of polyethylene and zinc ionomers of polyethylene methacrylic acid copolymers are reported. Charge distributions across the thickness were measured with the thermal pulse method. Charging variables included applied fields from 0 to 50 V/micro, temperatures from 20 to 80 C, times up to one week and electrode metal. Initial differences in the signs and distributions of charge between nominally similar films were related to differences in the mechanical and thermal histories. At long charging times all films approach a steady state, nearly uniform distribution of negative charge. In the steady state the field due to the space charge was found to be a significant fraction of the applied charging field. The results are compatible with a model of homopolar injection of negative charge at the cathode and conduction of the injected charge through the film.

General

900,579
PB89-180038 Not available NTIS
National Bureau of Standards (NML), Gaithersburg, MD. Office of Standard Reference Data.
Chemical and Spectral Databases: A Look into the Future.
Final rept.
J. R. Rumble, and D. R. Lide. 1985; 5p
Pub. in Jnl. of Chemical Information and Computer Sciences 25, p231-235 1985.

Keywords: *Chemistry, Chemical engineering, Information systems, Spectra, Reprints, *Data bases, Expert systems.

Over 50 databases of chemical and spectral information are now available, and in the coming years many more will be built. The paper discusses some of the current trends in the use of these databases and how such databases might affect chemistry.

CIVIL ENGINEERING

Construction Equipment, Materials, & Supplies

900,580
PB89-146971 Not available NTIS
National Bureau of Standards (NEL), Gaithersburg, MD. Building Materials Div.

CIVIL ENGINEERING

Construction Equipment, Materials, & Supplies

Interpretation of the Effects of Retarding Admixtures on Pastes of C3S, C3A plus Gypsum, and Portland Cement.
Final rept.
H. M. Jennings, H. Taleb, G, Frohnsdorff, and J. R. Clifton. 1986, 5p
Pub. in Proceedings of International Congress on the Chemistry of Cement (8th), Rio de Janeiro, Brazil, September 22-27, 1986, p239-243.

Keywords: *Cements, *Hydration, *Retardants, Portland cements, Gypsum cements, Coagulation, Setting time, Reprints, Induction period, Tricalcium aluminate, Tricalcium silicate hydration.

From studies of the effects of a range of organic retarders on pastes of C3S and of C3A-gypsum, it may be concluded that the two systems have different, possibly independent, influences on the concentrations of retarders in the aqueous phase of a portland cement paste. In the case of C3S, some of the retarder is incorporated into the layer of product which forms around the grains during the early stages of reaction. It appears that a process which controls the induction period occurs within this C-S-H layer and that the incorporation of a retarder in it somehow slows the process. It may be that a phase change, or other reaction which ends the induction period by making the layer more permeable, is poisoned by the retarder which is incorporated into the layer. The C3A-gypsum-water system removes a large amount of retarder from the aqueous phase during the early stages of hydration, with the rate of removal falling with time. Delaying retarder addition, therefore, decreases the amount of retarder which can be incorporated into a C-S-H layer. Thus these systems affect, in separate ways, a retarder's ability to extend the induction period of a portland cement. There is a delay time after initial mixing, when it is most efficient to add a retarder. It seems probable that, at least for some organic retarders, the effects on the portland cement reaction may be considered to be a combination of separate processes occuring in the C3S-water and C3A-gypsum-water subsystems.

900,581
PB89-146969 Not available NTIS
National Bureau of Standards (NEL), Gaithersburg, MD. Building Materials Div.
Implications of Computer-Based Simulation Models, Expert Systems, Databases, and Networks for Cement Research.
Final rept.
G. Frohnsdorff, J. Clifton, H. Jennings, P. Brown, L. Struble, and J. Pommersheim. 1986, 5p
Pub. in Proceedings of International Congress on the Chemistry of Cement (8th), Rio de Janeiro, Brazil, September 22-27, 1986, p598-602.

Keywords: *Cements, *Computer networks, *Research management, Chemical properties, Concretes, Hydration, Mathematical models, Decision making, Expenses, Physical management, *Data bases.

The simulation models, expert systems, and databases will complement each other to make possible improved predictions of performance of cements and concretes under diverse conditions, and improved decisions on selection of concrete materials; they will also facilitate identification of research needed to fill gaps in the knowledge base. More sharply focused statements of research needs should help improve decisions on research expenditures and lead to accelerated progress in cement research. Also, cooperative research efforts should become easier to carry out, thereby further contributing to the rate of progress. Examples from the authors' laboratory are used to illustrate some areas where computers are changing the direction of research related to cement hydration. They are: (1) modeling of microstructure development, (2) use of knowledge-based expert systems, and (3) storage, retrieval and transmission of knowledge needed for modeling and for expert system development. The implications of the growing ease of collection, storage, retrieval, and sharing of knowledge, and the possible use of knowledge, probably by teams, in the development of interconnected computer-based knowledge systems are discussed. It is concluded that these developments provide new opportunities for collaboration among cement researchers. If exploited, these opportunities will accelerate progress in cement and concrete science and technology.

900,582
PB89-176119 Not available NTIS
National Bureau of Standards (NEL), Gaithersburg, MD. Building Materials Div.

Integrated Knowledge Systems for Concrete and Other Materials.
Final rept.
G. Frohnsdorff, H. Jennings, L. Struble, P. Brown, and J. Clifton. 1986, 4p
Pub. in Communications on the Materials Science and Engineering Study, p57-60 1986.

Keywords: *Concretes, *Materials, Information systems, Mathematical models, Simulation, Materials tests, Reprints, *Expert systems, *Data bases, Knowledge bases(Artificial intelligence).

Advances in understanding and the ability to predict the performance of concrete and other materials will result from the development of integrated materials knowledge systems composed of expert systems, simulations models (and other mathematical models), and databases.

900,583
PB90-124306 PC A03/MF A01
National Inst. of Standards and Technology (NEL), Gaithersburg, MD. Center for Building Technology.
Adsorption of High-Range Water-Reducing Agents on Selected Portland Cement Phases and Related Materials.
D. R. Rossington, and L. J. Struble. Sep 89, 27p
NISTIR-89/4172
Prepared in cooperation with New York State Coll. of Ceramics, Alfred.

Keywords: *Portland cements, *Plasticizers, *Water reducing agents, *Adsorption, Concretes, Silicate cements, Chemical reactions, Silica materials, Slumping, Hydration, Chemical analysis.

The quantities of high-range water-reducing agents (superplasticizer) absorbed from their aqueous solutions by portland cement, tricalcium silicate, silica fume, wollastonite, and calcium silicate hydrate gel were determined using an ultraviolet spectrophotometer. The two superplasticizers (sulfonated melamine formaldehyde and sulfonated naphthalene formaldehyde) produced generally similar results. Wollastonite, a calcium silicate that does not react with water, produced no measurable adsorption of superplasticizer. The adsorption of superplasticizer by tricalcium silicate was similar to the adsorption by portland cement. From these results, it appears that superplasticizer is not absorbed by anhydrous silica or anhydrous calcium silicate, but rather by calcium silicate hydrate gel. Additional studies are recommended for improving the understanding of the absorption of superplasticizer by portland cement.

Highway Engineering

900,584
PB89-174924 PC A12/MF A01
National Inst. of Standards and Technology (NEL), Gaithersburg, MD. Center for Building Technology.
Inelastic Behavior of Full-Scale Bridge Columns Subjected to Cyclic Loading.
Final rept.
W. C. Stone, and G. S. Cheok. Jan 89, 265p NIST/BSS-166
Also available from Supt. of Docs. as SN003-003-02925-4. Library of Congress catalog card no. 88-600600. Errata sheet inserted. on National Science Foundation, Washington, DC., Federal Highway Administration, Washington, DC., and California State Dept. of Transportation, Sacramento.

Keywords: *Bridges(Structures), *Columns(Supports), *Cyclic loads, *Concrete structures, Elastic properties, Energy absorption, Failure, Dynamic structural analysis, Loading rate, Mechanical tests.

Circular, spirally reinforced concrete bridge columns were subjected to cyclic loading (representing seismic loads) in the laboratory. The test articles were prototype columns designed in accordance with recent (1983) California Department of Transportation (CALTRANS) specifications. Two full-scale columns, each measuring 5 feet (1.52 m) in diameter with aspect ratios (height/width) of 3 and 6, were subjected to slow reversed cyclic lateral load with constant axial load to simulate the gravity weight of the bridge superstructure. Details are presented concerning the design of the special test apparatus required to conduct the project as well as recommendations for future im-

provements in test procedures. Results from the tests are presented in the form of energy absorption graphs and bar charts, load-displacement hysteresis curves, longitudinal and confining steel strains, and curvature profiles. Comparisons of the ultimate moment capacities, measured displacement ductilities, plastic hinge lengths, cyclic energy absorption capacity and the failure modes of the full-scale specimens are made with those observed from 1/6-scale model tests.

Soil & Rock Mechanics

900,585
PB89-147045 Not available NTIS
National Bureau of Standards (IMSE), Gaithersburg, MD. Ceramics Div.
Prediction of Tensile Behavior of Strain Softened Composites by Flexural Test Methods.
Final rept.
T. J. Chuang, and Y. W. Mai. 1987, 11p
Contract DE-AI05-80OR20679
Sponsored by Department of Energy, Oak Ridge, TN. Oak Ridge Operations Office.
Pub. in Advanced Composite Materials and Structures, p647-657 1987.

Keywords: *Composite materials, *Tensile properties, *Fiber reinforced concretes, Bend test, Flexural strength, Loads(Forces), Tensile strength, Bending, Fiber composites, Polymer concrete, Strain softening.

A simple scheme of predicting tensile properties of a composite exhibiting strain-softening behavior from flexural tests is developed based on establishment of a relationship between bending and tensile properties. It is shown that strain-softening materials give higher bending strengths than tensile strengths, a result consistent with experimental observation. Also, given the load-displacement diagram obtained from a flexural test, it is possible to predict the entire tensile response. Bending data of polymer concrete are used to demonstrate the principles.

COMBUSTION, ENGINES, & PROPELLANTS

Combustion & Ignition

900,586
PB89-147482 Not available NTIS
National Bureau of Standards (NEL), Gaithersburg, MD. Fire Science and Engineering Div.
Aerodynamics of Agglomerated Soot Particles.
Final rept.
H. R. Baum, D. M. Corley, and A. F. Rabb. 1987, 4p
Sponsored by Defense Nuclear Agency, Washington, DC.
Pub. in Proceedings of Fall Technical Meeting--Chemical and Physical Processes in Combustion, San Juan, PR., December 15-17, 1986, p52.1-52.4 1987.

Keywords: *Combustion, *Soot, *Aerodynamic forces, Agglomeration, Kinetic theory, Numerical analysis, Particles.

Most studies of the effects of soot particles on combustion phenomena assume that the aerodynamic forces on an individual particle can be calculated as if the particle were a sphere. However, for particles of sizes significantly larger than 10(sup -2) microM, most of the growth comes from agglomeration of individual spheres of approximately this diameter into long irregular chains. These chains of reasonably distinct spheres form the characteristic open shapes of large soot particles which range up to several microns in overall length. Under combustion conditions it is apparent that any realistic analysis must recognize the chainlike shapes of the soot particles, and must be based on the kinetic theory of gases. The present study describes an analysis which satisfies these requirements.

900,587
PB89-149173 Not available NTIS
National Bureau of Standards (NEL), Gaithersburg, MD. Fire Science and Engineering Div.
Combustion of Oil on Water.
Final rept.
D. Evans, H. Baum, B. McCaffrey, G. Mulholland, M. Harkleroad, and W. Manders. 1986, 36p
Pub. in Proceedings of Annual Arctic and Marine Oil-spill Program (9th), Edmonton, Alberta, Canada, Jun 10-12, 1986, p 301-336.

Keywords: *Fires, *Oil spills, *Combustion products, *Water, *Crude oil, *Air pollution, Water pollution, Aerosols, Carbon monoxide, Carbon dioxide, Particle size distribution, Xylenes, Toluene, Smoke, Electron microscopy, Aromatic compounds, Alkanes, Benzene, *Environmental monitoring.

The report contains the results of measurements performed on both 0.4 m and 0.6 m diameter pool fires produced by burning a layer of Prudhoe Bay crude oil supported by a thermally deep layer of water. Both steady and vigorous burning caused by boiling of the water sublayer were observed. The measured energy release rate for steady burning was about 640 kW/sq m. The emission rate, the size distribution, and specific extinction coefficient were measured for the smoke aerosol produced by the fires. Data were also obtained on the structure of the smoke aerosol by electron microscopy and on emission of CO and CO2. Analysis of the crude oil burn residue indicated selected depletion of the short chain alkanes and cycloalkanes when compared to the fresh oil. Mono-ring aromatics including benzene, toluene, and xylenes present in the fresh crude were absent in the burn residue. Calculations of the induced air flow into a stimulated distribution of 20 fires over a 100 m x 100 m area showed that the maximum inflow velocity near the largest size fire (2.5 m diameter, 3.2 MW) was 1.1 m/s.

900,588
PB89-157234 Not available NTIS
National Bureau of Standards (NEL), Gaithersburg, MD. Thermophysics Div.
Light Scattering from Simulated Smoke Agglomerates.
Final rept.
R. D. Mountain, and G. W. Mulholland. 1988, 6p
Pub. in Langmuir 4, n6 p1321-1326 Nov/Dec 88.

Keywords: *Smoke, *Light scattering, Agglomerates, Computerized simulation, Reprints, Langevin formula, Fractal dimensions.

The computer simulation technique of Langevin dynamics is used to simulate the growth of smoke agglomerates. Clusters consisting of between 10 and 700 primary particles are generated, and the light scattering from these clusters is calculated in the Rayleigh-Debye limit. The results of these calculations are then used to illustrate how light-scattering measurements can be used to infer the concentration, the size, the radius of gyration (R sub g), and the fractal dimension of the agglomerates. The pair distribution function for these agglomerates is shown to be a scaled function of r/R sub g, and the cutoff function describing large separations of particles in the clusters has the form exp(-c(r/R sub g) to the 2.5 power).

900,589
PB89-157572 Not available NTIS
National Bureau of Standards (NEL), Gaithersburg, MD. Fire Science and Engineering Div.
Structure and Radiation Properties of Large-Scale Natural Gas/Air Diffusion Flames.
Final rept.
J. P. Gore, G. M. Faeth, D. Evans, and D. B. Pfenning. 1986, 9p
Pub. in Fire and Materials 10, n3-4 p161-169 Sep-Dec 86.

Keywords: *Diffusion flames, Mathematical models, Natural gas, Heat flux, Heat transfer, Combustion, Temperature, Thermal measurements, Soot, Reprints, *Flame radiation, *Flame structure, Radiative heat transfer.

Recent data from large-scale turbulent natural gas/air diffusion flames (135-210 MW) were used to evaluate analysis of flame structure and radiation properties. The conserved-scalar formalism, in conjunction with the laminar flamelet concept, was used to estimate flame structure. The discrete-transfer method, in conjunction with a narrow-band radiations model, was used to predict radiative heat fluxes. The narrow-band model considered the nonluminous gas bands of water vapor, carbon dioxide, methane and carbon monoxide in the 1000 - 6000 nm wavelength range. Structure predictions were encouraging, with discrepancies for mean temperatures comparable to experimental uncertainties (ca. 200 K in the hottest portions of the flames). Radiative heat flux predictions were also reasonably good, e.g., predictions based on mean scalar properties were generally 15% lower than the measurements. The findings also suggest that continuum radiation from soot is negligible for these flames.

900,590
PB89-171904 Not available NTIS
National Bureau of Standards (NEL), Gaithersburg, MD. Fire Measurement and Research Div.
Chemical Structure of Methane/Air Diffusion Flames: Concentrations and Production Rates of Intermediate Hydrocarbons.
Final rept.
J. H. Miller, and K. C. Smyth. 1986, 7p
Pub. in Preprints of Papers, American Chemical Society, Division of Fuel Chemistry 31, n2 p105-111 1986.

Keywords: *Molecular structure, *Combustion products, *Diffusion flames, *Methane, Aromatic polycyclic hydrocarbons, Concentration(Composition), Public health, Acetylene, Hydrocarbons, Soot, Benzene, *Air pollution effects(Humans).

The production of intermediate and large hydrocarbon species is common to most combustion systems. These products range in size from acetylene, benzene and polynuclear aromatic hydrocarbons (PAH) to very large soot particles. Radiation from particles is the dominant mode of heat transfer in large fires. In addition, sampled particles often have PAH adsorbed onto them. Many of these molecules are known carcinogens and their presence on inhalable soot particles poses an obvious long-term health hazard. Despite the important role that such species play in flames and the danger they present as combustion byproducts, the mechanism for their formation is as yet unknown.

900,591
PB89-171912 Not available NTIS
National Bureau of Standards (NEL), Gaithersburg, MD. Fire Measurement and Research Div.
Methyl Radical Concentrations and Production Rates in a Laminar Methane/Air Diffusion Flame.
Final rept.
J. H. Miller, and P. M. Taylor. 1987, 11p
Pub. in Combustion Science and Technology 52, n1-3 p139-149 1987.

Keywords: *Diffusion flames, *Reaction kinetics, *Combustion products, *Concentration(Composition), Combustion, Mass spectroscopy, Temperature, Velocity, Ionization, Reprints, *Methyl radicals.

Methyl radicals have been observed in a laminar methane/air diffusion flame via an application of the scavenger probe technique. In these experiments, a quartz microprobe was modified such that iodine vapor was pumped from a storage side arm into the inside tip of the probe. Sampled methyl radicals react quantitatively with iodine to product methyl iodide which was detected by an on-line mass spectrometer. These quantitative profiles are compared to profiles of stable intermediate hydrocarbons which have been observed in this flame, as well as profile signals which are due to methyl radical ionization by laser radiation. The concentration of methyl is combined with velocity and temperature data to calculate the net rates of chemical reactions of methyl in the flame. The use of methyl concentration and rate data to estimate the concentrations of other reactive species is discussed.

900,592
PB89-173850 Not available NTIS
National Bureau of Standards (NEL), Boulder, CO. Chemical Engineering Science Div.
Development of Combustion from Quasi-Stable Temperatures for the Iron Based Alloy UNS S66286.
Final rept.
J. W. Bransford, P. A. Billiard, J. A. Hurley, K. M. McDermott, and I. Vazquez. 1988, 12p
Sponsored by National Aeronautics and Space Administration, Huntsville, AL. George C. Marshall Space Flight Center.
Pub. in Flammability and Sensitivity of Materials in Oxygen-Enriched Atmospheres, ASTM STP 986, v3 p146-157 1988.

Keywords: *Iron alloys, *Combustion, Ignition, Exothermic reactions, Endothermic reactions, Temperature, Carbon dioxide lasers, Ignition temperature, Oxidation, Reprints.

The development of ignition and subsequent combustion from quasi-stable temperatures was studied for several iron, nickel, and cobalt-based alloys. The quasi-stable temperature was produced by heating a specimen with a continuous wave carbon dioxide laser. Endothermic and exothermic transitions appear to play an important role in the development of thermal runaway, ignition, and combustion. The apparent effect of the endothermic transitions was to accelerate the rate of oxidation of the alloy, which produced abrupt changes in surface temperature as well as increasing the rate of increase in surface temperature. In the final stages of the thermal runaway phase, endothermic and exothermic events forced the alloy surface rapidly into combustion. Total destruction of the specimen followed immediately. The results for the iron-based alloy UNS S66286, which represent the phenomena observed, are presented. A spontaneous ignition temperature, enhanced oxidation temperature and ignition temperature for the solid alloy have been defined. Data are presented for the oxygen pressure range of 1.7 to 13.8 MPa.

900,593
PB89-175913 Not available NTIS
National Bureau of Standards (NEL), Gaithersburg, MD. Fire Safety Technology Div.
Very Large Methane Jet Diffusion Flames.
Final rept.
B. J. McCaffrey, and D. D. Evans. 1988, 7p
Pub. in Proceedings of International Symposium on Combustion 21, p25-31 1988.

Keywords: *Diffusion flames, *Methane, *Jet flow, Combustion, Flame propagation, Subsonic flow, Temperature, Radiation, Heat transfer.

Methane jet diffusion flames with heat release approaching to 350 MW in both subsonic and supercritical configurations have been studied regarding lift-off height and flame height, absolute flame stability, radiative characteristics, and for an evaluation of their propensity for extinguishment using water sprays. Flames from orifices up to 38 mmD could be blown off with sufficient gas pressure. For D = 51 mm the flame could not be blown off even for stagnation pressures to 2300 kPa. The data point at 38 mm allows a more accurate extrapolation, in the manner of Kalghatgi, leading to a predicted critical orifice size of 48 mm for absolute flame stability for CH4. Failure to ignite gas from a 1 mm diameter aperture in a reservoir at 12,000 kPa is consistent with the shape of the upper portion of the locus of the derived stability curve. Wide-band radiation measurements coupled with flame temperature measurements confirm small scale absorptivity determinations for use in simple flame radiation models.

900,594
PB89-176622 Not available NTIS
National Bureau of Standards (NEL), Gaithersburg, MD. Mathematical Analysis Div.
Solution for Diffusion-Controlled Reaction in a Vortex Field.
Final rept.
R. G. Rehm, H. R. Baum, and D. W. Lozier. 1987, 4p
Pub. in Chemical and Physical Processes in Combustion 1986, p3.1-3.4 1987.

Keywords: *Combustion, *Diffusion, *Vortices, Convection, Mixing, Mathematical models, Reaction kinetics, Turbulence, Navier-Stokes equations, Reprints.

A mathematical model of local, transient, constant-density diffusion-controlled reaction between unmixed species initially occupying adjacent half-spaces is analyzed. An axisymmetric viscous vortex field satisfying the Navier-Stokes equations winds up the interface between the species as they diffuse together and react. A flame-sheet approximation of the rapid reaction is made using Shvab-Zeldovich dependent variables. The model was originally proposed by F. Marble, who performed a local analysis of consumption rates along the flame sheet. The present paper describes a global similarity solution to the problem which is Fourier analyzed in a Lagrangian coordinate system.

COMBUSTION, ENGINES, & PROPELLANTS

Combustion & Ignition

900,595
PB89-176762 Not available NTIS
National Bureau of Standards (NEL), Gaithersburg,
MD. Fire Measurement and Research Div.
**Cigarette as a Heat Source for Smolder Initiation in
Upholstery Materials.**
Final rept.
T. Ohlemiller, and R. Breese. 1987, 4p
Pub. in Chemical and Physical Processes in.Combustion 1986, p6.1-6.4 1987.

Keywords: *Flammability testing, *Upholstery, Ignition,
Fire resistant materials, Flammability, Fabrics, Reprints, *Cigarettes.

Cigarettes, which themselves undergo smoldering
combustion, are prone to induce smoldering in upholstery materials with which they make accidental contact. A set of 32 experimental cigarettes with five
varied design parameters is being tested for lesser
tendency to ignite such substrates. Heat flux scans
show cigarettes with lessened ignition tendencies do
yield a lesser heat input.

900,596
PB89-179261 Not available NTIS
National Bureau of Standards (NEL), Gaithersburg, .
MD. Fire Science and Engineering Div.
Fire Safety Science-Proceedings of the First International Symposium.
Final rept.
B. J. McCaffrey. 1987, 1p
Pub. in Combustion and Flame 69, p369 1987.

Keywords: *Meetings, *Fire safety, Fires, Fire protection, Fire tests, Fire prevention, Reprints.

The proceedings of the first meeting of the International Association for Fire Safety Science, dedicated to the
improvement of man's understanding of fire phenomena, is presented. By bringing together practitioners of
fire research -- physicists, chemists, engineers, architects, codes-and-standards, and insurance people -- it
is hoped that effective information exchange might
result, leading to more focused fire problem definitions
and ultimate solutions. This multidisciplinary group with
over 300 registrants from 17 countries met at NBS in
October 1985.

900,597
PB89-188577 PC A03/MF A01
California Univ., Berkeley. Dept. of Mechanical Engineering.
Fire Propagation in Concurrent Flows.
Final rept. 1 Aug 87-31 Jul 88.
A. C. Fernandez-Pello, Aug 88, 42p NIST/GCR-89/
560
See also PB87-140190. Sponsored by National Inst. of
Standards and Technology (NEL), Gaithersburg, MD.
Center for Fire Research.

Keywords: *Flame propagation, *Fire tests, Air flow,
Combustion, Fuels, Flow rate, Gas flow, Ignition, Turbulence, Diffusion, Pyrolysis, Numerical analysis,
*Flame spread.

The research tasks completed during the reporting
period include an experimental study of the effect on
the spread of flames of the turbulence intensity of an
opposed air flow, and a theoretical analysis of the concurrent spread of flames over thin fuels. Both studies
are, in the author's opinion, important contributions in
the study of the flame spread process. The results of
the experimental study show that the flame spread
process is significantly affected by the flow turbulence
intensity for flames spreading over both thin and thick
fuels. The results of the theoretical analysis, which are
in good agreement with previous experimental measurements, give detailed information about the flame
structure and mechanisms of flame spread.

900,598
PB89-201172 Not available NTIS
National Bureau of Standards (NEL), Gaithersburg,
MD. Fire Measurement and Research Div.
**Importance of Isothermal Mixing Processes to the
Understanding of Lift-Off and Blow-out of Turbulent Jet Diffusion Flames.**
Final rept.
W. M. Pitts. 1989, 16p
Sponsored by Air Force Office of Scientific Research,
Bolling AFB, DC.
Pub. in Combustion and Flame 76, p197-212 1989.

Keywords: *Diffusion flames, *Jet flow, *Fuel sprays,
*Turbulent flow, *Mixing, Two phase flow, Flame propagation, Reprints.

Many different theoretical analyses have been developed to predict the lift-off and blowout behaviors of
axisymmetric, turbulent jet diffusion flames. Past theoretical and experimental work is summarized. It is then
shown that these flame stability properties can be predicted using the known time-averaged concentration
and velocity profiles of the corresponding nonreacting
jet flows of the fuels into air. In contrast to past theoretical treatments of these processes, it is not necessary
to consider interactions of the turbulent concentration
and/or velocity fluctuations with the combustion process. Possible physical mechanisms that can lead to
such a finding are discussed.

900,599
PB89-201966 Not available NTIS
National Bureau of Standards (NEL), Gaithersburg,
MD. Fire Measurement and Research Div.
Soot Inception in Hydrocarbon Diffusion Flames.
Final rept.
K. C. Smyth, and J. H. Miller. 1987, 13p
Pub. in Chemical and Physical Processes in Combustion 1986, pC.1-C.13 1987.

Keywords: *Diffusion flames, *Soot, *Hydrocarbons,
Combustion, Spectrometers, Chemical reactions, Reaction kinetics, Flame propagation, Combustion products, Reprints.

In the last two years detailed species concentration
measurements have been made in hydrocarbon diffusion flames for the first time. Far from being hopelessly
complex (as many had thought), the interplay between
chemical and transport processes can now be unravelled. In addition, questions about the chemical growth
processes leading to polycyclic aromatic hydrocarbon
and soot formation can be addressed. The results thus
far reveal striking parallels with premixed flame studies.

900,600
PB89-211866 Not available NTIS
National Bureau of Standards (NEL), Gaithersburg,
MD. Chemical Process Metrology Div.
**FT-IR (Fourier Transform-Infrared) Emission/
Transmission Spectroscopy for in situ Combustion Diagnostics.**
Final rept.
P. R. Solomon, P. E. Best, R. M. Carangelo, J. R.
Markham, P. Chien, R. J. Santoro, and H. G.
Semerjian. 1988, 9p
Pub. in Proceedings on International Symposium on
Combustion (21st), p1763-1771 1988.

Keywords: *Combustion products, Infrared spectroscopy, Temperature, Soot, Optical measurement, Particles, Diffusion flames, Gases, Emission, Transmission,
*Fourier transform spectrometers, *In-situ combustion.

An infrared emission/transmission (E/T) technique (a
method previously used as an in-situ diagnostic for
gases and soot using dispersive infrared) has been implemented with a Fourier Transform Infrared (FT-IR)
spectrometer and extended to include measurements
on particles, as well as gases and soot. The method
can measure both the concentration and temperature
of each of the phases. The paper presents the applications of FT-IR E/T spectroscopy to gases, soot and
particles in reacting flows. Several examples are presented including a coannular laminar diffusion flame.
The results for soot temperature and concentration in
this flame are in good agreement with measurements
reported by other investigators.

900,601
PB89-212096 Not available NTIS
National Bureau of Standards (NML), Gaithersburg,
MD. Chemical Kinetics Div.
**Evaluated Kinetics Data Base for Combustion
Chemistry.**
Final rept.
W. Tsang, and J. T. Herron. 1987, 6p
Pub. in Proceedings of the International CODATA Conference on Comput. Handl. Dissemination Data (10th),
p229-234 1987.

Keywords: *Reaction kinetics, *Combustion, *Methane, Computerized simulation, Gases, Hydrocarbons,
Chemical reactions.

A continuing effort at developing a chemical kinetic
data base for the computer simulation of hydrocarbon
combustion is described. Because of the complexity of
the problem present efforts are directed towards providing data related to the initial stages of methane

combustion are considered. The reactions are a key
subset of all hydrocarbon combustion processes and
the rate expressions form the basis for predictive
methods for larger compounds. A key factor in the
work is the necessity of intercalating and extrapolating
experimental data so as to derive rate constant expressions valid over temperature and pressure ranges
of 300-2500K and 10(sup 16)-10(sup21) particles/cu
cm, respectively. Unimolecular reactions pose special
problems, and the methodology for treating and presenting the data is discussed.

900,602
PB89-216366 PC A07/MF A01
National Inst. of Standards and Technology (NEL),
Gaithersburg, MD. Center for Fire Research.
**Technical Reference Guide for FAST (Fire and
Smoke Transport) Version 18.**
Technical note.
W. W. Jones, and R. D. Peacock. May 89, 130p
NIST/TN-1262
Also available from Supt. of Docs: as SN003-003-
02944-1..

Keywords: *Mathematical models, *Computer programs, *Fires, Algorithms, Differential equations, Compartment analysis, Heating, Radiation, Convection,
Conduction, Venting, Smoke, Plumes.

FAST is a model to describe fire growth and smoke
transport in multi-compartment structures. The implementation consists of a set of programs to describe
the structure to be modeled, run the model and
produce usable output. The reference guide describes
the equations which constitute the model, data which
are used by the model and explains how to operate the
model. The physical basis of zone models, their limitations, and development of the predictive equations are
described elsewhere and therefore are only summarized in the reference guide. The intent of this guide is
to provide a complete description of the way the model
is structured. In particular the relationship between the
equations and the numerical implementation of the
equations is laid out. It is intended as a complete description of the parameters and key words available to
control various aspects of a simulation. It is hoped that
there is sufficient information provided to adapt the
model for specialized applications.

900,603
PB89-231096 Not available NTIS
National Inst. of Standards and Technology (NEL),
Gaithersburg, MD. Fire Measurement and Research
Div.
**Assessment of Theories for the Behavior and
Blowout of Lifted Turbulent Jet Diffusion Flames.**
Final rept.
W. M. Pitts. 1989, 8p
Pub. in Proceedings of International Symposium on
Combustion (22nd), Seattle, WA., August 14-19, 1988,
p809-816 1989.

Keywords: *Diffusion flames, *Turbulent flow, *Jet
flow, Mixing, Combustion stability, Flame photometry.

Many competing theories have been published to describe the characteristics and blowout of lifted turbulent jet diffusion flames. The assumptions which are
made as to the physical processes responsible for
these behaviors vary widely. In the paper these assumptions are summarized for each model and compared with the actual turbulent behaviors of unignited
fuel jets. As part of this discussion, recent unpublished
measurements of real-time concentration fluctuations
along a line in a turbulent fuel jet are introduced. To the
extent possible, each theory is also assessed as to its
capabilities to accurately predict experimentally observed lift off and blowout behaviors. The conclusion
of these analyses is that none of the currently available theories for flame stabilization are satisfactory.
Further experimentation is required before the actual
physical processes responsible for flame stabilization
can be identified and models which are capable of accurate prediction of lift off heights and blowout velocities developed.

900,604
PB89-231179 Not available NTIS
National Bureau of Standards (NEL), Gaithersburg,
MD. Fire Safety Technology Div.

Combustion Efficiency, Radiation, CO and Soot Yield from a Variety of Gaseous, Liquid, and Solid Fueled Buoyant Diffusion Flames.
Final rept.
B. J. McCaffrey, and M. F. Harkleroad. 1988, 11p
Sponsored by Defense Nuclear Agency, Washington, DC., and Department of the Interior, Washington, DC.
Pub. in Proceedings of International Symposium on Combustion (22nd), Seattle, WA., August 14-19, 1988, p1251-1261 1989.

Keywords: *Combustion efficiency, *Diffusion flames, *Combustion products, *Radiation, Carbon monoxide, Soot, Thermal measurements, Heptane, Propane, Wood, Crude oil.

The following compilation of data describes combustion conditions for the free burning of propane, heptane, 3 crude oils, wood and polyurethane cribs in fuel package sizes less than a meter, having heat release rates up to 1/2 MW. Measurements include heat release rate from O2 depletion calorimetry, mass loss rate, incident radiative flux to nearby targets, flame and liquid interface temperatures and overall major gaseous species yield including soot. The information contained herein provides a data base for buoyant diffusion flame modelling regarding radiation and elementary chemical characterization. There appears to be a one-to-one correspondence between a fuels radiative behavior and its soot and CO yield. Urethanes and crude oil have significantly higher values for all these parameters as compared to propane, heptane and wood. No correlation was found between these measurements and time-averaged thermocouple readings of temperature.

900,605
PB90-127101 PC A10/MF A02
National Inst. of Standards and Technology (NEL), Gaithersburg, MD. Center for Fire Research.
Summaries of Center for Fire Research In-House Projects and Grants: 1989.
Final rept.
S. M. Cherry. Oct 89, 214p NISTIR-89/4188
See also PB89-127302.

Keywords: *Combustion, Fire prevention, Research projects, Grants, Soot, Polymers, Carbon monoxide, Flame propagation, Toxicity, Furniture, Charring, Models, Ignition, Hazards, *Fire research, Fire models, Building fires, National Institute of Standards and Technology.

The report describes the research projects performed in the Center for Fire Research and under its grants program during FY1989. Topics considered include the following: Turbulent combustion, Soot formation, CO prediction, Polymer gasification, Flame spread, Toxic potency, Furniture flammability, Building fire modelling and smoke transport, Fire hazard assessment, Engineering analysis system and fire reconstruction, Suppression, Cone calorimeter, Technology transfer, Fire/modeling interactions, and Fire protection technology.

900,606
PB90-136821 Not available NTIS
National Inst. of Standards and Technology (NEL), Gaithersburg, MD. Fire Science and Engineering Div.
Summary of the Assumptions and Limitations in Hazard I.
Final rept.
R. W. Bukowski. 1989, 15p
Pub. in Proceedings of Fire Retardant Chemicals Association Spring Conference, Baltimore, MD., June 7-8, 1989, p1-15.

Keywords: *Fire extinguishing agents, Toxicity, Computer systems programs, Documentation, *Risk assessment.

A brief description of the function, major assumptions and limitations of the programs and procedures which comprise the prototype Hazard Assessment Method-Hazard I is presented. The method consists of a large software package and three volume report.

Reciprocation & Rotating Combustion Engines

900,607
PB89-147094 Not available NTIS

National Bureau of Standards (NEL), Gaithersburg, MD. Chemical Process Metrology Div.
Thin Film Thermocouples for Internal Combustion Engines.
Final rept.
K. G. Kreider. 1986, 6p
Contract DE-AI05-83OR21375
Sponsored by Oak Ridge National Lab., TN.
Pub. in Jnl. of Vacuum Science and Technology A 4, n6 p2618-2623 Nov/Dec 86.

Keywords: *Thermocouples, *Internal combustion engines, *Fabrication, *Measurement, *Temperature, *Coatings, Thin films, Valves, Cylinders, Pistons, Combustion chambers, Aluminum oxide, Iron alloys, Sputtering, Adhesion, Platinum alloys, Aluminum alloys, Chromium alloys, Microscopy, Stainless steels, Electrical properties, Reprints.

The feasibility of fabricating thin film thermocouples on internal combustion engine hardware was investigated. The goal was to find a procedure that would be useful for the measurement of the metal temperature of valve, valve seats, combustion chamber surface, cylinder walls, and piston heads during engine operation. The approach pursued was to coat the engine hardware material with an aluminum-containing, oxidation-resistance ferrous alloy (FeCrAlY) which forms an oxide layer with good electrical resistance. This thermal oxide was coated with a thin layer of reactively sputtered aluminum oxide and sputtered thin film type S thermocouple legs of platinum and platinum plus rhodium. This project was used to investigate the materials problems related to obtaining good adhesion in the metal metal-oxide-metal laminate and the electrical insulating properties of the oxide. Thermal oxidation, reactive sputtering of Al2O3 and platinum alloy sputtering were investigated using optical microscopy, x-ray photoemission spectroscopy (XPS), laminar adhesion testing, and the evaluation of high temperature electrical properties.

Rocket Engines & Motors

900,608
PB90-128661 Not available NTIS
National Bureau of Standards (NEL), Boulder, CO. Chemical Engineering Science Div.
Vortex Shedding Flowmeter for Fluids at High Flow Velocities.
Rept. for 1984-86.
J. D. Siegwarth. 1986, 15p
Sponsored by National Aeronautics and Space Administration, Huntsville, AL. George C. Marshall Space Flight Center.
Pub. in Advanced Earth-to-Orbit Propulsion Technology, NASA Conference Publication 2437, v2 p139-153 1986.

Keywords: *Flowmeters, Liquid oxygen, Hydrogen, Reprints, *Space shuttle main engine, Vortex shedding, Cryogenic fluids.

Vortex shedding flowmeter designs developed for this project are capable of measuring water flowrates at velocities above the liquid oxygen (LOX) flow velocities of the space shuttle main engines (SSME). These meters have been tested in two sections of actual SSME ducts with high velocity water flow. The results show that a meter vane fitted through a pair of appropriately located standard 11.2 mm diameter instrument ports can be used to measure flow without any upstream conditioning. The vortex shedding flowmeter with a 41 mm (1.61 in) bore has undergone preliminary testing with air at SSME fuel-flow densities. In spite of the much lower sound velocity of air, a strong vortex-generated spectrum line has been observed up to 180 m/s (590 ft/s). Successful meter designs suitable for high-pressure cryogenic flow measurement have been built and tested.

Rocket Propellants

900,609
PB89-146278 PC A03/MF A01
National Inst. of Standards and Technology (IMSE), Gaithersburg, MD. Polymers Div.

In Situ Fluorescence Monitoring of the Viscosities of Particle-Filled Polymers in Flow.
Annual rept.
A. J. Bur, F. W. Wang, A. Lee, R. E. Lowry, S. C. Roth, and T. K. Trout. Nov 88, 37p NISTIR-88/3892
Grant N00014-86-F-0115
See also FY 87, PB88-157698. Sponsored by Office of Naval Research, Arlington, VA.

Keywords: *Laminar flow, *Propellants, *Binders, *Polymers, Fluorometers, *Fluorescence, Monitoring, Spectroscopy, Mixing, Rheology, Viscosity, Polybutadiene, Aluminum oxide, Measuring instruments, Synthesis(Chemistry).

Work during FY 88 has focused on three areas: the chemical synthesis of a polymeric chromophore; the design of experiments to measure fluorescence anisotropy and non-Newtonian viscosity as a function of shear rate; and, measuring the quality-of-mix of a two component material using a fluorescence microscope. Significant results from these areas of work are: (a) a polymeric chromophore, consisting of anthracene covalently bonded to polybutadiene, has been synthesized and characterized by gel permeation chromatography and infrared observations. The number average molecular weight is 12,000 which is above the entanglement molecular weight for polybutadiene; (b) experiments using the polymeric chromophore as a dopant in a Newtonian fluid, very low molecular weight polybutadiene, show that fluorescence anisotropy correlates with the viscosity, i.e., it remains constant as a function of shear rate; and (c) using a fluorescence microscope, the authors have measured optical transmittance and near neighbor distances between particles in a matrix/particle mixing experiment and correlated these data with fluorescence intensity fluctuations.

COMMUNICATION

Common Carrier & Satellite

900,610
PB89-166086 PC A09/MF A01
National Inst. of Standards and Technology, Gaithersburg, MD.
Ongoing Implementation Agreements for Open Systems Interconnection Protocols: Continuing Agreements.
T. Boland. Dec 88, 176p NISTIR-88/3824-2
See also PB89-132351. Proceedings of the NIST/OSI Implementor's Workshop Plenary Assembly, held in Gaithersburg, MD. on December 16, 1988.

Keywords: Standards, Tests, *Open Systems Interconnections, *Protocol(Computers), Computer networks, Channels(Data transmission).

The document records current agreements on implementation details of Open Systems Interconnection Protocols among the organizations participating in the National Institute of Standards and Technology (NIST)/OSI Workshop Series for Implementors of OSI Protocols. These decisions are documented to facilitate organizations in their understanding of the status of agreements. The standing document is updated after each workshop (about 4 times a year).

900,611
PB89-171326 Not available NTIS
National Bureau of Standards (ICST), Gaithersburg, MD. Systems and Network Architecture Div.
Transport Layer Performance Tools and Measurement.
Final rept.
R. Aronoff, K. Mills, and M. Wheatley. 1987, 11p
Pub. in IEEE (Institute of Electrical and Electronics Engineers) Network 1, n3 p21-31 Jul 87.

Keywords: *Performance tests, Delay time, Reprints, *Computer networks, *Computer systems performance, Distributed computer systems, Throughput, INA-960 computers.

The paper describes the function, design and implementation of a Transport Experiment Control System (TECS) that enables transport layer performance ex-

COMMUNICATION

Common Carrier & Satellite

periments between stations on a local area network. The TECS is applied to evaluate the performance of Intel's iNA-960 transport service including profiles of throughput, delay and multiple applications. The results define an upper bound on throughput and a lower bound on delay. A problem is described involving flow control interactions between the transport and link layers. The effect of taut flow control on transport connections is investigated. The suitability of iNA-960 for real-time applications is evaluated.

900,612
PB89-171334   Not available NTIS
National Bureau of Standards (ICST), Gaithersburg, MD. Systems and Network Architecture Div.
Prediction of Transport Protocol Performance through Simulation.
Final rept.
K. Mills, M. Wheatley, and S. Heatley. 1986, 9p
Pub. in Computer Communication Review 16, n3 p75-83 Aug 86.

Keywords: *Data transmission, *Computerized simulation, Manufacturing, Automation, Reaction time, Reprints, *Protocol(Computers), *Computer networks.

A five-layer simulation model of OSI protocols is described and applied to predict transport user performance on a local area network (LAN). Emphasis is placed on time-critical applications typical of a small, flexible manufacturing system. The results suggest that, with current technology, OSI protocols can provide 1.5 Mbps throughputs, one-way delays between 6 and 10 ms, and response time between 15 and 25 ms. The results also demonstrate that CSMA/CD is a reasonable access method for time-critical applications on small factory LANS.

900,613
PB89-172357   Not available NTIS
National Bureau of Standards (NEL), Gaithersburg, MD. Building Environment Div.
Standardizing EMCS Communication Protocols.
Final rept.
S. T. Bushby, and H. M. Newman. 1989, 4p
Pub. in ASHRAE (American Society of Heating, Refrigerating and Air-Conditioning Engineers) Jnl. 31, n1 p33-36 Jan 89.

Keywords: *Telecommunication, *Standards, Reprints, *Protocols, *Communication networks, Computer architecture, Open System Interconnection.

The article summarizes the approach to standardizing EMCS communication protocols being taken by ASHRAE Standards Project Committee 135P. The advantages of a layered architecture based on the international standard for Open System Interconnection (OSI) and the possibility of a collapsed architecture are discussed. The structure of the Standards Project Committee along with the progress of its first year of activities are described.

900,614
PB89-174064   Not available NTIS
National Bureau of Standards (NML), Boulder, CO. Time and Frequency Div.
Dual Frequency P-Code Time Transfer Experiment.
Final rept.
J. R. Clynch, B. W. Tolman, M. A. Weiss, D. W. Allan, and D. Davis. 1987, 8p
Pub. in Proceedings of Annual Precise Time and Time Interval (PTTI) Applications and Planning Meeting (19th), Redondo Beach, CA., December 1-3, 1987, p25-32.

Keywords: *Atomic clocks, Performance evaluation, Time measuring instruments, Frequency standards, Experimental data, *Global positioning system, Autonomous spacecraft clocks, Cesium oscillators.

The Clock Evaluation and Time Keeping Experiment was the cooperative effort of Applied Research Laboratories, the National Bureau of Standards, and the United States Naval Observatory. It was designed to collect a dense Global Positioning System (GPS) data set for evaluating methods of monitoring a ground-based atomic clock with on-site data alone, and of investigating improved methods of time setting and time transfer using GPS data. The experiment collected two-frequency, P-code, pseudorange and Doppler data for five weeks at three sites: Austin, Texas; Boulder, Colorado; and Washington, D.C. All sites used two-frequency receivers on cesium oscillators. A time history of the cesium oscillators against hydrogen

masters also was recorded at two of the sites. The behavior of the cesium oscillators was well tracked by the range residuals over the five weeks of the experiment. Residuals of Doppler data were strongly correlated across stations. This implies that time transfer with accuracy approaching 1 ns looks promising over the time period of one satellite pass. With a full GPS constellation, continuous use of phase data could significantly improve time transfer via GPS.

900,615
PB89-177125   Not available NTIS
National Bureau of Standards (ICST), Gaithersburg, MD. Systems and Network Architecture Div.
Automatic Generation of Test Scenario (Skeletons) from Protocol-Specifications Written in Estelle.
Final rept.
J. P. Favreau, and R. J. Linn. 1987, 12p
Pub. in Proceedings of IFIP (International Federation for Information Processing) WG 6.1 International Workshop on Protocol Specification, Testing, and Verification (6th), Montreal, Quebec, Canada, June 10-13, 1986, p191-202 1987.

Keywords: *Standards, Tests, Automation, Proving, Specifications, *Protocols, *Computer communications, Protocol(Computers).

Estelle is a new formal description technique developed by the International Organization for Standardization for the description of computer communication protocols destined to become international standards. Specification, validation and testing computer communication protocols are all complex issues. The paper focuses on methods for generating test sequences for communication protocols and describes a model and method for partially automated generation of test skeletons for protocols written in Estelle.

900,616
PB89-177133   Not available NTIS
National Bureau of Standards (ICST), Gaithersburg, MD. Systems and Network Architecture Div.
Application of the ISO (International Standards Organization) Distributed Single Layer Testing Method to the Connectionless Network Protocol.
Final rept.
J. S. Nightingale. 1987, 12p
Pub. in Proceedings of IFIP (International Federation for Information Processing) WG 6.1 International Workshop on Protocol Specification, Testing, and Verification (6th), Montreal, Quebec, Canada, June 10-13, 1986, p123-134 1987.

Keywords: *Tests, Conformity, *Protocols, *Communication networks, Protocol(Computers), Computer networks, Connectionless Network Protocol.

An architecture for testing implementations of the ISO Connectionless Network Protocol is described. After a brief description of the Connectionless Network Protocol, or Internet, the architectures for both end system and intermediate system testing are motivated through a set of design principles. The design uses a test management protocol which is explained herein. Syntactic constructs within the test language are given. The final section highlights some reactions which have been received concerning the test system.

900,617
PB89-186629   Not available NTIS
National Bureau of Standards (ICST), Gaithersburg, MD. Systems and Network Architecture Div.
Simplified Discrete Event Simulation Model for an IEEE (Institute of Electrical and Electronics Engineers) 802.3 Local Area Network.
Final rept.
S. K. Heatley. 1986, 7p
Pub. in Proceedings of GLOBECOM '86: IEEE (Institute of Electrical and Electronics Engineers) Global Telecommunications Conference Communications Broadening Technology Horizons, Houston, TX., December 1-4, 1986, v1 p143-149.

Keywords: *Models, *Computerized simulation, *Computer networks, *Protocols, Computer performance evaluation, Computer communications, Throughput.

In the Protocol Performance Group at the National Bureau of Standards, the performance of multi-layered ISO protocols is under study. The paper describes a simplified model of the IEEE 802.3 Local Area Network which was developed to serve the lowest layer in a discrete simulation model of multilayered ISO protocols. In the paper, throughput and delay measure-

ments obtained from the simplified model are compared to measurements from a more detailed model of 802.3. Execution times of the simplified and detailed model are also compared. Sample execution times from the integrated, multi-layered model are presented.

900,618
PB89-193312   PC A22/MF A01
National Inst. of Standards and Technology, Gaithersburg, MD.
Stable Implementation Agreements for Open Systems Interconnection Protocols. Version 2, Edition 1. December 1988.
Final rept.
T. Boland. Feb 89, 511p NIST/SP-500/162
Also available from Supt. of Docs. as SN003-003-02921-1. See also PB88-168331. Library of Congress catalog card no. 89-600726. Based on the Proceedings of the NIST (National Institute of Standards and Technology) Workshop for Implementors of OSI Held at Gaithersburg, Maryland.

Keywords: *Protocols, *Open systems interconnection, *Computer networks, National Institute for Standards and Technology, Computer communications.

The document records current stable agreements for Open Systems Interconnection (OSI) Protocols among the organizations participating in the National Institute of Standards and Technology (NIST)/OSI Workshop Series for Implementors of OSI Protocols.

900,619
PB89-196158   PC A05/MF A01
National Inst. of Standards and Technology (NCSL), Gaithersburg, MD. Systems and Network Architecture Div.
User Guide for the NBS (National Bureau of Standards) Prototype Compiler for Estelle (Revised).
Technical rept. (Final).
J. P. Favreau, M. Hobbs, B. Strausser, and A. Weinstein. Feb 89, 79p ICST/SNA-87/3, NBS/SW/MT-89/004A
Supersedes PB88-124185. For system on magnetic tape, see PB89-196141.

Keywords: *Compilers, Prototypes, Computerized simulation, Syntax, Computer programs, Documentation, *Open Systems Interconnection, *Estelle, Software tools, Debugging(Computers), User manuals(Computer programs), Protocols, Computer communications.

The NBS Prototype Compiler for Estelle describes an implementation model for Estelle, the output of the compiler, the run-time library of support routines, and the syntax of Estelle used by the compiler. Instructions are provided for installing the compiler, executing it and providing the necessary implementation environment. Complete source is provided for a practical example of a simple protocol simulation; the report describes this example in some detail.

900,620
PB89-196166   PC A03/MF A01
National Inst. of Standards and Technology (NCSL), Gaithersburg, MD. Systems and Network Architecture Div.
User Guide for Wise: A Simulation Environment for Estelle.
Technical rept.
R. Sijelmassi. Feb 89, 26p NCSL/SNA-89/6, NBS/SW/MT-89/004B
For system on magnetic tape, see PB89-196141.

Keywords: *Computerized simulation, Semantics, Models, Documentation, *Computer communications, *Protocols, Wise system, Software tools, Estelle, User manuals (Computer programs).

The Wise software system provides a simulation environment for Estelle, which is a formal description technique for specifying computer communication protocols. The Wise tool is based on a model of Estelle semantics that is implemented as a collection of classes in the object-oriented language Smalltalk. For simulation, these classes are augmented with control and observation capabilities and the ability to simulate a distributed environment. The report describes how to use the Wise tool, particularly the observation and control features. A comparison document describes the underlying model. The Wise tool works together with the Wizard tool, a syntax-directed editor and translator, which also has a User Guide.

74

900,621
PB89-196174　　　　　　　　PC A03/MF A01
National Inst. of Standards and Technology (NCSL),
Gaithersburg, MD. Systems and Network Architecture
Div.
**User Guide for Wizard: A Syntax-Directed Editor
and Translator for Estelle.**
Technical rept.
B. Strausser. Feb 89, 13p NCSL/SNA-89/5, NBS/
SW/MT-89/004C
For system on magnetic tape, see PB89-16141.

Keywords: *Editing routines, *Translator routines,
Syntax, Documentation, *Computer communications,
*Protocols, Wizard system, Estelle, Software tools,
User manuals(Computer programs).

The Wizard editor and translator provides syntax-directed editing for Estelle, which is a formal description
technique for specifying computer communication protocols. The Wizard also translates an Estelle specification into code in the object-oriented language Smalltalk for use with the Wise tool. The report describes
some of the editor's features. The Wizard tool was developed with the Cornell Synthesizer Generator; documentation of editor operation from Cornell is desirable.

900,622
PB89-196182　　　　　　　　PC A04/MF A01
National Inst. of Standards and Technology (NCSL),
Gaithersburg, MD. Systems and Network Architecture
Div.
Free Value Tool for ASN.1.
Technical rept.
P. Gaudette, S. Trus, and S. Collins. Feb 89, 71p
NCSL/SNA-89/1, NBS/SW/MT-89/004D
For system on magnetic tape, see PB89-196141.

Keywords: *Syntax, Data, Models, Coding, Documentation, *Computer communications, *Protocols, Software tools, Open Systems Interconnection, Data
structures.

The Free Value Tool provides the ability to manipulate
data values described with Abstract Syntax Notation
One (ASN.1). ASN.1 is a non-procedural notation for
describing data objects at the application and presentation layers of computer communication protocols in
the Open Systems Interconnection model. An abstract
syntax in ASN.1 that does not use the macro feature is
converted into data structures that capture the type information from the input. Values of those types can
then be manipulated in an interpretive style, with transformations among printable value notation, encoded
transfer values, and other forms of values. The report
describes how to use the Free Value Tool and some of
the implementation details. It includes a tutorial on
ASN.1.

900,623
PB89-196190　　　　　　　　PC A05/MF A01
National Inst. of Standards and Technology (NCSL),
Gaithersburg, MD. Systems and Network Architecture
Div.
**Object-Oriented Model for Estelle and Its Smalltalk
Implementation.**
Technical rept.
R. Sijelmassi. Feb 89, 81p NCSL/SNA-89/7, NBS/
SW/MT-89/004E
For system on magnetic tape, see PB89-196141.

Keywords: *Computerized simulation, Semantics,
Models, Documentation, *Computer communications,
*Protocols, Wise system, Estelle, Software tools.

The Wise software system provides a simulation environment for Estelle, which is a formal description technique for specifying computer communication protocols. The Wise tool is based on a model of Estelle semantics that is implemented as a collection of classes
in the object-oriented language Smalltalk. The report
describes the model and the Smalltalk implementation
in some detail, which is useful background for extending or modifying the Wise tool. A companion report, the
User Guide for Wise, describes how to use the tool.

900,624
PB89-221196　　　　　　　　PC A10/MF A01
National Inst. of Standards and Technology (NCSL),
Gaithersburg, MD.
**Working Implementation Agreements for Open
Systems Interconnection Protocols.**
Final rept.
T. Boland. May 89, 213p NISTIR-89/4082
Proceedings of the NIST (National Institute of Stand-
ards and Technology) Workshop for Implementors of

OSI Plenary Assembly; Gaithersberg, MD, March 17,
1989.

Keywords: *Agreements, Security, Digital systems,
Specifications, Meetings, *Protocols, *Computer net-
works, *Communication networks, National Institute of
Standards and Technology.

The document records current agreements on implementation details for Open Systems Interconnection
Protocols among the organizations participating in the
NIST/OSI Workshop Series for Implementors of OSI
Protocols. These decisions are documented to facilitate organizations in their understanding of the status
of agreements. This is a standing document that is updated after each workshop (about 4 times a year).

900,625
PB89-235576　　　　　　　　PC A03/MF A01
National Inst. of Standards and Technology (NCSL),
Gaithersburg, MD.
**Trial of Open Systems Interconnection (OSI) Pro-
tocols Over Integrated Services Digital Network
(ISDN).**
C. A. Edgar. Aug 89, 21p NISTIR-89/4160

Keywords: Information retrieval, Compatibility, *Computer networks, *Computer communications, *Protocols, Electronic mail, Integrated systems, File management systems, Computer architecture, Workstations.

Presented are the results of the National Institute of
Standards and Technology Open Systems Intercon-
nection/Integrated Services Digital Network (OSI/
ISDN) Trial. The trial was organized to demonstrate
the use of ISDN as a lower layer technology for OSI
application. The document addresses the trial's topology, hardware/software configuration, parameters,
and results.

900,626
PB90-100736　　　　　　　　PC A03/MF A01
National Inst. of Standards and Technology (NEL),
Boulder, CO. Electromagnetic Fields Div.
**X-Band Atmospheric Attenuation for an Earth Ter-
minal Measurement System.**
M. H. Francis. Jul 89, 46p NISTIR-89/3918

Keywords: *Atmospheric attenuation, *Satellite antennas, Measurement, Antennas, Gain, Spacecraft communication, Errors, Sky brightness, Heat transfer, X-
Band, Tipping curve method.

The National Institute of Standards and Technology
has developed an Earth Terminal Measurement
System to be used by the Camp Parks Communications Annex in determining satellite effective isotropic
radiated power and antenna gain. In determining these
quantities the effect of atmospheric attenuation must
be taken into account. The paper provides an overview
of the methods used for determining atmospheric attenuation with emphasis on a tipping-curve method.
An error analysis is also provided.

Graphics

900,627
FIPS PUB 152　　　　　　　　PC E19
National Inst. of Standards and Technology (NCSL),
Gaithersburg, MD.
Standard Generalized Markup Language (SGML).
Federal information processing standards (Final).
L. A. Welsch. c26 Sep 88, 174p
Three ring vinyl FIPS binder also available, North
American Continent price $7.00; all others write for
quote.

Keywords: Documents, Publishing, *Federal information processing standards, *High level languages,
*Text processing.

The publication announces the adoption of the International Standards Organization Standard Generalized Markup Language (SGML), ISO 8879-1986, as a
Federal Information Processing Standard (FIPS). ISO
8879-1986 specifies a language for describing documents to be used in office document processing, interchange between authors and between authors and
publishers, and publishing. The language provides a
coherent and unambiguous syntax for describing the
elements within a document. (Copyright (c) International Standards Organization, 1986.)

COMMUNICATION

Common Carrier & Satellite

Policies, Regulations, & Studies

900,628
PB89-174072　　　　　　　　Not available NTIS
National Bureau of Standards (NML), Boulder, CO.
Time and Frequency Div.
**Comparison of Time Scales Generated with the
NBS (National Bureau of Standards) Ensembling
Algorithm.**
Final rept.
F. B. Varnum, D. R. Brown, D. W. Allan, and T. K.
Peppler. 1987, 11p
Pub. in Proceedings of Annual Precise Time and Time
Interval (PTTI) Applications and Planning Meeting
(19th), Redondo Beach, CA., December 1-3, 1987,
p13-23.

Keywords: *Time standards, *Algorithms, Time measurement, Time measuring instruments, Clocks,
Cesium, Weighting functions.

The National Bureau of Standards (NBS), Boulder,
Colorado, uses an algorithm which generates
UTC(NBS) from its ensemble of clocks and automatically, optimally, and dynamically weights each clock in
the ensemble. The same algorithm was used at the
Master Control Station (MCS) of the Global Positioning
System (GPS) to generate a time scale from a small
ensemble of cesium clocks. Time transfer employing
the GPS common view technique between NBS (Boulder) and MCS (Colorado Springs) was used to evaluate the stability of the MCS ensemble relative to
UTC(NBS). The results demonstrate the power of the
NBS algorithm in providing a stable time scale from a
small ensemble of clocks. The resulting scale is, in
principle, more stable than the best clock and a poor
clock need not degrade the ensemble.

900,629
PB89-185722　　　　　　　　Not available NTIS
National Bureau of Standards (NML), Boulder, CO.
Time and Frequency Div.
**Millisecond Pulsar Rivals Best Atomic Clock Stabil-
ity.**
Final rept.
D. W. Allan. 1987, 10p
See also PB86-138821.
Pub. in Proceedings of Annual Symposium on Fre-
quency Control (41st), Philadelphia, PA., May 27-29,
1987, p2-11.

Keywords: *Atomic clocks, *Stability, Comparison,
Orbits, Time measuring instruments, Signal to noise
ratios, Dispersions, Ephemerides, Interplanetary
medium, Interstellar matter, Gravity waves, *Millisecond pulsar.

The measurement time residuals between the millisecond pulsar PSR 1937+21 and atomic time have been
significantly reduced. Analysis of data for the most
recent 865 period indicates a fractional frequency stability (square root of the modified Allan variance) of
less than 2 x 10 to the (-14) power for integration times
of about 1/3 year. Analysis of the measurements
taken in two frequency bands revealed a random walk
behavior for dispersion along the 12,000 to 15,000
light year path from the pulsar to the earth. The
random walk accumulates to about 1,000 nanoseconds (ns) over 265 days. The final residuals are nominally characterized by a white phase noise at a level of
369 ns. Following improvement of the signal-to-noise
ratio, evidence was found for a residual modulation.
Possible explanations for the modulation include: a
binary companion to the pulsar with approximate
period(s) of 120 days and with a mass of the order of 1
x 10 to the (-9) power that of the pulsar; irregular magnetic drag in the pulsar; unaccounted delay variations
in the interstellar medium; modeling errors in the
earth's ephemeris; reference atomic clock variations in
excess of what are estimated; or gravity waves. For
gravity waves, the amplitude of the length modulation
would be about 5 parts in 10 to the (19) power. Further
study is needed to determine which is the most probable explanation.

900,630
PB89-212070　　　　　　　　Not available NTIS
National Inst. of Standards and Technology (NML),
Boulder, CO. Time and Frequency Div.
**Calibration of GPS (Global Positioning System)
Equipment in Japan.**
Final rept.
M. A. Weiss, and D. D. Davis. 1988, 10p
See also PB88-122023.

COMMUNICATION

Policies, Regulations, & Studies

Pub. in Proceedings of Annual Precise Time and Time Interval (PTTI) Applications and Planning Meeting (20th), Vienna, VA., November 29-December 1, 1988, p101-110.

Keywords: *Calibrating, *Time standards, *Frequency standards, Japan, Telecommunication, Time measurement, Frequency measurement, Receivers, Synchronism, *Global positioning systems, Time transfer.

With the development of common view time comparisons using Global Positioning System (GPS) satellites the Japanese time and frequency standards laboratories have been able to contribute with more weight to the international unification of time under the coordination of the Bureau International de Poids et Measures. During the period from June 1 through June 11, 1988, the differential delays of time transfer receivers of the GPS were calibrated at three different laboratories in Japan, linking them for absolute time transfer with previously calibrated labs of Europe and North America. The differential delay between two receivers was first calibrated at the National Institute of Standards and Technology (NIST)(formerly the National Bureau of Standards) in Boulder, Colorado, USA. Then one of these receivers was carried to each of the three laboratories: the Tokyo Astronomical Observatory, the Communications Research Laboratory, both in Tokyo, and the National Research Laboratory of Metrology in Tsukuba City. At each lab data was taken comparing receivers. Finally the traveling receiver was taken back to NIST for closure of the calibration. On the way back the GPS receiver at the WWVH radio station of NIST in Hawaii was also calibrated. The results of the calibration trip are reported, along with some interesting problems that developed concerning the technique.

900,631
PB89-212211   Not available NTIS
National Bureau of Standards (NML), Boulder, CO. Time and Frequency Div.
NBS (National Bureau of Standards) Calibration Service Providing Time and Frequency at a Remote Site by Weighting and Smoothing of GPS (Global Positioning System) Common View Data.
Final rept.
M. A. Weiss, and D. W. Allan. 1986, 2p
Pub. in Proceedings of Conference on Precision Electromagnetic Measurements, Gaithersburg, MD., June 23-27, 1986, p125-126.

Keywords: *Calibrating, *Time standards, *Frequency standards, Time measurement, Frequency measurement, Synchronism, Data smoothing, Data processing, *Global positioning systems, Kalman filtering.

The National Bureau of Standards Time and Frequency Division now performs precision time and frequency transfer using common view measurements of Global Positioning System (GPS) satellites as a calibration service. Using the service the authors have been able to transfer time with time stabilities of a few nanoseconds, time accuracies of the order of 10 ns, and frequency stabilities of a part in 10 to the 14th power and better for measurement times of about four days and longer. The paper describes the technique used for weighting and smoothing the data to produce this level of stability and accuracy.

900,632
PB90-117367   Not available NTIS
National Inst. of Standards and Technology (NML), Boulder, CO. Time and Frequency Div.
In Search of the Best Clock.
Final rept.
D. W. Allan, M. A. Weiss, and T. K. Peppler. 1989, 7p
Pub. in IEEE (Institute of Electrical and Electronics Engineers) Transactions on Instrumentation and Measurement 38, n2 p624-630 Apr 89.

Keywords: *Atomic clocks, *Stability, Time standards, Frequency meters, Time measuring instruments, Frequency standards, Time measurement, Chronometers, Timing devices, Gravitational fields, Reprints.

There is an increased need for better clock performance than is currently available. Past work has focused on developing better clocks to meet the increased need. Significant gains have been and can be obtained through the algorithms which optimize the clock readings and through international comparisons now available via satellite. Algorithms for processing are more important than the proportionate attention generally given them. In fact, to date, one of the main ways to investigate some of the long-term performance as-

pects of the millisecond pulsar, PSR 1937 + 21, is by using such optimization algorithms. Since there are indications in the pulsar data of variations which could be explained as arising from the influence of gravitational waves, these long-term stability studies take on a new importance. Improved long-term stability of earth-bound clock systems will significantly assist the study of the incredibly stable spin rates of these neutron stars.

Verbal

900,633
PB89-176713   Not available NTIS
National Bureau of Standards (ICST), Gaithersburg, MD. Systems Components Div.
PCM/VCR Speech Database Exchange Format.
Final rept.
D. S. Pallett. 1986, 4p
Pub. in Proceedings of IEEE-IECEJ-ASJ International Conference on Acoustics, Speech, and Signal Processing, Tokyo, Japan, April 7-11, 1986, p317-320.

Keywords: *Speech, Speech recognition, Data storage, Acoustics, *Data bases, Computer applications, Format, Information transfer.

The use of PCM/VCR technology is described for use as a storage and exchange medium for speech databases. In order to provide a limited amount of digital data, use is made of a recorded modem signal for ASCII character string headers associated with the speech tokens. The format can be used to store field recordings of speech material for subsequent digitization, for transfer of speech database material between laboratories, and in implementing automated testing of speech recognition devices.

900,634
PB89-176721   Not available NTIS
National Bureau of Standards (ICST), Gaithersburg, MD. Systems Components Div.
Compensating for Vowel Coarticulation in Continuous Speech Recognition.
Final rept.
J. L. Hieronymus, and W. J. Majurski. 1986, 4p
Pub. in Proceedings of ICASSP, IEEE (Institute of Electrical and Electronics Engineers) International Conference on Acoustics, Speech and Signal Processing, Tokyo, Japan, April 7-11, 1986, p2787-2790.

Keywords: *Speech recognition, Speech articulation, *Vowels, Formants.

Seven English monothong vowels were studied in continuous sentences. The study determined what methods are likely to be successful in compensating for coarticulation in all vowel and consonantal contexts. A method by Kuwahara is examined in detail. The Kuwahara compensation improves the separation of the vowel regions in a space composed of the first and second formant. Important issues are where to measure the formant target frequencies, how to obtain good formant tracks, measuring speaking rate accurately, and how to label vowels accurately.

COMPUTERS, CONTROL & INFORMATION THEORY

Computer Hardware

900,635
PB89-162614   PC A03/MF A01
National Inst. of Standards and Technology (NEL), Boulder, CO. Center for Computing and Applied Mathematics.

Analysis of Computer Performance Data.
J. C. M. Wang, J. M. Gary, and H. K. Iyer. Jan 89, 45p NISTIR-88/4019
Prepared in cooperation with Colorado State Univ., Fort Collins. Dept. of Statistics.

Keywords: *Data reduction, Statistics, Computer systems hardware, Computer systems programs, Tables(Data), Bench marks, *Computer performance evaluation, *Data analysis.

The paper is devoted to an analysis of the data from the Livermore loops benchmark. It will be shown that in a general predictive sense the dimension of this data is rather small; perhaps between two and five. Two techniques are used to reduce the 72 loop timings for each machine to a few scores which characterize the machine. The first is based on a principal component analysis, the second on a cluster analysis of the loops. The validity of the reduction of the data to a lesser dimension is checked by various methods.

900,636
PB89-173793   Not available NTIS
National Bureau of Standards (ICST), Gaithersburg, MD. Advanced Systems Div.
Performance Measurement of a Shared-Memory Multiprocessor Using Hardware Instrumentation.
Final rept.
A. Mink, and G. Nacht. 1989, 10p
Sponsored by Defense Advanced Research Projects Agency, Arlington, VA.
Pub. in Proceedings of Annual Hawaii International Conference on System Sciences (22nd), Kailua-Kona, HI., January 3-6, 1989, p267-276.

Keywords: Parallel processors, *Multiprocessors, *Computer performance evaluation, Microprocessors, Parallel computers.

A hardware approach is presented for the design of performance measurement instrumentation for a shared-memory, tightly coupled MIMD multiprocessor. The Resource Measurement System (REMS) is a non-intrusive, hardware measurement tool used to obtain both trace measurement and resource utilization information. This approach provides more detailed and extensive measurement information than alternative software or hybrid approaches without introducing artifacts into the test results. This is accomplished at a significantly higher tool cost than the alternative software or hybrid approaches. Certain features of today's microprocessors limit the applicability of such a hardware tool. Measurements obtained using this hardware tool on two kernel (small benchmark) routines are presented.

900,637
PB89-186845   Not available NTIS
National Bureau of Standards (ICST), Gaithersburg, MD. Systems Components Div.
Design Factors for Parallel Processing Benchmarks.
Final rept.
G. Lyon. 1988, 12p
Sponsored by Defense Advanced Research Projects Agency, Arlington, VA.
Pub. in High Performance Computer Systems, p103-114 1988.

Keywords: Models, Reprints, *Parallel processing(Computers), *Computer performance evaluation, Computer architecture, Computer systems design.

Performance benchmarks should be embedded in comprehensive frameworks that suitably set their context of use. One universal framework is beyond reach, since distinct clusters of use are emerging with separate emphases. Large application benchmarks are most successful when they run well on a machine, and thereby demonstrate concrete compatibility of job and architecture. The present value of smaller benchmarks is diagnostic, although sets of them would encourage the parametric study of architectures and applications.

900,638
PB89-186852   Not available NTIS
National Bureau of Standards (ICST), Gaithersburg, MD. Systems Components Div.

Hardware Instrumentation Approach for Performance Measurement of a Shared-Memory Multiprocessor.
Final rept.
G. Nacht, and A. Mink. 1988, 17p
Sponsored by Defense Advanced Research Projects Agency, Arlington, VA.
Pub. in Proceedings of International Conference on Modelling Techniques and Tools for Computer Performance Evaluation (4th), Palma de Mallorca, Spain, September 14-16, 1988, v2 p321-337.

Keywords: Computer systems hardware, *Computer performace evaluation, *Multiprocessors, Parallel computers, Hardware.

A hardware approach is presented for the design of performance measurement instrumentation for a shared memory, tightly coupled MIMD multiprocessor. The Resource Measurement System (REMS) is a nonintrusive, hardware measurement tool used to obtain both trace measurement and resource utilization information. The approach provides more detailed and extensive measurement information than alternative software or hybrid approaches without introducing artifacts into the test results. This is accomplished at a significantly higher cost than the alternative software or hybrid approaches. When access to pertinent signals is restricted, the applicability of such a hardware tool is limited.

900,639
PB89-189161 PC A03/MF A01
National Inst. of Standards and Technology (ICST), Gaithersburg, MD. Advanced Systems Div.
Hybrid Structures for Simple Computer Performance Estimates.
G. E. Lyon. Mar 89, 25p NISTIR-89/4063
Sponsored by Defense Advanced Research Projects Agency, Arlington, VA., Bureau of Export Administration, Washington, DC., and Department of Energy, Washington, DC.

Keywords: Computer systems hardware, Computer components, *Computer architecture, *Computer performance evaluation, Computer applications.

Even the coarsest performance estimators for a modern computer must account for architectural dependencies and variabilities. For example, average execution rate is rather sensitive to the match between machine and application workload. Computing can be viewed as computational components that are loaded by demands of an application, or alternately, as an application workload partitioned by system components. Models based upon these simple perspectives help organize simple performance measurements. Several examples demonstrate strengths of a straight-forward and flexible partitioning scheme based upon tree graphs. Quite explicit, the graphs promote a more critical view of measurements and support multiple interpretations.

900,640
PB89-216477 PC A03/MF A01
National Inst. of Standards and Technology (NCSL), Gaithersburg, MD. Advanced Systems Div.
Architecturally-Focused Benchmarks with a Communication Example.
G. E. Lyon, and R. D. Snelick. Mar 89, 39p NISTIR-89/4053
Sponsored by Defense Advanced Research Projects Agency, Arlington, VA.

Keywords: *Performance evaluation, Performance tests, Measurement, Computer systems hardware, Comparison, *Computer architecture, Computer communications, Parallel processing, Vector processing.

The discussion first sketches a framework of modalities for an architecturally-focused performance evaluation. The result is a hybrid of benchmarking and modeling: elements of capacity-and-use trees (CUTs) are explored as a simplified notation. There follows a description of the structure and preliminary results from a practical benchmark set for process communication. Argument is given that performance within a class of architecture is often dominated by unavoidable competitions within distinct machine modalities, such as scalar-vector. A k-alternative, forced choice defines a dimension of comparison equally well in SI- or MIMD architectures. Performance estimators are interpolations between values from basis benchmarks for modes; ideally in the two-alternative forced choice only two benchmark measurements are needed. Refinements in basis benchmarks support CUT-based estimates of performance.

900,641
PB89-229017 PC A03/MF A01
National Inst. of Standards and Technology (ICST), Gaithersburg, MD. Advanced Systems Div.
Processing Rate Sensitivities of a Heterogeneous Multiprocessor.
G. E. Lyon. Aug 89, 12p NISTIR-89/4128
Sponsored by Defense Advanced Research Projects Agency, Arlington, VA., and Department of Energy, Washington, DC.

Keywords: *Multiprocessors, Mathematical models, Performance evaluation, Parallel processing, Capacity, Computer architecture, Sensitivity.

A recent trend in multiprocessor evaluation has been to seek fundamental but easily parameterized performance characterizations. Freed of specialized detail, simplified models can convey good insight while applying easily and widely. A simple performance estimator and alternate scheduling schemes can, from very modest effort, highlight some first-order improvement tradeoffs in multiprocessors. An application example of a quicksort is used to demonstrate the approach.

900,642
PB89-235931 PC A16/MF A01
National Inst. of Standards and Technology (NCSL), Gaithersburg, MD.
Working Implementation Agreements for Open Systems Interconnection Protocols.
T. Boland. Aug 89, 374p NISTIR-89/4140
See also PB89-221196.

Keywords: *Computer networks, Tests, Standards, Data processing security, *National Institute of Standards and Technology, *Open systems interconnection, *Protocols, Local area networks, Message processing, File maintenance, Data base management.

The document records current agreements on implementation details of Open Systems Interconnection Protocols among the organizations participating in the NIST/OSI Workshop Series for Implementors of OSI Protocols. The decisions are documented to facilitate organizations in their understanding of the status of agreements: It is a standing document that is updated after each workshop (about 4 times a year).

900,643
PB90-112418 PC A07/MF A01
National Inst. of Standards and Technology (NEL), Gaithersburg, MD. Building Environment Div.
Guideline for Work Station Design.
A. Rubin, and G. Gillette. Sep 89, 150p NISTIR-89/4163
Sponsored by Public Buildings Service, Washington, DC.

Keywords: *Ergonomics, *Design, *Office equipment, Furniture, Human factors engineering, Acoustics, Illuminating, Wiring, *Workstations, Man machine systems.

The report describes a workstation design process, starting with an analysis of the activities performed, then deals with environmental, building design, space planning and furniture issues required for designing workstations suitable for a range of office activities. A limited number of generic workstations are presented, as examples of types of configurations that might meet specific office requirements. The examples are illustrative of the results of following the design process suggested and are not intended as recommended approaches for particular workstation designs. Technological, ergonomic, and organizational factors are considered from the standpoint of their design implications for automated workstations. Criteria and checklists are included as an aid to making workstation design decisions.

900,644
PB90-117672 Not available NTIS
National Inst. of Standards and Technology (NCSL), Gaithersburg, MD. Advanced Systems Div.
Design Factors for Parallel Processing Benchmarks.
Final rept.
G. Lyon. 1989, 15p
See also PB89-166845. Sponsored by Defense Advanced Research Projects Agency, Arlington, VA.
Pub. in Theoretical Computer Science 64, p175-189 1989.

Keywords: *Parallel processors, Models, Measurement, Reprints, *Computer performance evaluation, Computer architecture.

Performance benchmarks should be embedded in comprehensive frameworks that suitably set their context of use. One universal framework appears beyond reach, since distinct architectural clusters are emerging with separate emphases. Large application benchmarks are most successful when they run well on a machine, and thereby demonstrate the economic compatibility of job and architecture. The present value of smaller benchmarks is diagnostic, although sets of them would encourage the parametric study of architectures and applications; an extended example illustrates the last aspect.

900,645
PB90-129891 PC A03/MF A01
National Inst. of Standards and Technology (NEL), Boulder, CO. Center for Computing and Applied Mathematics.
Allocating Staff to Tax Facilities: A Graphics-Based Microcomputer Allocation Model.
P. D. Domich. Jun 89, 18p NISTIR-89/4059

Keywords: *Microcomputers, *Allocation models, Taxes, Mathematical models, Computer graphics, Facilities, Stuffing.

The paper presents a computer model to allocate Internal Revenue Service staff to field offices so to satisfy a user-defined level of workload. The system is microcomputer-based and uses menus and graphically displayed zip code maps of IRS districts for interactive inputs and solution outputs. The level of workload is interactively specified and a detailed report of workload and staff assignments is automatically generated.

Computer Software

900,646
PB89-151807 PC A04/MF A01
National Inst. of Standards and Technology (ICST), Gaithersburg, MD. Advanced Systems Div.
Wavefront Matrix Multiplication on a Distributed-Memory Multiprocessor.
J. R. Nechvatal. Jan 89, 71p NISTIR-88/4001
Sponsored by Defense Advanced Research Projects Agency, Arlington, VA.

Keywords: *Matrices(Mathematics), *Multiplication, Algorithms, *Multiprocessors, Distributed computer systems, Computer networks.

The report considers the problem of efficiently multiplying matrices on distributed-memory, message-passing computers. Block decomposition and wavefront computing are employed to yield a communication-efficient solution. It also explores interconnections between distributed data structures, physical networks of nodes and virtual networks of nodes and processes. The notion of wavefront computing is extended to include pipelining of data between nodes of networks of processes as well as physical networks of nodes. Algorithms are developed in layers, to facilitate porting between topologies and programming environments. The authors show how different topologies can be employed in a single application. A mathematical characterization of data-routing for efficient matrix multiplication on distributed-memory machines is developed, which exhibits a wavefront version of the Dekel/Nassimi/Sahni algorithm as a special case. The authors discuss the notion of granularity in this context and use it to distinguish between algorithms on grounds of communication complexity.

900,647
PB89-176226 Not available NTIS
National Bureau of Standards (NEL), Boulder, CO.
Electromagnetic Fields Div.
Creating CSUBs in BASIC.
Final rept.
E. J. Vanzura. 1988, 5p
Pub. in HP Design and Automation, pt1 p18-21 Oct 88 and HP Design and Automation, pt2 p25 Nov 88.

Keywords: *Subroutines, Operating systems(Computers), Compilers, Reprints, *Basic programming language, *Fortran programming language,

Computer Software

PASCAL programming language, Computer software, Precompilers, Input output processing.

CSUBs are compiled subprograms created using the Pascal operating system which run in the BASIC environment. A new technique is described in which programs written in FORTRAN can be turned into CSUBs. Thus, powerful, well-documented FORTRAN routines become accessible to the BASIC-language programmer. I/O and variable interfacing are discussed, and a comprehensive example is provided.

900,648
PB89-177000　　　　　　　　　Not available NTIS
National Bureau of Standards (NEL), Gaithersburg, MD. Factory Automation Systems Div.
Increment: A Graphical Technique for Manipulating Parameters.
Final rept.
S. Ressler. 1987, 5p
Pub. in ACM (Association for Computing Machinery) Transactions on Graphics 6, n1 p74-78 Jan 87.

Keywords: Reprints, *Interactive graphics, *Man computer interface, Man machine systems, Computer graphics, Input output processing.

A visually compact technique for manipulating variables is presented. The technique uses both a mouse and keyboard for redundant entry methods.

900,649
PB89-177117　　　　　　　　　Not available NTIS
National Inst. of Standards and Technology (ICST), Gaithersburg, MD. Systems and Network Architecture Div.
Object-Oriented Model for ASN.1 (Abstract Syntax Notation One).
Final rept.
P. Gaudette, S. Trus, and S. Collins. 1988, 14p
Pub. in Proceedings of International Conference on Formal Description Techniques (1st), Stirling, Scotland, September 6-9, 1988, p121-134.

Keywords: Models, *Data definition languages, *ASN.1 data definition language, High level languages, Abstract Syntax Notation One, Protocols, Protocol(Computers).

ASN.1 is a data definition language associated with the OSI Presentation Layer. The model of ASN.1 provides an object-oriented definition of types and values, the most basic concepts of ASN.1. The model is both a target for syntax-directed translation and an input to several tools for working with ASN.1. One tool allows evaluation of an abstract syntax in ASN.1 by providing a collection of transformations of values. Some difficult problems of ASN.1 tools, such as macros, are also discussed.

900,650
PB89-180012　　　　　　　　　Not available NTIS
National Bureau of Standards (ICST), Gaithersburg, MD. Systems Components Div.
Definitions of Granularity.
Final rept.
C. P. Kruskal, and C. H. Smith. 1986, 12p
Pub. in High Performance Computer Systems, p257-268 1988.

Keywords: Parallel processors, Reprints, *Granularity, *Parallel processing, Parallel computers, Computations, Balance.

'Granularity' is a well known concept in parallel processing. While intuitively, the distinction between 'coarse grain' and 'fine grain' parallelism is clear, there is no rigorous definition. The paper develops two notions of granularity, each defined formally and represented by a single rational number. The two notions are compared and contrasted with each other and with previously proposed definitions of granularity.

900,651
PB89-193833　　　　　　　　PC A03/MF A01
National Inst. of Standards and Technology (NCSL), Gaithersburg, MD.
Software Configuration Management: An Overview.
Final rept.
W. M. Osborne. Mar 89, 34p NIST/SP-500/161
Also available from Supt. of Docs. as SN003-003-02927-1. See also AD-A130 109. Library of Congress catalog card no. 89-600728.

Keywords: Revisions, Quality assurance, *Software configuration management, *Software engineering,

*Computer software management, *Computer program reliability, Reusable software, Computer software maintenance, Computer program verification, Software tools, Life cycles.

The guide provides an overview of software configuration management, a support function dedicated to making both the technical and managerial software activities more effective. It addresses the problems associated with managing software changes; the importance of implementing software configuration management (SCM) procedures early; and the application of those procedures throughout the software lifecycle. A brief summary of SCM tools and their applicable functionality is provided. SCM extends to more than just the code (source, relocatable, executable) and documentation (e.g., system and software requirements and design specifications). It also covers control files, test data, test suites, support tools, and other components used to develop and maintain the software product.

900,652
PB89-211908　　　　　　　　　Not available NTIS
National Inst. of Standards and Technology (ICST), Gaithersburg, MD. Systems and Network Architecture Div.
Application of Formal Description Techniques to Conformance Evaluation.
Final rept.
J. P. Favreau, R. J. Linn, and P. Gaudette. 1988, 15p
Pub. in Proceedings of International Conference on Formal Description Techniques (1st), Stirling, Scotland, September 6-9, 1988, p295-309.

Keywords: Tests, Systems engineering, *Protocols, *Communication networks, Software tools, Estelle, Abstract Syntax Notation One, Electronic mail.

Formal description techniques, and software tools based on them, are applied successfully in the development of test systems for OSI protocols. The test system architecture uses reference implementations augmented with test features in a manner that facilitates multi-layer testing and minimizes the effort necessary to develop test cases. The designs for test systems in general and for a specific gateway are described, focusing on the use of Estelle and ASN.1.

900,653
PB89-211916　　　　　　　　　Not available NTIS
National Inst. of Standards and Technology (ICST), Gaithersburg, MD. Systems and Network Architecture Div.
Object-Oriented Model for Estelle.
Final rept.
R. Sijelmassi, and P. Gaudette. 1988, 15p
See also PB89-196190.
Pub. in Proceedings of International Conference on Formal Description Techniques (1st), Stirling, Scotland, September 6-9, 1988, p91-105.

Keywords: Models, *Protocols, *Communication networks, Protocol(Computers), Estelle, Translators, Distributed computer systems, Finite state machines.

Estelle is a formal description technique for specifying OSI communication protocols. An object-oriented model for Estelle is presented that is concrete enough to allow execution but abstract enough to remain concise and workable. A synchronization protocol captures the subtleties of parallelism in the semantics of Estelle and allows module instances to execute as if they were asynchronous. The remainder of the basic model defines objects that are very close to the entities defined by Estelle semantics. The model constitutes an excellent specification of the output of a translator. It has been extended to allow appropriate observation and user control of the objects that simulate an Estelle formal description of a protocol. The combination of translation into an implementation of the model provides a powerful tool for specifying, studying, and debugging distributed systems.

900,654
PB90-111683　　　　　　　　PC A03/MF A01
National Inst. of Standards and Technology (NCSL), Gaithersburg, MD.
Computer Viruses and Related Threats: A Management Guide.
Special pub. (Final).
J. P. Wack, and L. J. Carnahan. Aug 89, 46p NIST/SP-500/166
Also available from Supt. of Docs. as SN003-003-02955-6. Library of Congress catalog card no. 89-600750.

Keywords: *Computer security, *Instructions, Computer software management, Computer networks, Contingency, Copyrights, Vulnerability, Trojan horses, Personal computers, *Computer viruses.

The document contains guidance for managing the threats of computer viruses and related software and unauthorized use. The document emphasizes that organizations cannot effectively reduce their vulnerabilities to viruses and related threats unless the organization commits to a virus prevention program, involving the mutual cooperation of all computer managers and users. The guidance is aimed at helping managers prevent and deter virus attacks, detect when they occur or are likely to occur, and then to contain and recover from any damage caused by the attack. The document contains an overview of viruses and related software, and several chapters of guidance for managers of multi-user computers, managers and users of personal computers, managers of wide and local area networks including personal computer networks, and managers of end-user groups.

900,655
PB90-111691　　　　　　　　PC A03/MF A01
National Inst. of Standards and Technology (NCSL), Gaithersburg, MD.
Software Verification and Validation: Its Role in Computer Assurance and Its Relationship with Software Project Management Standards.
Special pub. Jul 88-May 89.
D. R. Wallace, and R. U. Fujii. Sep 89, 40p NIST/SP-500/165
Also available from Supt. of Docs. as SN003-003-02959-9. Library of Congress catalog card no. 89-600754. Prepared in cooperation with Logicon, Inc., San Pedro, CA.

Keywords: *Computer software management, *Software engineering, Computer software maintenance, Computer program verification, Computer security, Computer performance evaluation, Software validation.

The report describes how the software verification and validation methodology and V&V standards provide a strong framework for developing quality software. First, the report describes software V&V; its objectives, recommended tasks, and guidance for selecting techniques to perform V&V. It explains the difference between V&V and quality assurance, development system engineering, and user organization functions. The report explains that V&V produces maximum benefits when it is performed independent of development functions and provides a brief discussion of how V&V benefits change when embedded in quality assurance, development systems engineering, and user organizations. An analysis of two studies of V&V's cost-effectiveness concludes that cost benefits of V&V's early error detection outweigh the cost of performing V&V.

900,656
PB90-128752　　　　　　　　　Not available NTIS
National Inst. of Standards and Technology (NEL), Boulder, CO. Electromagnetic Fields Div.
Creating CSUBs Written in FORTRAN That Run in BASIC.
Final rept.
E. J. Vanzura. 1988, 18p
Pub. in Proceedings of Conference on HP Technical Computer Users, Orlando, FL., August 7-12, 1988, p1-18.

Keywords: *Subroutines, Operating systems(Computers), Computer programming, *Basic programming language, *Fortran programming language, Pascal programming language.

CSUBs are compiled subprograms created using the Pascal operating system and which run in the BASIC environment. A new technique in which programs written in FORTRAN can be turned into CSUBs is described. Thus, powerful, well-documented FORTRAN routines become accessible to the BASIC-language programmer. I/O and variable interfacing are discussed, and a comprehensive example is provided.

900,657
PB90-130253　　　　　　　　PC A03/MF A01
National Inst. of Standards and Technology (NEL), Gaithersburg, MD. Applied and Computational Mathematics Div.

Supercomputers Need Super Arithmetic.
Final rept.
D. W. Lozier, and P. R. Turner. Oct 89, 31p NISTIR-89/4135
Prepared in cooperation with Naval Academy, Annapolis, MD. Dept. of Mathematics.

Keywords: Algorithms, Parallel processors, Error analysis, *Supercomputers, *Floating point arithmetic, Arithmetic units, Vector processing, Symmetric level-index system.

The paper discusses the parallel computation of vector norms and inner products in floating-point for vector and parallel computers. It concentrates on the vectorization of algorithms for the operations and proposes a new form of computer arithmetic, the symmetric level-index system.

900,658

PB90-133091   **PC A03/MF A01**
National Inst. of Standards and Technology (NCSL), Gaithersburg, MD.
Graphics Application Programmer's Interface Standards and CALS (Computer-Aided Acquisition and Logistic Support).
S. J. Kemmerer, and M. W. Skall. Oct 89, 17p
NISTIR-89/4199

Keywords: *Computer graphics, *Standards, interfaces, Computer aided design, Computer programs.

The principal purpose of a graphics Application Programmer's Interface (API) standard is to provide portability for an application program across a wide range of computers, operating systems, programming languages, and interactive graphics devices. The graphics API's are represented by four major standards' projects: Graphical Kernel System (GKS), GKS-3D, Programmer's Hierarchical Interactive Graphics System (PHIGS), and Programmer's Imaging Kernel (PIK).

Control Systems & Control Theory

900,659

PB89-151815   **PC A06/MF A01**
National Inst. of Standards and Technology (ICST), Gaithersburg, MD. Advanced Systems Div.
Small Computer System Interface (SCSI) Command System: Software Support for Control of Small Computer System Interface Devices.
J. Gorczyca, and E. S. Villagran. Jan 89, 125p
NISTIR-89/4023

Keywords: Systems engineering, Computer programs, *Command and control systems, *Computer software, *Small Computer System Interface (SCSI) devices, Microcomputers, C programming language, User manuals(Computer programs).

The Small Computer System Interface (SCSI) Command System was created by the National Computer and Telecommunications Laboratory of the National Institute of Standards and Technology (NCTL/NIST) personnel for the control of SCSI devices from a microcomputer equipped with a SCSI host adapter. The Command Interfacing Function permits all SCSI standard and manufacturer unique commands to be sent to external devices. The system allows two levels of user programming. The upper and lower levels offer the ability to utilize libraries of commands, and the ability to edit system parameters and commands directly by using the system's variables. Programming for the system is done in the 'C' language. Also included with the documentation are references, which provide additional information that may be of reader interest.

Information Processing Standards

900,660

FIPS PUB 134-1   **PC A03/MF A01**
National Inst. of Standards and Technology (NCSL), Gaithersburg, MD.

Coding and Modulation Requirements for 4,800 Bit/Second Modems, Category: Telecommunications Standard.
Federal information processing standards (Final).
S. M. Radack. 4 Nov 88, 19p
Supersedes FIPS PUB 134.
Three ring vinyl binder also available, North American Continent price $7.00; all others write for quote.

Keywords: *Standards, *Information processing, Coding, Modulation, Communication equipment, Computer communications, *Federal Information Processing Standards.

The standard establishes coding and modulation requirements for 4,800 bit/s modems owned or leased by the Federal government for use over analog transmission channels. It is based upon techniques described in CCITT Recommendations V.27 bis, V.27 ter, and V.32. The standard supersedes former Federal Standard (FED-STD) 1006 in its entirety.

900,661

FIPS PUB 149   **PC E08**
National Inst. of Standards and Technology (NCSL), Gaithersburg, MD.
General Aspects of Group 4 Facsimile Apparatus, Category: Telecommunications Standard.
Federal information processing standards (Final).
S. M. Radack. 4 Nov 88, 7p
Three ring vinyl binder also available, North American Continent price $7.00; all others write for quote.

Keywords: *Standards, *Information processing, Communication equipment, *Federal Information Processing Standards, Facsimile coding.

The standard adopts Electronic Industries Association (EIA) Standard EIA-536-1988, which defines the facsimile coding schemes and their control functions for Group 4 facsimile apparatus.

900,662

FIPS PUB 150   **PC E08**
National Inst. of Standards and Technology (NCSL), Gaithersburg, MD.
Facsimile Coding Schemes and Coding Control Functions for Group 4 Facsimile Apparatus, Category: Telecommunications Standard.
Federal information processing standards (Final).
S. M. Radack. 4 Nov 88, 7p
Three ring vinyl binder also available North American Continent price $7.00; all others write for quote.

Keywords: *Standards, *Information processing, Computer communication, Coding, Communication equipment, *Federal Information Processing Standards, Facsimile coding.

The standard adopts Electronic Industries Association (EIA) Standard EIA-538-1988, which defines the facsimile coding schemes and their control functions for Group 4 facsimile apparatus.

900,663

FIPS PUB 154   **PC E09**
National Inst. of Standards and Technology (NCSL), Gaithersburg, MD.
High Speed 25-Position Interface for Data Terminal Equipment and Data Circuit-Terminating Equipment, Category: Telecommunications Standard.
Federal information processing standards (Final).
S. M. Radack. 4 Nov 88, 8p
Three ring vinyl binder also available, North American Continent price $7.00; all others write for quote.

Keywords: *Standards, *Information processing, Computer communication, Communication equipment, Data processing terminals, Interfaces, Data transmission, *Federal Information Processing Standards.

The standard adopts Electronic Industries Association (EIA) Standard EIA-530-1987, which specifies the interconnection of data terminal equipment (DTE) and data circuit-terminating equipment (DCE) employing serial binary data interchange circuits with control information exchanged on separate control circuits. In particular, the standard defines the signal characteristics, interface mechanical characteristics, functional description of interchange circuits, and standard interfaces for selected communication system configurations.The electrical characteristics of the interchange circuits are specified by reference to Electronic Industries Association (EIA) standard EIA-422-A (FED-STD-1020A) and EIA-423-A (FED-STD-1030A).

900,664

FIPS PUB 155   **PC E11**
National Inst. of Standards and Technology (NCSL), Gaithersburg, MD.
Data Communication Systems and Services User-Oriented Performance Measurement Methods, Category: Telecommunications Standard.
Federal information processing standards (Final).
S. M. Radack. 4 Nov 88, 7p
Three ring vinyl binder also available, North American Continent price $7.00; all others write for quote.

Keywords: *Standards, *Information processing, Computer communications, Data transmission systems, Performance tests, Services, *Federal Information Processing Standards.

The standard adopts American National Standard X3.141-1987, which specifies uniform methods of measuring the performance of data communication services at digital interfaces between data communication systems and their users. These methods may be used to characterize the performance of any data communication service in accordance with the user-oriented performance parameters defined in a companion standard, American National Standard X3.102-1983, which has been adopted as FIPS 144 (former Federal Standard 1033).

900,665

PB89-149116   Not available NTIS
National Bureau of Standards (ICST), Gaithersburg, MD. Systems and Software Technology Div.
Case History: Development of a Software Engineering Standard.
Final rept.
T. Daughtrey, R. Fujii, and D. Wallace. 1986, 5p
Pub. in Proceedings of Computer Standards Conference: Striking a Balance between Technology, Economics, Politics, and Reality, San Francisco, CA., May 13-15, 1986, p21-25.

Keywords: *Standards, Verifying, Proving, Quality assurance, *Software engineering, Electronic mail.

The IEEE Standard for Verification and Validation Plans (SVVP) (1012) has been under development since March 1983. The paper presents the case history of the development of the SVVP Standard. The history characterizes the development based on the purpose of the Standard, the schedule for its development, the issues resolved for the content of the document, and the working group members. The history includes the results of using electronic mail to aid in the standards development process.

900,666

PB89-211965   Not available NTIS
National Bureau of Standards (ICST), Gaithersburg, MD. Systems and Software Technology Div.
Federal Software Engineering Standards Program.
Final rept.
D. R. Wallace. 1986, 11p
Pub. in Proceedings of Concepts and Principles for the Validation of Computer Systems Used in the Manufacture and Control of Drug Products, Chicago, IL., April 20-23, 1986, p103-113.1986.

Keywords: *Standards, Data processing, Tests, Proving, Verifying, *Software engineering, *National Institute of Standards and Technology, Federal information processing standards, Computer software maintenance, Computer software management.

The National Bureau of Standards, through its Institute for Computer Science and Technology (ICST), develops standards and guidelines to aid Government agencies in their effective use of automatic data processing resources. The paper describes the Federal computer software standards program at ICST in general and the program of its Software Engineering Group in detail.

900,667

PB90-111212   **PC A07/MF A01**
National Inst. of Standards and Technology (NCSL), Gaithersburg, MD.
Government Open Systems Interconnection Profile Users' Guide.
Final rept.
T. Boland. Aug 89, 149p NIST/SP-500/163
See also FIPS PUB 146. Also available from Supt. of Docs. Library of Congress catalog card no. 89-600749.

COMPUTERS, CONTROL & INFORMATION THEORY

Information Processing Standards

Keywords: Computer networks, Standardization, Government procurement, *Government Open Systems Interconnection Profile, *Federal information processing standards, *Computer communications, Computer architecture, Protocols.

The document provides guidance to users concerning implementation of the Government Open Systems Interconnection Profile (GOSIP) Federal Information Processing Standard (FIPS). Information in the document will help users to better understand and employ the GOSIP FIPS. The document will be updated annually.

Pattern Recognition & Image Processing

900,668
PB89-148415   **PC A04/MF A01**
National Inst. of Standards and Technology, Gaithersburg, MD.
Standards for the Interchange of Large Format Tiled Raster Documents.
F. E. Spielman. Dec 88, 72p NISTIR-88/4017

Keywords: *Standards, Sweep generators, Documents, *Image processing, *Computer graphics, Raster graphics, Information processing, Computer architecture.

The document is a compilation of five separately prepared, but interrelated, reports which discuss aspects of raster image processing, primarily as they relate to a standard tiling scheme being developed to support the interchange of large raster images. The first report provides the reader with a brief introduction to raster graphics and the current standards used to support raster graphics applications. The second provides the reader with a non-technical overview of the tiling scheme. The third report describes the user's requirements for tiling that were identified by an ad hoc Tiling Task Group (TTG). The fourth report, a Document Application Profile, presents information about all the attributes pertaining to a tiling application. The fifth report is a proposed addendum to an ANSI standard which is required to support the tiling scheme.

General

900,669
PB89-157366   Not available NTIS
National Bureau of Standards (NEL), Gaithersburg, MD. Robot Systems Div.
Automated Analysis of Operators on State Tables: A Technique for Intelligent Search.
Final rept.
T. R. Kramer. 1986, 27p
Pub. in Jnl. of Autom. Reasoning 2, n2 p127-153 1986.

Keywords: *Search structuring, *Searching, *Artificial intelligence, Selection, Tables(Data), Objectives, Operators(Mathematics), Reprints.

In one approach to the artificial intelligence problem of searching, the current situation is represented by a state table which gives the set of current conditions. Changes in the situation are brought about by operators which create a new state by adding some conditions to the table and deleting others. The search proceeds by applying operators one after another to the current state until the current state is identical to the goal state. The paper describes a technique for helping select operators. The technique is based on the fact that, during a search in which a partial sequence of operators has been selected, if the last operator or two are known, only a limited group of operators will be sensible to try next, and this group may be selected without prior knowledge of operators most recently tried. The output of the technique is a table that gives, for each feasible partial sequence of one or two operators, a list of operators that might be tried next.

900,670
PB89-168009   **PC A09/MF A01**
National Inst. of Standards and Technology (NCSL), Gaithersburg, MD. Computer Security Div.

Report of the Invitational Workshop on Integrity Policy in Computer Information Systems (WIPCIS).
Final rept.
S. W. Katzke, and Z. G. Ruthberg. Jan 89, 196p
NIST/SP-500/160
Also available from Supt. of Docs. as SN003-003-02904-1. Library of Congress catalog card no. 88-600608. Sponsored by Association for Computing Machinery, New York. Special Interest Group on Security, Audit and Control, Computer Society (IEEE), Washington, DC. Technical Committee on Security and Privacy, and National Computer Security Center, Fort George G. Meade, MD.

Keywords: *Meetings, Policies, Standards, *Information systems, *Computer information security, *Data integrity, Computer privacy.

Reported is the Invitational Workshop on Integrity Policy in Computer Information Systems. The workshop established a foundation for further progress in defining a model for information integrity. The workshop was held in response to the paper by David Clark of M.I.T. and David Wilson of Ernst and Whinney entitled 'A Comparison of Military and Commercial Data Security Policy.' The report's 10 sections contain an introduction, the composition of the organizing committee with a list of participants and a workshop agenda, a summary report by Don Parker and Peter Neumann of SRI International, the reports of the five working groups, a response by Clark and Wilson, and a proposal by the National Bureau of Standards for continuing the effort to define an integrity policy. The appendices include a copy of the original Clark-Wilson paper.

900,671
PB89-231021   Not available NTIS
National Bureau of Standards (ICST), Gaithersburg, MD. Systems Components Div.
National Bureau of Standards Message Authentication Code (MAC) Validation System.
Final rept.
M. E. Smid, E. B. Barker, and D. M. Balenson. 1986, 9p
Sponsored by National Computer Security Center, Fort George G. Meade, MD.
Pub. in Proceedings of National Computer Security Conference (9th), September 15-18, 1986, p99-107.

Keywords: *Standards, Authentication, Cryptology, Reprints, *National Institute of Standards and Technology, *Data encryption, *Computer information security, Computer privacy, Validation.

The paper describes the National Bureau of Standards Message Authentication Code (MAC) Validation System (MVS) for testing the conformance of vendor devices to Federal and commercial data authentication standards. Topics which are covered include the events which led to the development of the MVS, the standards it validates, its design philosophy, the requirements it places on vendors validating their devices, its performance characteristics, and the results of the validations performed to date.

900,672
PB89-235675   (Order as PB89-235634, PC A04)
National Inst. of Standards and Technology (NCSL), Gaithersburg, MD.
Conference Reports: National Computer Security Conference (11th). Held in Baltimore, MD. on October 17-20, 1988.
E. B. Lennon. 1989, 5p
Included in Jnl. of Research of the National Institute of Standards and Technology, v94 n4 p263-267 Jul/Aug 89.

Keywords: *Meetings, Auditing, Models, Information systems, Networks, *Computer security.

The Eleventh National Computer Security Conference held in Baltimore, Maryland on October 17-20, 1988 is summarized. The theme of the conference was the future of computer security. More than 1600 attendees from government, industry and academics partipated. Issues addressed included models and modeling integrity, risk management, audit and intrusion detection, security applications, verification, database management security, networking, system security requirements, automated tools and security architecture.

900,673
PB90-213687   **PC A04**
National Inst. of Standards and Technology, Gaithersburg, MD.

Journal of Research of the Institutes of Standards and Technology. September-October 1989.
Volume 94, Number 5.
1989, 70p
See also PB90-213695 through PB90-213711 and Volume 94, Number 4, PB89-235634. Also available from Supt. of Docs. as SN7003-027-00030-0.

Contents: Instrument-Independent MS/MS Database for XQQ Instruments: A Kinetics-Based Measurement Protocol; A Cotinine in Freeze-Dried Urine Reference Material; The NIST Automated Computer Time Service.

900,674
PB90-213695   (Order as PB90-213687, PC A04)
National Inst. of Standards and Technology, Gaithersburg, MD.
Instrument-Independent MS/MS Database for XQQ Instruments: A Kinetics-Based Measurement Protocol.
R. I. Martinez. 1989, 24p
Included in Jnl. of Research of the National Institute of Standards and Technology, v94 n5 p281-304 Sep/Oct 89.

Keywords: *CAD, CBRIS, Characteristic branching ratios of ionic substructures, Collisionally-activated dissociation, Database, Ion-molecule kinetics, Measurement protocol, MS/MS, NIST-EPA International Round Robin, Spectral library, Standardization, Tandem mass spectrometers, XQQ instruments (QQQ, BEQQ, etc).

A detailed kinetics-based measurement protocol is proposed for the development of a standardized MS/MS database for XQQ tandem mass spectrometers. The technical basis for the protocol is summarized. A CAD database format is proposed.

900,675
PB90-213703   (Order as PB90-213687, PC A04)
National Inst. of Standards and Technology, Gaithersburg, MD.
Cotinine in Freeze-Dried Urine Reference Material.
L. C. Sander, and G. D. Byrd. 1989, 5p
Included in Jnl. of Research of the National Institute of Standards and Technology, v94 n5 p305-309 Sep/Oct 89.

Keywords: *Cotinine, Cotinine - perchlorate, GC-MS, Passive smoking, Side stream smoke, Standards, Tobacco.

A cotinine in freeze-dried urine reference material (RM 8444) was prepared at three concentrations: (1) a 'blank' level typical of nonsmokers with no exposure to cigarette smoke, (2) a 'low' level corresponding to non-smokers with passive exposure to side-stream smoke, and (3) a 'high' level typical of smokers. Low- and high-level materials were prepared gravimetrically from pooled urine by the addition of appropriate amounts of continine perchlorate. Cotinine was determined by GC-MS using continine-d3 as an internal standard. No evidence for sample inhomogeneity was observed. This reference material will fulfill a need for a urine-based standard to assist in the validation of field methods used for assessing exposure to cigarette smoke.

900,676
PB90-213711   (Order as PB90-213687, PC A04)
National Inst. of Standards and Technology, Gaithersburg, MD.
NIST Automated Computer Time Service.
J. Levine, M. H. Weiss, D. D. Davis, D. W. Allan, and D. B. Sullivan. 1989, 11p
Included in Jnl. of Research of the National Institute of Standards and Technology, v94 n5 p311-321 Sep/Oct 89.

Keywords: *Automation, Computers, Delay, Digital systems, Frequency, Propagation delay, Telephone, Synchronization, Time.

The NIST Automated Computer Time Service (ACTS) is a telephone time service designed to provide computers with telephone access to time generated by the National Institute of Standards and Technology at accuracies approaching 1 ms. Features of the service include automated estimation by the transmitter of the telephone-line delay, advanced alert for changes to and from daylight saving time, and advanced notice of insertion of leap seconds. The ASCII-character time code-operates with most standard modems and computer systems. The system can be used to set computer clocks and simple hardware can also be developed to set non-computer clock systems.

900,677
PB90-780172 . . . . . . PC A03/MF A01
National Inst. of Standards and Technology (NCSL), Gaithersburg, MD.
Computer Security Training Guidelines.
Final rept.
M. A. Todd, and C. Guitian. Nov 89, 41p NIST/SP-500/172
Also available from Supt. of Docs. as SN003-003-02975-1. Library of Congress catalog card no. 89-600771.

Keywords: Standards, Objectives, Computer personnel, *Computer security, *Training, Computer information security, Computer privacy.

The guidelines describe what should be the learning objectives of agency security training programs. They focus on what the employee should know, and what they should be able to direct or perform. This allows agencies to design training programs that fit their environments and to clearly state the purpose of the training. Effectiveness can be measured by determining how many of the learning objectives were met.

DETECTION & COUNTERMEASURES

Acoustic Detection

900,678
PB89-173488 Not available NTIS
National Bureau of Standards (NEL), Gaithersburg, MD. Automated Production Technology Div.
Higher-Order Crossings: A New Acoustic Emission Signal Processing Method.
Final rept.
N. N. Hsu, and D. G. Eitzen. 1986, 8p
Pub. in Proceedings of International Acoustic Emission Symposium (9th), Kobe, Japan, November 14-17, p59-66 1988.

Keywords: Acoustic measurement, Signal processing, Spectrum analysis, Data reduction, Data smoothing, Fatigue tests, Crack propagation, *High order crossing, *Acoustic emission, Failure analysis, Stress waves.

A new signal processing technique has been developed called the higher-order crossings (HOC) technique. It is intuitively simple yet efficient and useful in many spectral analysis and data reduction applications. Some feasibility studies of the adaptation of HOC for acoustic emission (AE) signal discrimination are reported. First introduced are the mathematical concept and the physical significance of HOC, and then the experience on using the HOC technique to classify some simulated AE, AE during fatigue testing of pre-cracked aluminum specimens, impact-echo signals; and signals from machine tool monitoring is reported. It was found that the first few order crossings are sufficient to distinguish different types of AE, but specific pattern recognition schemes must be devised based on specific applications. To encourage others to experiment with these techniques, a scheme of modifying a conventional multi-channel AE system to do real time AE signal processing using higher-order crossings is presented.

900,679
PB89-211809 Not available NTIS
National Bureau of Standards (NEL), Gaithersburg, MD. Automated Production Technology Div.
Ultrasonic Sensor for Measuring Surface Roughness.
Final rept.
G. V. Blessing, and D. G. Eitzen. 1988, 9p
Pub. in Proceedings of SPIE (Society of Photo-Optical Instrumentation Engineers): Surface Measurement

and Characterization, Hamburg, FRG, September 19-21, 1988, v1009 p261-289.

Keywords: *Ultrasonic frequencies, *Detectors, *Surface roughness, Acoustics, Sound ranging, Surface properties, Topography, Metal plates, Measurement, Process control, Quality control.

Ultrasonic reflectance/scattering measurements have been made on metal samples possessing a large range of surface roughness values. The root-mean-square roughnesses $R(q)$ ranged from 0.3 to nearly 40 micrometers on the mostly periodic surfaces. The echo amplitude from short incident pulses of ultrasound in the frequency range of 1 to 30 MHz was used, in the manner of a comparator, to measure relative roughnesses with an area-averaging approach defined by the ultrasonic beam spot size. Ultrasonic wavelengths ranged from about 50 to 300 micrometers at these frequencies, and the beam spot sized varied from 0.2 to 5 mm in diameter. Both air and fluid coupling techniques were used between the sensor (transducer) and surface, on both static and rapidly (in excess of 5 m/sec surface speed) moving parts. On static surfaces, a resolution of better than 1.0 micrometers $R(q)$ was achieved at the higher ultrasonic frequencies. By focusing the ultrasonic beam at 30 MHz, a profilometry capability was demonstrated on a 1 micrometer $R(q)$ sinusoidal specimen of 800 micrometers wavelength.

Electromagnetic & Acoustic Countermeasures

900,680
PB89-161525 Not available NTIS
National Bureau of Standards (NEL), Boulder, CO. Electromagnetic Fields Div.
Techniques for Measuring the Electromagnetic Shielding Effectiveness of Materials. Part 1. Far-Field Source Simulation.
Final rept.
P. F. Wilson, M. T. Ma, and J. W. Adams. 1988, 12p
See also Part 2, PB89-161533.
Pub. in IEEE (Institute of Electrical and Electronics Engineers) Transactions on Electromagnetic Compatibility 30, n3 p239-250 Aug 88.

Keywords: *Materials tests, *Electromagnetic shielding, *Coaxial cables, Radiation shielding, Far field, Electromagnetic fields, Simulation, Effectiveness, Plane waves, Transmission lines, Time domain systems.

Shielding effectiveness relates to the ability of a material to reduce the transmission of propagating fields in order to electromagnetically isolate one region from another. Because the shielding capability of a complex material is difficult to predict, it often must be measured. A number of approaches to simulating a far-field source are studied, including the use of coaxial transmission-line holders and a time-domain system. In each case the authors consider the system frequency range, test sample requirements, test field type, dynamic range, measurement time required, and analytical background, and present data taken on a common set of materials.

900,681
PB89-161533 Not available NTIS
National Bureau of Standards (NEL), Boulder, CO. Electromagnetic Fields Div.
Techniques for Measuring the Electromagnetic Shielding Effectiveness of Materials. Part 2. Near-Field Source Simulation.
Final rept.
P. F. Wilson, and M. T. Ma. 1988, 9p
See also Part 1, PB89-161525.
Pub. in IEEE (Institute of Electrical and Electronics Engineers) Transactions on Electromagnetic Compatibility 30, n3 p251-259 Aug 88.

Keywords: *Materials tests, *Electromagnetic shielding, *Measurement, Radiation shielding, Electromagnetic fields, Antennas, Simulation, Effectiveness.

The paper continues to discuss the topic of measurements of electromagnetic shielding effectiveness of materials by simulating a near-field source. Two specific measurement approaches, the use of a dual TEM cell and the application of an apertured TEM cell in a reverberating chamber, are studied. In each case the

system frequency range, test sample requirements, test field type, dynamic range, measurement time required, analytical background, and present data are considered taken on a common set of materials.

900,682
PB89-173769 Not available NTIS
National Bureau of Standards (NEL), Boulder, CO. Electromagnetic Fields Div.
Automated TEM (Transverse Electromagnetic) Cell for Measuring Unintentional EM Emissions.
Final rept.
M. T. Ma, and W. D. Bensema. 1987, 12p
Pub. in Proceedings of International Conference on Electromagnetic Compatibility, EMC Expo 1987, San Diego, CA., May 19-21, 1987, pT11.1-T11.12.

Keywords: *Electromagnetic radiation detection, *Automatic control, Remote control, Leakage, Electromagnetic fields, Emission, Dipole moments, Electric moments, Magnetic moments, Adaptive systems, *Transverse electromagnetic cells.

The paper summarizes the basic electrical properties of a transverse electromagnetic (TEM) cell, and the underlying theoretical background, based on which a TEM cell is used to measure accurately the emission of an unknown, unintentional leakage source. The theory and measurements have been verified by the results of a simulated example and two experiments using a spherical dipole radiator and a small loop antenna. Recent development of an automated measurement system is also included.

900,683
PB89-176184 Not available NTIS
National Bureau of Standards (NEL), Boulder, CO. Electromagnetic Fields Div.
Photonic Electric Field Probe for Frequencies up to 2 GHz.
Final rept.
K. Masterson. 1987, 5p
Pub. in Proceedings of SPIE (Society of Photo-Optical Instrumentation Engineers), High Bandwidth Analog Applications of Photonics, September 23-24, 1986, Cambridge, MA., v720 p100-104 1987.

Keywords: *Electric fields, Electromagnetic radiation, Frequencies, Birefringence, Electrooptics, Measurement, Electromagnetic properties, Optical properties, *Photonic probes.

A photonic electric field probe using the Pockel's effect in bulk LiNbO3 is described. It was used to measure electromagnetic fields from 10 to 100 V/m. The observed frequency response was flat up to 1.6 GHz and extended beyond 2 GHz. Over the majority of the frequency range field strengths down to about 6 V/m would be detectable above the noise floor when using a 10 kHz detection bandwidth. Present experimental results indicate a linear dynamic range for the probe of approximately 30 dB. Increasing the optical carrier power and lowering the system noise floor is expected to improve the dynamic range to above 50 dB.

Infrared & Ultraviolet Detection

900,684
PB90-117458 Not available NTIS
National Inst. of Standards and Technology (NML), Boulder, CO. Time and Frequency Div.
Coherent Tunable Far Infrared Radiation.
Final rept.
D. A. Jennings. 1989, 3p
Contracts NASW-15, NASW-047
Sponsored by National Aeronautics and Space Administration, Washington, DC.
Pub. in Applied Physics B 48, p311-313 1989.

Keywords: *Far infrared radiation, *Coherent radiation, Electromagnetic radiation, Continuous radiation, Carbon dioxide lasers, Tuned circuits, Beams(Radiation), Reprints, Metal insulator metal diodes.

Tunable, cw, far infrared (FIR) radiation has been generated by nonlinear mixing of radiation from two CO2 lasers in a metal-insulator-metal (MIM) diode. The FIR difference-frequency power was radiated from the MIM diode antenna to a calibrated indium antimonide

81

DETECTION & COUNTERMEASURES

Infrared & Ultraviolet Detection

bolometer. Two-tenths of a microwatt of FIR power was generated by 250 mW from each of the CO_2 lasers. Using the combination of lines from a waveguide CO_2 laser, with its larger tuning range, with lines from CO_2, N_2O, and CO_2 isotope lasers promises complete coverage of the entire far infrared band from 100 to 5000 GHz (3-200/cm) with stepwise-tunable cw radiation.

Magnetic Detection

900,685
PB89-148365 PC A03/MF A01
National Inst. of Standards and Technology (NEL), Boulder, CO. Electromagnetic Fields Div.
Magnetostatic Measurements for Mine Detection.
R. G. Geyer. Oct 88, 35p NISTIR-88/3098
Sponsored by U.S. Army Belvoir Research and Development Center, Fort Belvoir, VA.

Keywords: *Magnetic detection, *Land mine detection, *Magnetic measurement, Magnetic permeability, Magnetostatics, Electromagnetic induction.

Magnetic susceptibility measurements are applied to the passive magnetometric detection problem of an arbitrarily shaped susceptible (metallic) mine buried in a magnetically permeable earth. For analysis purposes a conservative susceptibility contrast between a typical metallic mine and moist soil having the same measured magnetic characteristics as the U.S. Army Belvoir Research and Development Center (BRDC) magnetitesand mine lane mixture was assumed. Anomalous detection limits were then calculated for various total field intensity (Proton precession) sensor head heights and offset distances, given mine dimensions as small as 7.6 cm on a side.

Optical Detection

900,686
PB90-128281 Not available NTIS
National Inst. of Standards and Technology (NEL), Boulder, CO. Electromagnetic Fields Div.
Broadband, Isotropic, Photonic Electric-Field Meter for Measurements from 10 kHz to above 1 GHz.
Final rept.
K. D. Masterson, and L. D. Driver. 1989, 12p
Sponsored by Army Aviation Systems Command, St. Louis, MO.
Pub. in Proceedings of SPIE (Society of Photo-Optical Instrumentation Engineers) High Bandwidth Analog Applications of Photonics II, Boston, MA., September 8-9, 1988, v987 p107-118 1989.

Keywords: *Electric fields, *Electron probes, Field strength, Measurement, Electromagnetic fields, Detectors, Isotropy, Field emission, Fiber optics, Electrooptics, Crystals, Dipoles.

An electro-optic electric-field meter (PEFM-15) having 15 cm resistively tapered dipole elements and Pockels effect electro-optic modulators is used to measure electric fields of 10 to 100 V/m from 10 kHz to beyond 1 GHz. The probe's frequency response is flat within +1-dB from 30 kHz to 100 MHz except for a region between 1 and 10 MHz where acoustic resonances occur in the LiNbO3 modulator crystals. For a 3 kHz detection bandwidth, the noise equivalent field is approximately 7 V/m, thereby giving a calculated linear dynamic range of 68 dB in field power density. The probe's isotropic response is flat within +1-dB, and the response of each dipole closely follows the curve predicted by theory. An optical-beam switch that connects the individual dipoles to a laser source and optical receiver is also described.

Radiofrequency Detection

900,687
PB89-149272 Not available NTIS
National Bureau of Standards (NEL), Boulder, CO. Electromagnetic Fields Div.

Microwave Power Standards.
Final rept.
N. T. Larsen. 1987, 7p
Pub. in Proceedings of NCSL Workshop and Symposium 'Innovation: Key to the Future,' Denver, CO., July 12-16, 1987, p34-1-34-7.

Keywords: *Microwaves, *Standards, Calorimeters, Measuring instruments, Calibrating, National Institute for Standards and Technology.

A general review of the history and present status of the microwave power standards in use at the National Bureau of Standards (NBS) is presented. The standards are calorimeters, and the quantity measured is 'effective efficiency.' The calibration services are based on these standards. The design and evaluation of these standards are discussed.

900,688
PB89-229676 PC A03/MF A01
National Inst. of Standards and Technology (NEL), Boulder, CO. Electromagnetic Fields Div.
Clutter Models for Subsurface Electromagnetic Applications.
D. A. Hill. Feb 89, 46p NISTIR-89/3909
Sponsored by Army Belvoir Research Development and Engineering Center, Fort Belvoir, VA.

Keywords: *Subsurface investigations, *Tunnel detection, Magnetic dipoles, Remote sensing, Clutter, Electric dipoles, Born approximation.

Clutter models for subsurface electromagnetic applications are discussed with emphasis on tunnel detection applications. Random medium models are more versatile and require less detailed information than deterministic models. The Born approximation is used to derive expressions for the incoherent field, and electric and magnetic dipoles are treated in detail. When random inhomogeneities are located in the near field of the dipole source, an electric dipole radiates a larger incoherent field than a magnetic dipole because of its larger reactive electric field.

900,689
PB90-118167 Not available NTIS
National Inst. of Standards and Technology (NML), Gaithersburg, MD. Inorganic Analytical Research Div.
Safety Guidelines for Microwave Systems in the Analytical Laboratory.
Final rept.
H. M. Kingston, and L. B. Jassie. 1988, 13p
Pub. in Introduction to Microwave Sample Preparation-Theory and Practice, Chapter 11, p231-243 1988.

Keywords: *Microwave equipment, *Electronics laboratories, *Safety, Regulations, Laboratory equipment, Accident prevention, Electromagnetic radiation, Safe handling, Instructions, Reprints.

Safety considerations for working with microwave systems in the laboratory are discussed. The regulatory implications of modifying microwave equipment are presented. Guidelines for the selection and use of common laboratory equipment and materials are examined.

900,690
PB90-118183 Not available NTIS
National Inst. of Standards and Technology (NML), Gaithersburg, MD. Inorganic Analytical Research Div.
Monitoring and Predicting Parameters in Microwave Dissolution.
Final rept.
H. M. Kingston, and L. B. Jassie. 1988, 62p
Pub. in Introduction to Microwave Sample Preparation-Theory and Practice, Chapter 6, p93-154 1988.

Keywords: *Microwaves, *Decomposition reactions, *Pressure measurement, *Temperature measurement, Dissolving, Thermal measurement, Predictions, Numerical analysis, Catalysts, Reducing agents, Monitoring, Reprints, Standard reference materials.

Procedures are described for the real-time measurement of temperature and pressure during closed-vessel microwave sample decomposition. Pressure and temperature profiles of typical Standard Reference Materials and solitary as well as mixed acids are given to illustrate unique advantages that are available with the closed-vessel technique. A set of equations that permits prediction of target temperatures and times is derived from the fundamental heat capacity relationship for absorptive materials. From a series of fundamental measurements, original equations are in-

troduced that permit the power consumption of common mineral acids to be calculated. The method is proposed as a model to approximate the thermal behavior of reagents intended for microwave use. The fundamental degradation patterns of biological matrices are presented for model compounds.

900,691
PB90-128588 Not available NTIS
National Inst. of Standards and Technology (NEL), Boulder, CO. Electromagnetic Fields Div.
Thermo-Optic Designs for Microwave and Millimeter-Wave Electric-Field Probes.
Final rept.
J. Randa, M. Kanda, D. Melquist, and R. D. Orr.
1989, 5p
Sponsored by Naval Ocean Systems Center, San Diego, CA.
Pub. in Proceedings of IEEE (Institute of Electrical and Electronics Engineers) Symposium on Electromagnetic Compatibility, Denver, CO., May 23-25, 1989, p7-11.

Keywords: *Electron probes, *Microwaves, *Millimeter waves, Design, Electric fields, Electromagnetic radiation, Radio waves, Extremely high frequency, Temperature measuring instruments, *Thermo-optics.

The authors have considered various thermo-optic designs for electric-field probes for the approximate frequency range of 1-110 GHz. The designs are all based on using an optically sensed thermometer to measure the temperature rise of a resistive material in an electric field. The paper presents calculations of the sensitivities of the different designs, measurement results for the most easily fabricated design, and a discussion of possible improvements. The results indicate that a probe based on the design could detect a minimum electric field of about 30-50 V/m.

General

900,692
PB89-176705 Not available NTIS
National Bureau of Standards (ICST), Gaithersburg, MD. Systems Components Div.
Standard Format for the Exchange of Fingerprint Information.
Final rept.
R. T. Moore, and R. M. McCabe. 1986, 4p
Pub. in Proceedings of Carnahan Conference on Security Technology, Lexington, KY., May 14-16, 1986, p13-16.

Keywords: *Identification systems, *Standards, Data, Images, *Fingerprints, Format.

There are a number of automatic or semiautomatic identification systems which may be used to extract information from fingerprint images. The information extracted generally relates to discrete features, such as minutiae (ridge endings and bifurcations), cores, deltas, and ridge features, such as tracings or counts, and to their spatial and/or topological relationships with each other. The details of these features and the way that information about them is measured, described, and presented varies from system to system. As a consequence, fingerprint information read on one system cannot be searched directly against the files of fingerprint information read on another system. To cope with this problem, an American National Standard is being developed to define a format that the fingerprint information can be converted into, or from, to provide a mechanism for the exchange of data. This requires that each system have only a single set of data conversion programs to exchange data with any other system rather than a separate set of data conversion routines for each different system. An overview of the standard and the development process is presented.

ELECTROTECHNOLOGY

Antennas

900,693
PB89-149280　　　　　　　Not available NTIS
National Bureau of Standards (NEL), Boulder, CO.
Electromagnetic Fields Div.
Calibrating Antenna Standards Using CW and Pulsed-CW Measurements and the Planar Near-Field Method.
Final rept.
D. P. Kremer, and A. G. Repjar. 1988, 9p
Pub. in Proceedings of Annual Antenna Measurement Techniques Association (10th), Atlanta, GA., September 12-16, 1988, p13-21-13-29.

Keywords: *Antennas, *Continuous radiation, Coherent radiation, Near fields, Measurements, Gain, Calibrating, Cross polarization, Sidelobes.

The National Bureau of Standards (NBS) has calibrated an antenna to be used to evaluate both a near-field range and a compact range. These ranges are to be used to measure an electronically-steerable antenna which transmits only pulsed-CW signals. The antenna calibrated by NBS was chosen to be similar in physical size and frequency of operation to the array and was also calibrated with the antenna transmitting pulsed-CW. The calibration included determining the effects of using different power levels at the mixer, the accuracy of the receiver in making the amplitude and phase measurements, and the effective dynamic range of the receiver. Comparisons were made with calibration results obtained for the antenna transmitting CW and for the antenna receiving CW. The parameters compared include gain, sidelobe and cross polarization levels. The measurements are described and some results are presented.

900,694
PB89-150726　　　　　　　Not available NTIS
National Bureau of Standards (NEL), Boulder, CO.
Electromagnetic Fields Div.
Antenna Measurements for Millimeter Waves at the National Bureau of Standards.
Final rept.
M. H. Francis, A. G. Repjar, and D. P. Kremer. 1988, 5p
Pub. in Proceedings of Annual Antenna Measurement Techniques Association (10th), Atlanta, GA., September 12-16, 1988, p13-13-13-17.

Keywords: *Antennas, *Millimeter waves, *Error analysis, Amplification, Insertion loss, Extremely high frequencies, Polarized electromagnetic radiation, Power loss, Waveguides, Near fields, Radiation patterns.

For the past two years the National Bureau of Standards (NBS) has been developing the capability to perform on-axis gain and polarization measurements at millimeter wave frequencies from 33-65 GHz. The paper discusses the error analysis of antenna measurements at these frequencies. The largest source of error is insertion loss measurements. In order to make accurate insertion loss measurements, flanges on antennas need to be flat and perpendicular to the wave guide axis to within approximately 0.001 cm (0.0005 in). In addition, waveguide screws need to be tightened with a device that supplies constant torque. NBS is continuing development of its measurement capabilities, including measuring probe correction coefficients required in planar near-field processing, in order to provide accurate pattern measurements at these frequencies.

900,695
PB89-153886　　　　　　　PC A03/MF A01
National Inst. of Standards and Technology (NEL), Boulder, CO. Electromagnetic Fields Div.
Iterative Technique to Correct Probe Position Errors in Planar Near-Field to Far-Field Transformations.
Technical note.
L. A. Muth, and R. L. Lewis. Oct 88, 49p NIST/TN-1323
Also available from Supt. of Docs. as SN0003-003-02916-5.

Keywords: *Antennas, *Electrostatic probes, *Electromagnetic fields, Correction, Taylors series, Integral equations, Analysis(Mathematics), Electromagnetic radiation, Field strength, Measurement, *Position errors.

The authors have developed a general theoretical procedure to take into account probe position errors when planar near-field data are transformed to the far field. If the probe position errors are known, measured data can be represented as a Taylor series, whose terms contain the error function and the ideal spectrum of the antenna. Then the ideal spectrum can be solved in terms of the measured data and the measured position errors by inverting the Taylor series. This is complicated by the fact that the derivatives of the ideal data are unknown; that is, they can only be approximated by the derivatives of the measured data. This introduces additional computational errors, which must be properly taken into account. The authors have shown that the first few terms of the inversion can be easily obtained by simple approximation techniques, where the order of the approximation is easily specified. A more general solution can also be written by formulating the problem as an integral equation and using the method of successive approximations to obtain a general solution. An important criterion that emerges from the condition of convergence of the solution to the integral equation is that the total averaged position error must be less than some fraction of the sampling criterion for the antenna under test.

900,696
PB89-156796　　　　　　　Not available NTIS
National Bureau of Standards (NEL), Boulder, CO.
Electromagnetic Fields Div.
Spherical-Wave Source-Scattering Matrix Analysis of Coupled Antennas: A General System Two-Port Solution.
Final rept.
R. L. Lewis. 1987, 6p
Pub. in IEEE (Institute of Electrical and Electronics Engineers) Transactions on Antennas and Propagation AP-35, n12 p1375-1380 Dec 87.

Keywords: *Coupled antennas, *S matrix theory, Spherical waves, Elastic waves, Wave propagation, Backscattering, Waveguides, Reprints.

Expressions are given for the coupling between two antennas in terms of each antenna's spherical-wave source-scattering matrix. A comparison with the 'classical' scattering matrix representation is given in sufficient detail to permit conversion back and forth between the source-scattering matrix and the classical scattering matrix. Expressions for the transmission formulas, showing two different expressions corresponding to reversing the direction of propagation are given. However, if both antennas are reciprocal with equal characteristic waveguide impedances, then the two-port scattering matrix is a symmetric matrix.

900,697
PB89-156806　　　　　　　Not available NTIS
National Bureau of Standards (NEL), Boulder, CO.
Electromagnetic Fields Div.
Improved Spherical and Hemispherical Scanning Algorithms.
Final rept.
R. L. Lewis, and R. C. Wittmann. 1987, 8p
Pub. in IEEE (Institute of Electrical and Electronics Engineers) Transactions on Antennas and Propagation AP-35, n12 p1381-1388 Dec 87.

Keywords: *Algorithms, *Antennas, Electrostatic probes, Surveillance, Electromagnetic fields, Reprints, *Spherical scanning.

A probe-corrected hemispherical-scanning algorithm has been developed which is applicable when the antenna under test radiates negligibly into its rear hemisphere. For a hundred-wavelengths diameter antenna, hemispherical scanning would be about three times more efficient computationally than prior full-sphere scanning algorithms. Improvements have also been made to full-sphere scanning, significantly increasing that algorithm's computational efficiency.

900,698
PB89-156814　　　　　　　Not available NTIS
National Bureau of Standards (NEL), Boulder, CO.
Electromagnetic Fields Div.

900,699
Improved Polarization Measurements Using a Modified Three-Antenna Technique.
Final rept.
A. C. Newell. 1988, 3p
Pub. in IEEE (Institute of Electrical and Electronics Engineers) Transactions on Antennas and Propagation 36, n6 p852-854 Jun 88.

Keywords: *Antennas, *Electromagnetic fields, *Polarization(Waves), Measurement, Electromagnetic radiation, Calibrating, Standards, Electrostatic probes, Wave phases, Reprints.

An improved three-antenna measurement of polarization that greatly reduces the uncertainty due to phase measurement errors is described. The technique is used to calibrate polarization standards and probes used in near-field antenna measurements.

900,699
PB89-156822　　　　　　　Not available NTIS
National Bureau of Standards (NEL), Boulder, CO.
Electromagnetic Fields Div.
Gain and Power Parameter Measurements Using Planar Near-Field Techniques.
Final rept.
A. C. Newell, R. Ward, and E. McFarlane. 1988, 12p
Pub. in IEEE (Institute of Electrical and Electronics Engineers) Transactions on Antennas and Propagation 36, n6 p792-803 Jun 88.

Keywords: *Antennas, *Electromagnetic fields, Field strength, Measurement, Far field, Amplification, Flux density, Reprints, Effective radiated power, Intelsat satellites.

Equations are derived and measurement techniques described for obtaining gain, effective radiated power, and saturating flux density using planar near-field measurements. These are compared with conventional far-field techniques, and a number of parallels are evident. These give insight to the theory and help to identify the critical measurement parameters. Applications of the techniques to the INTELSAT VI satellite are described.

900,700
PB89-156830　　　　　　　Not available NTIS
National Bureau of Standards (NEL), Boulder, CO.
Electromagnetic Fields Div.
Fields of Horizontal Currents Located Above the Earth.
Final rept.
D. A. Hill. 1986, 7p
Pub. in IEEE (Institute of Electrical and Electronics Engineers) Transactions on Geoscience and Remote Sensing 26, n6 p726-732 Nov 88.

Keywords: *Dipole antennas, *Electromagnetic fields, Far field, Electromagnetic radiation, Field strength, Plane waves, Reprints, Fast fourier transforms, Integral transforms.

The plane-wave spectrum technique is used to derive the fields of horizontal currents located in a horizontal plane above the earth. The far field is derived asymptotically, and the near field is computed by two-dimensional fast Fourier transform. Specific numerical results are presented for a pair of oppositely directed dipoles, and the results have application to detection of buried objects. When the antenna is located at low heights, the field is enhanced in the earth and decreased in air.

900,701
PB89-156848　　　　　　　Not available NTIS
National Bureau of Standards (NEL), Boulder, CO.
Electromagnetic Fields Div.
Error Analysis Techniques for Planar Near-Field Measurements.
Final rept.
A. C. Newell. 1988, 15p
Pub. in IEEE (Institute of Electrical and Electronics Engineers) Transactions on Antennas and Propagation 36, n6 p754-768 Jun 88.

Keywords: *Antennas, *Electromagnetic fields, *Error analysis, Measurements, Numerical analysis, Electromagnetic radiation, Computerized simulation, Reprints.

The results of an extensive error analysis on planar near-field measurements are described. The analysis provides ways for estimating the magnitude of each individual source of error and then combining them to estimate the total uncertainty in the measurement. Mathematical analysis, computer simulation, and measurement tests are all used where appropriate.

900,702
PB89-156855 Not available NTIS
National Bureau of Standards (NEL), Boulder, CO.
Electromagnetic Fields Div.
Comparison of Measured and Calculated Antenna Sidelobe Coupling Loss in the Near Field Using Approximate Far-Field Data.
Final rept.
M. H. Francis, and C. F. Stubenrauch. 1988, 4p
Pub. in IEEE (Institute of Electrical and Electronics Engineers) Transactions on Antennas and Propagation 36, n3 p438-441 Mar 88.

Keywords: *Antennas, *Far field, *Computer systems programs, Transmission loss, Attenuation, Reprints, *Coupling(Interaction).

Computer programs are presently in existence to calculate the coupling loss between two antennas provided that the amplitude and phase of the far field are available. However, for many antennas the complex far field is not known accurately.

900,703
PB89-156863 Not available NTIS
National Bureau of Standards (NEL), Boulder, CO.
Electromagnetic Fields Div.
Brief History of Near-Field Measurements of Antennas at the National Bureau of Standards.
Final rept.
R. C. Baird, A. C. Newell, and C. F. Stubenrauch. 1988, 7p
Pub. in IEEE (Institute of Electrical and Electronics Engineers) Transactions on Antennas and Propagation 36, n6 p727-733 Jun 88.

Keywords: *Microwave antennas, *Electromagnetic fields, *Light(Visible radiation), Velocity measurement, Light pulses, Speed indicators, Solar radiation, Cylindrical antennas, Spherical antennas, Reprints.

The National Bureau of Standards (NBS) played a pioneering role in the development of practical planar near-field antenna measurement techniques. A brief history is presented of that role, which began with the theoretical studies to determine corrections for diffraction in a microwave measurement of the speed of light. NBS contributions to the development of nonplanar near-field measurement theory and practice are also described.

900,704
PB89-156871 Not available NTIS
National Bureau of Standards (NEL), Boulder, CO.
Electromagnetic Fields Div.
Accurate Determination of Planar Near-Field Correction Parameters for Linearly Polarized Probes.
Final rept.
A. G. Repjar, A. C. Newell, and M. H. Francis. 1988, 14p
Pub. in IEEE (Institute of Electrical and Electronics Engineers) Transactions on Antennas and Propagation 36, n6 p855-868 Jun 88.

Keywords: *Antenna radiation patterns, *Electromagnetic fields, *Electrostatic probes, Plane waves, Far field, Correction, Accuracy, Measurements, Reprints.

The receiving patterns (both amplitude and phase) of two probes must be known and utilized to determine accurately the complete far field of an antenna under test from near-field measurements. The process is called probe correction. When the antenna to be tested is nominally linearly polarized, the measurements are more accurate and efficient if nominally linearly polarized probes are used. Further efficiency is obtained if only one probe which is dual polarized is used instead of two probes to allow for simultaneous measurements of both components. It should be noted, however, that a single-port probe can be rotated by 90 deg (in effect, the second probe) to obtain the second component. A procedure used by the National Bureau of Standards (NBS) for accurately determining the plane-wave receiving parameters of both single- and dual-port linearly polarized probes is described. Examples are presented, and the effect of these probe receiving characteristics in the calculation of the parameters for the antenna under test is demonstrated using the required planar near-field theory.

900,705
PB89-157036 Not available NTIS

National Bureau of Standards (NEL), Boulder, CO.
Electromagnetic Technology Div.
SIS Quasiparticle Mixers with Bow-Tie Antennas.
Final rept.
L. Xizhi, P. L. Richards, and F. L. Lloyd. 1988, 33p
Grant AFOSR-85-0230
Sponsored by Air Force Office of Scientific Research, Bolling AFB, DC.
Pub. in International Jnl. of Infrared and Millimeter Waves 9, n2 p101-133 1988.

Keywords: *Biconical antennas, Elementary excitations, Antenna couplers, Gain, Electromagnetic noise, Reprints, *Quasiparticle mixers, Microstrip transmission lines.

The authors have designed and evaluated planar lithographed W-band SIS mixers with bow-tie antennas and several different RF coupling structures. Both Pb-In-Au/Pb-Bi and Nb/Pb-In-Au junctions were used, each with omega times (R sub N) times C> >1. Single junctions and series arrays of five junctions directly attached to bow-tie antennas with no additional coupling structure gave poor performance, as expected. Single junctions with inductive microstrips and five-junction arrays with parallel wire inductors gave good coupling over bandwidths of approximately 5 and 25% respectively. Good agreement was found between design calculations based on a simple equivalent circuit and measurements of the frequency dependence of the mixer gain. When good coupling was achieved, typical values of mixer gain G sub M(DSB) approximately equal to 0 dB, noise T sub M(DSB) approximately equal to 150 K, and receiver noise approximately 200 K were observed.

900,706
PB89-157051 Not available NTIS
National Bureau of Standards (NEL), Boulder, CO.
Electromagnetic Technology Div.
Measurement of Integrated Tuning Elements for SIS Mixers with a Fourier Transform Spectrometer.
Final rept.
Q. Hu, C. A. Mears, P. L. Richards, and F. L. Lloyd. 1988, 18p
Grant AFOSR-85-0230
Sponsored by Air Force Office of Scientific Research, Bolling AFB, DC.
Pub. in International Jnl. of Infrared and Millimeter Waves 9, n4 p303-320 1988.

Keywords: *Biconical antennas, *Spectrometers, *Mixing circuits, *Tuners, Optical measuring instruments, Antenna couplers, Lithography, Reprints, Fast fourier transforms.

Planar lithographed quasioptical mixers can profit from the use of integrated tuning elements to improve the coupling between the antenna and the SIS mixer junctions. The authors have used a Fourier transform spectrometer to measure the frequency response of such integrated tuning elements. The SIS junction connected to the tuning element served as the direct detector for the spectrometer. This relatively quick, easy experiment can give enough information over a broad range of millimeter and submillimeter wavelengths to test both design concepts and success in fabrication. One type of tuning element, an inductive wire connected in parallel with a series array of 5 SIS junctions across the terminals of a bow-tie antenna, shows a resonant response peak at 100 GHz with a 30% bandwidth.

900,707
PB89-157457 Not available NTIS
National Bureau of Standards (NEL), Boulder, CO.
Electromagnetic Fields Div.
Efficient and Accurate Method for Calculating and Representing Power Density in the Near Zone of Microwave Antennas.
Final rept.
R. L. Lewis, and A. C. Newell. 1988, 12p
See also PB86-181963.
Pub. in IEEE (Institute of Electrical and Electronics Engineers) Transactions on Antennas and Propagation 36, n6 p890-901 Jun 88.

Keywords: *Microwave antennas, Electromagnetic fields, Plane waves, Traveling waves, Wave diffraction, *Fresnel integrals, Power density, Fast fourier transforms.

An efficient and reliable method has been developed for computing and exhibiting Fresnel-region fields radiated by microwave antennas using plane-wave scattering matrix analysis. That is, near fields are calculat-

ed by numerically integrating the complex far-field antenna pattern. The predicted near-fields are exhibited as relative power-density contours lying in a longitudinal plane bisecting the antenna's aperture. The crux of the analysis consists of handling a numerical instability which arises from integrating discrete data. A criterion is developed for excluding highly oscillatory regions of the integrand. In turn, this leads to restricting the output domain where the near field computations are valid. With the numerical instability problem resolved, the fast Fourier transform is used for efficient numerical integration. The predicted near fields have been compared against both measured and theoretical data, confirming that the authors' near-field computation algorithm is capable of extremely high accuracy.

900,708
PB89-171839 Not available NTIS
National Bureau of Standards (NEL), Boulder, CO.
Electromagnetic Fields Div.
Effect of Random Errors in Planar Near-Field Measurement.
Final rept.
A. C. Newell, and C. F. Stubenrauch. 1988, 5p
See also PB87-233896.
Pub. in IEEE (Institute of Electrical and Electronics Engineers) Transactions on Antennas and Propagation 36, n6 p769-773 Jun 88.

Keywords: *Antenna radiation patterns, *Electromagnetic fields, *Signal to noise ratio, Far field, Attenuation, Electromagnetic interference, Electromagnetic noise, Transmitter characteristics, Random error, Reprints.

Expressions which relate the signal-to-noise ratio in the near field to the signal-to-noise ratio in the far field are developed. The expressions are then used to predict errors in far-field patterns obtained from near-field data. A technique to measure the noise in the far-field pattern is also given.

900,709
PB89-176218 Not available NTIS
National Bureau of Standards (NEL), Boulder, CO.
Electromagnetic Fields Div.
Experimental Study of Interpanel Interactions at 3.3 GHz.
Final rept.
L. A. Muth. 1987, 5p
Sponsored by Rome Air Development Center, Hanscom AFB, MA. Electromagnetics Directorate.
Pub. in Proceedings of Antenna Measurement Technique Association (AMTA) Annual Meeting and Symposium (9th), Seattle, WA., September 28-October 2, 1987, p25-29.

Keywords: *Antenna arrays, *Electric fields, *Electromagnetic scattering, Reflections, Strip transmission lines, Backscattering, Scattering loss, Deflection, Electromagnetic radiation.

A general theoretical approach is formulated to describe the complex electromagnetic environment of an N-element array. The theory reveals the element-to-element interactions and multiple reflections within the array. To experimentally verify some features of the theory, measurements on experimental array panels in various configurations were made. The array panels consisted of 256 microstrip radiating elements, in each of the configurations both the near-field and portside signals were measured to study the interactions between these panels. In particular, the effects of open-circuited array panels on the radiation pattern of a single panel are observed both in the near field and in the far field. It is found that internal scattering is the main mechanism of interaction between panels, rather than reradiation of signals received from adjacent panels. The effects of scattering are observable at the -50 dB level.

900,710
PB89-179857 Not available NTIS
National Bureau of Standards (NEL), Boulder, CO.
Electromagnetic Fields Div.
Antennas for Geophysical Applications.
Final rept.
D. A. Hill. 1988, 26p
Pub. in Antenna Handbook: Theory, Applications, and Design, Chapter 23, p23-1-23-26 1988.

Keywords: *Loop antennas, *Subsurface investigations, Extremely low radio frequencies, Geophysical prospecting, Geologic investigations, Excitation,

Surges, Stimulation, Reprints, Grounded wire antennas.

The book chapter is one of approximately forty chapters which will appear in the handbook. It is in the section on Applications, and the other handbook sections are Fundamentals, Antenna Theory, and Related Topics. The chapter discusses a number of antennas which are used for subsurface probing of the earth. The two most commonly used antennas are grounded wires and loops, and they are covered in detail for both time-harmonic and transient excitations. Emphasis is placed on the extremely low frequency (ELF) portion of the spectrum where it is possible to probe the earth to depths of several hundred meters.

900,711
PB89-185623   PC A03/MF **A01**
National Bureau of Standards (NEL), Gaithersburg, MD. Center for Electronics and Electrical Engineering.
Center for Electronics and Electrical Engineering Technical Publication Announcements Covering Center Programs, April-June 1986 with 1987 CEEE Events Calendar.
E. J. Walters. Jun 87, 20p NBSIR-87/3578
See also PB89-136311.

Keywords: *Electronics, *Electrical engineering, *Research, *Abstracts, Semiconductor devices, Semiconductors(Materials), Metrology, Waveforms, Antennas, Microwaves, Lasers, Fiber optics, Electric power, Electromagnetic interference, Superconductors.

The report is the ninth issue of a quarterly publication providing information on the technical work of the National bureau of Standards Center for Electronics and Electrical Engineering. The Center for Electronics and Electrical Engineering Technical Publication Announcement covers the second quarter of calendar year 1986. Abstracts are provided by technical area for papers published this quarter.

900,712
PB89-187595   Not available NTIS
National Bureau of Standards (NEL), Boulder, CO. Electromagnetic Fields Div.
Optically Linked Electric and Magnetic Field Sensor for Poynting Vector Measurements in the Near Fields of Radiating Sources.
Final rept.
L. Driver, and M. Kanda. 1988, 9p
See also PB89-132773.
Pub. in IEEE (Institute of Electrical and Electronics Engineers) Transactions on Electromagnetic Compatibility 30, n4 p495-503 Nov 88.

Keywords: *Electromagnetic fields, *Radiation patterns, Antennas, Vector analysis, Differential equations, Near fields, Measurement, Detectors, Electric fields, Magnetic fields, Electrooptics, Reprints, *Poynting theorem.

A unique, single-element antenna measurement scheme that can simultaneously measure the electric, magnetic, and time-dependent Poynting vectors of electromagnetic (EM) fields is described. The electric and magnetic responses of the antenna sensor are separated by a 0 degree/180 degree hybrid junction. The resulting two radio frequency (RF) voltages, along with relative phase and frequency information, are transmitted to a remotely located vector analyzer by a pair of well-matched phase and frequency downlinks. The remote receiver measures and displays the electric dipole response, the magnetic loop response, and the time phase difference between the two. The information is sufficient to determine the time-dependent Poynting vector. Both a theoretical analysis and a discussion of experimental measurements performed, which describe the capabilities and performance of a working prototype of the antenna measurement scheme, are presented. The results demonstrate that a three-axis (isotropic) version of the system could be used to measure the near fields of EM sources, as well as to completely describe the resultant flow of energy.

900,713
PB90-126208   Not available NTIS
National Inst. of Standards and Technology (NEL), Boulder, CO. Electromagnetic Fields Div.
Near-Field Detection of Buried Dielectric Objects.
Final rept.
D. A. Hill. 1989, 5p
Pub. in IEEE (Institute of Electrical and Electronics Engineers) Transactions on Geoscience and Remote Sensing 27, n4 p364-368 Jul 89.

Keywords: *Antennas, *Electromagnetic fields, *Dielectrics, *Detection, Electric fields, Lossy materials, Dipoles, S matrix theory, Scattering cross sections, Subsurface investigations, Reprints.

The plane-wave scattering-matrix method is used to compute the response of a detector to a buried dielectric scatterer. The Born approximation is used to derive the scattering matrix for scatterers of small dielectric contrast, but the general theory is not limited to such cases. Specific numerical results are generated for a UHF dipole detector swept over a buried dielectric cube. The maximum response is obtained when the detector is located at the air-earth interface, and the response decays rapidly with detector height. The sweep curves are symmetrical in the horizontal direction and have a null when the detector is directly over the object. An experimental curve for a free-space environment has the same qualitative features.

Circuits

900,714
PB89-171649   Not available NTIS
National Bureau of Standards (NML), Boulder, CO. Time and Frequency Div.
New Cavity Configuration for Cesium Beam Primary Frequency Standards.
Final rept.
A. DeMarchi, J. Shirley, D. Glaze, and R. Drullinger. 1988, 6p
Pub. in IEEE (Institute of Electrical and Electronics Engineers) Transactions on Instrumentation and Measurement 37, n2 p185-190 Jun 88.

Keywords: *Cesium frequency standards, *Cavity resonators, Oscillators, Resonant frequency, Microwaves, Waveguides, Mathematical models, Reprints.

In the design of cesium beam frequency standards, the presence of distributed cavity-phase-shifts (associated with residual running waves) in the microwave cavity, due to the small losses in the cavity walls, can become a significant source of error. To minimize such errors in future standards, it has been proposed that the long Ramsey excitation structure be terminated with ring-shaped cavities in place of the conventional shorted waveguide. The ring cavity will minimize distributed cavity-phase-variations at the position of the atomic beam, provided only that the two sides of the ring and the T-junction feeding the ring are symmetric. In the paper, a model is developed to investigate the validity of the concept in the presence of the small asymmetries that inevitably accompany the fabrication of such a cavity. The model, partially verified by laboratory tests, predicts that normal tolerances will allow the frequency shifts due to distributed cavity-phase-variations to be held at the 10 to the negative 15th power level for a beam tube with a Q of 10 to the 8th power.

900,715
PB89-173777   Not available NTIS
National Bureau of Standards (NEL), Boulder, CO. Electromagnetic Fields Div.
ANA (Automatic Network Analyzer) Measurement Results on the ARFTG (Automatic RF Techniques Group) Traveling Experiment.
Final rept.
L. F. Saulsbery, and R. T. Adair. 1987, 14p
Pub. in ARFTG (Automatic RF Techniques Group) Conference Digest (28th), St. Petersburg, FL., December 4-5, 1986, p65-78 1987.

Keywords: *Measuring instruments, Reflectivity, Attenuation, Standing wave ratios, Assessments, Phase shift, *Automatic network analyzers.

The Automatic RF Techniques Group (ARFTG) Executive Committee has assembled two traveling measurement assessment kits. Each of these kits consists of: 1-dB, 20-dB, 40-dB, and 60-dB attenuators; a 50-ohm termination; a 10-centimeter air line; 1.2-VSWR and 2.0-VSWR mismatched terminations; and a short circuit termination. These devices are equipped with precision 7-mm coaxial connectors. The traveling kits are being circulated among measurement laboratories that wish to assess their ability to measure reflection coefficient, attenuation, and phase shift from 300 MHz to 17 GHz. The results obtained on ten different automated measurement systems are presented.

900,716
PB89-174056   Not available NTIS

National Bureau of Standards (NML), Boulder, CO. Time and Frequency Div.
Low Noise Frequency Synthesis.
Final rept.
F. L. Walls, and C. M. Felton. 1987, 7p
Pub. in Proceedings of Annual Symposium on Frequency Control (41st), Philadelphia, PA., May 27-29, 1987, p512-518.

Keywords: *Frequency synthesizers, *Electromagnetic noise, Microwave oscillators, Crystal oscillators, Signal generators, Frequency converters, Frequency multipliers, Frequency dividers.

The paper reviews the various definitions of phase noise and changes in the phase noise of a signal under noiseless multiplication, division, and translation. Next the phase noise in selected noncryogenic rf and microwave oscillators is reviewed. Using a systems approach one can synthesize a microwave signal where the close in phase noise is controlled by a low frequency crystal oscillator while the high frequency phase noise is controlled by a microwave source. The approach yields a phase noise performance that is superior to that possible with a single source. Finally the phase noise of various amplifiers, multipliers, and dividers is compared. The phase noise of dividers while generally inferior to that of the best multipliers, is often sufficient for most applications.

900,717
PB89-201537   Not available NTIS
National Bureau of Standards (NEL), Gaithersburg, MD. Electrosystems Div.
Audio-Frequency Current Comparator Power Bridge: Development and Design Considerations.
Final rept.
N. M. Oldham, O. Petersons, and B. C. Waltrip. 1989, 5p
See also PB88-239561.
Pub. in IEEE (Institute of Electrial and Electronics Engineers) Transactions on Instrumentation and Measurement 38, n2 p390-394 Apr 89.

Keywords: *Comparator circuits, *Power measurement, Design, Calibrating, Wattmeters, Electric power meters, Electric measurement, Electric bridges, Reprints.

The development, design, construction, and partial evaluation of a system for performing active and reactive power measurement from 50 to 20 kHz is described. The technique is an extension of a power bridge based on a current comparator capacitance bridge that was originally restricted to power frequencies. The design features and component characteristics for wide-band operations are emphasized. A digitally synthesized, dual-channel signal source provides the required voltage and current signals.

900,718
PB89-201545   Not available NTIS
National Bureau of Standards (NEL), Gaithersburg, MD. Electrosystems Div.
International Comparison of Power Meter Calibrations Conducted in 1987.
Final rept.
W. J. M. Moore, E. So, N. M. Oldham, P. N. Miljanic, and R. Bergeest. 1989, 7p
See also PB88-239546.
Pub. in IEEE (Institute of Electrical and Electronics Engineers) Transactions on Instrumentation and Measurement 38, n2 p395-401 Apr 89.

Keywords: *Power meters, *Calibrating, Comparison, Power measurement, Watt meters, Electrical measurement, Evaluation, International trade, Reprints.

The results of an intercomparison of power meter calibrations conducted during 1987 among the National Research Council(Ottawa), the National Bureau of Standards(Gaithersburg), The Physikalisch-Technische Bundesanstalt(Braunschweig), and the Institut Mihailo Pupin(Belgrade), are described. The comparison was implemented by a transfer standard consisting of a time-division multiplier watt-converter developed at the Institut Mihailo Pupin. The measurements were made at 120 V, 5 A, 50 and 60 Hz, at power factors of 1.0, 0.5 lead and lag, and 0.0 lead and lag. An agreement between the laboratories of better than 20 ppm is indicated.

900,719
PB89-228597   Not available NTIS

ELECTROTECHNOLOGY

Circuits

National Inst. of Standards and Technology (NEL),
Gaithersburg, MD.
**Determination of AC-DC Difference in the 0.1 - 100
MHz Frequency Range.**
Final rept.
J. R. Kinard, and T. X. Cai. 1989, 8p
Pub. in IEEE (Institute of Electrical and Electronics En-
gineers) Transactions on Instrumentation and Meas-
urement 38, n2 p360-367 Apr 89.

Keywords: *AC to DC converters, Differences, Frequencies, Reprints, Voltage converters(AC to AC),
Voltage converters(DC to DC).

Thermal voltage converter structures have been modeled theoretically and studied experimentally to determine their ac-dc differences in the 0:1-100 MHz frequency range. Estimated uncertainties, corresponding
to the one standard deviation confidence level, for
these ac-dc differences vary from 20 ppm at 1 MHz to
2000 ppm at 100 MHz.

900,720
PB89-230452   Not available NTIS
National Bureau of Standards (NML), Gaithersburg,
MD. Electricity Div.
**Recharacterization of Thermal Voltage Converters
After Thermoelement Replacement.**
Final rept.
J. R. Kinard, and T. E. Lipe. 1989, 6p
Pub. in IEEE (Institute of Electrical and Electronics En-
gineers) Transactions on Instrumentation and Meas-
urement 38, n2 p351-356, Apr 89.

Keywords: *Electrical measurement, *Electric converters, AC to DC converters, Standards, Impedance, Analyzing, Reprints, *Thermal voltage converters, *Thermoelements, Replacing.

The relationship between the characteristics of various
thermoelements (TEs) as voltage or current converters and the overall ac-dc differences of a voltage
range in a coaxial thermal voltage converter (TVC) set
is described. An algorithm to predict the relationships
between the ac-dc differences of individual voltage
ranges with different TEs is presented, and a method
for recharacterizing a TVC containing a replacement
TE is given. The measured results show that for most
applications a complete recharacterization of the TVC
set is unnecessary.

900,721
PB90-117854   Not available NTIS
National Bureau of Standards (NEL), Boulder, CO.
Electrosystems Div.
**Calculable, Transportable Audio-Frequency AC
Reference Standard.**
Final rept.
N. M. Oldham, P. S. Hetrick, and X. Zeng. 1989, 4p
Sponsored by Instrumentation and Measurement So-
ciety (IEEE), New York.
Pub. in IEEE (Institute of Electrical and Electronics En-
gineers) Transactions on Instrumentation and Meas-
urement 38 n2 p368-371 Apr 89.

Keywords: *AC generators, *Electric potential, Alternating current, Digital techniques, Periodic functions,
Electric power generation, Power supply circuits, Time
standards, Synthesis, Reprints.

A transportable ac voltage source is described, in
which sinusoidal signals are synthesized digitally in the
audio-frequency range. The rms value of the output
waveform may be calculated by measuring the dc level
of the individual steps used to generate the waveform.
The uncertainty of the calculation at the 7-V level is
typically less than +/-5 ppm from 60 Hz to 2 kHz and
less than +/-10 ppm from 30 Hz to 15 kHz.

900,722
PB90-128703   Not available NTIS
National Inst. of Standards and Technology (NEL),
Gaithersburg, MD. Electrosystems Div.
Ambiguity Groups and Testability.
Final rept.
G. N. Stenbakken, T. M. Souders, and G. W.
Stewart. 1989, 7p
Pub. in IEEE (Institute of Electrical and Electronics En-
gineers) Transactions on Instrumentation and Meas-
urement 38, n5 p941-947 Oct 89.

Keywords: *Sensitivity, *Tests, Analog systems, Performance evaluation, Precision, Transient response,
Acuity, Accuracy, Reprints, *Circuit analysis.

An efficient method has been developed for determining component ambiguity groups which arise in analog

circuit testing. The method makes use of the sensitivity
model of the circuit. The ambiguity groupings are
shown to depend on the test points selected and the
measurement accuracy, and is, therefore, a useful tool
for determining where to add or delete test points. The
concept of ambiguity groups can be used to refine the
testability measure of a circuit.

Electromechanical Devices

900,723
PB89-146930   Not available NTIS
National Bureau of Standards (NEL), Boulder, CO.
Electromagnetic Fields Div.
**Simple Technique for Investigating Defects in Co-
axial Connectors.**
Final rept.
W. C. Daywitt. 1987, 5p
Pub. in IEEE (Institute of Electrical and Electronics En-
gineers) Transactions on Microwave Theory and Tech-
niques MTT-35, n4 p460-464 Apr 87.

Keywords: *Electric connectors, *Coaxial cables,
*Error analysis, Frequency analyzers, Reprints, Sweep
frequency, Automatic Network Analyzer.

The paper describes a technique that uses swept-frequency automatic network analyzer (ANA) data for investigating electrical defects in coaxial connectors.
The technique will be useful to connector and ANA
manufacturers and to engineers interested in determining connector characteristics for error analyses. A
simplified theory is presented and the technique is illustrated by applying it to perturbations caused by the
center conductor gap in a 7-mm connector pair.

Optoelectronic Devices & Systems

900,724
PB89-171722   Not available NTIS
National Bureau of Standards (NEL), Boulder, CO.
Electromagnetic Technology Div.
**Effect of Multiple Internal Reflections on the Sta-
bility of Electrooptic and Magnetooptic Sensors.**
Final rept.
K. S. Lee, and G. W. Day. 1988, 3p
Pub. in Applied Optics 27, n22 p4609-4611, 15 Nov 88.

Keywords: *Reflection, *Electrooptics, *Magnetoop-
tics, *Detectors, Optical measurement, Reflectance,
Electromagnetic radiation, Stability, Transient response, Reprints.

The effects of multiple internal reflections are evaluated analytically. Response functions showing changes
in shape as a function of optical path length are computed. The variation in sensitivity is obtained as a function of the reflectance of the sensing element and is
found to be significant (several tenths of a percent)
even when the reflectance is reduced to 0.1 percent.

900,725
PB89-173967   Not available NTIS
National Bureau of Standards (NEL), Boulder, CO.
Electromagnetic Technology Div.
**Optical Fiber Sensors for Electromagnetic Quanti-
ties.**
Final rept.
G. W. Day, K. S. Lee, A. H. Rose, L. R. Veeser, B. J.
Papatheofanis, and H. K. Whitesel. 1988, 3p
Sponsored by Department of Defense, Washington,
DC., and Department of Energy, Washington, DC.
Pub. in Proceedings of International Instrumentation
Symposium (34th), Albuquerque, NM., May 2-6, 1988,
p205-207.

Keywords: *Optical measuring instruments, *Fiber
optics, Alternating current, Magnetic fields, Electrical
measurement, *Optical fibers, Sensors, Voltage.

Several sensors used for the measurement of both
pulsed and ac current, voltage, and magnetic field are
described. Design considerations, including the choice
of components and configurations, and performance
achievements, are discussed.

900,726
PB89-176200   Not available NTIS

National Bureau of Standards (NEL), Boulder, CO.
Electromagnetic Technology Div.
**NBS (National Bureau of Standards) Standards for
Optical Power Meter Calibration.**
Final rept.
T. R. Scott. 1988, 14p
Pub. in Proceedings of DOD/ANSI/EIA Fiber Optics
Standardization Symposium, Arlington, VA., December
7-10, 1987, p224-237 1988.

Keywords: *Calibrating, *Power meters, *Lasers,
Measuring instruments, Power measurement, Calorimeters, Frequency standards, Performance standards, Beam splitters.

The measurement of optical power in the microwatt to
milliwatt power range at NBS is based upon a standard
reference laser calorimeter called the C-series calorimeter. The C-series calorimeter, which is used as a
national standard for the measurement of laser power
or energy, was designed to be rugged, easy to use,
and capable of measuring energy over a wide range of
laser wavelengths. The calorimeter, in conjunction
with various laser sources and a calibrated beamspli-
ter measurement system, is used to calibrate transfer
standards which are, in turn, used to calibrate other
optical power meters. The paper will review the operation and capabilities of the standard calorimeter and
associated measurement system, and will summarize
the uncertainties associated with these energy calibration measurements.

900,727
PB89-176689   Not available NTIS
National Bureau of Standards (NEL), Boulder, CO.
Electromagnetic Technology Div.
**Picosecond Pulse Response from Hydrogenated
Amorphous Silicon (a-Si:H) Optical Detectors on
Channel Waveguides.**
Final rept.
D. R. Larson, and R. J. Phelan. 1987, 5p
Pub. in Proceedings of SPIE (Society of Photo-Optical
Instrumentation Engineers) Integrated Optical Circuit
Engineering V, San Diego, CA., August 17-20, 1987,
v835 p59-63.

Keywords: Photodiodes, *Optical detectors, Integrated optics, Amorphous silicon, Optical waveguides,
Lithium niobates, Picosecond pulses.

The authors have fabricated high speed optical detectors on channel waveguides formed by both potassium
ion-exchange in glass and titanium diffusion in lithium
niobate. These new waveguide detectors show response times of 200 ps full width at half maximum amplitude (FWHM) when illuminated with subpicosecond
optical pulses. The detectors consist of back-to-back
Schottky photodiodes formed by chromium-gold metal
contacts on hydrogenated amorphous silicon (a-Si:H).
When interdigitated metal contacts with the contact
separation and semiconductor film thickness dimensions close to one micrometer are used, the detectors
are both fast and efficient.

900,728
PB89-176697   Not available NTIS
National Bureau of Standards (NEL), Boulder, CO.
Electromagnetic Technology Div.
**Potential Errors in the Use of Optical Fiber Power
Meters.**
Final rept.
X. Li, and R. L. Gallawa. 1988, 3p
Pub. in Proceedings of SPIE (Society of Photo-Optical
Instrumentation Engineers) Fiber Optic Networks and
Coherent Technology in Fiber Optic Systems II, San
Diego, CA., August 17-19, 1987, v841 p231-233 1988.

Keywords: *Power measurement, Power meters,
Errors, Fiber optics, Optical fibers, Optical connectors,
Calibration.

The authors discuss the potential errors associated
with the measurement of optical power in a field environment. The potential errors arise because field use
is often inconsistent with the calibration method.
Errors may be due to the use of connectors of different
types and due to variation among vendors for a given
connector type. The authors consider potential errors
in power measurements due to the variation in a connector type among vendors.

900,729
PB89-176796   Not available NTIS
National Bureau of Standards (NML), Gaithersburg,
MD. Radiometric Physics Div.

86

Characteristics of Ge and InGaAs Photodiodes.
Final rept.
E. Zalewski. 1989, 8p
Pub. in Proceedings of International Conference on Optical Radiometry, NPL (National Physical Laboratory), London, April 12-13, 1988, p47-54 1989.

Keywords: *Photodiodes, *Infrared detectors, Quantum efficiency, Indium phosphides, Near infrared radiation, Radiometry, *Ge semiconductor detectors, Gallium indium arsenides.

Measurements of the internal quantum efficiency of recently developed near infrared (1 to 1.6 micrometers) photodiodes show that considerable improvement has been made in the radiometric quality of these devices. Among commercially available devices, the newer InGaAs/InP photodiodes exhibit better characteristics than the Ge devices that have been traditionally used for near infrared radiometry. However, experimental induced junction Ge photodiodes produced at Purdue University have been observed to have nearly ideal internal quantum efficiency.

900,730
PB89-212161   Not available NTIS
National Bureau of Standards (NEL), Gaithersburg, MD. Semiconductor Electronics Div.
Blocked Impurity Band and Superlattice Detectors: Prospects for Radiometry.
Final rept.
J. Geist. 1989, 12p
Pub. in Proceedings of International Conference on Optical Radiometry, London, UK, April 12-13, 1988, p99-110 1989.

Keywords: *Radiometry, Infrared detectors, Ultraviolet detectors, Gallium arsenides, Cadmium tellurides, Mercury tellurides, Standards, *Optical detectors, Quantum wells, Superlattices, Aluminum gallium arsenides, Blocked impurity band detectors.

Blocked impurity band detectors and photomultipliers, which have been described by Petroff and Stapelbroek, may be suitable for use as high-accuracy standards for low background optical radiation measurements extending from the near ultraviolet to beyond 25 micrometers in the infrared. The current status of their development from the point of view of standards applications is reviewed. Superlattice technology offers new materials properties, new degrees of freedom, and new possibilities for optical radiation detectors displaying a large range of tailorability and tunability. GaAs/AlGaAs superlattices are used to illustrate new properties, HgTe/CdTe superlattices are used to illustrate new degrees of freedom, and GaAs-doping superlattices are used to illustrate tailorability and tunability.

900,731
PB89-228498   Not available NTIS
National Inst. of Standards and Technology (NEL), Gaithersburg, MD. Semiconductor Electronics Div.
Silicon Photodiode Detectors for EXAFS (Extended X-ray Absorption Fine Structure).
Final rept.
C. E. Bouldin, and A. C. Carter. 1989, 3p
Pub. in Physica B 158, p339-341 1989.

Keywords: *Photodiodes, *Silicon, X ray absorption, Fluorescence, Vacuum, Cryogenics, Linearity, Photons, Flux, Reprints.

Results are shown of using a large-area silicon diode as a fluorescence detector for EXAFS measurements. A direct comparison of this diode detector relative to a gas ionization fluorescence detector is made. Advantages of the diode detector include: higher signal for a given photon flux (due to higher quantum efficiency), vacuum and cryogenic compatibility, freedom from microphonic noise, good linearity, extremely wide dynamic range, operation without high voltage or gas connections, very simple electronics, and low cost. Use of photodiodes for transmission EXAFS is discussed.

900,732
PB89-229207   Not available NTIS
National Bureau of Standards (NEL), Boulder, CO. Electromagnetic Fields Div.
Photonic Electric-Field Probe for Frequencies up to 2 GHz.
Final rept.
K. D. Masterson. 1986, 5p
See also PB89-176184. Sponsored by Army Aviation Systems Command, St. Louis, MO.

Pub. in Proceedings of SPIE (Society of Photo-Optical Instrumentation Engineers) High Bandwidth Analog Applications of Photonics, v720 p100-104 1986.

Keywords: *Electric fields, Microwave equipment, Birefringence, Measurement, *Optoelectronic devices, Lithium niobates, Probes(Electromagnetic).

A photonic electric field probe using the Pockels effect in bulk LiNbO3 is used to measure electromagnetic fields from 10 to 100 V/m. The observed frequency response is flat up to 1.6 GHz and extends beyond 2 GHz. Over the majority of the frequency range, field strengths down to about 6 V/m would be detectable above the noise floor when using a 3 kHz detection bandwidth. Present experimental results indicate a linear dynamic range of 30 dB for the probe. Increasing the optical carrier power and lowering the noise floor are expected to improve the dynamic range to above 50 dB.

900,733
PB90-117599   Not available NTIS
National Inst. of Standards and Technology (NEL), Gaithersburg, MD. Semiconductor Electronics Div.
High Accuracy Modeling of Photodiode Quantum Efficiency.
Final rept.
J. Geist, and H. Baltes. 1989, 11p
Pub. in Applied Optics 28, n18 p3929-3939, 15 Sep 89.

Keywords: *Photodiodes, *Quantum efficiency, *Mathematical models, Diffusion theory, Silicon, Accuracy, Reprints.

A new silicon photodiode model is proposed which is optimized for high-accuracy measurement usage. The new model differs from previous models in that the contribution to the quantum efficiency from the diode front region is described by an integral transform of the equilibrium minority carrier concentration. The description is accurate as long as the recombination of excess minority carriers in the front region occurs only at the front surface and the diode is operating linearly.

900,734
PB90-118159   Not available NTIS
National Inst. of Standards and Technology (NEL), Gaithersburg, MD. Semiconductor Electronics Div.
Silicon Photodiode Self-Calibration.
Final rept.
J. C. Geist. 1988, 16p
Pub. in Theory and Practice of Radiation Thermometry, Chapter 14, p821-836 1988.

Keywords: *Photodiodes, *Calibrating, *Silicon, Thermal radiation, Temperature measurement, Blackbody radiation, Reprints.

A new and rapidly evolving technique for radiometric calibrations that is of considerable potential interest to radiation thermometry is considered. Although the method is not yet fully developed, it has already demonstrated an accuracy that is superior to that of all other methods in the spectral region traditionally used for high-precision radiation thermometry (650 nm). Through a simple experimental procedure using relatively inexpensive equipment that is widely available, the quantum efficiency and the spectral reflectance of a high-quality shallow-junction silicon photodiode can be readily determined.

900,735
PB90-128059   Not available NTIS
National Inst. of Standards and Technology (NML), Gaithersburg, MD. Radiation Physics Div.
Stability and Quantum Efficiency Performance of Silicon Photodiode Detectors in the Far Ultraviolet.
Final rept.
L. R. Canfield, J. Kerner, and R. Korde. 1989, 4p
Pub. in Applied Optics 28, n18 p3940-3943, 15 Sep 89.

Keywords: *Ultraviolet detectors, *Photodiodes, Far ultraviolet radiation, Quantum efficiency, Silicon dioxide, Radiometry, Stability, Reprints.

Recent improvements in silicon photodiode fabrication technology have resulted in the production of photodiodes which are stable after prolonged exposure to short wavelength radiation and which have efficiencies in the far ultraviolet close to those predicted using a value of 3.63 eV for electron-hole pair production in Si. Quantum efficiency and stability data are presented in the 6-124-eV region for several variations on the basic successful design and on devices with extremely thin silicon dioxide antireflecting/passivating layers. The

results indicate that the oxide is dominant in determining many of the performance parameters and that a stable efficient far ultraviolet diode can be fabricated by careful control of the Si-SiO2 interface quality.

900,736
PB90-130303   PC A03/MF A01
National Inst. of Standards and Technology (NEL), Boulder, CO. Electromagnetic Technology Div.
Improved Low-Level Silicon-Avalanche-Photodiode Transfer Standards at 1.064 Micrometers.
A. L. Rasmussen, P. A. Simpson, and A. A. Sanders. Aug 89, 40p NISTIR-89/3917
Sponsored by Aerospace Guidance and Metrology Center, Newark AFS, OH.

Keywords: *Photodiodes, *Avalanche diodes, *Standards, Semiconductor devices, Voltage regulators, Detectors, Lasers, Numerical analysis, Calibrating, Impulse response.

Three silicon-avalanche-photodiode (APD) transfer standards were calibrated from approximately 10(-8) to approximately 10(-5) W/sq cm peak power density at approximately 10 percent uncertainty. The calibrations are for 1.064 micrometer wavelength pulses of 10 to 100 ns duration. For the calibration, an acousto-optically modulated laser beam generated alternately equal levels of pulsed power and cw power into a low-level beam splitter. The cw power measured by a transfer standard in the transmitted beam of the splitter was used to determine the pulsed power into the APD transfer standard in one of the low-level reflected beams of the splitter. To increase the sensitivity, one or two 20 dB, 500 MHz bandwidth amplifiers followed the preamplifier. With very low pulsed power, a 30 MHz low-pass filter with Gaussian roll-off was attached to the amplifier output to reduce the noise. A transient digitizer recorded the impulse responses of the APD detectors at 1.064 micrometer. The data were read into computer programs that convolved the unit-area impulse response with unit-height Gaussian pulses. From the data, correction factors of the pulse peak for observed pulse durations from 10 to 100 ns were determined. Instructions, calibrations, error budgets, and system descriptions are included.

Power & Signal Transmission Devices

900,737
PB89-149264   Not available NTIS
National Inst. of Standards and Technology (NEL), Boulder, CO. Electromagnetic Fields Div.
Mode-Stirred Chamber for Measuring Shielding Effectiveness of Cables and Connectors: An Assessment of MIL-STD-1344A Method 3008.
Final rept.
M. L. Crawford, and J. M. Ladbury. 1988, 7p
Pub. in Proceedings of IEEE (Institute of Electrical and Electronics Engineers) International Symposium on Electromagnetic Compatibility, Seattle, WA., August 2-4, 1988, p30-36.

Keywords: *Electromagnetic shielding, *Communication cables, Standing wave ratios, Measurements, Specifications, Electric wire, Electric connectors, Antennas.

The mode-stirred method for measuring the shielding effectiveness (SE) of cables and connectors as specified in MIL-STD-1344A Method 3008 is examined. Problems encountered in applying the method are identified and recommendations to improve the measurement results are provided. These include chamber design, type and placement of transmitting and reference receiving antenna, determination and correction for VSWR of the reference antenna and equipment under test (EUT), and the measurement approach to use at specified test frequencies. Design and measurement setups for a small mode-stirred chamber suitable for performing SE measurements in the frequency range 1 - 18 GHz with dynamic ranges up to 130 dB are given along with SE measurement results of some sample EUTs.

900,738
PB89-156780   Not available NTIS
National Bureau of Standards (NEL), Boulder, CO. Electromagnetic Fields Div.

ELECTROTECHNOLOGY

Power & Signal Transmission Devices

Two-Layer Dielectric Microstrip Line Structure: SiO2 on Si and GaAs on Si; Modeling and Measurement.
Final rept.
R. A. Lawton, and W. T. Anderson. 1988, 5p
Sponsored by Naval Research Lab., Washington, DC.
Pub. in IEEE (Institute of Electrical and Electronics Engineers) Transactions on Microwave Theory and Techniques 36, n4 p785-789 Apr 88.

Keywords: *Backscattering, *Dielectrics, Characteristic impedance, Gallium arsenides, Lossy materials, Silicon dioxide, Sensitivity, Reflection, Measurement, Reprints, *Microstrip transmission lines.

Further development is reported of the modeling of the two-layer dielectric microstrip line structure by computing the scattering parameter S sub 21 derived from the model and comparing the computed value with the measured value over the frequency range from 90 MHz to 18 GHz. The sensitivity of the phase of S sub 21 and the magnitude of the characteristic impedance to various parameters of the equivalent circuit is determined.

900,739
PB89-157028   Not available NTIS
National Bureau of Standards (NEL), Boulder, CO. Electromagnetic Technology Div.
Waveguide Loss Measurement Using Photothermal Deflection.
Final rept.
R. K. Hickernell, D. R. Larson, R. J. Phelan, and L. E. Larson. 1988, 3p
Pub. in Applied Optics 27, n13 p2636-2638, 1 Jul 88.

Keywords: *Optical communication, *Transmission loss, *Waveguides, Transmission lines, Backscattering, Telecommunication, Lasers, Refractivity, Reprints, *Photothermal deflection.

Photothermal deflection (PTD) is introduced as a technique for measuring propagation loss in optical channel waveguides. A probe laser beam is deflected by the thermally-induced refractive-index gradient due to the absorption of guided pump light. The technique is non-contact and is applicable to a wide range of channel waveguide geometries and materials, including buried guides. Scattering centers and unguided background light affect the measurement only indirectly, since the PTD signal depends on the gradient of the local temperature and not the light intensity directly. Scans of the PTD signal as a function of distance along the waveguide yielded propagation loss measurements with lower uncertainty than scans of the scattered light intensity. The PTD technique should be useful in the study of waveguide loss mechanisms.

900,740
PB89-157861   Not available NTIS
National Bureau of Standards (NEL), Boulder, CO. Electrosystems Div.
Tiger Tempering Tampers Transmissions.
Final rept.
F. D. Martzloff. 1988, 2p
Pub. in BICSI Newsletter, p3 and p10 Dec 88.

Keywords: *Surges, *Wire lines, *Transmission lines, Buildings, Overcurrent, Overvoltage, Retarding, Reprints.

The article is the second of a two-part update on progress at the National Institute of Standards and Technology in a study on the propagation of surges in building wiring systems. In the first part, the problems associated with surge propagation in building wiring systems were described and classified as pussycats (they can just purr or they can scratch you and make you bleed) or tigers (they can eat you alive). Fast transients were shown to be pussycats. In the article, tests on the propagation of slower surges are cited; an important finding is that attempts at suppressing surges (tempering or taming tigers) on the power lines can have unexpected effects on the data transmission level of an electronic data processing system.

900,741
PB89-162572   PC A04/MF A01
National Inst. of Standards and Technology (NEL), Gaithersburg, MD. Center for Fire Research.
Flammability Characteristics of Electrical Cables Using the Cone Calorimeter.
E. Braun, J. R. Shields, and R. H. Harris. Jan 89, 63p
NISTIR-88/4003
Sponsored by Naval Sea Systems Command, Washington, DC.

Keywords: *Electric wire, *Flammability testing, Ignition, Combustion, Fire safety, Fire tests, Ignition delay, Smoking, Thermal measurements, *Cone calorimeters, Heat release rate.

Cone calorimeter tests were performed on eight multiconductor electrical cables. Measurements of ignition delay time, heat release rate, mass loss rate, and gas and smoke generation rates were made in the vertical (2 irradiance levels) and horizontal (3 irradiance levels) orientations. It was found that comparable ignition delay times were observed for all of the cross-linked polyolefin jacketed cables. The PVC jacketed cable had a substantially lower ignition delay time. All of the cables exhibited an ignition delay time dependence on external irradiance proportional to 1/q2. Sample orientation did not significantly affect the ignition delay time. Heat release rate measurements showed that cables burned in multiple stages. Each stage of burning was associated with the decomposition of a different layer of the cable assembly. For some cables, at low external irradiances (25.kW/m2) only the outer jacket burst open exposing the interior cable materials and secondary heat release rate peaks resulted. Changes in the cable components actually burning were reflected in variations in mass loss, gas, and smoke generation rates as well as small changes in the effective heat of combustion.

900,742
PB89-171664   Not available NTIS
National Bureau of Standards (NEL), Boulder, CO. Electromagnetic Fields Div.
Magnetic Dipole Excitation of a Long Conductor in a Lossy Medium.
Final rept.
D. A. Hill. 1988, 6p
Pub. in IEEE (Institute of Electrical and Electronics Engineers) Transactions on Geoscience and Remote Sensing 26, n6 p720-725 Nov 88.

Keywords: *Trahsmission lines, *Magnetic dipoles, *Excitation, Electric current, Polarized electromagnetic radiation, Stimulation, Emission, Electric discharges, Reprints, *Electric dipoles.

Formulations for the excitation of currents on an infinitely long conductor by electric or magnetic dipoles of arbitrary orientation are presented. The conductor can be either insulated or bare to model ungrounded or grounded conductors. Specific calculations are presented for a vertical magnetic dipole source because the source produces the appropriate horizontal polarization and could be used in a borehole-to-borehole configuration. Numerical results for the induced current and secondary magnetic field indicate that long conductors produce a strong anomaly over a broad frequency range. The secondary magnetic field decays slowly in the direction of the conductor and eventually becomes larger than the dipole source field.

900,743
PB89-171706   Not available NTIS
National Bureau of Standards (NEL), Boulder, CO. Electromagnetic Technology Div.
Fast Optical Detector Deposited on Dielectric Channel Waveguides.
Final rept.
D. R. Larson, and R. J. Phelan. 1988, 4p
Pub. in Optical Engineering 27, n6 p503-506 Jun 88.

Keywords: *Waveguides, *Optical communication, *Detectors, Thin films, Transmission lines, Dielectric films, Silicon, Hydrogeneration, Amorphous materials, Reprints.

A thin-film optical detector has been fabricated for detecting short optical pulses propagating in channel waveguides. The detectors show response times of 200 ps full width at half maximum amplitude when illuminated by guided, subpicosecond optical pulses. The detectors are formed by depositing hydrogenated amorphous silicon (a-Si:H) directly on the dielectric channel waveguide. Back-to-back Schottky photodiodes are then formed when interdigitated chrome-gold metal contacts are deposited on the a-Si:H.

900,744
PB89-173413   Not available NTIS
National Bureau of Standards (NEL), Gaithersburg, MD. Electrosystems Div.
Measuring Fast-Rise Impulses by Use of E-Dot Sensors.
Final rept.
R. H. McKnight. 1988, 3p
Sponsored by Department of Energy, Washington, DC.

Pub. in Proceedings of International Symposium on High Voltage Engineering (5th), Braunschweig, Federal Republic of Germany, August 24-28, 1987, v2 p1-3 1988.

Keywords: *Transmission lines, *High voltage, *Electromagnetic pulses, Calibrating, Frequencies, Pulsation, Electromagnetic radiation, Measurement, Power lines, *E-dot detectors.

Field coupled sensors such as capacitive dividers, derivative (E-dot or B-dot) sensors and Rogowski coils are commonly used in pulse power applications. Measurement devices using E-dot sensors in combination with passive or active integrators provide broadband capability, but with limited sensitivity. The use of this category of sensor in measurements of fast rise pulses, such as a nuclear electromagnetic pulse (EMP), in power system equipment offers some advantages, such as ease of construction and versatility in installation.

900,745
PB89-173439   Not available NTIS
National Bureau of Standards (NEL), Gaithersburg, MD. Electrosystems Div.
Measurement of Electrical Parameters Near AC and DC Power Lines.
Final rept.
M. Misakian. 1988, 1p
Pub. in Proceedings of IEEE (Institute of Electrical and Electronics Engineers) Instrumentation and Measurement Technology Conference, San Diego, CA., April 20-22, 1988, p114.

Keywords: *Power lines, *Power measurement, *Electric fields, Transmission lines, Electric wire, Power meters, Calibrating, Charge density, Ion currents, Current density, Magnetic fields.

The presentation surveys the instrumentation, calibration procedures, measurement techniques, and measurement standards which can be used for characterizing fields near AC power lines, and the electric field strength, ion current density, monopolar charge density, and net space charge near DC power lines.

900,746
PB89-173470   Not available NTIS
National Bureau of Standards (NEL), Gaithersburg, MD. Electrosystems Div.
AC Electric and Magnetic Field Meter Fundamentals.
Final rept.
M. Misakian. 1988, 23p
Pub. in Proceedings EPRI (Electric Power Research Institute) Utility Seminar: Power-Frequency Electric and Magnetic Field Exposure Assessment, Colorado Springs, CO, October 12-14, 1988, p1-23.

Keywords: *Power lines, *Electric fields, *Magnetic fields, Measurement, Calibrating, Standards, Power meters, Transmission lines.

Questions raised in the early 1970s regarding possible adverse environmental effects due to high-voltage AC transmission line fields focused attention on the need for accurate measurements of the power-frequency electric and magnetic fields. Following a brief description of the fields near AC power lines, the paper will survey the instrumentation, calibration procedures, measurement techniques and standards that have been developed since the early 1970s to characterize the electric and magnetic fields near AC power lines.

900,747
PB89-176176   Not available NTIS
National Bureau of Standards (NEL), Boulder, CO. Electromagnetic Fields Div.
Some Questions and Answers Concerning Air Lines as Impedance Standards.
Final rept.
C. A. Hoer. 1987, 13p
Pub. in Proceedings of ARFTG (Automatic RF Techniques Group) Conference (29th), Las Vegas, NV., June 12-13, 1987, p161-173.

Keywords: *Characteristic impedance, *Transmission lines, Calibrating, Standards, Electric conductors, Reflection, Measurement, *Automatic network analyzers, Scattering parameters.

The paper attempts to answer a number of questions that arise when using one or more lengths of precision coaxial transmission line to calibrate a dual 6-port automatic network analyzer, questions such as: how important is the quality of the test port relative to that of the line; what type connectors should the line standards have; what are the advantages of using two lines instead of one line and a through connection when test port imperfections are considered; and how many lines are optimum from a quality control point of view and what should the lengths be. The answers to these questions appear to be: (1) The quality of the line is much more important than that of the test port. A perfect line will calibrate out most imperfections in the test port. An example is given where 75-omega test ports are calibrated with 50-omega lines, and then used to measure reflection coefficient relative to 50 omega with very little error. (2) Greatest accuracy is achieved with line standards having male connectors. (3) Two lines get rid of many test port imperfections that one line cannot. Three lines will show up a problem if one line is bad. Five lines will identify which line is bad. Five is probably optimum. (4) There may not be an optimum for the actual lengths of a set of lines, but there does appear to be an optimum difference in the lengths.

900,748
PB89-176671   Not available NTIS
National Bureau of Standards (NEL), Boulder, CO. Electromagnetic Technology Div.
Optical Fiber Sensors for the Measurement of Electromagnetic Quantities.
Final rept.
A. H. Rose, G. W. Day, K. S. Lee, D. Tang, L. R. Vesser, B. J. Paptheofanis, and H. R. Whitesel. 1988, 3p
Pub. in Proceedings of Sensors Expo, Chicago, IL., September 13-15, 1988, p209A-1-209A-3.

Keywords: *Optical detection, *Electromagnetic interferences, Fiber optics, Electromagnetic compatibility, Magnetic fields, Electric current, Electric potential, Measurement.

Sensors used for the measurement of pulsed and AC current, voltage, and magnetic fields are described. Design considerations, including the choice of components and configurations, and performance achievements are discussed. The paper describes several sensor configurations presently being used to measure current, voltage, and magnetic fields in environments where electromagnetic interference is a problem. The current and magnetic field sensors are based on the Faraday effect either in single mode optical fiber or in bulk glass or polycrystalline materials. The voltage sensors are based on the linear electro-optic (Pockels) effect in cubic crystalline materials.

900,749
PB89-179816   Not available NTIS
National Bureau of Standards (NEL), Boulder, CO. Electromagnetic Technology Div.
Profile Inhomogeneity in Multimode Graded-Index Fibers.
Final rept.
C. W. Oates, and M. Young. 1989, 3p
Pub. in Jnl. of Lightwave Technology 7, n3 p530-532 Mar 89.

Keywords: *Fiber optics, *Heterogeneity, *Profiles, Optical communication, Impurities, Inclusions, Gradients, Optical materials, Reprints, *Multimode fiber, *Graded index fibers.

The authors have measured the profile parameters (g) of several multimode graded index fibers and found that g may vary azimuthally by + or - 0.15 or more in fibers for which the average value is between 1.8 and 2.2.

900,750
PB89-184121
                    (Order as PB89-184069, PC A04)
National Inst. of Standards and Technology, Boulder, CO.
Scattering Parameters Representing Imperfections in Precision Coaxial Air Lines.
Bi-monthly rept.
D. R. Holt. 1989, 17p
Included in Jnl. of Research of the National Institute of Standards and Technology, v94 n2 p117-133 Mar-Apr 89.

Keywords: *Coaxial cables, *Pneumatic lines, *Skin effect, *Surface roughness, Scattering, Measurement, Conformal mapping.

Scattering parameter expressions are developed for the principal mode of a coaxial air line. The model

allows for skin-effect loss and dimensional variations in the inner and outer conductors. Small deviations from conductor circular cross sections are conformally mapped by the Bergman kernel technique. Numerical results are illustrated for a 7 mm air line. An error analysis reveals that the accuracy of the scattering parameters is limited primarily by the conductor radii measurement precision.

900,751
PB89-188593   PC A04/MF A01
National Bureau of Standards (NEL), Boulder, CO. Center for Electronics and Electrical Engineering.
System for Measuring Optical Waveguide Intensity Profiles.
L. E. Larson, D. R. Larson, and R. J. Phelan. Aug 88, 67p NBSIR-88/3092

Keywords: *Waveguides, *Optical communication, *Luminous intensity, Directional couplers, Fiber optics, Telecommunication, Measurement, Profiles, Gradients, Radiance, Brightness, *Computer control.

A computer controlled system has been developed to measure the intensity profile of optical waveguides. Knowledge of the intensity profile provides an indication of the shape of the waveguide, and therefore the degree to which light can be coupled to the guide from an optical fiber. The report describes the construction and operation of the system.

900,752
PB89-189179   PC A03/MF A01
National Bureau of Standards (NEL), Boulder, CO. Electromagnetic Technology Div.
Group Index and Time Delay Measurements of a Standard Reference Fiber.
B. L. Danielson, and C. D. Whittenberg. Jul 88, 20p NBSIR-88/3091
Sponsored by Naval Weapons Center Corona Annex, CA.

Keywords: *Optical materials, *Standards, *Calibrating, *Fiber optics, Measurement, Interferometers, Reflectometers, Time lag, Length.

Measurement techniques for establishing a standard reference fiber with well characterized group index and time or group delay are described. Evaluation of an interferometric method indicates that fiber group index can be determined with a total estimated uncertainty of about 0.03% in small samples. Group delay of the reference fiber was measured with an overall uncertainty less than 0.004% in a 7 km waveguide. The application of a standard reference fiber to calibration of the distance measurement accuracy of an optical time-domain reflectometer (OTDR) is discussed.

900,753
PB89-201057   Not available NTIS
National Bureau of Standards (NEL), Boulder, CO. Electromagnetic Fields Div.
Reflection Coefficient of a Waveguide with Slightly Uneven Walls.
Final rept.
D. A. Hill. 1989, 9p
Pub. in IEEE (Institute of Electrical and Electronics Engineers) Transactions on Microwave Theory and Techniques 37, n1 p244-252 Jan 89.

Keywords: *Waveguides, Shape, Asymmetry, Numerical analysis, Transmission lines, Coaxial cables, Telecommunication, Electromagnetic radiation, Reflectivity, Reprints, *Reflection coefficient.

First-order results are derived for the reflection coefficient of a waveguide with slightly uneven walls. Exact analytical and numerical results are given for rectangular waveguides and coaxial transmission lines. Simple upper bounds are given for reflection coefficients in terms of the maximum deviation of the waveguide. For typical tolerances the reflection coefficients are very small (less than .01), but the results are important in precise six-port measurements.

900,754
PB89-237986   PC A02
National Inst. of Standards and Technology, Gaithersburg, MD.
National Institute of Standards and Technology (NIST) Information Poster on Power Quality.
Special rept.
B. F. Field. Jul 89, 12p NIST/SP-768
See also PB83-245068. Also available from Supt. of Docs. as SN003-003-02957-2.

Keywords: *Power lines, *Disturbances, Electric power failures, Outages, Protection, Reliability(Electronics), Cost analysis, Voltage regulators, Dictionaries.

The poster answers seven questions about power quality that should help one pinpoint problems and solutions related to power disturbances; describes the types of power disturbances, the equipment affected, the types of protection equipment that is effective against the disturbance; contains a glossary of common power terms.

900,755
PB90-117326   Not available NTIS
National Inst. of Standards and Technology (NEL), Gaithersburg, MD. Electrosystems Div.
Discussion of 'Steep-Front Short-Duration Voltage Surge Tests of Power Line Filters and Transient Voltage Suppressors'.
Final rept.
P. R. Barnes, and T. L. Hudson. 1989, 2p
Pub. in IEEE (Institute of Electrical and Electronics Engineers) Transactions on Power Delivery 4, n2 p1035-1036 Apr 89.

Keywords: *Power lines, *Electric filters, *Surges, *Suppressors, Overvoltage, Transmission lines, Electric wire, Voltage regulators, Retarding, Damping, Inhibitors, Reprints.

The authors report interesting results of their tests on commercial filters (presumably consisting of linear elements), enhanced by two types of nonlinear surge-protective devices. While there is no problem with the reported performance per se, the wording of the report summary is likely to cause ambiguities when the paper appears in literature abstracts.

900,756
PB90-117474   Not available NTIS
National Inst. of Standards and Technology (NEL), Boulder, CO. Electromagnetic Technology Div.
Comparison of Far-Field Methods for Determining Mode Field Diameter of Single-Mode Fibers Using Both Gaussian and Petermann Definitions.
Final rept.
T. J. Drapela, D. L. Franzen, A. H. Cherin, and R. J. Smith. 1989, 5p
Pub. in Jnl. of Lightwave Technology 7, n8 p1153-1157 Aug 89.

Keywords: *Far field, *Fiber optics, Electromagnetic fields, Electromagnetic radiation, Normal density functions, Optical communication, Transmission lines, Reprints, *Single mode fibers, *Mode field diameters.

An interlaboratory comparison of far-field measurement methods to determine mode field diameter of single-mode fibers was conducted among members of the Electronic Industries Association. Measurements were made on dispersion unshifted and shifted fibers at 1300 and 1550 nm. Results were calculated using both Petermann and Gaussian definitions. The Petermann definition gave better agreement than the Gaussian in all cases. A systematic offset of 0.52 micro m was observed between methods when applied to dispersion shifted fibers. Such an offset may be caused by limited angular collection.

900,757
PB90-117482   Not available NTIS
National Inst. of Standards and Technology (NEL), Boulder, CO. Electromagnetic Technology Div.
Numerical Aperture of Multimode Fibers by Several Methods: Resolving Differences.
Final rept.
D. Franzen, M. Young, A. Cherin, E. Head, M. Hackert, K. Raine, and J. Baines. 1989, 6p
Pub. in Jnl. of Lightwave Technology 7, n6 p896-901 Jun 89.

Keywords: *Fiber optics, *Apertures, Electric fields, Optical communication, Far field, Electromagnetic fields, *Reprints, *Multimode fibers, Gradient index optics, Index profiles.

An industry-wide study among members of the Electronic Industries Association was conducted to document differences among three numerical aperture measurement methods. Results on 12 multimode graded index fibers indicate systematic differences exist among commonly used far-field and index profile techniques. Differences can be explained by a wavelength dependent factor and choice of definitions.

Conversion factors may be used to relate the various methods.

Resistive, Capacitive, & Inductive Components

900,758
PB89-171805 Not available NTIS
National Bureau of Standards (IMSE), Gaithersburg, MD. Ceramics Div.
Fracture Behavior of Ceramics Used in Multilayer Capacitors.
Final rept.
T. L. Baker, and S. W. Freiman. 1986; 10p
Pub. in Materials Research Society Symposium Proceedings on Electronic Packaging Materials Science 72, n2 p81-90 1986.

Keywords: *Ceramics, *Crack propagation, *Aging tests(Materials), *Capacitors, *Dielectric breakdown, Microstructure, Chemical composition, Toughness, Fracturing, Stress corrosion, Indentation, Fatigue(Materials), Dynamic loads, Strength, Reprints.

The study involved the determination of the effects of composition and microstructure on the fracture toughness and susceptibility to environmentally enhanced crack growth of several ceramic materials used in multilayer capacitors. Indentation-fracture procedures were used to measure K(IC) as well as to assess the possible effects of internal stresses on the fracture behavior of these materials and to correlate dielectric aging phenomena with strength. The environmentally enhanced crack growth behavior of these materials was determined by conducting dynamic fatigue tests in water.

900,759
PB89-173785 Not available NTIS
National Bureau of Standards (NEL), Boulder, CO. Electromagnetic Fields Div.
Transient Response Error in Microwave Power Meters Using Thermistor Detectors.
Final rept.
F. R. Clague, and N. T. Larsen. 1987, 11p
Pub. in ARFTG (Automatic RF Techniques Group) Conference Digest (28th), St. Petersburg, FL., December 4-5, 1986, p79-89 1987.

Keywords: *Thermistors, *Transient response, Dynamic response, Amplification, Power measurement, Semiconductor devices, Variable resistors, Microwave frequencies, Bolometers, Electric power meters.

Broadband coaxial thermistor mounts are commonly used in automated precision microwave measurement systems such as six-port networks. To reduce the effect of temperature drift and to decrease the total measurement time, it is desirable to measure the DC bias voltage on the thermistor mount very quickly after turning the rf on or off. However, investigation has revealed that a coaxial mount may take much longer to settle to a stable DC bias voltage than the thermistor element time constant or the associated power meter servo bandwidth would indicate. If the bias voltage is measured before this transient ends, the error in the calculated rf power can be very large; as much as 1.4% has been observed. The paper describes these transients and gives measured durations and maximum error for a number of different bolometer mounts.

900,760
PB89-176648 Not available NTIS
National Bureau of Standards (NEL), Gaithersburg, MD. Electrosystems Div.
Selecting Varistor Clamping Voltage: Lower is Not Better.
Final rept.
F. D. Martzloff, and T. F. Leedy. 1989, 6p
Pub. in Proceedings of Zurich International EMC (Electromagnetic Compatibility) Symposium, Zurich, Switzerland, March 7-9, 1989, p137-142.

Keywords: *Protectors, *Varistors, *Surges, Overcurrent, Semiconductor devices, Thermistors, Variable resistors, Overvoltage, Power lines, Clamping circuits, Electric potential, Premature aging.

Surge protective devices, such as varistors, are applied to protect sensitive load equipment against power-line surges. The need to provide low clamping

voltage for protection of equipment with low inherent immunity must be balanced against the risk of premature aging of the protective device. Lower clamping voltage causes more frequent interventions of the protective device, accelerating its aging. The paper describes four possible causes of such premature aging, calling for a more careful and thus more reliable application of protective devices.

900,761
PB89-193890 PC A03/MF A01
National Inst. of Standards and Technology (NEL), Boulder, CO. Electromagnetic Fields Div.
Theory and Measurements of Radiated Emissions Using a TEM (Transverse Electromagnetic) Cell.
Technical note.
G. H. Koepke, M. T. Ma, and W. D. Bensema. Jan 89, 40p NIST/TN-1326
Also available from Supt. of Docs. as SN003-003-02932-1.

Keywords: *Electromagnetic radiation, Measurement, Antennas, Emission, Test equipment, *Transverse electromagnetic cells.

The transverse electromagnetic cell is widely used to evaluate the electromagnetic characteristics of electrically small devices. The paper reviews the theoretical basis for a technique to quantify the radiated emissions from any such device in the cell. The technique is well suited to an automated test system provided that the mechanical motions required can be controlled by a computer. The difficulties associated with these mechanical motions are discussed and possible solutions are proposed. The measurement technique is also expanded to include multiple-frequency sources in addition to single-frequency sources,

900,762
PB89-200505 Not available NTIS
National Inst. of Standards and Technology (NEL), Boulder, CO. Electromagnetic Technology Div.
Superconducting Kinetic Inductance Bolometer.
Final rept.
J. E. Sauvageau, and D. G. McDonald. 1989, 4p
Sponsored by Redstone Arsenal, AL., and Aerospace Guidance and Metrology Center, Newark AFS, OH.
Pub. in IEEE (Institute of Electrical and Electronics Engineers) Transactions on Magnetics 25, n2 p1331-1334 Mar 89.

Keywords: *Bolometers, Temperature measuring instruments, Inductance, Niobium, Silicon, Substrates, Reprints, *Superconducting devices, *Thermometers, SQUID (Detectors).

The authors are developing a bolometer with a temperature sensor based on the temperature dependence of the inductance of a superconducting microstrip line. As a first step in exploring this idea, they have designed experiments to test only the temperature sensor. The experimental devices are all-niobium inductance thermometers fabricated on silicon substrates which have been deeply etched to provide areas of relative thermal isolation. The ground plane superconductor is thin enough that its kinetic inductance dominates the audio frequency impedance of the stripline near its critical temperature, at 0.09(T sub c). This differential thermometer uses a commercial SQUID as the preamplifier. Preliminary results demonstrate a proof-of-principle for the thermometer design.

900,763
PB89-201032 Not available NTIS
National Bureau of Standards (NEL), Boulder, CO. Electromagnetic Technology Div.
Noise in DC SQUIDS with Nb/Al-Oxide/Nb Josephson Junctions.
Final rept.
M. W. Cromar, J. A. Beall, D. Go, K. A. Masarie, and R. H. Ono. 1989, 3p
Pub. in IEEE (Institute of Electrical and Electronics Engineers) Transactions on Magnetics 25, n2 p1005-1007 Mar 89.

Keywords: *Electromagnetic noise, Josephson junctions, Direct current, Aluminum oxide, Niobium, Reprints, *SQUID devices.

The authors have developed a process which incorporates very high quality Nb/Al-oxide/Nb Josephson junctions. The junctions had low subgap conductance yielding (V sub m) greater than 50 mV for critical current densities of 1000 A/sq cm. Low inductance SQUIDs made with these junctions were apparently free from junction conductance fluctuations, at least

for frequencies above 1 Hz. The SQUIDs exhibited flux noise of currently unknown origin.

900,764
PB89-201719 Not available NTIS
National Bureau of Standards (NML), Gaithersburg, MD. Temperature and Pressure Div.
Impedance of Radio-Frequency Biased Resistive Superconducting Quantum Interference Devices.
Final rept.
R. J. Soulen, and D. Van Vechten. 1987, 27p
Pub. in Physical Review B-Condensed Matter 36, n1 p239-265 1987.

Keywords: *Electrical impedance, Josephson junctions, Electrical measurement, Radio frequencies, Reprints, *SQUID devices, Noise thermometers.

The authors have measured with high accuracy (100 ppm) and high precision (5-10 ppm) the impedance of an rf-driven resistive SQUID as a function of the amplitude and frequency of the radio-frequency (rf) bias, as a function of the current bias, and as a function of several other circuit parameters. They have developed the coupled differential equations for the resistive SQUID and for the rf tank circuit. The fit of the solutions of these equations to the data for non-hysteretic junctions is excellent. The authors were unable to develop solutions for the hysteretic case and thus compare them to the data obtained. The relationship of the study to noise thermometry using resistive SQUIDs is indicated.

900,765
PB89-202014 Not available NTIS
National Bureau of Standards (IMSE), Gaithersburg, MD. Metallurgy Div.
Vector Calibration of Ultrasonic and Acoustic Emission Transducers.
Final rept.
J. A. Simmons, C. D. Turner, and H. N. G. Wadley. 1987, 9p
Pub. in Jnl. of the Acoustical Society of America 82, n4 p1122-1130 1987.

Keywords: *Sound transducers, *Transient response, Electric converters, Dynamic responses, Damping, Impedance, Vectors(Mathematics), Vector analysis, Calibrating, Reprints.

The independent transient response of a transducer to displacements polarized in each of the three orthogonal directions have been determined using a new method which assumes only linearity of transducer response and circular symmetry of both the transducer and the source. The method forms the basis for vector calibration of acoustic emission and ultrasonic transducers.

900,766
PB89-211114 (Order as PB89-211106, PC A04)
National Inst. of Standards and Technology, Gaithersburg, MD.
Calibration of Voltage Transformers and High-Voltage Capacitors at NIST.
W. E. Anderson. 1989, 17p
Included in Jnl. of Research of the National Institute of Standards and Technology, v94 n3-4 p179-195 May-Jun 89.

Keywords: *Calibration, Capacitors, Dissipation factor, Electric power, Electrical standards, NIST services, Voltage transformers.

The National Institute of Standards and Technology (NIST) calibration service for voltage transformers and high-voltage capacitors is described. The service for voltage transformers provides measurements of ratio correction factors and phase angles at primary voltages up to 170 kV and secondary voltages as low as 10 V at 60 Hz. Calibrations at frequencies from 50-400 Hz are available over a more limited voltage range. The service for high-voltage capacitors provides measurements of capacitance and dissipation factor at applied voltages ranging from 100 V to 170 kV at 60 Hz depending on the nominal capacitance. Calibration over a reduced voltage range at other frequencies are also available. As in the case with voltage transformers, these voltage constraints are determined by the facilities at NIST.

90

Semiconductor Devices

900,767
PB89-146880 Not available NTIS
National Bureau of Standards (NEL), Gaithersburg,
MD. Semiconductor Electronics Div.
**Analytical Model for the Steady-State and Tran-
sient Characteristics of the Power Insulated-Gate
Bipolar Transistor.**
Final rept.
A. R. Hefner, and D. L. Blackburn. 1988, 20p
Pub. in Solid-State Electronics 31, n10 p1513-1532
1988.

Keywords: *Field effect transistors, *Models, Wave-
forms, Electric potential, Unsteady flow, Steady state,
Electric current, Reprints, Ambipolar diffusion,
MOSFET.

An analytical model for the power Insulated-Gate Bipolar Transistor (IGBT) is developed. The model consistently describes the IGBT steady-state current-voltage
characteristics and switching transient current and
voltage waveforms for all loading conditions. The
model is based on the equivalent circuit of a MOSFET
which supplies the base current to a low-gain, high-
level injection, bipolar transistor with its base virtual
contact at the collector end of the base. The basic element of the model is a detailed analysis of the bipolar
transistor which uses ambipolar transport theory and
does not assume the quasi-static condition for the
transient analysis. This analysis differs from the previous bipolar transistor theory in that (1) the relatively
large base current which flows from the collector end
of the base is properly accounted for, and (2) the component of current due to the changing carrier distribution under the condition of a moving collector-base depletion edge during anode voltage transitions is accounted for. Experimental verification of the model
using devices with different base lifetimes is presented
for the on-state current-voltage characteristics, the
steady-state saturation current, and the current and
voltage waveforms for the constant voltage transient,
the inductive load transient, and the series resistor-inductor load transient.

900,768
PB89-146955 Not available NTIS
National Bureau of Standards (NEL), Gaithersburg,
MD. Semiconductor Electronics Div.
**Use of Artificial Intelligence and Microelectronic
Test Structures for Evaluation and Yield Enhance-
ment of Microelectronic Interconnect Systems.**
Final rept.
M. W. Cresswell, N. Pessall, L. W. Linholm, and D. J.
Radack. 1986, 10p
Pub. in Proceedings of International IEEE (Institute of
Electrical and Electronics Engineers) VLSI (Very Large
Scale Integration) Multilevel Interconnection Confer-
ence (3rd), Santa Clara, CA., June 9-10, 1986, p331-
340.

Keywords: *Circuit interconnections, *Integrated circuits, *Test equipment, Tests, Artificial intelligence,
Very large scale integration, Expert systems.

A major factor limiting the production and performance
of high-density VLSI integrated circuits is the fabrication of reliable interconnect systems. Properly designed microelectronic test structures and appropriate
test methods can be used to characterize the processes used to fabricate these systems. However, the
computer-controlled testing of comprehensive process evaluation and diagnosis structures often results in
large quantities of data which cannot be readily or effectively interpreted by the user. As a result, important
features of the data are often overlooked or not considered in the evaluation of the fabrication processes.
The paper describes an expert system for assisting the
user to interpret test results associated with fabricating
selected aspects of VLSI interconnect systems.

900,769
PB89-146997 Not available NTIS
National Bureau of Standards (IMSE), Gaithersburg,
MD.
Stress Effects on III-V Solid-Liquid Equilibria.
Final rept.
F. C. Larche, and J. W. Cahn. 1987, 8p
Pub. in Jnl. of Applied Physics 62, n4, p1232-1239, 15
Aug 87.

Keywords: *Epitaxy, *Thermodynamic equilibrium,
Phase diagrams, Compositions, Stresses, Reprints,
*III-V compounds.

The equilibrium of an epitaxial layer grown from the
melt is analyzed in detail. The stresses produced by
lattice mismatch with the substrate interact with the
chemistry of the process. The general equilibrium
equations are derived for 3-5 compounds. The stress
field, and its approximations for thin layers are obtained. An expression for the composition of the epi-
layer as a function of liquid composition, valid in the
vicinity of the lattice matching composition, is derived.
Very good agreement is found between calculated
values and measured results from LPE growth. Properties of systems that would present lattice latching phenomena are quantitatively discussed. It is shown that
epitaxial equilibrium also affects the results of several
methods of phase diagram determination, which are
discussed in light of the preceeding results.

900,770
PB89-150825 Not available NTIS
National Bureau of Standards (NEL), Gaithersburg,
MD. Semiconductor Electronics Div.
**Review of: Thermal Characterization of Power
Transistors.**
Final rept.
D. L. Blackburn. 1988, 7p

Keywords: *Thermal measurements, *Transistors,
Thermal environments, Power supplies, Surges,
Switching circuits, Field effect transistors, Darlington
transistors.

The thermal characteristics of power transistors and
their measurement are discussed. Topic areas addressed include general methods for measuring
device temperature, control of the thermal environment, selection of a temperature-sensitive electrical
parameter, measurement of temperature-sensitive
electrical parameters, reasons for measuring temperature, and temperature measurement of integrated
power devices. Procedures for detecting nonthermal
switching transients, extrapolation of the measured
temperature to the instant of switching, and for measuring the temperature of Darlington transistors are included.

900,771
PB89-150973 Not available NTIS
National Bureau of Standards (NEL), Gaithersburg,
MD. Precision Engineering Div.
Submicrometer Optical Metrology.
Final rept.
R. D. Larrabee. 1988, 2p
See also PB87-201646.
Pub. in Proceedings of Annual Meeting of the Electron
Microscopy Society of America (46th), Milwaukee, WI.,
August 7-12, 1988, p50-51.

Keywords: *Standards, *Metrology, *Optical measurements, *Line width, *Integrated circuits, Performance
evaluation, Pitch(Inclination), Critical dimensions.

The National Bureau of Standards (NBS) has developed optical linewidth standards for the integrated circuit industry for over 10 years. The past work has concentrated on the development and the certification of
photomask linewidth and pitch standards. The recent
work is directed at extending the feature sizes on
these standards to cover the range from 0.5 to 30 micrometers, and at doubling the certification accuracy
to 0.025 micrometers. Features with heights larger
than approximately 1/4 wavelength of light cannot be
modeled as zero-thickness layers as is done for photomasks. The development of models to handle this
thick-layer case and to develop practical edge-detec-
tion criteria are currently under development at NBS. It
is generally not possible to interpret the image profiles
of thick features and thereby measure an accurate
linewidth. The basic obstacles that must be overcome
to achieve accurate submicron feature size measurements are reviewed and the prospects for future NBS
optical standards for features such as photoresist lines
on silicon wafers are assessed. Some suggestions
about what to do until these standards become available are given.

900,772
PB89-151831 PC A04/MF A01
National Inst. of Standards and Technology (NEL),
Gaithersburg, MD. Semiconductor Electronics Div.

**Semiconductor Measurement Technology: Auto-
matic Determination of the Interstitial Oxygen
Content of Silicon Wafers Polished on Both Sides.**
Final rept.
W. K. Gladden, S. R. Slaughter, W. M. Duncan, and
A. Baghdadi. Nov 88, 73p NIST/SP-400/81
Also available from Supt. of Docs. as SN003-003-
02915-7. Library of Congress catalog card no. 88-
600601. Prepared in cooperation with Texas Instru-
ments, Inc., Dallas.

Keywords: *Semiconductor devices, *Interstitials, *Silicon, *Computer systems programs, *Infrared spectra,
*Oxygen, Computerized simulation, Tests, Automatic
control, Wafers, Algorithms.

The Special Publication contains FORTRAN and
PASCAL computer programs which implement an
ASTM test method for the automatic determination of
the interstitial oxygen content of silicon. The programs
are to be used as illustrative examples by programmers wishing to implement the ASTM algorithm on
their computers. The Publication also includes sample
data that can be used to test the computer programs.
The sample data are included in two forms: in print,
and on an MS-DOS floppy disk.

900,773
PB89-157655 Not available NTIS
National Bureau of Standards (NEL), Gaithersburg,
MD. Semiconductor Electronics Div.
**Effect of Neutrons on the Characteristics of the In-
sulated Gate Bipolar Transistor (IGBT).**
Final rept.
A. R. Hefner, D. L. Blackburn, and K. F. Galloway.
1986, 7p
Pub. in IEEE (Institute of Electrical and Electronics En-
gineers) Transactions on Nuclear Science NS-33, n6
p1428-1434 Dec 86.

Keywords: *Neutron irradiation, Mathematical models,
Field effect transistors, Reprints, *Bipolar transistors,
*Insulated gate bipolar transistors, *Physical radiation
effects, MOSFET, Carrier lifetime.

The effects of neutrons on the operating characteristics of Insulated Gate Bipolar Transistors (IGBT) are
described. Experimental results are presented for devices that have been irradiated up to a fluence of 10 to
the 13th power neutrons/c sq cm, and an analytical
model is presented which explains the observed effects. The effects of neutrons on the IGBT are compared with the known effects on power MOSFETs, and
it is shown that the IGBT characteristics begin to degrade at a fluence that is an order of magnitude less
than the fluence at which the power MOSFET begins
to degrade. At high fluences, the IGBT takes on the
characteristics of a power MOSFET.

900,774
PB89-158042 Not available NTIS
National Bureau of Standards (NEL), Gaithersburg,
MD. Center for Electronics and Electrical Engineering.
**Standards and Test Methods for VLSI (Very Large
Scale Integration) Materials.**
Final rept.
R. I. Scace. 1985, 5p
Pub. in Technical Program: Proceedings of Semicon-
ductor Technology Symposium, Tokyo, Japan, De-
cember 7-8, 1984, p4-1-4-5 1985.

Keywords: *Semiconductors(Materials), *Standards,
Integrated circuits, Specifications, Tests, Standardization, Reprints, *Very large scale integration.

Standard measurement methods and specifications
for the semiconductor industry will be reviewed and
discussed with emphasis on applications to VLSI processes. These standards are well accepted in the U.S.
and in Europe, but are not so well known in Japan. The
standards development process is an excellent way
for material producers and users to develop good
working relations and to solve their shared measurement problems; this process will be described in some
detail. Because the semiconductor industry is an international one, serious efforts have been made for a
number of years to rationalize the technical differences between test method standards in Europe and
the U.S. with considerable success. The present state
of such cooperative activity with Japan, which as a
more recent origin, will also be reported.

900,775
PB89-168033 PC A03/MF A01

91

ELECTROTECHNOLOGY

Semiconductor Devices

National Inst. of Standards and Technology (NEL), Gaithersburg, MD. Center for Electronics and Electrical Engineering.
Center for Electronics and Electrical Engineering Technical Progress Bulletin Covering Center Programs, July to September 1988, with 1989 CEEE Events Calendar.
E. J. Walters. Jan 89, 40p NISTIR-88/4020
See also PB88-130315.

Keywords: *Semiconductor devices, *Electric devices, *Electromagnetic interference, Silicon, Photodetectors, Packaging, Metrology, Superconductors, Millimeter waves, Microwaves, Fiber optics, Electrooptics, Antennas, *Signal acquisition.

The report is the twenty-fourth issue of a quarterly publication providing information on the technical work of the National Institute of Standards and Technology (formerly the National Bureau of Standards) Center for Electronics and Electrical Engineering. The topics discussed are semiconductor technology; first signal acquisition, processing and transmission; electrical systems; and electromagnetic interference.

900,776
PB89-172555 PC A03/MF A01
National Inst. of Standards and Technology (NEL), Gaithersburg, MD. Precision Engineering Div.
Approach to Accurate X-Ray Mask Measurements in a Scanning Electron Microscope.
M. T. Postek, R. D. Larrabee, and W. J. Keely. Jan 89, 14p NISTIR-89/4047
Sponsored by Naval Research Lab., Washington, DC.

Keywords: *Dimensional measurement, *Integrated circuits, *Masking, *Lithography, Metrology, Accuracy, *X ray lithography, Scanning electron microscopes.

The paper presents the concept and some preliminary experimental data on a new method for measuring critical dimensions on masks used for x-ray lithography. The method uses a scanning electron microscope (SEM) in a transmitted-scanning electron microscope (TSEM) imaging mode and can achieve nanometer precision. Use of this technique in conjunction with measurement algorithms derived from electron beam interaction modeling may ultimately enable measurements of these masks to be made to nanometer accuracy.

900,777
PB89-176259 Not available NTIS
National Bureau of Standards (NEL), Gaithersburg, MD. Semiconductor Electronics Div.
Analytical Modeling of Device-Circuit Interactions for the Power Insulated Gate Bipolar Transistor (IGBT).
Final rept.
A. R. Hefner. 1988, 9p
Pub. in Conference Record 1988, IEEE (Institute of Electrical and Electronics Engineers) Industry Applications Society Annual Meeting, Pittsburgh, PA., October 2-7, 1988, p605-614.

Keywords: *Field effect transistors, *Mathematical models, Simulation, Semiconductor devices, Transient response, Impedance, Surges, Protection, Ratings, Performance standards, *Bipolar transistors, Snubbers.

The device-circuit interactions of the power Insulated Gate Bipolar Transistor (IGBT) for a series resistor-inductor load, both with and without a snubber, are simulated. An analytical model for the transient operation of the IGBT is used in conjunction with the load circuit state equations for the simulations. The simulated results are compared with experimental results for all conditions. Devices with a variety of base lifetimes are studied. For the fastest devices studied, the voltage overshoot of the series resistor-inductor load circuit approaches the device voltage rating for load inductances greater than 1 microHenrys. For slower devices, the voltage overshoot is much less and a larger inductance can therefore be switched without a snubber circuit. In the study, the simulations are used to determine the conditions for which the different devices can be switched safely without a snubber protection circuit. Simulations are also used to determine the required values and ratings for protection circuit components when protection circuits are necessary.

900,778
PB89-179675 Not available NTIS
National Bureau of Standards (NML), Gaithersburg, MD. Electricity Div.

92

Quantized Hall Resistance Measurement at the NML (National Measurement Laboratory).
Final rept.
B. W. Ricketts, and M. E. Cage. 1987, 4p
Pub. in IEEE (Institute of Electrical and Electronics Engineers) Transactions on Instrumentation and Measurement 36, n2 p245-248 1987.

Keywords: *Gallium arsenides, *Electrical resistance, Semiconductors, Quantum interactions, Impedance, Measurement, Standards, Reprints, *Hall effect, Heterostructures.

An automatic measurement system has been developed to determine the values of quantized Hall resistances of two GaAs/AlGaAs heterostructures. The n=2 step of one heterostructure and the n=4 step of the other were measured over a seven-month period. A weighted mean of these determinations gives a Si value for the quantity h/e squared of 0.47 ppm (0.11 ppm 1 sigma uncertainty) above the nominal 25812.80 ohm value.

900,779
PB89-180020 Not available NTIS
National Bureau of Standards (NEL), Gaithersburg, MD. Semiconductor Electronics Div.
Correlation between CMOS (Complementary Metal Oxide Semiconductor) Transistor and Capacitor Measurements of Interface Trap Spectra.
Final rept.
T. J. Russell, H. S. Bennett, M. Gaitan, J. S. Suehle, and P. Roitman. 1986, 6p
Sponsored by Defense Nuclear Agency, Washington, DC.
Pub. in IEEE (Institute of Electrical and Electronics Engineers) Transactions on Nuclear Science NS-33, n6 p1228-1233 Dec 86.

Keywords: *Silicon dioxide, *Radiation effects, *Capacitors, *Transistors, Interfacial tension, Energy, Spectra, Semiconductor devices, Charging, Reprints.

The radiation induced change in the energy spectra of SiO2-Si interface traps as determined using the charge-pumping and weak-inversion techniques on complementary metal oxide semiconductors (CMOS) transistors and using the quasi-static capacitance voltage (C-V) and detailed model techniques on CMOS capacitors are compared. Over the range of approximately 10(sup 10) to 10(sup 12)/cm/ev, good quantitative agreement is obtained between these methods.

900,780
PB89-180426 Not available NTIS
National Bureau of Standards (NEL), Boulder, CO. Electromagnetic Fields Div.
Electromagnetic Fields in Loaded Shielded Rooms.
Final rept.
E. Vanzura, and J. W. Adams. 1987, 7p
Pub. in Test and Measurement World, p72, 74, 76, 78, 80, 82, 83 Nov 87.

Keywords: *Electromagnetic fields, *Field strength, *Feedback amplifiers, Electromagnetic shielding, Radiation shielding, Feedback control, Automatic control, Electromagnetic interference, Cybernetics, Magnetic permeability, Reprints, *Shielded rooms.

The paper describes a computer-controlled feedback system that can maintain field strength levels within moderate bounds inside a partially-loaded shielded room. The levels are relatively uniform over a large enough volume to allow radiated immunity testing of moderate-sized objects. The frequency range depends on the characteristics of the transmit antenna; 50 to 200 MHz was used, which is a difficult range to cover because of limitations of other EMC susceptibility test facilities. The measurement system consists of a computer, signal generator, amplifier, biconical antenna and an isotropic probe system.

900,781
PB89-166637 Not available NTIS
National Bureau of Standards (NEL), Gaithersburg, MD. Semiconductor Electronics Div.
Numerical Analysis for the Small-Signal Response of the MOS (Metal Oxide Semiconductors) Capacitor.
Final rept.
M. Gaitan, and I. Mayergoyz. 1989, 7p
Pub. in Solid-State Electronics 32, n3 p207-213 1989.

Keywords: *Metal oxide semiconductors, *Capacitors, *Frequency response, Quantum statistics, Numerical

---

analysis, Continuity equation, Surface properties, Poisson density functions, Reprints, Bulk trap dynamics.

Simulation results for the small-signal sinusoidal steady-state response of the MOS capacitor using time perturbation analysis of the basic semiconductor equations are presented. The effects of interface and bulk trap dynamics are included. The model uses Fermi-Dirac statistics and Shockley-Read-Hall recombination to describe the traps. The analysis is an improvement over previous techniques since it can simulate the effect of trap dynamics on the small-signal sinusoidal steady-state response of a semiconductor device with arbitrary geometry, doping and trap distributions.

900,782
PB89-189344 PC A06/MF A01
CD Metrology, Inc., Germantown, MD.
Narrow-Angle Laser Scanning Microscope System for Linewidth Measurement on Wafers.
D. Nyyssonen. Apr 89, 111p NISTIR-88/3808
Sponsored by National Inst. of Standards and Technology (NEL), Gaithersburg, MD. Precision Engineering Div.

Keywords: *Measuring instruments, *Semiconductors(Materials), *Integrated circuits, *Coherence, Scanning, Lasers, Line width, Computer systems programs, Thin films, Optical properties, Graphs(Charts), Microscopy.

The integrated-circuit industry in its push to finer and finer line geometries approaching submicrometer dimensions has created a need for ever more accurate and precise feature-size measurements to establish fighter control of fabrication processes. In conjunction with the NBS Semiconductor Linewidth Metrology Program, a unique narrow-angle laser measurement system was developed. The report describes the theory, optical design, and operation of the system and includes computer software useful for characterizing the pertinent optical parameters and images for patterned thin layers. For thick layers, the physics is more complex and only elements of the theory are included. For more detail the reader is referred to several related reports listed in the references.

900,783
PB89-201156 Not available NTIS
National Bureau of Standards (NEL), Gaithersburg, MD. Statistical Engineering Div.
Graphical Analyses Related to the Linewidth Calibration Problem.
Final rept.
M. C. Croarkin. 1986, 8p
Pub. in Proceedings of Measurement Science Conference, Irvine, CA., January 23-24, 1986, p159-166.

Keywords: *Graphic methods, *Calibrating, *Integrated circuits, Metrology, Line width, Statistical tests, Measurement, Control charts.

The paper demonstrates that graphical analyses, when properly understood, can often supplant more formal statistical analyses, and in all cases enhance such analyses. Statistical tests, which reduce the information in the data to a one line 'yes' or 'no' finding, are referenced to support the graphical findings. The power of the approach is illustrated with a measurement problem that led to the development of Standard Reference Material SRM-475. Assessment of the extent of the measurement problem in the integrated circuit industry and procedures for evaluating system performance are discussed.

900,784
PB89-201974 Not available NTIS
National Bureau of Standards (NEL), Gaithersburg, MD. Semiconductor Electronics Div.
Radiation-Induced Interface Traps in Power MOSFETs.
Final rept.
G. Singh, K. F. Galloway, and T. J. Russell. 1986, 6p
Pub. in IEEE (Institute of Electrical and Electronics Engineers) Transactions on Nuclear Science NS-33, n6 p1454-1459 Dec 86.

Keywords: *Field effect transistors, *Radiation effects, Irradiation, Gamma rays, Transconductance, Reprints, *MOSFET.

Methods for estimating values of radiation-induced interface trapped charge from the current-voltage (I-V) characteristics of MOSFETs are described and applied to commercially available power MOSFETs. The power MOSFETs show severe degradation on radiation exposure with the effects of positive oxide trapped charge dominating; however, interface trap buildup is significant. The results are compared to experimental measurements available on other technologies.

900,785
PB89-202964   Not available NTIS
National Bureau of Standards (NEL), Gaithersburg, MD. Chemical Process Metrology Div.
Fundamental Characterization of Clean and Gas-Dosed Tin Oxide.
Final rept.
S. Semancik, and D. F. Cox. 1987, 6p
Pub. in Sensors and Actuators 12, n2 p101-106 1987.

Keywords: *Tin oxides, *Single crystals, *Water, *Oxygen, *Adsorption, Surface chemistry, High pressure tests, Semiconductors(Materials), Defects, Sensors, Photoelectric emission, Reprints.

The chemical, electronic and structural properties of a SnO2(110) crystal have been studied before and after exposure to H2O and O2. The results show that the initial stages of water adsorption on SnO2(110) can be influenced by structural factors and/or the concentration level of surface defects. The degree of surface oxidation has also been demonstrated to have an observable effect on the nature of water adsorption at higher pressures. These results relate both to basic gas sensing mechanisms for SnO2 and to the influence of humidity on tin oxide.

900,786
PB89-209225   PC A03/MF A01
National Inst. of Standards and Technology (NEL), Gaithersburg, MD. Center for Electronics and Electrical Engineering.
Center for Electronics and Electrical Engineering: Technical Progress Bulletin Covering Center Programs, January to March 1989, with 1989 CEEE Events Calendar.
E. J. Walters. Jun 89, 27p NISTIR-89/4095
See also PB89-168033.

Keywords: *Semiconductor devices, *Signal processing, *Metrology, Photodetectors, Electric devices, Radiation effects, Waveforms, Lasers, Fiber optics, Electrooptics, Electric power, Superconductors, Electromagnetic interference, Signals and systems, Cryoelectronics.

The report is the twenty-sixth issue of a quarterly publication providing information on the technical work of the National Institute of Standards and Technology (formerly the National Bureau of Standards) Center for Electronics and Electrical Engineering. The issue of the CEEE Technical Progress Bulletin covers the first quarter of calendar year 1989.

900,787
PB89-209241   PC A03/MF A01
National Inst. of Standards and Technology (NEL), Gaithersburg, MD. Center for Electronics and Electrical Engineering.
Center for Electronics and Electrical Engineering Technical Publication Announcements. Covering Center Programs, October/December 1988, with 1989 CEEE Events Calendar.
E. J. Walters. May 89, 30p NISTIR-89/4096
See also PB89-189302.

Keywords: *Semiconductor devices, *Signal processing, *Metrology, Silicon, Photodetectors, Electric devices, Waveforms, Lasers, Antennas, Microwaves, Millimeter waves, Fiber optics, Electrooptics, Electric power, Superconductors, Electromagnetic interference, Signals and systems, Cryoelectronics.

The report is the nineteenth issue of a quarterly publication providing information on the technical work of the National Institute of Standards and Technology (formerly the National Bureau of Standards) Center for Electronics and Electrical Engineering. The issue of the Center for Electronics and Electrical Engineering Technical Publication Announcements covers the fourth quarter of calendar year 1988.

900,788
PB89-212187   Not available NTIS
National Bureau of Standards (NEL), Gaithersburg, MD. Semiconductor Electronics Div.

Neural Network Approach for Classifying Test Structure Results.
Final rept.
D. Khera, M. E. Zaghloul, L. W. Linholm, and C. L. Wilson. 1989, 4p
Pub. in Proceedings of IEEE (Institute of Electrical and Electronics Engineers) International Conference on Microelectronic Test Structures, Edinburgh, Scotland, March 13-14, 1989, v2 n1 p201-204.

Keywords: *Integrated circuits, *Manufacturing, *Neural nets, Artificial intelligence, Algorithms, Learning machines, Tests, *Semiconductors, Chips(Electronics).

The paper describes a new approach for identifying and classifying semiconductor manufacturing process variations using test structure data. The technique described employs a machine-learning algorithm based on neural networks to train computers to detect patterns associated with test structure results. The objective of the work is to develop more reliable machine-learning classification procedures using test structure data from a semiconductor manufacturing environment. An example based on characterizing the performance of a 1 micro m lithography process is presented as well as description of the test chip.

900,789
PB89-228308   PC A03/MF A01
National Inst. of Standards and Technology (NEL), Gaithersburg, MD. Center for Electronics and Electrical Engineering.
Center for Electronics and Electrical Engineering Technical Publication Announcements. Covering Center Programs, January-March 1989, with 1989 CEEE Events Calendar.
E. J. Walters. Jul 89, 20p NISTIR-89/4118
See also PB89-209241.

Keywords: *Semiconductor devices, *Metrology, Electrical engineering, Integrated circuits, Photodetectors, Electrical insulators, Waveforms, Lasers, Fiber optics, Electrooptics, Superconductors, Electromagnetic interference, Abstracts, *Signals and systems, Cryoelectronics.

The report is the twentieth issue of a quarterly publication providing information on the technical work of the National Institute of Standards and Technology (formerly the National Bureau of Standards) Center for Electronics and Electrical Engineering. The issue covers the first quarter of calendar year 1989. Abstracts are provided by technical area for papers published during the quarter.

900,790
PB89-228530   Not available NTIS
National Inst. of Standards and Technology (NEL), Gaithersburg, MD. Semiconductor Electronics Div.
Machine-Learning Classification Approach for IC Manufacturing Control Based on Test Structure Measurements.
Final rept.
M. E. Zaghloul, D. Khera, L. W. Linholm, and C. P. Reeve. 1989, 7p
Pub. in IEEE (Institute of Electrical and Electronics Engineers) Transactions on Semiconductor Manufacturing 2, n2 p47-53 May 89.

Keywords: *Integrated circuits, *Manufacturing, Semiconductor devices, Learning machines, Classifying, Artificial intelligence, Reprints, *Chips(Electronics), Expert systems.

The paper describes the use of a machine-learning method for classifying electrical measurement results from a custom-designed test chip. The techniques are used for characterizing the performance of a 1 micrometer integrated circuit lithographic process. The focus of the work is to develop a method for producing reliable classification rules from data bases containing large samples of measurement data. The paper describes a test chip, data-handling methods, rule generation techniques, and statistical data reduction and parameter extraction techniques. An analysis of error introduced by noise in the rule formation process is presented.

900,791
PB89-230460   Not available NTIS
National Inst. of Standards and Technology (NEL), Gaithersburg, MD. Center for Mfg. Engineering.

High-Mobility CMOS (Complementary Metal Oxide Semiconductor) Transistors Fabricated on Very Thin SOS Films.
Final rept.
D. J. Dumin, S. Dabral, M. Freytag, P. J. Robertson, G. P. Carver, and D. B. Novotny. 1989, 3p
Pub. in IEEE (Institute of Electrical and Electronics Engineers) Transactions on Electron Devices 36, n3, p596-598, Mar 89.

Keywords: *Metal oxide transistors, *Thin films, Semiconducting films, Semiconductor devices, Carrier mobility, Reprints, *CMOS, *Complementary metal oxide semiconductors, Silicon films.

The increased emphasis on submicrometer geometry CMOS/SOI devices has created a need for high-mobility CMOS transistors fabricated on high-quality SOI films with thicknesses of the order of 0.1 to 0.2 micro m. To date, the only demonstrated way of producing high-mobility transistors on very thin high-quality SOS films in this thickness range has been to apply recrystallizations and regrowths to the films prior to transistor fabrication. It has been found that the mobility of CMOS transistors fabricated on very thin SOS films is a function of the film growth rate. Transistors with mobilities nearly as high as those obtained on 1.0-micro m-thick films have been fabricated on SOS films 0.2 micro m thick that have been grown at growth rates above 4 micro m/min.

900,792
PB89-231195   Not available NTIS
National Inst. of Standards and Technology (NEL), Gaithersburg, MD. Semiconductor Electronics Div.
IEEE (Institute of Electrical and Electronics Engineers) IRPS (International Reliability Physics Symposium) Tutorial Thermal Resistance Measurements, 1989.
Final rept.
F. F. Oettinger. 1989, 33p
Pub. in Proceedings of International Reliability Physics Symposium, Phoenix, AZ., April 10, 1989, p7.1-7.33.

Keywords: *Thermal resistance, *Integrated circuits, *Transistors, Thermal conductivity, Thermal stability, Temperature, Thermal measuring instruments, Junctions, Measurement.

The tutorial reviews the thermal properties of power transistors and to discuss methods for characterizing these properties. The devices discussed include bipolar transistors and metal-oxide-semiconductor field-effect-transistors (MOSFETs). Measurement problems common to these devices, such as deciding the reason a particular measurement is required, adequate reference temperature control, selection of a temperature-sensitive electrical parameter, and separation of electrical and thermal effects during measurement are addressed. The thermal characterization of the packaged integrated circuit chip surface/junction for the new generation of VLSI devices generally takes one of three forms: indirect (i.e., electrical) measurements, direct (e.g., infrared), or computer simulations of the surface/junction temperatures. Due to the inherent difficulties in measuring and analyzing the thermal properties of active integrated circuits, an approach using specifically designed thermal test chips for evaluation of new die attachment and packaging schemes is finding wide acceptance in the industry.

900,793
PB89-231203   Not available NTIS
National Inst. of Standards and Technology (NEL), Gaithersburg, MD. Semiconductor Electronics Div.
AC Impedance Method for High-Resistivity Measurements of Silicon.
Final rept.
W. R. Thurber, J. R. Lowney, R. D. Larrabee, P. Talwar, and J. R. Ehrstein. 1989, 2p
Sponsored by Air Force Systems Command, Washington, DC.
Pub. in Proceedings of Electrochemical Society Meeting, Los Angeles, CA., May 7-12, 1989, p365-366.

Keywords: *Impedance, *Alternating current, *Silicon, Semiconductors(Materials), Electrical resistance, Electron probes, Frequencies.

An AC impedance method for measuring the average bulk resistivity of ingots and slices of high-resistivity silicon is described. Easily removable contacts, such as silver paste, are applied to the end faces of the sample and the impedance of the resulting capacitive sandwich is measured as a function of frequency. The resis-

ELECTROTECHNOLOGY
Semiconductor Devices

tivity can be calculated from the frequency of the negative peak in the imaginary part of the impedance and model consistency can be checked by comparison of values of resistance obtained from real and imaginary parts at this peak. Comparisons with van der Pauw and four-probe measurements are consistent with the impedance method.

900,794
PB89-231211 Not available NTIS
National Inst. of Standards and Technology (NEL), Gaithersburg, MD. Semiconductor Electronics Div.
Experimental Verification of the Relation between Two-Probe and Four-Probe Resistances.
Final rept.
J. J. Kopanski, J. H. Albers, and G. P. Carver. 1989, 2p
Pub. in Proceedings of Electrochemical Society Meeting, Los Angeles, CA., May 7-12, 1989, p367-368.

Keywords: *Electron probes, *Electrical resistance, Measurement, Wafers, Test equipment, Semiconductors(Materials).

Recent innovations in the measurement of two-probe (spreading) resistance and four-probe resistance using an array of lithographically fabricated, geometrically well-defined contacts have enabled the measurement of the quantities with high accuracy and reproducibility. It has permitted experimental verification of the relationship between the two-probe resistances and the four-probe resistance. Verification was also made of the predicted dependence of the four-probe resistance on the ratio of wafer thickness to probe spacing for in-line and square probe configurations.

900,795
PB89-231229 Not available NTIS
National Inst. of Standards and Technology (NEL), Gaithersburg, MD. Semiconductor Electronics Div.
Improved Understanding for the Transient Operation of the Power Insulated Gate Bipolar Transistor (IGBT).
Final rept.
A. R. Hefner. 1989, 11p
Pub. in Proceedings of IEEE (Institute of Electrical and Electronics Engineers) Power Electronics Specialists Conference (20th) - PESC '89- Milwaukee, WI., June 26-29, 1989, p303-313.

Keywords: *Field effect transistors, *Electric current, *Waveforms, Models, Comparison, Capacitance, Electric potential, Analyzing.

It is shown that a non-quasi-static analysis must be used to describe the transient current and voltage waveforms of the Insulated Gate Bipolar Transistor (IGBT). The non-quasi-static analysis is necessary because the transport of electrons and holes are coupled for the low-gain, high-level injection conditions, and because the quasi-neutral base width changes faster than the base transit speed for typical load circuit conditions. To verify that both of these non-quasi-static effects must be included, the predictions of the quasi-static and non-quasi-static models are compared with measured current and voltage switching waveforms. The comparisons are performed for different load circuit conditions and for different device base lifetimes.

900,796
PB89-231237 Not available NTIS
National Inst. of Standards and Technology (NEL), Gaithersburg, MD. Semiconductor Electronics Div.
Power MOSFET Failure Revisited.
Final rept.
D. L. Blackburn. 1988, 8p
Pub. in IEEE (Institute of Electrical and Electronics Engineers) Power Electronics Specialists Conference.- PESC '88- Kyoto, Japan, April 11-14, 1988, p681-688.

Keywords: *Field effect transistors, *Failure, Avalanche breakdown, Electrical faults, Nondestructive tests, Energy dissipation, *MOSFET semiconductors.

The failure of power MOSFETs during avalanche breakdown is discussed. A theory is presented that relates the failure to the temperature rise of the chip during the avalanche breakdown and to a critical current for failure. It is shown that the energy that can be safely dissipated during avalanche breakdown decreases as the starting current increases or as the case temperature increases. Thus, if power MOSFETs are to be rated for their energy dissipation capability during avalanche breakdown, both the starting current and temperature must be specified as it is these two parameters that determine the failure limits and not the energy.

900,797
PB90-123589 Not available NTIS
National Inst. of Standards and Technology (NEL), Gaithersburg, MD. Semiconductor Electronics Div.
Numerical Simulations of Neutron Effects on Bipolar Transistors.
Final rept.
H. S. Bennett. 1987, 4p
Sponsored by Defense Nuclear Agency, Washington, DC.
Pub. in IEEE (Institute of Electrical and Electronics Engineers) Transactions on Nuclear Science NS-34, n6 p1372-1375 Dec 87.

Keywords: *Neutron irradiation, *Radiation effects, Carrier mobility, Concentration(Composition), Mathematical models, Reprints, *Bipolar transistors, Carrier lifetime, Numerical solution.

A detailed device model that has been verified by comparisons with experimental measurements on unirradiated, state-of-the-art bipolar devices has been modified to include the effects of neutron radiation on carrier lifetimes, concentrations, and mobilities. Numerical experiments on the degradation due to neutron fluences in the dc common emitter gains for bipolar transistors with submicrometer emitter and base widths are given and compared in general terms with the few published measurements.

900,798
PB90-128109 Not available NTIS
National Inst. of Standards and Technology (NEL), Gaithersburg, MD. Semiconductor Electronics Div.
Growth and Properties of High-Quality Very-Thin SOS (Silicon-on-Sapphire) Films.
Final rept.
D. J. Dumin, S. Dabral, M. Freytag, P. J. Robertson, G. P. Carver, and D. B. Novotny. 1989, 5p
Pub. in Jnl. of Electronic Materials 18, n1 p53-57 1989.

Keywords: *Thin films, *Crystal growth, Thickness, Quality control, Reprints, *SOS(Semiconductors), CMOS.

The increased emphasis on submicron- geometry CMOS/SOS devices has created a need for high quality silicon-on-sapphire films with thicknesses of the order of 0.1 to 0.2 microns. To date the only viable way of producing high quality SOS films with the thicknesses has been through the application of recrystallization and regrowth techniques. The need for as-grown, high-quality, very-thin SOS films has prompted a study of film quality growth rate for films with thicknesses in the 0.1 to 0.2 micron range as a possible way of producing thin high-quality SOS films. It has been found that film quality increased as the growth rate increased. It was possible to produce films as thin as 0.1 micron with mobilities nearly as high as 1 micron films, if the film growth rate was higher than 4 micron/min.

900,799
PB90-128182 Not available NTIS
National Inst. of Standards and Technology (NEL), Gaithersburg, MD. Semiconductor Electronics Div.
Silicon and GaAs Wire-Bond Cratering Problem.
Final rept.
G. G. Harman. 1989, 6p
Pub. in Proceedings of VLSI and GaAs Chip Packaging Workshop, Santa Clara, CA., September 12-14, 1988, p65-70 1989.

Keywords: *Integrated circuits, *Cratering, Electric wires, Bonding, Gallium arsenides, Silicon, Microelectronics, *Very large scale integration, Fracture mechanics.

The complex synergystic cratering effects of the VLSI era involve not only bonding parameters, but also Au-Al compound-induced stress, silicon nodules in the metallization, plastic package stress, and surface mount stress. The situation is even worse in GaAs.

900,800
PB90-128224 Not available NTIS
National Inst. of Standards and Technology (NML), Boulder, CO. Time and Frequency Div.
Very Low-Noise FET Input Amplifier.
Final rept.
S. R. Jefferts, and F. L. Walls. 1989, 3p
Grant NSF-PHY86-04504
Sponsored by National Science Foundation, Washington, DC.
Pub. in Review of Scientific Instruments 60, n6 p1194-1196 Jun 89.

Keywords: *Field effect transistors, *Low noise amplifiers, Semiconductor devices, Transistor amplifiers, Design, Schematic diagrams, Performance evaluation, Reprints, *Cascade amplifiers.

The design, schematics, and performance of a very low-noise FET cascode input amplifier are described. The amplifier has noise performance of less than 1.2 nV/square root(Hz) and 0.25 fA/square root(Hz) over the 500 Hz to 500 kHz frequency range. With modest changes it could be extended to a wide variety of uses requiring low-noise gain in the 1-Hz to 30 MHz frequency range.

General

900,801
PB89-149058 Not available NTIS
National Bureau of Standards (NML), Gaithersburg, MD. Electricity Div.
Possible Quantum Hall Effect Resistance Standard.
Final rept.
M. E. Cage, R. F. Dziuba, and B. F. Field. 1982, 15p
Sponsored by Naval Research Lab., Washington, DC., and Department of Defense, Washington, DC.
Pub. in Proceedings of NCSL Workshop and Symposium Metrology Management and Technology-A Scientific Approach, Gaithersburg, MD., October 4-7, 1982, pB1-1-B1-15.

Keywords: *Hall effect, *Standards, Semiconductor devices, Metrology, Fundamental constants, Cryogenics, *Quantum Hall effect, *Resistance standards.

The discovery of the quantum Hall effect by K. v. Klitzing, using semiconductor devices that are cryogenically cooled in large applied magnetic fields, has opened up the exciting possibility that this effect could stimulate the discipline of electrical metrology to an extent analogous to that of the Josephson effect. The paper describes the quantum Hall effect, and explains how it is being used in experiments at NBS in an attempt to achieve a new resistance standard accurate to a few parts in 100 million, in which the resistance is defined in terms of fundamental constants of nature.

900,802
PB89-149066 Not available NTIS
National Bureau of Standards (NML), Gaithersburg, MD. Electricity Div.
NBS (National Bureau of Standards) Ohm: Past-Present-Future.
Final rept.
R. F. Dziuba. 1987, 13p
Pub. in Proceedings of Measurement Science Conference, Irvine, CA., January 29-30, 1987, pVI-A(15)-VI-A(27).

Keywords: *Electrical measurement, Electrical resistance, Units of measurement, Hall effect, Measuring instruments, *Absolute Ohm, *Resistance standards.

A brief history is given of the NBS Ohm commencing from the establishment of the NBS in 1901 to the present. It includes a description of the resistance standards and measurement methods used to maintain the NBS Ohm during its 85-year history. Indications of the drift of the NBS Ohm based on absolute-Ohm determinations and quantized-Hall effect measurements are presented. The results of these measurements may lead to an adjustment of the value of the NBS Ohm in 1990.

900,803
PB89-149074 Not available NTIS
National Bureau of Standards (NEL), Gaithersburg, MD. Electrosystems Div.
Thermal-Expansive Growth of Prebreakdown Streamers in Liquids.
Final rept.
C. Fenimore. 1988, 4p
Sponsored by Office of Naval Research, Arlington, VA.
Pub. in Proceedings of Conference Record IEEE (Institute of Electrical and Electronics Engineers) International Symposium on Electrical Insulation, Boston, MA., June 5-8, 1988, p27-30.

Keywords: *Dielectric breakdown, *Electric discharges, *Liquids, Electrical insulation, Mathematical models, Bubbles, Pressure effects.

94

·The growth of electrically conductive, low-density regions has been observed in dielectric breakdown in a variety of liquids. This phenomenon motivates the development of the present theory for coupling thermal effects with fluid mechanical effects in the dynamics of elongated, impulsively-driven bubbles. The model explicitly describes the time-dependent dimensions of a growing ellipsoidal bubble. In previous work, a model for such effects associated with the growth of a bubble about an arc in a liquid was developed. In the case of the prebreakdown streamer, the geometry is not as simple as for an arc, and the evolution of the bubble is found in ellipsoidal coordinates. The effect of pressure is to shorten the time scale of the bubble dynamics.

900,804
PB89-170872 PC A04/MF A01
National Inst. of Standards and Technology (NEL), Gaithersburg, MD. Center for Electronics and Electrical Engineering.
High-Current Measurement Techniques. Part II. 100-kA Source Characteristics and Preliminary Shunt and Rogowski Coil Evaluations.
J. D. Ramboz. Mar 89, 52p NISTIR-89/4040
See also PB85-100444. Sponsored by Sandia National Labs., Albuquerque, NM.

Keywords: *Electrical measurement, Electric current, Calibrating, Bypasses, Circuits, Alternating currents, Electric coils, Rogowski coils.

The characterization of a 100-kA current source is discussed. The source is intended for use in the calibration of high-current sensors such as shunts and Rogowski coils commonly employed in resistance welders. The output current from the source is derived from SCR-gated signals in the form of bursts of 'chopped' 60-Hz sinusoidal waveforms. These waveforms and their spectral content were investigated. The near-field magnetic field strength was mapped. Initial calibrations were performed on a 30-kA, 10 microohms shunt. Preliminary results indicate a temperature coefficient of about 130 ppm/deg which is thought to be related to a thermally induced strain. Several Rogowski coil type current sensors were evaluated and calibrated. Each of the coils measured had outputs which were sensitive to the rotational position about the current carrying conductor. The calibration philosophy and approach is discussed and estimates of measurement uncertainty are given. Suggested improvements for the measurement process are offered. Planned efforts are outlined.

900,805
PB89-171656 Not available NTIS
National Bureau of Standards (NEL), Gaithersburg, MD. Electrosystems Div.
Power Quality Site Surveys: Facts, Fiction, and Fallacies.
Final rept.
F. D. Martzloff, and T. M. Gruzs. 1988, 14p
See also PB87-201679.
Pub. in IEEE (Institute of Electrical and Electronics Engineers) Transactions on Industry Applications 24, n6 p1005-1018 Nov/Dec 88.

Keywords: *Site surveys, *Power supplies, Quality assurance, Reliability, Reprints, *Electronic equipment, Power losses.

The quality of the power supplied to sensitive electronic equipment is an important issue. Monitoring disturbances of the power supply has been the objective of various site surveys, but results often appear to be instrument-dependent or site-dependent, making comparisons difficult. After a review of the origins and types of disturbances, the types of monitoring instruments are described. A summary of nine published surveys reported in the last 20 years is presented, and a close examination of underlying assumptions allows meaningful comparisons which can reconcile some of the differences. Finally, the paper makes an appeal for improved definitions and applications in the use of monitoring instruments.

900,806
PB89-173447 Not available NTIS
National Bureau of Standards (NEL), Gaithersburg, MD. Electrosystems Div.
Estimates of Confidence Intervals for Divider Distorted Waveforms.
Final rept.
R. H. McKnight, and J. Lagnese. 1986, 4p
Pub. in Proceedings of International Symposium on High Voltage Engineering (5th), Braunschweig, Federal Republic of Germany, August 24-28, 1987, v3 p1-4 1988.

Keywords: *Confidence limits, *High voltage, *Frequencies, Statistical analysis, Pulsation, Measurement, Stochastic processes, Errors.

The paper describes a method for computing confidence intervals for a high voltage impulse distorted by a divider system. The technique is based on a recent algorithm designed to calculate confidence intervals for solutions to ill-posed problems subject to inequality constraints. Applications of the method to measurements made with a resistive divider illustrate its value for obtaining useful stochastic error bounds for high voltage impulse restoration.

900,807
PB89-176192 Not available NTIS
National Inst. of Standards and Technology (NEL), Boulder, CO. Electromagnetic Fields Div.
NIST (National Institute of Standards and Technology) Automated Coaxial Microwave Power Standard.
Final rept.
F. R. Clague. 1989, 14p
Pub. in Proceedings of Measurement Science Conference, Anaheim, CA., January 26-27, 1989, p1C-1-1C-14.

Keywords: *Microwaves, *Frequency standards, *Calorimeters, *Bolometers, Measuring instruments, Performance standards, Automatic control, Electrical measurement, Adaptive systems, Electronic control.

The national microwave power standards consist of two parts: a microcalorimeter and a bolometer mount used as the transfer standard. In the past, operation of the microcalorimeter has been slow and complicated, and required skilled personnel. The paper details the automation of the 0.1 to 18 GHz coaxial microcalorimeter and the design of a new coaxial transfer standard. Together, these have reduced measurement time by a factor of 10. A highly skilled operator is no longer required and largely unattended operation 24 hours a day is possible. The basic theory of operation of both devices, design considerations, some error evaluation problems, and performance results are included.

900,808
PB89-184097 .· r.
(Order as PB89-184089, PC A04)
National Inst. of Standards and Technology, Gaithersburg, MD.
New Internationally Adopted Reference Standards of Voltage and Resistance.
Bi-monthly rept.
B. N. Taylor. 1989, 9p
Included in Jnl. of Research of the National Institute of Standards and Technology, v94 n2 p95-103 Mar-Apr 89.

Keywords: *Electrical measurement, *Standards, Voltage measurement, Electrical resistance, Voltmeters, Ohmmeters, Hall effect, Units of measurement, Calibrating, Consultative Committee on Electricity(CCE), International Committee of Weights and Measures(CIPM), International System of Units(SI).

The report provides the background for and summarizes the main results of the 18th mee ng of the Consultative Committee on Electricity (CCE) of the International Committee of Weights and Measures (CIPM) held in September 1988. The principal recommendations originating from the meeting, which were subsequently adopted by the CIPM, establish new international reference standards of voltage and resistance based on the Josephson effect and the quantum Hall effect, respectively. The new standards, which are to come into effect starting January 1, 1990, will result in improved uniformity of electrical measurements worldwide and their consistency with the International System of Units or SI. The CCE also recommended a particular method, affirmed by the CIPM, of reporting calibration results obtained with the new reference standards that is to be used by all national standards laboratories.

900,809
PB89-186423 Not available NTIS
National Bureau of Standards (NEL), Gaithersburg, MD. Electrosystems Div.

International Comparison of HV Impulse Measuring Systems.
Final rept.
T. R. McComb, R. C. Hughes, H. A. Lightfoot, K. Schon, R. Schulte, R. McKnight, and Y. X. Zhang. 1989, 10p
Pub. in IEEE (Institute of Electrical and Electronics Engineers) Transactions on Power Delivery 4, n2 p906-915 Apr 89.

Keywords: *High voltage, Comparison, Measurement, Standards, Reprints, *Impulses.

Present standards for qualifying high voltage (HV) impulse measuring systems by unit-step-response parameters are complex and difficult to apply and some systems which have response parameters within the limits of the standards have unacceptable errors. The paper takes the first step in providing a simplified method based on simultaneous measurements of an HV impulse by a reference system and the system under test. Comparative measurements have been made in four National Laboratories and the relative differences are reported. The results are discussed and the further work which is required is outlined.

900,810
PB89-189211 PC A06/MF A01
National Inst. of Standards and Technology (NEL), Boulder, CO. Electromagnetic Fields Div.
Bibliography of the NIST (National Institute of Standards and Technology) Electromagnetic Fields Division Publications.
A. M. Reidy, and K. A. Gibson. Sep 88, 112p NISTIR-88/3900
Supersedes PB86-191947. See also PB89-147847.

Keywords: *Bibliographies, *Electromagnetic fields, Antennas, Dielectrics, Electromagnetic interference, Electrical measurement, Microwaves, Metrology, Electromagnetic noise, Remote sensing, Wave forms, Time domain.

The bibliography lists the publications by the staff of the Electromagnetic Fields Division of the National Institute of Standards and Technology for the period January 1970 through August 1988. Selected earlier publications from the Division's predecessor organizations are included.

900,811
PB89-189245 PC A05/MF A01
National Inst. of Standards and Technology (NEL), Gaithersburg, MD. Center for Electronics and Electrical Engineering.
Emerging Technologies in Electronics and Their Measurement Needs.
Mar 89, 79p NISTIR-89/4057

Keywords: *Electric security, *Electrical measurement, *Metrology, Semiconductor devices, Microwaves, Magnetic measurement, Superconductors, Fiber optics, Optical communication, Television systems, Bioelectricity, *Emerging technologies, Smart systems, Light waves.

The report identifies emerging technologies in electronics that the Center for Electronics and Electrical Engineering (CEEE) believes will require increased measurement support from CEEE in coming years. The emerging technologies described here are new to the marketplace or are experiencing major technological advances. The document is designed to stimulate feedback that CEEE needs to refine its plans for developing measurement capability to support emerging electronic technologies that are important to the national interest.

900,812
PB89-189302 PC A03/MF A01
National Inst. of Standards and Technology (NEL), Gaithersburg, MD. Center for Electronics and Electrical Engineering.
Center for Electronics and Electrical Engineering Technical Publication Announcements: Covering Center Programs, July/September 1988, with 1989 CEEE Events Calendar.
E. J. Walters. Mar 89, 24p NISTIR-89/4067
See also PB89-136311.

Keywords: *Electrical engineering, *Bibliographies, Semiconductors(Materials), Metrology, Integrated circuits, Signal processing, Transmission, Waveforms, Antennas, Microwaves, Fiber optics, Superconductors, Electric power, Electromagnetic interference,

ELECTROTECHNOLOGY

General

Electronics, Dimensional measurement, Cryoelectronics.

The eighteenth issue of a quarterly publication providing information on the technical work of the National Institute of Standards and Technology (formerly the National Bureau of Standards) Center for Electronics and Electrical Engineering. The issue of the Center for Electronics and Electrical Engineering Technical Publication Announcements covers the third quarter of calendar year 1988.

900,813
PB89-193270   **PC A03/MF A01**
National Inst. of Standards and Technology (NEL), Gaithersburg, MD. Center for Electronics and Electrical Engineering.
Center for Electronics and Electrical Engineering Technical Progress Bulletin Covering Center Programs, October to December 1988, with 1989 CEEE Events Calendar.
E. J. Walters. May 89, 44p NISTIR-89/4076
See also PB89-168033.

Keywords: *Electrical engineering, *Semiconductor devices, *Signal processing, *Systems, Metrology, Silicon, Photodetectors, Interfaces, Lasers, Microwaves, Millimeter waves, Fiber optics, Electrooptics, Superconductors, Electromagnetic interference, Signal acquisition, Cryoelectronics.

The report is the twenty-fifth issue of a quarterly publication providing information on the technical work of the National Institute of Standards and Technology (formerly The National Bureau of Standards) Center for Electronics and Elecrical Engineering. The issue of the CEEE Technical Progress Bulletin covers the fourth quarter of calendar year 1988. Abstracts are provided by technical area for both published papers and papers approved by NIST for publication.

900,814
PB89-193916   **PC A07/MF A01**
National Inst. of Standards and Technology (NEL), Boulder, CO. Electromagnetic Fields Div.
Performance Evaluation of Radiofrequency, Microwave, and Millimeter Wave Power Meters.
Technical note.
E. M. Livingston, and R. T. Adair. Dec 88, 149p NIST/TN-1310
Also available from Supt. of Docs. as SN003-003-02931-9.

Keywords: *Power measurement, *Electrical measurement, *Power meters, Thermistors, Wattmeters, Measuring instruments, Radio waves, Performance evaluation, Microwaves, Millimeter waves, Electromagnetic radiation, Calibrating, Temperature.

Measurement techniques are described for the evaluation of the electrical performance of commercially available radiofrequency (rf), microwave (mw), and millimeter wave (mmw) power meters which use bolometric power sensors and typically operate from 10 MHz to 26.5 GHz for an average power range of 10 microW to 10 mW with appropriate attenuation for higher power ranges. Techniques are described for analysis of: ranges of frequency and power, operating temperature, stability, response time, calibration factor, extended power measurement, overload protection, and characteristics of the internal power reference source. Some automated methods are discussed. Block diagrams of test setups are presented. Sources of uncertainty in the bolometric method are analyzed.

900,815
PB89-201552   Not available NTIS
National Bureau of Standards (NEL), Gaithersburg, MD. Electrosystems Div.
Accurate RF Voltage Measurements Using a Sampling Voltage Tracker.
Final rept.
T. M. Souders, and P. S. Hetrick. 1989, 6p
See also PB88-239579.
Pub. in IEEE (Institute of Electrical and Electronics Engineers) Transactions on Instrumentation and Measurement 38, n2 p451-456 Apr 89.

Keywords: *Electrical measurement, *Electric potential, Electrical properties, Voltmeters, Calibrating, Radio frequencies, Step response, Ramp response, Reprints.

The RF voltage measurement capability of an equivalent-time sampling system has been investigated over the frequency range of 1-100 MHz. The system is easily calibrated from step response measurements, independent of thermal transfer standards. Comparison measurements made with NBS-calibrated thermal converters show agreement generally within the stated uncertainties presently provided by NBS for such calibrations. The system offers several advantages over conventional thermal transfer techniques; ac/dc transfers are not required, loading and transmission line problems are reduced, and direct measurement of voltages from 2 V to as low as 10 mV are possible. In addition, other waveform characteristics are readily obtained, e.g., average and peak values, harmonic distortion, etc.

900,816
PB89-201560   Not available NTIS
National Bureau of Standards (NML), Gaithersburg, MD. Electricity Div.
AC-DC Difference Calibrations at NBS (National Bureau of Standards).
Final rept.
J. R. Kinard. 1986, 6p
Pub. in Proceedings of Measurement Science Conference, Irvine, CA., January 23-24, 1986, p3-8.

Keywords: *Calibrating, *AC to DC converters, Electric potential, Electric current, Electric converters, Rectifiers, Standards, National Institute of Standards and Technology.

The NBS calibration service for thermal voltage and current converters relies on a group of primary multijunction thermal converters and sets of reference and working standards for extending their ranges and frequencies. The converter sets which constitute the NBS standards--primary, reference and working--as well as the build-up and bootstrap techniques used in their characterization over the full ranges of voltage, current, and frequency are described briefly. Routine NBS uncertainties for ac-dc difference calibrations are given as well as a summary of current activities and plans.

900,817
PB89-214761   **PC A05/MF A01**
National Inst. of Standards and Technology (NML), Gaithersburg, MD. Electricity Div.
Guidelines for Implementing the New Representations of the Volt and Ohm Effective January 1, 1990.
Technical note (Final).
N. B. Belecki, R. F. Dziuba, B. F. Field, and B. N. Taylor. Jun 89, 76p NIST/TN-1263
Also available from Supt. of Docs as SN003-003-02941-6.

Keywords: *Electrical measurement, *Electrical resistance, *Electric potential, Standards, Voltage measuring instruments, Impedance, Electrical measuring instruments, *Calibration standards.

The document provides general guidelines and detailed instructions on how to bring laboratory reference standards of voltage and resistance and related instrumentation into conformity with newly established and internationally adopted representations of the volt and ohm. Based on the Josephson and quantum Hall effects, respectively, the new representations are to come into effect worldwide starting on January 1, 1990. Their implementation in the U.S. will result in increases in the values of the national volt and ohm representations maintained at the National Institute of Standards and Technology (NIST, formerly the National Bureau of Standards or NBS) of 9.264 parts per million (ppm) and 1.69 ppm, respectively. The resulting increase in the value of the U.S. representation of the ampere will be 7.57 ppm and in the U.S. electrical representation of the watt, 16.84 ppm. Also discussed are the effects on electrical standards of the January 1, 1990, replacement of the International Practical Temperature Scale of 1968 by the International Temperature Scale of 1990, and of the January 1, 1990, approximate 0.14 ppm decrease in the U.S. representation of the farad.

900,818
PB89-222616   **PC A14/MF A01**
National Inst. of Standards and Technology (NML), Gaithersburg, MD. Center for Basic Standards.

NIST (National Institute of Standards and Technology) Measurement Services: AC-DC Difference Calibrations.
Final rept.
J. R. Kinard, J. R. Hastings, T. E. Lipe, and C. B. Childers. May 89, 311p NIST/SP-250/27
Also available from Supt. of Docs. as SN003-003-02950-5. See also PB89-201560. Library of Congress catalog card no. 89-600736.

Keywords: *Electrical measurement, *Test facilities, Electric current, Direct current, Alternating current, Electric potential, Mathematical models, *Calibration standards.

The publication collects and summarizes the specialized information needed to operate the ac-dc difference laboratory and calibration service at the National Institute of Standards and Technology (NIST) in Gaithersburg. It also serves as a convenient reference source for the users of this calibration service and other interested people by documenting the service and its underlying background in considerable detail. It contains the following: an annotated table of contents, a topical index, and a glossary of common ac-dc acronyms; an overview of the service; selected published papers; instructions for the operation of the comparator systems; a schedule for the recalibration and periodic checks of the NIST thermal converters; and a sample report of calibration.

900,819
PB89-230387   Not available NTIS
National Bureau of Standards (NML), Gaithersburg, MD. Electricity Div.
Determination of the Time-Dependence of ohm NBS (National Bureau of Standards) Using the Quantized Hall Resistance.
Final rept.
M. E. Cage, R. F. Dziuba, C. T. Van Degrift, and D. Yu. 1989, 7p
Pub. in IEEE (Institute of Electrical and Electronics Engineers) Transactions on Instrumentation and Measurement 38, n2, p263-269, Apr 89.

Keywords: *Electrical resistance, *Standards, Electrical measurement, Precision, Reprints, *Resistance standards, *Quantum Hall effect, *Ohm, Temperature dependence.

The Quantum Hall effect is being used to monitor the U.S. legal representation of the ohm, or as-maintained ohm, ohm(NBS). Measurements have been made on a regular basis since August 1983. Individual transfers between the quantized Hall resistance, R(H) and the five 1-ohm resistors which comprise ohm(NBS) can now be made with a total one standard deviation (1 sigma) uncertainty of + or - 0.014 ppm. This uncertainty is the root-sum-square of 32 individual components. The time-dependent expression for T(H) in terms of ohm(NBS) is: R(H) = 25 812.8 (1 + (1.842 + or - 0.012) x 10 to the -6th power + (0.0529 + or - 0.0040) (t - 0.7785) x 10 to the -6th power/year) ohm(NBS), where t is measured in years from January 1, 1987. The value of ohm(NBS) is, therefore, decreasing at the rate of (0.0529 + or - 0.0040) ppm/year.

900,820
PB89-230403   Not available NTIS
National Bureau of Standards (NML), Gaithersburg, MD. Electricity Div.
Josephson Array Voltage Calibration System: Operational Use and Verification.
Final rept.
R. L. Steiner, and B. F. Field. 1989, 6p
Pub. in IEEE (Institute of Electrical and Electronics Engineers) Transactions on Instrumentation and Measurement 38, n2, p296-301, Apr 89.

Keywords: *Electric potential, *Standards, Josephson junctions, Arrays, Reprints, *Voltage standards, *Calibration.

A new Josephson array system now maintains the U.S. Legal Volt. This system is almost fully automated, operates with a typical precision of 0.009 microV, and readily allows U.S. Legal Volt measurements weekly, or more frequently if desired. The system was compared to the previous volt maintenance system, and agreement was made to within 0.03 ppm. This verification is limited by uncertainties in the resistive divider instruments of the previous system.

900,821
PB89-230429   Not available NTIS

National Bureau of Standards (NML), Gaithersburg, MD: Electricity Div.
Measurement of the NBS (National Bureau of Standards) Electrical Watt in SI Units.
Final rept.
P. T. Olsen, R. E. Elmquist, W. D. Phillips, E. R. Williams, G. R. Jones, and V. E. Bower. 1989, 7p
Pub. in IEEE (Institute of Electrical and Electronics Engineers) Transactions on Instrumentation and Measurement 38, n2, p238-244, Apr 89.

Keywords: *Electric power, *Standards, Reprints, *Watt, Ampere, Josephson effect.

The authors have measured the NBS electric watt in SI units to be: W(NBS)/W = K(w) = 1 - (16.69 + or - 1.33) ppm. The uncertainty of 1.33 ppm has the significance of a standard deviation and includes their best estimate of random and known or suspected systematic uncertainties. The mean time of the measurement is May 15, 1988. Combined with the recent measurement of the NBS ohm in SI units: ohm(NBS)/ohm = K(ohm) = 1 - (1.593 + or - 0.022) ppm, this leads to a Josephson frequency/voltage quotient of E(J) = ((0)(1 + (7.94 + or - 0.67) ppm) where E(0) = 483 594 GHz/V.

900,822
PB89-231039   Not available NTIS
National Inst. of Standards and Technology (NEL), Gaithersburg, MD. Electrosystems Div.
Production and Stability of S2F10 in SF6 Corona Discharges.
Final rept.
I. Sauers, M. C. Siddagangappa, G. Harman, R. J. Van Brunt, and J. T. Herron. 1989, 4p
Sponsored by Department of Energy, Washington, DC. Office of Energy Storage and Distribution.
Pub. in Proceedings of International Symposium on High Voltage Engineering (6th), New Orleans, LA., August 28-September 1, 1989, p1-4.

Keywords: *Sulfur hexafluoride, Decomposition reactions, Gas chromatography, Mass spectroscopy, Water vapor, Production, Stability, *Corona discharges.

The authors report the yield of S2F10 produced in corona discharges in SF6 and the dependence of the S2F10 yield on various parameters. The data were obtained from two experimental systems, both employing point-to-plane geometry, a small corona cell (200-ml in volume) with a gas chromatograph-thermal conductivity analyzer (GC-TCD) at ORNL and a large (3.7-1) corona cell with a gas chromatograph-mass spectrometer (GC-MS) analyzer at NIST (formerly NBS). The GC-MS technique was found to be quite sensitive to S2F10 when the mass analyzer was tuned to mass 86 (SOF2(1+)).

900,823
PB89-231153   Not available NTIS
National Bureau of Standards (NEL), Boulder, CO. Electromagnetic Technology Div.
Faraday Effect Sensors: The State of the Art.
Final rept.
G. W. Day, and A. H. Rose. 1988, 13p
Sponsored by Department of Energy, Washington, DC., Electric Power Research Inst., Palo Alto, CA., Empire State Electric Energy Research Corp., New York, and Department of Defense, Washington, DC.
Pub. in Proceedings of SPIE (Society of Photo-Optical Instrumentation Engineers), Fiber Optic and Laser Sensors VI, v985, p138-150 1988.

Keywords: *Faraday effect, Electric current, Magnetic fields, Electrical measurement, Magnetic measurement, Optical measurement, Sensitivity, Stability, Detectors, Reviews, Reprints, State of the art, Sensors.

The Faraday effect is becoming widely used as an optical method of measuring electric current or magnetic field. It is particularly advantageous where the measurements must be made at high voltage or in the presence of electromagnetic interference, and where speed or stability are considerations. The paper reviews the development of the technology over the last twenty years, with an emphasis on the basic principles, design considerations, and performance capabilities of sensors that represent the latest achievements. Faraday effect current sensors are now used routinely in the measurement of large current pulses and are starting to become available for ac current measurements in the power industry. Recent developments include their extension to the measurement of currents in the milliampere range and substantial reductions in size.

Similar devices, in slightly different configurations, can be used for magnetic field measurements. Further improvements, based on new fiber types and new materials, are projected.

900,824
PB90-116195   PC A03/MF A01
National Bureau of Standards (NEL), Gaithersburg, MD. Center for Electronics and Electrical Engineering.
Center for Electronics and Electrical Engineering Technical Publication Announcements Covering Center Programs, October to December 1986, with 1987 CEEE Events Calendar.
E. J. Walters. Aug 87, 17p NBSIR-87/3620
See also PB87-226890.

Keywords: *Electronics, *Electrical engineering, Abstracts, Metrology, Signal processing, Electromagnetic interference, Semiconductor devices, Catalogs(Publications), *Center for Electronics and Electrical Engineering.

The report is a quarterly publication providing information on the technical work of the National Bureau of Standards Center for Electronics and Electrical Engineering. The issue of the CEEE Technical Publication Announcements covers the fourth quarter of calendar year 1986. The issue contains citations and abstracts for Center papers published in the quarter. Entries are arranged by technical topic as identified in the table of contents and alphabetically by first author within each topic. Following each abstract is the name and telephone number of the individual to contact for more information on the topic (usually the first author).

900,825
PB90-117680   Not available NTIS
National Inst. of Standards and Technology (NEL), Boulder, CO. Electromagnetic Fields Div.
Calibrating Network Analyzers with Imperfect Test Ports.
Final rept.
J. R. Juroshek, C. A. Hoer, and R. F. Kaiser. 1989, 4p
Pub. in IEEE (Institute of Electrical and Electronics Engineers) Transactions on Instrumentation and Measurement 38, n4 p898-901 Aug 89.

Keywords: *Calibrating, *Network analyzers, Transmission lines, Test equipment, Frequency analyzers, Electric analyzers, Measuring instruments, Test sets, Reprints, *Test ports.

The test ports on automatic network analyzers are generally built with an impedance that matches the impedance of the calibration standards. The paper gives experimental evidence that substantial impedance discontinuities can be tolerated at the test port interface if proper calibration procedures are observed. The 50-Omega test port on one of the six-ports in a dual six-port network analyzer was replaced with a 75-Omega test port. The test port was then calibrated to look like a 50-Omega test port. Measurements on various devices showed that indeed it was possible to make a 75-Omega test port indistinguishable from a 50-Omega test port.

900,826
PB90-117953   Not available NTIS
National Inst. of Standards and Technology (NEL), Boulder, CO. Electromagnetic Fields Div.
Hybrid Representation of the Green's Function in an Overmoded Rectangular Cavity.
Final rept.
D. I. Wu, and D. C. Chang. 1988, 9p
Pub. in IEEE (Institute of Electrical and Electronics Engineers) Transactions on Microwave Theory and Techniques 36, n9 p1334-1342 Sep 88.

Keywords: *Electromagnetic waves, *Greens function, Cavities, Computation, Reprints, Point sources, Rectangular configuration, Numerical solution, Modes.

A hybrid ray-mode representation of the Green's function in a rectangular cavity is developed using the finite Poisson summation formula. In order to obtain a numerically efficient scheme for computing the field generated by a point source in a large rectangular cavity, the conventional modal representation of the Green's function is modified in such a way that all the modes near resonance are retained while the truncated remainder of the mode series is expressed in terms of a weighted contribution of rays. For an electrically large cavity, the contribution of rays from distant images becomes small; therefore the ray sum can be approximated by one or two dominant terms without a loss of

numerical accuracy. To illustrate the accuracy and the computational simplification of this ray-mode representation, numerical examples are included with the conventional mode series (summed at the expense of long computation time) serving as a reference.

900,827
PB90-128190   Not available NTIS
National Inst. of Standards and Technology (NEL), Boulder, CO. Electromagnetic Fields Div.
Electromagnetic Detection of Long Conductors in Tunnels.
Final rept.
D. A. Hill. 1988, 20p
Pub. in Proceedings of Technical Symposium on Tunnel Detection (3rd), Golden, CO., January 12-15, 1988, p518-537.

Keywords: *Tunnel detection, *Electromagnetic induction, Magnetic dipoles, Magnetic fields, Transmission lines, Electric dipoles.

Formulations for the excitation of currents on an infinitely long conductor by electric or magnetic dipoles of arbitrary orientation are presented. The conductor can be either insulated or bare to model ungrounded or grounded conductors. Specific calculations are presented for a vertical magnetic dipole source because this source produces the appropriate horizontal polarization and could be used in a borehole-to-borehole configuration. Numerical results for the induced current and secondary magnetic field indicate that long conductors produce a strong anomaly over a broad frequency range. The secondary magnetic field decays slowly in the direction of the conductor and eventually becomes larger than the dipole source field.

900,828
PB90-128315   Not available NTIS
National Inst. of Standards and Technology (NEL), Gaithersburg, MD. Electrosystems Div.
Effect of Pressure on the Development of Pre-breakdown Streamers.
Final rept.
P. J. McKenny, E. O. Forster, E. F. Kelley, and R. E. Hebner. 1988, 6p
Pub. in Annual Report of Conference on Electrical Insulation and Dielectric Phenomena, Ottawa, Canada, October 16-20, 1988, p263-268.

Keywords: *Liquids, *Breakdown(Electronic threshold), *Prebreakdown, Pressure effects.

The initiation of streamers in a liquid under the application of negative impulse voltages applied to a needle-sphere gap is investigated. A square pulse is applied so that the prebreakdown streamer will not grow to breakdown. With the application of pressure the initial streamer is observed to collapse and disappear while the voltage remains on the tip. When the voltage is chopped, a new streamer appears which resembles the structure of the new streamers. The branches of the new streamer do not strictly follow the previous branches of the cathode streamer which injected the charge in the liquid. Using a simple model, approximately 11 nC is estimated to be injected into the liquid producing a charge density of 49 microC/cc.

900,829
PB90-128745   Not available NTIS
National Inst. of Standards and Technology (NEL), Gaithersburg, MD. Electrosystems Div.
Method for Measuring the Stochastic Properties of Corona and Partial-Discharge Pulses.
Final rept.
R. J. Van Brunt, and S. V. Kulkarni. 1989, 12p
Sponsored by Department of Energy, Washington, DC. Div. of Electric Energy Systems.
Pub. in Review of Scientific Instruments 60, n9 p3012-3023 Sep 89.

Keywords: *Electric corona, *Gas discharges, Electrical measurement, Stochastic processes, Nitrogen, Oxygen, Mixtures, Pulse height analyzers, Markov processes, Reprints.

A new method is described for measuring the stochastic behavior of corona and partial-discharge pulses which uses a pulse selection and sorting circuit in conjunction with a computer-controlled multichannel analyzer to directly measure various conditional and unconditional pulse-height and pulse-time-separation distributions. From these measured distributions it is possible to determine the degree of correlation between successive discharge pulses. Examples are

ELECTROTECHNOLOGY

General

given of results obtained from measurements on negative, point-to-plane (Trichel-type) corona pulses in a N2/O2 gas mixture which clearly demonstrate that the phenomenon is inherently stochastic in the sense that development of a discharge pulse is significantly affected by the amplitude of and time separation from the preceding pulse. It is found, for example, that corona discharge pulse amplitude and time separation from an earlier pulse are not independent random variables. Discussions are given about the limitations of the method, sources of error, and data analysis procedures required to determine self-consistency of the various measured distributions.

900,830
PB90-128794   Not available NTIS
National Inst. of Standards and Technology (NEL), Gaithersburg, MD. Electrosystems Div.
Method for Fitting and Smoothing Digital Data.
Final rept.
Y. X. Zhang, R. H. McKnight, and C. Fenimore. 1989, 4p
Pub. in Proceedings of International Symposium on High Voltage Engineering (6th), New Orleans, LA., August 28-September 1, 1989, p1-4.

Keywords: *Lightning, *Digital systems, *Data smoothing, *Splines, Cubic equations, Data processing, Data reduction, Numerical analysis, Electric discharges, Static electricity, Step response.

Measurements of the full-lightning or chopped-lightning waveforms or of the step response are confounded by noise, radiated electro-magnetic interference, and high frequency oscillations. Before the advent of digitizers, these effects were minimized by filtering in hardware and fitting photographic data by hand. Unfortunately, unacceptably large errors may be introduced by the process. A method has been developed and evaluated to fit and smooth digital data using cubic splines. These are piecewise cubic polynomials. Their advantage is that the spline and its first two derivatives are continuous, including at the knots (where two polynomials are joined). It is important because the first derivative is used in calculating the unit-step response-time. The location of the knots may be freely chosen. In the case of oscillatory data the choice is critical. A selection criterion is given and evaluated.

900,831
PB90-130699   PC A03/MF A01
National Inst. of Standards and Technology (NEL), Gaithersburg, MD.
Report on Interactions between the National Institute of Standards and Technology and the Institute of Electrical and Electronic Engineers.
G. K. Ehrlich. Feb 89, 43p NISTIR-89/4037

Keywords: Technology transfer, Standards, Computers, Electrical engineering, Reliability, Ultrasonic frequencies, *National Institute of Standards and Technology, *Institute of Electrical and Electronic Engineers, Robotics.

The report highlights examples of interactions between the National Institute of Standards and Technology (NIST) and the Institute of Electrical and Electronic Engineers (IEEE) since October 1, 1987. It is meant to be representative, not all-inclusive. The interactions are organized by discipline in the following categories: IEEE honors and awards, editors, committee memberships and contribution to standards, conferences and workshops, publications, and other interactions. The report illustrates many activities which are designed to disseminate NIST's most recent technical advances and to learn of the technical challenges facing engineers in industry.

900,832
PB90-130907   PC A03/MF A01
National Inst. of Standards and Technology (NEL), Boulder, CO. Electromagnetic Fields Div.
Radiometer Equation and Analysis of Systematic Errors for the NIST (National Institute of Standards and Technology) Automated Radiometers.
Technical note.
W. C. Daywitt. Mar 89, 27p NIST/TN-1327
Also available from Supt. of Docs. as SN003-003-02965-3.

Keywords: *Radiometers, *Microwaves, *Millimeter waves, Noise analyzers, Radiation measuring instruments, Waveguides, Coaxial cables, Calibrating, Temperature measurement, Error analysis.

Equations used in the NIST coaxial and waveguide automated radiometers to estimate the noise tempera-

ture and associated errors of a single-port noise source are derived in the report. The equations form the foundation upon which the microwave and millimeter wave noise calibration and special test services are performed. Results from the 1-12 GHz coaxial radiometer are presented.

ENERGY

Batteries & Components

900,833
PB89-173421   Not available NTIS
National Bureau of Standards (NEL), Gaithersburg, MD. Electrosystems Div.
Measurement of Partial Discharges in Hexane Under DC Voltage.
Final rept.
E. F. Kelley, M. Nehmadi, R. E. Hebber, M. O. Pace, A. L. Wintenberg, T. V. Blalock, and J. V. Foust. 1986, 9p
Sponsored by Department of Energy, Washington, DC.
Pub. in Proceedings of Annual Report Conference on Electrical Insulation and Dielectric Phenomena, Ottawa, Canada, October 16-20, 1986, p394-402.

Keywords: *Electrical phenomena, *Measurement, *Electrodes, *High speed photography, Hexanes, Timing devices, Direct current, Liquids, Dielectric properties, Electronic photography.

The very first stages of the low density region (LDR) at an electrode in hexane are recorded by high-magnification, high-speed photography and low-noise electronics to determine the relationship between the inception and growth of the LDR and the current. Observations of the initiation phase of the LDR are made by an image-preserving optical delay which permits photography of unpredictable events after their occurrence. The current is monitored by a special low-noise preamplifier which also provides timing signals to the photographic system. Here with DC, the current producing the LDR is in the form of a growing pulse train which is in contrast to the linear growth observed in the LDR.

Electric Power Transmission

900,834
PB89-173462   Not available NTIS
National Bureau of Standards (NEL), Gaithersburg, MD. Electrosystems Div.
Characterizing Electrical Parameters Near AC and DC Power Lines.
Final rept.
M. Misakian. 1988, 12p
Sponsored by Department of Energy, Washington, DC.
Pub. in Proceedings of United States-Japan Seminar on Electromagnetic Interference in Highly Advanced Social Systems, Honolulu, HI., August 1-4, 1986, p6-32-6-43.

Keywords: *High voltage, *Power transmission lines, *Electric fields, *Magnetic fields, Measuring instruments, Ion currents, Ion density(Concentration), Charge density.

In the mid-1970s there were no standards which provided guidance for the measurement of fields near power lines or for the calibration of instrumentation used for such measurements. Today an ANSI/IEEE standard exists for measurements of electric and magnetic fields near AC power lines. An IEC standard exists for measuring power-frequency electric fields. In addition, an IEEE standard is currently being prepared for the measurement of DC electric fields and ion related parameters near DC power lines. The paper briefly surveys the instrumentation currently in use for characterizing fields near AC power lines and the electric field, and ion current density and monopolar charge density near DC power lines.

900,835
PB89-176630   Not available NTIS

National Bureau of Standards (NEL), Gaithersburg, MD. Electrosystems Div.
Power Frequency Electric and Magnetic Field Measurements: Recent History and Measurement Standards.
Final rept.
M. Misakian. 1986, 16p
Sponsored by Department of Energy, Washington, DC.
Pub. in Proceedings of Workshop and Symposium of the National Conference of Standards Laboratories, Gaithersburg, MD., October 6-9, 1986, p5-1-5-16.

Keywords: *Electric fields, *Magnetic fields, *Standards, *Power transmission lines, Calibrating, Measurement, Wirelines, Electric wire, Substations, Electric switches.

During the early 1970s reports appeared in the literature which raised questions regarding possible biological effects from exposure to power frequency electric and magnetic fields in the vicinity of high voltage transmission lines and in substations. Today a U.S. (IEEE) standard exists for measurement of AC power line fields and an NBS technical note is available which describes the measurement of electrical parameters in biological exposure systems. The paper focuses on selected results of NBS studies which have been incorporated into the U.S. standard for measurement of power frequency electric and magnetic fields or included in the NBS technical note for measuring electrical parameters in bioeffects exposure systems.

Fuel Conversion Processes

900,836
PB90-110065   PC A07/MF A01
National Bureau of Standards (IMSE), Gaithersburg, MD. Ceramics Div.
Effect of Slag Penetration on the Mechanical Properties of Refractories: Final Report.
S. M. Wiederhorn, and R. F. Krause. Jun 87, 145p
NBSIR-87/3584
Contract DE-AI05-83OR21349
Sponsored by Oak Ridge National Lab, TN.

Keywords: *Refractories, *Slags, *Coal gasification, *Corrosion, Ceramics, Mechanical properties.

The problem of selecting refractory insulation is the result of refractory lining exposure to the gasifier environment and to slag present in that environment. The lifetime of the refractory depends on the temperature of operation and the composition of slag at the hot face of the refractory. Higher temperatures (1500 C to 1700 C) and more corrosive slags (high alkali; high Fe) reduce the lifetime of refractories. Since slag is a universal solvent for refractories, all refractories eventually dissolve in coal slag. However, by appropriate materials design, the resistance of the refractory to dissolution (and hence lifetime) can be increased.

Fuels

900,837
PB89-148100   PC A03/MF A01
National Bureau of Standards (NEL), Boulder, CO. Chemical Engineering Science Div.
Development of Standard Measurement Techniques and Standard Reference Materials for Heat Capacity and Heat of Vaporization of Jet Fuels.
Final rept. Aug 87-May 88.
J. E. Callanan. Aug 88, 33p NBSIR-88/3093
Contract F33615-85-C-2508
Sponsored by Air Force Wright Aeronautical Labs., Wright-Patterson AFB, OH. Aero-Propulsion Lab.

Keywords: *Jet engine fuels, *Specific heat, *Heat of vaporization, Measurement, Standards, Thermogravimetry, Temperature, Heat measurement, Calorimeters, JP-4, JP-7, JP-8x.

Procedures developed in the NBS-Boulder Laboratory for heat capacity measurements of solids have been adapted successfully for the determination of liquid heat capacities. Heptane was used as a standard in the measurement of the heat capacity of three jet fuels, JP-4, JP-7, and JP-8x, from 220 to 360 K. Transitions were observed in these fuels as follows (fuel, temperature): JP-4, 320 K; JP-7, 220-230 K; JP-8x, 220-240 K. Combined thermogravimetric and scanning calorimetric techniques were used successfully to measure heats of vaporization of cyclohexane, cis-decalin, and heptane. The precision of these measurements was better than 0.9 percent; agreement with literature values was satisfactory in view of small differences between the measuring temperature and the literature values for the boiling points.

900,838
PB89-173868 Not available NTIS
National Bureau of Standards (NEL), Boulder, CO.
Chemical Engineering Science Div.
Enthalpies of Desorption of Water from Coal Surfaces.
Final rept.
J. E. Callanan, B. J. Filla, K. M. McDermott, and S. A. Sullivan. 1987, 4p
Pub. in Proceedings of ACS (American Chemical Society) Symposium, Division of Fuel Chemistry, Denver, CO., April 5-10, 1987, p185-188.

Keywords: *Enthalpy, *Desorption, *Water, *Coal, *Calorimeters, Surface properties, Exothermic reactions, Oxidizers, Porosity, Measurement, Outgassing.

An exotherm observed on the initial heating of coal, but not on subsequent heating, has been attributed to the interaction of moisture with the oxidized coal surface. This enthalpy difference has been evaluated as a function of specimen size, coal oxidation, and atmosphere within the cell. A simple procedure has been devised which uses a commercially available calorimeter, for measurement of enthalpies of desorption of water on coal. The method is useful, not only for coal-water interactions, but also for solvent-porous solid interactions in general.

900,839
PB89-173900 Not available NTIS
National Bureau of Standards (NEL), Boulder, CO.
Chemical Engineering Science Div.
Specific Heat Measurements of Two Premium Coals.
Final rept.
J. E. Callanan, and K. M. McDermott. 1988, 6p
Pub. in Proceedings of ACS (American Chemical Society) National Meeting (194th), New Orleans, LA., August 31-September 4, 1987, p237-242 1988.

Keywords: *Coal, *Specific heat, Volatility, Thermophysical properties, Exothermic reactions, Measurement.

Specific heats of high volatile and medium volatile coals from the Argonne Premium Coal Sample Program were measured from 300 to 520 K. The results for the premium coals did not manifest the exothermic behavior on initial runs shown by weathered coals. Qualitative and quantitative differences in the behavior of the premium coals, by comparison with oxidized coals, will be presented.

900,840
PB89-174031 Not available NTIS
National Bureau of Standards (NEL), Boulder, CO.
Thermophysics Div.
Speed of Sound in Natural Gas Mixtures.
Final rept.
R. D. McCarty. 1986, 7p
Sponsored by Gas Research Inst., Chicago, IL.
Pub. in Proceedings of International Symposium on Fluid Flow Measurement, November 16-19, 1986, 7p.

Keywords: *Natural gas, *Acoustic velocity, *Sound transmission, Flow measurement, Equations of state, Mathematical models, Density(Mass/volume), Pressure, Temperature, Sound waves, Mixtures.

Accurate values for the speed of sound in natural gas mixtures are important in the application of sonic metering devices and in many design applications. In the case of mixtures, it is not possible to obtain experimentally determined speed of sound data for all possible compositions of the pure components found in natural gases. The alternative is a mathematical model of acceptable accuracy which allows the prediction of the speed of sound at an arbitrary state point and composition. The paper describes the 'state of the art' for the prediction of the speed of sound for natural gases.

900,841
PB89-176747 Not available NTIS

National Bureau of Standards (NEL), Boulder, CO.
Thermophysics Div.
Comprehensive Study of Methane + Ethane System.
Final rept.
W. M. Haynes, and R. D. McCarty. 1986, 16p
Pub. in Proceedings of International Symposium on Fluid Flow Measurement, Washington, DC., November 16-19, 1986, p14-27.

Keywords: *Methane, *Ethane, Mixtures, Thermodynamic properties, Transport properties, Thermal conductivity Specific heat, Viscosity, Acoustic velocity.

The paper reports on the use of the methane-ethane system as a model for developing and testing predictive techniques for mixtures. Comprehensive data have been obtained for both thermodynamic (PVT, heat capacity, sound velocity) and transport (viscosity, thermal conductivity) properties of three mixtures of methane and ethane, as well as for both components. The sound velocity and heat capacity data serve as extremely stringent tests for evaluating the performance of calculational techniques, as well as providing key information essential for optimizing the models. A critical enhancement in the thermal conductivity was observed for the system; current theory does not predict such behavior for mixtures.

900,842
PB89-195663 PC A10/MF A01
National Inst. of Standards and Technology (NEL), Gaithersburg, MD. Center for Fire Research.
Alaska Arctic Offshore Oil Spill Response Technology Workshop Proceedings.
Special pub.
N. H. Jason. Apr 89, 211p NIST/SP-762
Also available from Supt. of Docs. Library of Congress catalog card no. 89-600731. Sponsored by Minerals Management Service, Herndon, VA. Held at Anchorage, Alaska on November 29-December 1, 1988.

Keywords: *Offshore drilling, *Oil pollution, *Proceedings, *Alaska, Combination, Ignition, Artic regions, Water pollution, *Oil spills, *Alaska, *Emergency planning, Chemical treatments, In-situ burning, Mechanical containment, Mechanical recovery.

The Proceedings of the Alaska Arctic Offshore Oil Spill Response Technology Workshop contains papers by keynote speakers on the following topics: Mechanical Containment and Recovery; Chemical Treatment; In-Situ Burning; Readiness; the Technology Assessment and Research Program and the OHMSETT Program; the Arctic and Marine Oil Spill Program; the Canadian Environmental Science Revolving Fund; the Alaskan Clean Seas Research and Development Program; the NOFO Program. These papers served as a stimulus to the discussions that followed in the various Panel sessions. The Panels were organized into broad research areas: Mechanical Containment; Mechanical Recovery; Chemical Treatment; In-Situ Burning; Readiness. Each Panel summary is included in the Proceedings and these recommendations reflect the combined input from experts in the field. The Proceedings will serve as a working document to the Minerals Management Service (MMS) to identify their future research program.

900,843
PB89-222608 PC A21/MF A01
National Inst. of Standards and Technology (NEL), Boulder, CO. Thermophysics Div.
Tables for the Thermophysical Properties of Methane.
Technical note.
D. G. Friend, J. F. Ely, and H. Ingham. Apr 89, 469p NIST/TN-1325
Also available from Supt. of Docs. as SN003-003-02947-5.

Keywords: *Methane, *High pressure tests, *Low temperature tests, *Equations of state, Thermodynamic properties, Tables(Data), Specific heat, Viscosity, Thermal conductivity, Computer programs.

The thermophysical properties of methane are tabulated for a large range of fluid states based on recently formulated correlations. For the thermodynamic properties, temperatures from 91 to 600 K at pressures less than 100 MPa are included. For the viscosity, the corresponding range is 91 - 400 K with pressures to 55MPa, while for the thermal conductivity the range is 91 - 600 K with pressures to 100 MPa. In addition to the tables of properties, algebraic expressions and associated tables of coefficients are given to allow addi-

tional property calculations. Tables of comparisons between experimental property determinations and the correlations are also given both for primary data used in the formulation of the correlations and for additional data. A listing of a FORTRAN program for the evaluation of methane thermophysical properties using the Schmidt-Wagner equation of state is included.

900,844
PB90-117896 Not available NTIS
National Inst. of Standards and Technology (NEL), Gaithersburg, MD. Thermophysics Div.
Measurements of Molar Heat Capacity at Constant Volume: Cv,m(xCH4 + (1-x)C2H6' T = 100 to 320 K, p < or = 35 MPa).
Final rept.
J. E. Mayrath, and J. W. Magee. 1989, 15p
Sponsored by Gas Research Inst., Chicago, IL.
Pub. in Jnl. of Chemical Thermodynamics 21, p499-513 1989.

Keywords: *Specific heat, *Thermal measurement, *Thermophysical properties, *Methane, *Ethane, Mixtures, Temperature, Density, Volume, Calorimeters, Reprints.

Measurements of the molar heat capacity at constant volume Cv,m(xCH4 + (1-X)C2H6, X=0.35, 0.50, 0.68) at temperatures from 100 K to 320 K and at pressures to 35 MPa are reported. Heat capacities have been measured for 626 state conditions and these measurements complement thermodynamic and transport-property measurements previously reported for the same mixtures. The measurements were made on samples of constant mass contained by a calorimeter vessel of nearly constant volume. Uncertainties of the heat-capacity measurements are estimated to be less than 2.0%. Critical enhancement of heat capacity for each mixture is apparent in the critical-temperature region for constant density and similarly in the critical-density region for constant temperature. Liquid heat capacities at saturation, obtained by extrapolation at constant density, have a minimum for each composition occurring at twice the estimated critical density.

900,845
PB90-136839 Not available NTIS
National Bureau of Standards (NML), Gaithersburg, MD. Chemical Thermodynamics Div.
Evaluation of Data on Higher Heating Values and Elemental Analysis for Refuse-Derived Fuels.
Final rept.
T. J. Buckley, and E. S. Domalski. 1988, 8p
Sponsored by Department of Energy, Washington, DC. Biofuels and Municipal Waste Technology Div.
Pub. in Proceedings of National Waste Processing Conference, Philadelphia, PA., May 1-4, 1988, p77-84.

Keywords: Experimental design, Numerical analysis, Error analysis, *Refuse derived fuels, *Heating rate, *Multi-element analysis, Waste utilization, Data bases, Interlaboratory comparisons.

Elemental analyses and higher heating values taken from ASTM Round Robin Testing of RDF-3 have been evaluated. The data base was composed of five rounds of tests with eight to twelve laboratories performing four tests each. The authors found that established formulas can be used to calculate higher heating values on a moisture free basis (HHV2) from elemental analysis. A comparison is made between several methods of calculating HHV2. The Dulong formula and Institute of Gas Technology formula can predict HHV2 to within 3%.

Heating & Cooling Systems

900,846
PB89-150924 Not available NTIS
National Bureau of Standards (NEL), Washington, DC.
Building Equipment Div.
Combustion Testing Methods for Catalytic Heaters.
Final rept.
E. R. Kweller. 1986, 19p
Sponsored by Department of Energy, Washington, DC.
Pub. in ASHRAE (American Society of Heating, Refrigerating and Air-Conditioning Engineers) Transactions, v92 pt2B p239-257 1986.

ENERGY

Heating & Cooling Systems

Keywords: *Test facilities, Natural gas, Propane, Fuel consumption, Standards, Tests, Thermal efficiency, Reprints, *Catalytic heaters, *Combustion tests, Appliance Efficiency Test Procedures.

Both vented and unvented designs of catalytic heaters using natural gas and propane as the fuel were tested by three test methods. The objective of the study was to determine an appropriate laboratory test method for determining the percentage of unreacted fuel during typical heater use. Results of the study are expected to be useful for manufacturers who are currently developing an ANSI standard for catalytic heaters and by the Department of Energy for their Appliance Efficiency Test Procedures. Results showed that an analytical approach of calculating unreacted fuel using a carbon balance was equivalent to an empirical method. Results of tests in two open-room test methods and one closed room test method showed that the methods to use will depend on whether the heater is vented or unvented. A closed room test method was shown to yield results consistent with those obtained by open-room methods for the unvented heater, and may be used in place of an open-room method if the room oxygen is not depleted below 19% as an end-point of the test. A completely sealed test room was found to be unnecessary for this closed room test.

900,847
PB89-188619   PC A04/MF A01
National Bureau of Standards (NEL), Gaithersburg, MD.
Thermal and Economic Analysis of Three HVAC (Heating, Ventilating, and Air Conditioning) System Types in a Typical VA (Veterans Administration) Patient Facility.
G. N. Walton, and S. R. Petersen. Aug 87, 59p
NBSIR-87/3619
Sponsored by Veterans Administration, Washington, DC. Office of Facilities.

Keywords: *Thermal analysis, *Economic analysis, *Hospitals, *Military facilities, Computerized simulation, Radiant heating, Environmental engineering, Tables(Data), *Energy analysis, *HVAC systems, Energy demand, Variable air volume systems, Fan coil systems, Life cycle costs.

Thermal and economic analyses were performed for three different types of heating, ventilating, and cooling systems for a patient room in a typical VA patient facility in each of four locations. Thermal analysis was done with the U.S. Army's Building Loads Analysis and System Thermodynamics (BLAST) energy analysis program. Radiant panel, variable air volume (VAV), and fan coil systems were simulated. Some subroutines were developed and added to the BLAST program in order to simulate the radiant panel system. The predicted energy requirements, energy cost projections, and system costs were then evaluated using the NBS Federal Building Life-Cycle Cost (FBLCC) program to determine the 20-year life-cycle cost of each system in each location.

900,848
PB89-212146   Not available NTIS
National Bureau of Standards (NEL), Gaithersburg, MD. Building Equipment Div.
Part Load, Seasonal Efficiency Test Procedure Evaluation of Furnace Cycle Controllers.
Final rept.
R. A. Wise, and E. R. Kweller. 1986, 11p
Sponsored by Department of Energy, Washington, DC. Pub. in ASHRAE (American Society of Heating, Refrigerating and Air-Conditioning Engineers) Transactions, v92 p2B p674-684 1986.

Keywords: *Furnaces, *Performance standards, Thermostats, Thermal efficiency, Heating load, Tests, Temperature control, Reprints, Energy consumption, Energy conservation.

Claims of energy savings of 20 to 40% for space heating have been made by marketers of rapid cycling furnace controllers (cyclers). The present testing method of ANSI/ASHRAE 103-1982 can not be used to evaluate the heating seasonal efficiency of furnaces or boilers should they be installed using cyclers. To develop a test method that will reflect possible changes in efficiency, NBS has tested a furnace under various cycling conditions. A supply air temperature based cycler was tested, time based unit was simulated, and the test results were compared to results using simulated standard thermostat control. The comparisons were made using cycles corresponding to the average outdoor air temperatures of four weather condition bins of

approximately the same total heating hours. A test method has been developed which can evaluate the effect of cyclers on furnace efficiency. No proposal is made to use this method because: the test is more difficult to perform and replicate, change would require retesting of all furnaces and destroy the present data base of past test results, and changes in furnace efficiency attributable to cyclers were too small to justify a change.

Policies, Regulations & Studies

900,849
PB89-151211   CP D99
National Center for Health Statistics, Hyattsville, MD.
NBS (National Bureau of Standards) Life-Cycle Cost (NBSLCC) Program (for Microcomputers).
Software.
S. R. Petersen, and R. T. Ruegg. 21 Dec 88, 1 diskette NBS/SW/DK-89/001
The software is contained on 5 1/4-inch diskettes, double density (360K), compatible with the IBM PC microcomputer. The diskettes are in the ASCII format. Price includes documentation, PB87-180253.

Keywords: *Software, *Public buildings, Economic analysis, Benefit cost analysis, Diskettes, *National Institute of Standards and Technology, *Life-cycle cost, Energy conservation, L=BASIC, H=IBM PC/XT/AT.

The diskette provides the National Bureau of Standards Life-Cycle Cost (BNSLCC) programs and related files referenced in NBS SP 709, Comprehensive Guide for Least-Cost Energy Decisions. The NBSLCC programs perform economic analysis of buildings, building systems and components with special emphasis on energy conservation projects. Software description: The software is written in BASIC for implementation on IBM-PC/XT/AT microcomputers using the MS-DOS operating system. A minimum of 128 K of memory is required.

900,850
PB89-153960   PC A04/MF A01
National Inst. of Standards and Technology (NEL), Gaithersburg, MD. Center for Computing and Applied Mathematics.
Energy Prices and Discount Factors for Life-Cycle Cost Analysis 1988: Annual Supplement to NBS (National Bureau of Standards) Handbook 135 and NBS Special Publication 709.
Annual rept.
B. C. Lippiatt, and R. T. Ruegg. Nov 88, 61p NISTIR-85/3273-3
See also PB88-109913. Sponsored by Department of Energy, Washington, DC. Assistant Secretary for Conservation and Renewable Energy.

Keywords: *Cost analysis, *Energy management systems, *Federal buildings, Prices, Heating fuels, Investments, Energy conservation, *Life cycle costs, Federal Energy Management and Planning Program, Discount factors.

The 1988 energy prices and discount factors for life-cycle cost analysis are reported as established by the U.S. Department of Energy. The data are provided as an aid to implementing life-cycle cost evaluations of potential energy conservation and renewable energy investments in existing and new federally owned and leased buildings.

Solar Energy

900,851
PB89-229058   PC A05/MF A01
National Inst. of Standards and Technology (NML), Gaithersburg, MD. Electricity Div.
Lightning and Surge Protection of Photovoltaic Installations. Two Case Histories: Vulcano and Kythnos.
F. D. Martzloff. Jun 89, 78p NISTIR-89/4113
Sponsored by Sandia National Labs., Albuquerque, NM.

Keywords: *Photovoltaic cells, *Lightning, *Surges, *Circuit protection, Overvoltage, Overcurrent, Lightning protection, *Solar cell arrays.

Two installations of photovoltaic systems were damaged during lightning storms. The two sites were visited and the damaged equipment that was still available on the site was examined for analysis of the suspected occurrence. The evidence, however, is insufficient to conclude that all the observed damage was caused by the direct effect of a lightning flash. A possible scenario may be that lightning-induced overvoltages caused insulation breakdown at the edges of the photovoltaic modules, with subsequent damage done by the dc current of the array. Other surge protection considerations are also addressed, and suggestions presented for further investigations.

General

900,852
PB89-175897   Not available NTIS
National Bureau of Standards (NEL), Gaithersburg, MD. Office of Energy-Related Inventions.
Periodic Heat Conduction in Energy Storage Cylinders with Change of Phase.
Final rept.
M. E. McCabe. 1986, 9p
Sponsored by Department of Energy, Washington, DC. Office of Solar Heat Technologies.
Pub. in Proceedings of Joint AIAA/ASME (American Institute of Aeronautics and Astronautics/American Society of Mechanical Engineers) Thermophysics and Heat Transfer Conference, Boston, MA., June 2-4, 1986, 9p.

Keywords: *Conduction, *Heat transmission, *Phase transformations, *Cylinders, Flux(Rate), Thermophysical properties, Heat of fusion, Diurnal variations, Enthalpy, Specific heat, Solar heating, *Thermal energy storage equipment.

The thermal performance of an energy storage system subjected to a time dependent flux representing the 24-hour diurnal cycle of the sun is discussed. The governing mathematical equations are presented for a cylindrical configuration and a numerical solution based on the enthalpy model is used to obtain the temperature distribution and phase interface location in a phase change energy storage system. The steady-periodic solution is investigated to determine the effects of cylinder size, fusion temperature and thermophysical properties on the thermal performance of a radiatively charged/convectively discharged energy storage system in a passive solar heated building. The relationship between fusion temperature and the mean surface temperature is shown to be critical for storage of energy. The numerical results show that maximum energy storage and minimum surface temperature variation occur when the cyclic mean surface temperature equals the fusion temperature.

900,853
PB90-112442   PC A06/MF A01
National Inst. of Standards and Technology (NEL), Gaithersburg, MD. Electrosystems Div.
Research for Electric Energy Systems: An Annual Report.
R. J. Van Brunt. Oct 89, 109p NISTIR-89/4167
See also rept. for 1987, PB89-132310. Sponsored by Department of Energy, Washington, DC. Div. of Electric Energy Systems.

Keywords: *Potential energy, *Ions, Research projects, Electric fields, Activity coefficients, Charged particles, Transmission lines, Atmospheric pressure, Measurements, Charge density, Gas cylinders, Stochastic processes, Dielectrics, Hydrostatic pressure, Transients.

Reported is technical progress in four investigations conducted at NIST and supported by the U.S. Department of Energy under Task Order Number 137. The first investigation is concerned with the measurements of electric fields and ions in the vicinity of high-voltage transmission lines and biological exposure facilities. Results are reported on evaluations of two methods for measuring monopolar charge densities in air. The second investigation is concerned with development of advanced diagnostics for compressed gas-insulated power systems. Results are reported on measurements of collisional electron detachment and negative ion conversion reactions in SF6 and on a new technique for measuring the stochastic behavior of partial discharges. The third investigation is concerned with measurement of prebreakdown phenomena at solid-liquid dielectric interfaces. Results are presented here from optical observations of the influence of hydrostatic pressure on prebreakdown pa tia discharge development and measurement of rené-second impulse breakdown at liquid-solid interfaces. The fourth area of research is concerned with electrical measurement of fast transient phenomena. Results are presented from an investigation into the interactions between two dividers used simultaneously to measure fast impulse voltages.

ENVIRONMENTAL POLLUTION & CONTROL

Air Pollution & Control

900,854
PB89-148134 PC A03/MF A01
National Bureau of Standards (NEL), Gaithersburg, MD. Center for Building Technology.
Predicting Formaldehyde Concentrations in Manufactured Housing Resulting from Medium-Density Fiberboard.
S. Silberstein. Apr 88, 19p NBSIR-88/3761
Sponsored by Department of Housing and Urban Development, Washington, DC.

Keywords: *Formaldehyde, *Fumes, *Air pollution, Fiberboards, Houses, Emission, Particle boards, Plywood, Panels, Temperature, Humidity, Urea-formaldehyde resins, Forecasting.

HUD previously issued Manufactured Home Construction and Safety Standards limiting formaldehyde emissions of particleboard and plywood paneling that were manufactured using urea-formaldehyde resins for use in manufactured homes. The report uses indoor air quality models to predict how much medium-density fiberboard(mdf) may be added to manufactured homes already containing formaldehyde emitting particleboard and plywood paneling, without raising the formaldehyde concentration beyond 400 ppb. It was found that any combination of mdf that results in a chamber-test concentration of 300 ppm may be added to such a home. A sensitivity analysis was done to predict how this formaldehyde concentration limit is affected by variations in temperature, relative humidity, and air exchange rate. It was concluded that limiting chamber concentration to 200 ppb would allow for small errors in temperature, relative humidity, and air exchange rate that might be expected to arise in practice.

900,855
PB89-150775 Not available NTIS
National Bureau of Standards, Gaithersburg, MD. Office of Standards Management.
Draft International Document on Guide to Portable Instruments for Assessing Airborne Pollutants Arising from Hazardous Wastes.
Draft rept.
S. Chappell, J. Driscoll, G. Flanagan, S. Levine, K. Leichnitz, P. Lilienfeld, and R. Turpin. 1988, 26p
Pub. in Draft International Document on Guide to Portable Instruments for Assessing Airborne Pollutants Arising from Hazardous Wastes, p1-26 1988.

Keywords: *Hazardous materials, *Guidelines, *Portable equipment, *Waste disposal, Assessments, Sites, Concentration(Composition), Performance equipment, Sampling, Calibrating, Reprints, *Environmental monitoring, *Air pollution sampling.

The document provides definitions and guidelines for selecting portable monitoring instruments to measure airborne pollutants at hazardous waste sites. A brief description of six types of instruments, including some important metrological and technical characteristics, is given. The document also provides background and literature references on the application of these instruments. Emphasis is placed on methods and requirements for testing and calibrating instruments. Specifics of sampling methods and of instrument type evaluation are not addressed.

900,856
PB89-214779 PC A03/MF A01
National Inst. of Standards and Technology (NEL), Gaithersburg, MD. Center for Fire Research.
Synergistic Effects of Nitrogen Dioxide and Carbon Dioxide Following Acute Inhalation Exposures in Rats.
B. C. Levin, M. Paabo, L. Highbarger, and N. Eller. May 89, 45p NISTIR-89/4105
Sponsored by Society of the Plastics Industry, Inc., New York.

Keywords: *Air pollution, *Carbon dioxide, *Nitrogen dioxide, *Rats, Toxicology, Respiratory diseases, Oxidizers, Hemoglobins, Exposure, Blood, Mortality, Air pollution control, Standards, Methemoglobins, Fire gases.

All fires occurring in air produce carbon dioxide (CO2). Fire involving nitrogen-containing products will also generate nitrogen dioxide (NO2), a pulmonary irritant. In Fischer 344 male rats, the LC50 (30 minute exposure plus 14 day post-exposure observation period) for NO2 was 200 ppm (with 95% confidence limits of 43 to 51%); whereas, the LC50 for NO2 in the presence of 5% CO2 was 90 ppm (with 90% confidence limits ranging from 70-120 ppm). Exposure to NO2 increased the methemoglobin (MetHb) levels in the arterial blood. At the end of the 30 minute exposures, the MetHb levels were 2-3 times higher in the animals exposed to the combination of NO2 (200 ppm) and CO2 (5%) than in those exposed to NO2 only. Deaths from NO2 were all post-exposure and occurred earlier in the presence of NO2 plus 5% CO2 than in the absence of the CO2. The time of death was concentration-dependent when both gases were present. At death, evidence of hemorrage and extensive edema was observed in the lungs. The mean lung wet weight/body weight ratio from rats exposed to 200 ppm NO2 with and without 5% CO2 was 3-4 times that of non-exposed rats. More edema was noted with NO2 and CO2 than with NO2 alone.

900,857
PB89-229686 PC A03/MF A01
National Inst. of Standards and Technology (NEL), Gaithersburg, MD. Building Environment Div.
Ventilation and Air Quality Investigation of the U.S. Geological Survey Building.
W. S. Dols, and A. Persily. Jul 89, 44p NISTIR-89/4126
Sponsored by Geological Survey, Reston, VA.

Keywords: *Ventilation, *Buildings, *Air quality, *Laboratories, Odors, Air flow, Heating, Measurement, Carbon dioxide, Carbon monoxide, Radon, Concentration(Composition), Formaldehyde, Standards, Recommendations, Improvement, Graphs(Charts), *Indoor air pollution, Tracer gas tests.

The National Center of the U.S. Geological Survey in suburban Washington, DC is a seven story building containing both office and laboratory space. Based on a history of occupant complaints regarding the air quality within the building, an investigation was conducted by the National Institute of Standards and Technology to quantify the ventilation characteristics of the building and to determine the indoor levels of selected indoor pollutants. The investigation of the building included measurements of air exchange rates using the tracer gas decay technique and measurements of indoor concentrations of carbon dioxide, carbon monoxide, radon, formaldehyde and particulates. The measurement results are compared to appropriate standards and guidelines in order to investigate the role of ventilation and pollutant concentrations in the indoor air quality complaints. Based on the investigation, several recommendations are made to improve the environmental conditions within the building.

900,858
PB89-235899 PC A04/MF A01
National Inst. of Standards and Technology (NEL), Gaithersburg, MD. Building Environment Div.
Method for Measuring the Effectiveness of Gaseous Contaminant Removal Filters.
B. M. Mahajan. Aug 89, 54p NISTIR-89/4119
See also PB88-155882. Sponsored by Consumer Product Safety Commission, Washington, DC.

Keywords: *Air pollution abatement, *Air filters, Ventilation, Air conditioning equipment, Effectiveness, Gases, Activated carbon, Adsorbents, Performance evaluation, Contaminants, *Indoor air pollution.

The report presents a brief review of the gas adsorption kinetics theory applicable to adsorption of gaseous contaminants by filter media, and an algorithm for assessing the effectiveness of filtering devices with flow bypass. It briefly describes the selected testing technique for measuring the effectiveness of filter media, and presents experimental data for adsorption of n-butane, toluene, and carbon monoxide.

900,859
PB90-123571 Not available NTIS
National Inst. of Standards and Technology (NML), Gaithersburg, MD. Organic Analytical Research Div.
Mobile Sources of Atmospheric Polycyclic Aromatic Hydrocarbons: A Roadway Tunnel Study.
Final rept.
B. A. Benner, G. E. Gordon, and S. A. Wise. 1989, 10p
Pub. in Environmental Science and Technology 23, n10 p1269-1278 1989.

Keywords: *Aromatic polycyclic hydrocarbons, *Exhaust emissions, *Chemical analysis, *Particles, *Motor vehicles, Chromatographic analysis, Gas analysis, Gasoline engines, Diesel engines, Tunnels, Reprints, *Air pollution sampling, *Air pollution detection, Heavy duty vehicles, Light duty vehicles.

Suspended particulate matter samples were collected by high-volume samplers in a heavily traveled roadway tunnel to characterize the mobile source emissions (diesel- and gasoline-fueled vehicles) for polycyclic aromatic hydrocarbons (PAHs). Liquid and gas chromatographic techniques were employed to isolate and quantify individual PAHs in dichloromethane extracts of Teflon and glass-fiber filters. Mobile-source PAH emission estimates were generated from the particle- and vapor-phase samples collected in the tunnel and agree with emission estimates reported previously for roadway tunnels in Japan. Factor analysis of PAH concentrations from 47 filter samples yielded two factors; one which possibly represented the diesel-fueled vehicles (heavy-duty trucks) and the other, either gasoline-fueled vehicles or a composite of the gasoline and diesel sources.

900,860
PB90-126166 Not available NTIS
National Inst. of Standards and Technology (NML), Gaithersburg, MD. Organic Analytical Research Div.
Residential Wood Combustion: A Source of Atmospheric Polycyclic Aromatic Hydrocarbons.
Final rept.
F. R. Guenther, S. N. Chesler, G. E. Gordon, and W. H. Zoller. 1986, 6p
Pub. in Jnl. of High Resolution Chromatography and Chromatography Communications 11, p761-766 Nov 88.

Keywords: *Aromatic polycyclic hydrocarbons, *Particles, *Combustion products, *Stoves, Sources, Residential building, Exhaust emissions, Automobiles, Chemical analysis, Chromatographic analysis, Atmospheric composition, Reprints, *Air pollution detection, *Air pollution sampling, *Wood burning appliances, Coal combustion, Fairbanks(Alaska).

Samples were taken of the atmospheric particulate matter over Fairbanks Alaska in the winter of 1985, and from wood stoves burning the major wood types locally available. These samples were then analyzed for polycyclic aromatic hydrocarbon (PAH). A PAH emission profile was determined from the wood stove samples and applied to the atmospheric samples to determine the residential wood combustion contribution to the local atmospheric particulate burden. Emission profiles for coal burning and automobile emissions for PAH were also used to estimate their relative contributions.

Solid Wastes Pollution & Control

900,861
PB89-212104 Not available NTIS
National Bureau of Standards (NML), Gaithersburg, MD. Chemical Kinetics Div.

101

ENVIRONMENTAL POLLUTION & CONTROL

Solid Wastes Pollution & Control

Fundamental Aspects of Key Issues in Hazardous Waste Incineration.
Final rept.
W. Tsang. 1986, 6p
Pub. in Proceedings of ASME (American Society of Mechanical Engineers) Winter Annual Meeting, Anaheim, CA., December 7-12, 1986, 6p. .

Keywords: *Hazardous materials, *Waste disposal, *Incineration, *Pyrolysis, Reaction kinetics, Oxidation, Chemical reactions, Performance evaluation.

The incineration of hazardous waste is considered from the point of view of its important chemical reactions. The authors begin with a brief review of the earlier work on destruction mechanisms and expand it by allowing for regions where pyrolytic reactions are important. The authors then discuss the quantitative relationship between destruction efficiency and the appropriate rate constants for hazardous waste destruction. This leads into considerations of the products of incomplete combustion and the author describe in qualitative terms the type of compounds that can be expected. The concept of surrogates is discussed. The authors consider recent results on OH attack on organics near levels of CO usually observed in various types of incinerators, and conclude that there is a high probability that the organics in the incineration effluent stream are formed by pyrolysis mechanism. Finally the authors suggest a number of measures for validating proper incinerator operation.

Water Pollution & Control

900,862
PB89-173827   Not available NTIS
National Bureau of Standards (NEL), Gaithersburg, MD. Statistical Engineering Div.
Designs for Assessment of Measurement Uncertainty: Experience in the Eastern Lake Survey.
Final rept.
W. S. Liggett. 1986, 15p
Sponsored by Corvallis Environmental Research Lab., OR.
Pub. in Proceedings of International Biometric Conference: Invited Papers (13th), Seattle, WA., July 27-August 1, 1986, p1-15.

Keywords: *Lakes, *Error analysis, *Environmental surveys, Quality assurance, Statistical analysis, Assessments, Probability theory, Chromatographic analysis, Inorganic nitrates, Sulfates, Performance evaluation, United States, *Water pollution sampling.

As is typical of environmental studies, the Eastern Lake Survey generated data that reflect both properties of the environment and properties of the sampling and measurement procedures. Thus, environmental conclusions cannot be drawn from these data without assessment of the sampling and measurement error. The paper shows how the properties of ion chromatography affect the nitrate and sulfate measurements in the survey. The investigation, which is based on measurements of routine-duplicate samples, field blanks, and field audit samples, is interesting because of its complexity. The measurement error has a within-day component and a day-to-day component. The variance of the within-day component increases linearly with concentration except near zero where the algorithm used to interpret the chromatographs seems to behave poorly. The day-to-day component depends on the calibration procedure. As an example, this investigation illustrates the strengths and weaknesses of some common designs for uncertainty assessment.

900,863
PB89-185581   PC A04/MF A01
National Bureau of Standards (NEL), Gaithersburg, MD. Center for Fire Research.
Combustion of Oil on Water. November 1987.
D. Evans, H. Baum, B. McCaffrey, G. Mulholland, M. Harkleroad, and W. Manders. Nov 87, 56p NBSIR-86/3420
See also PB89-149173. Sponsored by Minerals Management Service, Reston, VA.

Keywords: *Fuel oil, *Combustion products, *Water pollution, *Fires, *Aerosols, Tanker ships, Measurement, Smoke, Aromatic hydrocarbons, Air flow, Offshore drilling, Graphs(Charts).

The report contains the results of measurements performed on both 0.4 m and 0.6 m diameter pool fires

produced by burning a layer of Prudhoe Bay crude oil supported by a thermally deep layer of water. Both steady and vigorous burning caused by boiling of the water sublayer were observed. The measured energy release rate for steady burning was about 640 kW per sq meter. The emission rate, the size distribution, and specific extinction coefficient were measured for the smoke aerosol produced by the fires. Data were also obtained on the structure of the smoke aerosol by electron microscopy and on emission of CO and CO2. Analysis of the crude oil burn residue indicated selected depletion of the short chain alkanes and cycloalkanes when compared to the fresh oil. Mono-ring aromatics including benzene, toluene, and xylenes present in the fresh crude were absent in the burn residue. Calculations of the induced air flow into a simulated distribution of 20 fires over a 100 m x 100 m area showed that the maximum inflow velocity near the largest size fire (2.5 m diameter, 3.2 MW) was 1.1 m/s.

900,864
PB89-229280   Not available NTIS
National Inst. of Standards and Technology (IMSE), Gaithersburg, MD. Polymers Div.
Biotransformation of Mercury by Bacteria Isolated from a River Collecting Cinnabar Mine Waters.
Final rept.
F. Baldi, M. Filippelli, and G. J. Olson. 1989, 12p
Pub. in Microbial Ecology 17, p263-274 1989.

Keywords: *Aerobic bacteria, *Mercury inorganic compounds, *Water pollution, Mercury ore deposits, Rivers, Reprints, *Mine waters, *Biotransformation, Resistance(Biology).

One hundred six strains of aerobic bacteria were isolated from the Fiora River which drains an area of cinnabar deposits in southern Tuscany, Italy. Thirty-seven of the strains grew on an agar medium containing 10 microg/ml Hg (as HgCl2) with all of these strains producing elemental mercury. Seven of the 37 strains also degraded methylmercury. None of 106 sensitive and resistant strains produced detectable monomethylmercury although 15 strains produced a benzene-soluble mercury species. Two strains of alkylmercury (methyl-, ethyl- and phenylmercury) degrading bacteria were tested for the ability to degrade several other analogous organometals and organic compounds, but no activity was detected toward these compounds. Mercury methylation is not a mechanism of Hg resistance in aerobic bacteria from this environment. Growth of bacteria on the agar medium containing 10 microg/ml HgCl2 was diagnostic for Hg detoxification based on reduction.

General

900,865
PB89-150940   Not available NTIS
National Bureau of Standards (NML), Gaithersburg, MD. Office of Standard Reference Data.
Environmental Standard Reference Materials - Present and Future Issues.
Final rept.
S. D. Rasberry. 1988, 9p
Pub. in Proceedings of International Symposium on Trace Analysis in Environmental Samples and Standard Reference Materials, Honolulu, HI., January 6-8, 1988, p62-66-4.

Keywords: *Environmental surveys, *Chemical analysis, Measurement, Chemical properties, Physical properties, Quality assurance, Forecasting, *Standard reference materials, Certified reference materials.

Accurate measurements are an important consideration in environmental analysis. The National Bureau of Standards (NBS) provides several types of services to aid analysts in obtaining accurate measurements and in validating the accuracy of measurement methods and measurement systems. The most well known of these services is Standard Reference Materials (SRMs), which are well-characterized, homogeneous materials or simple artifacts with specific chemical or physical properties certified by NBS. There are several hundred SRMs in support of environmental chemists covering inorganic and organic constituents in water, air, sediments, rocks, plant and animal materials. Attention is given to providing SRMs which closely approximate the matrix being analyzed in a specific environmental application. The extremely wide range of

environmental matrices makes it economically unfeasible for the NBS to offer nearly exact matches to every matrix. For this reason, analysts frequently should use more than one SRM in a 'benchmark-bracketing' way to validate their analytical methods. Non-matrix-specific SRMs, such as single element spectrometric solutions, are also available for instrument calibration or validation.

900,866
PB90-123969   Not available NTIS
National Inst. of Standards and Technology (NML), Gaithersburg, MD. Organic Analytical Research Div.
Experiences in Environmental Specimen Banking.
Final rept.
S. A. Wise, B. J. Koster, R. M. Parris, M. M. Schantz, S. F. Stone, and R. Zeisler. 1989, 16p
Pub. in International Jnl. of Environmental Analytical Chemistry 37, p91-106 1989.

Keywords: *Chemical analysis, Shelf life, Reprints, *National Institute of Standards and Technology, *Environmental monitoring, *Sample preparation, *Environmental specimen banking.

For the past 10 years, the National Institute of Standards and Technology has been involved in environmental specimen banking activities. These activities have resulted in the development of collection, storage, processing, and analysis procedures for long-term archiving of a variety of environmental specimens including human liver, fish muscle and liver, oysters, mussels, sediment, and seal tissues. In the paper, the authors describe some of the experiences and results from these efforts as related to environmental trend monitoring and the potential value of a specimen bank for future retrospective analyses.

INDUSTRIAL & MECHANICAL ENGINEERING

Industrial Safety Engineering

900,867
PB89-151781   PC A04/MF A01
California Univ., Berkeley. Dept. of Mechanical Engineering.
Fire Propagation in Concurrent Flows.
Final rept.
A. C. Fernandez-Pello. Jan 89, 67p NIST/GCR-89/557
Grant NANB-7-D0737
See also PB86-181849. Sponsored by National Inst. of Standards and Technology (NEL), Gaithersburg, MD. Center for Fire Research.

Keywords: *Air flow, *Flame propagation, *Fire tests, Flames, Heat transfer, Combustion, Turbulence, Fuels, Gases, Flammability, *Rooms.

The research tasks completed during this reporting period include an experimental study of the effect on the spread of flames of the turbulence intensity of an opposed air flow, and a theoretical analysis of the concurrent spread of flames over thin fuels. Both studies are in the authors opinion important contributions in the study of the flame spread process. The results of the experimental study show that the flame spread process is significantly affected by the flow turbulence intensity for flame spreading over both thin and thick fuels. For a fixed flow velocity, the spread rate decreases as the turbulent intensity is increased. This appears to be due to the turbulent convective cooling of the fuel surface and gas in the vicinity of the flame front. Also observed is that extinction of the flames occurs at lower velocities as the turbulent intensity increases. The results of the theoretical analysis, which are in good agreement with previous experimental measurements, give detailed information about the flame structure and mechanisms of flame spread. It is shown that the flame length and consequently the rate

900,868
PB89-174890 Not available NTIS
National Bureau of Standards (NEL), Gaithersburg, MD. Fire Measurement and Research Div.
Smoke and Gas Evolution Rate Measurements on Fire-Retarded Plastics with the Cone Calorimeter.
Final rept.
V. Babrauskas. 1989, 8p
Pub. in Fire Safety Jnl. 14, p135-142 1989.

Keywords: *Plastics, *Flammability testing, *Fire resistant materials, Smoke, Reprints, *Cone calorimeter.

The cone calorimeter was developed originally for making improved rate of heat release measurements. The basic design has proved to be highly versatile, allowing it to become a suitable test bed for making smoke and gas evolution studies. The smoke measurement procedures with the cone calorimeter have recently been published by ASTM as a proposed method. Gas evolution measurements are still at the research laboratory stage, but are being actively developed, with a strong role being seen in the near future for both fire modeling and combustion toxicity. The techniques evolved are seen to be especially useful for comparing fire-retarded materials to non-retarded ones.

900,869
PB89-212047 Not available NTIS
National Bureau of Standards (NEL), Gaithersburg, MD. Fire Measurement and Research Div.
Toxicity of Mixed Gases Found in Fires.
Final rept.
B. C. Levin, M. Paabo, J. L. Gurman, and S. E. Harris. 1986, 18p
Sponsored by Army Medical Research Inst. of Chemical Defense, Aberdeen Proving Ground, MD., and Society of the Plastics Industry, Washington, DC.
Pub. in Proceedings of SPI and FRCA Joint Meeting, Washington, DC., March 19-21, 1986, p71-88.

Keywords: *Toxicity, *Fires, *Gases, Fire hazards, Carbon dioxide, Carbon monoxide, Hydrogen cyanide, Concentration(Composition), Lethal dose 50.

The Center for Fire Research is developing a computer model to predict the toxic hazard that people will experience under various fire scenarios. The toxicity of single and multiple fire gases is being studied to determine whether the toxic effects of a material's combustion products can be explained by the biological interactions of the primary fire gases or, if minor, more obscure gases need to be considered. LC50 values for Fisher 344 rats have been calculated for carbon monoxide (CO) in air for 2, 5, 10, 30, and 60 minute exposures using the NBS Toxicity Test Method. LC50 values have also been calculated for hydrogen cyanide (HCN) in air for similar time periods, plus one minute exposures. Combination experiments with CO and HCN indicate that they act in an addictive manner. A preliminary comparison of the concentrations of the major combustion products generated from some materials tested at their LC50 values with the combined gas results indicates that the observed toxicity for these materials appear to be explained by the biological interactions of the primary toxic fire gases.

Laboratory & Test Facility Design & Operation

900,870
PB89-147086 Not available NTIS
National Bureau of Standards (NEL), Gaithersburg, MD. Chemical Process Metrology Div.
Economical Ultrahigh Vacuum Four-Point Resistivity Probe.
Final rept.
J. Erickson, and S. Semancik. 1987, 3p
Pub. in Jnl. of Vacuum Science and Technology A 5, n1 p115-117 Jan/Feb 87.

Keywords: *Electric measuring instruments, *Ultrahigh vacuum, Semiconductors(Materials), Oxides, Measurement, Reprints, *Resistivity probes, Metal oxides.

An economical four-point probe has been built to measure changes in the surface resistivity of metal

oxide semiconductors. Its fabrication from ultra-high-vacuum-compatible parts is described.

900,871
PB89-147847 PC A04/MF A01
National Bureau of Standards (NEL), Boulder, CO. Electromagnetic Technology Div.
Metrology for Electromagnetic Technology: A Bibliography of NBS (National Bureau of Standards) Publications.
M. E. DeWeese. Aug 88, 63p NBSIR-88/3097
Supersedes PB88-123682.

Keywords: *Metrology, *Electromagnetic properties, *Bibliographies, Superconductors, Magnetic measurement, Lasers, Optical fibers, Cryoelectronics.

The bibliography lists the publications of the personnel of the Electromagnetic Technology Division of NBS in the period from January 1970 through publication of the report. A few earlier references that are directly related to the present work of the Division are also included.

900,872
PB89-149108 Not available NTIS
National Bureau of Standards (NML), Gaithersburg, MD. Length and Mass Div.
Interpretation of a between-Time Component of Error in Mass Measurements.
Final rept.
R. S. Davis. 1987, 7p
Pub. in Proceedings of Measurement Science Conference, Irvine, CA., January 29-30, 1987, pIII-B(1)-III-B(7) 1987.

Keywords: *Error analysis, *Reliability, Measuring instruments, Metrology, Random error, Mathematical models, *Between-time error, *Calibration errors, *Mass measurement.

Detection of a between-time component of error in a calibration process involves analysis of a lengthy record of data. An individual object or device which is being calibrated must be returned to the customer in a timely way. Therefore, evidence of a between-time component of error can only be inferred from studies of standards which are internal to the calibration facility. If study of internal standards reveals such an error component, it then becomes necessary to propagate this error to the reported calibration uncertainty. Because sophisticated calibrations usually involve measurement designs which are solved by least squares techniques, the propagation of a between-time error component is obscured. In order to examine such propagation it is necessary to choose a mathematical model which can account for the observed effects. Since choice of this model has a profound effect on the analysis, the metrologist must be aware of its implications. These considerations are illustrated by a real example from precision mass metrology.

900,873
PB89-149215 Not available NTIS
National Bureau of Standards (NML), Gaithersburg, MD. Gas and Particulate Science Div.
Special Calibration Systems for Reactive Gases and Other Difficult Measurements.
Final rept.
W. D. Dorko, and E. E. Hughes. 1987, 6p.
Pub. in ASTM (American Society for Testing and Materials) Special Technical Publication 957, p132-137 1987.

Keywords: *Measuring instruments, *Gas analysis, Gas chromatography, Measurement, Reliability, *Reactive gases, *Calibration standards.

The most popular method for analyzing a component of interest in a gas mixture is to use a detector which responds to the analyte and is calibrated directly with a cylinder gas mixture containing the analyte at the approximate concentration of interest. In many instances, however, this cannot be achieved; cylinder calibration mixtures may be unstable or the direct response detector may not be sensitive enough. The unstable cylinder mixture can be replaced by a dynamic dilution gas calibration system where the analyte can be introduced to any one of several means including permeation devices, or a higher concentration gas mixture which can be more easily stabilized. If the detector for the analyte is insensitive or difficult to calibrate then the analyte can be converted to another compound for which there is a detector which is sufficiently sensitive and easy to calibrate. Any system which will produce a mixture of the required concentra-

tion, of acceptable uncertainty, and of stability sufficient for the analysis at hand can be a special calibration system.

900,874
PB89-153894 PC A05/MF A01
National Bureau of Standards (NEL), Gaithersburg, MD. Precision Engineering Div.
NIST (National Institute of Standards and Technology) Measurement Services: Mass Calibrations:
Final rept.
R. S. Davis. Jan 89, 76p NIST/SP-69/31
Also available from Supt. of Docs. as SN003-003-02919-0. Library of Congress catalog card no. 88-600608.

Keywords: Calibrating, Density, Mass, Least squares method, Measurement, Primary standards, Setting(Adjusting), *Mass calibration, National Institute for Standards and Technology.

The NIST calibration service for standard masses is described. Weights which are accepted for calibration range in nominal values from 1 mg to 13,600 kg (30,000 pounds). Weights used to generate standard pressures in piston gages are also accepted. Cleaning procedures used on weights prior to calibration are described. The measurement algorithms (including density determinations of single-piece kilogram weights) and the uncertainties assigned to calibrated weights are discussed. The system now in place to monitor the quality of calibrations is described. Finally, the limitations of the present controls on measurement quality and outline improvements which are underway are assessed.

900,875
PB89-157184 Not available NTIS
National Bureau of Standards (NEL), Gaithersburg, MD. Thermophysics Div.
Semi-Automated PVT Facility for Fluids and Fluid Mixtures.
Final rept.
D. Linsky, J. M. H. L. Sengers, and H. A. Davis. 1987, 5p
Pub. in Review of Scientific Instruments 58, n5 p817-821 1987.

Keywords: *Automatic control equipment, *Mixtures, *Fluids, *Supercritical flow, Geothermal prospecting, Boiling points, Capacitance bridges, Dew point, Gas laws, Measurement, Pressure, Volume, Temperature, Bubbles, Density(Mass/volume), Reprints.

A manually-operated Burnett PVT apparatus has been converted into a semi-automated Burnett-isochoric facility. An automated pressure injector with dedicated control logic nulls a sensitive differential pressure indicator. A microcomputer is used in setting the control temperature, monitoring equilibration, measuring temperature and pressure, processing the raw data and storing the information. The quality of the apparatus is demonstrated by, means of a low-density and a high-density isochore obtained for a geothermal working fluids mixture in one-phase, two-phase and supercritical regimes, including a dew and a bubble point.

900,876
PB89-162234 PC A13/MF A01
National Inst. of Standards and Technology, Gaithersburg, MD.
Electrical Performance Tests for Hand-Held Digital Multimeters.
Final rept.
T. F. Leedy, K. J. Lentner, O. B. Laug, and B. A. Bell. Jan 89, 290p NISTIR-88/4021
Sponsored by Army Communications-Electronics Command, Fort Monmouth, NJ.

Keywords: *Performance tests, *Standards, *Multimeters, Electric measuring instruments, Electrical measurement, Reliability, Voltmeters, Ammeters, Digital systems.

Electrical performance test procedures for battery-powered, hand-held digital multimeters were developed for the purpose of evaluating samples submitted by electronic instrument manufacturers in response to specifications issued by the U.S. Army Communications-Electronics Command. The detailed, step-by-step test procedures are based on the Army specifications and include sample data sheets and tables for the recording of interim data and final test results. The report discusses the measurement principles and techniques underlying each of the procedures. In addi-

103

tion, the sources of measurement uncertainty are discussed.

900,877
PB89-173496 Not available NTIS
National Bureau of Standards (NML), Gaithersburg, MD. Ionizing Radiation Physics Div.
Computer Program for Instrument Calibration at State-Sector Secondary-Level Laboratories.
Final rept.
H. T. Heaton. 1988, 13p
Pub. in Proceedings of Midyear Health Physics Society Topical Meeting on Instrumentation (22nd), San Antonio, TX., December 4-8, 1988, 13p.

Keywords: *Calibrating, *Measuring instruments, Standardization, X rays, Gamma rays, Quality control, *Instrument compensation, Computer software. .

To provide the quality assurance necessary for secondary-level laboratories, a computer program has been developed for use by state-sector laboratories calibrating instruments used in x-ray and gamma-ray fields. Although specific to the equipment in state-sector laboratories, the basic procedures are also applicable to secondary-level laboratories in the private and federal sectors. The program is written in a compiled BASIC. After the initial program selection, options are menu driven. There are specific programs for each of the classes of instruments calibrated by these laboratories. They perform the routine quality control on all of the pertinent equipment used for calibrations and make the measurements to characterize the laboratory, which are necessary for the uncertainty analysis and generation of the final report.

900,878
PB89-173934 Not available NTIS
National Bureau of Standards, Gaithersburg, MD. Associate Director of Industry and Standards.
NBS (National Bureau of Standards) Calibration Services: A Status Report.
Final rept.
G. A. Uriano. 1984, 24p
Pub. in Proceedings of Workshop and Symposium of the National Conference of Standards Laboratories, Gaithersburg, MD., October 1-4, 1984, p35-1-35-24.

Keywords: *Standards, *Metrology, Documentation, National government, *National Institute of Standards and Technology, *Calibration, Standard reference materials, Federal agencies.

A status report is given for NBS calibration service and related activities. Discussed are recent and pending administrative changes in the program, and the budget/resource outlook for the next several years. A description is given of a number of recent NBS publications aimed at improving the quality and quantity of information available to users of NBS measurement services. Highlights of new NBS technical services recently initiated are also described. These include (1) a new MAP for the measurement coefficient of luminous intensity of retroreflectors, (2) the reestablishment of the calibration services for odd decade electrical resistors, (3) new NBS capabilities in x-ray dosimetry, (4) new NBS capabilities and services for microwave and antenna measurements, (5) improvements in dimensional metrology and pressure/vacuum services, (6) several new SRMs for use in measurement of temperature, the dimensions of fine particles, and for the evaluation of the performance of coordinate measuring machines.

900,879
PB89-173975 Not available NTIS
National Bureau of Standards (NML), Gaithersburg, MD. Electron Physics Div.
Precision Weight Calibration with a Specialized Robot.
Final rept.
R. D. Cutkosky. 1988, 6p
Pub. in Proceedings of Measurement Science Conference, Long Beach, CA., January 29-30, 1988, p1-6.

Keywords: *Weight measurement, *Automation, *Robots, Calibrating, Weight indicators, Automatic control, Balance.

A selected commercial top-loading balance with a range of 200 grams and a resolution of 10 micrograms has been adapted for use in conjunction with a specially designed robot arm configured to load and unload the balance in accordance with established weighing designs. The complete system includes a personal computer for control of the robot and the balance, a 6-

axis stepper motor controller, and a system for maintaining the balance, the robot, and the stored weights at a constant temperature.

900,880
PB89-175699 Not available NTIS
National Bureau of Standards (NML), Gaithersburg, MD. Temperature and Pressure Div.
NBS (National Bureau of Standards) Orifice-Flow Primary High Vacuum Standard.
Final rept.
K. E. McCulloh, C. R. Tilford, S. D. Wood, and D. F. Martin. 1986, 1p
Pub. in Jnl. of Vacuum Science and Technology A 4, n3 pt1 p362 May/Jun 86.

Keywords: *Standards, Orifice flow, Vacuum gages, Pressure, Reprints, *High vacuum, Spinning rotor gages, Molecular drag gages.

The National Bureau of Standards has constructed a primary high vacuum standard of the orifice-flow type that currently covers the range 10 to the -6 power to 10 to the -1 power Pa. The uncertainty of the standard is 3.4% at 10 to the -6 power pa, 1.4% at 10 to the -4 power pa, and 2.5% at 10 to the -1 power pa.

900,881
PB89-176135 Not available NTIS
National Bureau of Standards (NEL), Gaithersburg, MD. Building Physics Div.
Circular and Square Edge Effect Study for Guarded-Hot-Plate and Heat-Flow-Meter Apparatuses.
Final rept.
B. Peavy, and B. Rennex. 1986, 7p
Pub. in Jnl. of Thermal Insulation 9, p254-300 Apr 86.

Keywords: *Heat flow meters, *Size determination, *Shape, *Heat transmission, Circles(Geometry), Squares(Geometry), Thermal insulation, Thermal resistance, Thermal conductivity, Mathematical models, Reprints, *Edge effect, *Guarded hot plate.

The guarded-hot-plate apparatus measures thermal resistance across thermal insulation specimens. The measurement of very thick specimens results in an edge error due to the loss of some of the centrally metered heat to the peripheral ambient surrounding the specimen. To choose the appropriately sized apparatus for a particular test, it is important to know the order of magnitude of this edge error as a function of specimen thickness. The report defines the edge effect, derives a model for its calculation for both circular and square geometries, and indicates graphically the sensitivity of the edge-effect curves (as a function of thickness) with respect to the following parameters: the ratio of the guard to metered sizes, the metered size itself, the ratio of the surface heat transfer coefficient to the specimen apparent conductivity, and the apparent-thermal-conductivity anisotropy. The results of the square metered area are compared with those of the circular metered area, and the theoretical results for a circular geometry (using a 1-m guarded hot plate).

900,882
PB89-176432 Not available NTIS
National Bureau of Standards, Gaithersburg, MD. Office of Product Standards Policy.
Laboratory Accreditation Systems in the United States, 1984.
Final rept.
J. W. Locke. 1984, 13p
Pub. in Proceedings of Workshop and Symposium of National Conference of Standards Laboratories, Gaithersburg, MD., October 1-4, 1984, p29-1-29-13.

Keywords: *Laboratories, *Test facilities, Measurement, Standards, *Accreditation, Calibration.

Laboratory accreditation in the United States is growing as is demonstrated by the comparison of the newest survey conducted for the NBS Office of Product Standards Policy by the Marley Organization with the original study completed in 1979. The largest interest in accreditation comes from administrators of government funded programs (e.g. health care and defense) where accurate measurement is needed to justify payment of fees or contracts. There is a growing interest in accreditation by some product marketeers since the use of test data from accredited testing laboratories in specific areas can help sell products. There is little thrust toward developing a coordinated accreditation system to serve all users, although the continuing increase in the number of accreditation systems and laboratories seeking accreditation and the

expansion of export markets will tend to encourage such coordination. Verification of calibrations is important in most accreditation systems and is emphasized in defense systems, but in no case is the identification of uncertainties throughout the chain of measurement required.

900,883
PB89-176580 Not available NTIS
National Bureau of Standards (NML), Gaithersburg, MD. Gas and Particulate Science Div.
Trace Gas Calibration Systems Using Permeation Devices.
Final rept.
G. D. Mitchell. 1987, 11p
Pub. in ASTM (American Society for Testing and Materials) Special Technical Publication 957, p110-120 1987.

Keywords: *Permeameters, *Flow measurement, Temperature control, Stability, Gas analysis, Diffusion, Solubility, Reprints, *Calibration, *Trace gases.

The paper describes trace calibration systems and the factors that influence their operation. COGAS, a dynamic trace calibration system that serves as both a permeation device and an instrument calibration system, is presented. Such a system improves precision, reduces manpower requirements, and is versatile in its application.

900,884
PB89-176606 Not available NTIS
National Bureau of Standards (NEL), Gaithersburg, MD. Building Physics Div.
Summary of Circular and Square Edge Effect Study for Guarded-Hot-Plate and Heat-Flow-Meter Apparatuses.
Final rept.
B. Peavy, and B. Rennex. 1988, 26p
Pub. in Thermal Conductivity 19, p173-198 1988.

Keywords: *Heat flow meters, *Size determination, *Shape, *Thermal measuring instruments, *Heat transmission, Circles(Geometry), Squares(Geometry), Thermal insulation, Thermal resistance, Thermal conductivity, Mathematical models, Reprints, *Edge effect.

The report provides the information necessary to choose the appropriate apparatus size for a measurement of thermal resistance of a specimen of insulation material of a particular thickness. The information consists of the order of magnitude of apparatus edge error as a function of specimen thickness. The report defines the edge effect; derives a model for its calculation for both circular and square geometries, and indicates graphically the sensitivity of the edge-effect curves (as a function of thickness) with respect to the following parameters: the ratio of the guard to metered sizes, the metered size itself, the ratio of the surface heat transfer coefficient to the specimen apparent conductivity, and the apparent-thermal-conductivity anisotropy. The results of the square metered area are compared with those of the circular metered area, and the theoretical results are compared with the experimental results for a circular geometry (using a 1-m guarded hot plate).

900,885
PB89-177190 Not available NTIS
National Bureau of Standards (NEL), Gaithersburg, MD. Automated Production Technology Div.
Automated Fringe Counting Laser Interferometer for Low Frequency Vibration Measurements.
Final rept.
B. F. Payne. 1986, 7p
Pub. in Proceedings of International Instrumentation Symposium on Instrumentation in the Aerospace Industry (32nd), Seattle, WA., May 5-8, 1986, p1-7.

Keywords: *Accelerometers, *Vibration meters, *Transducers, Automation, *Laser interferometers, *Calibration, Computerized control systems.

Low frequency accelerometers and velocity transducers are widely used for vibration investigations on structures such as buildings, bridges, aircraft, ships, power plant equipment, and seismic applications. Previous work in this area has focused on the development of accurate calibration methods for transducers by optical methods in the frequency range of 2-100 Hz. The paper describes a computer-controlled fringe counting system for transducer calibration. The calibration system uses digital signal analysis for accurate low frequency voltage measurements. The measurement procedures are fully automated, with menu driven programs using the computer soft keys for controlling the test frequencies, acceleration levels, setting test parameters, collecting and storing data and producing reports and graphs. An error analysis is given and experimental data are presented for transducer calibrated on this system.

900,886
PB89-179162   Not available NTIS
National Bureau of Standards (NML), Gaithersburg, MD. Gas and Particulate Science Div.
Precision and Accuracy Assessment Derived from Calibration Data.
Final rept.
H. L. Rook. 1987, 6p
Pub. in ASTM (American Society for Testing and Materials) Special Technical Publication 957, p81-86 1987.

Keywords: *Precision, *Accuracy, *Assessments, Data processing, Error analysis, Bias, Standards, Reprints, *Calibration.

During the past ten years, the importance of assessing the precision and the accuracy of data derived from standard methods has become recognized. Both ASTM and EPA have developed policies requiring precision and accuracy assessment for methods before they are designated as standard methods or reference methods. The difficulty in implementing these policies is not in developing a meaningful error estimate for a given method, it is in separating the random from the systematic components of the error. In the paper, a method of separating error components into precision and bias is given. The method uses calibration data taken repeatedly at a fixed point using a standard whose total uncertainty is small compared to the uncertainty of the measurement method.

900,887
PB89-184089   PC A04
National Inst. of Standards and Technology, Gaithersburg, MD.
Journal of Research of the National Institute of Standards and Technology. Volume 94, Number 2, March-April 1989.
Bi-monthly rept.
1989, 58p
Also available from Supt. of Docs. as SN703-027-00027-0. See also PB89-184097 through PB89-184121 and PB89-133367.

Keywords: *Measurement, *Standards, Electrical measurement, Chromatography, Termal conductivity, Coaxial cables, Pneumatic lines, Spectroscopy, Supercritical fluids, Hot-wire flowmeters.

The journal contains the following articles: New international adopted reference standards of voltage and resistance; A supercritical fluid chromatograph for physicochemical studies; Relation between wire resistance and fluid pressure in the transient hot-wire method; and Scattering parameters representing imperfections in precision coaxial air lines.

900,888
PB89-186381   Not available NTIS
National Bureau of Standards (NEL), Boulder, CO. Statistical Engineering Div.
Calibration with Randomly Changing Standard Curves.
Final rept.
D. F. Vecchia, H. K. Iyer, and P. L. Chapman. 1989, 8p
Pub. in Technometrics 31, n1 p83-90 Feb 89.

Keywords: *Measuring instruments, Statistical samples, Regression analysis, Calibrating, Curves(Graphs), Reprints, *Calibration standards.

Changes in calibration curves from one time to the next caused by drift often require measuring devices to be recalibrated at frequent intervals. In such situations, the usual practice is to estimate the unknown values of test samples using only data from the corresponding calibration period. Under a random coefficient regression model for the different calibration curves, however, it can be shown that it is more efficient to combine the data from all calibration periods to estimate the unknowns. The authors consider a particular class of point estimators obtained by inverting suitable prediction functions and show that the estimator obtained from a best prediction function is optimal in a sense defined by Godambe and Durbin in the context of unbiased estimating equations. The small sample perform-

ance of the estimator is compared with the usual estimator using the Pitman closeness criterion.

900,889
PB89-186787   Not available NTIS
National Bureau of Standards (NEL), Boulder, CO. Chemical Engineering Science Div.
NBS (National Bureau of Standards)-Boulder Gas Flow Facility Performance.
Final rept.
S. E. McFaddin, J. A. Brennan, and C. F. Sindt. 1988, 4p
Sponsored by Gas Research Inst., Chicago, IL.
Pub. in American Gas Association Operating Section Proceedings, Toronto, Canada, May 16-18, 1988, p492-495.

Keywords: *Gas flow, *Test facilities, *Flow measurement, Performance, Efficiency, Gas pipelines, Temperature, Flowmeters, Precision, Reprints.

Major modifications have been made to the gas flow facility at the National Bureau of Standards in Boulder, Colorado. Significant improvements in steady state operation and overall efficiency have resulted. Variability in the gas temperature has been decreased by a factor of five and the precision of performance data on flowmeters has increased by a factor of two. The mass-based facility provides the gas industry with an accurate, efficient, and precise research facility operating at pipeline conditions.

900,890
PB89-189278   PC A04/MF A01
National Inst. of Standards and Technology, Gaithersburg, MD.
NVLAP (National Voluntary Laboratory Accreditation Program) Directory of Accredited Laboratories.
J. L. Donaldson, and J. Horlick. 1 Apr 89, 57p
NISTIR-89/4056
See also PB88-169529.

Keywords: *Directories, *Laboratories, Fields, Test facilities, Tests, Accreditation.

The 1989 NVLAP Directory of Accredited Laboratories provides information on the activities of the National Institute of Standards and Technology in administering the National Voluntary Laboratory Accreditation Program (NVLAP) during calendar year 1989. The status of current programs is briefly described and a summary of laboratory participation is provided. Indexes cross reference the laboratories by name, NVLAP Lab Code Number, accreditation program, and geographical location and cross reference NVLAP code numbers with test method designations. The scope of accreditation of each laboratory, listing the test methods for which it is accredited, is provided.

900,891
PB89-193841   PC A04/MF A01
National Inst. of Standards and Technology (NML), Gaithersburg, MD. Center for Basic Standards.
NIST (National Institute of Standards and Technology) Measurement Services: High Vacuum Standard and Its Use.
Interim rept.
S. Dittmann. Mar 89, 71p NIST/SP-250/34
Also available from Supt. of Docs as SN003-003-02934-3. Library of Congress catalog card no. 88-600605.

Keywords: *Standards, *High vacuum, *Pressure measurement, Vacuum gages, Test facilities, Flowmeters, Ionization, Resistance, Pressure, Orifices, Calibration standards.

The document presents an in-depth discussion of the National Institute of Standards and Technology primary high vacuum standard, used between 10 to the minus 6th power and .01 Pa. Included are discussions of the theory, design, and construction of the standard. The systematic and random errors in the standard and the methods used to check the accuracy of the standard are presented. Also included is a brief discussion of the molecular drag gage and its use as a transfer standard between .0001 and .1 Pa.

900,892
PB89-201164   Not available NTIS
National Bureau of Standards (NML), Gaithersburg, MD. Office of Standard Reference Data.

Standard Reference Materials for Dimensional and Physical Property Measurements.
Final rept.
L. J. Kieffer. 1986, 6p
Pub. in Proceedings of Measurement Science Conference, Irvine, CA., January 23-24, 1986, p167-172.

Keywords: *Standards, *Mechanical properties, *Measurement, Measuring instruments, Test facilities, *Standard Reference Materials.

An overview of Standard Reference Materials (SRMs) distributed by the National Bureau of Standards for use in standardizing physical property measurements is presented. In addition some reference materials (artifacts) to be used in standardizing dimensional metrology measurements are discussed in some detail.

900,893
PB89-201180   Not available NTIS
National Bureau of Standards (NML), Gaithersburg, MD. Temperature and Pressure Div.
Development of New Standard Reference Materials for Use in Thermometry.
Final rept.
B. W. Mangum. 1986, 11p
Pub. in Proceedings of Measurement Science Conference, Irvine, CA., January 23-24, 1986, p148-158.

Keywords: *Temperature measurement, *Standards, *Thermodynamic properties, Temperature measuring instruments, *Standard Reference Materials.

In recent years, several SRMs have been developed for use in thermometry. These cover the range from 0.015 K to 2326 K. The article will review the use and importance of thermometric fixed points in precision thermometry and discuss new developments in SRMs related to those fixed points.

900,894
PB89-201818   Not available NTIS
National Bureau of Standards (NML), Gaithersburg, MD. Length and Mass Div.
Stability of the SI (International System) Unit of Mass as Determined from Electrical Measurements.
Final rept.
R. S. Davis. 1989, 2p
Pub. in Metrologia 26, p75-76 1989.

Keywords: *Mass, *Standards, *Electrical measurement, Precision, Test facilities, Stability, Reprints, Calibration standards.

The apparatus developed by Kibble at the National Physical Laboratory has recently achieved an uncertainty smaller than 1 x 10 to the minus seventh power in measuring the ratio between the electrical watt as maintained and the watt as defined in the International System (SI). Since there is reason to anticipate further experimental improvements, the possibility of monitoring the stability of the SI unit of mass (the kilogram) through such an apparatus is being seriously discussed. The kilogram is the only remaining base unit of the SI still defined by an artifact. The letter points out that electrical measurements even now provide the most critical test of the long-term stability of the SI unit of mass.

900,895
PB89-209266   PC A04/MF A01
National Bureau of Standards (NEL), Gaithersburg, MD. Precision Engineering Div.
Length Scale Measurement Procedures at the National Bureau of Standards.
J. S. Beers. Sep 87, 72p NBSIR-87/3625

Keywords: *Interferometers, *Measurement, *Length, Optical measuring instruments, Performance, Calibrating, Precision, Test facilities.

Precision graduated length scales have been measured by interferometry at NBS since 1965. An instrument called the Line Scale Interferometer was designed for this purpose. The history, development, improvement, operation and evaluation of the line scale interferometer are described. Special emphasis is given to detailed operating procedures to provide guidance in the use of the instrument. Evaluating performance through a measurement assurance program is also emphasized.

900,896
PB89-209324   PC A07/MF A01

Laboratory & Test Facility Design & Operation

Virginia Polytechnic Inst. and State Univ., Blacksburg.
Dept. of Mechanical Engineering.
Development of an Automated Probe for Thermal Conductivity Measurements.
Final rept.
B. P. Dougherty, and W. C. Thomas. May 89, 131p
NIST/GCR-89/563
Grant NSNB-6-D0642
Sponsored by National Inst. of Standards and Technology (NEL), Gaithersburg, MD. Center for Building Technology.

Keywords: *Temperature measuring instruments, *Thermistors, *Thermal conductivity, Thermal resistance, Thermophysical properties, Liquids, Thermal measurements, Mathematical models, Calibrating.

A transient technique was validated for making thermal conductivity measurements. The technique incorporated a small, effectively spherical, heat source and temperature sensing probe. The actual thermal conductivity measurements lasted 30 seconds. After approximately 15 minutes of data reduction, a value for thermal conductivity was obtained. The probe yielded local thermal conductivity measurements. Spherical sample volumes less than 8 cu cm were required for the materials tested. Thermal conductivity (and moisture) distributors can be measured for relatively dry or wetted samples.

900,897
PB89-209340　　　　　　　　　PC A10/MF A01
National Inst. of Standards and Technology (NML), Gaithersburg, MD. Temperature and Pressure Div.
NIST (National Institute of Standards and Technology) Measurement Services: The Calibration of Thermocouples and Thermocouple Materials.
Final rept.
G. W. Burns, and M. G. Scroger. Apr 89, 204p NIST/SP-250/35
Also available from Supt. of Docs. as. SN003-003-02939-4. Library of Congress catalog card no. 89-600732.

Keywords: *Thermocouples, Test facilities, Platinum, Temperature measurement, Thermal measurements, Quality control, Mathematical models, Thermophysical properties, Rhodium, Statistics, Accuracy, *Calibration standards.

The document describes the calibration services for thermocouples and thermocouple materials presently provided by the NIST at temperatures from -210 to 2100 C. Three general calibration methods used at NIST, which provide traceability to the International Practical Temperature Scale of 1968, are outlined in detail. Consideration is given primarily to the calibration of type S (platinum-10% rhodium versus platinum) thermocouples at defining fixed points of the IPTS-68 as standard instruments, and to the calibration of other letter-designated type thermocouples (types B, R, S, E, J, K, N, and T) by comparison with a calibrated reference thermocouple and by comparison with a standard platinum resistance thermometer. The procedures followed to maintain internal quality control of apparatus and standards in the NIST thermocouple calibration laboratories are covered, and statistical assessments of the uncertainties involved in the calibrations of type S thermocouples by the fixed-point method and by the comparison method are presented.

900,898
PB89-211874　　　　　　Not available NTIS
National Inst. of Standards and Technology (NEL), Gaithersburg, MD. Chemical Process Metrology Div.
Gas Flow Measurement Standards.
Final rept.
G. E. Mattingly. 1989, 20p
Pub. in Proceedings of AIChE (American Institute of Chemical Engineers) Meeting, Houston, TX., April 1989, 20p.

Keywords: *Gas flow, *Flow measurement, *Standards, Air flow, Test facilities, Errors, Pressure, Temperature, Precision.

The National Institute of Standards and Technology (NIST), formerly the National Bureau of Standards, maintains the U.S. standards for gas flow measurement. These consist of a number of flow facilities that enable the arrangements and measurement of a wide range of fluid and flow conditions. In the paper, descriptions are given for the air flow-rate facilities at NIST-Gaithersburg, MD. The air flow measurement facilities maintained at NIST-Gaithersburg include a variety of timed-collection of fluid techniques that are used

to calibrate flowmeters and transfer standards. As well, these facilities are available for establishing and maintaining the different types of traceability needed by U.S. indus r and other government agencies, etc. These needs tare based upon the increased concerns for improved gas flow measurements because of the increasing values of the materials involved. Higher accuracies and precision levels are being sought not only in custody transfer between buyer and seller but also in the continuous process industries where optimal control can only be attained via precise measurements.

900,899
PB89-211882　　　　　　Not available NTIS
National Bureau of Standards (NEL), Gaithersburg, MD. Chemical Process Metrology Div.
Prediction of Flowmeter Installation Effects.
Final rept.
T. T. Yeh, and G. E. Mattingly. 1989, 32p
Pub. in Proceedings of AIChE (American Institute of Chemical Engineers) Spring National Meeting, Houston, TX., April 1989, p1-32.

Keywords: *Flowmeters, *Pipe flow, Fluid flow, Turbulent flow, Flow measurement, Pipes(Tubes), Velocity, Orifice flow, Laser doppler velocimeters.

The research program described is intended to improve flow meter performance in 'non-ideal' installation conditions. The program involves a procedure through which a strategy is proposed and evaluated for predicting flowmeter performance when the meter is installed too near elbows, reducers, flow conditioning elements, etc. The strategy is based upon understanding and parameterizing the salient features of the pipe flows produced by selected piping configurations and understanding how specific meters perform in these flows. The results described include: laser Doppler velocimetry (LDV) measurements of the mean and the turbulence velocities for the pipe flows produced by single and double elbow configurations -- the elbows-out-of-plane with different spacings between the elbows, the quantification of these secondary flows in the downstream piping, and the demonstration that the flowmeter prediction strategy works for selected turbine-type and orifice-type flowmeters in these flows.

900,900
PB89-211890　　　　　　Not available NTIS
National Bureau of Standards (NEL), Gaithersburg, MD. Chemical Process Metrology Div.
Prediction of Flowmeter Installation Effects.
Final rept.
T. T. Yeh, and G. E. Mattingly. 1989, 32p
Pub. in Proceedings of Institute of Gas Technology Symposium, Chicago, IL., June 1989, p1-32.

Keywords: *Flowmeters, *Pipe flow, Fluid flow, Turbulent flow, Flow measurement, Pipes(Tubes), Velocity, Orifice flow, Laser doppler velocimeters.

The research program described is intended to improve flow meter performance in 'non-ideal' installation conditions. The program involves a procedure through which a strategy is proposed and evaluated for predicting flowmeter performance when the meter is installed too near elbows, reducers, flow conditioning elements, etc. The strategy is based upon understanding and parameterizing the salient features of the pipe flows produced by selected piping configurations and understanding how specific meters perform in these flows. The results described include: laser Doppler velocimetry (LDV) measurements of the mean and the turbulence velocities for the pipe flows produced by single and double elbow configurations -- the elbows-out-of-plane with different spacings between the elbows, the quantification of these secondary flows in the downstream piping, and the demonstration that the flowmeter prediction strategy works for selected turbine-type and orifice-type flowmeters in these flows.

900,901
PB89-215370　　　　　　PC A03/MF A01
National Inst. of Standards and Technology (NEL), Gaithersburg, MD. Applied and Computational Mathematics Div.
Mechanism for Shear Band Formation in the High Strain Rate Torsion Test.
T. J. Burns. Jun 89, 46p NISTIR-89/4121

Keywords: *Torsion tests, Plastic flow, Shear flow, Shear properties, Strains, Conduction, Numerical analysis, Mathematical models.

An asymptotics argument is given, which shows that rigid unloading from the ends of the thin-walled tubular

specimen, enhanced by conductive heat transfer, is a plausible mechanism for adiabatic shear band formation during the high strain rate torsion test. The argument assumes that thickness variations, as well as elastic and dynamic effects in the tube, can be ignored, but that heat conduction and heat-sink thermal boundary conditions must be included. The proposed mechanism is supported by a numerical analysis of a mathematical model of the torsion test, which is based on recent torsional Kolsky bar experimental work of Marchand and Duffy (1988), on a physical model of thermoelastic-plastic flow due to Wallace (1985), and on a phenomenological Arrhenius model of the plastic flow surface. The numerical technique used is the semi-discretization method of lines.

900,902
PB89-228282　　　　　　PC A03/MF A01
National Bureau of Standards, Gaithersburg, MD. Associate Director of Industry and Standards.
Update of U.S. Participation in International Standards Activities.
P. W. Cooke. Jul 89, 25p NISTIR-89/4124
See also PB88-164165.

Keywords: *Standards, *United States, Standardization, Industrial engineering, *International Standards Organization, *International Electrotechnical Commission, International organizations.

The report presents updated information on the current level of U.S. participation in the two major international standardization bodies, International Standards Organization and International Electrotechnical Commission. Data on the new ISO/IEC Joint Technical Committee 1 on Information Technology are also presented.

900,903
PB89-228324　　　　　　PC A03/MF A01
National Inst. of Standards and Technology, Gaithersburg, MD. National Voluntary Lab. Accreditation Program.
NVLAP (National Voluntary Laboratory Accreditation Program) Assessment and Evaluation Manual.
R. L. Gladhill. Jul 89, 20p NISTIR-86/3853
See also PB85-200079.

Keywords: *Laboratories, Test facilities, Standards, Evaluation, Test facilities, Manuals, *Accreditation, *National Voluntary Laboratory Accreditation Program.

The National Voluntary Laboratory Accreditation Program (NVLAP), established in 1976, is administered by the National Institute of Standards and Technology (NIST) formerly the National Bureau of Standards (NBS). NVLAP is a voluntary system for assessing and evaluating testing laboratories and accrediting those found competent to perform specific test methods or types of test methods. Laboratory accreditation programs are established for specified product or service areas in response to requests and demonstrated need. The publication, intended for the NVLAP technical experts who serve as assessors and evaluators, describes general policies and practices of NVLAP assessment and evaluation. The specific technical criteria for assessing and evaluating laboratories are provided elsewhere, in NVLAP Handbooks and checklists for each technical area.

900,904
PB89-231104　　　　　　Not available NTIS
National Inst. of Standards and Technology (NML), Gaithersburg, MD. Temperature and Pressure Div.
InSb as a Pressure Sensor.
Final rept.
V. E. Bean. 1986, 7p
Pub. in Proceedings of Seminar on High Pressure Standards, Paris, France, May 24-25, 1988, p125-131.

Keywords: *Indium antimonides, Pressure sensors, Pressure measurement, Semiconductor devices, Sensitivity, Performance, Standards, *Pressure transducers.

The resistivity of InSb increases exponentially with pressure. A pressure transducer was made, based on InSb, that has nearly 70 times the sensitivity at 645 MPa of a manganin pressure transducer. The transducer has adequate sensitivity and short term stability to be used to measure the change of pressure generated by a controlled-clearance piston gage resulting from a change of jacket pressure from which the cylinder distortion coefficient can be determined. The long term performance data necessary to evaluate the suit-

900,905
PB89-231112 Not available NTIS
National Inst. of Standards and Technology (NML),
Gaithersburg, MD. Temperature and Pressure Div.
**Non-Geometric Dependencies of Gas-Operated
Piston Gage Effective Areas.**
Final rept.
C. R. Tilford, R. W. Hyland, and S. Yi-Tang. 1988, 9p
Pub. in Proceedings of Seminar on High Pressure Me-
trology, Paris, France, May 24-25, 1988, p105-113.

Keywords: *Pressure measurement, Manomet$_{e}$r$_{s}$,
Standards, Gases, *P$_{is}$on gages, Effective area.

Using a mercury manometer, the authors determined
the effective areas of different gas-operated piston
gages as a function of pressure, mode of operation
(absolute or differential, and gas species. They have
observed changes in the effective area of individual
gages that vary from zero to 25 ppm as these parameters are changed. Over the 5-160 kPa range of these
experiments, changes in the geometry of the pistons
and cylinders cannot explain these effects. These results demonstrate the need for a more refined theory
of the interaction of the pressure fluid and the piston/
cylinder. Until that is available, effective areas of primary standard piston gages calculated on the basis of
geometric factors alone can have significant uncertainties.

900,906
PB89-231120 Not available NTIS
National Inst. of Standards and Technology (NML),
Gaithersburg, MD. Temperature and Pressure Div.
**Observations of Gas Species and Mode of Oper-
ation Effects on Effective Areas of Gas-Operated
Piston Gages.**
Final rept.
B. E. Welch, R. E. Edsinger, V. E. Bean, and C. D.
Ehrlich. 1988, 14p
Pub. in Proceedings of Seminar on High Pressure
Standards, Paris, France, May 24-25, 1988, p81-94.

Keywords: *Pressure measurement, Helium, Neon, Nitrogen, Argon, Krypton, Carbon dioxide, *Piston
gages, Effective area.

The effective areas of four gas-operated piston gages
have been determined by the pressure calibration
technique with a state-of-the-art manometer in both
both helium and nitrogen in the absolute mode. For all
four gages, the effective areas with nitrogen are greater than the effective areas using helium. The differences range from 4 to 28 parts-per-million. Pairs of
these gages have been cross-floated in both the gage
and the absolute modes with helium, neon, nitrogen,
argon, carbon dioxide, and krypton. For a given gas,
the effective area in the absolute mode is greater than
that for the gage mode. The magnitude of the difference is dependent upon the species of gas.

900,907
PB89-231146 Not available NTIS
National Inst. of Standards and Technology (NEL),
Boulder, CO. Electromagnetic Fields Div.
**Advances in NIST (National Institute of Standards
and Technology) Dielectric Measurement Capabil-
ity Using a Mode-Filtered Cylindrical Cavity.**
Final rept.
E. J. Vanzura, and W. A. Kissick. 1989, 4p
Pub. in Proceedings of IEEE (Institute of Electrical and
Electronics Engineers) MTT-S International Micro-
wave Symposium, Long Beach, CA., p901-904 Jun 89.

Keywords: *Measuring instruments, *Cavity resonators, Microwave frequencies, Design, Performance
evaluation.

A 60-mm diameter cylindrical cavity resonator has
been constructed for performing high-accuracy permittivity measurements on low-loss materials at microwave frequencies. The cavity's design and evaluation
are described. Estimated errors in seven parameters
result in approximately 0.2% uncertainty in permittivity
and 6% uncertainty in loss tangent for a fused silica
measurement.

900,908
PB89-235634 PC A04
National Inst. of Standards and Technology, Gaithers-
burg, MD.

**Journal of Research of the National Institute of
Standards and Technology, Volume 94, Number 4,
July-August 1989.**
Bi-monthly rept.
1989, 73p
See also PB89-235642 through PB89-235675 and
Volume 94, Number 3; PB89-211106. Also available
from Supt. of Docs. as SN703-027-00029-6.

Keywords: *Spectroscopy, *Standards, *Measurement, Iodine, Tungsten, Fluids, Computer security.

Contents: Determination of trace level iodine in biological and botanical reference materials by isotope dilution mass spectrometry; The spectrum of doubly ionized tungsten (W III); Apparatus for neturon scattering
measurements on sheared fluids; Eleventh National
Computer Security Conference.

900,909
PB89-235915 PC A06/MF A01
National Inst. of Standards and Technology (NEL),
Gaithersburg, MD. Automated Production Technology
Div.
**Intercomparison of Load Cell Verification Tests
Performed by National Laboratories of Five Coun-
tries.**
Final rept.
R. A. Mitchell, S. L. Yaniv, K. Yee, and O. K.
Warnlof. Aug 89, 124p NISTIR-89/4101

Keywords: *Load cells, *Verifying, Measuring instruments, Temperature, Weight indicators, Mass, Test facilities, Metrology, *Calibration standards, International cooperation.

A round-robin intercomparison of OILM IR 60 load cell
verification tests, as performed by national laboratories of five countries, is reported. The five participating
countries were Australia, the Federal Republic of Germany, the Netherlands, the United Kingdom, and the
United States. Six OIML Class C load cells, ranging in
capacity from 18 kg to 25000 kg, were tested by the
five laboratories. The objective was to determine the
comparability of the results from the verification test
processes of the five laboratories, so that the laboratories could accept the results from any one laboratory
and avoid the cost of retesting. Overall, the test results
indicate reasonably good agreement among the five
laboratories in the measurement of most of the characteristics of the six load cells. The degree and pattern
of the differences in the results can serve as a guide to
making refinements in the verification test processes.

900,910
PB90-111220 PC A05/MF A01
National Inst. of Standards and Technology (NEL),
Gaithersburg, MD. Fluid Flow Group.
**NBS' (National Bureau of Standards) Industry;
Government Consortium Research Program on
Flowmeter Installation Effects: Summary Report
with Emphasis on Research July-December 1987.**
Summary rept.
G. E. Mattingly, and T. T. Yeh. Nov 88, 82p NISTIR-
88/3698
See also PB89-189120.

Keywords: *Flowmeters, *Pipe flow, *Turbulence,
Flow distribution, Fluid flow, Pipe bends, Secondary
flow, Flow measurement, Research projects, *Installation.

The objective of the research program is to produce
improved flowmeter performance when meters are installed in 'non-ideal' conditions. This objective is being
attained via a strategy to measure, understand, and
quantify the 'non-ideal' pipeflows from such pipeline
elements as elbows, reducers, valves, or combinations
of these; for selected types of flowmeters, correlate
meter factor 'shifts' relative to the features of these
'non-ideal' installations; and disseminate the resulting
technology through appropriate channels such as publishing results in pertinent journals and upgrading
'paper' standards for flow measurement.

900,911
PB90-111675 PC A05/MF A01
National Inst. of Standards and Technology (NEL),
Boulder, CO. Chemical Engineering Science Div.

**Optimum Location of Flow Conditioners in a 4-inch
Orifice Meter.**
Technical note.
S. E. McFaddin, C. F. Sindt, and J. A. Brennan. Jul
89, 84p NIST/TN-1330
Contract GRI-5088-271-1680
Also available from Supt. of Docs. as SN003-003-
02961-1. Sponsored by Gas Research Inst., Chicago,
IL.

Keywords: *Flow measurement, *Orifice meters,
*Position(Location), *Standards, Flow rate, Flow-
meters, Pressure measurement, Flow distribution, Experimental data, Test facilities.

Two orifice flow measurement standards are presently
used, one in the United States (ANSI/API 2530) and
another in Europe (ISO 5167). These two standards
have significantly different specifications for installations. One important specification is the location of the
flow conditioner relative to the orifice plate.

900,912
PB90-117938 Not available NTIS
National Inst. of Standards and Technology (NML),
Gaithersburg, MD. Radiation Physics Div.
**Use of Thorium as a Target in Electron-Spin Ana-
lyzers.**
Final rept.
J. J. McClelland, M. R. Scheinfein, and D. T. Pierce.
1989, 5p
Sponsored by Department of Energy, Washington, DC.
Pub. in Review of Scientific Instruments 60, n4 p683-
687 Apr 89.

Keywords: *Electron spin, *Analyzers, Electron scattering, Gold, Thin films, Reprints, *Thorium 230 target,
Mott scattering, Sherman tables.

Measurements of the effective Sherman function have
been carried out for 10-100-keV spin-polarized electrons scattering from a thick thorium target in a retarding Mott analyzer. At 20 and 100 keV the dependence
on the maximum energy loss accepted by the detector
has been measured. Comparison is made with scattering from a 1250 A gold film. Thorium is seen to have a
S(eff) up to 30% higher than gold. The higher S(eff)
can not only improve the figure of merit of a spin detector, but also lessen its sensitivity to instrumental asymmetries. Comparison is also made with preliminary theoretical results. Good agreement between theory and
experiment is seen in the thorium Sherman function
relative to that of gold.

900,913
PB90-127820 PC A04/MF A01
National Inst. of Standards and Technology (NML),
Gaithersburg, MD. Office of Physical Measurement
Services.
**NIST (National Institute of Standards and Technol-
ogy) Calibration Services, Users Guide: Fee
Schedule.**
Special pub.
Mar 89, 70p NIST/SP-250-APP/89ED

Keywords: *Fees, *Metrology, Measuring instruments,
Test facilities, Measurement, Setting(Adjusting),
Standards, *Calibration, *National Institute of Stand-
ards and Technology.

The physical measurement services of the National Institute of Standards and Technology are designed to
help the makers and users of precision instruments
achieve the highest possible levels of measurement
quality and productivity. The hundreds of individual
services listed in the Fee Schedule constitute the high-
est-order calibration services available in the United
States. They directly link a customer's precision equipment or transfer standards to national measurement
standards. These services are offered to public and
private organizations and individuals alike.

900,914
PB90-128273 Not available NTIS
National Inst. of Standards and Technology (NEL),
Gaithersburg, MD. Statistical Engineering Div.
Bootstrap Inference for Replicated Experiments.
Final rept.
W. Liggett. 1988, 6p
Pub. in Computing Science and Statistics, Proceedings
of Symposium on the Interface (20th), Fairfax, VA.,
p68-73 Apr 88.

Laboratory & Test Facility Design & Operation

Keywords: *Experimental design, *Replicating, *Statistical inference, Error analysis, Estimating, Robustness(Mathematics).

Inference methods valid for nonnormal error are proposed for experiments in which each design point is replicated three or more times. Differences between the replicates provide the data needed for a pooled estimate of the error density, and the density forms the basis for the bootstrap. The density estimator is specified for symmetric error, and the symmetric estimator has been generalized to asymmetric error. In the paper, the application of the density estimator to designed experiments is considered. The lack-of-fit test is of particular interest. The extension of the density estimator to data requiring a blocking variable and to data with dispersion effects is discussed. The bootstrap based on the density estimator is shown to be valid for smaller sample sizes when the test statistics are robust. Estimation of the error density is illustrated with measurements replicated at different laboratories.

Manufacturing Processes & Materials Handling

900,915
PB89-147003 Not available NTIS
National Bureau of Standards (ICST), Gaithersburg, MD. Advanced Systems Div.
Notion of Granularity.
Final rept.
C. P. Kruskal, and C. H. Smith. 1988, 14p
Pub. in Jnl. of Superconducting 1, p395-408 1988.

Keywords: *Grain size, Definitions, Fineness, Comparison, Metal industry, Reprints, Parallel processing.

Granularity is a well known concept in parallel processing. While intuitively, the distinction between coarse-grain and fine-grain parallelism is clear, there is no rigorous definition. The paper develops two notions of granularity, each defined formally and represented by a single rational number. The two notions are compared and contrasted with each other and with previously proposed definitions of granularity.

900,916
PB89-156384 PC A03/MF A01
National Bureau of Standards (NEL), Gaithersburg, MD. Factory Automation Systems Div.
Integrated Manufacturing Data Administration System (IMDAS) Operations Manual.
D. A. Nickerson. 21 Apr 89, 21p NBSIR-88/3743
See also PB88-177290.

Keywords: *Data administration, Manufacturing, Planning, Manuals, *Automated manufacturing.

The report is an operator's manual for the Integrated Manufacturing Data Administration System (IMDAS) of the Automated Manufacturing Research Facility (AMRF). The IMDAS is designed to provide the control systems of the AMRF access to the data necessary to support the design, planning, manufacturing, and inspection of parts. The manual describes the necessary steps to place the IMDAS into service within the AMRF.

Nondestructive Testing

900,917
PB89-151625 PC A04/MF A01
National Inst. of Standards and Technology, Gaithersburg, MD.
Institute for Materials Science and Engineering, Nondestructive Evaluation: Technical Activities 1988.
Annual rept.
H. T. Yolken. Oct 88, 72p NISTIR-88/3839
See also PB88-153655.

Keywords: *Nondestructive tests, *Ceramics, *Metals, *Powder(Particles), *Reviews, Fabrication, Consolidation, Formability, Composite materials, Processing, Interfaces, Standards, Graphs(Charts), Ultrasonics, Thermal properties, Radiography.

The report provides brief reviews of technical activities in nondestructive evaluation (NDE) that were carried out by or for the National Institute of Standards and Technology (NIST--formerly the National Bureau of Standards) in fiscal year 1988 (October 1, 1987, through September 30, 1988). The reviews in the annual report are arranged in the following sections that reflect the NDE Program's four major activity areas: NDE for Ceramic and Metal Powder Production and Consolidation; NDE for Formability of Metals; NDE for Composites Processing and Interfaces; and NDE Standards and Methods. Each of the sections is preceded by an introduction.

900,918
PB89-187579 Not available NTIS
National Inst. of Standards and Technology (NEL), Boulder, CO. Electromagnetic Technology Div.
Optical Power Measurements at the National Institute of Standards and Technology.
Final rept.
T. R. Scott. 1989, 11p
Pub. in Proceedings of Measurement Science Conference, Anaheim, CA., January 26-27, 1989, p3C-19-3C-29.

Keywords: *Fiber optics, *Power measurement, *Calorimeters, Lasers, Thermal measurements, Standards, Detectors, Calibrating, Telecommunications, Light transmission, Optical communication.

The measurement of optical power (that is, laser power or energy at wavelengths and power levels of interest to the fiber optic community) at NIST is based upon a standard reference calorimeter called the C-series calorimeter. The C-series calorimeter is a national reference standard for measuring absolute energy and power levels of cw laser sources over a wide range of wavelengths. Various infrared laser sources and a calibrated beamsplitter measurement system are used to compare an electrically calibrated pyroelectric radiometer (ECPR) to the C-series calorimeter. The calibrated ECPR is then used as a laboratory standard for the calibration of measurement of optical power. The measurement of optical power at NIST is reviewed starting with a discussion of the primary reference standard and the associated measurement system. The system used for calibrating optical power detectors is discussed and the associated uncertainties are identified.

900,919
PB89-193866 PC A03/MF A01
National Inst. of Standards and Technology (NEL), Gaithersburg, MD. Center for Mfg. Engineering.
Acoustical Technique for Evaluation of Thermal Insulation.
D. R. Flynn, D. J. Evans, and T. W. Bartel. Apr 89, 47p NISTIR-88/3882
Sponsored by Department of Energy, Washington, DC. Building Systems Div., and Mineral Insulation Mfrs. Association, Alexandria, VA.

Keywords: *Thermal insulation, *Acoustic measurement, Thermal resistance, Sound transmission, Heat transfer, Evaluation, Cellulose, Houses, Thermal efficiency, Fiberglass, Rockwool, Blown-in-place attic insulation.

A laboratory apparatus has been constructed that enables rapid measurement of the sound insertion loss of a sample of insulation as a function of frequency. An extensive series of measurements of the sound insertion losses associated with blown samples of fiberglass, rockwool, and cellulose has been completed. The results of these acoustical measurements are highly correlated with coverage (mass per unit area) and thermal resistance (R-value). An investigation is planned to extend the acoustical techniques used in the laboratory apparatus to in-situ determination of the sound transmission loss through thermal insulation installed in attics. Two possible approaches to such field measurements are described.

900,920
PB89-202006 Not available NTIS
National Bureau of Standards (IMSE), Gaithersburg, MD. Metallurgy Div.
Sensors for Intelligent Processing of Materials.
Final rept.
H. N. G. Wadley. 1986, 5p
Pub. in Jnl. of Metals 38, n10 p49-53 1986.

Keywords: *Process control, *Sensors, *Nondestructive tests, Ultrasonic frequencies, Eddy current tests, Acoustics, Mathematical models, Reprints.

To implement new process control strategies including intelligent Processing of Materials strategies advanced sensors are required to nonintrusively evaluate process and microstructure variables. Examples of emerging sensors based upon ultrasonics, eddy currents and acoustic emission and other new nondestructive evaluation methods are described. In general, it is becoming evident that sophisticated sensor can reduce the dependence upon quantitative process models and vice versa. It is advisable therefore to assess sensor needs and process models together when developing the control scenario for a new process.

900,921
PB89-211924 Not available NTIS
National Bureau of Standards (IMSE), Gaithersburg, MD. Metallurgy Div.
Acoustic Emission: A Quantitative NDE Technique for the Study of Fracture.
Final rept.
H. N. G. Wadley. 1987, 16p
Pub. in Proceedings of ONR Symposium on Solid Mechanics Research for QNDE, Evanston, IL., September 18-20, 1985, p25-40 1987.

Keywords: *Nondestructive tests, *Acoustic absorption, *Emission, *Crack propagation, Structural forms, Wave propagation, Mechanics, Dislocations(Materials).

Acoustic emission, an NDE technique that shows promise for detecting and locating cracks in engineering structures, has been used as an experimental technique for the basic study of fracture. Examples of the use of acoustic emission for the latter purpose are reviewed, and future research opportunities are identified.

900,922
PB90-123415 Not available NTIS
National Inst. of Standards and Technology (IMSE), Gaithersburg, MD. Metallurgy Div.
Mossbauer Imaging: Experimental Results.
Final rept.
U. Atzmony, S. J. Norton, L. J. Swartzendruber, and L. H. Bennett. 1987, 2p
Pub. in Nature 330, n6144 p153-154 1987.

Keywords: *Mossbauer effect, Nondestructive tests, Reprints, *Imaging techniques, Iron 57.

The letter reports the first experimental demonstration of the concept of Mossbauer imaging. For simplicity, a one-dimensional imaging experiment is described; however, the fundamental imaging principle thus demonstrated has obvious extensions to high dimensions. Finally, speculations on some possible applications of the technique in materials science are made.

900,923
PB90-128679 Not available NTIS
National Inst. of Standards and Technology (IMSE), Boulder, CO. Fracture and Deformation Div.
Improved Standards for Real-Time Radioscopy.
Final rept.
T. A. Siewert. 1989, 3p
Pub. in Proceedings of Nondestructive Evaluation NDE Planning and Application Conference, Honolulu, HI., July 23-27, 1989, p95-97.

Keywords: *Standards, *Radiology, Questionnaires, Nondestructive tests, Fracture tests, Deformation methods, *Real-time radioscopy.

The National Institute of Standards and Technology is assisting in the development of a radiation transfer standard for real-time radioscopy (RTR). The report describes and discusses the replies to a questionnaire which was developed to quantify the parameters with which RTR systems are used, and to identify the appropriate features of such a transfer standard.

900,924
PB90-128687 Not available NTIS
National Inst. of Standards and Technology (IMSE), Boulder, CO. Fracture and Deformation Div.
Standards for Real-Time Radioscopy.
Final rept.
T. Siewert. 1988, 5p
Pub. in Proceedings of Defense Conference on Nondestructive Testing (37th), Jacksonville, FL., November 1-3, 1988, p161-165.

Keywords: *Radiology, *Standards, Questionnaires, Nondestructive tests, Fracture tests, Deformation methods, *Real-time radioscopy.

Standards developed to measure the quality of film radiographs are unable to evaluate all the features of real-time systems. Particularly, problems exist when the specimen is rotated to a degree where the image quality indicators are no longer orthogonal to the beam or when the evaluation is performed while the specimen is in motion. The report describes the development of new standards that will allow quantitative evaluation of real-time systems under these conditions. Other NBS activities, such as a survey of real-time usage and the development of a 150 kV radiation transfer standard, will also be presented.

900,925
PB90-132739   PC A05/MF A01
National Inst. of Standards and Technology (IMSE), Gaithersburg, MD.
Institute for Materials Science and Engineering, Nondestructive Evaluation: Technical Activities, 1989.
Annual rept.
H. T. Yolken. Nov 89, 77p NISTIR-89/4147
See also PB89-151625.

Keywords: *Nondestructive tests, Ceramics, Powder metallurgy, Sintering, Machined, Formability, Ultrasonic tests, Surface finishing, Polymers, Composite materials, Standards, Eddy currents, Research, Radiography, Thermal analysis, National Institute of Standards and Technology, Acoustic emission.

A review of the Nondestructive Evaluation Program at NIST for fiscal year 1989 is presented in the annual report. Topics include the following: Intelligent processing of rapidly solidified metal powders; Nondestructive characterization of ceramic sintering; Monitoring of machined ceramic surfaces by thermal waves; Eddy current temperature sensing; Ultrasonic sensor for sheet metal formability; Ultrasonic metrology for surface finish and part thickness; Measurement and control of polymer processing parameters using fluorescence spectroscopy; Nondestructive evaluation of diamond films; Transient elastic waves in laminates; Intelligent processing of solder joint connections for printed wiring assemblies; Ultrasonics and acoustic emission; Real-time x-ray radioscopy; Magnetic methods and standards for NDE; Eddy current techniques; New standard test methods for characterizing performance of thermal imaging systems; and Capacitive array research for characterization of ceramics.

Quality Control & Reliability

900,926
PB89-200216   PC A10/MF A01
National Inst. of Standards and Technology (NML), Gaithersburg, MD. Office of Physical Measurement Services.
NIST (National Institute of Standards and Technology) Calibration Services Users Guide. 1989 Edition.
Special pub.
J. D. Simmons. Jan 89, 212p NIST/SP-250/89ED
Also available from Supt. of Docs. as SN003-003-02909-2. See also PB87-174041.Color illustrations reproduced in black and white.

Keywords: *Calibration, *Measurement, Standards, Services, Quality assurance, Tests, Fees.

The National Institute of Standards and Technology (NIST) Calibration Services Guide provides detailed descriptions of currently available NIST calibration services, measurement assurance programs, and special-test services. The following measurement areas are covered: (1) dimensional; (2) mechanical, including flow, acoustic, and ultrasonic; (3) thermodynamic; (4) optical radiation; (5) ionizing radiation; and (6) electromagnetic, including dc, ac, rf, and microwave. A separate Fee Schedule is issued as required, providing current prices for the services offered, updates on points-of-contact, and information on measurement seminars.

LIBRARY & INFORMATION SCIENCES

Information Systems

900,927
PB89-170864   PC A04/MF A01
National Inst. of Standards and Technology (NEL), Gaithersburg, MD. Center for Computing and Applied Mathematics.
Internal Structure of the Guide to Available Mathematical Software.
R. F. Boisvert, S. E. Howe, and J. L. Springmann. Mar 89, 55p NISTIR-89/4042
See also PB84-171305.

Keywords: *Catalogs(Publications), *Mathematics, *Statistical analysis, *Applications of mathematics, *Computer software, *Computer software catalog, *Scientific data, *On line systems, *Classification, *Data base management systems.

The purpose of the NIST Guide to Available Mathematical Software (GAMS) project is to provide convenient documentation tools for users and maintainers of scientific computer software. The main components of this effort are a detailed tree-structured, problem-oriented classification scheme for mathematical and statistical software, a printed catalog based upon this classification scheme which integrates information about all available software, an on-line interactive version of this catalog, and a relational database containing all information upon which the on-line and off-line catalogs rely, along with associated maintenance programs. The report presents a detailed specification of the internal structure of the GAMS database and the programs used to manipulate it. The information is useful to those who wish to implement the GAMS systems on their own computer systems.

900,928
PB89-193874   PC A03/MF A01
National Inst. of Standards and Technology (NCTL), Gaithersburg, MD. Office Systems Engineering Group.
Document Interchange Standards: Description and Status of Major Document and Graphics Standards.
J. Moline. Sep 88, 37p NISTIR-88/3851

Keywords: *Documents, *Standards, *Document circulation, Information retrieval, Document storage, Computer graphics.

Document interchange standards have emerged in response to two distinct needs. First, there is the need to interchange documents among workstations and tools in the office environment. Second, there is the need to exchange versions of a document between an author and a publisher. The document describes standards which attempt to satisfy those needs. Each relevant standard is presented in summary form and includes the following information: standard name, standard number, status, scope, description, use, and references.

900,929
PB89-228993   PC A04/MF A01
National Inst. of Standards and Technology (NCSL), Gaithersburg, MD. Information Systems Engineering Div.
Detailed Description of the Knowledge-Based System for Physical Database Design. Volume 1.
Internal rept. Jul 85-Dec 88.
C. E. Dabrowski. Aug 89, 63p NISTIR-89/4139/1
See also Volume 2, PB89-229033.

Keywords: *Data base management systems, Factor analysis, Data structures, Information systems, Artificial intelligence, Mathematical models, Design, *Knowledge-based systems.

A knowledge-based system for physical database design has been developed at the National Computer Systems Laboratory. The system was previously de-

scribed in NIST Special Publication 500-151. The report is a follow-up report to that publication which describes the knowledge base for the system in detail. The description includes a complete explanation of each component of the knowledge base together with the actual rules used by the system.

900,930
PB89-229033   PC A09/MF A01
National Inst. of Standards and Technology (NCSL), Gaithersburg, MD. Information Systems Engineering Div.
Detailed Description of the Knowledge-Based System for Physical Database Design. Volume 2.
Internal rept. Jul 85-Dec 88.
C. E. Dabrowski. Aug 89, 177p NISTIR-89/4139/2
See also Volume 1, PB89-228993.

Keywords: *Data base management systems, Instructions, Information systems, Artificial intelligence, Mathematical models, *Knowledge-based systems.

A knowledge-based system for physical database design has been developed at the National Computer Systems Laboratory. The system was previously described in NIST Special Publication 500-151. The report is a follow-up report to that publication which describes the knowledge base for the system in detail. The description includes a complete explanation of each component of the knowledge base together with the actual rules used by the system.

900,931
PB90-112467   PC A03/MF A01
National Inst. of Standards and Technology (NCSL), Gaithersburg, MD.
Use of the IRDS (Information Resource Dictionary System) Standard in CALS (Computer-Aided Acquisition and Logistic Support).
D. K. Jefferson, and C. M. Furlani. Sep 89, 16p NISTIR-89/4169

Keywords: *Information systems, *Data management systems, Dictionaries, Architecture(Computers), Distributed data bases, Standards, Models, Information processing languages, *Computer-aided Acquisition and Logistic Support.

The objective of the point paper is to show how the Information Resource Dictionary System (IRDS) can fulfill critical design and operational requirements for CALS Phase II. First, a series of assumptions are made about the data management services which are needed by CALS Phase II. Next, these assumptions are used to develop a series of requirements for a dictionary system. The structure of the IRDS family of standards is then described. Examples are provided to illustrate how the IRDS could meet the requirements. A schedule is presented to show that the IRDS and other data management standards will be available when needed to meet the immediate requirements of CALS. An architecture is presented to illustrate additional standards required to achieve longer-range goals of distributed database. Finally, development tasks are recommended.

Reference Materials

900,932
PB89-160014   PC A06/MF A01
National Inst. of Standards and Technology, Gaithersburg, MD. Information Resources and Services Div.
Data Bases Available at the National Institute of Standards and Technology Research Information Center.
Special pub. (Final).
D. Cunningham. Nov 88, 117p NIST/SP-753
Supersedes PB88-153754. Also available from Supt. of Docs. as SN003-003-02903-3. Library of Congress catalog card no. 88-600602.

Keywords: *Information systems, *Directories, Information centers, *Bibliographic data bases, *National Institute of Standards and Technology, *Data bases.

Data bases available online at the National Institute of Standards and Technology (NIST) Research Information Center are listed by acronym and by full title. In addition, descriptions of the data bases, periods of coverage, producers, corresponding hard copy titles and principal sources and vendors are listed. A general subject index and a cross reference index are also supplied.

900,933
PB89-185599   PC A05/MF A01
National Bureau of Standards, Gaithersburg, MD.
Directory of NVLAP (National Voluntary Laboratory Accreditation Program) Accredited Laboratories, 1986-87.
H. W. Berger. Jan 87, 99p NBSIR-87/3519
See also PB88-169529.

Keywords: *Directories, *Laboratories, Tests, Projects, Position(Location), Accreditation.

The report lists laboratories accredited under the procedures of the National Voluntary Laboratory Accreditation Program (NVLAP) as of January 1, 1987. Indexes cross reference the laboratories by name, NVLAP Lab Code Number, test method, accreditation program, and geographical location. The scope of accreditation of each laboratory, listing the test methods for which it is accredited, is provided along with a tabulation of test methods and the laboratories accredited for those test methods.

General

900,934
PB89-147052   Not available NTIS
National Bureau of Standards (NEL), Gaithersburg, MD. Statistical Engineering Div.
Theory and Practice of Paper Preservation for Archives.
Final rept.
A. Calmes, R. Schofer, and K. Eberhardt. 1988, 16p
Pub. in Restaurator: International Jnl. for the Preservation of Library and Archival Material 9, p96-111 1988.

Keywords: *Documents, Papers, *Archives, Operations research, Reprints, *Preservation.

The task of preserving huge quantities of paper records may appear so overwhelming that an archivist may not know where to begin or how best to use his limited resources. The paper addresses those difficulties and offers suggestions and a model for reducing the overall paper preservation problem into manageable and efficient subtasks. Based on a National Archives/National Bureau of Standards study, the archival/conservation principles employed are: sample surveying, careful planning, protective packaging, use of copies instead of originals, monitoring the condition of records as they are being used, and professional conservation treatment.

900,935
PB89-214753   PC A03/MF A01
National Inst. of Standards and Technology (NCSL), Gaithersburg, MD.
Electronic Publishing: Guide to Selection.
Final rept.
L. S. Rosenthal. Jun 89, 39p NIST/SP-500/164
Also available from Supt. of Docs. as SN003-003-02938-6. Library of Congress catalog card no. 89-600734.

Keywords: *Publishing, *Composition, *Documentation, Computer applications, Printing, Instructions, Documents, Fonts, Typography, Data processing, *Electronic publishing.

The purpose of the report is to assist managers and users in making informed decisions on which systems are best for them. The report presents the technical and managerial choices and implications associated with selecting and using electronic publishing systems. A matrix of publishing capabilities and features is presented in the appendix to illustrate one method of comparing and selecting a publishing system.

MANUFACTURING TECHNOLOGY

Computer Aided Design (CAD)

900,936
PB89-151799   PC A06/MF A01
Catholic Univ. of America, Washington, DC.
Design Protocol, Part Design Editor, and Geometry Library of the Vertical Workstation of the Automated Manufacturing Research Facility at the National Bureau of Standards.
T. R. Kramer, and J. S. Jun. 28 Jan 88, 113p
NISTIR-88/3717
Grant NANB-5-D0522
Sponsored by National Bureau of Standards, Gaithersburg, MD.

Keywords: Design, Machining, Planning, *Computer aided design, *Computer aided manufacturing, interactive systems, Workstations, Automated Manufacturing Research Facility.

In the Vertical Workstation (VWS) of the NBS Automated Manufacturing Research Facility, metal parts are machined automatically from a feature-based design. A simple two-and-a-half dimensional part may be designed and machined within an hour, allowing half the time for design input. Workstation activity may be divided into design, process planning, data execution, and physical execution stages. The design of a part is expressed as a list of features on a block-shaped workpiece. Each feature is a removed volume. A feature is expressed by giving the name of the feature type and values for several parameters appropriate to that feature type. The design editor is an interactive system that runs on a Sun computer which is used to create or change designs. The system engages the user in a dialog to determine what the user wants to do, and prepares a design according to the user's instructions. The editor draws a three-view picture of the part being edited. The geometry library is a set of LISP functions that do geometric calculations to support the operation of the design editor and other modules of the VWS software.

900,937
PB89-166102   PC A06/MF A01
National Inst. of Standards and Technology (NEL), Gaithersburg, MD. Center for Building Technology.
Guidelines for the Specification and Validation of IGES (Initial Graphics Exchange Specification) Application Protocols.
R. J. Harrison, and M. E. Palmer. Jan 89, 111p
NISTIR-88/3846
Prepared in cooperation with Sandia National Labs., Albuquerque, NM.

Keywords: *Data converters, *Standards, Proving, *Computer aided design, *Initial Graphics Exchange Specification, Data management, Computer applications.

The document provides a background discussion of product data, describes the concept of IGES (Initial Graphics Exchange Specification) application protocols, specifies the technical content of an IGES application protocol, describes a validation methodology for these application protocols, and provides guidelines for the implementation of an IGES application protocol. A key conclusion of the background discussion is that IGES application protocols must be developed in order to achieve consistent and reliable exchanges of product data within specified application areas. The technical content of an IGES application protocol includes a conceptual information model for the application area with its supporting documentation, an application protocol format specification with a protocol usage guide, and a set of application protocol format test cases. These test cases must be used in concert with a well-defined testing methodology. Since no complete IGES application protocols currently exist, the document describes a current implementation of an application protocol process that is based on a partially complete application protocol.

900,938
PB90-112426   PC A03/MF A01
National Inst. of Standards and Technology (NEL), Gaithersburg, MD. Factory Automation Systems Div.
Product Data Exchange: The PDES Project-Status and Objectives.
B. M. Smith. Sep 89, 12p NISTIR-89/4165
See also PB89-144794.

Keywords: *Data, *Standardization, Product development, Standards, Mathematical models, *Computer aided design, *Product Data Exchange Specification, Digital data.

The paper details the strategy behind the development of the Product Data Exchange Specification (PDES) project, identifies the various technical resources that have been brought together to develop, standardize and use PDES technology, gives the status of the effort as of early 1989 and enumerates project plans for the balance of the year.

900,939
PB90-112434   PC A03/MF A01
National Inst. of Standards and Technology (NEL), Gaithersburg, MD. Factory Automation Systems Div.
External Representation of Product Definition Data.
B. M. Smith. Sep 89, 11p NISTIR-89/4166

Keywords: *Data, *Standards, Standardization, Product development, *Computer aided design, Computer aided manufacturing, International Organization for Standardization.

The ability to exchange product data files among a variety of different vendor CAD/CAM systems is critical to both a company's internal plans for integration and its external relationships with contractors and customers. Therefore, several national projects are being coordinated through the International Organization for Standardization (ISO) to develop a single world standard for data exchange. In addition to geometry, the standard will support a wide range of non-geometry data such as features, tolerance specifications, material properties and surface finish specifications. The geometry model will include solid representations for both boundary and constructive solid geometry forms. The geometry model coupled with the non-geometric data and the relationship information preserved from the sending system will enable the standard to communicate a complete product model. The paper details the strategy behind the development of the ISO project, identifies the various technical resources that have been brought together to develop, standardize and uses product data technology, gives the status of the effort as of early 1989 and enumerates project plans for the balance of 1989.

Computer Aided Manufacturing (CAM)

900,940
PB89-144794   PC A99/MF E16
National Inst. of Standards and Technology (NEL), Gaithersburg, MD. Factory Automation Systems Div.
Product Data Exchange Specification: First Working Draft.
Interim rept.
B. Smith, and G. Rinaudot. Dec 88, 2513p NISTIR-88/4004
Portions of this document are not fully legible.

Keywords: *Product development, *Data, *Standardization, Standards, *Product Data Exchange Specification, Computer aided manufacturing.

The document contains a neutral format for the representation and communication of product data. Known as the Product Data Exchange Specification (PDES), the document has been developed by the IGES/PDES Organization with active cooperation from the Working Group 1 of ISO/TC184/SC4. It represents the first working draft of PDES for presentation to the international community.

900,941
PB89-150809   Not available NTIS
National Bureau of Standards (NEL), Gaithersburg, MD. Factory Automation Systems Div.

Automated Documentation System for a Large Scale Manufacturing Engineering Research Project.
Final rept.
H. M. Bloom, and C. E. Wenger. 1983, 10p.
Pub. in Proceedings of International Conference on Systems Documentation (2nd), Seattle, WA., April 29-30, 1983, 10p.

Keywords: Data storage, *Automated Manufacturing Research Facility, *Computer aided manufacturing, *Software engineering, Information processing, Computer communications, Data management systems.

The Automated Manufacturing Research Facility (AMRF) being implemented at the National Bureau of Standards (NBS) will involve the development of a software manufacturing system integrating the various information processing, communications and data storage functions required in a totally automated environment. For such a research environment, there exists a need to develop an automated system for generating documentation for the software life cycle that could be used for the following purposes: (1) tracking progress of individual module development; (2) allowing for the availability of up-to-date information on module description to other members of the project who need to interface to a given module; (3) developing a cross reference of module and data element relationships; and (4) generating working level documentation that can be easily modified and serve as information to be given to any one with interest in the project. The paper describes the structure of a software development system that will function in a research environment.

900,942
PB89-151823   PC A04/MF A01
National Inst. of Standards and Technology (IMSE), Gaithersburg, MD. Office of Nondestructive Evaluation.
Intelligent Processing of Materials: Report of an Industrial Workshop Conducted by the National Institute of Standards and Technology.
H. T. Yolken, and L. Mordfin. Jan 89, 51p NISTIR-89/4024

Keywords: *Materials, *Processing, *Artificial intelligence, Plastics processing, Detectors, Nondestructive tests, Polymers, Ceramics, Metallurgy, Isostatic pressing, *Expert systems, Thermomechanical treatment, National institute of Standards and Technology.

Intelligent processing of materials has been established as a major new program area in the Institute for Materials Science and Engineering, National Institute of Standards and Technology (NIST). The goal of the program is to develop some of the generic scientific and technological bases for intelligent processing and, by means of selected pilot or demonstration projects, to encourage American industry to pursue and to adopt this powerful new approach to materials processing. In developing the new program, NIST cooperated in organizing two national workshops in 1985-86 to help identify the principal industrial needs in this field of technology and to solicit guidance for program planning activities. On August 30 and September 1, 1988, NIST convened a third workshop in this series. This one was comprised primarily of selected industrial specialists and was designed to define the specific materials processes upon which the NIST program should focus, and to discuss suitable approaches for accomplishing the work. The report documents the results of that workshop.

900,943
PB89-159636   PC A04/MF A01
National Bureau of Standards, Gaithersburg, MD.
Data Handling in the Vertical Workstation of the Automated Manufacturing Research Facility at the National Bureau of Standards.
T. R. Kramer. 21 Apr 88, 70p NBSIR-88/3763
Grants NANB-5-D0522, NANB-7-H0716
Sponsored by Catholic Univ. of America, Washington, DC.

Keywords: Machining, Tooling, Tables(Data), *Automated Manufacturing Research Facility, *Data management systems, Computer aided manufacturing, Workstations, Data base management, Automatic programming.

In the Vertical Workstation (VWS) of the NBS Automated Manufacturing Research Facility, metal parts are machined automatically from a feature-based design. A simple two-and-a-half dimensional part may be de-

signed and machined within an hour, allowing half the time for design input. Workstation activity may be divided into design, process planning, data execution, and physical execution stages: The VWS requires data for: features, machining operations, tooling, fixturing, workpieces, trays and other items. Hierarchical LISP property lists are used to store most of this data. The VWS also exchanges data with a global database. A local database manager is used to handle all exchanges with the global database and also to emulate the global database. Data exchanges take place through structured reports. The VWS system includes a subsystem which automatically writes LISP functions to read and write reports.

900,944
PB89-160634   PC A03/MF A01
Catholic Univ. of America, Washington, DC.
Parser That Converts a Boundary Representation into a Features Representation.
T. R. Kramer. Feb 89, 22p NISTIR 88/3864
Grant NANB7H0716
Sponsored by National Inst. of Standards and Technology (NEL), Gaithersburg, MD. Center for Mfg. Engineering.

Keywords: Pattern recognition, Boundaries, *Automated Manufacturing Research Facility, *Computer aided manufacturing, Parsing algorithms, Computer aided design, Workstations, Feature extraction.

The VWS2 B-rep Parser is a computer program written in LISP that takes a file giving the boundary representation of a part as input and produces a file giving a feature-based representation of the part as output. The format of the input file is a Product Data Exchange Specification (PDES)/Standard for The Exchange of Product data (STEP) boundary representation, and the format of the output file is that required by the VWS2 system of the National Institute of Standards and Technology (NIST) Automated Manufacturing Research Facility (AMRF). The parser deals with a limited range of two-and-a-half dimensional parts. The general approach to parsing is to expect that the part is parsable and look for arrangements of faces which are the signatures of features. The initial implementation of the approach recognizes five feature types. The approach is extendible to a wider range of feature and subfeature types, and to parts which have features made from several sides. Parts having features which intersect in a complex manner are likely to test the limits of this approach, or be beyond the limits. With the addition of this parser, the AMRF Vertical Workstation is capable of making a part from a PDES/STEP file without human intervention.

900,945
PB89-172571   PC A03/MF A01
National Inst. of Standards and Technology (NEL), Gaithersburg, MD. Center for Mfg. Engineering.
Artificial Intelligence Techniques in Real-Time Production Scheduling.
W. J. Davis, and A. T. Jones. Feb 89, 20p NISTIR-88/3891
Prepared in cooperation with Illinois Univ. at Urbana-Champaign. Dept. of General Engineering.

Keywords: *Production planning, *Scheduling, *Artificial intelligence, Decision making, Automation, *Computer aided manufacturing, Real time.

The paper addresses the real-time production scheduling problem as a special case of a much larger class of real-time decision-making/control problems. The paper first reviews the definition of the scheduling problem, and then reviews an earlier algorithm proposed by the authors to address this problem. It then concentrates on the possible application of various AI techniques to many of the functions that make up that algorithm.

900,946
PB89-172589   PC A03/MF A01
National Inst. of Standards and Technology (NEL), Gaithersburg, MD. Center for Mfg. Engineering.
Functional Approach to Designing Architectures for Computer Integrated Manufacturing.
W. J. Davis, and A. T. Jones. Feb 89, 20p NISTIR-88/3872
Prepared in cooperation with Illinois Univ. at Urbana-Champaign. Dept. of General Engineering.

Keywords: *Production planning, *Scheduling, *Manufacturing, *Automation, Mathematical models, Design, Hierarchies, *Computer integrated manufacturing, Computer systems design.

Developing effective CIM architecture is hampered by automation and integration problems. The key to resolving these problems lies in a better understanding of each manufacturing function and how it is related to other manufacturing functions. The authors view is that mathematical models can provide this understanding. The paper presents the results of their initial efforts to develop such models. They can be used to guide the development of the technology needed for automation. They also specify the inputs, outputs, and interrelations needed for integration, regardless of the specific CIM architecture used.

900,947
PB89-172597   PC A03/MF A01
National Inst. of Standards and Technology (NEL), Gaithersburg, MD. Center for Mfg. Engineering.
Real-Time Optimization in the Automated Manufacturing Research Facility.
W. J. Davis, R. H. F. Jackson, and A. T. Jones. Feb 89, 27p NISTIR-88/3865
Prepared in cooperation with Illinois Univ. at Urbana-Champaign. Dept. of General Engineering.

Keywords: *Scheduling, *Routing, *Hierarchical control, Optimization, *Factory automation, *Automated Manufacturing Research Facility, *Flexible manufacturing, Real time.

A major manufacturing research facility has been established at the National Institute of Standards and Technology. The Automated Manufacturing Research Facility has been designed to address the standards and measurement needs for the factory of the future. A five-layer hierarchical planning/control architecture is under development to manage production and support activities. A three layer architecture is being developed to manage the data requirements of the modules within that hierarchy. Each of these architectures contain functions that require the solution to one or more optimization problems. The paper describes both the production planning/control and the data management architectures being developed at NBS. It emphasizes the optimization problems contained within those architectures. It also discusses the work underway at NBS to address some of those problems.

900,948
PB89-172605   PC A03/MF A01
National Inst. of Standards and Technology (NEL), Gaithersburg, MD. Center for Mfg. Engineering.
On-Line Concurrent Simulation in Production Scheduling.
W. J. Davis, and A. T. Jones. Feb 89, 27p NISTIR-88/3837
Prepared in cooperation with Illinois Univ. at Urbana-Champaign. Dept. of General Engineering.

Keywords: *Production control, *Computerized simulation, Scheduling, Industrial engineering, Computer programming, Numerical control, *Computer aided manufacturing, *On-line programming, *Flexible manufacturing systems, *Automated Manufacturing Research Facility.

Flexible manufacturing systems (FMS) have been installed in many factories around the world. Production scheduling is the function responsible for assigning FMS resources to various manufacturing tasks. On-line simulation is being used as an analysis tool to choose among several candidate scheduling rules. The paper defines on-line simulation, and describes the inputs to and outputs from the on-line simulation trails. It also addresses the statistical analysis of those outputs to determine the 'best' compromise scheduling rule. Finally, it presents results from some preliminary scheduling experiments on the Automated Manufacturing Research Facility (AMRF) at the National Bureau of Standards.

900,949
PB89-172613   PC A03/MF A01
National Inst. of Standards and Technology (NEL), Gaithersburg, MD. Center for Mfg. Engineering.
Hierarchies for Computer-Integrated Manufacturing: A Functional Description.
W. J. Davis, and A. T. Jones. Feb 89, 30p NISTIR-88/3744
Prepared in cooperation with Illinois Univ. at Urbana-Champaign. Dept. of General Engineering.

Keywords: *Automatic control, *Hierarchical control, Decision making, Mathematical models, *Computer in-

111

MANUFACTURING TECHNOLOGY

Computer Aided Manufacturing (CAM)

tegrated manufacturing, *Hierarchies, Organizational structure.

In the recent past, several hierchies have been proposed as candidate models for the integration of decision and control functions within a Computer-Integrated Manufacturing environment. A common theme in the definition of these models is to construct an analog to managerial hierarchies that are currently employed in many corporate settings. The paper will adopt an alternate approach. Rather than defining a hierarchy, the paper will discuss the manufacturing functions that a CIM hierarchy must address. Whenever possible, mathematical formulations for the functions will be given with consideration for the stochastic environment in which they will function. The conclusion outlines several concerns arising in the definition of a generic CIM hierarchy and associated research topics that must be addressed.

900,950
PB89-176663 Not available NTIS
National Bureau of Standards (NEL), Gaithersburg, MD. Precision Engineering Div.
Vertical Machining Workstation of the AMRF (Automated Manufacturing Research Facility): Equipment Integration.
Final rept.
E. B. Magrab. 1986, 18p
See also PB87-218368.
Pub. in Proceedings of ASME (American Society of Mechanical Engineers) Winter Annual Meeting on Integrated and Intelligent Manufacturing, Anaheim, CA., December 7-12, 1986, p83-100.

Keywords: *Machining, Automation, Robots, Manipulators, *Automated Manufacturing Research Facility, *Workstations.

The integration and automation of the equipment comprising the vertical machining workstation (VMW) of the Automated Manufacturing Research Facility are presented. The workstation consists of a CNC vertical machining center, robot, vacuum chip removal system, robot cart delivery system, NBS designed grippers and modified pneumatic vise, and a hydraulic clamping system. In conjunction with the workstation controller, this VMW has attained a high level of sophistication and flexibility. Described in detail are the rules and assumptions governing the workstation's equipment operation, its equipment control software structure, the operations of the various pieces of equipment and the placement and type of sensors used to ensure proper execution of its command sets.

900,951
PB89-177083 Not available NTIS
National Bureau of Standards (NEL), Gaithersburg, MD. Robot Systems Div.
Real-Time Control System Software: Some Problems and an Approach.
Final rept.
L. S. Haynes, and A. J. Wavering. 1986, 12p
Pub. in Proceedings of IEEE (Institute of Electrical and Electronics Engineers) International Conference on Robotics and Automation, San Francisco, CA., p1705-1716 1986.

Keywords: Computer systems programs, Robots, *Automated Manufacturing Research Facility, *Computer aided control systems, *Computer software, Real time systems, Robotics, Software engineering.

The Automated Manufacturing Research Facility (AMRF) is currently composed of six workstations, six robots, an automated material handling system, buffer storage, a networking system for communication, a database system to support the needs of the facility and over 100 sensors throughout the facility to provide continued monitoring of all processes. One of the major challenges of the facility is to implement a real-time control system which continually reads the sensors within portions of the facility, interprets those readings, makes decisions as to required actions or changes, and then effects the required change via commands to actuators, manipulators, or subsystems. During development of the AMRF it became clear that real-time control software is fundamentally different, and more complex than scientific or business type systems. The paper discusses the reasons why this is true and describes the NBS Real-Time Control System, a system designed to help deal with these problems.

900,952
PB89-177091 Not available NTIS

National Bureau of Standards (NEL), Gaithersburg, MD. Factory Automation Systems Div.
AMRF (Automated Manufacturing Research Facility) Material Handling System Architecture.
Final rept.
C. R. McLean, and C. E. Wenger. 1986, 10p
Pub. in Proceedings of Annual Control Engineering Conference (5th), Rosemont, IL., May 6-8, 1986, p40-49.

Keywords: *Materials handling, Materials handling equipment, Computer systems hardware, *Automated Manufacturing Research Facility, *Computer aided manufacturing, Workstations, Computer software, Data bases.

The paper describes the architecture of the Automated Manufacturing Research Facility (AMRF) Material Handling Workstation and interface techniques that are used to integrate the system with other factory components. The material handling system (MHS) is comprised of two automatically guided vehicles, tray roller tables, a storage and retrieval system, control computers, and a tender terminal to coordinate manual support services. These services include: kitting, tray loading, tool setup, and raw material preparation. The architectural aspects of the system that are presented include: the hardware components of the system, a description of major software modules, material handling work element definitions, the programming of handling operations via process plans, the execution of these plans by the workstation controller, database structures and communications interfaces.

900,953
PB89-177109 Not available NTIS
National Bureau of Standards (NEL), Gaithersburg, MD. Factory Automation Systems Div.
Software for an Automated Machining Workstation.
Final rept.
T. R. Kramer, and J. S. Jun. 1986, 36p
Pub. in Proceedings of Biennial International Machine Tool Technical Conference (3rd), McLean, VA., September 3-10, 1986, p12-9-12-44.

Keywords: *Machining, *Automation, Computer systems programs, Manufacturing, *Workstations, *Computer software, Computer aided manufacturing, Computer aided design, Control systems.

Software written in LISP is used to control the activities of a workstation developed at the National Bureau of Standards. The controlled activities include the design-using parametric programming, process planning, and data and physical execution.

900,954
PB89-185607 PC A10/MF A01
National Bureau of Standards, Gaithersburg, MD.
Turning Workstation in the AMRF (Automated Manufacturing Research Facility).
A. Donmez, R. Gavin, L. Greenspan, K. Lee, V. Lee, J. Peris, E. Reisenauer, C. Shoemaker, and C. Yang. 20 Apr 88, 207p NBSIR-88/3749
See also PB87-218368. Sponsored by Naval Research Lab., Washington, DC. Navy Manufacturing Technology Program.

Keywords: *Manufacturing, *Automatic control equipment, *Turning(Machining), Components, Loading, Revisions, Robots, Detection, Failure, Controllers, Manipulators, *Flexible manufacturing workstation, Collet changing.

The Turning Workstation is a flexible manufacturing workstation developed in the Automated Manufacturing Research Facility (A.M.R.F.) at the National Bureau of Standards. The development of the workstation addressed some of the problems associated with an unattended turning operation which include tool changing, collet loading, collet changing, flexible robot end-effectors, and machine malfunction detection. The document describes the components of the Turning Workstation and its relationship to the A.M.R.F.

900,955
PB89-189286 PC A04/MF A01
National Inst. of Standards and Technology, Gaithersburg, MD.
Workstation Controller of the Cleaning and Deburring Workstation.
R. J. Norcross. 16 Feb 89, 70p NISTIR-89/4046
See also PB88-194279.

Keywords: Metal cleaning, Deburring, Robots, Automation, *Automated Manufacturing Research Facility, *Workstations, *Control systems.

The Cleaning and Deburring Workstation at NIST's Automated Manufacturing Research Facility employs two robots and numerous supporting equipment to wash, buff, and deburr discreet metal workpieces. The manual describes the workstation controller for perspective users and for researchers interested in expanding the workstation's capabilities. The manual describes the general control problem and provides the theoretical foundation of the solution along with specific implementation details. These details include important data structures, programming formats, interfaces, and the current operation of the controller and workstation.

900,956
PB89-193882 PC A06/MF A01
National Inst. of Standards and Technology, Gaithersburg, MD.
NBS AMRF (National Bureau of Standards) (Automated Manufacturing Research Facility) Process Planning System: System Architecture.
P. F. Brown, and S. R. Ray. Mar 89, 117p NISTIR-88/3826
See also PB87-234050.

Keywords: Process control, Automation, Research projects, Control systems, Architecture, Research facilities, *Process planning, *Automated Manufacturing Research Facility, Computerized control systems.

The purpose of the document is to provide a general description of design and implementation of the Automated Manufacturing Research Facility (AMRF) Process Planning System. The document should provide the reader with an understanding of the concepts behind the work in the process planning project as well as on the approach adopted. Details on system implementation are provided.

900,957
PB89-201727 Not available NTIS
National Bureau of Standards (IMSE), Gaithersburg, MD. Office of Nondestructive Evaluation.
Automated Processing of Advanced Materials. The Path to Maintaining U.S. Industrial Competitiveness in Materials.
Final rept.
H. T. Yolken, and L. Mordfin. 1986, 4p
Pub. in ASTM (American Society for Testing and Materials) Standardization News 14, n10 p32-35 Oct 86.

Keywords: *Process control, Powder metals, Production controls, Mechanical properties, Process variables, Reprints, *Computer-aided manufacturing, Nondestructive analysis.

Automation of materials processing offers a path to maintaining U.S. industrial competitiveness in materials. The key components of a systems approach to automating materials processing are discussed. These components included: the process, a process model, NDE sensors for materials properties and process variable sensors, process controllers, and an expert computer system to integrate and control the operation of the system. An example is present of a model systems approach at the National Bureau of Standards (NBS) involving the production of rapidly solidified metal powders by high pressure gas atomization. NBS invites joint sponsorship and active participation by a consortium of industrial firms.

900,958
PB89-209233 PC A03/MF A01
National Inst. of Standards and Technology, Gaithersburg, MD.
Inventory of Equipment in the Cleaning and Deburring Workstation.
F. M. Proctor, and R. Russell. 11 May 89, 16p
NISTIR-89/4092
See also PB89-176663.

Keywords: *Machining, *Inventories, *Control equipment, Cleaning, Deburring, Computer software, Robots, *Workstations, *Automated Manufacturing Research Facility, Computer hardware.

The manual provides a complete inventory of equipment in the Cleaning and Deburring Workstation of the Automated Manufacturing Research Facility of the National Institute of Standards and Technology.

900,959
PB89-209258 PC A03/MF A01
National Inst. of Standards and Technology (NEL),
Gaithersburg, MD. Center for Mfg. Engineering.
Data Management Strategies for Computer Integrated Manufacturing Systems.
W. Davis, A. Jones, and S. Ram. Jun 89, 25p
NISTIR-88/4002
Prepared in cooperation with Illinois Univ. at Urbana-
Champaign. Dept. of General Engineering, and Arizona Univ., Tucson. Dept. of Management Information Systems.

Keywords: Automation, Marketing, Sales, Production engineering, Scheduling, Planning, Inventory control, Quality assurance, Models, *Computer aided manufacturing, *Data management, Data bases, Data base administrators.

A recent worldwide trend to improve productivity in manufacturing has centered around the adoption of computer technology. Efforts are underway in many plants to use that technology to automate and integrate all manufacturing functions. This is transforming those plants into computer integrated manufacturing (CIM) systems. The paper addresses some of the special problems that have been and will be encountered in designing data management strategies for CIM. It describes both the major manufacturing functions themselves and the data required to carry out those functions. It also includes discussions on the various alternatives for data placement, data modeling, data administration, and data communications for CIM.

900,960
PB89-215198 PC A03/MF A01
National Inst. of Standards and Technology (NEL),
Gaithersburg, MD. Factory Automation Systems Div.
Use of GMAP (Geometric Modeling Applications Interface Program) Software as a PDES (Product Data Exchange Specification) Environment in the National PDES Testbed Project.
K. L. Perlotto. Jun 89, 26p NISTIR-89/4117
Prepared in cooperation with Pratt and Whitney Aircraft Group, East Hartford, CT. Sponsored by Department of Energy, Washington, DC. Office of Buildings and Community Systems.

Keywords: *Production models, *Test facilities, Specifications, Manufacturing, Computer programs, *Computer aided manufacturing, GMAP programming language, Computer architecture, Computer applications, National Institute of Standards and Technology, Automated Manufacturing Research Facility.

The report is a basic guide to the use of the GMAP System Architecture as installed on the NIST AMRF VAX as part of the National PDES Testbed Project. An overview of the GMAP System Architecture is provided. The use of the GMAP software to create an implementation environment for the PDES Draft Proposal Specification (February 1989) is outlined. The software organization on the NIST AMRF VAX and the development of test and validation applications are described. The GMAP System Architecture consists of software system components which meet the three basic requirements of an automated product data environment. The requirements are data definition, application support, and data exchange. The system components are defined and the role they play is described.

900,961
PB89-215339 PC A03/MF A01
National Inst. of Standards and Technology (NEL),
Gaithersburg, MD. Automated Production Technology Div.
Inventory of Equipment in the Turning Workstation of the AMRF (Automated Manufacturing Research Facility).
K. Lee. Jun 89, 12p NISTIR-88/3810·
See also PB89-185607 and PB89-215347. Sponsored by Naval Research Lab., Washington, DC. Navy Manufacturing Technology Program.

Keywords: *Computer aided Manufacturing, *Turning(Machining), *Automatic control equipment, *Robots, Manipulators, Components, Detectors, Controllers, Loading, Microcomputers, *Flexible manufacturing workstation, Collect changing.

The manual serves as an inventory guide to all electronic and mechanical systems in the Automated Turning Workstation at the Automated Manufacturing Research Facility (AMRF).

900,962
PB89-215347 PC A03/MF A01
National Inst. of Standards and Technology (NEL),
Gaithersburg, MD. Automated Production Technology Div.
Recommended Technical Specifications for Procurement of Equipment for a Turning Workstation.
K. Lee. Jun 89, 49p NISTIR-88/3611
See also PB89-215339. Sponsored by Naval Research Lab., Washington, DC. Navy Manufacturing Technology Program.

Keywords: *Turning(Machinery), *Specifications, *Robots, Machine tools, Procurement, Nuclear powered ships, Submarines, *Workstations, *Computer aided manufacturing, Automated Manufacturing Research Facility.

The manual serves as a technical guide to the specifications required to procure commercially available, major components such as a turning center and a robot system for an automated turning workstation.

900,963
PB89-221873 PC A03/MF A01
National Inst. of Standards and Technology (NEL),
Gaithersburg, MD. Center for Computing and Applied Mathematics.
AutoMan: Decision Support Software for Automated Manufacturing Investments. User's Manual.
Final rept.
S. F. Weber, B. C. Lippiatt, and K. S. Johnson. Aug 89, 45p NISTIR-89-4116, NBS/SW/DK-89/006A
For system on diskette, see PB89-221741. Sponsored by Office of the Assistant Secretary of the Navy, Washington, DC. Mfg. Technology Program.

Keywords: *Investments, *Fixed investment, Cost effectiveness, Economic analysis, Benefit cost analysis, Automation, Performance evaluation, Documentation, *Computer aided manufacturing, *Decision support systems, Microcomputers, User manuals(Computer programs).

The manual documents AutoMan, a microcomputer program designed to support multi-criteria decisions about automated manufacturing investments: The program permits users to combine quantitative and qualitative criteria in evaluating investment alternatives. Quantitative criteria could include such traditional financial measures as Life-Cycle Cost and Net Present Value as well as such engineering performance measures as throughput and setup time. Qualitative criteria could include flexibility and product quality. AutoMan combines ratings with criteria weights into an overall rating for each investment alternative and then ranks alternatives. AutoMan comes with sample decision models and a manual that includes a detailed tutorial, a glossary of evaluation criteria, a bibliography, and an index.

900,964
PB90-112350 PC A03/MF A01
National Inst. of Standards and Technology (NEL),
Gaithersburg, MD. Factory Automation Systems Div.
Experience with IMDAS (Integrated Manufacturing Data Administration System) in the Automated Manufacturing Research Facility.
E. Barkmeyer, and J. Lo. Sep 89, 23p NISTIR-89/4132

Keywords: Automation, *Automated Manufacturing Research Facility, *Distributed computer systems, *Data base management systems, Computer aided manufacturing, National Institute of Standards and Technology.

The National Institute of Standards and Technology has been operating a locally-developed distributed data management system in its Automated Manufacturing Research Facility (AMRF) since mid-1987. The system, called the Integrated Manufacturing Data Administration System (IMDAS), front-ends existing databases and database management systems with a unified conceptual model and a common data manipulation language and interface. The paper first describes the AMRF, a totally automated manufacturing plant in microcosm, using commercially available equipment for the most part, and its data and data repositories. It then describes the operation and performance of IMDAS in the AMRF. Finally, it looks at areas for performance improvement in the IMDAS and draws conclusions about the usage of IMDAS for production manufacturing.

900,965
PB90-112459 PC A03/MF A01
National Inst. of Standards and Technology (NEL),
Gaithersburg, MD. Factory Automation Systems Div.
Generic Architecture for Computer Integrated Manufacturing Software Based on the Product Data Exchange Specification.
J. E. Fowler. Sep 89, 29p NISTIR-89/4168

Keywords: *Computer aided manufacturing, *Computer graphics, *Architecture(Computers), Computer software, Programming languages, Data bases, Data processing systems, Interfaces, Specifications, *Computer integrated manufacturing.

The Product Data Exchange Specification (PDES) is an emerging standard that is intended to address the problems of data exchange and representation for a variety of manufacturing enterprises. The National Institute of Standards and Technology (NIST) has a long-standing research program that addresses the problems of integration and development of automated manufacturing systems. The document presents a software architecture that forms the basis for the incorporation of PDES into the software applications that are part of NIST's work in Computer Integrated Manufacturing (CIM).

900,966
PB90-128596 Not available NTIS
National Inst. of Standards and Technology (NEL),
Gaithersburg, MD. Factory Automation Systems Div.
Modular Process Planning System Architecture.
Final rept.
S. R. Ray. 1989, 5p
Sponsored by Department of Defense, Washington, DC.
Pub. in Proceedings of IIE Integrated Systems Conference, Atlanta, GA., November 12-15, 1989, p1-5.

Keywords: *Planning, Systems engineering, Automation, *Computer aided manufacturing, *Computer architecture, Integrated systems.

A general purpose architecture for a modular process planning system is presented. Based upon emerging national standards in manufacturing, it offers easy integration among planning subsystems.

900,967
PB90-129446 PC A03/MF A01
National Inst. of Standards and Technology, Gaithersburg, MD.
AMRF Part Model Extensions.
A. Barnard. 5 Oct 89, 32p NISTIR-89/4189

Keywords: Standards, Specifications, Mathematical models, Topology, *Automated Manufacturing Research Facility, *Data management, *Computer aided manufacturing, Product Data Exchange Specification, Format, Data base management systems.

The document specifies the addition of ellipses, hyperbolas, parabolas, and b-splines to the AMRF Part Model Report Format. The reports are used throughout the AMRF to communicate part model data between application processes and the global AMRF database. Part model data consists of geometry, topology, features, and tolerances. The document is intended to be used by programmers implementing systems that make use of AMRF part model data. While the report is a complete description of the extensions, the document 'AMRF Database Report Format: Part Model' by T.H. Hopp is also required to complete the grammar.

900,968
PB90-132713 PC A04/MF A01
Catholic Univ. of America, Washington, DC.
Enhancements to the VWS2 (Vertical Workstation 2) Data Preparation Software.
T. R. Kramer. 1 Nov 89, 66p NISTIR-89/4201
Sponsored by National Inst. of Standards and Technology, Gaithersburg, MD.

Keywords: *Computer software, *Data processing, *Manufacturing, Automatic control, Computer aided manufacturing, Automatic programming, Algorithms, Vertical workstation.

In the Vertical Workstation (VWS) of the NIST Automated Manufacturing Research Facility, metal parts are machined automatically from a feature-based design. Workstation activity may be divided into

113

MANUFACTURING TECHNOLOGY

Computer Aided Manufacturing (CAM)

design, process planning, data execution, and physical execution stages. The first three of these are data preparation stages. Major enhancements to the VWS2 data preparation software included: Development of new methods for creating contour outlines; Complete rebuilding of the text system; Introduction of tolerance information; A parser that extracts features from a boundary representation; Improvements in the Part Design Editor; An algorithm for the three-dimensional sculpting of contour grooves.

Engineering Materials

900,969
PB89-201768 Not available NTIS
National Bureau of Standards (IMSE), Gaithersburg, MD.
Versailles Project on Advanced Materials and Standards Evolution to Permanent Status.
Final rept.
L. Schwartz, and B. Steiner. 1986, 5p
Pub. in ASTM (American Society for Testing and Materials) Standardization News 14, n10 p40-44 Oct 86.

Keywords: *Ceramics, *Standardization, *Polymers, Cryogenics, Superconductors, Composite materials, International relations, Corrosion, Creep properties, Reprints, *Versailles Project on Advanced Materials and Standards.

The international activities in advanced materials started at the Versailles Heads of State meeting is described. The motivation for the collaboration, the various technical activities under way, and the organizational structure are summarized.

900,970
PB90-136672 Not available NTIS
National Inst. of Standards and Technology (IMSE), Boulder, CO. Fracture and Deformation Div.
Measuring In-Plane Elastic Moduli of Composites with Arrays of Phase-Insensitive Ultrasound Receivers.
Final rept.
D. W. Fitting, R. D. Kriz, and A. V. Clark. 1989, 8p
Sponsored by National Research Council, Washington, DC.
Pub. in Review of Progress in Quantitative Nondestructive Evaluation, v8B p1497-1504 1989.

Keywords: *Composite materials, *Phased arrays, *Modulus of elasticity, *Ultrasonic tests, Nondestructive tests, Measurement, Wave phases, Oriented fiber composites, Velocity, Anisotropy, Reprints, Graphite-epoxy composites.

Ultrasonic measurements of elastic moduli of composite materials have traditionally been made on either small specimens cut from a larger component or by using a scanning technique with the specimen immersed in a water bath. A phase-insensitive array has been developed for rapid nondestructive measurement of in-plane elastic moduli. The array characteristics allow for determination of energy flux deviation angle in the composite, as well as measurement of group and phase velocity. Techniques for computing the phase velocity from array measurements are described. Energy flux deviation in unidirectional graphite-epoxy specimens has been determined with the array. Comparisons are made herein of results with an analytical (bulk wave) predictions.

Job Environment

900,971
PB89-161697 PC A04/MF A01
Maryland Univ., College Park. Dept. of Mechanical Engineering.
Transient Cooling of a Hot Surface by Droplets Evaporation.
Final rept.
M. di Marzo, F. Kavoosi, and M. Klassen. Nov 88, 65p NIST/GCR-89/559
See also PB87-145421. Sponsored by National Inst. of Standards and Technology (NEL), Gaithersburg, MD. Center for Fire Research.

Keywords: *Evaporative cooling, *Drops(Liquids), *Extinguishing, Fire protection, Metal plates, Fire safety, Aluminum, Computerized simulation, Interfaces, Vaporizing.

The report describes the research performed during the period March 1987 - July 1988 under a joint research program between the Mechanical Engineering Department of the University of Maryland and the Center for Fire Research of the National Bureau of Standards. The formulation of a model for the prediction of the cooling induced by an evaporating droplet impinging a semi-infinite solid is the subject of the report. The thermal interactions during the evaporation of a liquid droplet deposited on a low conductivity semi-infinite solid are complex because the evaporative process is coupled to the solid intense local cooling. Numerical techniques based on finite difference methods have failed to provide meaningful results. This is due to the sharp temperature gradients in the proximity of the droplet edge which cause instabilities in the solution for reasonable time steps due to the explicit coupling of the liquid-vapor regions. An integral method was proposed by Dr. Baum (CFR-NBS) in order to overcome difficulties. The methodology and its application to this specific problem is described in detail.

Joining

900,972
PB90-117391 Not available NTIS
National Inst. of Standards and Technology (IMSE), Boulder, CO. Fracture and Deformation Div.
On-Line Arc Welding: Data Acquisition and Analysis Using a High Level Scientific Language.
Final rept.
G. Adam, and T. A. Siewert. 1989, 7p
Pub. in Intelligent Instruments and Computers, p14-20 Jan/Feb 89.

Keywords: *Gas metal arc welding, *Welding, Data acquisition, Electrical measurement, Electric current, Electric potential, Reprints, *Data analysis, Personal computers.

A personal computer was used to monitor and record the current and voltage during gas metal arc welding experiments. These data were then evaluated by signal analysis techniques to characterize the welding transfer modes. The necessary hardware for high acquisition rates (up to 90kHz), as well as the software which was written to drive the equipment and to analyze the results, are described.

Manufacturing, Planning, Processing & Control

900,973
PB89-183214 PC A03/MF A01
National Inst. of Standards and Technology, Gaithersburg, MD.
Operations Manual for the Automatic Operation of the Vertical Workstation.
F. F. Rudder. 12 Jan 89, 38p NISTIR-89/4031

Keywords: *Automation, *Manufacturing, Operations, Manuals, *Automated Manufacturing Research Facility, Workstations, Man machine systems.

The Vertical Workstation (VWS) is located in the Automated Manufacturing Research Facility at the National Institute of Standards and Technology. The manual is for those individuals who wish to use the VWS equipment and control it from the workstation controller. Anyone possessing the ability to flip a switch, press a button, turn a valve, and operate a personal computer can follow these steps and place the VWS under command of the workstation controller. Only the steps required to configure the VWS for automatic operation are presented. The hardware details are contained in the VWS Operator's Manual.

900,974
PB89-183230 PC A03/MF A01
National Inst. of Standards and Technology (NEL), Gaithersburg, MD. Center for Mfg. Engineering.

Real-Time Simulation and Production Scheduling Systems.
W. J. Davis, and A. T. Jones. Apr 89, 20p NISTIR-89/4070
Prepared in cooperation with Illinois Univ. at Urbana-Champaign.

Keywords: *Production control, *Scheduling, *Simulation, *Real time operations, Production planning, Automation, Manufacturing, Statistical analysis, Flexible manufacturing systems, Concurrent processing.

The efficient scheduling of resources in a flexible manufacturing system (FMS) has a direct impact on the company's goal of increased profits. Many techniques, including mathematical programming, expert systems, and discrete event simulation have been used to solve these scheduling problems. However, they have all been ineffective in dealing with the unexpected delays that occur on the shop floor. The paper deals with a new approach to address production scheduling problems in an FMS - real-time, concurrent simulations. These simulations can be initialized to the current system state and run any time a new schedule is needed.

900,975
PB89-201495 Not available NTIS
National Bureau of Standards (IMSE), Gaithersburg, MD. Polymers Div.
Necking Phenomena and Cold Drawing.
Final rept.
L. J. Zapas, and J. M. Crissman. 1985, 23p
Pub. in Viscoelasticity and Rheology, Chapter 4, p81-103 1985.

Keywords: *Cold drawing, *Necking, *Tensile properties, Stresses, Strains, Loading rate, Elastic properties, Reprints.

The various experimental results presented suggest strongly that the instability leading to neck formation is associated with the nonlinearity of the mechanical behavior coupled with the time dependence. In experiments involving constant rate of clamp separation, necking occurs well beyond the point where the stress-strain curve goes through a maximum, and the amount beyond depends on the rate of strain. For polypropylene, at the lowest rate of clamp separation shown, the strain at which the neck became visible was about three times the value at which the maximum in the stress occurred. At constant rate of loading, the stress-strain behavior is monotonically increasing up to the point of necking where immediately thereafter the specimens break. To describe the instability behavior in a hard device one might in some way use elastic behavior. However for a soft device it becomes very clear that one needs to consider the viscoelastic behavior of the material.

Quality Control & Reliability

900,976
PB89-146740 Not available NTIS
National Bureau of Standards (IMSE), Gaithersburg, MD. Office of Nondestructive Evaluation.
Optical Nondestructive Evaluation at the National Bureau of Standards.
Final rept.
G. Birnbaum, D. Nyyssonen, C. M. Vest, and T. Vorburger. 1986, 17p
Pub. in Proceedings of SPIE (Society of Photo-Optical Instrumentation Engineers), v604 p1-17 1986.

Keywords: *Nondestructive tests, *Optical measurement, Light scattering, Holography, Standards, Surfaces, Microscopy, Scratches, Inspection, Reviews, Reprints, Optical fibers, Calibration.

The report reviews recent and current work on a variety of optical techniques applied to nondestructive evaluation (NDE) carried out by the National Bureau of Standards. The optical methods discussed include holography, scattering from surfaces, microscopy, scattering from particles, and methods employing optical fibers. Much of this work is aimed at the development of accurate measurement methods for in-service inspection and process monitoring in manufacturing, and the development of standards and calibration procedures.

114

900,977
PB89-150874 Not available NTIS
National Bureau of Standards (NEL), Gaithersburg,
MD. Automated Production Technology Div.
Generalized Mathematical Model for Machine Tool Errors.
Final rept.
M. A. Donmez, C. R. Liu, and M. M. Barash. 1986,
13p
Pub. in Proceedings of ASME (American Society of
Mechanical Engineers) Winter Annual Meeting - Modeling, Sensing, and Control of Manufacturing Processes, Anaheim, CA., December 7-12, 1986, p231-243.

Keywords: *Mathematical models, *Machine tools,
*Errors, Accuracy, Compensation,
Turning(Machining), Deflections, Wear, Positioning
error, Coordinate transformations, Turning center,
Thermally-induced errors.

In the paper, the authors describe a general mathematical model, which is able to incorporate the error
sources to determine the positional error vector of the
cutting tool with respect to the workpiece. In the process of the development of the model, the authors first
represent individual elements in the machine tool-fixture-workpiece system by assigning a homogeneous
coordinate transformation matrix to each element.
This matrix describes the position and orientation of a
body in space, and can incorporate the error motions
of the body in six degrees of freedom. The machine
tool-fixture-workpiece system is then considered as a
chain of linkages, and the relationships between these
linkages are determined by matrix multiplications.
Based on this idea, a matrix equation corresponding to
the structural loop of the machine tool-fixture-workpiece system is constructed, and solved for the error
vector. Although this model is applied to a turning
center, it can easily be modified for any type of machine tools, coordinate measuring machines, and
robots.

900,978
PB89-154322 PC E06/MF E01
National Inst. of Standards and Technology, Gaithersburg, MD. Office of Standards Code and Information.
U.S. Organizations Represented in the Collection of Voluntary Standards.
Jan 89, 46p
Supersedes PB88-145560.
Contains (Ten sheets of 48X reduction microfiche). A
supplementary document, U.S. Organizations Represented in the Collection of Voluntary Standards, accompanies this index.

Keywords: *Standards, Indexes(Documentation), Subjec; index terms, Specifications, Tests, United States,
Engineering standards, Product standards.

The list names the names of U.S. organizations
which develop standards or provide technical information on standards-related activities. The standards-developing organizations are represented in the National
Center for Standards and Certification Information
(NCSCI) reference collection (in microform or hard
copy) and are listed in the Key-Word-In-Context
(KWIC) index of U.S. voluntary industry standards. The
acronym list (Contents) is listed in alphabetical order
by acronym. The organization names, address and
telephone numbers are listed in alphabetical order by
organization name.

900,979
PB89-176655 Not available NTIS
National Bureau of Standards (NEL), Gaithersburg,
MD. Precision Engineering Div.
Optical Roughness Measurements for Industrial Surfaces.
Final rept.
D. Gilsinn, T. Vorburger, L. X. Cao, C. Giauque, F.
Scire, and E. C. Teague. 1986, 9p
Pub. in Proceedings of SPIE (Society of Photo-Optical
Instrumentation Engineers) Optical Techniques for Industrial Inspection, Quebec City, Canada, June 4-6,
1986, p8-16.

Keywords: *Optical measurement, *Surface roughness, *Nondestructive tests, Optical measuring instruments, Light scattering, Diffraction, Mathematical
models.

The paper reviews the effort to develop the theory and
instrumentation needed to measure surface roughness of manufactured surfaces by optical scattering
methods. Three key problems are addressed: devel-

oping a valid and sufficient optical scattering theory for
the roughness range; applying appropriate mathematical inversion techniques so that practical roughness
parameters can be calculated from scattering distributions; and evaluating a compact commercial instrument that is easy to align on a wide variety of part geometries. Recent results suggest that the simple
phase screen approximation model of optical scattering theory validly describes light scattering from machined metal surfaces with a predominant surface lay
in the 0.01 Ra to 3.0 Ra range. New measurements of
light scattered out of the plane-of-incidence are discussed. A model for scattering in the entire far-field
hemisphere and observations on the inverse problem
is given.

900,980
PB89-177018 Not available NTIS
National Bureau of Standards (NEL), Gaithersburg,
MD. Factory Automation Systems Div.
CAD (Computer Aided Design)-Directed Inspection.
Final rept.
T. H. Hopp. 1984, 15p
Pub. in Annals of the CIRP 33, n1 p1-15 1984.

Keywords: *Inspection, Automation, Measuring instruments, Artificial intelligence, Quality assurance, Reprints, *Computer aided design, *Computer aided
manufacturing, Control systems.

The paper describes a control system architecture,
based on hierarchical task-decomposition techniques,
for multi-axis coordinate measuring machines. An inspection program consists of a series of high level
goals to be satisfied. Goals are satisfied when specified information has been obtained regarding the part.
There is a decision hierarchy, each level of which provides logic for partially decomposing goals into simpler
goals. The control system executes the inspection program by interpreting the decision hierarchy logic. A
world model hierarchy executing in parallel with the
task decomposition hierarchy provides information
which aids in the decomposition decisions at each
level. Artificial intelligence techniques will allow the
convenient incorporation of quality assurance standards into inspection tasks.

900,981
PB89-187587 Not available NTIS
National Bureau of Standards (NEL), Boulder, CO.
Electromagnetic Technology Div.
New Standard Test Method for Eddy Current Probes.
Final rept.
L. L. Dulcie, and T. E. Capobianco. 1987, 7p
Sponsored by Army Materials Technology Lab., Watertown, MA.
Pub. in Proceedings of Defense Conference on Nondestructive Testing (36th), St. Louis, MO., October 27-29, 1987, p154-160.

Keywords: *Eddy current tests, *Standards, Test
equipment, Electrical impedance, Electrical properties,
Eddy currents, Nondestructive tests, Electrical measurement, Equipment specifications, *Military equipment.

Recently, a draft military standard for the characterization of eddy current probes was submitted to the U.S.
Army Materials Technology Laboratory by the National
Bureau of Standards. The development of a standard
test set and future plans for a round robin study for
evaluating the draft standard in a controlled study are
discussed. The test set will be used to determine impedance measurement capability and consists of two
parts; a prototype test block set as specified by the
draft standard and a specially designed and characterized probe set. A round robin survey will be conducted
to determine ease of use, repeatability of characterization measurements, and impedance measurement
precision when using the test blocks as specified in the
standard.

900,982
PB89-193296 PC A06/MF A01
National Inst. of Standards and Technology, Gaithersburg, MD.
Progress Report of the Quality in Automation Project for FY88.
C. D. Lovett. Apr 89, 108p NISTIR-89/4045
Prepared in cooperation with Department of the Navy,
Washington, DC., and Department of Energy, Washington, DC.

Keywords: *In-process quality control, *Automation,
Verification inspection, Process control, Monitors,
Sensors.

The document describes a quality control architecture
that uses real-time sensing, deterministic metrology
methods, machine tool characterization, process-intermittent gauging, and process certification to control
the machining process with the objective of reducing
the reliance on traditional post-process inspections
methods and moving toward greater reliance on in-process verification. The 'Quality In Automation'
project is developing a computer-based system that includes real-time sensors, a quality database, quality
monitors, quality controllers, an inspection station, and
a communication network which allows data to flow
among components of the quality system. The quality
architecture includes control loops such as post-processing characterization, pre-processing characterization, process intermittent gauging, and real-time sensing. The sensors for temperature, force and ultrasonic
may result in a command signal to the machine tool
controller to alter its pre-programmed positions, to
change its manipulative variables (speed and feed),
and to modify the NC code. The corrective action will
depend on the time varying nature of the errors and
the predictions derived from the analysis models.

900,983
PB89-228290 PC A03/MF A01
National Inst. of Standards and Technology (IMSE),
Gaithersburg, MD.
Computer-Controlled Test System for Operating Different Wear Test Machines.
E. P. Whitenton, and A. W. Ruff. Jul 89, 50p NISTIR-89/4107
Grant N00014-89-F-0021
Sponsored by Office of Naval Research, Arlington, VA.

Keywords: *Wear tests, Computer systems hardware,
Computer programs, *Tribology, *Computer aided
control systems, Computer architecture, File structures.

The report discusses a wear tester control system,
where the same computer and software runs three different wear test machines; a commercial crossed-cylinder, a commercial block-on-ring, and an in-house designed controlled-atmosphere tribometer. The computer hardware, the interface to the wear test machines, and the aspects that make the machines functionally similar are examined. The program itself, its
use, and the data file structure are also explored.

900,984
PB89-229025 PC A03/MF A01
National Inst. of Standards and Technology (IMSE),
Gaithersburg, MD. Office of Nondestructive Evaluation.
NDE (Nondestructive Evaluation) Publications, 1985.
L. Mordfin. Aug 89, 43p NISTIR-89/4131
See also PB87-201406.

Keywords: *Nondestructive tests, *Bibliographies, Radiography, Thermography, Ultrasonic tests, Acoustic
measurement, Magnetic tests, Optical measurement,
Leak detectors, Nondestructive analysis.

The report provides bibliographic citations, with selected abstracts, for 131 publications that appeared in the
open literature, primarily during calendar year 1985. A
detailed subject index is included as well as information on how copies of many of the publications may be
obtained.

900,985
PB89-229199 Not available NTIS
National Inst. of Standards and Technology (IMSE),
Gaithersburg, MD. Metallurgy Div.
Quantitative Problems in Magnetic Particle Inspection.
L. J. Swartzendruber. 1989, 8p
Pub. in Review of Progress in Quantitative Nondestructive Evaluation, v8B p2133-2140 1989.

Keywords: *Magnetic particle tests, *Reproducibility,
*Quantitative analysis, Magnetic properties, Steels,
Quality control, Nondestructive tests, Defects, Inspection, Reprints.

Although long considered a mature technology, a
number of questions remain on how to best control the
magnetic particle inspection process to obtain repro-

MANUFACTURING TECHNOLOGY

Quality Control & Reliability

ducible, quantitative results. The primary factors that must be controlled to obtain reproducible and predictable results are briefly discussed, followed by a detailed discussion of the magnetic leakage field from defects and a proposed method for determining the applied field necessary to detect defects of a given geometry.

900,986
PB90-130246   PC A03/MF A01
National Inst. of Standards and Technology, Gaithersburg, MD. Office of Standards Code and Information.
Glossary of Standards-Related Terminology.
Final rept.
D. R. Mackay. Oct 89, 29p NISTIR-89/4194
Library of Congress catalog card no. 89-600767.

Keywords: *Dictionaries, *Standards, Definitions, Terminology, Quality control, Tests, *Standardization, Accreditation, Certification.

The glossary provides definitions of 95 terms that are commonly used in standardization, certification, laboratory accreditation, and quality control activities. Multiple definitions are provided in some cases to identify organizational differences in the use of terms. In each case the source of the definition is indicated. The terms are presented in a logically structured format, beginning with general terms and moving to more specific terms.

Robotics/Robots

900,987
PB89-157358   Not available NTIS
National Bureau of Standards (NEL), Gaithersburg, MD. Robot Systems Div.
Optical Sensors for Robot Performance Testing and Calibration.
Final rept.
N. Dagalakis, and K. C. Lau. 1988, 5p
Pub. in Vision 5, n3 p10-14 1988.

Keywords: *Robots, *Optical measuring instruments, *Metrology, Performance evaluation, Performance tests, Calibrating, Reprints.

The article presents a brief review of robot performance measurements which may use optical metrology sensors. The desired optical metrology sensors performance characteristics are also discussed. Finally, a brief description of the available optical robot metrology sensors is provided.

900,988
PB89-159644   PC A04/MF A01
National Bureau of Standards, Gaithersburg, MD.
Material Handling Workstation Implementation.
C. E. Wenger. 19 May 88, 55p NBSIR-88/3784

Keywords: *Materials handling, *Robots, *Manufacturing, *Workstations, Automation, Work place layout, Operations, Instructions, Machining, Buffing, Hydraulic power pumps, Computers, *Automated Manufacturing Research Facilities, *AMRF material handling workstations, Roller tables, Automatic storage and retrieval systems, Automatic guided vehicles.

The purpose of the document is to provide a general description of design and implementation of the AMRF Material Handling Workstation (MHWS). The MHWS equipment includes two Automatic Guided Vehicles (AGVs), an Automatic Storage and Retrieval System (ASRS), and roller tables at other workstations. The document should provide the reader with an understanding of concepts used to implement the MHWS.

900,989
PB89-159651   PC A03/MF A01
National Bureau of Standards, Gaithersburg, MD.
Material Handling Workstation: Operator Manual.
C. E. Wenger. 19 May 88, 18p NBSIR-88/3785
See also PB89-159644.

Keywords: *Materials handling, *Robots, *Manufacturing, *Workstations, Automation, Work place layout, Operations, Instructions, Machining, Buffing, Hydraulic power pumps, Computers, *Automated Manufacturing Research Facilities, *AMRF material handling workstations, Roller tables, Automatic storage and retrieval systems, Automatic guided vehicles.

The purpose of the document is to provide operating instructions for the AMRF Material Handling Workstation (MHWS). The material handling equipment includes Automatic Guided Vehicles (AGV) and an Automatic Storage and Retrieval System (ASRS). The document includes operating instructions to startup and shutdown the material handling system. Also included are instructions to operate the AGV in its manual operating mode.

900,990
PB89-159669   PC A03/MF A01
National Bureau of Standards, Gaithersburg, MD.
Real-Time Control System Modifications for a Deburring Robot. User Reference Manual.
K. N. Murphy. 4 Aug 86, 43p NBSIR-88/3832

Keywords: *Control equipment, *Real time operations, *Robots, Deburring, Cleaning, *Automated Manufacturing Research Facility, Computer aided manufacturing, Computer aided design, Workstations, User manuals(Computer programs).

At the National Bureau of Standards' Automated Manufacturing Research Facility a PUMA 760 robot deburrs metal parts at the Cleaning and Deburring Workstation. The robot is controlled by the NBS developed Real-Time Control System (RCS). The basic RCS was extended to meet the needs of the workstation and the manual explains these additions.

900,991
PB89-177059   Not available NTIS
National Bureau of Standards (NEL), Gaithersburg, MD. Robot Systems Div.
Building Representations from Fusions of Multiple Views.
Final rept.
E. W. Kent, M. O. Shneier, and T. H. Hong. 1986, 6p
Pub. in Proceedings of IEEE (Institute of Electrical and Electronics Engineers) International Conference on Robotics and Automation, San Francisco, CA., April 7-10, 1986, p1634-1639.

Keywords: *Robots, Pattern recognition, Models, *Computer vision, Position sensing, Spatial resolution, Sensors.

A robot sensing system is described that uses multiple sources of information to construct an internal representation of its environment. Initially, object models are used to form the basic representations. These are modified by processes that operate on sequences of sensory information which is obtained from sensors that move about in the environment. Two representations are constructed. One is a description of the spatial layout of the environment, represented as an octree. The other is an object- and feature-based representation. The system handles both expected and unexpected objects, and attempts to register its internal representation with the external world using a variety of predictive, sensory-processing, and matching procedures.

900,992
PB89-177067   Not available NTIS
National Bureau of Standards (NEL), Gaithersburg, MD. Robot Systems Div.
Fast Path Planning in Unstructured, Dynamic, 3-D Worlds.
Final rept.
M. Herman. 1986, 8p
Pub. in Proceedings of SPIE (Society of Photo-Optical Instrumentation Engineers), Applications of Artificial Intelligence III, v635 p505-512 1986.

Keywords: *Paths, *Planning, Collision avoidance, Search structuring, Algorithms, *Robotics, Obstacle avoidance.

Issues dealing with fast motion planning in unstructured, dynamic 3-D worlds are discussed and a fast path planning system under development at NBS is described. It is argued that an octree representation of the obstacles in the world leads to fast path planning algorithms. The system performs the path search in an octree space and uses a hybrid search technique that combines hypothesis and test, hill climbing, A*, and multiresolution grid search.

900,993
PB89-181739   PC A03/MF A01
National Inst. of Standards and Technology (NEL), Gaithersburg, MD. Robot Systems Div.

Interfaces to Teleoperation Devices.
Technical note (Final).
J. C. Fiala. Oct 88, 16p NIST/TN-1254
Also available from Supt. of Docs. as SN003-003-02924-6.

Keywords: *Control equipment, *Robots, Servomechanisms, Interfaces, *Teleoperators, Active control.

The document describes a basic logical architecture for teleoperation control devices and interfaces for integrating these devices with a telerobot control system architecture. The interfaces described are for manipulator control only. The discussion will consider teleoperation devices as divided into two classes. Section 2 describes the interfaces for joint-space teleoperation devices. The interface requirements for Cartesian teleoperation devices are detailed in section 3.

900,994
PB89-221186   PC A04/MF A01
National Inst. of Standards and Technology (NEL), Gaithersburg, MD. Center for Mfg. Engineering.
Visual Perception Processing in a Hierarchical Control System: Level 1.
Technical note (Final).
K. Chaconas, and M. Nashman. Jun 89, 53p NIST/TN-1260
Also available from Supt. of Docs. as SN003-003-02949-1.

Keywords: *Robotics, *Control systems, *Visual perception, Image processing, Interfaces, Models, Algorithms, *Architecture(Computers).

The document describes the interfaces and functionality of the first level of the visual perception branch of a real time hierarchical manipulator control system. It includes a description of the scope of the processing performed and the outputs generated. It defines the interfaces and the information exchanged between the modules at this level, as well as interfaces to a camera, a human operator, and to higher levels of the system.

Tooling, Machinery, & Tools

900,995
PATENT-4 836 145   Not available NTIS
Department of Commerce, Washington, DC.
Multiple Actuator Hydraulic System and Rotary Control Valve Therefor.
Patent.
A. H. Slocum, and J. P. Peris. Filed 18 Jun 87, patented 13 Jun 89, 11p PB89-222368, PAT-APPL-7-063 558
Supersedes PB87-218384. Sponsored by National Inst. of Standards and Technology, Gaithersburg, MD. This Government-owned invention available for U.S. licensing and, possibly, for foreign licensing. Copy of patent available Commissioner of Patents, Washington, DC 20231 $1.50.

Keywords: *Hydraulic equipment, *Rotary valves, *Actuators, *Patents, Control equipment, Fluidic control devices, Creep strength, Phased arrays, PAT-CL-91-536.

The present invention provides a rotary hydraulic valve and a multiple hydraulic actuator system which, together, offer a truly practical solution to the problem of automating machine-tool parts fixtures. The control valve of the invention is capable of carrying out the complex logic required for multiple actuator control but is nonetheless both simple and compact. The actuator system is designed to incorporate a rotary valve but avoids the problem of creep which is characteristic of prior rotary-valve controlled actuators. The valve design of the invention also prevents thrust loads on the rotary valve member and thus avoids the need for heavy duty thrust bearings even in high pressure applications.

900,996
PB89-146781   Not available NTIS
National Bureau of Standards (NEL), Gaithersburg, MD. Automated Production Technology Div.

116

General Methodology for Machine Tool Accuracy Enhancement by Error Compensation.
Final rept.
A. Donmez, D. S. Blomquist, R. J. Hocken, C. R. Liu, and M. M. Barash. 1986, 10p
Pub. in Precision Engineering 8, n4 p187-196 Oct 86.

Keywords: *Machine tools, *Accuracy, Calibrating, Tolerances(Mechanics), Temperature, Errors, Least squares method, Compensation, Reprints.

The methodology introduces a general mathematical model, which relates the error in the position of the cutting tool with respect to the workpiece to the errors of the individual machine elements are then decomposed into geometric and thermally-induced components. A predictive machine calibration procedure to predict the geometric and thermally-induced errors of the machine slides is described. Based on the calibration data, empirical models for the error components are generated and valves for the parameters in these models are obtained using least-squares curve fitting techniques. A flexible, modular, and structured software system compensates for the predicted errors in real-time. The compensation system monitors the temperatures on the machine tool structure and the nominal axes positions, and then uses this information to predict the errors. The software is written in high level language and implemented in a dedicated, low cost, single-board microcomputer. The cutting tests carried out under transient thermal conditions have shown that the accuracy enhancement of up to 20 times is achievable using the methodology described in the paper, without a machine warm-up period.

900,997
PB89-150841   Not available NTIS
National Bureau of Standards (NEL), Gaithersburg, MD. Precision Engineering Div.
Preliminary Experiments with Three Identical Ultraprecision Machine Tools.
Final rept.
C. Evans, P. Hannah, and R. Rhorer. 1989, 10p
Pub. in Proceedings of SPIE (Society of Photo-Optical Instrumentation Engineers), v996 10p.

Keywords: *Machine tools, Precision, Mirrors, Optical equipment, Comparison, *Ultraprecision machine tools, *Precision machining, Diamond turning machines.

The paper outlines the underlying philosophy of, and reports preliminary results from, experiments with three nominally identical diamond turning machines. Identical tools, specially fabricated from the same diamond, were used to machine mirrors from blanks cut from the same base material using previously specified process parameters; parts produced are compared.

900,998
PB89-162564   PC A03/MF A01
National Bureau of Standards, Gaithersburg, MD.
Material Handling Workstation, Recommended Technical Specifications for Procurement of Commercially Available Equipment.
C. E. Wenger. 19 May 88, 32p NBSIR-88/3786

Keywords: *Materials handling equipment, *Procurement, Specifications, Manufacturing, Requirements, Inspection, *Automated Manufacturing Research Facility, *Workstations.

The purpose of the document is to provide specifications to be used in the procurement of material handling equipment to be used in an automated facility. The equipment specified includes Automatic Guided Vehicles (AGV) and an Automatic Storage and Retrieval System (ASRS).

900,999
PB89-201198   Not available NTIS
National Bureau of Standards (NML), Gaithersburg, MD. Temperature and Pressure Div.
Progress in Vacuum Standards at NBS (National Bureau of Standards).
Final rept.
C. R. Tilford. 1986, 14p
Pub. in Proceedings of Measurement Science Conference, Irvine, CA., January 23-24, 1986, p94-107.

Keywords: *Vacuum apparatus, *Standards, Vacuum gages, Pressure measurement, Ionization gages, Leakage, Test facilities, Calibration, National Institute of Standards and Technology.

The increasing reliance of American industry and science on vacuum technology has generated a continu-

ing demand for improved vacuum measurement accuracy. The National Bureau of Standards has responded with a vacuum and leak standards program. The article describes the goals of the NBS program, the current state of vacuum standards and calibration services at NBS, and the operation of the U.S. voluntary standards program. Information is contained on the performance of vacuum instruments that might be used in an industrial vacuum calibration laboratory.

901,000
PB89-216469   PC A04/MF A01
National Inst. of Standards and Technology (NEL), Gaithersburg, MD. Center for Building Technology.
Static Tests of One-third Scale Impact Limiters.
L. T. Phan, and H. S. Lew. May 89, 56p NISTIR-89/4089

Keywords: *Impact tests, *Shock absorbers, Energy absorption, Test facilities, Mechanical shock, Stresses, Vibration damping, Shock resistance.

The National Institute of Standards and Technology carried out four tests of one-third scale impact limiters for Transnuclear, Inc. The impact limiters were tested under static load in a 12-million pound capacity universal testing machine. Energy absorbed by the impact limiters, as indicated by the area under the load-deformation curve, was computed and compared with the required value which was specified for each specimen by Transnuclear, Inc. The testing was terminated when the absorbed energy value exceeded the required value.

Tribology

901,001
PB89-147391   Not available NTIS
National Bureau of Standards (IMSE), Gaithersburg, MD. Metallurgy Div.
Metallographic Evidence for the Nucleation of Subsurface Microcracks during Unlubricated Sliding of Metals.
Final rept.
P. J. Blau, and E. D. Doyle. 1987, 7p
Pub. in Wear 117, n3 p361-387 1987.

Keywords: *Wear, *Crack propagation, Sliding, Crack initiation, Nucleation, Coalescing, Metallography, Reprints.

The location, surface or subsurface, for nucleation and propagation of microcracks during sliding wear has been a continuing source of discussion and controversy in the tribology community. Rather than to suggest that there is only one correct answer to this question, the authors propose that depending on the contact conditions and materials involved, several preferred sites for microcrack nucleation can exist. The present paper provides metallographic evidence in support of a model which involves nucleation ahead of the slider, crack closure, and reopening after the slider passes. Once microcracks nucleate, subsequent slider passes can cause crack propagation, coalescence, and the eventual formation of wear debris particles. Observations of uplifted flakes are similar to those of the 'delamination theory of wear,' but the mechanism proposed for crack nucleation differs from that originally proposed based on the subsurface stress analysis from the 'delamination theory.'

901,002
PB89-228274   PC A14/MF A01
National Bureau of Standards (NEL), Gaithersburg, MD.
Development and Use of a Tribology Research-in-Progress Database.
S. Jahanmir, and M. B. Peterson. Jul 89, 316p NISTIR-89/4112
Portions of this document are not fully legible.

Keywords: *Databases, *Management information systems, Data processing, Mechanical engineering, Research management, Lubricants, Components, Systems engineering, Friction, Wear, *Tribology.

Preliminary efforts leading to the development of a research-in-progress database on tribology are described. The database contains brief abstracts of current tribology research being conducted by industry, universities, research institutes and government lab-

oratories based on a survey of active researchers. It also contains information on the types of activities, general areas of interest, program objectives, and tribology applications. The primary program objectives cited in connection with the tribology activities include long life, low maintenance, failure-free machinery, fundamental understanding, and materials development for improved performance.

901,003
PB90-130295   PC A05/MF A01
National Inst. of Standards and Technology, Gaithersburg, MD. Tribology Group.
Measurements of Tribological Behavior of Advanced Materials: Summary of U.S. Results on VAMAS (Versailles Advanced Materials and Standards) Round-Robin No. 2.
A. W. Ruff, and S. Jahanmir. Sep 89, 76p NISTIR-89/4170

Keywords: *Wear, *Sliding friction, *Materials tests, Measurement, Tables(Data), Standards, Ceramics, *Tribology, Interlaboratory comparisons, International cooperation, Coordinated research programs.

An interlaboratory comparison of tribological measurements was carried out among 16 U.S. laboratories as part of a large world effort involving six countries within the VAMAS (Versailles Advanced Materials and Standards) activity. Results for friction and wear of five material pairs are described in the report, along with a statistical analysis of the data, and interpretation of some of the findings.

General

901,004
PB89-172563   PC A03/MF A01
National Inst. of Standards and Technology (NEL), Gaithersburg, MD.
Report on Interactions between the National Institute of Standards and Technology and the American Society of Mechanical Engineers.
G. K. Ehrlich. Feb 89, 32p NISTIR-89/4036

Keywords: *Technology transfer, *Mechanical engineering, *Industrial engineering, *Standards, Computers, Heat transfer, Pressure vessels, Solar energy, Tribology, Meetings, *American Society of Mechanical Engineers(ASME), *National Institute of Standards and Technology (NIST).

The report highlights examples of interactions between the National Institute of Standards and Technology (NIST) and the American Society of Mechanical Engineers (ASME) over the past several years. It is meant to be representative, not all-inclusive. The interactions are organized by discipline in the following categories: Conferences, Committee memberships and contribution to standards, Editors, Publications which are designed to disseminate NIST's most recent technical advances and to learn of the technical challenges facing engineers in industry.

901,005
PB89-173876   Not available NTIS
National Bureau of Standards (NEL), Boulder, CO.
Chemical Engineering Science Div.
Ineffectiveness of Powder Regenerators in the 10 K Temperature Range.
Final rept.
R. Radebaugh. 1987, 19p
Sponsored by Air Force Flight Dynamics Lab., Wright-Patterson AFB, OH.
Pub. in Proceedings of Interagency Meeting on Cryocoolers (2nd), Easton, MD., September 24, 1986, p145-163 1987.

Keywords: *Regenerators, *Low temperature tests, *Effectiveness, Mechanical refrigeration, Cryogenics, Performance tests, Powder(Particles), Specific heat, Phase angle, Test facilities, Measurement.

Regenerators for temperatures around 10 K usually have rather high values of ineffectiveness because of a lack of matrix heat capacity. The paper describes various models used to predict regenerator performance and shows how the pressure oscillation and flow through a temperature gradient affect the overall heat transfer in the regenerator. These two effects can partially cancel each other and reduce the regenerator

MANUFACTURING TECHNOLOGY

General

loss when the phase angle between the displacer and compressor is less than 90 degrees. The paper also describes an apparatus used to measure the ineffectiveness under realistic operating conditions and gives results of measurements on two powder materials with different heat capacities - Pb and GdRh. The effect of the phase angle between mass flow rate and pressure are examined and are shown to have a strong effect.

901,006
PB89-173884 Not available NTIS
National Bureau of Standards (NEL), Boulder, CO.
Chemical Engineering Science Div.
Measurement of Regenerator Ineffectiveness at Low Temperatures.
Final rept.
R. Radebaugh, B. Louie, and D. Linenberger. 1987, 25p
Sponsored by Air Force Flight Dynamics Lab., Wright-Patterson AFB, OH.
Pub. in Proceedings of Interagency Meeting on Cryocoolers (2nd), Easton, MD., September 24, 1986, p89-93 1987.

Keywords: *Regenerators, *Low temperature tests, *Measurement, *Effectiveness, Heat exchangers, Cryogenics, Helium, Test facilities, Pressure, Temperature, Phase angle, Flow rate, Mass flow.

The low temperature limit of regenerative-cycle cryocoolers is usually determined by the ineffectiveness of the regenerator for temperatures below approximately 20 K. It is also this temperature range that the energy stored in the void volume gas becomes significant, which makes regenerator modeling much more difficult compared with regenerators operating at higher temperatures. The paper describes an experimental apparatus for measuring the ineffectiveness of regenerators in the temperature range 4-40 K. All operating parameters, such as mass flow, pressure, temperature, and phase angle can be measured and controlled independently of each other. A discussion of materials and geometries to be tested in the apparatus is given and estimates of performance of refrigerators with improved regenerators are made.

901,007
PB89-173892 Not available NTIS
National Bureau of Standards (NEL), Boulder, CO.
Chemical Engineering Science Div.
Refrigeration Efficiency of Pulse-Tube Refrigerators.
Final rept.
R. Radebaugh, and S. Herrmann. 1987, 15p
Sponsored by Department of the Navy, Washington, D.C., and National Aeronautics and Space Administration, Moffett Field, CA. Ames Research Center.
Pub. in Proceedings of International Cryocoolers Conference (4th), Easton, MD., September 25-26, 1986, p119-133 1987.

Keywords: *Low temperature tests, *Thermal efficiency, Cryogenics, Refrigerating machinery, Mass flow, Regenerators, Heat transfer, Performance tests, Power, Test facilities, *Pulse tube refrigerators.

The report describes measurements of the refrigeration capacity per unit mass flow as well as the thermodynamic efficiency of the cooling process which occurs within these pulse tubes. The effect of tube diameter, tube length, orifice setting and frequency were investigated. Efficiencies as high as 90% of Carnot efficiency were measured in some cases when compressor and regenerator losses were neglected. Gross refrigeration power at the optimum orifice setting was as high as 10 W at 80 K for a tube 12.7 mm O.D. by 240 mm long. It is shown that the performance of orifice pulse tubes is dependent on the tube volume and not on diameter and length, and heat transfer to the tube walls is detrimental to the performance. The regenerator losses can be relatively large since high mass-flow rates occur in these devices.

901,008
PB89-165748 Not available NTIS
National Bureau of Standards (NML), Boulder, CO.
Time and Frequency Div.
Analysis of High Performance Compensated Thermal Enclosures.
Final rept.
F. L. Walls. 1987, 5p
Pub. in Proceedings of Annual Symposium on Frequency Control (41st), Philadelphia, PA., May 27-29, 1987, p439-443.

Keywords: *Cryostats, *Ovens, *Thermal measurements, *Temperature control, Low temperature tests, Thermostats, Errors, Thermal analysis.

Approximate analysis of the conventional thermal enclosures such as ovens and cryostats reveals that the limitation to achievable thermal regulation is in many cases not the gain of the thermal servo loop, but rather the fact that the experiment under observation within the thermal enclosure is still coupled to the outside temperature. So, even if the thermal enclosure is perfectly stable in temperature, the experiment is not. A new configuration is suggested which uses an additional sensor to measure changes in the outside temperature and compensate the temperature set point of the thermal enclosure in order to just correct for the temperature error induced by the coupling to the outside.

901,009
PB89-186720 Not available NTIS
National Bureau of Standards (NEL), Gaithersburg, MD. Chemical Process Metrology Div.
Measurement of Shear Rate on an Agitator in a Fermentation Broth.
Final rept.
B. Robertson, and J. J. Ulbrecht. 1987, 5p.
Pub. in Biotechnology Processes, p31-35 1987.

Keywords: *Shear rate, *Turbine blades, *Fermentation, *Vats, Strain rate, Shear properties, Food processing, Electrochemistry, Reprints.

The shear rate was measured on the front face of a Rushton turbine blade rotating in an aqueous polyox solution that models a fermentation broth. The measurements agree with the theory of stagnation flow on the blade. A formula is given for use in scaling the results up to larger-diameter fermentation vats.

901,010
PB89-189120 PC A04/MF A01
National Inst. of Standards and Technology (NEL), Gaithersburg, MD. Fluid Flow Group.
NBS' (National Bureau of Standards) Industry; Government Consortium Research Program on Flowmeter Installation Effects: Summary Report with Emphasis on Research January-July 1988.
Summary rept.
G. E. Mattingly, and T. T. Yeh. Apr 89, 75p NISTIR-89/4080

Keywords: *Fluid flow, *Pipe flow, *Flowmeters, *Pipe bends, *Installing, Turbulent flow, Pressure, Measurement, Graphs(Charts), Vortices.

The report presents results produced in a consortium-sponsored research program on Flowmeter Installation Effects. The project is a collaborative one that has been underway for three years; it is supported by an industry-government consortium that meets twice yearly to review and discuss results and to plan subsequent phases of the work. The report contains the results and conclusions of the recent meeting of the consortium at NIST-Gaithersburg, MD in August 1988. Specific results included in the report include the following research results for the pipe flow from a conventional, long radius elbow: the distributions of the mean and the turbulence velocities in the axial and vertical directions; the pressure loss measurements; the performance of selected types of flowmeter installed downstream of the elbow; and the demonstration that satisfactory performance for the selected meters can be predicted using the research results of the study.

901,011
PB89-218341 PC A03/MF A01
National Inst. of Standards and Technology (NEL), Boulder, CO. Chemical Engineering Science Div.
Interlaboratory Comparison of the Guarded Horizontal Pipe-Test Apparatus: Precision of ASTM (American Society for Testing and Materials) Standard Test Method C-335 Applied to Mineral-Fiber Pipe Insulation.
D. R. Smith. Apr 89, 47p NISTIR-89/3913
Sponsored by Oak Ridge National Lab., TN.

Keywords: *Thermal conductivity, *Pipes(Tubes), *Mineral wool, Comparison, Measurement, Methodology, Accuracy, Insulation, Heat transfer, Refractory materials, Graphs(Charts).

Apparent thermal conductivity of refractory pipe insulation from the same production lot was measured by seven different laboratories. The comparison as-

sessed the precision and bias of the ASTM Test for Measurement of Steady-State Heat-Transfer Properties of Horizontal Pipe Insulation (C 335). For all test results from all seven participants, the standard deviation was 5%. This value is offered as the estimated precision of the horizontal pipe-test method. The accuracy of the pipe-test method cannot be estimated from the data obtained in the intercomparison.

901,012
PB90-130568 PC A07/MF A01
National Inst. of Standards and Technology (NEL), Gaithersburg, MD. Center for Mfg. Engineering.
Publications of the Center for Manufacturing Engineering Covering the Period January 1978-December 1988.
P. Nanzetta, A. Weaver, J. Wellington, and L. Wood. Sep 89, 141p NISTIR-89/4180

Keywords: *Manufacturing, *Production engineering, *Bibliographies, Automation, Artificial intelligence, Operations research, Robots, Information systems.

A list of publications by staff of the Center for Manufacturing Engineering for the period 1978-1988, indexed by subject area. Publications cover research done by the Center in the areas of high precision dimensional measurement and precision engineering; robotics and intelligent machines; manufacturing data description; data administration, and information processing; and sensors for manufacturing processes.

901,013
PB90-132747 PC A04/MF A01
National Inst. of Standards and Technology (NEL), Gaithersburg, MD. Center for Mfg. Engineering.
Emerging Technologies in Manufacturing Engineering.
Oct 89, 51p NISTIR-89/4187

Keywords: *Production engineering, *Manufacturing, *Technology assessment, *Industrial engineering, Production methods, Robots, Artificial intelligence, Industrial plants, *Computer aided manufacturing, National Institute for Standards and Technology.

This is an internal report produced by the managers and staff of the Center for Manufacturing Engineering for planning purposes only. It represents current best thinking about emerging technologies in manufacturing engineering, the impact these technologies will have on the Center's programs, and the directions the Center's programs will go if sufficient resources are available. The emerging technologies discussed are those that the Center believes, will require increased support and leadership in coming years.

MATERIALS SCIENCES

Ceramics, Refractories, & Glass

901,014
PB89-146823 Not available NTIS
National Bureau of Standards (IMSE), Gaithersburg, MD. Ceramics Div.
Defect Intergrowths in Barium Polytitanates. 1. Ba2Ti9O20.
Final rept.
P. K. Davies, and R. S. Roth. 1987, 13p
See also PB89-146831.
Pub. in Jnl. of Solid State Chemistry 71, n2 p490-502 1987.

Keywords: *Barium titanates, *Defects, *Electron microscopy, Microwaves, Dielectrics, Crystal structure, Stoichiometry, Reprints.

Using high-resolution transmission electron microscopy, the mechanisms of defect formation in samples of the microwave dielectric material, Ba(sub 2)Ti(sub 9)O(sub 20) were investigated. Materials prepared by a variety of different techniques show considerable structural disorder. The most prevalent intergrowth involved formation of a new triclinic polytype with an ionic arrangement closely related to that in the accepted structure. Defects also resulted from considerable microtwinning and were observed mainly in the samples prepared from a vanadate flux. The degree of nonstoichiometric defect formation was small in comparison to the stoichiometric intergrowths. In this case defects appeared to result from the incorporation of excess vacancies into the close-packed layers of the structure. Barium-deficient surface phases were also formed via a similar mechanism.

901,015
PB89-146831 Not available NTIS
National Bureau of Standards (IMSE), Gaithersburg, MD. Ceramics Div.
Defect Intergrowths in Barium Polytitanates. 2. Ba7Ti6O11.
Final rept.
P. K. Davies, and R. S. Roth. 1987, 10p
See also PB89-146823.
Pub. in Jnl. of Solid State Chemistry 71, n2 p503-512 1987.

Keywords: *Barium titanates, *Defects, *Electron microscopy, Crystal lattices, Diffraction, Reprints.

High resolution electron microscopy has been used to investigate the structure of Ba7Ti(sub 5)O(sub 11). Single phase materials were prepared from alkoxide precursors and studied using lattice imaging and microdiffraction techniques. Considerable structural disorder was observed in all the samples investigated. In general, isolated defects were observed. These resulted from a displacement of the close-packed layers of the structure giving some face-sharing of the Ti octahedra. However, in several regions of the samples, systematic stacking faults lead to the formation of a new polytypic structure. The paper describes the defect mechanisms, and relates these to the structure of the new polytype.

901,016
PB89-146849 Not available NTIS
National Bureau of Standards (IMSE), Gaithersburg, MD. Ceramics Div.
Critical Assessment of Requirements for Ceramic Powder Characterization.
Final rept.
A. L. Drago, C. R. Robbins, and S. M. Hsu. 1987, 4p
Pub. in Advances in Ceramics 21, p711-720 1987.

Keywords: *Ceramics, *Powder(Particles), *Microstructure, *Particle size distribution, *Quality control, Measurement, Standards, Sampling, Reprints.

The detailed characterization of ceramic powders is very important for the reproducible manufacture of advanced ceramics which will perform reliably in service. The basic issues are: what to measure, how to measure it and how to assure quality in analytical measurements by all laboratories. 'What to measure' involves understanding the relationship between powder characteristics and ceramic microstructures. 'How to measure it' requires the development of measurement methods and the determination of repeatability and reproducibility. Standard Reference Materials (SRMs) are required to assure the quality of measurements and the comparability of measurements between different laboratories and techniques. From considerations of the distributed nature of powder and ceramic properties, new SRMs are proposed which will certify distributed properties. As an example of such SRMs, technical requirements are developed for the production of an SRM with a certified particle size distribution for ceramic powders. Factors which enter the certification of a particle size distribution for a ceramic powder include the approximate mathematical representation of the distribution, weighting of the distribution by the measurement technique, the particle shape distribution, and statistical variances introduced by the powder, the sampling methods, and the methods of measurement.

901,017
PB89-146856 Not available NTIS
National Bureau of Standards (IMSE), Gaithersburg, MD. Ceramics Div.
Application of SANS (Small-Angle Neutron Scattering) to Ceramic Characterization.
Final rept.
K. G. Frase, K. A. Hardman-Rhyne, and N. F. Berk. 1986, 8p
Pub. in Materials Research Society Symposia Proceedings Better Ceram. Chem. 73, n2 p179-186 1986.

Keywords: *Neutron scattering, *Ceramics, *Defects, *Nondestructive tests, Spinel, Porosity, Particle size,

Agglomeration, Sintering, Precipitation(Chemistry), Microstructure, Reprints.

Traditionally, small angle neutron scattering (SANS) has been used to study dilute concentrations of defects 1 - 100 nm in size. Recent extensions of the scattering theory have allowed the expansion of the technique to include larger sizes through the use of multiple scattering. With multiple small angle neutron scattering, defects (pores, microcracks, precipitates) up to 10 micro m in size can be studied. SANS is inherently a non-destructive, bulk probe of microstructure, with wide applications in the characterization of materials. A number of studies of ceramic materials using multiple and traditional (single particle diffraction) small angle neutron scattering will be discussed. The emphasis will be on the strength of the technique in the characterization of materials. Particular examples will include: the assessment of pore size distributions in spinel compacts as a function of sintering and agglomeration, the characterization of primary and secondary particle sizes in precipitated aggregates, and the determination of microporosity in MDF cements.

901,018
PB89-148373 PC A03/MF A01
National Inst. of Standards and Technology (NEL), Boulder, CO. Chemical Engineering Science Div.
Microporous Fumed-Silica Insulation Board as a Candidate Standard Reference Material of Thermal Resistance.
D. R. Smith, and J. G. Hust. Oct 88, 25p NISTIR-88/3901
Sponsored by Oak Ridge National Lab., TN.

Keywords: *Thermal conductivity, *Insulating boards, *Silicon dioxide, Microporosity, Vapor deposition, Temperature, Standards, Pressure, Tables(Data), *Standard Reference Materials.

Measurements of apparent thermal conductivity of microporous fumed-silica insulation board are reported in order to provide a basis for certifying it as a Standard Reference Material (SRM) of thermal resistance. These data, for a pair of specimens having a mean density of 301 kg/m sup 3, encompass a range of temperature from 321 to 723 K and environmental gas pressures at and below ambient atmospheric pressure (40 to 83.7 kPa). Detailed analyses and intercomparisons of previously published data are given. Correlations of thermal conductivity with temperature and with pressure are given which represent the data within a standard deviation of 0.2%. This fumed-silica material has a thermal conductivity of 19.66 mW/(mK) at 300 K and is suitable for use as an SRM of very low conductivity from room temperature up to temperatures beyond 720 K (450 deg C). Great care in handling this material is necessary because of its fragility.

901,019
PB89-148381 PC A05/MF A01
National Inst. of Standards and Technology (IMSE), Gaithersburg, MD. Ceramics Div.
Institute for Materials Science and Engineering, Ceramics: Technical Activities 1988.
S. M. Hsu. Feb 88, 92p NISTIR-88/3840
See also PB86-196771.

Keywords: *Ceramics, *Superconductors, *Phase diagrams, *Images, Composite materials, Tensile properties, Lubrication, Grain boundaries, Raman spectroscopy, Synchrotron radiation, Topography.

Current programs of the Ceramics Division are reviewed. Programs include: grain boundary chemistry and structure of YBaCuO superconductors; compilation of phase diagrams for salt systems; diamond film synthesis; tribology; composites; wear maps; tensile tests; time resolved micro RAMAN analysis; typographic imaging (synchrotron radiation analysis).

901,020
PB89-149231 Not available NTIS
National Bureau of Standards (NML), Gaithersburg, MD. Gas and Particulate Science Div.
Stokes and Anti-Stokes Fluorescence of Er(3+) in the Raman Spectra of Erbium Oxide and Erbium Glasses.
Final rept.
E. Etz, and J. Travis. 1986, 2p
Pub. in Proceedings of International Conference on Raman Spectroscopy (10th), Eugene, OR., August 31-September 5, 1986, p11/67-11/68.

Keywords: *Erbium oxides, *Glass, *Fluorescence, *Raman spectra, *Lasers, *Excitation, *Energy levels, Silicate minerals, Energy absorption.

The Abstract reports new observations on the laser-excited visible fluorescence in the micro-Raman spectra of erbium oxide and erbium-bearing silicate glasses. These results relate to the unusual optical behavior of rare-earth ions in solid phase matrices attributed to certain energy transfer reactions, proceeding by various upconversion mechanisms, when stimulated by infrared or visible radiation. The results reported here include the observation of anti-Stokes fluorescence from Er3+ when excited with 568.2 nm laser light.

901,021
PB89-150742 Not available NTIS
National Bureau of Standards (NEL), Gaithersburg, MD. Building Materials Div.
Standard Specifications for Cements and the Role in Their Development of Quality Assurance Systems for Laboratories.
Final rept.
G. Frohnsdorff, P. W. Brown, and J. H. Pielert. 1986, 5p
Pub. in Proceedings of International Congress on the Chemistry of Cement (8th), Rio de Janeiro, Brazil, September 22-27, 1986, p316-320.

Keywords: *Cements, *Specifications, *Standards, Portland cements, Quality assurance, Performance standards, Durability, Laboratories, Test facilities, Performance evaluation.

Standard specifications for portland and related cements contain both prescriptive and performance criteria. Prescriptive criteria may hinder innovation by placing unnecessary limits on composition. Performance specifications define the performance required of a cement in terms which are closely related to the performance characteristics required in service. Performance tests may take too long to carry out to be suitable as acceptance tests. For this reason cement specifications must continue to include both prescriptive and performance components. It should be practical for a specification to be structured to consist of a complete performance specification supplemented by prescriptive criteria to be developed by the manufacturer according to specified rules. New factors affecting cement specifications may include computer-aided checks for logical consistency, and the needs for precise chemical and physical information for use in mathematical models of cement performance and in integrated project information systems.

901,022
PB89-150759 Not available NTIS
National Bureau of Standards (NEL), Gaithersburg, MD. Building Materials Div.
Implications of Phase Equilibria on Hydration in the Tricalcium Silicate-Water and the Tricalcium Aluminate-Gypsum-Water Systems.
Final rept.
P. W. Brown. 1986, 8p
Pub. in Proceedings of International Congress on the Chemistry of Cement (8th), Rio de Janeiro, Brazil, September 22-27, 1986, p231-238.

Keywords: *Cements, *Phase transformations, *Reaction kinetics, *Setting time, Hydration, Calcium silicates, Aluminates, Gypsum, Morphology, Hardening(Materials).

The mechanisms and kinetics of hydration in the C3S-H2O and C3A-gypsum-H2O systems are discussed within the context of the relevant phase equilibria. The development of a layer of C-S-H of variable composition results in the onset of the induction period. The composition gradient in this layer causes a morphological transformation leading to the onset of the acceleratory period. The first-formed aluminate-containing phase in the C3A-gypsum-H2O system may be AH3 or ettringite, depending on the initial solution composition. The rate of early C3A hydration is lowest when ettringite is the first-formed phase. This suggests that initial AH3 formation does not have a major influence on retardation.

901,023
PB89-156350 PC A03/MF A01
National Bureau of Standards, Gaithersburg, MD. Ceramics Div.

MATERIALS SCIENCES

Ceramics, Refractories, & Glass

Structural Reliability and Damage Tolerance of Ceramic Composites for High-Temperature Applications. Semi-Annual Progress Report for the Period Ending September 30, 1987.
E. R. Fuller, T. W. Coyle, and R. F. Krause. Feb 88, 27p NBSIR-88/3710
See also PB87-208310. Sponsored by Department of Energy, Oak Ridge, TN. Advanced Research and Technology Fossil Energy Materials Program.

Keywords: *Engines, *Thermal efficiency, *Ceramics, High temperature tests, Composite materials, Microstructure, Mechanical properties, Corrosion resistance, Heat recovery, Structural members, Compressing, Silicon carbides, Strength, Fibers, Shear strength, Damage, Graphs(Charts), Whisker composites, Crack propagation, Creep strength.

The achievement of higher efficiency heat engines and heat recovery systems requires the availability of high-temperature, high-performance structural materials. Structural ceramics, and more recently, ceramic matrix composites have received particular attention for these applications due to their high strength and resistance to corrosion and thermal shock. Even with these positive attributes, improved reliability and extended lifetime under service conditions are necessary for structural ceramics to gain industrial acceptance. The problems with these materials are mechanical and chemical in nature and are enhanced by the fact that they are subjected to high temperatures, reactive environments and extreme thermal gradients. With an objective of improved performance for heat engine/heat recovery applications, the NBS program on structural ceramics and ceramic composites addresses these problems through the determination of critical factors that influence mechanical and microstructural behavior. The activities of the program are grouped under two major subtasks, each designed to develop key data, associated test methods and companion predictive models. The status of the subtasks for the period ending September 30, 1987 are provided.

901,024
PB89-156368   PC A03/MF A01
National Inst. of Standards and Technology (IMSE), Gaithersburg, MD. Ceramics Div.
Structural Reliability and Damage Tolerance of Ceramic Composites for High-Temperature Applications. Semi-Annual Progress Report for the Period Ending March 31, 1988.
E. R. Fuller, T. W. Coyle, T. R. Palamides, and R. F. Krause. Aug 88, 18p NISTIR-88/3817
See also PB89-156350. Sponsored by Department of Energy, Oak Ridge, TN. Advanced Research and Technology Fossil Energy Materials Program.

Keywords: *Engines, *Thermal efficiency, *Ceramics, High temperature tests, Composite materials, Microstructure, Mechanical properties, Corrosion resistance, Heat recovery, Structural members, Compressing, Silicon carbides, Fibers, Shear strength, Damage, Graphs(Charts), Whisker composites, Crack propagation, Creep strength.

The achievement of higher efficiency heat engines and heat recovery systems requires the availability of high-temperature, high-performance structural materials. Structural ceramics, and more recently, ceramic matrix composites have received particular attention for these applications due to their high strength, and resistance to corrosion and thermal shock. Even with these positive attributes, improved reliability and extended lifetime under service conditions are necessary for structural ceramics to gain industrial acceptance. The problems with these materials are mechanical and chemical in nature and are enhanced by the fact that they are subjected to high temperatures, reactive environments and extreme thermal gradients. With an objective of improved performance for heat engine/heat recovery applications, the NBS program on structural ceramics and ceramic composites addresses these problems through the determination of critical factors that influence mechanical and microstructural behavior. The activities of the program are grouped under two major subtasks, each designed to develop key data, associated test methods and companion predictive models. The status of the subtasks for the period ending March 31, 1988 are provided.

901,025
PB89-157044   Not available NTIS
National Bureau of Standards (NEL), Boulder, CO. Electromagnetic Technology Div.

Oxygen Isotope Effect in the Superconducting Bi-Sr-Ca-Cu-O System.
Final rept.
H. Katayama-Yoshida, T. Hirooka, A. Oyamada, Y. Okabe, T. Takashashi, T. Sasaki, A. Ochiai, T. Suzuki, A. J. Mascarennas, J. I. Pankove, T. F. Ciszek, S. K. Deb, R. B. Goldfarb, and Y. Li. 1988, 4p
Contract DE-AC02-83CH10093
Sponsored by Department of Energy, Washington, DC.
Pub. in Physica C 156, p481-484 1988.

Keywords: *Superconductors, *Oxygen, *Isotopic labeling, *Copper oxides, Bismuth, Strontium, Calcium, Electrical resistivity, Magnetic permeability, Phonons, Reprints.

An oxygen isotope effect is observed in mixed-phase Bi-Sr-Ca-Cu-O superconductors when O18 is substituted for O16. The superconducting transition temperature Tc, measured by electrical resistivity and magnetic susceptibility, is lowered by 0.32 K for the higher-Tc (110 K) phase and by about 0.34 K for the lower Tc (75 K) phase. These results suggest a measurable contribution to the superconductivity from phonons.

901,026
PB89-157564   Not available NTIS
National Bureau of Standards (IMSE), Gaithersburg, MD. Ceramics Div.
Small Angle Neutron Scattering from Porosity in Sintered Alumina.
Final rept.
N. F. Berk, A. Hardman-Rhyne, and N. F. Berk. 1986, 3p
Pub. in Jnl. of the American Ceramic Society 69, n11 pC,285-C.287 Nov 86.

Keywords: *Aluminum oxide, *Neutron scattering, *Sintering, *Porosity, Microstructure, Nondestructive tests, Ceramics, Voids, Reprints.

Large voids (approximately 0.2 micrometers and larger) can remain in the ceramic material after the sintering process is over. Often sintered ceramic materials are quite thick as well. Larger volume fraction and pore sizes are not seen with traditional single particle diffraction techniques in small angle neutron scattering. Multiple scattering methods were employed to elucidate microstructural information relating to pore size and porosity.

901,027
PB89-158034   Not available NTIS
National Bureau of Standards (IMSE), Gaithersburg, MD. Ceramics Div.
Effect of Coal Slag on the Microstructure and Creep Behavior of a Magnesium-Chromite Refractory.
Final rept.
S. M. Wiederhorn, R. F: Krause, and J. Sun. 1988, 10p
Sponsored by Department of Energy, Washington, DC.
Pub. in American Ceramic Society Bulletin 67, n7 p1201-1210 Jul 88.

Keywords: *Slags, *Refractory materials, *Ceramics, Microstructure, Creep tests, Magnesium, Chromites, Bricks, Aggregates, Viscosity, Penetration, Reprints.

As slag penetrates magnesium chromite refractory, ion exchange between the refractory and the slag changes the composition of the slag from anorthite-like to diopside-like. This modification of the slag composition is believed to reduce its viscosity and enhance its reactivity. The creep rate of the refractory brick is increased by a factor of about 3 when slag penetrates into the brick. Creep behavior is rationalized in terms of the spacing and size of the aggregate particles within the refractory, and the structure and composition of the material bonding the aggregate particles together. It is suggested that refractory disintegration involves creep deformation at the hot face of the brick caused by mechanical loading and stresses due to ion exchange within the brick. The results of the present paper suggest that refractory performance can be improved by chemical modification of the brick to prevent a reduction slag viscosity during penetration.

901,028
PB89-162606   PC A03/MF A01
National Inst. of Standards and Technology (IMSE), Gaithersburg, MD. Ceramics Div.

Toughening Mechanisms in Ceramic Composites: Semi-Annual Progress Report for the Period Ending September 30, 1988.
Interim rept.
E. R. Fuller, R. F. Krause, M. D. Vaudin, and T. R. Palamides. Feb 89, 24p NISTIR-88/4018
Sponsored by Department of Energy, Oak Ridge, TN. Advanced Research and Technology Fossil Energy Materials Program.

Keywords: *Ceramics, *Compression tests, *Fractures(Materials), *Crack propagation, Silicon carbides, Borosilicate glass, Toughness, Fiber composites, Aluminum oxide, Creep properties, Bursting, Graphs(Charts).

A fracture mechanics specimen known as the double-cleavage drilled-compression (DCDC) specimen has been used to study crack-fiber interactions and toughening increments in a model composite system of SiC monofilaments in a borosilicate glass matrix. The toughening increments were measured from changes in applied stress intensity factor as a function of crack length and number of monofilament fibers. Both the fiber-matrix debond strength and the interfacial frictional shear resistance, which influence these toughening increments, were measured independently by a single fiber pull-out test and were correlated with the toughness increases measured by the DCDC specimen. A ceramic composite material that has received much attention because of its increased toughness and creep resistance compared with alumina is aluminum oxide reinforced with silicon carbide whiskers. In the study, the creep and creep rupture behavior of a 25 wt% SiC whisker-reinforced alumina ceramic with 4.9% porosity were measured at temperature between 100 deg C and 1300 deg C and at applied stresses between 55 and 306 MPa, although the applied stresses at each temperature varied over a much narrower range. Creep strains were determined from loading-point displacement measurements in four-point flexure.

901,029
PB89-165427   Not available NTIS
National Bureau of Standards (NEL), Gaithersburg, MD. Chemical Process Metrology Div.
Sputtered Thin Film Ba2YCu3On.
Final rept.
K. G. Kreider. 1986, 3p
Pub. in Proceedings of Colloquium High Temperature Superconductivity: Prospects and Challenges, Washington, DC., October 9, 1987, p51-53 1988.

Keywords: *Thin films, *Sputtering, *Barium oxides, *Oxidation, *Perovskites, *Superconductors, Yttrium, Copper, Zirconium, Aluminum, Sapphire, Magnesium, Silver, Silicon, Crystallization, High temperature tests, Targets, Ceramics, Reprints.

Thin sputtered films were produced from premixed targets of Ba2Y Cu3 O with a planar magnetron. The films were deposited on ZrO2(Y), Al2O3, sapphire, MgO, Ag, and Si. The ZrO2(Y) and MgO enabled crystallization and oxidation of the thin films at 800C into perovskite. Copper loss was observed in films produced at elevated temperatures and with target heating. The sputtering and thermal processing parameters are discussed.

901,030
PB89-171730   Not available NTIS
National Bureau of Standards (IMSE), Gaithersburg, MD. Reactor Radiation Div.
Crystal Chemistry of Superconductors: A Guide to the Tailoring of New Compounds.
Final rept.
A. Santoro, F. Beech, M. Marezio, and R. J. Cava. 1988, 8p
Pub. in Physica C 156, p693-700 1988.

Keywords: *Superconductivity, *Copper oxides, *Crystal defects, Perovskites, Lanthanum, Calcium, Thallium, Bismuth, Barium, Strontium, Yttrium, Shearing, Reprints.

The crystal structures of the known superconducting copper oxides can be described in terms of two basic structural types. The series La2Ca(n-1)CunO(2n+2), (Tl,Bi)2(Ba,Sr)2Ca(n-1)CunO(2n+4) and TlBa2Ca(n-1)CunO(2n+3) can be viewed as made of alternating slices having the rock salt and perovskite structure. The compounds Ba2YCu4O8 and Ba4Y2Cu7O(14+x), on the other hand, are comprised of perovskite blocks alternating with blocks in which a

120

crystallographic shear is present. The effect of this shear is that of forming double chains of edge sharing squares with oxygen atoms at the corners and copper atoms at the center. The superconductor Ba2YCu3O7 can be described in terms of both structural types and may be considered as an intermediate type between the other two. The basic building blocks of these superconducting materials can be further broken down into constituent nets (or meshes). This description allows one to envisage new structures built from these meshes containing the key structural elements present in the currently known superconductors. As such, the structural schemes used in this description may be used as a guide in the preparation of new materials with interesting electronic properties.

901,031
PB89-171771 Not available NTIS
National Bureau of Standards (IMSE), Gaithersburg, MD. Ceramics Div.
Rising Fracture Toughness from the Bending Strength of Indented Alumina Beams.
Final rept.
R. F. Krause. 1988, 6p
Pub. in Jnl. of the American Ceramic Society 71, n5 p338-343 May 88.

Keywords: *Crack propagation, *Ceramics, *Fracturing, *Aluminum oxide, *Indentation, Defects, Tensile properties, Optical microscopes, Surface properties, Vickers hardness, Impact tests, Loads(Forces), Bending, Toughness, Annealing, Residual stress, Polishing, Beams(Supports), Reprints.

The analytical function of crack extension to a fractional power is used to represent the fracture resistance of a vitreous-bonded 96% alumina ceramic. A varying flaw size, controlled by Vickers indentation loading between 3 and 300 N, was placed on the prospective tensile surfaces of four-point bend specimens, previously polished and annealed. The lengths of surface cracks were measured by optical microscopy. Straight lines were fitted to the logarithmic functions of observed bending strength versus indentation load in two series of experiments, including the residual stress due to indentation and having the residual stress annealed out at an elevated temperature. Within the precision of measurement these lines have the same slope, being about 32% less than the slope which a fracture toughness independent of crack extension would indicate.

901,032
PB89-171789 Not available NTIS
National Bureau of Standards (IMSE), Gaithersburg, MD. Ceramics Div.
Phase Relations between the Polytitanates of Barium and the Barium Borates, Vanadates and Molybdates.
Final rept.
J. M. Millet, R. S. Roth, and H. S. Parker. 1986, 4p
Pub. in Jnl. of the American Ceramic Society 69, n11 p811-814 Nov 86.

Keywords: *Ceramics, *Barium oxides, *Phase diagrams, Barium titanates, Vanadates, Molybdates, Chemical equilibrium, Borates, Ternary systems, x ray diffraction, Liquids, Melting, Reprints.

Phase equilibria in ternary systems involving BaO, TiO2, and the low melting oxides B2O3, V2O5 and MoO3 are reported. Alternative ways to synthesize BaTi4O9 have been found. These compounds can be obtained by heating at relatively low temperature appropriate compositions in the system BaO-TiO2-V2O5 or BaO-TiO2-MoO3 and dissolving the excess phases in dilute HCl. Ba2Ti9O20 can also be prepared at low temperature in the BaO-TiO2-B2O3 system, in which it may show a limited solid solution, offering the possibility of obtaining dense ceramics at low temperature. The low temperature phase BaTi5O11 has not been obtained in equilibrium in any of the studied systems. It has been found that the compound reported with the composition Ba3Ti3V4O15 has in fact the composition Ba3TiV4O14. It decomposes at 995 + or - 5 C (Ba3TiV4O13 yields liq + Ba2V2O7). A ternary compound Ba2Ti2B2O9 which decomposes from a BaTiO3 and a liquid at approximately 950 C was also found.

901,033
PB89-171797 Not available NTIS
National Bureau of Standards (IMSE), Gaithersburg, MD. Ceramics Div.

Phase Equilibria and Crystal Chemistry in the Ternary System BaO-TiO2-Nb2O5: Part 1.
Final rept.
J. M. Millet, L. D. Ettlinger, H. S. Parker, and R. S. Roth. 1987, 12p
Pub. in Jnl. of Solid State Chemistry 67, n2 p259-270 1987.

Keywords: *Phase diagrams, *Ternary system, *Barium titanates, Single crystals, Barium oxides, Titanium, Niobium, Solid solutions, Diffraction, Ceramics, Reprints.

A partial subsolidus phase diagram is presented of the BaO-TiO2-Nb2O5 ternary system. Eight new compounds or solid solutions in the ternary system are described and the existence of three solid solutions previously reported are confirmed and further studied.

901,034
PB89-171813 Not available NTIS
National Bureau of Standards (IMSE), Gaithersburg, MD. Ceramics Div.
Effect of Lateral Crack Growth on the Strength of Contact Flaws in Brittle Materials.
Final rept.
R. F. Cook, and D. H. Roach. 1986, 12p
Sponsored by IBM Thomas J. Watson Research Center, Yorktown Heights, NY.
Pub. in Jnl. of Materials Research 1, n4 p589-600 Jul/Aug 86.

Keywords: *Ceramics, *Glass, *Brittle fracturing, *Crack propagation, *Impact tests, Defects, Single crystals, Contacting, Elastic properties, Plastic properties, Residual stress, Indentation, Radial stress, Lateral pressure, Failure, Fatigue(Materials), Strength, Loads(Forces), Reprints, Fracture mechanics.

The effect of lateral cracks on strength controlling contact flaws in brittle materials is examined. Inert strength studies using controlled indentation flaws on a range of ceramic, glass and single crystal materials reveal significant increases in strength at large contact loads, above the predicted load dependence extrapolated from strength measurements at low indentation loads. The increases are explained by the growth of lateral cracks decohesing the plastic deformation zone associated with the contact from the elastically restraining matrix, thereby reducing the residual stress field driving the strength controlling radial cracks. A strength formulation is developed from indentation fracture mechanics which permits inert strengths to be described over the full range of contact loads. The formulation takes account of the decreased constraint of the plastic deformation zone by lateral crack growth as well as post-contact non-equilibrium growth of the radial cracks. Simple extensions permit the strengths of specimens controlled by impact flaws to be described, as well as those falling under non-equilibrium (fatigue) conditions. The work reinforces the conclusion that a full understanding of the residual stress field at dominant contact flaws is necessary to describe the strength of brittle materials.

901,035
PB89-171821 Not available NTIS
National Bureau of Standards (IMSE), Gaithersburg, MD. Ceramics Div.
Densification, Susceptibility and Microstructure of Ba2YCu3O(6+x).
Final rept.
J. E. Blendell, and L. C. Stearns. 1988, 10p
Pub. in Ceramic Transactions 1, ptB p1146-1155 1988.

Keywords: *Ceramics, *Microstructure, *Densification, *Magnetic permeability, *Sintering, *Superconductivity, Measurement, Pressure, Temperature, Porosity, Chemical composition, Grain boundaries, Barium oxides, Yttrium, Copper, Compacting, Atmospheres, Reprints.

The densification of Ba2YCu3O(6+x) for different sintering conditions has been measured. The effects of compaction pressure, atmosphere and temperature have been investigated. Measurements of the AC magnetic susceptibility have been used to determine the superconducting properties. Dense samples were found to have less superconducting material as compared to porous samples. Compositional mapping of the samples showed variations in composition in the samples, corresponding to Cu-rich regions and Y-poor regions. Sinter-forging was done in order to align the grains.

901,036
PB89-175244

(Order as PB89-175194, PC **A06**)
National Inst. of Standards and Technology (IMSE), Gaithersburg, MD. Ceramics Div.
Structural Ceramics Database: Technical Foundations.
Bi-monthly rept.
R. G. Munro, F. Y. Hwang, and C. R. Hubbard. 1989, 11p
Included in Jnl. of Research of the National Institute of Standards and Technology, v94 n1 p37-46 Jan-Feb 89.

Keywords: *Ceramics, Thermodynamic properties, Materials specifications, Thermal expansion, Thermal conductivity, Thermal diffusivity, Specific heat, Shock resistance, Thermal resistance, Bibliographics, Systems engineering, *Numerical data bases, Structural ceramics data base system, User requirements.

The development of a computerized database on advanced structural ceramics can play a critical role in fostering the widespread use of ceramics in industry and in advanced technologies. A preliminary system has been completed as phase one of an ongoing program to establish the Structural Ceramics Database system. The system is designed to be used on personal computers. Developed in a modular design, the preliminary system is focused on the thermal properties of monolithic ceramics. The initial modules consist of materials specification, thermal expansion, thermal conductivity, thermal diffusivity, specific heat, thermal shock resistance, and a bibliography of data references. Query and output programs also have been developed for use with these modules. Three primary considerations provide the guidelines to the system's development: (1) The user's needs; (2) The nature of materials properties; and (3) The requirements of the programming language. The report discusses the manner and rationale by which each of these considerations leads to specific features in the design of the system.

901,037
PB89-175939 Not available NTIS
National Bureau of Standards (IMSE), Gaithersburg, MD. Ceramics Div.
Pore Morphology Analysis Using Small Angle Neutron Scattering Techniques.
Final rept.
K. A. Hardman-Rhyne. 1987, 12p
Pub. in Advanced Ceramics 21, p767-778 1987.

Keywords: *Ceramics, *Porosity, *Compacting, *Powder(Particles), *Neutron scattering, Aluminum oxide, Microstructure, Green strength, Reprints.

Unfired ceramics are very difficult to characterize due to the often fragile nature of the compacted powder and the high concentration of pores. Furthermore, these pores can have a wide distribution in sizes depending on the size and nature of the particles and the compaction procedure used to form the compacts. Small angle neutron scattering techniques are used to study alumina compacts. Various microstructural phenomena, generally in the size range of 1 to 10,000 nm, are discussed including pore size and distribution, pore fraction and surface area.

901,038
PB89-175954 Not available NTIS
National Bureau of Standards (IMSE), Gaithersburg, MD. Ceramics Div.
Creep Cavitation in Liquid-Phase Sintered Alumina.
Final rept.
R. A. Page, K. S. Chan, K. Hardman-Rhyne, J. Lankford, and S. Spooner. 1987, 9p
Pub. in Jnl. of the American Ceramic Society 70, n3 p137-145 1987.

Keywords: *Aluminum oxide, *Ceramics, *Creep tests, *Cavitation, *Grain boundaries, *Sintering, Phases, Liquids, Neutron scattering, Stress analysis, Strain rate, Nucleation, Reprints.

The early stages of creep cavitation in a liquid-phase sintered alumina have been characterized using small-angle neutron scattering. Grain boundary cavities were found to nucleate throughout creep, although at a steadily decreasing rate. The cavities were located on two-grain junctions as well as triple points and were spaced approximately 100 to 200 nm apart. Cavity nucleation was also found to be relatively independent of the applied stress. The behavior has been rationalized based on the decreasing ratio of epsilon(gbs)/

epsilon(t), where epsilon(gbs) is the strain due to grain boundary sliding and epsilon(t) is the total strain, at increasing stresses. Cavity growth, on the other hand, was highly stress dependent. Above a certain 'threshold' stress cavity growth was observed. In all cases, however, the observed growth was transient, i.e., the cavity growth rate decreased with time. Lowering the stress below the 'threshold' resulted in a condition in which cavities nucleated but continued growth of the cavities did not occur. In all cases the cavities nucleated and grew, when growth did occur, with relatively equiaxed shapes.

901,039
PB89-177182 Not available NTIS
National Bureau of Standards (NEL), Gaithersburg, MD. Automated Production Technology Div.
Dynamic Poisson's Ratio of a Ceramic Powder during Compaction.
Final rept.
M. P. Jones, and G. V. Blessing. 1987, 4p
Pub. in Proceedings of IEEE (Institute of Electrical and Electronics Engineers) Ultrasonics Symposium, Denver, CO., October 14-16, 1987 p587-590.

Keywords: *Ceramics, *Powder testing, *Elastic properties, *Ultrasonic tests, Compacting, Poisson ratio, Pressure, Secondary waves, Longitudinal waves, Green strength.

Shear and longitudinal wave transit times were measured in a ceramic powder during its compaction. These transit times were used to calculate the dynamic Poisson's ratio of the powder. The goal of the work was to better understand the particles' interaction and rearrangement under pressure. It was shown that Poisson's ratio depended upon the type of powder evaluated and the loading path. Information obtained from this technique could be used as real-time feedback to control and monitor the quality of ceramic parts.

901,040
PB89-179717 Not available NTIS
National Bureau of Standards (IMSE), Gaithersburg, MD. Ceramics Div.
Syntheses and Unit Cell Determination of Ba3V4O13 and Low- and High-Temperature Ba3P4O13.
Final rept.
J. M. Millet, H. S. Parker, and R. S. Roth. 1986, 3p
Pub. in Jnl. of the American Ceramic Society 69, n5 pC103-C105 May 86.

Keywords: *Ceramics, *Barium oxides, *Crystal structure, *X ray diffraction, Vanadates, Phosphate deposits, Crystal lattices, Triclinic lattices, Monoclinic lattices, Orthorhombic lattices, Reprints.

Syntheses and unit cell determination of Ba3V4O13 and the two forms (low and high) of Ba3P4O13 are presented. Ba3V4O13 crystallizes in the monoclinic system, space group Cc or C2/c with unit cell dimensions a = 16.087, b = 8.948, c = 10.159 nm x 10, beta = 114.52 degrees. Low-Ba3P4O13 crystallizes in the triclinic system, space group P1 or P-1 with unit cell dimensions, a = 7.240, b = 8.011, c = 5.689 nm x 10, alpha = 104.02 degrees, beta = 109.510, gamma = 83.62 degrees. Low Ba3V4O13 transforms at 870 degrees C into high Ba3P4O13 which crystallizes in the orthorhombic system, space group Pbcm (No. 57) (or Pbc2, No. 29) with unit cell dimensions, a = 7.107, b = 13.883, c = 19.219 nm x 10. No relations have been found between the structures of the tri-barium tetravanadate and the tri-barium tetraphosphate.

901,041
PB89-179741 Not available NTIS
National Bureau of Standards (IMSE), Gaithersburg, MD. Ceramics Div.
Synthesis, Stability, and Crystal Chemistry of Dibarium Pentatitanate.
Final rept.
R. S. Roth, J. J. Ritter, H. S. Parker, and D. B. Minor. 1986, 5p
Pub. in Jnl. of the American Ceramic Society 69, n12 p858-862 Dec 86.

Keywords: *Ceramics, *Synthesis(Chemistry), *Barium oxides, *Titanates, *Phase transformations, Niobium, Tin, Zirconium, Stability, Triclinic lattices, Orthorhombic lattices, X ray diffraction, Single crystals, Metastable state, Reprints.

A metastable phase corresponding to the previously reported Ba2Ti5O12 was found to form between 650-675 C from hydrolized ethoxide precursors. The stability was increased to approximately 850 C by the addition of 1-2 mole% Nb2O5 to the precursor solutions. Addition of 5 mole% SnO2 failed to yield any sign of this phase in solid state preparations, contrary to the previous report. Addition of 8 mole% ZrO2, however, did produce the desired phase as reported, both with and without additional Nb2O5, apparently stable to greater than 1300 C. Small single crystals, picked from a ZrO2 stabilized specimen with one mole% Nb2O5, showed the compound Ba2Ti(5-x)ZrxO12 to be a 10-layer structure, triclinic, pseudo-orthorhombic with A-centered symmetry and a = 9.941(5), b = 11.482(4), c = 23.528(10) nm x 10. The corresponding reduced triclinic unit cell has a equal 9.941, b = 11.482, c = 13.090 nm x 10, alpha = 116.01 degrees, beta = 90.0 degrees, gamma = 90.0 degrees.

901,042
PB89-185573 PC **A03**/MF **A01**
National Bureau of Standards (NEL), Gaithersburg, MD. Center for Building Technology.
Epoxy Impregnation of Hardened Cement Pastes for Characterization of Microstructure.
Interim rept.
L. Struble, and E. Byrd. Nov 86, 22p NBSIR-87/3504
Sponsored by Air Force Office of Scientific Research, Bolling AFB, DC.

Keywords: *Cements, *Hardening(Materials), *Polishing, *Impregnating, *Epoxy resins, *Ethanols, Tests, Microstructure, Crack propagation, Drying, Methodology, Electron microscopy.

Methods were explored for drying of hardened cement paste prior to impregnation with epoxy, in order to polish for microscopic examination. All were shown to cause microcracking of the paste. An alternative method involves replacing pore water with ethanol, then replacing ethanol with epoxy. The method appears to minimize the occurrence of microcracks associated with drying.

901,043
PB89-186290 Not available NTIS
National Bureau of Standards (IMSE), Gaithersburg, MD. Ceramics Div.
Computer Graphics for Ceramic Phase Diagrams.
Final rept.
P. K. Schenck, and J. R. Dennis. 1987, 5p
Pub. in Computer Handling and Dissemination of Data, p184-188 1987.

Keywords: *Ceramics, *Phase diagrams, Reprints, *Interactive graphics, Computer graphics, Computer applications, Data bases, Computer software.

Specialized graphics software has been written in support of the NBS-ACerS Phase Diagrams for Ceramists Data Base Program. The software runs on a stand alone desk top computer and allows both binary and ternary ceramic phase diagrams to be entered into a graphics data base. The phase diagrams may be entered by keyboard, direct digitization of a published diagram, or from properly formatted data files. The software allows for dynamic on-screen editing of the diagrams and includes many special features to enhance the appearance and the accuracy of the figures. The rapid generation of uniform, camera-ready copy for inclusion in subsequent volumes of Phase Diagrams for Ceramists has allowed the Data Center to accelerate publication of critically evaluated phase diagrams. In addition, the graphics data base will eventually be integrated with other data bases under development in the Data Center.

901,044
PB89-186308 Not available NTIS
National Bureau of Standards (IMSE), Gaithersburg, MD. Ceramics Div.
Phase Diagrams for High Tech Ceramics.
Final rept.
S. J. Schneider, J. W. Hastie, and W. P. Holbrook. 1987, 16p
Pub. in Mater. Sci. Monogr. 38A, p59-74 1987.

Keywords: *Ceramics, *Phase diagrams, *Data processing, Mathematical models, Reprints.

Chemical behavior is the foundation from which advanced ceramics are designed, processed and used; it sets properties, dictates reliable manufacture and ultimately determines performance and lifetimes. Predictive and descriptive chemistry is required and the basis for this is phase equilibria and its graphical expression -- the phase diagram. Ceramic phase equilibria research is literally exploding with literature growth outstripping the capacity of any single company, or even nation, to compile, evaluate and disseminate the available phase diagram information. Cooperative efforts worldwide are needed and in recognition of this, the American Ceramic Society (ACerS) and the National Bureau of Standards (NBS) have initiated a program to expand their previous data effort to develop a comprehensive Phase Diagram Data System. The paper summarizes the status of the development program to achieve more frequent and up-to-date compilations of critically evaluated phase diagrams, expanded coverage for high technology ceramics and computer services for data base management, graphics and phase diagram modeling.

901,045
PB89-188569 PC **A06**/MF **A01**
National Inst. of Standards and Technology (IMSE), Gaithersburg, MD. Ceramics Div.
Advanced Ceramics: A Critical Assessment of Wear and Lubrication.
R. G. Munro, and S. M. Hsu. Jan 89, 109p NISTIR-88/3722, GRI-88/0290, CAM-8901
Contract GRI-5084-238-1302
See also PB88-215447. Sponsored by Gas Research Inst., Chicago, IL., and Pennsylvania State Univ., University Park. Center for Advanced Materials.

Keywords: *Wear, *Ceramics, *Gas engines, *Surveys, Thermal properties, Mechanical properties, Lubrication, Reciprocating engines, Gas turbine engines, Rotary combustion engines, Methane, Engine primers, Graphs(Charts), *Tribology.

A critical assessment of the state of the art of the tribology of ceramics is made. To identify the critical technical barriers confronting the utilization of advanced gas-fired engines, data were gathered specifically on the tribology of materials for gas-fired engine applications. Site visits and discussions with a number of GRI contractors in industry were conducted as the first step in identifying critical issues. Then, an extensive review of the technical literature was made to determine what information was available to resolve those issues, and, more importantly, what critical information was not yet available. These data were used to examine the issues for each of the principal engine types (rotary, reciprocating, and turbine). Materials property data for ceramics were then reviewed in the context of the operating environments and conditions for these engines. Thermal, mechanical, and tribological properties were examined, along with the important considerations for lubricating ceramics in engine applications. The analysis of these data considered the impact and relative merits of using various advanced materials and resulted in recommendations for research activities that could have a significant impact on the development of gas-fired prime movers.

901,046
PB89-193221 PC **A03**/MF **A01**
National Inst. of Standards and Technology (NEL), Gaithersburg, MD. Center for Building Technology.
Standard Aggregate Materials for Alkali-Silica Reaction Studies.
L. Struble, and M. Brockman. May 89, 39p NISTIR-89/4058

Keywords: *Alkali aggregate reactions, *Silicon oxides, *Expansion, *Concretes, Cements, Mortars(Materials), Opal, Quartzites, Rhyolite, Flint glass, Limestone, Graphs(Charts).

Preliminary studies have been carried out to identify candidate materials for use as a standard reactive aggregate in alkali-silica investigations. The materials studied included several commercial glasses, an opal, a quartzite, a rhyolite and a calcined flint. Candidate materials were tested for their expansion in mortars prepared using either a high-alkali or a low-alkali cement, a nonreactive limestone sand, and some proportion of reactive material. Tests were carried out according to ASTM C441-81, Standard Test Method for Effectiveness of Mineral Admixtures in Preventing Excessive Expansion of Concrete Due to the Alkali-Aggregate Reaction, and ASTM C227-87, Standard Test Method for Potential Alkali Reactivity of Cement-Aggregate Combinations (Mortar-Bar Method). The proportion of limestone replaced by each reactive material was varied so as to bracket the pessimum proportion (i.e., the proportion of reactive material producing the highest level of expansion). Mortar-bar expansion levels were measured throughout reaction periods of approximately 6 months to 1 year. Expansion results are presented and discussed. Based on the studies, the Vycor, fused quartz, fused silica, and calcined flint appear suitable as standard reactive materials; the calcined flint appears especially promising.

901,047
PB89-201636   Not available NTIS
National Bureau of Standards (NML), Gaithersburg, MD. Gas and Particulate Science Div.
Methods for the Production of Particle Standards.
Final rept.
J. A. Small, J. J. Ritter, P. J. Sheridan, and T. R. Pereles. 1986, 21p
Pub. in Jnl. of Trace Microprobe Tech. 4, n3 p163-183 1986.

Keywords: *Metal particle composites, *Standards, *Ceramics, *Glass particle composites, Particle size, Production, Fiber composites, Spheres, Microstructure, Granular materials, Reprints, National Institute of Standards and Technology.

The microanalysis group at the National Bureau of Standards (NBS) has been studying three different methods of fabricating microscopic and submicroscopic particle standards. These methods include the manufacture of ground glass shards, fibers, and microspheres from the NBS analytical glass standards, the manufacture of metal and metal alloy particles in the spark source emission spectrometer, and the manufacture of doped ceramic particles by alkoxide synthesis. Preliminary studies of these methods indicate that they can be used to produce particle standards which have a wide variety of elemental compositions and sizes.

901,048
PB89-201776   Not available NTIS
National Bureau of Standards (IMSE), Gaithersburg, MD.
Commercial Advanced Ceramics.
Final rept.
S. J. Schneider. 1986, 4p
Pub. in ASTM (American Society for Testing and Materials) Standardization News, p36-39 Oct 86.

Keywords: *Ceramics, Standards, Construction materials, Electronics, Engine blocks, Reprints.

Advanced ceramics represent a newer generation of high-performance materials that collectively are viewed as an enabling technology, development and exploitation of which are critical to advances in a host of high-technology applications ranging between modern microelectronic components to futuristic auto engines. New products are in the offing that could impact whole industrial segments including transportation, communications, computers, energy conversion and major defense systems. Even though products are being developed, diffusion to the marketplace is severely inhibited by the lack of commercial standards - the test methods, the classification system, the standard reference materials and the like. The standards needs have been identified and include requirements in the areas of: processing, characterization, properties, performance, statistical procedures and terminology.

901,049
PB89-202097   Not available NTIS
National Bureau of Standards (IMSE), Gaithersburg, MD. Ceramics Div.
Electron Microscopy Studies of Diffusion-Induced Grain Boundary Migration in Ceramics.
Final rept.
M. D. Vaudin, C. A. Handwerker, and J. E. Blendell. 1988, 8p
Pub. in Jnl. de Physique 49, n10 pC5-687-C5-692 Oct 88.

Keywords: *Ceramics, *Grain boundaries, Electron microscopy, Diffusion, Sintering, Crystal structure, Magnesium oxides, Annealing, Nickel oxides, Microstructure, Temperature, Surface properties, Interfaces, Strains, Reprints.

The stability of grain boundaries during the uptake of solute into polycrystalline MgO has been investigated by exposing sintered, well-annealed MgO specimens to NiO at various temperatures for different times, and observing their microstructure both before and after this exposure. Electron and optical microscopy techniques have been used to characterize the specimen surface morphology, and the variation in composition and elastic and plastic strains in the interface regions. The spatial distribution of nickel at various depths below the original specimen surface was determined

using electron microprobe analysis. The observations are discussed in terms of the coherency strain theory.

901,050
PB89-202105   Not available NTIS
National Bureau of Standards (IMSE), Gaithersburg, MD. Ceramics Div.
Green Function Method for Calculation of Atomistic Structure of Grain Boundary Interfaces in Ionic Crystals.
Final rept.
V. K. Tewary, E. R. Fuller, and R. M. Thomson. 1986, 10p
Pub. in Ceramic Microstructures '86 Role of interfaces, p167-176.

Keywords: *Atomic structure, *Crystal lattices, *Ceramics, Greens function, Microstructure, Grain boundaries, Interfaces, Ions, Polarity, Coulomb interactions, Crystal defects, Reprints.

A lattice statics Green function method is described for calculating the atomistic structure of grain boundary interfaces in ionic crystals. The grain boundary is taken along coincidence lattice sites of the two crystallites. The periodicity of the coincidence lattice is exploited by taking a partial Fourier transform of the Green function. It is also shown that the Coulomb interaction between ions across the grain boundary line can be represented in terms of effective dipolar and higher order polar interactions which makes them relatively short range.

901,051
PB89-211817   Not available NTIS
National Bureau of Standards (IMSE), Gaithersburg, MD. Ceramics Div.
Crack-Interface Traction: A Fracture-Resistance Mechanism in Brittle Polycrystals.
Final rept.
P. L. Swanson. 1988, 21p
Pub. in Proceedings of Conference on the Fractography of Glasses and Ceramics, Alfred, NY., August 3-6, 1986, p135-155 1988.

Keywords: *Ceramics, *Glass, *Interfaces, *Crack propagation, Fractures(Materials), Microscopy, Traction, Aluminum oxide, Resistance.

Crack-interface tractions have been identified as a source of increasing resistance to fracture with crack extension, or rising R-curve behavior, in previous studies on coarse-grained alumina. Real time in situ microscopy observations are used in the present study to investigate the generality of crack-interface tractions as a crack-resistance mechanism in three alumina and three glass-ceramics with varying R-curve characteristics. Interface tractions are found to operate to varying degrees in each material. Ligamentary-bridge formation is compared with the development of twist hackle, inclusion/wake hackle and cleavage hackle in simple material systems. Both sources of interface traction remain active as far as 100 particle dimensions behind the primary crack tip and, with sufficient crack-opening displacement, are eventually overcome by interface-localized microfracturing. Simple analytical fracture mechanics concepts are used to assess the influence of interface tractions on macroscopic fracture behavior. Because of the observed crack-history dependence of the interface-traction crack-tip shielding, it is suggested that neither R-curve behavior nor applied-K(sub I)/subcritical crack velocity relationships are unique properties of these and similar materials.

901,052
PB89-211833   Not available NTIS
National Bureau of Standards (IMSE), Gaithersburg, MD. Ceramics Div.
Design Criteria for High Temperature Structural Applications.
Final rept.
S. M. Wiederhorn, and E. R. Fuller. 1986, 19p
Sponsored by Department of Energy, Oak Ridge, TN.
Pub. in Proceedings of International Symposium on Ceramic Materials and Components for Engines (2nd), Luebeck-Travemuende, West Germany, April 14-17, 1986, p911-929.

Keywords: *Structural forms, *High temperature tests, *Fractures(Materials), *Crack propagation, *Ceramics, Creep rupture tests, Nucleation, Initiation, Damage, Engines.

Current methods of assessing high temperature, structural reliability of ceramics are fracture mechanics based, in the sense that failure is assumed to originate

from pre-existing flaws that grow until they reach a critical size for rupture. Recent data on structural ceramics at elevated temperatures suggest that the view may be over-simplified and that flaw initiation and damage accumulation may be important factors to consider in the design of ceramic materials for high temperature applications. The techniques that are currently being used for structural design are summarized and the limits of their application at elevated temperatures are defined. The need for incorporating information on flaw nucleation and damage accumulation into methods of structural design is emphasized.

901,053
PB89-211841   Not available NTIS
National Bureau of Standards (IMSE), Gaithersburg, MD. Ceramics Div.
PC-Access to Ceramic Phase Diagrams.
Final rept.
P. K. Schenck, and J. R. Dennis. 1989, 12p
Pub. in ASTM (American Society for Testing and Materials) Special Technical Publication 1017, p292-303 1989.

Keywords: *Phase diagrams, *Ceramics, Computer systems programs, Data retrieval, Reprints, Computer graphics.

A personal computer (PC)-based version of Phase Diagrams for Ceramists has been demonstrated by the joint National Bureau of Standards/American Ceramic Society Phase Diagrams for the Ceramists Data Center. A selection of phase diagrams from the nearly 1100 diagrams of Volume 6 has been transferred from the Data Center's graphics workstations to a PC. Demonstration software has been developed for retrieving and plotting the phase diagrams on the PC's monitor from the PC-based Phase Diagram Data Base. In addition, the software allows the operator to retrieve data from the diagram by means of an interactive graphics cursor whose location is displayed digitally on the monitor in a choice of user units (for example, C, F, K). Areas of the phase diagram can be magnified and replotted to clarify features. The operator can also overlay a second diagram for comparison purposes, reverse the diagram, magnify or rescale the diagram, and apply an electronic lever rule or curve-tracking mode. Future refinements include conversions between weight and mole percent and retrieval of ternary or higher-order phase diagrams.

901,054
PB89-221170   PC A09/MF A01
National Inst. of Standards and Technology (NEL), Gaithersburg, MD. Semiconductor Electronics Div.
Semiconductor Measurement Technology: Database for and Statistical Analysis of the Interlaboratory Determination of the Conversion Coefficient for the Measurement of the Interstitial Oxygen Content of Silicon by Infrared Absorption.
Special pub. (Final).
A. Baghdadi, R. I. Scace, and E. J. Walters. Jul 89, 183p NIST/SP-400/82
Also available from Supt. of Docs. as SN003-003-02943-2. Library of Congress catalog card no. 89-600739.

Keywords: *Semiconductors(Materials), *Oxygen, *Interstitials, *Silicon, *Infrared radiation, Quantitative analysis, Statistical analysis, Radioactivation analysis, Charged particles, Photons, Conversion, Tables(Data), Graphs(Charts), Electromagnetic absorption, Round robin tests.

The Special Publication contains the data collected for the worldwide, double-round-robin determination of the conversion coefficient used to calculate the interstitial oxygen content of silicon from infrared absorption measurements. It also contains detailed statistical analyses of the data. The approach taken to determine the conversion coefficient was to conduct inter-laboratory round robins for both the infrared measurements and the absolute measurements. The infrared measurements were carried out at 18 laboratories in China, Europe, Japan, and the United States, using either dispersive infrared or Fourier transform infrared spectrometers. The absolute measurements were carried out at eight laboratories in Europe, Japan, and the United States, using either charged-particle activation analysis, photon activation analysis, or inert gas fusion analysis.

901,055
PB89-229074   PC A06/MF A01

Ceramics, Refractories, & Glass

National Inst. of Standards and Technology (IMSE), Gaithersburg, MD. Ceramics Div.
NBS/BAM (National Bureau of Standards/Bundesanstalt fur Materialprufung) 1986 Symposium on Advanced Ceramics.
Special pub. (Final).
S. M. Hsu, and H. Czichos. May 89, 171p NIST/SP-766
Also available from Supt. of Docs. as SN003-003-02936-0. Library of Congress catalog card no. 89-600737. Prepared in cooperation with Bundesanstalt fuer Materialpruefung, Berlin (Germany, F.R.).

Keywords: *Meetings, *Ceramics, Fabrication, Standards, Defects, Quality control, Sintering, International relations, Graphs(Charts), Manufacturing, Reliability, Microstructure, Wear, Fretting, Nondestructive tests.

dvanced ceramics offer many advantages that other materials do not possess. They have high strength, dimensional stability, are chemically inert, lightweight, wear resistant, and have desirable properties in electrical, optical, and thermal applications. At high temperatures, they are the only class of material with reasonable properties. Worldwide production of advanced ceramics is growing rapidly. Since ceramics are based on alumina and silica, the most abundant minerals on earth, effective utilization of ceramics carries implications into the next several centuries. One of the major technical barriers to widespread use of ceramics is the inability of industry to manufacture reliable ceramics reproducibly and economically. Advanced ceramics are sensitive to small defects introduced during processing and generated during use. The symposium provides timely exchange of technical information on a very significant subject area of achieving reliable manufacturing through standards.

901,056
PB89-229231 Not available NTIS
National Inst. of Standards and Technology (IMSE), Gaithersburg, MD. Ceramics Div.
Effect of Heat Treatment on Crack-Resistance Curves in a Liquid-Phase-Sintered Alumina.
Final rept.
S. J. Bennison, H. M. Chan, and B. R. Lawn. 1989, 3p
Sponsored by Air Force Office of Scientific Research, Bolling AFB, DC.
Pub. in Jnl. of the American Ceramic Society 72, n4 p677-679 Apr 89.

Keywords: *Aluminum oxide, *Crack propagation, *Heat treatment, *Liquid phases, *Sintering, *Indentation hardness tests, Microstructure, Defects, Design criteria, Ceramics, Reprints.

The effects of heat treatment on the R-curve (crack-resistance) behavior of a commercial liquid-phase-sintered (LPS) alumina have been studied using the indentation-strength test. An enhancement of the R-curve characteristic of this LPS alumina is obtained by a treatment that increases the scale of the microstructure. The enhanced R-curve characteristics leads to the desirable property of flaw tolerance, albeit at the expense of a diminished strength at small crack sizes. The implications of the findings are discussed with reference to processing and design strategy.

901,057
PB90-111238 PC **A09**/MF **A02**
National Inst. of Standards and Technology (NEL), Gaithersburg, MD. Center for Building Technology.
Set Time Control Studies of Polymer Concrete.
R. G. Mathey, and J. M. Pommersheim. Sep 89, 195p NISTIR-89/4026
Prepared in cooperation with Bucknell Univ., Lewisburg, PA. Sponsored by Air Force Engineering and Services Center, Tyndall AFB, FL.

Keywords: *Concretes, *Polymers, *Setting time, Polyurethane resins, Temperature, Catalysts, Water, Pilot plants, Flexural strength, Mathematical models.

Set time data were obtained for polymer concrete made with a proprietary polyurethane resin for a wide range of aggregate and resin temperatures. Catalyst concentrations were adjusted so that setting occurred within a required time range. The effects of the presence of water and ice on set time were also studied. Set time data were also obtained from pilot tests using another polyurethane and catalyst for various aggregate and resin temperatures and moisture conditions. Considerably more catalyst was required in the pilot tests to obtain comparable set times. The impact of temperature variations on flexural strength was investigated. The flexural strength and failure mechanism at early ages depended on the temperature of aggregate and resin at the time of casting the polymer concrete. Using the model, a series of design charts were prepared which can be used to predict set time when catalyst concentration and initial aggregate and resin temperatures are given, or to determine the catalyst concentration needed to assure set time corresponding to specified aggregate and resin temperatures.

901,058
PB90-117383 Not available NTIS
National Inst. of Standards and Technology (IMSE), Gaithersburg, MD. Ceramics Div.
Grain-Size and R-Curve Effects in the Abrasive Wear of Alumina.
Final rept.
S. J. Cho, B. J. Hockey, B. R. Lawn, and S. J. Bennison. 1989, 4p
Sponsored by Air Force Office of Scientific Research, Bolling AFB, DC., Gas Research Inst., Chicago, IL., and Korea Science and Engineering Foundation.
Pub. in Jnl. of the American Ceramic Society 72, n7 p1249-1252 1989.

Keywords: *Ceramics, *Aluminum oxides, *Wear, *Abrasion tests, *Grain size, Crack propagation, Brittleness, Sliding friction, Fracture strength, Microstructure, Reprints.

Results of sliding wear tests on three alumina ceramics with different grain sizes are discussed in the light of crack-resistance (R-curve, or T-curve) characteristics. The degree of wear increases abruptly after a critical sliding period, reflecting a transition from deformation-controlled to fracture-controlled surface removal. The transition occurs at earlier sliding times for the aluminas with the coarser-grained microstructures, indicative of an inherent size effect in the wear process. A simplistic fracture mechanics model, incorporating the role of internal thermal expansion mismatch stresses in the crack-resistance characteristic, is developed. The results suggest an inverse relation between wear resistance and large-crack toughness for ceramics with pronounced R-curve behavior.

901,059
PB90-117722 Not available NTIS
National Inst. of Standards and Technology (IMSE), Gaithersburg, MD. Ceramics Div.
Tribochemical Mechanism of Alumina with Water.
Final rept.
R. S. Gates, S. M. Hsu, and E. E. Klaus. 1989, 7p
Sponsored by Department of Energy, Washington, DC.
Pub. in Tribology Transactions 32, n3 p357-363 1989.

Keywords: *Aluminum oxide, *Water, Reaction kinetics, Wear tests, X ray diffraction, Thermal analysis, Surface chemistry, Phase transformations, Aluminum hydroxides, Rubbing, Reprints, *Tribology.

Water has been found to exhibit significant effects on the tribological behavior of alumina. A film-like substance was found on the surfaces of water lubricated alumina wear surfaces, suggesting the possibility of tribochemical reaction between water and alumina in the contact junction. The paper describes an investigation of the alumina/water tribosystem to determine the chemical interaction between the two materials under rubbing conditions. A combination of x-ray powder diffraction and thermogravimetric analysis (TGA) has been used to investigate the kinetics of aluminal/water reactions. The experiments have determined that transition (gamma) alumina reacts with water to form hydroxides of aluminum. At high temperature (approximately 200 C) aluminum oxide hydroxide (boehmite - AlO(OH) is formed, while at lower temperature (approximately 100 C) the formation of aluminum trihydroxide (bayerite-Al(OH)3) is favored. A mechanism for lubrication of alumina with water is proposed whereby stresses and temperatures in the contact junction case phase transformation from alpha alumina to a transition alumina. The transition alumina subsequently reacts with water to form a lubricious hydroxide layer and reduce friction and wear.

901,060
PB90-128026 Not available NTIS
National Inst. of Standards and Technology (IMSE), Gaithersburg, MD. Ceramics Div.
Flaw Tolerance in Ceramics with Rising Crack Resistance Characteristics.
Final rept.
S. J. Bennison, and B. R. Lawn. 1989, 7p
Sponsored by Air Force Office of Scientific Research, Washington, DC.
Pub. in Jnl. of Materials Science 24, p3169-3175 1989.

Keywords: *Ceramics, *Crack propagation, *Defects, Toughness, Aluminum oxide, Polycrystalline, Brittleness, Stresses, Reprints, Closures.

The stabilizing influence of increasing toughness with crack size associated with a cumulative closure-stress process (R-curve, or T-curve) on the strength properties of brittle ceramic materials is analyzed. Three strength-controlling flaw types are examined in quantitative detail: microcracks with closure-stress history through both the initial formation and the extension in subsequent strength testing; microcracks with closure stresses active only during the subsequent extension; spherical pores. Using a polycrystalline alumina with pronounced T-curve behavior as a case study, it is demonstrated that the strength is insensitive to a greater or lesser extent on the initial size of the flaw, i.e. the material exhibits the quality of 'flaw tolerance'. The insensitivity is particularly striking for the flaws with full closure-stress history, with virtually total independence on initial size up to some 100 microns, for the flaws with only post-evolutionary exposure to the closure elements the effect is less dramatic, but the strength characteristics are nevertheless significantly more insensitive to initial flaw size than the counterparts for materials with single-value toughnesses. The implications of the results to engineering design methodologies, as expressed in conventional R-curve constructions, and to processing strategies for tailoring materials with optimal crack resistance properties, are discussed.

901,061
PB90-128638 Not available NTIS
National Inst. of Standards and Technology (IMSE), Boulder, CO. Fracture and Deformation Div.
Low Temperature Mechanical Property Measurements of Silica Aerogel Foam.
Final rept.
L. L. Scull, and J. M. Arvidson. 1988, 6p
Sponsored by Lawrence Livermore National Lab., CA.
Pub. in Advances in Cryogenic Engineering Materials, v34 p413-418 1988.

Keywords: *Cryogenics, *Mechanical properties, *Silicon oxides, *Aerogels, *Foam, *Measurement, Colloids, Ultimate strength, Thermal conductivity, Insulation, Reprints.

Silica aerogel is a low-density foam material produced by supercritically extracting the solvent from a colloidal suspension of silica in solution. The process produces a transparent, open-cell foam structure that possesses an extremely small cell size (approximately 60 nm) with a total porosity in excess of 90% by volume. The morphology gives the aerogel a large strength to density ratio with excellent insulating properties (thermal conductivity of 0.02 - 0.05 w/m/k in N2 gas). Also, the material has refractive index less than 1.10. The properties measured in the study include Young's modulus, proportional limit, yield strength and strain in compression. Other properties measured were Poisson's ratio and ultimate tensile strength.

901,062
PB90-136847 Not available NTIS
National Bureau of Standards (NEL), Gaithersburg, MD. Building Materials Div.
Model for Particle Size and Phase Distributions in Ground Cement Clinker.
Final rept.
P. W. Brown, and K. G. Galuk. 1987, 8p
Pub. in Materials Research Society Symposia Proceedings 85, p83-90 1987.

Keywords: *Mathematical models, *Particle size distribution, *Phase transformations, *Portland cements, *Grinding(Comminution), *Clinker, Hydration, Chemical analysis, Physical properties.

Relationships between the phase distributions in portland cement clinker and the phase and particle size distributions in the ground clinker are not well understood. Little experimental work to characterize the distribution of phases in ground clinker seems to have been done. As a result of the factors, it is difficult to develop physically significant, predictive mathematical models for cement hydration. It is the objective of the paper to describe a computer model that relates the phase and particle size distributions of the ground cement clinker to physical and chemical composition of unground clinker.

Coatings, Colorants, & Finishes

901,063
PB89-162598 PC A04/MF A01
National Inst. of Standards and Technology (NEL), Gaithersburg, MD. Building Materials Div.
Relationship between Appearance and Protective Durability of Coatings: A Literature Review.
T. Nguyen, B. Collins, L. Kaetzel, J. Martin, and M. McKnight. Dec 88· 57p NISTIR-88/4010
Sponsored by Civil Engineering Lab. (Navy), Port Hueneme, CA.

Keywords: *Protective coatings, *Forecasting, *Service life, *Discoloration, *Paints, Reviews, Construction materials, Fading, Weathering.

For coatings, improved service life prediction aids in the effective selection and use of materials and in the development of cost-effective maintenance strategies. However, quantitative measures of degradation are essential in predicting service life. Standard procedures are available to quantitatively measure small changes in the appearance properties of coatings, one of the two primary· functions of coatings, the other being protection · However, quantitative measurements of early changes associated with the protective function usually are not possible. Hence, the objective of the report is to ascertain, based upon the literature, whether changes in appearance properties of coatings can be used to predict changes in the protective properties of the film. It was concluded, that for the most part, changes in appearance properties are not related to changes in the protective properties of a coating film.

901,064
PB89-201040 Not available NTIS
National Bureau of Standards (IMSE), Gaithersburg, MD. Metallurgy Div.
Structural Study of a Metastable BCC Phase in Al-Mn Alloys Electrodeposited from Molten Salts.
Final rept.
B. Grushko, and G. R. Stafford. 1989, 6p
Pub, in Scripta Metallurgica 23, n4 p557-562 1989.

Keywords: *Electrodeposition, *Aluminum alloys, *Manganese, *Crystal structure, *Molten salt electrolytes, Body centered cubic lattices, Metastable state, Electron microscopy, X ray diffraction, Copper, Zinc, Reprints.

The structure of aluminum-manganese alloys electrodeposited from molten salt electrolytes at 150 C were characterized ·by transmission electron microscopy and x-ray diffraction. Electrodeposits containing 24-27 at .% Mn contain a crystalline phase co-existing with an amorphous phase. Ring electron diffraction data from the crystalline phase is very similar to that reported for the 'F' phase in this system. Selected area electron diffraction confirms the bcc structure but the appearance of additional weak reflections indicates that the structure should be indexed to a lattice which is 3 times larger. Such a unit cell is consistent with the .Cu5Zn8 structure and suggests that the crystalline phase is Al8Mn5 with a cubic gamma-brass structure.

901,065
PB89-209290 PC A03/MF A01
National Inst. of Standards and Technology (NEL), Gaithersburg, MD. Chemical Process Metrology Div.
Thin Film Thermocouples for High Temperature Measurement.
K. G. Kreider. May 89, 26p NISTIR-89/4087

Keywords: *Thin films, *Thermocouples, *Vacuum deposited coatings, High temperature tests, Graphs(Charts), Platinum, Rhodium, Measurement, Engines, Turbines.

Thin film thermocouples have unique capabilities for measuring surface temperatures at high temperatures (above 800K) under harsh conditions. Their low mass, approximately 2x10(sup-5) g/mm permits very rapid response and very little disturbance of heat transfer to the surface being measured. This has led to applications inside gas turbine engines and diesel engines measuring the surface temperature of first stage turbine blades and vanes and ceramic liners in diesel cylinders. The most successful high temperature (up to 1300K) thin film thermocouples are sputter deposited from platinum and platinum-10% rhodium targets although results using base metal alloys, gold, and platinel will also be presented. The paper reviews the fabrication techniques used to form the thermocouples.

approaches used to solve the high temperature insulation and adherence problems, current applications, and test results using the thin film thermocouples. In addition a discussion will be presented on the current problems and future trends related to applications of thin film thermocouples at higher temperatures up to 1900K.

901,066
PB89-212112 Not available NTIS
National Bureau of Standards (NEL), Gaithersburg, MD. Building Materials Div.
Quantitative Studies of Coatings on Steel Using Reflection/Absorption Fourier Transform Infrared Spectroscopy.
Final rept.
T. Nguyen, D. Bentz, and E. Byrd. 1986, 1p
Pub. in Abstracts of Papers, 1p 1986.

Keywords: *Infrared spectroscopy, *Thin films, *Steels, *Absorption, Reflection, Epoxy resins, Coatings. Quantitative analysis, Nondestructive tests, Reprints.

Reflection/absorption infrared spectroscopy (RAS), commonly referred to as external reflection infrared spectroscopy, has become a powerful, nondestructive technique for studies of thin and thick films on metal surfaces. The intensity and shape of the absorption bands obtained by the technique are different from those obtained by the conventional transmission technique and are a complex function of numerous parameters. The theory of RAS is valid only for very thin films and for a particular substrate/film system; for thick films, the theory deviates from experimental data. In the paper, the relationships between band intensities, angles of incidence, and film thicknesses of an amine-cured epoxy coating on cold-rolled steel are examined using reflection/absorption Fourier transform infrared spectroscopy. Mathematical expressions are developed to describe the relationships for thick films of a steel/epoxy system.

901,067
PB89-228571 Not available NTIS
National Inst. of Standards and Technology (IMSE), Gaithersburg, MD. Ceramics Div.
Fiber Coating and Characterization.
Final rept.
D. C. Cranmer. 1989, 5p
Sponsored by Strategic Defense Initiative Organization, Washington, DC. Innovative Science and Technology.
Pub. In American Ceramic Society Bulletin 68, n2 p415-419·Feb 89.

Keywords: *Coatings, *Ceramic fibers, *Carbon fibers, *Thickness, Monitors, Dimensional measurement, Reprints.

A variety of techniques exist for depositing coatings on ceramic and carbon fibers. The paper reviews several of the techniques and discusses the advantages and disadvantages. It also points out several deficiencies in the ability to uniformly and reproducibly coat fibers, especially multifilament tows. One of the most significant problems is the inability to monitor the coating composition and thickness in real-time, i.e., as it is deposited. Characterization techniques are currently limited to the examination of small amounts of fiber at a time and can not readily be adapted to continuous processing.

901,068
PB90-112343 PC A03/MF A01
National Inst. of Standards and Technology (IMSE), Gaithersburg, MD. Polymers Div.
Design and Synthesis of Prototype Air-Dry Resins for Use in BEP (Bureau of Engraving and Printing) Intaglio Ink Vehicles.
Annual rept.
B. Dickens, B. J. Bauer, W. R. Blair, and E. J. Parks. Sep 89, 44p NISTIR-89/4110
Sponsored by Bureau of. Engraving and Printing, Washington, DC.

Keywords: *Alkyd resins, *Synthesis(Chemistry), *Inks, Polymers, Drying oils, Pentaerythritol esters, Air, Dispersing, Water, Oxidation, *Synthetic resins, *Intaglio.

Over 60 air-dry resins were designed and synthesized to provide prototype resins for intaglio inks used in the Bureau of Engraving and Printing for printing currency. Most of the resins contain linseed oil fatty acids as the air-dry part. The polyols used are trimethylol propane,

pentaerythritol, dipentaerythritol, and tripentaerythritol. In other cases, 'super drying oil' resins we,e synthesized from tripentaerythritol and linseed oil fatty acids. Inks made from the resins must disperse in 1% aqueous alkali. Acids groups were introduced into the resins using trimellitic anhydride, phthalic anhydride or succinic anhydride. The inks must pass two preliminary tests, one for dispersion in aqueous alkali before curing and one for resistance to the same solution after curing. Two resins, one a super-drying-oil type molecule based on tripentaerythritol and one a more typical alkyd based on pentaerythritol, passed these tests.

901,069
PB90-117961 Not available NTIS
National Inst. of Standards and Technology (IMSE), Gaithersburg, MD. Ceramics Div.
Cathodoluminescence of Defects in Diamond Films and Particles Grown by Hot-Filament Chemical-Vapor Deposition.
Final rept.
L. H. Robins, L. P. Cook, E. N. Farabaugh, and A. Feldman. 1989, 11p
Pub. in Physical Review B 39, n18 p13 367-13 377, 15 Jun 89.

Keywords: *Cathodoluminescence, *Diamonds, *Thin films, *Defects, Crystal structure, Surface properties, Microstructure, Reprints, *Chemical vapor deposition, Scanning electron microscopy.

Point defects, impurities, and defect-impurity complexes in diamond particles and polycrystalline films were investigated by cathodoluminescence (CL) imaging and spectroscopy in a scanning electron microscope. The diamond films and particles were grown by hot-filament methane-hydrogen chemical-vapor deposition at several different temperatures; the nominal deposition temperature (Td) ranged from 600 to 850 C. Electron-beam energies used to excite the CL were 10-30 keV. By comparing the CL spectra to spectra of known defects in natural and synthetic diamond, the following luminescence centers were identified: 2.156-eV center attributed to a nitrogen-vacancy complex; 2.326-eV center, also thought to be a nitrogen-vacancy complex; violet-emitting center (observed peak position at 2.82 eV), associated with dislocation line defects, whose atomic structure is uncertain; 3.188-eV center, attributed to interstitial nitrogen or a nitrogen-(carbon-interstitial) complex; isolated neutral vacancy (denoted the general radiation center) with principal zero-phonon line at 1.673 eV. The luminescence from each center displayed a different dependence on Td and film morphology. ·

Composite Materials

901,070
PB89-147078 Not available NTIS
National Bureau of Standards (NEL), Boulder, CO. Chemical Engineering Science Div.
Performance of Alumina/Epoxy Thermal Isolation Straps.
Final rept.
R. D. Kriz, and L. L. Sparks. 1988, 8p
Sponsored by Ball Aerospace Systems Div., Boulder, CO.
Pub. in Advances in Cryogenic Engineering, v34 p107-114 1988.

Keywords: *Composite materials, *Aluminum oxide, *Epoxy resins, *Straps, *Mechanical properties, *Fibers, Reinforced plastics, Glass fibers, Thermal conductivity, Loads(Forces), Fatigue(Materials), Fractures(Materials), Reprints.

A study of advanced fiber-reinforced composites indicates improved thermal-mechanical performance for straps fabricated with alumina fiber over conventional fiber/epoxy systems. In particular, the study compared identical thermal-isolation strap configurations but with different fiber-reinforcement: S2-glass and alumina. Static and cyclic mechanical tests and thermal conductivity measurements indicate superior performance of straps with alumina fibers. Here a popular cryogenic grade resin was used in both configurations. Results of the study indicate that failure initiates in a region where the load is transferred by shear and compression.

MATERIALS SCIENCES

Composite Materials

901,071
PB89-148399   PC A05/MF A01
National Inst. of Standards and Technology (IMSE),
Boulder, CO. Fracture and Deformation Div.
**Institute for Materials Science and Engineering,
Fracture and Deformation: Technical Activities
1988.**
H. I. McHenry. Feb 89, 85p NISTIR-88/3841
See also report for 1987, PB88-153622.

Keywords: *Fractures(Materials), *Deformation,
*Composite materials, *Metals, *Ceramics, Nonde-
structive tests, Cryogenics, Service life, Mathematical
models, Superconductors, Microstructure, Welding.

The report describes the 1988 fiscal-year programs of
the Fracture and Deformation Division of the Institute
for Materials Science and Engineering. It summarizes
the principal accomplishments in three general research areas: materials performance, properties, and
processing. The Fracture Mechanics, Fracture Physics, Nondestructive Evaluation, and Composite Materials Groups work together to detect damage in metals
and composite materials and to assess the significance of the damage with respect to service performance. The Cryogenic Materials and Physical Properties
Groups investigate the behavior of materials at low
temperature and measure and model the physical
properties of advanced materials, including composites, ceramics and the new high-critical-temperature
superconductors. The Welding and Thermomechanical Processing Groups investigate the nonequilibrium
metallurgical changes that occur during processing
and affect the quality, microstructure, properties and
performance of metals. The report lists the division's
professional staff, their research areas, publications,
leadership in professional societies, and collaboration
in research programs with industries and universities.

901,072
PB89-149165   Not available NTIS
National Bureau of Standards (NEL), Gaithersburg,
MD. Fire Science and Engineering Div.
**Cone Calorimeter Method for Determining the
Flammability of Composite Materials.**
Final rept.
J. E. Brown. 1988, 10p
Sponsored by Naval Sea Systems Command, Wash-
ington, DC.
Pub. in Proceedings of Annual Conference on Ad-
vanced Composites 'How to Apply Advanced Com-
posites Technology' (4th), Dearborn, MI., September
13-15, 1988, p141-150.

Keywords: *Composite materials, *Calorimeters,
*Flammability testing, Heat measurement, Ignition,
Radiance, Fiberglass reinforced plastics, Epoxy resins,
Combustion, Sensitivity, Phenylene sulfide resins.

A study was undertaken to evaluate the fire performance of composite materials using the cone calorimeter as the bench-scale method of test simulating the
thermal irradiance from fires of various magnitudes.
Parameters were derived from the calorimetry measurements to characterize the ignitability and flammability of the composite materials. The parameters are, to
some extent, empirical since radiative heat losses
from the samples were unknown. These parameters
are: (1) minimum external radiant flux (MERF) required
to produce pilot ignition in a predetermined exposure
time; (2) thermal sensitivity index (TSI) which indicates
the burning intensity dependence on external heat
flux; and (3) extinction sensitivity index (ESI) which indicates the propensity for continued flaming combustion without an external heat flux. MERF values at 300
s for 3 mm composite panels of a FR epoxy resin and
of a poly(phenylene sulfide) (PPS) resin were about 18
and 28 kW/sq m, respectively. The TSI of the PPS
resin composite revealed that it had the greatest dependency on external flux. Additionally, the ESI of the
PPS composites was the only one to indicate an external flux requirement to sustain combustion during the
first 60 s after ignition.

901,073
PB89-157754   Not available NTIS
National Bureau of Standards (NML), Gaithersburg,
MD. Gas and Particulate Science Div.
**Computer-Aided Imaging: Quantitative Composi-
tional Mapping with the Electron Probe Microana-
lyzer.**
Final rept.
D. E. Newbury, R. B. Marinenko, D. S. Bright, and R.
L. Myklebust. 1988, 13p
Pub. in Scanning 10, p213-225 1988.

Keywords: *Electron probes, *Microanalysis, Composite materials, Quantitative analysis, Mapping, Reprints,
Image processing.

X-ray area scanning ('dot mapping') is a technique
widely used in electron probe microanalysis for determining the spatial distribution of elemental constituents. Although powerful, this technique is subject to
significant limitations on concentration sensitivity and
flexibility for subsequent processing. The new technique of compositional mapping overcomes these limitations. In compositional mapping, a complete quantitative electron probe analysis is carried out at each
point of a matrix scan. The resulting matrices of concentration values can be assembled into images in a
digital image processor by assigning gray or color intensities to the actual concentrations rather than the
raw spectral intensities. Digital compositional maps
can be readily manipulated by a wide variety of image
processing techniques to improve the visibility of features of interest.

901,074
PB89-179733   Not available NTIS
National Bureau of Standards (IMSE), Gaithersburg,
MD. Ceramics Div.
**Novel Process for the Preparation of Fiber-Rein-
forced Ceramic-Matrix Composites.**
Final rept.
W. Haller, U. V. Deshmukh, and S. W. Freiman.
1988, 3p
Grant N00014-86-F-0046
Sponsored by Strategic Defense Initiative Organiza-
tion, Washington, DC.
Pub. in Jnl. of the American Ceramic Society 71, n12
pC-498-C-500 Dec 88.

Keywords: *Ceramics, *Coatings, *Silicon carbides,
*Glass Fibers, Monofilaments, Powder(Particles),
Composite materials, Ultraviolet radiation, Curing, Re-
inforcing materials, Reprints.

A procedure for the reproducible production of monofi-
lament/powder composites has been developed. The
process consists of making a slurry of the powder in a
solventless ultraviolet-curing resin, and coating the
fiber with this slurry in a continuous process whereby
the coating solidifies immediately after leaving the
coater. The fast curing prevents the breakup of coating
into globules, which usually occurs with monofilaments. This technique can be applied to any composite using continuous filaments and matrices available
in the form of particulate precursors. The application of
the technique for preparing a silicon carbide monofila-
ment/glass composite is demonstrated.

901,075
PB89-180376   PC A04/MF A01
National Inst. of Standards and Technology (IMSE),
Gaithersburg, MD. Polymers Div.
Composites Databases for the 1990's.
D. H. Reneker, J. M. Crissman, and D. L. Hunston.
Feb 89, 62p NISTIR-88/4016

Keywords: *Composite materials, *Standards, *Polymers, *Tables(Data), Matrix methods, Tests, Data,
Composite structures, Surveys.

The report contains a draft standard for identification
of polymer matrix composite materials and for reporting test results. The draft standard is based on a comprehensive description of the flow of data through the
polymer matrix composites community. Two essentially different kinds of data bases are required, one oriented toward a particular group of data users, and one
designed to make the collection of all kinds of data
straightforward. An interactive dictionary to serve as a
tool in the development of the best names is described. Relationships of the draft standard to various
groups concerned with composites are noted.

901,076
PB89-199138   PC A05/MF A01
National Inst. of Standards and Technology (IMSE),
Gaithersburg, MD. Ceramics Div.
**Mechanical Property Enhancement in Ceramic
Matrix Composites.**
Interim rept. 1 Jan-31 Dec 88.
S. W. Freiman, D. C. Cranmer, E. R. Fuller, W.
Haller, M. J. Koczak, M. Barsoum, T. Palamides, and
U. V. Deshmukh. Apr 89, 80p NISTIR-89/4073
Contract N00014-86-F-0096
See also PB88-232863. Prepared in cooperation with
Drexel Univ., Philadelphia, PA. Dept. of Materials Engi-
neering. Sponsored by Office of Naval Research, Ar-
lington, VA.

Keywords: *Ceramics, *Composite materials, *Monofilaments, *Interfaces, *Borosilicate.glass, Silicon carbides, Fibers, Indentation hardness tests, Bonding
strength, Carbon, Equipment, Mechanical properties,
Graphs(Charts), Surface chemistry.

The fiber-matrix interfacial properties of several glass
and ceramic matrix composites have been determined
using two indentation techniques and a single fiber
pull-out technique. An instrumented indenter was developed to improve the acquisition and analysis of the
data. The effects of thermal expansion mismatch were
determined from three model composite systems containing large SiC monofilaments using the single fiber
pull-out test. An indentation push-out test was successfully used to determine the debond strength of a
borosilicate matrix/SiC monofilament/carbon core
material. A comparison of the various techniques for
determining the fiber-matrix interfacial properties was
conducted.

901,077
PB89-211825   Not available NTIS
National Bureau of Standards (IMSE), Gaithersburg,
MD. Ceramics Div.
**Creep Rupture of a Metal-Ceramic Particulate
Composite.**
Final rept.
T. J. Chuang, D. F. Carroll, and S. M. Wiederhorn.
1989, 12p
Contract DE-A105-85OR21569
Sponsored by Department of Energy, Oak Ridge, TN.
Pub. in Proceedings of International Conference on
Fracture (ICF7) (7th), Houston, TX., March 20-24,
1989, v1 p2965-2976.

Keywords: *Creep rupture tests, *Composite materi-
als, *Particulate composites, *Silicon carbides, Crack
propagation, Ceramics, Metals, Tensile properties,
Flexural strength, Loads(Forces), Deformation,
Strains, Coalescing, Fractures(Materials), Failure,
Bending, Service life.

The creep rupture behavior of a ceramic particulate
composite system was studied under tensile and flexural loading. The rupture process commences from
the heterogeneous formation of cavities at particle
interfaces in the tensile stress field. As deformation
proceeds, a critical strain is reached whereupon cavity
coalescence takes place forming large microcracks.
Ultimately, rupture occurs when one of these microcracks reaches a critical length and the remaining ligament cannot sustain the applied load. On the assumption that microcracks grow to a critical size through the
coalescence of cavities, a new rupture criterion is proposed based upon a critical strain concept of failure.
Using the criterion for fracture, together with a detailed
creep mechanics analysis, theoretical predictions are
made of lifetime under both bending and simple tension. Creep and creep rupture data for a grade of sili-
conized silicon carbide tested at 1300 C are collected
and compared with the proposed theory. Reasonable
agreement between theory and experiment were obtained in both modes of loading.

901,078
PB89-218358   PC A04/MF A01
National Inst. of Standards and Technology (NEL),
Boulder, CO. Chemical Engineering Science Div.
**Low-Temperature Thermal Conductivity of Com-
posites: Alumina Fiber/Epoxy and Alumina Fiber/
PEEK.**
D. L. Rule, and L. L. Sparks. May 89, 58p NISTIR-
89/3914
Sponsored by National Aeronautics and Space Admin-
istration, Moffett Field, CA. Ames Research Center.

Keywords: *Composite materials, *Thermal conductivity, *Thermal cycling tests, Plastics, Polymers, Aluminum oxide, Epoxy resins, Ceramic fibers, Orientation,
Low temperature tests, Graphs(Charts), Tables(Data),
Fiber laminates, Polyetheretherketone.

The thermal conductivities of poly-ether-ether-ketone
(PEEK), of alumina fiber in a matrix of PEEK, and of
alumina fiber in a matrix of epoxy, were determined
along with the effects of fiber orientation and thermal
cycling. Thermal conductivity was measured over the
temperature range of 4.2 to 310 K using a steady-state
apparatus. Data are presented and discussed relative
to specimen characteristics. It appears that after accounting for different fiber fractions in the specimens,
the thermal conductivity of the PEEK composite material is less than that of the epoxy composite material in
particular temperature ranges.

901,079
PB89-234223 Not available NTIS
National Inst. of Standards and Technology (IMSE),
Gaithersburg, MD. Polymers Div.
**Comparison of Microleakage of Experimental and
Selected Commercially Available Bonding Systems.**
Final rept.
A. A. Chohayeb, and N. W. Rupp. 1989, 3p
Sponsored by American Dental Association. Health
Foundation, Chicago, IL.
Pub. in Dental Materials 5, p241-243 Jul 89.

Keywords: *Dentistry, *Dental materials, *Bonding,
*Composites, Cavities, Leakage, Aluminum oxide,
Ferric compounds, Reprints, Permanent dental restorations.

The study observed microleakages of composite restorations bonded with two commercial and two experimental systems. A high-viscosity condensable composite and a low-viscosity composite were used as the restorative materials. The bonding systems used were two widely accepted commercial brands and two experimental systems, one containing ferric oxalate and the other aluminum oxalate. Restorations were placed in cavities prepared in extracted human teeth, then stored in 37 C water for 24 hours and then polished. The restored teeth were subjected to seven days of thermocycling (5 C-55 C for 540 cycles per day). Microleakage was detected and scored from 0-4 according to the degree of stain penetration. The experimental systems had lower scores than the commercial systems. The high-viscosity composite restorations had microleakage scores higher than those of the lower-viscosity composite restorations.

901,080
PB89-235907 PC A03/MF A01
National Inst. of Standards and Technology (IMSE),
Gaithersburg, MD. Ceramics Div.
**Toughening Mechanisms in Ceramic Composites.
Semi-Annual Progress Report for the Period
Ending March 31, 1989.**
Interim rept.
E. R. Fuller, E. P. Butler, R. F. Krause, and M. D.
Vaudin. Jul 89, 29p NISTIR-89/4111
Contract DE-AI05-800R20679
See also PB89-162606. Sponsored by Department of
Energy, Oak Ridge, TN. Advanced Research and
Technology Fossil Energy Materials Program.

Keywords: *Silicon carbides, *Fiber composites, *Borosilicate glass, *Fracture strength, Aluminum oxides,
Coatings(Materials), Creep tests, Strain rate, Stresses,
Microstructure, Ceramics, Whisker composites.

A silicon carbide fiber used as reinforcement in a borosilicate glass matrix has been shown to enhance the fracture toughness of the glass by as much as 22%. A ductile nickel coating on the fiber was found to reduce the interfacial shear strength and the frictional sliding between the fibers and the glass matrix, but any influence that the thickness of the nickel coating has on toughening was not conclusive. Time functions of creep strain and creep time to failure were measured for a 25 wt% SiC whisker-reinforced Al203 composite with 4.9% porosity. Beam specimens were used in four-point flexure with variously fixed bending moments and fixed temperatures between 1100 and 1300 C. The secondary creep-strain rates of specimens tested at lower stresses did not follow a power-law function of stress which was fitted to the high stress data. The microstructure of as-received specimens and specimens crept to failure were studied using optical and electron microscopy to determine the mechanisms of creep in the two stress regimes. Both kind of specimens contained cavitation at the whisker/Al203 grain boundary intersections to varying degrees. However, high stress specimens contained broken whiskers whereas those from the low stress regime did not.

901,081
PB90-112996 PC A03/MF A01
National Inst. of Standards and Technology (NEL),
Gaithersburg, MD. Center for Fire Research.
Assessing the Flammability of Composite Materials.
T. Ohlemiller. Jan 89, 27p NISTIR-89/4032
Sponsored by David W. Taylor Naval Ship Research
and Development Center, Annapolis, MD.

Keywords: *Flammability testing, *Composite materials, Ignition, Flame propagation, Heat measurement,
Polymers, Test equipment, Ship structural components, Honeycomb laminates.

A comprehensive approach to properly characterizing the flammability of composite materials is outlined. Laboratory-scale tests are described which provide measures of material ignitability, flame spread rate and heat release rate. Rather than expressing the measures as arbitrary indices, they are interpreted in terms of models of the controlling phenomena designed to provide information that can be generalized to full scale contexts, particularly compartment fires.

901,082
PB90-128265 Not available NTIS
National Inst. of Standards and Technology (IMSE),
Boulder, CO. Fracture and Deformation Div.
Edge Stresses in Woven Laminates at Low Temperatures.
Final rept.
R. D. Kriz. 1989, 12p
Sponsored by Department of Energy, Washington, DC.
Office of Fusion Energy.
Pub. in Composite Materials: Fatigue and Fracture,
ASTM STP 1012, p150-161 1989.

Keywords: *Laminates, *Woven fiber composites,
*Stress analysis, Epoxy resins, Composite materials,
Delaminating, Glass fibers, Cryogenics, Thermonuclear energy, Reprints, *Edges, Glass fiber reinforced
plastics.

Woven glass-epoxy laminates are used as nonmetallic components at low temperatures in magnetic fusion energy structures. Previous damage studies on G-10CR and G-11CR cryogenic grade woven laminates revealed that most of the damage occurred in the laminated interior. An existing, generalized plane strain, finite element model was modified to predict stress states at the laminate interior and free edges. Finite element results demonstrated that the weave geometry reduces edge stresses at low temperatures. Delamination edge stresses in woven laminates are more sensitive to small changes in temperature than those in nonwoven laminates.

901,083
PB90-128646 Not available NTIS
National Inst. of Standards and Technology (IMSE),
Boulder, CO. Fracture and Deformation Div.
Tensile and Fatigue-Creep Properties of a Copper-Stainless Steel Laminate.
Final rept.
L. L. Sculi, and R. P. Reed. 1988, 7p
Sponsored by Department of Energy, Washington, DC.
Pub. in Advances in Cryogenic Engineering Materials,
v34 p397-403 1988.

Keywords: *Tensile strength, *Fatigue(Materials),
*Creep strength, *Copper, *Stainless steels, *Laminates, Composite materials, Electrical resistivity, Thermal conductivity, Magnets, Reprints.

The design of compact ignition magnets uses a high-conductivity copper alloy. However, the large magnetic fields cause large stresses in the coil. The application may require a conductor with higher strength than that of the copper alloys and equally high electrical and thermal conductivity. A candidate material was produced by reinforcing the copper alloy with a stainless-steel alloy. The steel is roll-bonded as the midplane between two copper sheets. The material has the high thermal and electrical conductivity of the copper alloy and, possibly, sufficient strength to be used in compact ignition magnets. Tests were conducted at 295 and 76 K to characterize the tensile and creep-fatigue behavior of the laminated composite material in three roll-bonded conditions. The conditions correspond to a bulk reduction of 40, 50, and 60% cold work in the laminate as a whole. A mixing rule was used to predict the tensile behavior of the composite on the basis of the individual tensile properties of copper and stainless steel.

Corrosion & Corrosion Inhibition

901,084
PB89-176291 Not available NTIS
National Bureau of Standards (NEL), Gaithersburg,
MD. Building Materials Div.

**Corrosion Induced Degradation of Amine-Cured
Epoxy Coatings on Steel.**
Final rept.
T. Nguyen, and E. Byrd. 1987, 1p
Pub. in Abstracts of Papers of the American Chemical
Society 193, p133 Apr 87.

Keywords: *Protective coatings, *Degradation, *Epoxy
resins, *Amines, *Corrosion, *Curing agents, Interfaces, Steels, Aging tests(Materials), Oxidation, Alkalinity, Infrared spectroscopy, Reprints.

Organic protective coatings on metal can undergo physical and chemical changes under service conditions. The paper differentiates the interfacial degradation due to corrosion processes from that due to thermal oxidative reactions of an amine-cured epoxy coating on steel substrate exposed to corrosive environment. Relatively thin films of 40 and 400 nm of an amine cured-epoxy coating on well-prepared cold-rolled steel substrate aged in corrosive and thermal oxidative environments were studied by reflection/absorption Fourier Transform Infrared Spectroscopy. The results obtained showed degradation of the coating exposed to corrosive environment but not that exposed to thermal oxidative environment, suggesting that the corrosion reaction products, which are highly alkaline, are responsible for the degradation of the coating.

901,085
PB89-235345 PC A03/MF A01
National Inst. of Standards and Technology (NEL),
Gaithersburg, MD. Center for Building Technology.
Development of a Method to Measure in situ Chloride at the Coating/Metal Interface.
Technical note (Final).
T. Nguyen, and C. Lin. Jul 89, 20p NIST/TN-1266
Also available from Supt. of Docs. as SN003-003-02960-2. Prepared in cooperation with Xiamen Univ.
(China).

Keywords: *Chlorides, *Coatings, *Metals, *Corrosion, *Blistering, *Electrodes, Methodology, Interfaces, Electric potential, Microelectronics, Measurement, Graphs(Charts).

One of the main reasons for the lack of a complete understanding of corrosion and adhesion failures of a coated metal is the lack of analytical instrumentation to probe the behaviors of corrosive agents at the coating/steel interface. A procedure has been developed based in microelectrodes for studying in situ the behaviors of potential and chloride ions in blister and at a coating/metal interface. The procedure requires an attachment of a double-barred Cl(-1) selective microelectrode at the coating/metal interface, thus allowing direct measurements of Cl(-1) concentration and corrosion potential changes at localized areas under a coating. Although it is intricate to prepare the microelectrodes, the procedure provided very useful information for mechanistic studies of corrosion under coatings, as well as for transport studies of Cl(-1) ions through a coating on metal. The procedure should also be very useful for studying the roles of Cl(-1) in localized corrosion. The utility of an inverted electrode microsampling method for studies of Cl(-1) in very small volumes such as blisters was also demonstrated.

901,086
PB90-131152 PC A03/MF A01
National Inst. of Standards and Technology (IMSE),
Gaithersburg, MD. Metallurgy Div.
Corrosion Behavior of Mild Steel in High pH Aqueous Media.
A. C. Fraker, and J. S. Harris. Sep 89, 20p NISTIR-89/4173
Sponsored by Nuclear Regulatory Commission, Washington, DC. Office of Nuclear Material Safety and Safeguards.

Keywords: *Corrosion resistance, *Alkalinity, *Low
alloy steels, Basalt, Ground water, Pitting tests, Passivity, Ferrite, Pearlite, Packaging materials, Aqueous
electrolytes, Steel-ASTM-A27, Radioactive waste
management, Underground disposal.

The paper reports on a study of the corrosion behavior and localized corrosion susceptibility of mild steel in a simulated ground water with a pH of 9.75 and a temperature of 95C. The steel used in the study was A27, ASTM Grade 60-30. The steel did not passivate in the aqueous environment used. The corrosion rate decreased with exposure time. Corrosion occurred in an uneven form over the surface, and although some pit-

MATERIALS SCIENCES

Corrosion & Corrosion Inhibition

ting may have been present, no deep pits were observed. The amount and distribution of the areas of ferrite and pearlite as well as the impurities were determined to be important as related to uneven corrosion and to localized attack.

Elastomers

901,087
PB89-148118 PC A03/MF A01
National Bureau of Standards (IMSE), Gaithersburg, MD. Polymers Div.
Flow of Molecules Through Condoms.
Annual rept. 1 Mar-1 May 87 (Final).
C. M. Guttman. Oct 87, 47p NBSIR-88/3721
Contract FDA-224-79-5023
Sponsored by Food and Drug Administration, Rockville, MD. Office of Science and Technology.

Keywords: *Elastomers, *Contraceptives, *Diffusion, Latex, Pin holes, Molecular flow, *Condoms.

An apparatus for the measurement of flux of small molecules through whole condoms has been developed. It is shown that the experiment can measure diffusion constants as low as 10(sup -13) cm(sup 2)/s or a single pinhole as small as .4 micrometers in the condom. For pinhole measurements this is shown to be a factor of 10 better than current ASTM testing methods on the basis of flow considerations only. Analysis of the experimental data show the difficulties in making unambiguous determinations on the mechanisms of flow. Further experiments are necessary to distinguish between large holes and small holes and fluxes due to diffusion and those due to pinholes. The results suggest a more careful study of fluxes through condoms is necessary to assure that a particle about .1 micrometer cannot pass through the condom.

901,088
PB89-175830 Not available NTIS
National Bureau of Standards (IMSE), Gaithersburg, MD. Polymers Div.
Uniaxial Deformation of Rubber Network Chains by Small Angle Neutron Scattering.
Final rept.
H. Yu, T. Kitano, C. Y. Kim, E. J. Amis, T. Chang, M. R. Landry, J. A. Wesson, C. C. Han, T. P. Lodge, and C. J. Glinka. 1986, 14p
Pub. in Advanced Elastomers Rubber Elasticity, p407-420 1986.

Keywords: *Polyisoprene, *Neutron scattering, Deformation, Elastomers, Natural rubber, Isocyanates, Crosslinking, Strains, Molecular weight, Extensibility, Contraction, Radius of gyration, Elastic properties, Reprints.

Small angle neutron scattering (SANS) measurements were performed on poly(isoprene) networks at different uniaxial strains, i.e., 1.0 less than lambda (extension ratio) less than 2.1. The networks were prepared from anionically polymerized, alpha, mu-dihydroxy-poly(isoprene) precursors (H-chains) and the corresponding poly(isoprene-d8) isotopic counterparts (D-chains). Two molecular weights of D-chains, 26,000 and 64,000, crosslinked with approximately the same molecular weight H-chains (29,000 and 68,000 respectively) were examined for the deformation behaviors. The chain extensive deformation is found to follow a behavior intermediate between the junction affine model and the phantom network model which allows unrestricted fluctuations of network junctions. The chain contractive deformation follows closely the chain affine model, indicating an asymmetry between extensive and contractive chain deformation. In either case, the deformation behavior is found to be the same for the two molecular weights.

901,089
PB89-209308 PC A03/MF A01
National Inst. of Standards and Technology (IMSE), Gaithersburg, MD. Polymers Div.
Studies on Some Failure Modes in Latex Barrier Films.
Annual rept. (Final).
C. M. Guttman, G. B. McKenna, K. M. Flynn, and T. K. Trout. May 88, 50p NISTIR-89/4084
Contract FDA-224-79-5023
Sponsored by Food and Drug Administration, Rockville, MD. Office of Science and Technology.

Keywords: *Latex, *Thin films, *Barrier materials, *Liquid permeability, *Failure, *Crosslinking, Body fluids, Simulation, Prophylaxis, Natural rubber, Surgical gloves, Swelling, Graphs(Charts), Condoms.

The report covers work on a 1988 contract with the FDA to study failure modes of latex barrier films in their use as condoms or medical gloves. Two areas are reported on: The change in the failure of latex barrier films as a result of swelling in bodily fluid simulants and the cross-link density variation in condoms on the 0.1 mm scale.

901,090
PB89-228316 PC A07/MF A01
Brown Univ., Providence, RI. Div. of Engineering.
Experimental Study of the Pyrolysis of Pure and Fire Retarded Cellulose.
Doctoral thesis.
Y. Chen. Jun 89, 147p NIST/GCR-89/566
Grants NANB-8-D0851, NANB-6-D0629
Sponsored by National Inst. of Standards and Technology (NEL), Gaithersburg, MD. Center for Fire Research.

Keywords: *Pyrolysis, *Cellulose, *Fire resistant coatings, *Sodium hydroxide, Catalysts, Combustion products, Heat of vaporization, Nitrogen, Graphs(Charts), Vapors, Volatility, Theses.

The pyrolysis of pure and fire retarded bulk cellulose samples in a nitrogen atmosphere is studied. The study is directed toward determining the effects of the solid phase fire retardant (sodium hydroxide) on the burning of cellulose. The material property of pure and fire retarded cellulose which most directly affects its burning behavior, the heat of gasification, is measured by using a specially designed pyrolyzing chamber. Theoretical results are obtained for a one-dimensional pyrolysis wave propagating into cellulose by using a finite-difference calculation. The experimental data show that sodium hydroxide acts as a catalyst in the pyrolysis of cellulose. Its addition leads to decreases in the heat of gasification and the mass fraction of non-combustible volatiles in the total volatiles and to increases in stoichiometric ratio and the heat of combustion of combustible volatiles, and thereby has a dual effect on cellulose burning.

Fibers & Textiles

901,091
PB89-174122 Not available NTIS
National Bureau of Standards (NEL), Gaithersburg, MD. Fire Science and Engineering Div.
Flammability Tests for Industrial Fabrics: Relevance and Limitations.
Final rept.
K. M. Villa, and J. F. Krasny. 1988, 16p
Pub. in Proceedings of Annual Conference of the Industrial Fabrics Association International (76th), Chicago, IL., November 9-12, 1988, p119-134.

Keywords: *Flammability testing, *Industrial fabrics, *Fire resistant textiles, Criteria.

Flammability tests applicable to industrial fabrics, including tents and other outdoor equipment, inflatable structures, etc., are discussed briefly. These tests were designed to assure self-extinguishment after exposure to small flames. The specimens are generally held vertically in a U-shaped steel frame and ignited at the bottom. Most of these tests were developed when only char-forming materials like flame retardant cotton were available. The criteria chosen for the tests were char length, and afterflame and afterglow time. The applicability of these tests and criteria to both char-forming and thermoplastic fabrics is critically reviewed.

Iron & Iron Alloys

901,092
PB89-149062 Not available NTIS
National Bureau of Standards (IMSE), Boulder, CO. Fracture and Deformation Div.

Local Brittle Zones in Steel Weldments: An Assessment of Test Methods.
Final rept.
R. Denys, and H. I. McHenry. 1988, 7p
Pub. in Proceedings of International Conference on Offshore Mechanics and Arctic Engineering (7th), Houston, TX., February 7-12, 1988, p379-385.

Keywords: *Steels, *Weldments, *Fractures(Materials), *Offshore structures, *Plates(Structural member), *Nondestructive tests, Brittleness, Microstructure, Testing, Quality control, Toughness, Structural analysis.

Local brittle zones (LBZs) are regions of brittle microstructure within the heat affected zones (HAZs) of steel weldments that can initiate brittle fracture at low toughness levels. The paper describes the metallurgical nature of LBZs, reviews the various test methods used to detect and evaluate LBZs, and recommends test methods for controlling LBZs in offshore structures. The recommended tests and their specific functions are: (1) For pre-production qualification of steel plates, CTOD tests evaluate the susceptibility of steels to the formation of LBZs; (2) For quality control of steel plates, drop weight NDT tests evaluate the tolerance for LBZs of the steels used for offshore structures; (3) For qualification of welding materials and procedures, Charpy V-notch tests verify that the HAZ toughness exceeds the minimum toughness specified for the steel; (4) For fitness for purpose evaluations, wide plate tests assess the significance of LBZs in existing structures.

901,093
PB89-149090 Not available NTIS
National Bureau of Standards (IMSE), Boulder, CO. Fracture and Deformation Div.
J-Integral Values for Small Cracks in Steel Panels.
Final rept.
D. T. Read. 1986, 13p
Sponsored by Naval Sea Systems Command, Washington, DC., and David W. Taylor Naval Ship Research and Development Center, Annapolis, MD.
Pub. in Fracture Mechanics: Eighteenth Symposium, ASTM STP 945, p151-163 1988.

Keywords: *Steels, *Fractures(Materials), Toughness, Defects, Loads(Forces), Strains, Crack propagation, Plates(Structural members), Tearing, Reprints, *J-integral, Steel A710.

For a quantitative relationship between fracture toughness, flaw size, and applied loading for small flaws, to be used for fitness-for-service assessment, applied J-integral was measured as a function of applied strain in eight 14-mm-thick specimens of ASTM A710 Grade A Class 3 steel plate. All the edge cracks had lengths less than 3% of the specimen width of 82 mm. Six specimens were tested in tension; two were loaded by four-point-bending in the plane of the plate. One single-edge-cracked, transversely oriented specimen was tested at -30 deg C. Electrical resistance strain gage and clip-gage crack mouth opening displacement measurements were used to obtain quantities inside the J-integral. The J-integral was calculated by trapezoidal rule integration. Unloading crack mouth compliance measurements were used to obtain crack length values so that tearing effects could be observed. Lueder's strains occurring right after yield caused rapid increases in the applied J-integral values for the tension specimens. Except for the Lueder's strain effect, the behavior of the applied J-integral in bending was similar to that in tension. Tearing caused a smooth exponential rise in applied J as strain increased beyond the point of initiation. The initiation toughness and tearing resistance of the panels with short cracks were equal to or greater than those of conventional three-point-bend specimens of the same thickness.

901,094
PB89-156160 PC A04/MF A01
National Bureau of Standards (IMSE), Gaithersburg, MD. Fracture and Deformation Div.
Postweld Heat Treatment Criteria for Repair Welds in 2-1/4Cr-1Mo Superheater Headers: An Experimental Study.
D. T. Read, and H. I. McHenry. Aug 88, 60p NBSIR-87/3075
Sponsored by Naval Sea Systems Command, Washington, DC.

Keywords: *Weldments, *Maintenance, *Superheater headers, Boilers, Ships, Steels, Heat Treatment, Shielded metal arc welding, Bending,

Cracking(Fracturing), Toughness, Residual stress, Strain gages, Pressure vessels, Hydrostatics, Chromium, Molybdenum.

Wide-plate and standard-size specimens cut from repair welds in 2-1/4Cr-1Mo plate were tested as-welded and after post-weld heat treatment (PWHT). Three-point-bend specimens with cracks oriented in the TS direction were used to measure weld-metal and heat-affected-zone (HAZ) toughness values. Results of direct measurements of the applied J-integral on the wide plates were compared with critical J-value measurements of three-point-bend specimens. The comparison indicated that PWHT was highly beneficial, because it reduced the crack-driving force from residual stresses and increased the weld-metal and HAZ toughness. In the as-welded condition, very low toughness values were measured at the HAZ. These low toughness values, together with the measured crack-driving forces, indicated critical crack depths of a few millimeters. To extend the usefulness of these results, a new approach to the problem of the applied J-integral produced by residual stresses is being explored: strain-gage measurements made during notching are analyzed to obtain an applied J-integral as a function of crack depth. The preliminary results are encouraging. The residual-stress-produced J-value is roughly equivalent to that produced by a remote elastic loading to the same stress level.

901,095
PB89-157796 Not available NTIS
National Bureau of Standards (IMSE), Boulder, CO.
Fracture and Deformation Div.
Molybdenum Effect on Volume In Fe-Cr-Ni Alloys.
Final rept.
H. M. Ledbetter, and M. W. Austin. 1988, 5p
Sponsored by Department of Energy, Washington, DC.
Pub. in Jnl. of Materials Science 23, p3120-3124 1988.

Keywords: *Face centered cubic lattices, *Ray diffraction, *Iron alloys, *Compressibility, Crystallography, Chromium, Nickel, Molybdenum, Atomic properties, Atomic orbitals, Reprints.

The unit-cell size for six face-centered-cubic Fe-Cr-Ni alloys, nominally Fe-19Cr-12Ni (at%) were determined by x-ray diffraction on powder specimens. In these alloys, the molybdenum content ranged up to 2.4 at%. Molybdenum increases volume: 0.45% per at%. Usual models based on atomic volumes and elastic compressibilities fail to explain the large volume increase. The discrepancy was ascribed to changes in interatomic bonding, which are described in terms of 3d-electron models.

901,096
PB89-158018 Not available NTIS
National Bureau of Standards (NEL), Gaithersburg, MD. Building Materials Div.
Fractal-Based Description of the Roughness of Blasted Steel Panels.
Final rept.
J. W. Martin, and D. P. Bentz. 1987, 7p
Sponsored by Federal Highway Administration, Washington, DC.
Pub. in Jnl. of Coatings Technology 59, n745 p35-41 1987.

Keywords: *Steels, *Roughness, *Thermography, *Images, Panels, Blasting, Fractography, Cameras, Surface properties, Reprints.

The fractal dimensions of a standard series of blasted steel panels are shown to correlate very well with their perceived roughness. This occurs because the roughness of a blasted panel dictates the roughness of its image, and hence, its fractal dimension. The blasted steel panels are imaged with a thermographic camera, as opposed to a visual camera, because a thermographic image better delineates the peak-to-valley heights of the crater-like structures; it minimizes imaging problems due to light reflectance; and it eliminates most of the imaging problems associated with surface discoloration. It is concluded that the fractal dimension of a blasted steel surface captures most of the perceptually relevant shape structures on an abraded surface, and thus, provides a good quantitative representation of surface roughness.

901,097
PB89-171896 Not available NTIS
National Bureau of Standards (IMSE), Boulder, CO.
Fracture and Deformation Div.

Loading Rate Effects on Discontinuous Deformation in Load-Control Tensile Tests.
Final rept.
T. Ogata, K. Ishikawa, R. P. Reed, and R. P. Walsh.
1988, 8p
Sponsored by Department of Energy, Washington, DC.
Office of Fusion Energy.
Pub. in Advances in Cryogenic Engineering Materials, v34 p233-240 1988.

Keywords: *Tension tests, *Loading rate, *Load control, *Austenitic steels, *Deformation, Tensile properties, Strains, Ultimate strength, Low temperature tests, Cryogenics, Loads(Forces), Reprints.

In load-control tensile tests at liquid helium temperature, an abrupt and large discontinuous deformation occurs, which differs from the discontinuous deformation obtained from displacement-control tests. We investigated the effects of loading rate, varied from 0.5 to 5000 N/s, on the tensile properties of AISI 304L, 310, and 316LN steels at 4 K. A large deformation, near 40% strain, occurred in AISI 310. At the high loading rates, the ultimate strength of these materials was 65% of the strength obtained in displacement-control tests; the initiation strength of discontinuous deformation was also less.

901,098
PB89-172621 PC A06/MF A01
National Inst. of Standards and Technology (IMSE), Gaithersburg, MD. Metallurgy Div.
Elevated Temperature Deformation of Structural Steel.
B. A. Fields, and R. J. Fields. Mar 89, 121p NISTIR-88/3899
Sponsored by American Iron and Steel Inst., Washington, DC.

Keywords: *High temperature tests, *Creep tests, *Tensile strength, *Computation, *Deformation, *Construction materials, *Steels, Elastic properties, Plastic properties, Equations, Graphs(Charts), Steel ASTM A36.

The results of tensile and creep tests on steels close to the American specification for ASTM A36 have been used to formulate an equation from which elastic, plastic, creep and total strains can be calculated. Correlations between measured and predicted strains for Australian AS A149 and Japanese SS41 steels, both close to the A36 specification, are shown and good agreement is found. The above mentioned equation is also used to construct deformation mechanism (i.e., elastic, plastic, or creep) maps for times of 2 minutes to 4 hours at temperature. From these maps the deformation mechanisms operating at a given temperature and stress can be seen. The dominant mechanism for each set of conditions is given. In addition the maps show contours of total strain values 1, 2, and 5%.

901,099
PB89-173504 Not available NTIS
National Bureau of Standards (IMSE), Boulder, CO.
Fracture and Deformation Div.
Role of Inclusions in the Fracture of Austenitic Stainless Steel Welds at 4 K.
Final rept.
T. A. Siewert, and C. N. McCowan. 1987, 11p
Sponsored by Department of Energy, Washington, DC.
Pub. in Welding Metallurgy of Structural Steels, p415-425 1987.

Keywords: *Inclusions, *Austenitic stainless steels, *Weldments, *Impact tests, *Toughness, *Cryogenics, Yield strength, Shielded metal arc welding, Gas metal arc welding, Fractures(Materials), Surface properties, Morphology, Ductility, Reprints.

Inclusion densities were measured for three types of austenitic stainless steel welds and compared to the 4-K yield strengths, 76-K Charpy V-notch absorbed energies, and the ductile dimple densities on the respective fracture surfaces. The welds included shielded metal arc (SMA) welds and gas metal arc (GMA) welds. The inclusion density was consistently a factor of 8 to 10 less than the fracture surface dimple density. Inclusion and dimple densities ranged from 3.9 x 10 sup 4 inclusions per sq mm and 3.2 x 10 sup 5 dimples per sq mm for one SMA specimen to 1.1 x 10 sup 4 inclusions per sq mm and 1.2 x 10 sup 5 dimples per sq mm for the fully austenitic GMA specimen. Both ductile dimple density and dimple morphology varied with specimen type. The inclusion data for the welds agreed with a linear relationship between fracture toughness and inclusion spacing that had been developed for base metals.

901,100
PB89-173512 Not available NTIS
National Bureau of Standards (IMSE), Boulder, CO.
Fracture and Deformation Div.
Influence of Molybdenum on the Strength and Toughness of Stainless Steel Welds for Cryogenic Service.
Final rept.
C. N. McCowan, T. A. Siewert, and E. Kivineva.
1987, 12p
Sponsored by Department of Energy, Washington, DC.
Pub. in Proceedings of International Symposium on Welding Metallurgy of Structural Steels, Denver, CO., February 22-26, 1987, p427-438.

Keywords: *Molybdenum, *Austenitic stainless steels, *Additives, *Toughness, *Impact tests, *Cryogenics, Yield strength, Chromium alloys, Nickel, Manganese, Nitrogen, Weldments, Inclusions.

Molybdenum additions to austenitic stainless welds were found to increase the 4-K yield strength by approximately 30 MPa per weight percent. Molybdenum additions had little effect on the 76-k Charpy V-Notch impact energy, with one exception: when the molybdenum content was raised from 1.7 to 3.8 wt. % in an otherwise equivalent stainless steel composition of approximately 17Cr-9Ni-6.6Mn-0.17N, the absorbed energy decreased from 33 to 16 J. At a nickel content of 14 wt. %, the higher molybdenum contents did not reduce the impact toughness. The loss in impact toughness for the 9 wt.% nickel, 3.8 wt. % molybdenum composition was linked to an increased number of both small and large inclusion sizes in the weld. The effects of nickel and manganese on the cryogenic strength and toughness are also reported.

901,101
PB89-173835 Not available NTIS
National Bureau of Standards (IMSE), Boulder, CO.
Fracture and Deformation Div.
Failure Analysis of an Amine-Absorber Pressure Vessel.
Final rept.
H. I. McHenry, and D. T. Read. 1986, 17p
See also PB89-126783. Sponsored by Occupational Safety and Health Administration, Chicago, IL.
Pub. in Proceedings of International Conference on Structural Failure, Product Liability and Technical Insurance (2nd), Vienna, Austria, July 1-3, 1986, p141-157.

Keywords: *Pressure vessels, *Bursting, *Refineries, *Stress corrosion tests, Failure, Fractures(Materials), Microstructure, Hydrogen, Crack propagation, Pressure, Embrittlement, Petroleum products, Toughness, Weldments, Maintenance, Structural forms, Steels, Reprints.

In 1984, a pressure vessel ruptured at a petroleum refinery causing an explosion and fire. It fractured along a path that was weakened by extensive cracking adjacent to a repair weld joining a replacement section to the vessel. These pre-existing cracks initiated in areas of a hard microstructure due to hydrogen stress cracking. The cracks grew through the vessel wall due to hydrogen pressure cracking. When the depth of the largest of these cracks exceeded 90% of the wall thickness, the remaining ligament ruptured resulting in a through crack about 800 mm long. This crack triggered final fracture at the operating stress level of 35 MPa because the toughness of the vessel steel was reduced nearly 3-fold by hydrogen embrittlement.

901,102
PB89-174882 Not available NTIS
National Bureau of Standards (IMSE), Boulder, CO.
Fracture and Deformation Div.
Tensile Strain-Rate Effects in Liquid Helium.
Final rept.
R. P. Reed, and R. P. Walsh. 1988, 10p
Pub. in Advances in Cryogenic Engineering: Materials, v34 p199-208 1988.

Keywords: *Tensile strength, *Strain rate, *Liquids, *Helium, *Austenitic stainless steels, Cryogenics, Heat transfer, Dislocations(Materials), Surface properties, Adiabatic conditions, Reprints.

The effects of strain rate on tensile properties of three austenitic stainless steels at 4 K were examined. Strain rates ranged from 4.4 x 10 sup(-6) s sup(-1) to 8.8 x 10 sup(-3) s sup(-1). Strain rates less than 2.2 x 10 sup(-3) s sup(-1) had no effect on tensile properties.

MATERIALS SCIENCES

Iron & Iron Alloys

rates of 4.4×10 sup(-3) s sup(-1) or larger reduced the ultimate tensile strength, and stress-strain curves and temperature measurements indicated specimen warming to 100 K. Calculations are presented to estimate the work put into the specimen during deformation, stored energy in terms of dislocations and dislocation interactions, and dissipated heat. The reduction of tensile strength was associated with specimen warming which was caused by the transition from nucleate to film-boiling heat transfer on the specimen surface.

901,103

PB89-189195 PC A03/MF A01
National Inst. of Standards and Technology (IMSE), Boulder, CO. Fracture and Deformation Div.
Fracture Behavior of a Pressure Vessel Steel in the Ductile-to-Brittle Transition Region.
J. Heerens, and D. T. Read. Dec 88, 44p NISTIR-88/3099
Prepared in cooperation with GKSS - Forschungszentrum Geesthacht G.m.b.H., Geesthacht-Tesperhude (Germany, F.R.).

Keywords: *Fractures(Materials), *Steels, *Pressure vessels, *Ductile brittle transition, *Cleavage, Manganese, Initiation, Crack propagation, Nickel, Molybdenum, Metal alloys.

The reasons for the scatter of fracture toughness in the ductile-to-brittle transition region, as well as the mechanisms leading to cleavage fracture, have been investigated for a quenched and tempered pressure vessel steel, DIN 20 MnMoNi 55. The fracture surfaces indicate that cleavage fracture starts at one small area in the ligament, the cleavage initiation site. Cleavage initiation occurs ahead of the crack tip at the location of the maximum normal stresses. Fractography and metallography show four different types of initiation sites. The mechanisms which may trigger cleavage fracture at these initiation sites are discussed. The results indicate that the scatter of fracture toughness is due to the scatter in the distance between the cleavage initiation site and the fatigue crack tip.

901,104

PB89-189336 PC A03/MF A01
National Inst. of Standards and Technology (NEL), Boulder, CO. Chemical Engineering Science Div.
Ignition Characteristics of the Iron-Based Alloy UNS S66286 in Pressurized Oxygen.
J. W. Bransford, P. A. Billiard, J. A. Hurley, K. M. McDermott, and I. Vazquez. Nov 88, 50p NISTIR-88/3904
Sponsored by National Aeronautics and Space Administration, Huntsville, AL. George C. Marshall Space Flight Center.

Keywords: *Metal alloys, *Ignition, *Combustion, *Oxygen, *High pressure tests, *Iron alloys, Carbon dioxide lasers, Heat treatment, Solidus, Endothermic reactions, Surface properties, Oxidation, Tables(Data), Graphs(Charts), Alloy UNS S66286.

The development of ignition and combustion in pressurized oxygen atmospheres was studied for the iron based alloy UNS S66286. Ignition of the alloy was achieved by heating the top surface of a cylindrical specimen with a continuous-wave CO2 laser. Two heating procedures were used. In the first, laser power was adjusted to maintain an approximately linear increase in surface temperature. In the second, laser power was periodically increased until autoheating (self-heating) was established. It was found that the alloy would autoheat to destruction from temperatures below the solidus temperature. In addition, endothermic events occurred as the alloy was heated, many at reproducible temperatures. Many endothermic events occurred prior to abrupt increases in surface temperature and appeared to accelerate the rate of increase in specimen temperature to rates greater than what would be expected from increased temperature alone. It is suggested that the source of these endotherms may increase the oxidation rate of the alloy. Ignition parameters are defined and the temperatures at which these parameters occur are given for the oxygen pressure range of 1.72 to 13.8 MPa (25 to 2000 psia).

901,105

PB89-193262 PC A03/MF A01
National Inst. of Standards and Technology (IMSE), Gaithersburg, MD. Metallurgy Div.

Metallurgical Evaluation of 17-4 PH Stainless Steel Castings.

G. E. Hicho, and J. H. Smith. May 89, 35p NISTIR-89/4075
Sponsored by Naval Ordnance Station, Indian Head, MD.

Keywords: *Castings, *Stainless steels, *Heat treatment, *Microscopy, *Hardness, *Missile warheads, *Microstructure, Specifications, Metallography, Measurement, Temperature, Solution annealing, Aging tests(Materials), Tables(Data), Homogenizing, Graphs(Charts), Weapons, Head Caps.

A metallurgical evaluation was conducted to determine if selected castings of 17-4 PH stainless steel used in head caps on missile weapon systems had been properly heat treated as required by SAE specification AMS-5355D. Optical metallographic analysis and hardness measurements were made on four samples of as-received castings and on selected samples that were homogenized, solution annealed and aged at various temperatures.

901,106

PB89-201586 Not available NTIS
National Bureau of Standards (IMSE), Boulder, CO. Fracture and Deformation Div.
Ferrite Number Prediction to 100 FN in Stainless Steel Weld Metal.
Final rept.
T. A. Siewert, C. N. McCowan, and D. L. Olson. 1988, 10p
Pub. in Welding Research Supplement, p289-s-298-s Dec 88.

Keywords: *Weldments, *Stainless steels, *Ferrite, Manganese, Molybdenum, Nitrogen, Silicon, Solidification, Surveys, Reprints, Stainless steel 309, Data bases.

To improve the accuracy of ferrite number (FN) prediction in stainless steel weld metal, a new diagram has been developed using a database containing more than 950 alloy compositions from worldwide sources. In accuracy, the diagram surpasses the DeLong diagram for the low-FN austenitic stainless steel compositions of the 300 series, and it corrects a 2 FN bias detected for Type 309 stainless steel. The diagram is more accurate than the Schaeffler diagram for duplex stainless steel alloys and ferrite contents to 100 FN. It is most accurate when the Mn content is restricted to 10 wt-%, Mo content is restricted to 3 wt-%, N content is restricted to 0.2 wt-%, and Si content is restricted to 1 wt-%. Changes in the primary solidification mode are indicated on the diagram, and they appear to affect the FN response. Transitions in iso-FN line spacings may be caused by these mode changes.

901,107

PB89-231260 Not available NTIS
National Inst. of Standards and Technology (IMSE), Boulder, CO. Fracture and Deformation Div.
Stainless Steel Weld Metal: Prediction of Ferrite Content.
Final rept. Jan-Oct 88.
C. N. McCowan, T. A. Siewert, and D. L. Olson. 1989, 36p
Sponsored by Welding Research Council, New York, Colorado School of Mines, Golden, and Department of Energy, Washington, DC. Office of Fusion Energy.
Pub. in Welding Research Council Bulletin 342, p1-36, Apr 89.

Keywords: *Stainless steels, *Weldments, Diagrams, Ferrite, Solidification, Data, Solid phases, Reprints.

A new diagram to predict the ferrite number (FN) in stainless steel welds is proposed. The diagram has a range from 0 to 100 FN, and the primary solidification zones are indicated. The diagram more accurately predicts the ferrite content for welds having FN less than 18 than existing diagrams. It corrects overestimates made by the DeLong diagram for AWS type 309 stainless steel and it predicts the FN of duplex stainless steel more accurately than the Schaeffler diagram. Weld compositions used to develop the diagram ranged (in wt.%) from 0.01 to 0.2 C, 0.4 to 12 Mn, 0.1 to 1.3 Si, 15 to 32 Cr, 5 to 25 Ni, 0 to 7 Mo, 0.02 to 0.3 N, 0 to 0.9 Nb, and 0 to 0.1 Ti. The database contained over 950 welds and is included in Appendix I and II to the report.

901,108

PB90-117623 Not available NTIS
National Inst. of Standards and Technology (IMSE), Boulder, CO. Fracture and Deformation Div.

Linear-Elastic Fracture of High-Nitrogen Austenitic Stainless Steels at Liquid Helium Temperature.

Final rept.
R. L. Tobler, R. P. Reed, and P. T. Purtscher. 1989, 6p
Sponsored by Department of Energy, Washington, DC. Office of Fusion Energy.
Pub. in Jnl. of Testing and Evaluation 17, n1 p54-59 Jan 89.

Keywords: *Austenitic stainless steels, *Chromium nickel alloys, *Cryogenics, *Fractography, Fracture strength, Brittleness, Reprints, *Nitrogen additions, *Manganese additions, Thermonuclear reactors.

Four commercial Fe-Cr-Ni-Mn austenitic stainless steels containing 0.14, 0.26, and 0.37 wt% N were fractured in liquid helium at 4 K, and measurements of the linear-elastic plane-strain stress-intensity factor, K(sub 1c), were made. Interstitial nitrogen significantly strengthens these steels at low temperatures so that brittle fractures occur under plane strain conditions at 4 K despite moderate ductility in uniaxial tension. The brittle fracture mechanism at 4 K involves a form of cleavage or slip-band cracking as evidenced by the formation of transgranular facets on (111) planes.

901,109

PB90-117649 Not available NTIS
National Inst. of Standards and Technology (IMSE), Boulder, CO. Fracture and Deformation Div.
Nitrogen in Austenitic Stainless Steels.
Final rept.
R. P. Reed. 1989, 8p
Pub. in Jnl. of Metals 41, n3 p16-21 Mar 89.

Keywords: *Austenitic stainless steels, Corrosion resistant steels, Austenite, Mechanical properties, Toughness, Reprints, *Nitrogen additions.

Nitrogen alloyed in austenitic stainless steels improves austenite stability, mechanical properties and corrosion resistance. Steels supersaturated with nitrogen ('super-nitrogen steels') have been investigated, which rival the latest ferritic steels in strength but have potentially greater toughness.

901,110

PB90-128554 Not available NTIS
National Inst. of Standards and Technology (IMSE), Boulder, CO. Fracture and Deformation Div.
Effect of Chemical Composition on the 4 K Mechanical Properties of 316LN-Type Alloys.
Final rept.
P. T. Purtscher, R. P. Walsh, and R. P. Reed. 1988, 8p
Pub. in Advances in Cryogenic Engineering Materials, v34 p191-198 1988.

Keywords: *Chemical composition, *Cryogenics, *Mechanical properties, *Austenitic stainless steels, Liquid helium, Toughness, Yield strength, Crack propagation, Molybdenum containing alloys, Nickel containing alloys, Microstructure, Annealing, Reprints, *Stainless-steel-316LN, Molybdenum additions, Nickel additions.

A series of eight austenitic stainless steels was tested in liquid helium to determine the effect of Mo and Ni variations on the strength and toughness. The Mo content ranged from 0 to 4 wt.%; the Ni content varied from 11 and 14 wt.%. The microstructure of the alloys depended upon the composition and annealing temperature. Higher alloy content and lower annealing temperatures, 1000 to 1050 C, resulted in a nonuniform structure. The higher temperature, 1150 C, produced a uniform austenitic structure. The mechanical test results showed that Mo additions increased the yield strength (to a maximum at 3 wt.%) and decreased the K sub (Ic)(J) values, so that there was no improvement in the strength-toughness relationship. Increasing the Ni content decreased the strengthening effect of Mo and increased K sub (Ic)(J). It supports earlier work that showed that Ni does improve the strength-toughness relationship.

901,111

PB90-128562 Not available NTIS
National Inst. of Standards and Technology (IMSE), Boulder, CO. Fracture and Deformation Div.
Fracture Behavior of 316LN Alloy in Uniaxial Tension at Cryogenic Temperatures.
Final rept.
P. T. Purtscher, R. P. Walsh, and R. P. Reed. 1988, 8p
Sponsored by Department of Energy, Washington, DC.

Pub. in Advances in Cryogenic Engineering Materials, v34 p379-386 1988.

Keywords: *Fractures(Materials), *Cryogenics, *Austenitic stainless steels, *Tension tests, Axial stress, Inclusions, Surface properties, Microstructure, Nucleation; Reprints, *Stainless- steel-316LN,, Maganese sulfides.

The fracture behavior of an austenitic stainless steel, commercial-grade 316LN alloy, at cryogenic temperatures is studied by careful examination of the fracture surfaces and polished cross-sections through the fracture surfaces of round tensile specimens. The fracture is ductile (a dimpled rupture process) and is controlled by inclusions in the structure, MnS stringers and smaller spherical particles. The main effect of the MnS stringers was to decrease the percent reduction in area in the tensile test at 4K. The true stress and strain at fracture vary as a function of test temperature and are shown to be related to the nucleation of microvoids around the spherical particles.

901,112
PB90-130297   PC A03/MF A01
National Inst. of Standards and Technology (IMSE), Gaithersburg, MD. Metallurgy Div.
Tensile Tests of Type 305 Stainless Steel Mine Sweeping Wire Rope.
T. R. Shives, and S. R. Low. Oct 89, 29p NISTIR-89/4174

Keywords: *Tension tests, *Minesweepers(Ships), *Wire rope, Graphs(Charts), Tensile stress, Naval mine detection, *Stainless steel-305.

The Naval Coastal Systems Center submitted to the National Institute of Standards and Technology (NIST) approximately 360 feet of each of two different AISI 305 stainless steel wire ropes for testing. Both wire ropes were nominally 5/8 inch in diameter. One was stated as having a 6 x 19 configuration and the other a 7 x 7 configuration. The first number in such a designation indicates the number of strands in the wire rope and the second number indicates how many wires there are per strand. For example, the wire rope with a 6 x 19 configuration consists of six strands of 19 wires each. As shown later, the wire rope stated to have a 7 x 7 configuration actually had a 6 x 7 configuration with an independent wire rope core (IWRC). The core is one of the three basic parts of a wire rope. The other two are the wires and the strands. The core may be comprised of steel or fiber. In the case of both of the submitted wire rope samples, the core consisted of an independent wire rope.

901,113
PB90-136771   Not available NTIS
National Inst. of Standards and Technology (IMSE), Boulder, CO. Fracture and Deformation Div.
Low-Temperature Phase and Magnetic Interactions in fcc Fe-Cr-Ni Alloys.
Final rept.
C. Almasan, T. Datta, R. D. Edge, E. R. Jones, J. W. Cable, and H. M. Ledbetter. 1989, 10p
Pub. in Jnl. of Magnetism and Magnetic Materials 80, p329-338 1989.

Keywords: *Low temperature tests, *Phase transformations, *Antiferromagnetism, *Face centered cubic lattices, *Nickel chromium steels, *Iron alloys, Neutron diffraction, Ultrasonic tests, Elastic scattering, Magnetic measurement, Interactions, Cryogenics, Neel temperature, Crystal structure, Reprints, SQUID(Detectors).

The low-temperature (5 K $<$ T $<$ 300 K) magnetic properties of a set of nine isostructural fcc Fe-Cr-Ni (Fe approximately 66 at %, Cr approximately 20 at %, Ni approximately 9 at %) alloys were studied by SQUID magnetometry, neutron diffraction and ultrasonic techniques. Type-1 antiferromagnetic (AF) ordering was observed below the Neel temperature, T (sub N). The dc susceptibility, X(T), did not exhibit a simple Curie-Weiss dependence. Above T (sub N), a temperature independent component was observed. T (sub N) was systematically influenced by the lattice parameter, 'a', decreasing from (47.9 + or - 0.5) K to (35.0 + or - 0.5) K as a increased by only 0.25%. The average magnetic moment of approximately 0.6 obtained from neutron scattering was lower than the approximately 1 obtained from the SQUID data. Mean field estimates of antiferromagnetic nearest-neighbors exchange interaction (J1) and ferromagnetic second-nearest-neighbors interaction (J2) indicate that (J2/J1) is approximately 1.5, evidence of the RKKY interaction. Only the

external d electrons are responsible for the localized average moment. It may mean that s-d hybridization of the external electrons is weak in the alloys.

Lubricants & Hydraulic Fluids

901,114
PB89-175921   Not available NTIS
National Bureau of Standards (IMSE), Gaithersburg, MD. Ceramics Div.
Preparative Liquid Chromatographic Method for Pre Characterization of Minor Constituents of Lubricating Base Oils.
Final rept.
P. Pei, and S. M. Hsu. 1986, 35p
Pub. in Jnl. of Liquid Chromatography 9, n15 p3311-3345 Nov 86.

Keywords: *Lubricating oils, *Chromatography, *Polarity, *Hydrocarbons, Methodology, Liquids, Separation, Synthesis(Chemistry), Friction, Wear, Oxidation, Molecular structure, Reprints.

In an effort to isolate, identify, and measure the properties of the active ingredients in a lubricating base oil, a high performance liquid chromatographic (HPLC) separation scheme has been developed. The preparative mode of production is necessary to yield sufficient amounts of minor constituents for property measurements in terms of friction, wear, and oxidation characteristics. In friction and wear control, the polarity of the molecular species is more important than the functional groups in the species. Therefore the design of the separation scheme is based on the relative polarity of various functional groupings. Because the effort is directed towards identifying key components rather than analysis of the major compositions, mass recovery requirement is critical. The separation scheme is divided into two stages. The base oil first undergoes a clay-gel separation to yield the saturates, aromatics, and the polar fractions. The polar fraction then is separated further using a neutral alumina column and the sequential solvent extractions into molecular compound classes of varying polarity. The paper describes the separation scheme and the detailed chemical characterization of the fractions.

Materials Degradation & Fouling

901,115
PB89-147409   Not available NTIS
National Bureau of Standards (IMSE), Gaithersburg, MD. Metallurgy Div.
Ultrasonic Characterization of Surface Modified Layers.
Final rept.
B. J. Elkind, M. Rosen, and H. N. G. Wadley. 1987, 8p
Contract DARPA Order-4275
Sponsored by Defense Advanced Research Projects Agency, Arlington, VA.
Pub. in Metallurgical Transactions A-Physical Metallurgy and Materials Science 18, n3 p473-480 1987.

Keywords: *Nondestructive tests, *Solidification, *Steels, *Radiation effects, Surface properties, Martensite, Pearlite, Rayleigh waves, Penetration, Hardness, Depth, Electron beams, Ultrasonic tests, Process control, Monitors, Quench hardening, Reprints.

Nondestructive techniques are required for the inprocess characterization of rapidly solidified and surface modified layers to fulfill the role of sensors in emerging intelligent materials processing technologies. In steels, where surface modification via directed high energy sources is being investigated for surface hardening, it has been found that a difference exists in the Rayleigh wave velocity of martensite and pearlite. The difference in velocity can be used to characterize the hardness of a surface modified layer on a pearlite substrate. By varying the Rayleigh wave frequency (and thus the depth of wave penetration) and measuring velocity dispersion, it has also been possible to non-destructively determine the depth of modified surface layers on both AISI 1053 and 1044 steels produced by electron beam melting.

901,116
PB89-157960   Not available NTIS
National Bureau of Standards (NML), Gaithersburg, MD. Surface Science Div.
Electron and Photon Stimulated Desorption: Probes of Structure and Bonding at Surfaces.
Final rept.
T. E. Madey. 1986, 7p.
Pub. in Science 234, n4774 p316-322 1986.

Keywords: *Surface properties, *Solids, *Radiation damage, *Molecular structure, Photons, Quantitative analysis, Simulation, Chemical bonds, Sorption, Reprints.

Techniques for analyzing the structure and composition of solid surfaces using electron and photon beams often suffer from interferences due to radiation damage. Damage-producing processes compete with information-producing events during measurements, and beam damage can be a serious perturbation in quantitative surface analysis. However, there are also substantial benefits of electron and photon stimulated damage processes for studying molecules on surfaces. Direct information about the geometrical structure of adsorbed species can be obtained from measurements of the angular distributions of ions released by electron or photon stimulated desorption. The directions of ion emission are determined by orientations of the surface bonds which are ruptured by beam irradiation. The method of Electron Stimulated Desorption Ion Angular Distributions (ESDIAD) has proven particularly useful as a direct tool for characterizing local molecular structure at surfaces. Moreover, photon stimulated desorption studies using synchrotron radiation are revealing the fundamental electronic excitations which lead to bond-breaking processes at surfaces.

Miscellaneous Materials

901,117
PB89-186407   Not available NTIS
National Bureau of Standards (NEL), Gaithersburg, MD. Building Environment Div.
Experimental Determination of Forced Convection Evaporative Heat Transfer Coefficients for Non-Azeotropic Refrigerant Mixtures.
Final rept.
R. Radermacher, H. Ross, and D. Didion. 1983, 7p
Pub. in Proceedings of Winter Annual Meeting of the American Society of Mechanical Engineers, Boston, MA., November 13-18, 1983, p1-7.

Keywords: *Refrigerants, *Test facilities, *Heat transfer coefficient, Convection, Mixtures, Evaporators, Two phase flow.

Recently energy conservation requirements spurred interest in nonazeotropic refrigerant mixtures, because such mixtures can improve theoretically the COP of certain refrigerant cycles. The two phase heat transfer coefficient of such mixtures under forced convection conditions is virtually unknown. An experimental rig has been constructed to investigate whether it is possible to predict the heat transfer coefficient of the mixture based on the coefficients of the components. Initially data was taken on R-22 and compared to literature data and to existing predictive correlations. Good agreement was found with the literature's data on forced convection, single phase heat transfer correlations, and some two phase evaporative correlations.

901,118
PB89-229041   PC A06/MF A01
National Inst. of Standards and Technology (NEL), Gaithersburg, MD. Center for Building Technology.
Experimental Investigation and Modeling of the Flow Rate of Refrigerant 22 Through the Short Tube Restrictor.
D. A. Aaron, and P. A. Domanski. Jul 89, 102p NISTIR-89/4120
Sponsored by Department of Energy, Washington, DC. Office of Buildings and Community Systems.

Keywords: *Refrigerants, *Models, *Flow rate, *Tubes, Air conditioners, Heat pumps, Constrictions, Flow measurement, Experimental data, Pressure, Chamfering.

MATERIALS SCIENCES
Miscellaneous Materials

Refrigerant flow through the short tube expansion device was theoretically and experimentally investigated. The analysis was limited to initially subcooled R22 flowing through short tubes with 5 L/D 20. Flow dependency upon upstream subcooling, upstream pressure, downstream pressure, length, diameter, entrance chamfering and exit chamfering was determined. A flow model and flow charts were developed.

Nonferrous Metals & Alloys

901,119
PATENT-4 804 446   Not available NTIS
National Bureau of Standards, Gaithersburg, MD.
Electrodeposition of Chromium from a Trivalent Electrolyte.
Patent.
D. S. Lashmore, I. Weisshaus, and E. NamGoong.
Filed 19 Sep 86, patented 14 Feb 89, 14p PB89-160592, PAT-APPL-6-909 433
This Government-owned invention available for U.S. licensing and, possibly, for foreign licensing. Copy of patent available Commissioner of Patents, Washington, DC 20231 $1.50.

Keywords: *Chromium, *Coatings, *Patents, Electrodeposition, Electrodes, Electrolytes.

An electrodeposition process and a bath therefore are disclosed for performing the electrodeposition of hard smooth coatings of trivalent chromium. The electrodeposition process is accomplished energy efficiently. The bath includes chromium chloride as a source of chromium, citric acid to complex the chromium, and a wetting agent which is preferably Triton X-100. Preferably, bromide is also provided in the conductivity and also the current distribution in the bath. Boric acid is provided to advance the reaction kinetics. The pH of the bath is maintained at approximately 4.0 and the temperature is maintained at approximately 35 C. Either a direct current or pulsed current is used for the deposition process. Hard smooth coatings of trivalent chromium are deposited through use of the process and the bath of the claimed invention.

901,120
PB89-146690   Not available NTIS
National Bureau of Standards (IMSE), Gaithersburg, MD. Polymers Div.
Electronic, Magnetic, Superconducting and Amorphous-Forming Properties Versus Stability of the Ti-Fe, Zr-Ru and Hf-Os Ordered Alloys.
Final rept.
R. Kuentzler, and R. M. Waterstrat. 1986, 15p
Sponsored by American Dental Association Health Foundation, Chicago, IL.
Pub. in Jnl. of Less-Common Met. 125, p261-275 Nov 86.

Keywords: *Intermetallics, *Superconductivity, *Magnetic susceptibility, Titanium, Iron, Zirconium, Ruthenium, Hafnium, Osmium, Magnesium, Zinc, Phase transformations, Specific heat, Electron transitions, Reprints, Ordered alloys.

The electronic, magnetic, and superconducting properties of the Ti-Fe, Zr-Ru and Hf-Os ordered alloys of the B2-type and MgZn sub 2-type structures are described using original low temperature specific heat and susceptibility results and known magnetization data. The stability of the B2-type ordered alloys, including that of TiFe, ZrRu and HfOs is discussed. The high stability of the B2-type ordered alloys having an average number of 'd' electrons, N sub d, equal or nearly equal to 5 (which corresponds to a Fermi level lying in a deep valley of the band structure) is accompanied by a low electronic specific heat coefficient gamma, no magnetic order, no superconductivity and poor glass-forming ability. On the other hand, deviations from N sub d = 5 produce a decrease of the stability, high gamma values, appearance of magnetic order and martensitic transformations leading to superconductivity and good glass-forming ability.

901,121
PB89-146948   Not available NTIS
National Bureau of Standards (IMSE), Boulder, CO.
Fracture and Deformation Div.

132

Ultrasonic Texture Analysis for Polycrystalline Aggregates of Cubic Materials Displaying Orthotropic Symmetry.
Final rept.
P. P. Delsanto, R. B. Mignogna, and A. V. Clark.
1986, 9p
Sponsored by Naval Research Lab., Washington, DC.
Pub. in Nondestructive Characterization of Materials II, p535-543 1986.

Keywords: *Nondestructive tests, *Ultrasonic frequencies, *Steels, *Aluminum alloys, Texture, Crystal structure, Reprints, *Raleigh waves.

The general perturbation formalism for the propagation of Raleigh waves on the surface of initially deformed anisotropic material plates is applied to the investigation of material texture. The preferential alignment of crystallographic axes in a polycrystalline material can be conveniently described in terms of their orientation distribution function. The case of an orthotropic distribution of cubic crystallites, which occurs, for example, in aluminum and steel alloys is considered. It was shown that the measured values of the Raleigh wave phase velocity at these different angles on the material plate can be used for the determination of the three coefficients W sub 400; W sub 420 and W sub 440 which completely characterize the orientation distribution in the case considered.

901,122
PB89-147102   Not available NTIS
National Bureau of Standards (NEL), Gaithersburg, MD. Chemical Process Metrology Div.
Sputter Deposition of Icosahedral Al-Mn and Al-Mn-Si.
Final rept.
K. G. Kreider, F. S. Biancaniello, and M. J. Kaufman. 1987, 6p
Pub. in Scripta Metallurgica 21, n5 p657-662 May 87.

Keywords: *Thin films, *Sputtering, *Crystals, *Intermetallics, Aluminum, Manganese, Microscopy, X-ray analysis, Hexagonal lattices, Morphology, Fabrication, Reprints, Icosahedral.

Thin-film sputtered deposits of Al with 17.4% (atomic) Mn and Al with 20.2% Mn plus 4.7% Si were produced at temperatures ranging from -80 deg C to 420 deg C from prealloyed targets. These films were analyzed using electron microscopy and x-ray diffraction to determine their structures and compositions. The icosahedral quasicrystalline phase was observed in the films produced at the lower temperatures whereas a mixture of the quasicrystal and the hexagonal Al10Mn3 (Al9Mn3Si) was observed at the higher temperatures. The fabrication technique as well as the physical characterization of the films are described.

901,123
PB89-147383   Not available NTIS
National Bureau of Standards (IMSE), Gaithersburg, MD. Metallurgy Div.
Quasicrystals with 1-D Translational Periodicity and a Ten-Fold Rotation Axis.
Final rept.
L. Bendersky. 1986, 4p
Pub. in Rapidly Solidified Alloys and Their Mechanical and Magnetic Properties, p237-240 1986.

Keywords: *Quench hardening, *Intermetallics, *Aluminum alloys, *Manganese alloys, *Crystal lattices, Microscopy, Mechanical properties, Reprints, Icosahedral.

Studies of phase formation in rapidly solidified Al-Mn alloys (composition range 18-22 at % Mn) show that an icosahedral phase is replaced by another noncrystallographic phase, a decagonal phase. The decagonal phase is another example of quasicrystal: it has a noncrystallographic point group (10/m or 10/mmm) together with long-range orientational order and one-dimensional symmetry. The decagonal phase is an intermediate phase between an icosahedral phase and a crystal both from the symmetry and from the solidification condition points of view.

901,124
PB89-150957   Not available NTIS
National Bureau of Standards (IMSE), Boulder, CO.
Fracture and Deformation Div.

Ultrasonic Determination of Absolute Stresses in Aluminum and Steel Alloys.
Final rept.
A. V. Clark, J. C. Moulder, R. B. Mignogna, and P. P. Delsanto. 1986, 8p
Sponsored by Naval Sea Systems Command, Washington, DC.
Pub. in Residual Stresses in Science and Technology, Garmisch-Partenkirchen, Federal Republic of Germany, v1 p207-214 1986.

Keywords: *Ultrasonic tests, *Weldments, *Aluminum, *Steels, *Plates(Structural members), *Residual stress, Strain gages, Piezoelectricity, Electromagnetic fields, Transducers, Measurement, Acoustics, Reprints.

Components of plane stress have been measured with ultrasonic techniques for welded aluminum and steel alloy plates. Measurements of the difference of principal stresses were performed using both a piezoelectric transducer and an electromagnetic-acoustic transducer (EMAT) with good agreement. The EMAT was used to measure arrival times of ultrasonic shear waves along the centerline of baseplates before and after welding. Subject to certain assumptions, changes in arrival times, at a given location, are due to principal stresses at that location. For the aluminum alloy plates these EMAT measurements of principal stresses were within 20 MPa of strain gage values. For the steel plates, the difference between EMAT and strain gage results was about 20%. A second technique used to obtain the normal stress in welded aluminum alloy plates is described.

901,125
PB89-157432   Not available NTIS
National Bureau of Standards (IMSE), Gaithersburg, MD. Metallurgy Div.
Stable and Metastable Ti-Nb Phase Diagrams.
Final rept.
D. L. Moffat, and U. R. Kattner. 1988, 9p
Pub. in Metallurgical Transactions A 19A, p2389-2397 Oct 88.

Keywords: *Titanium alloys, *Niobium, *Phase transformations, Thermal degradation, Chemical composition, Thermodynamics, Metastable state, Equilibrium, Vanadium, Molybdenum, Zirconium, Reprints.

The phase transformations which occur in the Ti-Nb binary alloy system have been discussed in two recent papers. The phase relationships were investigated by varying alloy composition and thermal history. In the paper, these results are summarized in complete and thermodynamically consistent calculations of the stable and metastable phase diagrams. The calculations of the metastable equilibria are relevant to the Ti-V and Ti-Mo systems, as well as to several other titanium and zirconium-based transition metal alloy systems.

901,126
PB89-157598   Not available NTIS
National Bureau of Standards (IMSE), Gaithersburg, MD. Metallurgy Div.
Ostwald Ripening in a System with a High Volume Fraction of Coarsening Phase.
Final rept.
S. C. Hardy, and P. W. Voorhees. 1988, 9p
Sponsored by National Aeronautics and Space Administration, Washington, DC.
Pub. in Metallurgical Transactions A 19A, p2713-2721 Nov 88.

Keywords: *Lead alloys, *Tin, *Particle size distribution, *Sintering, Kinetics, Eutectics, Liquid phases, Reprints, Ostwald ripening.

Experiments on the coarsening behavior of two-phase mixtures in a model Pb-Sn system are reported. This system fulfills most of the assumptions of theory and has the particular advantage that all the materials parameters necessary for a comparison between the experimentally measured and theoretically predicted coarsening kinetics are known. The coarsening of Sn-rich and Pb-rich solid phases was examined in contact with eutectic liquid in the volume fraction solid range above approximately 0.6 where the development of a solid skeletal structure inhibits sedimentation. Particle intercept distributions are measured and found to be time independent when scaled by the average intercept. This invariance is interpreted as evidence that scale factor coarsening is present. The intercept distributions are in good agreement with the predictions of theory. Measurements of average intercept diameter as a function of time establish unambiguously that the coarsening follows the theoretically predicted t sup(1/3) kinetics. The coarsening rate constants are measured as a function of volume fraction solid and are found to exceed the values calculated from theory using the known thermophysical properties of the Pb-Sn system by factors ranging from approximately 2 to 5.

901,127
PB89-157606 Not available NTIS
National Bureau of Standards (IMSE), Gaithersburg, MD.
Observations on Crystal Defects Associated with Diffusion Induced Grain Boundary Migration in Cu-Zn.
Final rept.
S. A. Hackney, F. S. Biancaniello, D. N. Yoon, and C. A. Handwerker. 1986, 6p
Pub. in Scripta Metallurgica 20, n6 p937-942 Jun 86.

Keywords: *Copper, *Crystal defects, *Zinc, *Electron microscopy, Diffusion, Grain boundaries, Brasses, Transmission, Migrations, Dislocations(Materials), Gases, Reprints.

High purity copper foils (.025 mm thick) have been exposed to zinc vapor from an 11 at% zinc brass at 360 deg C. Standard transmission electron microscopy reveals that diffusion induced grain boundary migration (DIGM) has occurred at an average rate of 1.85 x 10(sup -11) m/s over the first 30 hours. The grain boundary structure, grain boundary morphology, and matrix dislocations associated with the DIGM phenomena have been studied in detail. The following general observations have been made: (1) the region over which the grain boundary has migrated is alloyed with zinc; (2) a 'wall' of dislocations marks the original position of the grain boundary; (3) the grain boundary structure includes a high density of defects with a step character having a maximum height of 3 nm; and (4) the matrix dislocation density is highest directly in front of the migrated grain boundary.

901,128
PB89-157614 Not available NTIS
National Bureau of Standards (IMSE), Gaithersburg, MD. Metallurgy Div.
Migration of Liquid Film and Grain Boundary in Mo-Ni Induced by W Diffusion.
Final rept.
H. K. Kang, S. Hackney, and D. N. Yoon. 1988, 5p
Pub. in Acta Metallurgica 36, n3 p695-699 1988.

Keywords: *Liquids, *Migrations, *Grain boundaries, *Nickel alloys, Molybdenum, Diffusion, Heat treatment, Sintering, Tungsten, Solid solutions, Coherence, Reprints.

The liquid films and grain boundaries in liquid phase sintered Mo-Ni alloy are observed to migrate during heat-treatment after adding W to the liquid matrix. Behind the migrating boundaries form Mo-Ni-W solid solution with the W concentration decreasing with the migration distance because of W depletion in the liquid matrix. The migration rate during the heat-treatment at 1540 deg C after adding W decreases with the decreasing pretreatment sintering temperature. When the sintering temperature is 1420 deg C, the migration rate is almost reduced to 0. Under this condition, the coherency strain due to the simultaneous diffusion of W and Ni into the grain surfaces is estimated to be almost 0. The results thus lead to the conclusion that the coherency strain due to lattice diffusion is the driving force for the liquid film and grain boundary migration.

901,129
PB89-157622 Not available NTIS
National Bureau of Standards (IMSE), Gaithersburg, MD. Metallurgy Div.
Metastable Phase Production and Transformation in Al-Ge Alloy Films by Rapid Crystallization and Annealing Treatments.
Final rept.
M. J. Kaufman, J. E. Cunningham, and H. L. Fraser. 1987, 12p
Pub. in Acta Metallurgica 35, n5 p1181-1192 1987.

Keywords: *Crystallization, *Aluminum alloys, *Germanium, *Quenching(Cooling), *Electron beam, Transmission, Electron microscopy, Films, Annealing, Metastable state, Nucleation, Kinetics, Reprints.

Metastable crystalline phases have been produced in Al-Ge alloy films which initially were either entirely or partially amorphous by rapid crystallization and subsequent annealing treatments. The studies were conducted directly in a transmission electron microscope where the electron beam was used as a local heating source to effect the reactions. The types and sequences of transformations are described and discussed in terms of competitive nucleation and growth kinetics. In addition, the results are related to those previously obtained on similar alloys subjected to rapid quenching and high undercooling treatments.

901,130
PB89-157630 Not available NTIS
National Bureau of Standards (IMSE), Gaithersburg, MD. Metallurgy Div.
Experimental Observations on the Initiation of DIGM (Diffusion Induced Grain Boundary Migration).
Final rept.
S. A. Hackney. 1986, 4p
Pub. in Scripta Metallurgica 20, n10 p1385-1388 Oct 86.

Keywords: *Microstructure, *Grain boundaries, *Diffusion, *Copper, *Zinc, Migrations, Kinetics, Crystal, Initiation, Dislocations(Materials), Reprints.

Diffusion induced grain boundary migration (DIGM) is now a well recognized phenomena which occurs during multi-component diffusion in many systems. Kinetic theories of diffusional mixing in crystalline solids will have to be modified to include the contributions of DIGM and the related process of diffusion induced recrystallization. The underlying driving force for this important process has been the topic of a great deal of speculation. The theoretical approach to the process has far outstripped the necessary experimental observations. One critical area of experimentation which has been neglected is the microstructural observations of the physical processes involved in the initiation of DIGM. The early observations from a study of DIGM initiation are presented for the Cu-Zn system.

901,131
PB89-157648 Not available NTIS
National Bureau of Standards (NEL), Gaithersburg, MD. Semiconductor Electronics Div.
Structural Unit in Icosahedral MnAlSi and MnAl.
Final rept.
Y. Ma, E. A. Stern, and C. E. Bouldin. 1986, 4p
Pub. in Physical Review Letters 57, n13 p1611-1614, 29 Sep 86.

Keywords: *x ray analysis, *Manganese, *Aluminum alloys, *Silicon, *Crystallization, Orthorhombic lattices, Quenching(Cooling), Reprints.

EXAFS measurements were made on icosahedral MnAl and MnSiAl and on the standards alpha-phase of MnSiAl and orthorhombic phase of MnAl6. Experimental evidence is presented that a cage of Mn atoms at the vertices of an icosahedron is the structural unit in the icosahedral MnSiAl and MnAl phases. The connections among these icosahedral units and between them and the Al atoms are different in the icosahedral phases and in the alpha-phase. As in the alpha-phase, the Mn icosahedra do not share vertices in the icosahedral phases; i.e., they are separated from one another. It is suggested that the i-phase grows by randomly nucleating together Mn icosahedra along their 20 threefold directions, as allowed by local steric constraints.

901,132
PB89-157671 Not available NTIS
National Bureau of Standards (NEL), Gaithersburg, MD. Automated Production Technology Div.
Dynamic Young's Modulus Measurements in Metallic Materials: Results of an Interlaboratory Testing Program.
Final rept.
A. Wolfenden, M. R. Harmouche, G. V. Blessing, Y. T. Chen, P. Terranova, V. Dayal, V. K. Kinra, J. W. Lemmens, R. R. Phillips, J. S. Smith, P. Mahmoodi, and R. J. Wann. 1989, 12p
Pub. in Jnl. of Testing and Evaluation 17, n1 p2-13 Jan 89.

Keywords: *Modulus of elasticity, *Test facilities, *Nickel alloys, Metals, Measurement, Comparison, Reprints.

The results of a round-robin testing study are presented for measurements of dynamic Young's modulus in two nickel-based alloys. The Interlaboratory Testing Program involved six types of apparatus, six different organizations, and specimens from a well-documented source. All the techniques yielded values of dynamic Young's modulus that agreed within 1.6% of each other. For Inconel alloy 600 the dynamic modulus was 213.5 GPa with a standard deviation of 3.6 GPa; for Incoloy alloy 907 the corresponding values were 156.6 and 2.2 GPa, respectively. No significant effect of frequency over the range 780 Hz to 15 MHz was found.

901,133
PB89-157804 Not available NTIS
National Bureau of Standards (IMSE), Boulder, CO. Fracture and Deformation Div.
Influence of Dislocation Density on the Ductile-Brittle Transition in bcc Metals.
Final rept.
I. H. Lin, and R. Thomson. 1986, 4p
Pub. in Scripta Metallurgica 20, n10 p1367-1370 Oct 86.

Keywords: *Body centered cubic lattices, *Ductile brittle transition, *Crack propagation, *Dislocations(Materials), *Metals, Emission, Fractures(Materials), Reprints.

The purpose of the paper is to show that the local k-field at the crack tip at which dislocation emission takes place is lowered by the action of external dislocation sources, and that this mechanism leads to the result that sufficiently high concentrations of dislocations and their sources in the influence field of the crack tip will limit the ability of the crack tip to cleave.

901,134
PB89-157911 Not available NTIS
National Bureau of Standards (IMSE), Gaithersburg, MD. Metallurgy Div.
Directional Invariance of Grain Boundary Migration in the Pb-Sn Cellular Transformation and the Tu-Turnbull Hysteresis.
Final rept.
S. A. Hackney, and F. S. Biancaniello. 1986, 6p
Pub. in Scripta Metallurgica 20, n10 p1417-1422 Oct 86.

Keywords: *Lead alloys, *Tin, *Grain boundaries, *Migrations, Solid solutions, Thermodynamic properties, Hysteresis, Cellular materials, Porosity, Diffusion, Dissolving, Reprints.

The cellular dissolution process in Pb-5.5% Sn first studied by Tu and Turnbull has been reexamined using instrumental techniques. It has been determined that cell dissolution by grain boundary migration does not recreate a homogeneous solid solution. This observation has been interpreted in terms of a thermodynamic hysteresis. The driving force for dissolution first proposed by Tu and Turnbull has been modified to include the macroscopic grain boundary curvature term. This allows a direct contrast between the forced oscillation of the grain boundary in the cellular transformation and that studied in DIGM. A simple thermodynamic evolution criteria reveals that the difference in behavior during dissolution between the two phenomena may be due to the presence of the precipitate phase rather than a difference in migration mechanism.

901,135
PB89-157986 Not available NTIS
National Bureau of Standards (NEL), Gaithersburg, MD. Mathematical Analysis Div.
ASM/NBS (American Society for Metals/National Bureau of Standards) Numerical and Graphical Database for Binary Alloy Phase Diagrams.
Final rept.
J. S. Sims, D. F. Redmiles, and J. B. Clark. 1988, 16p
Pub. in Computerized Metallurgical Databases, p119-134 1988.

Keywords: *Alloys, *Metals, *Data retrieval, *Phase diagrams, Surveys, Crystal structure, Reprints.

Under the ASM/NBS program on alloy phase diagrams, a comprehensive relational database of binary alloy phase diagrams has been developed. The phase diagrams, critical numerical data and crystal structure data of the phases of nearly 1600 binary alloy systems can be accessed by a user friendly database management program. Important features of the database program are: (a) Search and display of all the phase graphics prepared for the 'Bulletin of Alloy Phase Diagrams'; (b) Search and display of the critical numerical data summarizing the 'structure' of the phase diagram - the phase reactions, the reaction temperatures, the

133

MATERIALS SCIENCES

Nonferrous Metals & Alloys

compositions of the reacting phases, and the crystal structure data of the solid phases.

901,136
PB89-172324 Not available NTIS
National Bureau of Standards (IMSE), Gaithersburg, MD. Metallurgy Div.
Stable and Metastable Phase Equilibria in the Al-Mn System.
Final rept.
J. L. Murray, L. A. Bendersky, F. S. Biancaniello, A. J. McAlister, D. L. Moffat, and R. J. Schaefer. 1987, 8p
Pub. in Metallurgical Transactions A-Physical Metallurgy and Materials Science 18, n3 p365-392 1987.

Keywords: *Aluminum, *Manganese, *Phase diagrams, Thermodynamics, Mathematical models, Equilibrium, Stoichiometry, Liquidus, Quenching(Cooling), Stability, Metastable state, Thermal analysis, Reprints.

The aim of the present investigation was resolution of certain obscure features of the Al4Mn phase diagram. The experimental approach was guided by assessment of the previous literature and modeling of the thermodynamics of the system. It has been shown that two phases of approximate stoichiometry Al4Mn (lambda and mu) are present in stable equilibrium, lambda forming by a peritectoid reaction at 693 + or - C. The liquidus and invariant reactions as proposed by Godecke and Koester have been verified. A map has been made of the successive non-equilibrium phase transformations of as-splat-quenched alloys. Finally, the thermodynamic calculation of the phase diagram allows interpretation of complex reaction sequences during cooling in terms of a catalog of all the metastable invariant reactions involving (Al), Al6Mn, lambda, mu, theta, and Al11Mn4 phases.

901,137
PB89-172332 Not available NTIS
National Bureau of Standards (IMSE), Gaithersburg, MD. Metallurgy Div.
Solidification of Aluminum-Manganese Powders.
Final rept.
B. A. Mueller, R. J. Schaefer, and J. H. Perepezko. 1987, 9p
Pub. in Jnl. of Materials Research 2, n6 p809-817 Nov/Dec 87.

Keywords: *Aluminum, *Manganese, *Powder(Particles), *Quenching(Cooling), Phase diagrams, Metastable state, Nucleation, Temperature, Crystallization, Reprints.

The solidification behavior of Al-Mn powders was studied as a function of cooling rate and Mn content. It was found that the phases present in the powder differed from those expected at equilibrium. The Al6Mn phase was absent due to its failure to nucleate, and in the more concentrated and rapidly cooled powders the metastable quasicrystal phases were present. Nucleation temperatures measured in alloys cooled at 25 C/sec are believed to represent formation of the icosahedral phase, which subsequently transforms to the decagonal phase.

901,138
PB89-176457 Not available NTIS
National Bureau of Standards (IMSE), Gaithersburg, MD. Metallurgy Div.
Kinetics of Resolidification.
Final rept.
J. H. Perepezko, and W. J. Boettinger. 1987, 40p
Pub. in Proceedings of ASM (American Society for Metals) Materials Science Seminar: Surface Alloying by Ion, Electron, and Laser Beams, Toronto, Ontario, Canada, October 12-13, 1985, p51-90 1987.

Keywords: *Solidification, *Kinetics, *Surface finishing, *Metal alloys, *Lasers, *Ion beams, *Electron beams, *Microstructure, Solubility, Metastable state, Free energy, Morphology, Heat transfer, Phase diagrams.

While ion, laser and electron beam surface treatments involve a variety of experimental conditions, they share some important common kinetic and thermodynamic features. The alloy additions that are incorporated into the surface modified solid region by the various processes are often in metastable states of relatively high free energy. An examination of the metastable equilibrium features of phase diagrams can identify the possible choice of product structures depending on the particular constraints imposed on the system and the controlling phase selection kinetics during nucleation.

901,139
PB89-176465 Not available NTIS
National Bureau of Standards (IMSE), Gaithersburg, MD. Metallurgy Div.
Undercooling and Microstructural Evolution in Glass Forming Alloys.
Final rept.
M. J. Kaufman, and H. L. Fraser. 1987, 20p
Pub. in Undercooled Alloy Phases, Proceedings of Hume-Rothery Memorial Symposium, p249-268 1987.

Keywords: *Aluminum, *Amorphous materials, *Glass, *Solidification, *Nucleation, *Recalescence, Powder(Particles), Forecasting, Crystallization, Germanium, Metal alloys, Microstructure.

The microstructural evolution of highly undercooled submicron powders of a glass forming alloy (Al-30Ge) is considered. A simple calculation is developed for predicting the number of alpha-Al crystals that form during solidification using classical nucleation and growth equations appropriate for glass forming systems. In addition, a reasonable agreement between theory and experiment is achieved only when a solid-liquid temperature gradient, generated by recalescence effects, is considered.

901,140
PB89-176911 Not available NTIS
National Bureau of Standards (IMSE), Gaithersburg, MD. Metallurgy Div.
Dynamic Microindentation Apparatus for Materials Characterization.
Final rept.
R. S. Polvani, A. W. Ruff, and E. P. Whitenton. 1988, 5p
Pub. in Jnl. of Testing and Evaluation 16, n1 p12-16 Jan 88.

Keywords: *Indentation hardness tests, *Aluminum, *Composite materials, *Equipment, Dynamic tests, Loads(Forces), Penetration, Wear, Mechanical properties, Time, Reprints.

A microindentation system is described that provides a new approach to dynamic mechanical testing. The indentation is characterized in terms of continuous measurements of applied load and penetration depth. The indentation can be performed over a wide range of loading times from hours down to milliseconds. The shape of the loading waveform can also be selected. The deformation energy can be measured and partitioned into elastic, plastic, and anelastic components. The apparatus is also able to perform conventional hardness testing and can be used to determine conventional mechanical properties.

901,141
PB89-177026 Not available NTIS
National Bureau of Standards (NML), Gaithersburg, MD. Radiation Physics Div.
Electron Mean Free Path Calculations Using a Model Dielectric Function.
Final rept.
D. R. Penn. 1987, 5p
Pub. in Physical Review B-Condensed Matter 35, n2 p462-466 1987.

Keywords: *Electrons, *Mean free path, *Computation, Dielectric properties, Copper, Silver, Gold, Aluminum, Energy, Gas laws, Reprints.

The anelastic electron mean free path as a function of energy is calculated for Cu, Ag, Au, and Al. The calculations are based on a model dielectric function, epsilon(q,w), which is obtained from a modification of the statistical approximation. In this approach epsilon(o,w) is determined by the experimentally measured optical dielectric function. Calculated mean free paths are compared to experimental data and to other theories.

901,142
PB89-179170 Not available NTIS

National Bureau of Standards (IMSE), Gaithersburg, MD. Metallurgy Div.
Process Control during High Pressure Atomization.
Final rept.
S. D. Ridder, and F. S. Biancaniello. 1988, 5p
Pub. in Materials Science and Engineering 98, p47-51 Feb 88.

Keywords: *Liquid metals, *Powder(Particles), *Process control, *High pressure tests, *Atomizing, *Rare gases, Metal alloys, Solidification, Particle size, Drops(Liquids), High speed photography, Lasers, Tin alloys, Reprints.

High Pressure Inert Gas Atomization (HPIGA) has been studied using various metal alloy systems. The high yield of ultrafine (less than 45 micrometers) powder produced using HPIGA makes it an ideal test system for rapidly solidified metal powder. High speed photography and laser scattering techniques have been applied to study droplet formation and measure powder size with the intent of future feedback and control of particle size during atomization. Liquid metal droplet formation will be discussed as well as on-line particle size measurement and control.

901,143
PB89-179840 Not available NTIS
National Bureau of Standards (NEL), Boulder, CO. Electromagnetic Fields Div.
Transmission Loss through 6061 T-6 Aluminum Using a Pulsed Eddy Current Source.
Final rept.
K. H. Cavcey. 1989, 3p
Pub. in Materials Evaluation 47, p216-218 Feb 89.

Keywords: *Aluminum, *Aircraft, *Nondestructive tests, *Eddy currents, *Electric conductors, Pulse analyzers, Electromagnetic testing, Thickness, Frequency analyzers, Reprints, Aluminum T-6.

One method of nondestructive testing in conductors is that of pulsed eddy currents (PEC). The method involves the propagation of a modified electromagnetic field through the medium, resulting in attenuation and time delay of the pulse. The paper outlines work that was done to determine the frequency response for seven different thicknesses of aircraft-grade 6061 T-6 aluminum using a PEC source.

901,144
PB89-186316 Not available NTIS
National Bureau of Standards (IMSE), Gaithersburg, MD. Metallurgy Div.
Formation and Stability Range of the G Phase in the Aluminum-Manganese System.
Final rept.
R. J. Schaefer, F. S. Biancaniello, and J. W. Cahn. 1986, 6p
Pub. in Scripta Metallurgica 20, n10 p1439-1444 1986.

Keywords: *Aluminum alloys, *Manganese, *Phase transformation, *Solidification, Metastable state, Crystal structure, Nucleation, Eutectics, Stability, Microstructure, Quenching(Cooling), Reprints.

The G phase of Al-Mn, which has until now been considered to be metastable, is demonstrated to actually be a stable phase forming by a peritectoid reaction between 490 and 550 C. The growth of the G phase is extremely slow, but by rapid solidification the rate of nucleation of the G phase is greatly increased so that the transformation is almost complete after 1000 hours at 400 C. The G phase contains icosahedral clusters of atoms similar to those proposed in some models of the Al-Mn icosahedral phase.

901,145
PB89-186324 Not available NTIS
National Bureau of Standards (IMSE), Gaithersburg, MD. Metallurgy Div.
Nucleation and Growth of Aperiodic Crystals in Aluminum Alloys.
Final rept.
R. J. Schaefer, L. A. Bendersky, and F. S. Biancaniello. 1986, 10p
Pub. in Jnl. de Physique 47, nC-3 p311-320 1986.

Keywords: *Aluminum alloys, *Manganese, *Nucleation, *Solidification, Microstructure, Eutectics, Silicon, Crystal structure, Dendritic crystals, Phase transformation, Quenching(Cooling), Reprints.

Under rapid rates the analysis of resolidification involves the use of response functions to treat the liquid-solid interface conditions during solute trapping. These functions, when combined with solute redistribution and heat flow analysis, provide a basis for the prediction and analysis of microstructural evolution in surface treated layers in terms of size scale and morphology. The application of metastable equilibria and kinetics analysis offers an effective strategy for the generation of tailored microstructures to optimize the results of surface treatments.

The icosahedral and decagonal aperiodic phases dominate the microstructures of rapidly solidified Al-Mn alloys because of their nucleation and growth behavior, which differs substantially from that of the equilibrium phases. Electron beam surface melting can be used to produce a wide range of solidification conditions, in which the different stages of the nucleation and growth processes can be observed. It is found that the icosahedral phase nucleates abundantly in supercooled Al-Mn melts, and that the decagonal phase is subsequently nucleated by the icosahedral phase. Addition of Si to the Al-Mn alloys suppresses formation of the decagonal phase, but in these alloys the hexagonal beta phase can grow rapidly.

901,146
PB89-186332 Not available NTIS
National Bureau of Standards (IMSE), Gaithersburg, MD. Metallurgy Div.
Replacement of Icosahedral Al-Mn by Decagonal Phase.
Final rept.
R. J. Schaefer, and L. Bendersky. 1986, 6p
Pub. in Scripta Metallurgica 20, n5 p745-750 May 86.

Keywords: *Aluminum alloys, *Manganese, *Microstructure, *Solidification, *Crystal structure, Phase transformation, Nucleation, Quenching(Cooling), .Reprints.

It is concluded from microstructural evidence that the decagonal T phase is nucleated epitaxially by the icosahedral phase. At low cooling rates, the T phase grows and completely replaces the icosahedral phase, while at high cooling rates the icosahedral phase is preserved. Even in samples where none of the icosahedral phase is found, its presence at an early stage of solidification is revealed by the specific geometrical arrangement and orientational variants of the T phase crystals.

901,147
PB89-201321 PC A06/MF A01
National Inst. of Standards and Technology (IMSE), Gaithersburg, MD. Metallurgy Div.
Institute for Materials Science and Engineering: Metallurgy, Technical Activities 1988.
Annual rept.
G. N. Pugh, and J. H. Smith. Dec 88, 121p NISTIR-88/3843
See also report for 1987, PB88-157722.

Keywords: *Metallurgy, *Corrosion, Processing, Metals, Alloys, Mechanical properties, Chemical properties, Wear, Electrodeposition, Magnetic materials, Detectors, Technical activities, Metals processing.

The report summarizes the FY 1988 activities of the Metallurgy Division of the National Institute of Standards and Technology (NIST). The research centers upon the structure-processing-properties relations of metals and alloys and on the methods of their measurement. The activities also include the generation and evaluation of critical materials data. Efforts comprise studies of metallurgical processing, corrosion, mechanical properties, electrodeposition, process sensors, high temperature reactions and magnetic materials. The work described also includes four cooperative programs with American professional societies and industry: the National Association of Corrosion Engineers (NACE) - NIST Corrosion Data Program, the Aluminum Association - NIST Temperature Sensor Program, the American Iron and Steel Institute (AISI) - NIST Steel Sensor Program, and the ASM INTERNATIONAL (ASM) - NIST Alloy Phase Diagram Program.

901,148
PB89-201693 Not available NTIS
National Bureau of Standards (IMSE), Gaithersburg, MD. Reactor Radiation Div.
Magnetic Correlations in an Amorphous Gd-Al Spin Glass.
Final rept.
M. L. Spano, R. J. Gambino, S. K. Hasanain, T. R. McGuire, S. J. Pickart, and J. J. Rhyne. 1987, 3p
Pub. in Jnl. of Applied Physics 61, n8 p3639-3641 1987.

Keywords: *Neutron scattering, *Thin films, *Gadolinium, *Aluminum alloys, Sputtering, Lorentz transformations, Ferromagnetism, Spin orbit interactions, Temperature, Rare earth elements, Reprints.

Small angle neutron scattering (SANS) as well as magnetization measurements have been made on a sputtered film of Gd43Al57. The low field susceptibility

peaks at a freezing temperature, T sub f, of 33 K. It agrees well with the SANS data, which shows a peak in the intensity at this temperature for the lowest Qs measured. The SANS lineshapes are unusual in that they can be fitted with a Lorentzian-squared cross section with dissimilar correlation lengths. The Lorentzian correlation length peaks near 35 K at a value of approximately 14 angstroms, while the Lorentzian-squared correlation length exhibits a large, essentially resolution-limited value up to temperatures several times T sub f. These results are consistent with the coexistence of finite static spin clusters with relatively long range ferromagnetic correlations.

901,149
PB89-201701 Not available NTIS
National Bureau of Standards (IMSE), Gaithersburg, MD. Reactor Radiation Div.
Neutron Scattering Study of the Spin Ordering in Amorphous Tb45Fe55 and Tb25Fe75.
Final rept.
M. L. Spano, and J. J. Rhyne. 1987, 3p
Pub. in Jnl. of Applied Physics 61, n8 p4100-4102 1987.

Keywords: *Neutron scattering, *Spin orbit interactions, *Terbium, *Iron, *Ferromagnetism, Glass, Metal alloys, Temperature, Lorentz transformations, Rare earth elements, Phase transformations, Reprints.

Small angle neutron scattering measurements (SANS) have been made on Tb45Fe55 and Tb25Fe75 as a function of temperature. The SANS results show that long range ferromagnetic order is quenched in the alloys and is replaced by a spin glass-like state. For T greater than T(c), where T(c) is the transition temperature, both samples exhibit a conventional Lorentzian lineshape (in q), but they depart from the form below T(c). The low temperature lineshapes have been fitted with the Lorentzian plus Lorentzian-squared form appropriate for random field systems. As the temperature of the Tb45Fe55 alloy is lowered, the correlation length rises to a rounded maximum of 80 angstroms at 250 K (T(c) = 298 K) and decreases to about 60 angstroms at low T. In both alloys the coefficient of the Lorentzian term rises sharply as T approaches 0, whereas the Lorentzian-squared coefficient follows approximately the square of the order parameter divided by the correlation length. Both systems thus lend support to the suppression of long range order by the random anisotropy field.

901,150
PB89-201784 Not available NTIS
National Bureau of Standards (IMSE), Gaithersburg, MD. Ceramics Div.
Grain Boundary Structure in Ni3Al.
Final rept.
R. A. D. Mackenzie, M. D. Vaudin, and S. L. Sass. 1988, 6p
Contract DE-FG02-85ER45211 .
Sponsored by Department of Energy, Washington, DC.
Pub. in Jnl. de Physique 49, n10 pC5-227-C5-232 Oct 88.

Keywords: *Nickel alloys, *Aluminum, *Boron, *Grain boundaries, Separation, Stoichiometry, Electron microscopy, Single crystals, Additives, Diffraction, Crystal structure, Dislocations(Materials), Reprints.

The influence of boron segregation and non-stoichiometry on grain boundary structure in Ni3Al was studied by transmission and scanning electron microscopy techniques. Small angle twist and tilt boundaries were produced by hot pressing misoriented single crystals of both doped and undoped material. Dislocation structures were observed in both bicrystal and polycrystal grain boundaries. In most cases the grain boundary dislocations were found to have the expected a <100> lattice, however in one case dislocations with Burgers vector a/2 <110> have been observed. Using a SEM diffraction technique the frequency of occurrence of grain boundary types was examined and found to be unchanged by the addition of boron.

901,151
PB89-201982 Not available NTIS
National Bureau of Standards (IMSE), Gaithersburg, MD. Metallurgy Div.
In situ Observation of Particle Motion and Diffusion Interactions during Coarsening.
Final rept.
P. W. Voorhees, and R. J. Schaefer. 1987, 13p
Pub. in Acta Metallurgica 35, n2 p327-339 1987.

Keywords: *Metals, *Phase transformations, Drops(Liquids), Diffusion, Particles, Reprints, *Ostwald ripening, Coarsening.

In situ observation of the growth and Ostwald ripening of spherical second phase domains in a solid is reported. It was found that at relatively low (3%) volume fractions of coarsening phase, diffusional interactions between particles were sufficiently strong to alter significantly the individual particle coarsening rates from the theoretical predictions of Lifshitz and Slyozov and Wagner (LSW). As a result, the LSW theory was found to be an inadequate description of the coarsening behavior of the low volume fraction system. In addition, particle migration in the solid matrix during coarsening was observed. The experimental results were found to be qualitatively consistent with a theoretical analysis of particle migration due to interparticle diffusional interactions or nonuniform matrix concentration fields. The generality of the mechanism responsible for the particle migration implies that particle motion, and thus a time dependent spatial correlation function, during coarsening will occur to some extent in all systems undergoing first order phase transformations.

901,152
PB89-201990 Not available NTIS
National Bureau of Standards (IMSE), Gaithersburg, MD. Metallurgy Div.
Numerical Simulation of Morphological Development during Ostwald Ripening.
Final rept.
P. W. Voorhees, G. B. McFadden, R. F. Boisvert, and D. Meiron. 1988, 16p
Pub. in Acta Metallurgica 36, n1 p207-222 1988. .

Keywords: *Metals, *Microstructure, *Particles, *Spatial distribution, Diffusion, Separation, Interfaces, Sintering, Liquids, Reprints, *Ostwald ripening, Coarsening.

A boundary integral technique is employed to determine the morphological evolution of small number of particles during Ostwald ripening in two dimensions. The approach specifically allows the bodies to change shape consistent with interparticle diffusional interactions and the interfacial concentrations as given by the Gibbs-Thomson equation. It is shown that the strong interparticle diffusional interactions which occur at small interparticle separations can induce significant motions of the centers of mass of the particles. Such motion is shown to be a strong function of the spatial distribution of particles. The generality of the mechanism responsible for the particle migration suggests that particle motion is a generic aspect of the ripening process at high volume fractions of coarsening phase. It was found that significant shape distortions of particles during ripening requires particle arrangements which induce significant diffusional interactions. Through particle arrangements similar to those found in solid-liquid systems during liquid phase sintering, it is shown that the formation regions of flat interface between particles is completely consistent with an Ostwald ripening mechanism.

901,153
PB89-202089 Not available NTIS
National Bureau of Standards (IMSE), Gaithersburg, MD. Ceramics Div.
Diffraction Effects Along the Normal to a Grain Boundary.
Final rept.
J. M. Vitek, M. D. Vaudin, M. Ruhle, and S. L. Sass. 1989, 5p
Sponsored by Department of Energy, Washington, DC.
Pub. in Scripta Metallurgica 23, p349-353 1989.

Keywords: *Metals, *Grain boundaries, Interfaces, Diffraction, Crystal structure, Separation, Kinematics, Grain structure, Reprints.

Much work has been done in recent years on studying the structure and properties of grain boundaries and interfaces in general in metals. Included in the studies is work done by the present authors on diffraction effects from grain boundaries along the reciprocal lattice direction passing through the origin normal to a planar boundary. The studies considered the simplified case of kinematical diffraction effects caused by distortions normal to the planar interface in the boundary region. It has been noted both in the literature and in personal communications that discrepancies existed between the results of Vaudin, Sass et al., and Vitek and Ruhle. It was felt that the most appropriate action would be to discuss these differences in a joint paper. The

clarifies and resolves many of these differences, including some issues that have not been raised previously in the literature, and identifies the remaining areas of disagreement.

901,154
PB89-218333 PC A03/MF A01
National Inst. of Standards and Technology (NEL), Boulder, CO. Chemical Engineering Science Div.
Ignition Characteristics of the Nickel-Based Alloy UNS N07718 in Pressurized Oxygen.
J. W. Bransford, P. A. Billiard, J. A. Hurley, K. M. McDermott, and I. Vazquez. Apr 89, 50p NISTIR-89/3911.
Sponsored by National Aeronautics and Space Administration, Huntsville, AL. George C. Marshall Space Flight Center.

Keywords: *Nickel alloys, *Combustion, *Ignition, Critical temperature, Flammability tests, Graphs(Charts), Tables(Data), UNS N07718, Pressurized oxygen.

The development of ignition and combustion in pressurized oxygen atmospheres was studied for the nickel-based alloy UNS N07718. Ignition of the alloy was achieved by heating the top. It was found that the alloy would autoheat to destruction from temperatures below the solidus temperature. In addition, endothermic events occurred as the alloy was heated, many at reproducible temperatures. Many endothermic events occurred prior to abrupt increases in surface temperature and appeared to accelerate the rate of increase in specimen temperature. It appeared that the source of some endotherms may increase the oxidation rate of the alloy. Ignition parameters are defined and the temperatures at which these parameters occur are given for the oxygen pressure range of 1.72 to 13.8 MPa (250 to 2000 psia).

901,155
PB89-226985 PC A03/MF A01
National Inst. of Standards and Technology (NEL), Gaithersburg, MD. Applied and Computational Mathematics Div.
Effect of Anisotropic Thermal Conductivity on the Morphological Stability of a Binary Alloy.
S. R. Coriell, G. B. McFadden, and R. F. Sekerka.
Aug 89, 24p NISTIR-89/4143
Prepared in cooperation with Carnegie-Mellon Univ., Pittsburgh, PA.

Keywords: *Bismuth alloys, *Tin alloys, *Anisotropy, *Thermal conductivity, *Crystal growth, *Solidification, Microstructure, Dispersions, Stability, Orientation, Oscillations, Graphs(Charts), Equations.

A linear morphological stability analysis of a planar interface during unidirectional solidification of a planar alloy was performed for the case of a crystal having an anisotropic thermal conductivity. A dispersion relation was calculated which shows that the onset of instability depends on the orientation of the growth direction with respect to crystallographic axes and on the orientation of the wave vector of the perturbation. The onset of instability can be either oscillatory (travelling waves) or non-oscillatory in time. For growth along a principal axis of the crystal there is an exchange of stabilities, and the onset of instability is non-oscillatory. The dispersion relation for a uniaxial crystal was explored in detail. Numerical results for the case of an alloy of 0.78 at % bismuth in tin are given.

901,156
PB89-229306 Not available NTIS
National Bureau of Standards (IMSE), Gaithersburg, MD. Ceramics Div.
Grain Boundary Characterization in Ni3Al.
Final rept.
R. A. D. Mackenzie, M. D. Vaudin, and S. L. Sass.
1988, 6p
Contract DE-FG02-85ER45211
See also DE88009169. Sponsored by Department of Energy, Washington, DC.
Pub. in Materials Research Society Symposia Proceedings 122, p461-466 1988.

Keywords: *Nickel alloys, *Intermetallics, *Boron, *Additives, *Grain boundaries, *Electron microscopy, Polycrystals, Bicrystals, Aluminum, Crystal dislocations, Nickel aluminides.

Grain boundaries in both pure and boron doped Ni3Al have been studied using a variety of electron microscopy techniques. Small angle boundary structures were examined in both bicrystal and polycrystalline specimens. Dislocations with Burgers vector a/

2 <110> are observed in the presence of boron, while dislocations with Burgers vector a<100> are observed in the absence of boron. A possible explanation for the behavior is the presence of a disordered layer at the interface in boron doped Ni3Al. The addition of boron to Ni3Al was seen to induce faceting of grain boundaries, and to afford the boundaries some protection from etching. Using electron backscatter diffraction patterns, the frequency of occurrence of grain boundary types was found to be unchanged by the addition of boron.

901,157
PB89-229314 Not available NTIS·
National Bureau of Standards (IMSE), Gaithersburg, MD. Ceramics Div.
Grain Boundary Structure in Ni3Al.
1988, 2p
See also DE88009164.
Pub. in Proceedings of the Annual Meeting of the Electron Microscopy Society of America (46th), San Francisco, CA., p602-603 Aug 88.

Keywords: *Nickel alloys, *Intermetallics, *Boron, *Ductility, *Bicrystals, *Aluminum, Intergranular corrosion, Transgranular corrosion, Polycrystals, Grain boundaries, Additives, Nickel aluminides, Flat surfaces. .

Ni3Alk is a potentially useful high temperature alloy. In its single crystal form it exhibits good ductility; however in polycrystalline form the pure alloy is highly prone to intergranular failure. It has been seen that in slightly nickel-rich alloys the addition of small amounts of boron has the effect of dramatically increasing the material ductility and of changing the failure mode from intergranular to transgranular. In alloys which have been ductilitized by boron addition, atom probe investigation has shown the boron to be segregated to grain boundaries. The segregation may induce a change in the boundary structure as has been seen by Sickafus and Sass in gold doped iron bicrystals.

901,158
PB89-231302 Not available NTIS
National Inst. of Standards and Technology (NML), Gaithersburg, MD. Surface Science Div.
Interaction of Oxygen and Platinum on W(110).
Final rept.
R. A. Demmin, and T. E. Madey. 1989, 7p
Sponsored by Department of Energy, Washington, DC.
Pub. in Jnl. of Vacuum Science and Technology A7, n3 p1954-1960 May/Jun 89.

Keywords: *Tungsten, *Platinum, *Adsorption, *Oxygen, *Thin films, *Annealing, Desorption, Electron diffraction, Auger electrons, Spectroscopy, Monomolecular films, Reprints.

Low-energy electron diffraction (LEED) and Auger spectroscopy have been used to characterize Pt overlayers coadsorbed with oxygen on W(110). Previous work has shown that multilayers of Pt alone will cluster into three-dimensional crystallites of bulk Pt when annealed, leaving a pseudomorphic monolayer of Pt covering the W(110) surface. When oxygen is present on the surface, the monolayer of Pt is no longer stable. Annealing a W(110) surface on which both Pt and oxygen are adsorbed causes nearly all of the Pt to agglomerate into clusters as the oxygen largely replaces the monolayer on the surface. The presence of the Pt clusters reduces the temperature required for desorption of oxygen, however, and the Pt spreads across the W surface as the oxygen is removed. The complex LEED patterns observed when coadsorbed oxygen and Pt are annealed and the changes in the temperature required for desorption of oxygen suggest that there is an interaction between the two adsorbates and that there is not complete phase separation; i.e., some Pt remains dispersed on the W surface within the oxygen phase.

901,159
PB90-117409 Not available NTIS
National Inst. of Standards and Technology (IMSE), Boulder, CO. Fracture and Deformation Div.
Texture Monitoring in Aluminum Alloys: A Comparison of Ultrasonic and Neutron Diffraction Measurements.
Final rept.
A. V. Clark, R. C. Reno, R. B. Thompson, J. F. Smith, G. V. Blessing, R. J. Fields, P. P. Delsanto, and R. B. Mignogna. 1988, 9p
Sponsored by Office of Naval Research, Arlington, VA.

Pub. in Ultrasonics 26, p189-197 Jul 88.

Keywords: *Aluminum alloys, *Texture, *Monitors, Ultrasonic tests, Neutron diffraction, Metal sheets, Reprints, Piezoelectric transducers, Electroacoustic transducers.

Theories have been developed by several authors to calculate velocities of bulk, guided and surface waves in polycrystalline aggregates of cubic metals. The theories can be used to predict the effect of texture on ultrasonic velocity in rolled aluminum and steel sheet, provided that the effects of dislocations, second-phase particles, inclusions, etc. can be ignored. The theories predict that ultrasonic velocities will be influenced by three orientation distribution coefficients (ODCs). The ODCs are quantitative measures of the texture in the material. In the work, the texture of thin sheets of a commercial grade aluminum alloy was measured with both ultrasonics and neutron diffraction. Several ultrasonic techniques were employed, using bulk, guided and surface waves. Both piezoelectric and electromagnetic-acoustic transducers (EMATs) were used. Quantitative measurements of texture made with different ultrasonic techniques were in good agreement. The ultrasonic measurements also agreed with neutron diffraction measurements, indicating that the dominant features of the effect of texture on wave propagation have been modelled with sufficient accuracy.

901,160
PB90-117607 Not available NTIS
National Inst. of Standards and Technology (IMSE), Boulder, CO. Fracture and Deformation Div.
Fourth-Order Elastic Constants of beta-Brass.
Final rept.
R. R. Rao, and H. Ledbetter. 1989, 4p
Pub. in International Jnl. of Thermophysics 10, n4 p899-902 Jul 89.

Keywords: *Brasses, *Elastic properties, Strains, Cauchy problem, Partial differential equations, Elastic theory, Crystal structure, Reprints, *Brass-beta.

Combinations of the fourth-order elastic constants of beta-brass were calculated using the measured second-order and third-order elastic constants and the expressions for the effective elastic constants of a cubic crystal obtained from finite-strain theory. The present calculations show that the Cauchy relations for the fourth-order elastic constants in beta-brass are not satisfied. This implies that noncentral or many-body forces occur in this material. The authors considered two alloys. The higher-Zn alloy shows lower magnitudes of the fourth-order elastic constants and a larger Cauchy discrepancy.

901,161
PB90-117664 Not available NTIS
National Inst. of Standards and Technology (IMSE), Boulder, CO. Fracture and Deformation Div.
Typical Usage of Radioscopic Systems: Replies to a Survey.
Final rept.
T. A. Siewert. 1989, 5p
Pub. in Materials Evaluation 47, p701-705 Jun 89.

Keywords: Real time operations, Aluminum, Steels, Surveys, Reprints, *Radioscopy, Image quality.

A program has been initiated at the National Institute of Standards and Technology to develop image quality indicators for real-time radioscopic systems. A survey was conducted to obtain some general parameters to guide the development; the report presents the results of the survey.

901,162
PB90-117755 Not available NTIS
National Inst. of Standards and Technology (IMSE), Gaithersburg, MD. Metallurgy Div.
Equilibrium Crystal Shapes and Surface Phase Diagrams at Surfaces in Ceramics.
Final rept.
C. A. Handwerker, M. D. Vaudin, and J. E. Blendell.
1988, 7p
Sponsored by Office of Naval Research, Arlington, VA.
Pub. in Jnl. de Physique 49, n10 pC5-367-C5-373 Oct 88.

Keywords: *Ceramics, *Surface properties, *Crystal structure, *Shape, *Thermodynamic equilibrium, Free energy, Phase transformations, Magnesium oxides, Nickel oxides, Reprints.

The equilibrium shape of a crystal is the shape that minimizes the total surface free energy of the crystal and may contain any or all of the following: facets, sharp edges, smoothly curved surfaces, and corners. The relationship between the surface free energy per unit area, gamma, and the equilibrium crystal shape is seen straightforwardly from the Wulff plot, the polar plot of gamma as a function of orientation of the normal vectors, n. The boundaries between surfaces on the equilibrium crystal (for example, between adjacent facet planes) have been described as surface phase transitions. If gamma(n) changes with temperature, pressure, or chemical potentials of the components, the equilibrium shape of the crystal may change. The change in equilibrium shape can be represented by a surface phase diagram with axes of surface orientation, in terms of angle from an arbitrary orientation, and temperature, pressure, or chemical composition. The surface phase transitions of MgO have been determined as a function of NiO solute concentration and surface phase diagrams have been constructed.

901,163
PB90-118084   Not available NTIS
National Inst. of Standards and Technology (IMSE), Gaithersburg, MD. Metallurgy Div.
Magnetic Behavior of Compositionally Modulated Ni-Cu Thin Films.
Final rept.
L. H. Bennett, L. J. Swartzendruber, D. S. Lashmore, R. Oberle, U. Atzmony, M. P. Dariel, and R. E. Watson. 1989, 5p
Pub. in Physical Review B 40, n7 p4633-4637, 1 Sep 89.

Keywords: *Copper nickel alloys, *Thin films, *Magnetic properties, Composition(Property), Measurement, Electrodeposition, Reprints, Compositional modulation.

Magnetic measurements on Ni-Cu compositionally modulated multilayers prepared by electrodeposition indicate less diffusion of Cu into the Ni than previously obtained by either electrodeposition or other means. The samples exhibit magnetic behavior much more closely resembling that of bulk Ni than has been seen previously for Ni-Cu multilayers. No dead layer is found.

901,164
PB90-123423   Not available NTIS
National Inst. of Standards and Technology (IMSE), Gaithersburg, MD. Metallurgy Div.
Temperature Hysteresis in the Initial Susceptibility of Rapidly Solidified Monel.
Final rept.
U. Atzmony, L. J. Swartzendruber, and L. H. Bennett. 1988, 4p
Pub. in Scripta Metallurgica 22, n5 p721-724 May 88.

Keywords: *Hysteresis, *Solidification, *Monel, *Alternating current, Thermal cycling tests, Annealing, Temperature, Magnetic permeability, Reprints, *Rapid quenching(Metallurgy), AC losses.

The temperature dependence of the ac susceptibility for a Cu-Ni alloy near the monel composition (28 at.% Cu) has been measured over temperature cycles of cooling down and warming up, for both the as-spun and after-thermal-annealing condition. A time-dependent temperature hysteresis was observed. A transformation between the two values can be abruptly induced by external disturbance.

901,165
PB90-123431   Not available NTIS
National Inst. of Standards and Technology (IMSE), Gaithersburg, MD. Metallurgy Div.
Magnetization and Magnetic Aftereffect in Textured Ni/Cu Compositionally-Modulated Alloys.
Final rept.
U. Atzmony, L. J. Swartzendruber, L. H. Bennett, M. P. Dariel, D. Lashmore, M. Rubinstein, and P. Lubitz. 1987, 10p
Pub. in Jnl. of Magnetism and Magnetic Materials 69, n3 p237-246 1987.

Keywords: *Magnetization, *Copper nickel alloys, Resonance absorption, Ferromagnetism, Magnetic measurement, Texture, Reprints, Aftereffect.

The magnetic properties of Ni/Cu compositionally-modulated alloys with (100), (110) and (111) textures were measured by magnetometry and ferromagnetic resonance. Unexpectedly, a magnetic aftereffect was found.

901,166
PB90-123514   Not available NTIS
National Inst. of Standards and Technology (IMSE), Gaithersburg, MD. Metallurgy Div.
TEM Observation of Icosahedral, New Crystalline and Glassy Phases in Rapidly Quenched Cd-Cu Alloys.
Final rept.
L. A. Bendersky, and F. S. Biancaniello. 1987, 6p
Pub. in Scripta Metallurgica 21, n4 p531-536 Apr 87.

Keywords: *Cadmium alloys, *Copper containing alloys, *Solidification, *Phase diagrams, Intermetallics, Crystal structure, Microstructure, Metastable state, Reprints, *Rapid quenching(Metallurgy), *Transmission electron microscopy, *Icosahedrons, Metallic glasses.

Since the discovery of an icosahedral phase in rapidly solidified Al-Mn alloys, similar phases have been found in many other systems. Generally, a successful approach was to rapidly solidify alloys known to form equilibrium intermetallic compounds with structure exhibiting extensive icosahedral (or polytetrahedral) clustering. For example, Frank-Kasper phases have entirely tetrahedral bonding of atoms with coordination numbers 12, 14, 15 and 16. According to the criteria the Cd-Cu system seems to be appropriate for forming the icosahedral phase. The central portion of the Cd-Cu phase diagram consists of three intermetallic phases with unit cells containing the icosahedral clusters. In the present work the microstructure of the rapidly solidified Cu-Cd alloys was studied by means of transmission electron microscopy. The purpose was to explore: (1) a possibility of the icosahedral and other metastable phase formation; and (2) solidification of complex crystal structures under conditions of rapid freezing.

901,167
PB90-123522   Not available NTIS
National Inst. of Standards and Technology (IMSE), Gaithersburg, MD. Metallurgy Div.
Amorphous Phase Formation in Al70Si17Fe13 Alloy.
Final rept.
L. A. Bendersky, F. S. Biancaniello, and R. J. Schaefer. 1987, 4p
Pub. in Jnl. of Materials Research 2, n4 p427-430 Jul/Aug 87.

Keywords: *Aluminum alloys, *Phase transformations, *Silicon containing alloys, *Iron containing alloys, Microstructure, Solidification, Reprints, *Amorphous state, *Rapid quenching(Metallurgy).

The alloy Al70Si17Fe13 was subjected to a range of rapid solidification conditions and the resulting microstructures were evaluated. It was found that when solidification was sufficiently rapid to bypass the formation of primary intermetallic phases, the alloy consisted of spherical regions of amorphous (or micro-quasicrystalline) material surrounded by a crystalline phase(s). The microstructure is interpreted as the result of solidification of the amorphous phase from the melt by a first-order transformation. The structure of the amorphous phase is different from that of a liquid (or usual metallic glass).

901,168
PB90-123530   Not available NTIS
National Inst. of Standards and Technology (IMSE), Gaithersburg, MD. Metallurgy Div.
Solidification of an 'Amorphous' Phase in Rapidly Solidified Al-Fe-Si Alloys.
Final rept.
L. A. Bendersky, M. J. Kaufman, W. J. Boettinger, and F. S. Biancaniello. 1988, 4p
Pub. in Materials Science and Engineering 98, p213-216 Feb 88.

Keywords: *Solidification, *Aluminum alloys, *Iron containing alloys, *Silicon containing alloys, Crystal structure, Phase transformations, Microstructure, Metastable state, Reprints, *Amorphous state, *Rapid quenching(Metallurgy).

The focus of the work is the amorphous phase formation in Al-Fe-Si alloys. Depending on the concentration of Fe and Si, the phase appears either as an interdendritic constituent or as primary phase with globular morphology. Thus the amorphous phase acts like a normal crystalline phase. The globular morphology of the amorphous phase suggests the possibility of a metastable liquid miscibility gap. However there is no thermodynamic evidence to support a positive heat of

mixing for the liquid phase. Another possibility suggests crystallization of a phase structurally different from liquid but at the same time being neither crystalline nor quasicrystalline. A possible structural relationship among the amorphous phase, the cubic alpha(AlFeSi) and to quasicrystalline phases is discussed.

901,169
PB90-123548   Not available NTIS
National Inst. of Standards and Technology (IMSE), Gaithersburg, MD. Metallurgy Div.
Quasicrystals and Quasicrystal-Related Phases in the Al-Mn System.
Final rept.
L. A. Bendersky. 1988, 4p
Pub. in Materials Science and Engineering 99, p331-334 Mar 88.

Keywords: *Aluminum manganese alloys, *Crystal structure, *Phase transformations, Electron diffraction, Reprints, Icosahedrons, Amorphous state.

The Al-Mn system is very rich in phases with composition close to Al4Mn. As solidification conditions change from very slow (casting) to extremely fast (atomized submicron size droplets) the following phases will form: hexagonal mu phase (a = 1.995 nm, c = 2.452 nm), hexagonal lambda phase (a = 2.841 nm, c = 1.238 nm), decagonal quasicrystal, icosahedral quasicrystal, and microquasicrystalline or 'amorphous' phase. In the present work, the potential interrelationship between the structures of the crystalline, quasicrystalline and amorphous Al4Mn phases is investigated by a systematic study of electron diffraction intensities. Analysis of electron diffraction intensity modulations and the spatial relationships suggests that the phases have a structural skeleton of icosahedral units, possibly of Mackay icosahedron type. Different crystalline and quasiperiodic phases can be formed by different stackings of the same icosahedral clusters and they are not necessarily in a single orientation. The amorphous structure can be described as a network of randomly oriented clusters.

901,170
PB90-123621   Not available NTIS
National Inst. of Standards and Technology (IMSE), Gaithersburg, MD. Metallurgy Div.
Microstructural Variations in Rapidly Solidified Alloys.
Final rept.
W. J. Boettinger. 1988, 8p
Pub. in Materials Science and Engineering 98, p123-130 Feb 88.

Keywords: *Microstructure, *Alloys, *Solidification, Surface reactions, Powder metals, Atomizing, Reprints, *Rapid quenching(Metallurgy).

Depending on alloy composition and cooling conditions, rapid solidification can produce a wide variety of microstructures. Examples are presented from material produced by three common methods of rapid solidification: surface melting and resolidification, substrate quenching, and powder atomization. The examples illustrate the microstructural similarities and differences that can occur using the methods.

901,171
PB90-123639   Not available NTIS
National Inst. of Standards and Technology (IMSE), Gaithersburg, MD. Metallurgy Div.
Rapid Solidification and Ordering of B2 and L2 (sub 1) Phases in the NiAl-NiTi System.
Final rept.
W. J. Boettinger, L. A. Bendersky, F. S. Biancaniello, and J. W. Cahn. 1988, 4p
Pub. in Materials Science and Engineering 98, p273-276 1988.

Keywords: *Nickel alloys, *Aluminides, *Titanium intermetallics, *Phase transformations, Solidification, Thermodynamics, Metastable state, Reprints, Heusler alloys, Rapid quenching(Metallurgy).

Evidence is presented for the direct solidification of the B2 phase in the NiAl-NiTi system at some compositions where the L2sub1 phase is stable at the melting point. Subsequent continuous ordering produces the equilibrium phase. The metastable continuous ordering curve is used to discuss the result.

MATERIALS SCIENCES

Nonferrous Metals & Alloys

901,172
PB90-123647   Not available NTIS
National Inst. of Standards and Technology (IMSE),
Gaithersburg, MD. Metallurgy Div.
Formation of Dispersoids during Rapid Solidification of an Al-Fe-Ni Alloy.
Final rept.
W. J. Boettinger, L. A. Bendersky, R. J. Schaefer,
and F. S. Biancaniello. 1988, 7p
Pub. in Metallurgical Transactions A-Physical Metallurgy and Materials Science 19, n4 p1101-1107 1988.

Keywords: *Dispersions, *Aluminum alloys, *Iron containing alloys, *Nickel containing alloys, *Solidification, Microstructure, Eutectics, Cells, Velocity, Reprints, *Rapid quenching(Metallurgy).

Examination of the effect of rapid solidification velocity on the microstructure of Al-3.7 wt% Ni-1.5 wt% Fe has revealed a new mechanism for the formation of discrete second phase particles in rapidly solidified alloys. Cellular growth of alpha-Al occurs with the intercellular phase, Al9(Fe,Ni)2 in two distinct morphologies. At low velocity (<50 cm) the phase is continuous in the growth direction while at higher velocity discrete rounded particles are observed. Analysis of the orientation relationship and the number of variants that exists between phases leads to a mechanism where liquid droplets are deposited by the moving cellular interface. The droplets solidified subsequently to form the rounded second phase particles.

901,173
PB90-123779   Not available NTIS
National Inst. of Standards and Technology (IMSE),
Gaithersburg, MD. Metallurgy Div.
Pathways for Microstructural Development in TiAl.
Final rept.
J. A. Graves, L. A. Bendersky, F. S. Biancaniello, J. H. Perepezko, and W. J. Boettinger. 1988, 4p
Pub. in Materials Science and Engineering 98, p265-268 Feb 88.

Keywords: *Microstructure, *Titanium intermetallics, *Aluminides, Alloys, Solidification, Crystal structure, Phase diagrams, Transmission electron microscopy, Powder metallurgy, Metastable state, Reprints, *Rapid quenching(Metallurgy).

Rapid solidification processing (RSP) of intermetallic alloys can provide alternative solidification paths and lead to the formation of metastable products. For the intermetallic TiAl with an equilibrium Li sub 0 structure RSP has yielded a metastable hcp (alpha-Ti) phase in both fine powder and melt spun ribbon. Based upon transmission electron microscopy (TEM) observations of a fine anti-phase domain structure the hcp phase orders to the Ti3Al (DO sub 19) structure following solidification. A metastable phase diagram analysis indicates that a melt undercooling of about 100 C is required for partitionless formation of a hcp phase from an equiatomic melt and reveals other possible solidification pathways including the nucleation of metastable disordered bcc and fcc phases. Each of the potential pathways offers further opportunity for microstructural modification by solid state annealing treatments.

901,174
PB90-128125   Not available NTIS
National Inst. of Standards and Technology (NML),
Gaithersburg, MD. Surface Science Div.
Role of Adsorbed Gases in Metal on Metal Epitaxy.
Final rept.
W. F. Egelhoff, and D. A. Steigerwald. 1989, 7p
Pub. in Jnl. of Vacuum Science and Technology A 7,
n3 p2167-2173 May/Jun 89.

Keywords: *Adsorption, *Gases, *Transition metals, *Epitaxy, Surface properties, Thin films, Carbon monoxide, Water, Copper, Nickel, Iron, Silver, Monomolecular films, Reprints.

It is found that a variety of (deliberately) adsorbed gases influence in some manner the epitaxial growth of metals on metals. The adsorbed gases investigated included molecularly adsorbed CO and H2O and dissociatively adsorbed H, O, N, C, and S. The effects of adsorbed gases have been investigated in metal-on-metal epitaxial systems including Cu/Cu(100), Ni/Ni(100), Cu/Ni(100), Ni/Cu(100), Fe/Cu(100), Cu/Fe(100), and Fe/Fe(100). Although not all of the possible gas-metal combinations among the systems have been studied, enough have been to make the following generalizations. Around room temperature and above, the gases exhibit a strong tenden-

cy to 'float out' to the growing surface, hardly reducing the extent of epitaxial ordering. All but the most strongly bound, e.g., C or N, have a strong tendency to float out even during epitaxy at temperatures as low as 100 K. The most strongly bound, e.g., C, N, or O, tend to suppress the agglomeration or interdiffusion (which otherwise occurs) when a monolayer of a transition metal is deposited on a noble metal at room temperature or above, e.g., Fe/Cu(100) or Ni/Cu(100): In some cases, the effects may be useful as a new tool for gaining improved control over epitaxial growth.

901,175
PB90-128174   Not available NTIS
National Inst. of Standards and Technology (IMSE),
Gaithersburg, MD. Metallurgy Div.
Diffusion-Induced Grain Boundary Migration.
Final rept.
C. Handwerker. 1989, 4p
Pub. in McGraw-Hill Yearbook of Science and Technology, p155-158 1989.

Keywords: *Diffusion, *Grain boundaries, *Alloying, Metals, Solid solutions, Phase diagrams, Solutes, Polycrystalline, Thin films, Reprints.

When large grain-sized, well-annealed polycrystalline materials are exposed to a solute source and the solute diffuses down the grain boundaries, it is frequently observed that grain boundaries migrate, sometimes away from the centers of curvature, and aligns form in the regions swept by the moving grain boundaries. This newly recognized phenomenon is known as diffusion-induced grain boundary migration (DIGM) since the solute diffusion induces otherwise stable grain boundaries to move. Grain boundaries are placed in contact with solute sources in most engineering systems, for example, in thin film multi-layer contacts on computer circuit boards. The technological and economic impact of the unexpected breakdown of the systems by DIGM is enormous.

901,176
PB90-128604   Not available NTIS
National Inst. of Standards and Technology (IMSE),
Boulder, CO. Fracture and Deformation Div.
Effects of Grain Size and Cold Rolling on Cryogenic Properties of Copper.
Final rept.
R. P. Reed, R. P. Walsh, and F. R. Fickett. 1988,
10p
Sponsored by Department of Energy, Washington, DC.
Pub. in Advances in Cryogenic Engineering Materials,
v34 p299-308 1988.

Keywords: *Cryogenics, *Copper, *Grain size, *Cold rolling, Electrical resistivity, Yield strength, Tensile stress, Ultimate strength, Microstructure, Reprints.

The effects of grain size and cold rolling on the tensile properties and electrical resistivity at 295, 76 and 4 K were studied for oxygen-free, high-conductivity copper. Tensile-yield and ultimate strengths increase linearly with increasing 1/sq rt d (d = grain diameter), following the Hall-Petch relationship. At low temperatures, the dependence on grain size increases. Increasing grain size lowers resistivity slightly at all temperatures. Cold rolling to 10 percent reduction of area significantly increases the yield strength at all temperatures; subsequent rolling produces smaller strength increases. Resistivity increases with cold rolling.

901,177
PB90-128737   Not available NTIS
National Inst. of Standards and Technology (IMSE),
Boulder, CO. Fracture and Deformation Div.
Fatigue Resistance of a 2090-T8E41 Aluminum Alloy at Cryogenic Temperatures.
Final rept.
R. L. Tobler, J. K. Han, and R. P. Reed. 1989, 10p
Sponsored by Martin Marietta Aerospace, New Orleans, LA.; and Department of Energy, Washington, DC.
Pub. in Proceedings of International Cryogenic Materials Conference (1988), Los Angeles, CA., July 23-28,
1989, v2 p703-712.

Keywords: *Fatigue life, *Cryogenics, *Aluminum alloys, Crack propagation, Lithium containing alloys, Copper containing alloys, *Alloy 2090-T8E41, *Alloy 2014-T6.

Smooth-bar axial fatigue-life measurements were performed to evaluate the fatigue resistance of a 2090-T8E41 aluminum-copper-lithium alloy at cryogenic temperatures. Conventional S-N curves with failures

occuring between 10(sup 4) and 10(sup 6) cycles are presented with similar data for a 2014-T6 aluminum alloy. For specimens in the longitudinal orientation, the 2090 alloy performed better than 2014 in the low-cycle fatigue range at 295, 76, and 4 K. Factors contributing to the results are discussed.

901,178
PB90-136862   Not available NTIS
National Bureau of Standards (NML), Gaithersburg,
MD. Inorganic Analytical Research Div.
Technical Examination, Lead Isotope Determination, and Elemental Analysis of Some Shang and Zhou Dynasty Bronze Vessels.
Final rept.
I. L. Barnes, W. T. Chase, L. L. Holmes, E. C. Joel,
P. Meyers, and E. V. Sayre. 1986, 11p
Pub. in Proceedings of Conference on the Beginning of the Use of Metals and Alloys, Zhengzhou, China, October 21-26, 1986, p296-308 1988.

Keywords: *Lead isotopes, *Archaeology, *Mass spectroscopy, *Bronzes, *Containers, China, Microscopy, Age determination, Radiography, Nondestructive tests, Neutron activation analysis, Absorption spectroscopy, Antiquities, Artifacts.

The Arthur M. Sackler Collection of works of art contains several hundred fine early ritual bronze vessels. Among the vessels, 106 attributed to the Shang Dynasty, 127 attributed to the Western Zhou and 94 attributed to the Eastern Zhou Dynasty have been given a thorough technical examination. The examination has included x-ray radiography viewing under ultraviolet light, overall visual examination and examination of details under a binocular microscope. The metal of most of the vessels also has been analyzed for lead isotope ratios by means of mass spectrometry, for the major component by atomic absorption spectrometry, and for some minor and trace elemental concentrations by instrumental neutron activation analysis. The vessels may be divided into 11 groups by one particular statistical analysis of the data. The division may indicate the use of different sources of raw materials, although the locations of the different sources is not known at the present time. Data analysis using both chemical and isotopic information is continuing as is the search for samples from sources of ores which might have been used in ancient times.

Plastics

901,179
PB89-146706   Not available NTIS
National Bureau of Standards (IMSE), Gaithersburg,
MD. Polymers Div.
Effects of Space Charge on the Poling of Ferroelectric Polymers.
Final rept.
A. S. DeReggi, and M. G. Broadhurst. 1987, 11p
Pub. in Ferroelectrics 73, n3-4 p351-361 Jun 87.

Keywords: *Ferroelectric materials, *Polymers, *Thermal measurements, *Polarization, Poles, Electrical resistivity, Lasers, Charge density, Reprints, Vinylidene fluoride resins.

Thermal pulse measurements of the polarization distribution in ferroelectric polymers and copolymers after poling give distributions concentrated to one side when the electrical conductivity under poling conditions is significantly different from zero. These results could be explained if net charge were present in the material during poling. The direct observation of distributed charge in nonpolar polymers after exposure to poling conditions support this explanation. However, high resolution laser thermal pulse measurements on nominally well poled samples of polyvinylidene fluoride reveal a sharp drop of polarization to near zero value at a depth of about 1 micrometer from the surfaces which is not explained.

901,180
PB89-147821   PC A03/MF A01
National Bureau of Standards (NEL), Gaithersburg,
MD. Center for Building Technology.

138

Epoxy Impregnation Procedure for Hardened Cement Samples.
Progress rept.
L. Struble, and P. Stutzman. May 88, 19p NBSIR-88/3702
Sponsored by Air Force Office of Scientific Research, Bolling AFB, DC.

Keywords: *Epoxy resins, *Impregnating; *Cements, *Ethanols, *Curing, Viscosity, Cracks, Microscopy, Porosity, Crosslinking, Plastics processing.

A method was previously developed for epoxy impregnation of hydrated cementitious materials for microscopical examination without drying the samples, by sequentially replacing pore solution with ethanol, then the ethanol with epoxy. During subsequent application of the procedure, many specimens were cured. Studies were carried out to identify the cause of these problems and to modify the procedure for more reliable impregnation. Contamination with low levels (4%) of water or ethanol was found to prevent proper curing. Modifications in the procedure to prevent contamination, including monitoring the replacement of pore solution by ethanol, were shown to provide consistent and reliable impregnation.

901,181
PB89-157101 Not available NTIS
National Bureau of Standards (IMSE), Gaithersburg, MD. Polymers Div.
Polymers Bearing Intramolecular Photodimerizable Probes for Mass Diffusion Measurements by the Forced Rayleigh Scattering Technique: Synthesis and Characterization.
Final rept.
Q. Tran-Cong, T. Chang, and C. C. Han. 1988, 10p
Pub. in Polymer 29, p2261-2270 Dec 88.

Keywords: *Anthracene, *Photochemical reactions, *Synthesis, *Photochromism, *Probes, *Rayleigh scattering, *Polystyrene, Irradiation, Ultraviolet radiation, Solvents, Carbon tetrachloride, Temperature, Refractivity, Polymers, Mixtures, Diffusion, Dyes, Reprints.

A new photochromic probe, bis(9-anthryl methyl)ether(BAME) derivative, was synthesized and introduced as an effective probe for forced Rayleigh scattering (FRS) measurement. It is shown that BAME and polystyrene labelled with BAME exhibit a large change in refractive index under irradiation of u.v. light (363.8 nm). The self-diffusion of BAME and polystyrene labelled with BAME (PSA) were measured in various solvents. Results of PSA in good, marginal and poor solvents are consistent with those obtained from quasi-elastic light scattering (QELS). Since the photodimerization reaction of anthracene has been extensively studied in solution and crystal as well as in polymer matrix, unfavorable multistep photochemical reactions can be avoided. One such unfavorable case, which involves a fluorescence quenching solvent CCl4, is demonstrated. A temperature dependence study of polystyrene in semi-dilute theta solution by FRS has suggested that the non-exponential intensity decay is due to large concentration fluctuations in the solution. The well known photochemistry, large refractive index change and temperature stability have made BAME a very promising probe for FRS measurements.

901,182
PB89-157119 Not available NTIS
National Bureau of Standards (IMSE), Gaithersburg, MD. Polymers Div.
Phase Contrast Matching in Lamellar Structures Composed of Mixtures of Labeled and Unlabeled Block Copolymer for Small-Angle Neutron Scattering.
Final rept.
Y. Matsushita, Y. Nakao, R. Saguchi, K. Mori, H. Choshi, Y. Muroga, I. Noda, M. Nagasawa, T. Chang, C. J. Glinka, and C. C. Han. 1988, 5p
Pub. in Macromolecules 21, n6 p1802-1806 1988.

Keywords: *Polystyrene, *Mixtures, *Neutron scattering, *Lamellar structure, Isotopic labelling, Polymers, Deuterium, Molecular weight, Vinyl resins, Phase, Pyridines, Reprints.

To extract the single-chain scattering function of polystyrene block chain in lamellar structures of styrene-2-vinylpyridine diblock copolymers, the method of 'phase contrast matching' was studied for small-angle neutron scattering from blends of the deuterium-labeled and unlabeled block copolymers. The phase contrast matching is successfully applied for the samples with

the lowest molecular weights (3.4 x 10 sup 4 for the labeled portions) but not for the samples with the higher molecular weights (9.2 and 16.2 x 10 sup 4). It is concluded that the mismatching may be caused by concentration fluctuation in the mixture of hydrogenated and deuteriated polystyrenes in domains, as well as by nonuniform distribution of deuteriated species along the direction perpendicular to the lamellae due to the difference in lengths of the labeled and unlabeled blocks.

901,183
PB89-173942 Not available NTIS
National Bureau of Standards (IMSE), Gaithersburg, MD. Polymers Div.
Dynamics of Concentration Fluctuation on Both Sides of Phase Boundary.
Final rept.
C. C. Han, M. Okada, and T. Sato. 1988, 11p
Pub. in Dynamics of Ordering Processes in Condensed Matter, p433-443 1988.

Keywords: *Polymers, *Boundary layer flow, *Binary systems(Materials), Concentration(Composition), Temperature, Transition flow, Blends, Diffusion coefficient, Reprints.

The authors have shown for PSD/PVME polymer blend system that interfacial free energy is small and consistent with theoretical prediction, which does not play an important role in the dynamics of concentration fluctuations in either one phase or two phase regions. The interdiffusion coefficient is continuous at the phase separation boundary. Furthermore, the mobility M has been extracted and an Arrhenius type of temperature dependence has been found. To the best of their knowledge, this is the first time that M has been obtained on both sides of the phase boundary, and the static results have been incorporated into the kinetics study for a consistent evaluation of the Cahn-Hilliard-Cook theory.

Wood & Paper Products

901,184
PB89-172530 Not available NTIS
National Bureau of Standards (NEL), Gaithersburg, MD. Chemical Process Metrology Div.
Laser Induced Fluorescence for Measurement of Lignin Concentrations in Pulping Liquors.
Final rept.
J. J. Horvath, H. G. Semerjian, K. L. Biasca, and R. Attala. 1988, 10p
Pub. in Proceedings of SPIE (Society of Photo-Optical Instrumentation Engineers) Industrial Optical Sensing, Dearborn, MI., June 29-30, 1988, v961 p68-77.

Keywords: *Fluorescence, *Lasers, *Lignin, *Process control, Pulping, Dyes, Absorption spectra.

Laser excited fluorescence of pulping liquors was investigated for use in the pulp and paper industry for process measurement and control applications. A Nd-YAG pumped dye laser was used to generate the excitation wavelength of 280 nm; measurements were also performed using a commercially available fluorometer. Measurements on mill pulping liquors gave strong signals and showed changes in the fluorescence intensity during the cook. Absorption spectra of diluted mill liquor samples showed large changes during the cook. Samples from well controlled and characterized laboratory cooks showed fluorescence to be linear with concentration over two decades with an upper limit of approximately 1000 ppm dissolved lignin. At the end of these cooks, a possible chemical change was indicated by an increase in the observed fluorescence intensity. Results indicate that lignin concentrations in pulping liquors can be accurately determined with fluorescence in the linear optical region over a greater dynamic range than absorption spectroscopy.

901,185
PB89-212039 Not available NTIS
National Bureau of Standards (NEL), Gaithersburg, MD. Fire Measurement and Research Div.
Prediction of the Heat Release Rate of Douglas Fir.
Final rept.
Y. A. Parker. 1989, 10p
See also PB87-131819.
Pub. in Proceedings of International Symposium on Fire Safety Science (2nd), Tokyo, Japan, June 13-17, 1989, p337-346.

Keywords: *Douglas fir wood, *Thermal diffusivity, *Fire tests. Heat of combustion, Heat flux, Specific heat, Thermochemical properties, Combustion.

Measurements have been made on the thermal diffusivity of Douglas fir and its char up to 550 deg C. Its char contraction factors have also been determined. Interpretation of some data in the literature has resulted in the establishment of the specific heat as a function of temperature over this range. These thermophysical property data along with some data reported separately on the thermochemical properties of cellulose, mannan, xylan and lignin were used as input to a model for the heat release rate of wood in order to calculate the heat release rate and heat of combustion of Douglas fir exposed to an external radiant flux of 25 kW/sq m. These calculations were compared with measurements made in the Cone calorimeter. The agreement is reasonable at this stage of the model development.

General

901,186
PB89-175194 PC A06
National Inst. of Standards and Technology, Gaithersburg, MD.
Journal of Research of the National Institute of Standards and Technology, Volume 94, Number 1, January-February 1989. Special Issue: Numeric Databases in Materials and Biological Sciences.
Bi-monthly rept.
1989, 104p
Also available from Supt. of Docs. as SN703-027-00026-1. See also PB89-175202 through PB89-175293 and PB89-133367.

Keywords: *Biology, Thermodynamics, Chemical analysis, Ceramics, Crystallography, Proteins, Electron diffraction, Mass spectroscopy, Phase diagrams, *Materials science, *Numerical data bases.

The special issue on Numeric Databases in Materials and Biological Sciences includes the following articles: The importance of numeric databases to materials science; NIST/Sandia/ICDD electron diffraction database: A database for phase identification by electron diffraction; Numeric databases in chemical thermodynamics at the National Institute of Standards and Technology; Numeric databases for chemical analysis; The structural ceramics database: technical foundations; Applications of the crystallographic search and analysis system CRYSTDAT in materials science; New directions in bioinformatics; The use of structural templates in protein backbone modeling; Comparative modeling of protein structure—progress and prospects; and The computational analysis of protein structures: Sources, methods, systems, and results.

901,187
PB89-175202
(Order as PB89-175194, PC A06)
A.T. and T. Bell Labs., Murray Hill, NJ.
Importance of Numeric Databases to Materials Science.
Bi-monthly rept.
R. A. Matula. 1989, 4p
Included in Jnl. of Research of the National Institute of Standards and Technology, v94 n1 p9-14 Jan-Feb 89.

Keywords: *Industries, Fiber optics, Crystallography, Crystal lattices, Ferroelectricity, Superconductivity, *Materials science, *Numerical data bases.

Scientific numeric databases are important research tools for materials scientists. In distinction to bibliographic databases, these numeric databases are useful primarily to provide direct, immediate access to data, often evaluated data. Examples showing the application of crystallographic databases are given, including determining candidate materials for certain applications. Thermochemical data useful for optimizing optical fiber processing are discussed showing the importance of high-quality data. In addition, these databases are an important tool that can be utilized in the graduate education of the next generation of materials scientists.

901,188
PB89-179683 Not available NTIS

General

National Bureau of Standards .(NEL), Gaithersburg, MD. Thermophysics Div.
Computer Model of a Porous Medium.
Final rept.
R. A. MacDonald. 1988, 9p
Pub. in International Jnl. of Thermophysics 9, n6 p1061-1069 Nov 88.

Keywords: *Porous materials, *Computer systems programs, Diffusion, Channels, Porosity, Percolation, Random walk, Mathematical models, Reprints.

A computer model has been set up to represent a porous medium. The basis for this model is a two-dimensional square network (100 x 100) of channels that have randomly assigned widths between the value of zero (closed) and the value of one (open, unrestricted flow). The channel width assignments have been made by a random selection from five different distributions: f(q)=q, f(q)=si q, f(q)=erf(q), f(q)=1 - sin q, and f(q)=1-erf(q). Diffusion of particles in the network has been studied by a random-walk procedure for each realization of the channel width assignments. The diffusivity is quite sensitive to the distribution of channel widths. The percolation properties of the networks obtained from the three most restrictive distributions have been investigated and the independent, linked clusters within the network have been determined. For clusters sizes that are less than the full width of the network, the network does not percolate and either the flow is not diffusive or the diffusivity is severely reduced. An approximate value for the percolation threshold has been determined in each case and the fractal dimension has been calculated also.

901,189
PB89-211932 Not available NTIS
National Bureau of Standards (IMSE), Gaithersburg, MD. Metallurgy Div.
Mossbauer Spectroscopy.
Final rept.
L. J. Swartzendruber, and L. H. Bennett. 1986, 9p
Pub. in Metals Handbook, v10 p287-295 1986.

Keywords: *Mossbauer effect, *Metals, Spectroscopy, Scattering, Gamma rays, Fluorescence, Reprints.

A brief introduction to Mossbauer spectroscopy for the materials engineer is presented.

901,190
PB89-212237 Not available NTIS
National Bureau of Standards (IMSE), Gaithersburg, MD.
Materials Failure Prevention at the National Bureau of Standards.
Final rept.
L. H. Schwartz, and D. B. Butrymowicz. 1986, 14p
Pub. in Proceedings of International Conference and Exposition on Fatigue, Corrosion Cracking, Fracture Mechanics and Failure Analysis, Salt Lake City, UT., December 2-6, 1985, p1-14 1986.

Keywords: *Fractures(Materials), Measurement, Failure, Prevention, *Technology transfer, National Institute of Standards and Technology.

As a Commerce Department agency, the National Bureau of Standards (NBS) provides the measurement foundation that the national industrial economy needs. Crucial to the needs are the safe, efficient, and economical use of materials. The NBS programs that support generic technologies in materials and the mechanisms by which fundamental information is transferred are analyzed. Specific examples are drawn from recent developments in fracture of materials.

901,191
PB89-228332 PC A13/MF A01
National Inst. of Standards and Technology, Gaithersburg, MD.
International Cooperation and Competition in Materials Science and Engineering.
L. H. Schwartz, and S. J. Schneider. Jun 89, 280p
NISTIR-89/4041

Keywords: *Engineering, *Research management, Graphs(Charts), Competition, *Materials science, international cooperation, Coordinated research programs.

In 1986, the National Research Council commissioned a Committee on Materials Science and Engineering (COMMSE) to conduct a comprehensive study of the field, to define its progress, assess needs and opportunities and provide policy guidance at the national level.

A Summary Report of COMMSE was published in 1989; it was based primarily on informational inputs generated by five separate panels, each charged to investigate a different aspect of Materials Science and Engineering (MSE). The report documents the results of the individual study conducted by the COMMSE Panel 3 on International Cooperation and Competition in MSE. It deals with many facets of MSE, as practiced in other countries, and in the United States. It surveys national policies and programs for science and technology and MSE, elaborates on administrative structures to carry out R&D, and provides comparisons between the United States and the major industrial nations of the world. Much of the content revolves around the theme of industrial competitiveness as influenced by cooperative R&D.

MATHEMATICAL SCIENCES

Algebra, Analysis, Geometry, & Mathematical Logic

901,192
AD-A201 256/5 PC A03/MF A01
Maryland Univ., College Park.
Error Bounds for Linear Recurrence Relations.
F. W. Olver. Apr 88, 18p ARO-20806.9-MA
Contract DAAG29-84-K-0022, Grant NSF-DMS84-19820
Pub. in Mathematics of Computation, v50 n182 p481-499 Apr 88.

Keywords: Arithmetic, *Numerical analysis, Oscillation, *Statistical processes, Linearity, Monotone functions, *Linear recurrence relations.

Recurrence relations of a certain form are examined in two cases. In both cases, a posteriori methods are supplied for constructing strict and realistic error bounds in O (r) arithmetic operations. A priori bounds, also requiring O (r) arithmetic operations, are supplied in Case B. Several illustrative numerical examples are included. Keywords: Numerical analysis, Oscillatory systems, Monotonic systems.

901,193
PB89-143283 PC A03/MF A01
National Inst. of Standards and Technology (NEL), Gaithersburg, MD. Center for Computing and Applied Mathematics.
Finite Unions of Closed Subgroups of the n-Dimensional Torus.
J. Lawrence. Aug 88, 24p NISTIR-88/3777
Prepared in cooperation with George Mason Univ., Fairfax, VA.

Keywords: *Lattices(Mathematics), Analytic geometry, Optimization, *Toruses, Polytopes, Subgroups.

Let U be an open subset of the torus group T sup n. We show that the set of maximal subgroups of T sup n which miss U is of finite cardinality. This result is applied to show that the lattice of finite unions of closed subgroups of T sup n is a complete distributive lattice, and to show that, up to unimodular equivalence, there are only finitely many convex polytopes P R sup n having vertices in Z sup n but no interior points in Z sup n and such that each subgroup G of the additive group R sup n which properly contains Z sup n does have points in common with the interior of P.

901,194
PB89-147425 Not available NTIS
National Bureau of Standards (NEL), Gaithersburg, MD. Scientific Computing Div.
Mathematical Software: PLOD. Plotted Solutions of Differential Equations.
Final rept.
E. Agron, I. L. Chang, G. Gunaratna, D. K. Kahaner, and M. Reed. 1988, 6p
Pub. in IEEE (Institute of Electrical and Electronics Engineers) Micro 8, n4 p56-61 Aug 88.

Keywords: *Ordinary differential equations, *Problem solving, Plotting, Reprints, *Computer software, interactive systems, IBM-PC computers.

PLOD is an acronym for Plotted solutions of Differential Equations. PLOD can solve up to 25 ordinary differential equations with 10 parameters. It is entirely interactive, requiring very little programming experience. PLOD uses state-of-the-art numerical methods to perform the integration. The current version of PLOD has been implemented on the IBM Personal Computer family, including the XT, and AT. PLOD is in the public domain.

901,195
PB89-158166 Not available NTIS
National Bureau of Standards (NEL), Gaithersburg, MD. Scientific Computing Div.
Numerical Evaluation of Certain Multivariate Normal Integrals.
Final rept.
A. Genz, and D. K. Kahaner. 1986, 4p
Pub. in Jnl. of Computational and Applied Mathematics 16, n2 p255-258 Oct 86.

Keywords: Multivariate analysis, Numerical quadrature, Iteration, Computation, Numerical integration, Reprints, *Integrals.

It is shown that a multivariate normal integral with tridiagonal covariance matrix can be computed efficiently using iterated integration.

901,196
PB89-171623 Not available NTIS
National Bureau of Standards (NEL), Gaithersburg, MD. Robot Systems Div.
Real Time Generation of Smooth Curves Using Local Cubic Segments.
Final rept.
M. Roche, and W. Li. 1986, 9p
Pub. in ACCESS 7, n6 p23-31 Nov/Dec 88.

Keywords: *Curve fitting, *Real time operations, Algorithms, Interpolation, Coordinates, Reprints, Spline functions.

Computer listings of coded curve fitting algorithms for interpolation of coordinate points as they are generated are presented. As coordinate values are determined, they are added to an array and an interpolation procedure is applied to the new coordinate values of the array. In particular, as the sequence is being increased, it will be interpolated by a local cubic fitting procedure. The report exhibits a procedure which limits the cubic construction to be one segment behind the last segment of the sequence. The last input coordinate values are not the end points for the cubic segment being constructed. Another procedure will include these last input coordinate values as end coordinate values as well. As coordinate values and so coordinated. As will be seen, these methods lend themselves to real time curve generation.

901,197
PB89-172522 Not available NTIS
National Bureau of Standards (NEL), Gaithersburg, MD. Center for Applied Mathematics.
How to Estimate Capacity Dimension.
Final rept.
F. Sullivan, and F. Hunt. 1988, 4p
Pub. in Nuclear Physics B 5A, p125-128 1988.

Keywords: Monte Carlo method, Estimates, Algorithms, Reprints, *Euclidean space, Data compression, Fractal dimensions, Strange attractors, Robustness (Mathematics), Sorting, Chaos.

The authors describe a class of robust, computationally efficient algorithms for estimating capacity dimension and related quantities for compact subsets of (R sup n). The algorithms are based on Monte Carlo quadrature, data compression, and generalized distance functions.

901,198
PB89-175871 Not available NTIS
National Bureau of Standards (NEL), Gaithersburg, MD. Scientific Computing Div.
Efficient Algorithms for Computing Fractal Dimensions.
Final rept.
F. Hunt, and F. Sullivan. 1986, 8p
Pub. in Springer Series Synergetics 32, p74-81 1986.

Keywords: *Dimensions, Dimensional analysis, Algorithms, Monte Carlo method, Point set topology, Reprints, *Fractals, Data structures, Vector processing.

The purpose is to describe a new class of methods for computing the 'capacity dimension' and related quantities for point-sets. The techniques presented here build on existing work which has been described in the literature. The novelty of the methods lies first in the approach taken to the definition of computation of dimension (namely, via Monte Carlo calculation of the volume of an epsilon-cover of the point-set), and second in the use of data structures which result in extremely efficient codes for vector computers such as the Cyber 205 (the computation is reduced to the sorting and searching of one-dimensional arrays so that a calculation employing one million points requires less than 2 minutes).

901,199
PB89-177034   Not available NTIS
National Bureau of Standards (NEL), Gaithersburg, MD. Mathematical Analysis Div.
Note on the Capacitance Matrix Algorithm, Substructuring, and Mixed or Neumann Boundary Conditions.
Final rept.
D. P. O'Leary. 1987, 7p
Pub. in Applied Numerical Mathematics 3, n4 p339-345 1987.

Keywords: *Matrices(Mathematics), *Elliptic differential equations, Partial differential equations, Algorithms, Reprints.

The paper develops variants of the capacitance matrix algorithm which can be used to solve discretizations of elliptic partial differential equations when either the original system of equations or one which arises from substructuring has a rank deficient matrix.

901,200
PB89-209332   PC A03/MF A01
National Inst. of Standards and Technology (NEL), Gaithersburg, MD. Applied and Computational Mathematics Div.
Expected Complexity of the 3-Dimensional Voronoi Diagram.
J. Bernal. May 89, 23p NISTIR-89/4100

Keywords: *Euclidean geometry, Algorithms, Cubes(Mathematics), *Voronoi diagrams, *Polyhedrons, Complexity, Flat surfaces, Apexes.

Let S be a set of n sites chosen independently from a uniform distribution in a cube in three-dimensional Euclidean space. Work by Bentley, Weide and Yao is extended to show that the Voronoi diagram for S has an expected O(n) number of faces. A consequence of the proof of the result is that the Voronoi diagram for S can be constructed in expected O(n) time. Finally, it is shown that with the exception of at most an expected O(n sup(2/3)) number of polyhedra, each polyhedron in the Voronoi diagram for S has an expected constant number of faces.

901,201
PB90-123654   Not available NTIS
National Bureau of Standards (NEL), Gaithersburg, MD. Scientific Computing Div.
Guide to Available Mathematical Software Advisory System.
Final rept.
R. F. Boisvert. 1989, 11p
Pub. in Mathematics and Computers in Simulation 31, p453-463 1989.

Keywords: *Applications of mathematics, *Statistical analysis, Artificial intelligence, Documentation, Reprints, *Computer applications, *Computer software, Expert systems, Interactive systems, On line systems, Data bases.

The primary goal of the Guide to Available Mathematical Software (GAMS) project is to provide convenient access to information about mathematical and statistical software which is available to computer users at the National Institute for Standards and Technology. The principal vehicle through which the information is disseminated is an on-line advisory system called the GAMS Interactive Consultant. The paper describes the current status of the GAMS project. It then enumerates some of the weaknesses of the system and suggests knowledge engineering techniques which may alleviate them.

901,202
PB90-123688   Not available NTIS
National Bureau of Standards (NEL), Gaithersburg, MD. Scientific Computing Div.

Shortest Paths in Simply Connected Regions in R2.
Final rept.
R. D. Bourgin, and P. L. Renz. 1989, 36p
Pub. in Advances in Mathematics 76, n2 p260-295 1989.

Keywords: Topology, Reprints, *Jordan regions, Shortest paths.

Let R be a Jordan region in R2. A number of questions are settled concerning shortest paths in such regions: If there is a rectifiable path in R joining given points p and q then there is a unique shortest path in R joining these points. This path may be characterized by local geometric conditions. If another path joins p and q within R, a quantitative bound determined solely by the arclengths of this path and of the shortest path is presented which determines a tube about the shortest path guaranteed to contain the other path. When no rectifiable path in R exists joining p and q, there is, nevertheless, a unique locally shortest path in R joining them. Finally, the pathology-free nature of shortest paths is explicitly demonstrated by characterizing their connected sections within the boundary of R as well as within its interior.

901,203
PB90-129982   PC A03/MF A01
National Inst. of Standards and Technology (NEL), Gaithersburg, MD. Center for Computing and Applied Mathematics.
Polytope Volume Computation.
J. Lawrence. Oct 89, 27p NISTIR-89/4123
Prepared in cooperation with George Mason Univ., Fairfax, VA.

Keywords: *Volume, Algorithms, Simplex method, Linear programming, Linear inequalities, *Polytopes.

A combinatorial form of Gram's relation of convex polytopes can be adapted for use in computing polytope volume. The author presents an algorithm for volume computation based on the observation. The algorithm is useful in finding the volume of a polytope given as the solution set to a system of linear inequalities. As an example of the application of the method, the author computes a formula for the volume of a projective image of the n-cube.

Operations Research

901,204
PB89-174049   Not available NTIS
National Bureau of Standards (NML), Boulder, CO. Time and Frequency Div.
Variances Based on Data with Dead Time between the Measurements.
Final rept.
J. A. Barnes, and D. W. Allan. 1987, 8p
Pub. in Proceedings of Annual Precise Time and Time Interval (PTTI) Applications and Planning Meeting (19th), Redondo Beach, CA., December 1-3, 1987, p227-234.

Keywords: *Variance(Statistics), *Dead time, Bias, Displacement, Errors, Damping, Delay time, Resolution, Sensitivity, Data sampling, Frequency stability.

In 1974 a table of bias functions which related variance estimates with various configurations of number of samples and dead time to the two-sample (or Allan) variance was published. The tables were based on noises with pure power-law power spectral densities. Often situations recur which unavoidably have distributed dead time between measurements, but still the conventional variances are not convergent. Some of these applications are outside of the time and frequency field. Also, the dead times are often distributed throughout a given average, and this distributed dead time is not treated in the 1974 tables. The paper reviews the bias functions B1(N,r,mu) and B2(r,mu) and introduces a new bias function B3(2,r,mu), to handle the commonly occurring cases of the effect of distributed dead time on the computed variances. Some convenient and easy ways to interpret asymptotic limits are reported.

901,205
PB89-211957   Not available NTIS
National Bureau of Standards (NEL), Gaithersburg, MD. Mathematical Analysis Div.

Electronic Mail and the 'Locator's' Dilemma.
Final rept.
C. Witzgall, and P. B. Saunders. 1988, 20p
Pub. in Applications of Discrete Mathematics, p65-84 1988.

Keywords: *Linear programming, *Optimization, Telecommunication, Benefit cost analysis, Return on investment, Reprints, Electronic mail, Communication networks, Cost benefit analysis.

A general methodology for optimally selecting subconfigurations from a given universal configuration is discussed. It is based on a method for assessing the effects of shared fixed costs developed by J. M. W. Rhys and M. L. Balinski. A parametric minimum cost network flow algorithm for solving the resulting optimization problems is described. The problem arose in connection with a cost-benefit model called PAREC whose purpose was to analyze particular configurations of a contemplated electronic mail or message service system.

901,206
PB90-112335   PC A03/MF A01
National Inst. of Standards and Technology (NEL), Gaithersburg, MD. Factory Automation Systems Div.
Modeling Dynamic Surfaces with Octrees.
D. Libes. Sep 89, 14p NISTIR-89/4055

Keywords: *Mathematical models, Monte Carlo method, *Three dimensional models, *Octrees, Data structures, Dynamic models, Systems simulation.

Octrees in the past have been used to represent static objects. The paper discusses the extensions necessary to model dynamic surfaces. One particularly important aspect is the ability to represent expanding surfaces that grow to be arbitrarily large. The ability to represent dynamic surfaces allows one to apply octrees to new problems which could not previously have been modeled with static octrees. One such problem is the 'entropy of random surfaces.' Using dynamic octrees, a simulation of self-avoiding random surfaces using Monte Carlo techniques is produced.

901,207
PB90-112392   PC A03/MF A01
National Inst. of Standards and Technology (NEL), Gaithersburg, MD. Center for Mfg. Engineering.
FACTUNC: A User-Friendly System for Unconstrained Optimization.
R. H. F. Jackson, G. P. McCormick, and A. Sofer.
Aug 89, 37p NISTIR-89/4159
Prepared in cooperation with George Washington Univ., Washington, DC. Dept. of Operations Research, and George Mason Univ., Fairfax, VA. Dept. of Operations Research and Applied Statistics.

Keywords: *Nonlinear programming, Functions(Mathematics), Optimization, Constraints, Mathematical models, Regression analysis.

FACTUNC is a system for solving unconstrained minimization problems based on the concept of factorable programming. The concept enables the user to provide the problem function and data in a user friendly way and does not require user-supplied derivatives. The system utilizes the factorable function concept to obtain the first and second derivatives required for unconstrained optimization. As a system for nonlinear minimization, FACTUNG allows several options. First the user can solve regression (nonlinear least squares) problems by providing the regression equation and the data for the dependent and independent variables. The second option allows for the minimization of the sum of an indexed function. The user provides the function, and the indexed data. The third option is simply to minimize a function supplied by the user. Utilizing barrier function methodology, the third option can sometimes be used to solve constrained problems.

901,208
PB90-123944   Not available NTIS
National Bureau of Standards (NEL), Gaithersburg, MD. Scientific Computing Div.
Merit Functions and Nonlinear Programming.
Final rept.
J. W. Tolle, and P. T. Boggs. 1988, 8p
Pub. in Proceedings of International Conference on Operational Research (11th), Buenos Aires, Argentina, August 10-14, 1987, p882-889 1988.

141

MATHEMATICAL SCIENCES

Operations Research

Keywords: *Nonlinear programming, *Algorithms, Optimization, Operations research, Convergence.

In the paper, a merit function for inequality constrained nonlinear programs is proposed. A local convergence theorem is stated for an algorithm using the merit function in conjunction with a sequential quadratic programming procedure for generating steps. The algorithm is shown to permit unit step lengths in the presence of superlinear convergence.

Statistical Analysis

901.209
PB89-157812 Not available NTIS
National Bureau of Standards (NEL), Boulder, CO. Statistical Engineering Div.
Problems with Interval Estimation When Data Are Adjusted via Calibration.
Final rept.
J. M. Mulrow, D. F. Vecchia, J. P. Buonaccorsi, and H. K. Iyer. 1988, 15p
Pub. in Jnl. of Quality Technology 20, n4 p233-247 Oct 88.

Keywords: *Confidence limits, Error analysis, Estimates, Measurement, Reprints, *Tolerance limits, Calibration.

The analysis of adjusted data arising from a linear calibration curve is considered. Although it is obvious that adjusted values contain errors due to estimation of the calibration curve, some investigators may be tempted to analyze such data as if one applies 'naive' analyses to calibrated data. In particular, it is shown that standard one-sample confidence intervals have actual confidence levels that are always less than the nominal value. The authors also propose and evaluate two other methods for constructing confidence intervals. Tolerance intervals derived from adjusted data also, may yield deceiving results, having actual probability levels that are usually greater than the desired level but smaller in some cases.

901.210
PB89-157820 Not available NTIS
National Bureau of Standards (NEL), Boulder, CO. Statistical Engineering Div.
Estimation of the Error Probability Density from Replicate Measurements on Several Items.
Final rept.
W. Liggett. 1988, 11p
Pub. in Biometrika 75, n3 p557-567 1988.

Keywords: *Error analysis, Orthogonal functions, Monte Carlo method, Variance(Statistics), Probability density functions, Measurement, Computation, Estimates, Reprints, Robustness (Mathematics), Hermite polynomials.

Estimation of the measurement error probability density from data that consist of a few measurements on each of several dissimilar items is investigated. An estimator is proposed for independent and identically distributed measurement error with a symmetric density function. The estimator is based on an orthogonal function expansion. Computation begins with the differences between measurements on the same item and makes use of the fact that the characteristic function of these differences equals the square of the characteristic function of the measurement error. Application to robust inference for items measured in triplicate is considered. The M-estimates of the values of the items are compared on the basis of an estimated standard error computed from the density estimate. The circumstances under which this standard error estimator provides nearly valid inferences are delimited by Monte Carlo experiments.

901.211
PB89-171847 Not available NTIS
National Bureau of Standards (NEL), Gaithersburg, MD. Statistical Engineering Div.
Minimax Approach to Combining Means, with Practical Examples.
Final rept.
K. R. Eberhardt, C. P. Reeve, and C. H. Spiegelman. 1989, 20p
Grant N00014-86-F-0025
Sponsored by Office of Naval Research, Arlington, VA.
Pub. in Chemometrics and Intelligent Laboratory Systems 5, p129-148 1989.

Keywords: *Minimax technique, *Mean, Comparison, Estimating, Confidence limits, Statistics, Reprints.

The paper describes a method for combining sample means that accounts for bias in those means. It compares the unweighted mean, the weighted mean using reciprocal estimated variances for weights, and a minimax weighted mean. When the individual means are subject to nontrivial biases, the authors show that the minimax estimator can lead to important decreases in mean squared error and confidence interval width. The recommendations are based on statistical theory and on simulations based on three Standard Reference Material data sets.

901.212
PB89-201131 Not available NTIS
National Bureau of Standards (NEL), Gaithersburg, MD. Statistical Engineering Div.
Estimation of an Asymmetrical Density from Several Small Samples.
Final rept.
W. Liggett. 1989, 9p
Pub. in Biometrika 76, n1 p13-21 1989.

Keywords: *Skewed density functions, *Estimating, Probability density functions, Orthogonal functions, Statistical samples, Monte Carlo method, Least squares method, Reprints, Hermitian polynomial.

A method for estimation of the measurement error probability density from three or more measurements on each of several dissimilar items is presented. Differences between measurements on the same item provide estimates of the densities of the first and second differences between these densities and the error density, expressed in terms of characteristic functions and Hermite function expansions, are the basis for a nonlinear least-squares algorithm. Estimated percentiles of the error density are investigated by Monte Carlo experiments. The method is applied to measurements by several laboratories on an inhomogeneous reference material with asymmetric variability.

901.213
PB89-211130
(Order as PB89-211106, PC A04)
National Inst. of Standards and Technology, Gaithersburg, MD.
Consensus Values, Regressions, and Weighting Factors.
Bi-monthly rept.
R. C. Paule, and J. Mandel. 1989, 7p
Included in Jnl. of Research of the National Institute of Standards and Technology, v94 n3 p197-203 May-Jun 89.

Keywords: Variance(Statistics), Regression analysis, Taylor series, Computation, Convergence, *Consensus values, Weighted average, Iterative methods.

An extension to the theory of consensus values is presented. Consensus values are calculated from averages obtained from different sources of measurement. Each source may have its own variability. For each average a weighting factor is calculated, consisting of contributions from both the within- and the between-source variability. An iteration procedure is used and calculational details are presented. An outline of a proof for the convergence of the procedure is given. Consensus values are described for both the case of the weighted average and the weighted regression.

901.214
PB89-215321 PC A03/MF A01
National Inst. of Standards and Technology (NEL), Gaithersburg, MD. Applied and Computational Mathematics Div.
Computation and Use of the Asymptotic Covariance Matrix for Measurement Error Models.
P. T. Boggs, and J. R. Donaldson. 26 Jun 89, 30p
NISTIR-89/4102

Keywords: *Measurement, *Error analysis, Simultaneous equations, Covariance, Regression analysis, Confidence limits, Monte Carlo method, Matrices(Mathematics), Optimization.

The measurement error model assumes that errors occur in both the response variables and the predictor variables. In using this model, it is of interest to compute confidence regions and intervals for the estimators of the model parameters. An asymptotic form for the covariance matrix is used to construct approximate confidence regions and intervals. The solution of the minimization problem resulting from the use of the measurement error model is discussed, and a procedure for accurately computing the covariance matrix is developed. Then the quality of the confidence regions and intervals constructed from this matrix is assessed via a Monte Carlo study.

901.215
PB89-229066 PC A05/MF A01
National Inst. of Standards and Technology (NEL), Gaithersburg, MD. Applied and Computational Mathematics Div.
User's Reference Guide for ODRPACK: Software for Weighted Orthogonal Distance Regression Version 1.7.
Internal rept.
P. T. Boggs, R. H. Byrd, J. R. Donaldson, and R. B. Schnabel. Aug 89, 79p NISTIR-89/4103
Prepared in cooperation with Colorado Univ. at Boulder. Dept. of Computer Science.

Keywords: *Curve fitting, *Data smoothing, Orthogonality, Regression analysis, Algorithms, Mathematical models, Errors, Least squares method, Data, *User manuals(Computer programs), Computer applications, Independent variables, Dependent variables.

ORDPACK is a portable collection of NASI 77 Fortran subroutines for fitting a model to data. It is designed primarily for instances when the independent as well as the dependent variables have significant errors, implementing a high efficient algorithm for solving the weighted orthogonal distance regression problem, i.e., for minimizing the sum of the squares of the weighted orthogonal distances between each data point and the curve described by the model equation. It can also be used to solve the ordinary least squares problem where all of the errors are attributed to the observations of the dependent variable. A complete description of the orthogonal distance regression problem and the algorithm and implemented in ORDPACK is given by Boggs et al. ORDPACK is designed to handle many levels of user sophistication and problem difficulty.

MEDICINE & BIOLOGY

Anatomy

901.216
PB89-230478 Not available NTIS
National Inst. of Standards and Technology (ICST), Gaithersburg, MD. Advanced Systems Div.
Analysis of Ridge-to-Ridge Distance on Fingerprints.
Final rept.
R. T. Moore. 1989, 8p
Sponsored by Federal Bureau of Investigation, Washington, DC.
Pub. in Jnl. of Forensic Identification 39, n4, p231-238, Jul/Aug 89.

Keywords: Females, Males, Distance, Reprints, *Forensic medicine, *Fingerprints.

The distance from the center of one friction skin ridge to the center of the ridge next to it is quite variable in different regions of a given fingerprint. This distance has been measured on a small sample of fingerprints. The measured value ranged from 0.2 mm to 0.85 mm on fingerprints from male subjects, and from 0.2 mm to 0.75 mm on fingerprints from female subjects. The mean ridge-to-ridge distance for 731 measurements on the fingerprints of ten male subjects was 0.46 mm. For 1,046 measurements on the fingerprints of ten female subjects the mean value was 0.41 mm. A method is described for using these values to calculate ridge counts between near neighboring minutiae. Estimates are made of the errors likely to result from the use of calculated ridge counts.

Biochemistry

901,217
PB89-156897 Not available NTIS
National Bureau of Standards (NML), Gaithersburg,
MD. Inorganic Analytical Research Div.
**Sequential Determination of Biological and Pollut-
ant Elements in Marine Bivalves.**
Final rept.
R. Zeisler, S. Stone, and R. Sanders. 1988, 6p
Pub. in Analytical Chemistry 60, n24 p2760-2765, 15
Dec 88.

Keywords: *Marine biology, *Mollusca, *Radioassay,
*X ray fluorescence, Neutron absorption, Neutron acti-
vation analysis, Nondestructive analysis, Trace elements, Sequential analysis, Reprints, *Bivalves.

A unique sequence of instrumental methods has been
employed to obtain concentrations for 44 elements in
marine bivalves tissue. The techniques used were X-
ray fluorescence, prompt gamma activation analysis,
and neutron activation analysis. It is possible to use a
single subsample and follow it nondestructively
through the three instrumental analysis techniques. A
final radiochemical procedure for tin was also applied
after completing the instrumental analyses. Comparison of results for elements determined by more than
one technique in sequence showed good agreement,
as did results from certified reference material samples analyzed along with the samples. The concentrations found in the bivalve samples ranged from carbon
at more than 50% dry weight down to gold at several
microgram per kilogram.

901,218
PB89-156905 Not available NTIS
National Bureau of Standards (NML), Gaithersburg,
MD. Inorganic Analytical Research Div.
**Sample Validity in Biological Trace Element and
Organic Nutrient Research Studies.**
Final rept.
G. V. Iyengar. 1987, 10p
Sponsored by Department of Agriculture, Beltsville,
MD., and Food and Drug Administration, Washington,
DC.
Pub. in Jnl. of Radioanalytical and Nuclear Chemistry
112, n1 p151-160 1987.

Keywords: *Trace elements, *Sampling, *Bioassay,
Contamination, Acceptability, Selection, Validity,
Standards, Nutrients, Storage, Preservation, Organic
compounds, Tissues(Biology), Biological extracts, Reprints.

The complexities involved in dealing with the requirements of trace element research studies in the life sciences demand a comprehensive planning of the investigations and use of a variety of techniques. It also requires a combination of biological insight and analytical awareness on the part of the investigators in order
to obtain valid samples for analysis. Thus, the generation of meaningful conclusions from elemental composition studies on biological systems is vital for the overall success of the investigations. In addition, new initia-
tives are needed to produce multipurpose biological
reference standards to cope with the growing demands of this multifaceted area of research. These aspects are discussed.

901,219
PB89-157770 Not available NTIS
National Bureau of Standards (NML), Gaithersburg,
MD. Office of Standard Reference Materials.
**NBS (National Bureau of Standards) Activities in
Biological Reference Materials.**
Final rept.
S. D. Rasberry. 1988, 5p
Pub. in Fresenius' Zeitschrift fuer Analytische Chemie
332, p528-532 1988.

Keywords: *Biological products, *Cholesterol, *Vitamins, *Diet, *Nutrition, Reprints, *Standard Reference
Materials.

National Bureau of Standards (NBS) activities in biological reference materials during 1986 - 1988 are described with a preview of plans for future certifications
of reference material. During the period, work has
been completed or partially completed on about 40 reference materials of importance to health, nutrition, and
environmental quality. Some of the reference materials that have been completed during the period and
are described include: creatinine (SRM 914a), bovine
serum albumin (SRM 927a), cholesterol in human

serum (SRMs 1951 - 1952), aspartate aminotransfer-
ase (RM 8430), cholesterol and fat-soluble vitamins in
coconut oil (SRM 1563), wheat flour (SRM 1567a), rice
flour (SRM 1568a), mixed diet (RM 8431a), dinitropyr-
ene isomers and 1-nitropyrene (SRM 1596), and complex PAHs from coal tar (SRM 1597). Oyster tissue
(SRM 1566a) is being analyzed and should be available in 1988.

901,220
PB89-177216 Not available NTIS
National Bureau of Standards (IMSE), Gaithersburg,
MD. Ceramics Div.
**Element-Specific Epifluorescence Microscopy in
vivo Monitoring of Metal Biotransformations in En-
vironmental Matrices.**
Final rept.
T. K. Trout, G. J. Olson, F. E. Brinckman, J. M.
Bellama, and R. A. Faltynek. 1989, 14p
Pub. in ACS (American Chemical Society) Symposium
Series 383, Chapter 6, p84-97 1989.

Keywords: *In vivo analysis, *Quantitative analysis,
*Metals, *Absorption spectra, Bioaccumulation,
Chemical analysis, Fluorescence, *Epifluorescence
microscopy, Fluorescence spectroscopy.

Quantitative measurement of metal ion uptake in living
cells is accomplished via staining the biota with an appropriate fluorogenic ligand, determining the emission
photon flux by epifluorescence microscopy imaging
(EMI), and relating the latter quantity to absolute metal
ion concentration by atomic absorption analysis. By
thus combining the techniques of element-specific
fluorimetry and EMI, it is possible to observe the chemistry occurring during redox transformations of elements by colonies of living cells in reactions that have
significant economic potential. The strengths, weaknesses, and future directions to be taken in improving
this analytical method are discussed.

901,221
PB89-186761 Not available NTIS
National Bureau of Standards (NML), Gaithersburg,
MD. Chemical Thermodynamics Div.
Thermodynamics of Hydrolysis of Disaccharides.
Final rept.
Y. B. Tewari, and R. N. Goldberg. 1989, 8p
Pub. in Jnl. of Biological Chemistry 264, n7 p3966-
3971, 5 Mar 89.

Keywords: *Thermodynamics, *Hydrolysis, *Disaccharides, Carbohydrates, Cellobiose, Enzymes, Maltose, Chromatographic analysis. Calorimetry, Enthalpy,
Entropy, Reprints, Enzymatic hydrolysis, Gentiobiose,
Isomaltose.

The thermodynamics of the enzymatic hydrolysis of
cellobiose, gentiobiose, isomaltose, and maltose have
been studied using both high pressure liquid chromatography and microcalorimetry. The hydrolysis reactions were carried out in aqueous sodium acetate
buffer at a pH of 5.65 and over the temperature range
of 286 to 316 K using the enzymes Beta-glucosidase,
isomaltase, and maltase. The thermodynamic parameters which described for the hydrolysis reactions,
disaccharide(aq) + H2O(liq) = 2 glucose(aq), at
298.15 K. The standard state is the hypothetical ideal
solution of unit molality. Due to enzymatic inhibition by
glucose, it was not possible to obtain reliable values
for the equilibrium constants for the hydrolysis of either
cellobiose or maltose. The entropy changes for the hydrolysis reactions are in the range 32 to 43 J/mol K;
the heat capacity changes are approximately equal to
zero J/mol K. Additional pathways for calculating thermodynamic parameters for these hydrolysis reactions
are discussed.

901,222
PB89-186803 Not available NTIS
National Bureau of Standards (NML), Gaithersburg,
MD. Chemical Thermodynamics Div.
**Water Structure in Vitamin B12 Coenzyme Crys-
tals. 1. Analysis of the Neutron and X-ray Solvent
Densities.**
Final rept.
H. Savage. 1986, 19p
See also PB89-186811.
Pub. in Biophysical Jnl. 50, n5 p947-965 1986.

Keywords: *Water, *Molecular structure, Crystal structure, Neutron diffraction, Vitamin B12, X ray diffraction,
Reprints, *Vitamin B 12 coenzymes.

The disordered solvent distribution in crystals of vitamin B12 coenzyme was examined using the methods

of high resolution neutron and X-ray diffraction. One
set of neutron (O.95A) and two sets of X-ray (O.94A
and 1.1A) data were collected and the resulting
models were extensively refined using least squares
and Fourier syntheses. The solvent regions were ana-
lysed in two stages: firstly, 'main' sites were assigned
to the well defined regions of solvent density and refined using least squares; secondly, 'continuous' sites
were assigned representing the more disordered diffuse and elongated regions of solvent density around
and between the main sites. Water networks were formulated from the assigned sites in the above three
models and also from those assigned in the original
structure determination. The well established networks
extend throughout all the solvent regions of the crystal
with interesting orientational arrangements of the individual waters around both polar and apolar groups of
the coenzyme molecule. The networks were seen to
be consistent among each of the four models in terms
of occupying relatively similar positions, however, the
occupancy values of the individual networks varied between the models.

901,223
PB89-186811 Not available NTIS
National Bureau of Standards (NML), Gaithersburg,
MD. Chemical Thermodynamics Div.
**Water Structure in Vitamin B12 Coenzyme Crys-
tals. 2. Structural Characteristics of the Solvent
Networks.**
Final rept.
H. Savage. 1986, 14p
See also PB89-186803.
Pub. in Biophysical Jnl. 50, n5 p967-980 1986.

Keywords: *Water, *Molecular structure, Crystal structure, Vitamin B12, Reprints, *Vitamin B 12 coenzymes,
Biomolecules, Hydrogen bonding.

The geometrical details of the solvent structure in vitamin B12 coenzyme crystals with respect to hydrogen
bonding and non-bonded contacts are described. The
individual H-bond geometries varied over wide ranges,
similar to those observed in small molecule structures.
Large deviations from tetrahedral coordination were
found around a majority of the waters. The mutual positions and orientations of the water molecules could
not be adequately explained in terms of the H-bonding
relationships present in the structure. However, additional investigations which focused on the short range
non-bonded contacts around water positions in a variety of crystal hydrates, revealed several structural regularities. These features relate to the non-bonded
O...O, H...O and H...H interactions and give rise to a set
of repulsive restrictions that are seen to be very much
stronger stereochemical restraints than those associated with H-bonding. The repulsive restrictions can be
used as stereochemical restraints in the interpretation
and refinement of solvent structures within larger hydrate systems such as protein crystals. They may also
be included in potential functions used to simulate solvent structures in aqueous solutions and hydrate systems.

901,224
PB89-201594 Not available NTIS
National Bureau of Standards (NML), Gaithersburg,
MD. Chemical Thermodynamics Div.
**Crystal Structure of a Cyclic AMP (Adenosine Mon-
ophosphate)-Independent Mutant of Catabolite
Gene Activator Protein.**
Final rept.
I. T. Weber, G. L. Gilliland, J. G. Harman, and A.
Peterkofsky. 1987, 7p
Pub. in Jnl. of Biological Chemistry 262, n12 p5630-
5636 1987.

Keywords: *Crystal structure, Mutations, Reprints,
*Cyclic AMP receptors, *Catabolite gene activator pro-
teins, Genetic mapping, Conformation change.

E. coli NCR91 synthesizes a mutant form of catabolite
gene activator protein in which alanine 144 is replaced
by threonine. This mutant, which also lacks adenyl cy-
clase activity, has a crp* phenotype; in the absence of
cAMP it is able to express genes that normally require
cAMP. CRP91 has been purified and crystallized with
cAMP under the same conditions as crystals of the
wild type GAP-cAMP complex. X-ray diffraction data
were measured to 2.3A resolution on a Xentronics
area detector and the CAP91 structure was determined using initial model phase from the wild type
structure. A difference Fourier map calculated between CAP91 and wild type showed the two alanine to
threonine sequence changes in the dimer and also

MEDICINE & BIOLOGY

Biochemistry

change in the side chain of cysteine 178 in one of the subunits. The CAP91 coordinates were refined by restrained least-squares to an R factor of 0.186. Small changes in the atomic positions of the wild type and mutant protein structures were analyzed by a local vector average. The change included concerted motions of the small domains, the hinge between the two domains and in an adjacent loop between two beta strands. The mutation apparently caused changes in position of the protein atoms that are distal to the mutation site.

901,225
PB89-202576   Not available NTIS
National Bureau of Standards (NEL), Gaithersburg, MD. Fire Measurement and Research Div.
Spectroscopic Quantitative Analysis of Strongly Interacting Systems: Human Plasma Protein Mixtures.
Final rept.
M. R. Nyden, G. P. Forney, and K. Chittur. 1988, 7p
Sponsored by National Institutes of Health, Bethesda, MD.
Pub. in Applied Spectroscopy 42, n4 p588-594 1988.

Keywords: *Spectroscopic analysis, *Humans, *Infrared spectra, Proteins, Quantitative analysis, Reprints, *Plasma proteins, Blood proteins.

Blood plasma protein infrared spectra, while qualitatively very similar, display subtle differences in the frequencies and intensities of absorption bands. These small differences are sufficient to permit an accurate quantitative analysis of mixtures of these proteins. In the paper the authors examine the performance of some alternative methods of spectroscopic quantitative analysis in determining the concentrations of proteins in aqueous solutions. The widely-used K matrix method, using sloping baselines and intercept functions, was found to be inadequate for these determinations. In contrast, a method based on the little-known Q matrix approach, augmented by a robust equation solver, yielded results with a sufficient degree of accuracy to make it a viable tool for use in the study of proteins at solid interfaces and for more general applications in the field of protein chemistry.

901,226
PB89-227888   Not available NTIS
National Inst. of Standards and Technology (NML), Gaithersburg, MD. Chemical Thermodynamics Div.
Calorimetric and Equilibrium Investigation of the Hydrolysis of Lactose.
Final rept.
R. N. Goldberg, and Y. B. Tewari. 1989, 4p
Pub. in Jnl. of Biological Chemistry 264, n17 p9887-9900, 15 Jun 89.

Keywords: *Lactose, *Heat measurement, *Chemical equilibrium, Hydrolysis, Catalysis, Thermochemistry, Glucose, Galactose, Hypotheses, Reprints, Beta-galactosidase.

The thermodynamics of the hydrolysis of lactose to glucose and galactose have been investigated using both high pressure liquid chromatography and heat conduction microcalorimetry. The reaction was carried out over the temperature range 282-316 K and in 0.1 M sodium acetate buffer at a pH of 5.65 using the enzyme beta-galactosidase to catalyze the reaction. The standard state is the hypothetical ideal solution of unit molality. Thermochemical cycle calculations using enthalpies of combustion and solution, entropies, solubilities, activity coefficients, and apparent molar heat capacities have also been performed. These calculations indicate large discrepancies which are attributable primarily to errors in literature data on the enthalpies of combustion and/or third law entropies of the crystalline forms of the substrates.

901,227
PB89-227904   Not available NTIS
National Inst. of Standards and Technology (NML), Gaithersburg, MD. Chemical Thermodynamics Div.
Thermodynamics of the Hydrolysis of Sucrose.
Final rept.
R. N. Goldberg, Y. B. Tewari, and J. C. Ahluwalia. 1989, 4p
Pub. in Jnl. of Biological Chemistry 264, n17 p9901-9904, 15 Jun 89.

Keywords: *Sucrose, *Thermochemistry, Hydrolysis, Thermodynamics, Heat measurement, Liquid chromatography, Chemical equilibrium, Fructose, Glucose, Solutions, Reprints.

A thermodynamic investigation of the hydrolysis of sucrose to fructose and glucose has been performed using microcalorimetry and high-pressure liquid chromatography. The calorimetric measurements were carried out over the temperature range 298-316 K and in sodium acetate buffer (0.1 M, pH 5.65). Enthalpy and heat capacity changes were obtained for the hydrolysis of aqueous sucrose (process A). Equilibrium data was obtained from the literature, and was used to calculate a value of the equilibrium constant for the hydrolysis of aqueous sucrose. Additional thermochemical cycles that bear upon the accuracy of these results are examined.

901,228
PB90-117508   Not available NTIS
National Inst. of Standards and Technology (IMSE), Gaithersburg, MD. Polymers Div.
Biophysical Aspects of Lipid Interaction with Mineral: Liposome Model Studies.
Final rept.
E. D. Eanes. 1989, 6p
Sponsored by National Inst. of Dental Research, Bethesda, MD.
Pub. in Anatomical Record 224, p220-225 1989.

Keywords: *Lipids, *Minerals, *Biophysics, *Biochemistry, pH, Acidification, Calcification, Electrostatics, Reprints, *Liposomes, Phosphatidylcholines, Artificial membranes.

The paper reviews the use of liposomes as synthetic models for studying various biophysical aspects of matrix vesicle calcification, especially the involvement of acidic phospholipids in the nucleation and growth processes which occur during the initial stages of mineral formation in and around these membrane-bound structures. Recent results showed that acidic phospholipids incorporated into phosphatidylcholine-rich anionic liposome membranes were ineffective in initiating extraliposomal calcium phosphate precipitation from metastable solutions at physiological pH. On the contrary, certain acidic phospholipids such as phosphatidic acid and phosphatidylserine retarded the development of such precipitation when the latter was endogenously induced. The extent of inhibition correlated with the strength of the electrostatic interaction between the polar head group of the acidic phospholipid and the surface of the mineral phase. The results suggest that acidic phospholipids may play an important role in controlling the rate of early mineral development in matrix vesicle calcification.

901,229
PB90-123886   Not available NTIS
National Inst. of Standards and Technology (NML), Gaithersburg, MD. Organic Analytical Research Div.
Generic Liposome Reagent for Immunoassays.
Final rept.
A. L. Plant, M. V. Brizgys, L. Locasio-Brown, and R. A. Durst. 1989, 7p
Pub. in Analytical Biochemistry 176, p420-426 1989.

Keywords: *Biochemistry, *Antibodies, Lipids, Vitamin B complex, Antigens, Reprints, *Liposomes, *Immunoassay, Cell membranes, Fluorescence spectrometry, Cross-linking reagents, Avidin.

The report discusses the derivatization of liposomes with antibodies by using avidin to crosslink biotinylated phospholipid molecules in the liposome membranes with biotinylated antibody molecules. A comparison of the biotin binding activity of avidin in solution and avidin associated with liposomes shows that avidin bound to biotinylated phospholipid in liposome membranes retains full binding activity for additional biotin molecules. Changes in the fluorescence spectrum of avidin have been used to characterize the binding capacity of avidin for biotin in solution, and change in intensity of light scattered due to aggregation of liposomes was used to measure the biotin binding activity of avidin associated with liposomes. Relative amounts of the biotinylated phospholipid, avidin, and biotinylated antibody have been optimized to produce stable liposomes which are derivatized with up to 1.7 nmol of antibody/micromol of lipid. These derivatized liposomes are highly reactive to immunospecific aggregation in the presence of multivalent antigen. A linear increase in light scattering was recorded between 1 and 10 pmol of antigen. The work shows that liposomes containing biotinylated phospholipid can be successful generic reagent for immunoassays.

901,230
PB90-128117   Not available NTIS

National Inst. of Standards and Technology ,(IMSE), Gaithersburg, MD. Polymers Div.
Liposome Technology in Biomineralization Research.
Final rept.
E. D. Eanes, and B. R. Heywood. 1989, 22p
Sponsored by National Inst. of Dental Research, Bethesda, MD.
Pub. in New Biotechnology in Oral Research, Chapter 4, p54-75 1989.

Keywords: *Calcification, Calcium phosphate, Precipitation(Chemistry), Reprints, *Liposomes, *Mineralization, Membrane lipids, Lipid bilayers.

The chapter reviews the preparation and use of artificial lipid vesicles (liposomes) as synthetic models for studying membrane controlled precipitation of calcium phosphate salts in aqueous solutions. The impetus behind the development of liposomes for this purpose is the recognition that naturally occurring bilayer counterparts known as matrix vesicles have a primary role in initiating mineralization in many skeletal and dental tissues. Reviewed in the chapter are some of the lipid combinations, preparation procedures, and methods for characterizing liposomes that are currently used in mineralization studies. Also described is an endogenous reaction procedure which allows precipitations to be carried out in liposomal suspensions which parallel those postulated to occur during matrix vesicle calcification. With the use of the technology as currently developed, liposomes have proved to be a useful tool for elucidating in vitro the role membrane lipids possibly play in biocalcification processes.

901,231
PB90-136763   Not available NTIS
National Bureau of Standards (NEL), Boulder, CO. Thermophysics Div.
Bioseparations: Design and Engineering of Partitioning Systems.
Final rept.
M. H. Hariri, J. F. Ely, and G. A. Mansoori. 1989, 3p
Pub. in Biotechnology 7, p686-688 Jul.89.

Keywords: *Biochemistry, *Molecules, Extraction, Reprints, *Biotechnology, Equipment design, Separation processes.

Aqueous two-phase extraction offers a potential separation technique for separation of biomacromolecules. The manuscript briefly summarizes an approach to separating biomacromolecules using this technique and discusses requirements for industrial scale-up of this type of process.

901,232
PB90-136854   Not available NTIS
National Bureau of Standards (IMSE), Gaithersburg, MD. Ceramics Div.
Global Biomethylation of the Elements - Its Role in the Biosphere Translated to New Organometallic Chemistry and Biotechnology.
Final rept.
F. E. Brinckman, and G. J. Olson. 1988, 29p
Pub. in Special Publication - Royal Soc. of Chemistry Bio\. Alkylation Heavy Elem. 66, p168-196 1988.

Keywords: *Methylation, *Organometallic compounds, Metals, Environment, Microbiology, Reprints, *Biotechnology, *Biosphere.

The paper reviews important aspects of the field of bioorganometallic chemistry opened over 50 years ago by Professor Frederick Challenger. Considered are the scope and rate of organometallic biogenesis, mechanistic implications for environmental (chemical and biological) processes involving organometals and the prospects for control and manipulation of these processes for new biotechnical approaches to materials acquisition and environmental protection. Methylmetal(loid) species are microbially generated endo- and exocellularly by a variety of enzymatic and non-enzymatic processes. Large quantities of microbial metabolites which can act as metal methylating agents are generated in the global environment. Examples include methyl halides, methyl corrinoids and methylated sulfur compounds. Interesting prospects for materials processing arise in considering bioorganometallic chemistry with novel routes to metal dissolution precipitation and reduction being possible.

Clinical Chemistry

901,233
PB89-146773 Not available NTIS
National Bureau of Standards (NML), Gaithersburg, MD. Organic Analytical Research Div.
Developing Definitive Methods for Human Serum Analytes.
Final rept.
P. Ellerbe. 1986, 5p
Pub. in Pathologist 40, n9 p22, 24-27 1986.

Keywords: *Cholesterol, *Glucose, *Urea, *Uric acid, Reprints, *Mass spectroscopy, *Isotope dilution, *Blood serum, *Creatinine, Tracer techniques, Qualitative chemical analysis, Blood chemistry, Radioassay.

The National Bureau of Standards is actively involved in a program with the CAP to develop definitive methods for constituents of human serum. The author is involved in the effort of the Organic Analytical Research Division of the Center for Analytical Chemistry to develop isotope dilution mass spectrometric (IDMS) definitive methods for organic serum constituents. Analytes that have been examined include cholesterol, glucose, uric acid, urea, and creatinine.

901,234
PB89-149223 Not available NTIS
National Bureau of Standards (NML), Gaithersburg, MD. Gas and Particulate Science Div.
Micro-Raman Characterization of Atherosclerotic and Bioprosthetic Calcification.
Final rept.
E. S. Etz, B. B. Tomazic, and W. E. Brown. 1986, 8p
Pub. in Microbeam Analysis 1986, p39-46.

Keywords: *Raman spectra, *Arteriosclerosis, *Calcium metabolism disorders, *Chemical analysis, Vibrational spectra, Calcium phosphates, Carbonates, Deposits, Microanalysis, Minerals, Reprints, *Bioprosthesis, Aortic diseases, Calcinosis, Plaque formation.

Described is the application of Raman microprobe spectroscopy to the characterization of the mineral phase present in atherosclerotic and bioprosthetic calcified deposits. Examined are human aortic plaque and calcific deposits removed from a heart aseat device implanted in sheep. The vibrational Raman spectra of these mineralized deposits are interpreted relative to the various types of calcium phosphates known to participate in the formation of the biological mineral. A specific goal is the spatial tracking, in the microscopic domain, of carbonate species associated with the mineral phosphate to determine the type of carbonate substitution (Type A or Type B sites) in the biological apatite. Results are presented from the quantitation of carbonate contents based on an empirical procedure using relative scattering intensities of carbonate and phosphate species. These studies are part of a larger research program on the comprehensive physicochemical characterization of these pathologic, calcified deposits in animals and in man.

901,235
PB89-151922 Not available NTIS
National Bureau of Standards (NML), Gaithersburg, MD. Ionizing Radiation Physics Div.
Chemical Characterization of Ionizing Radiation-Induced Damage to DNA.
Final rept.
M. Dizdaroglu. 1986, 3p
Pub. in Biotechniques 4, n6 p536-538 1986.

Keywords: *Ionizing radiation, *Radiation effects, *Gas chromatography, *Mass spectroscopy, *Deoxyribonucleic acids, Spectroscopic analysis, Chromatographic analysis, Damage assessment, Sugars, Nucleosides, Microanalysis, Reprints, *Genetic effects, Base composition, DNA.

The report reviews the application of the capillary gas chromatography-mass spectrometry (GC-MS) technique to chemical characterization of radiation-induced damage to DNA. Damage to both sugar and base moieties of DNA exposed to ionizing radiation in aqueous solution can be unequivocally characterized by GC-MS. Sugar products released from DNA are reduced by NaBD, to corresponding polyalcohols and analyzed by GC-MS following trimethylsilylation. For those sugar products still bound to the DNA backbone, an additional step, alkaline treatment, following NaBD-reduction is required to release them. Incorporation of deuterium atoms into the polyalcohols permits the assessment of the presence and the position of the car-

bonyl and deoxy groups in the precursor sugar molecules. The t-butyldimethylsilyl derivatives provide a typical (M-57) plus ion, which appears in most instances as the base peak in the mass spectra, and is very useful for diagnostic purposes. The technique of selected-ion monitoring (SIM) permits the unequivocal characterization of the products at very low radiation doses that are considered biologically relevant. The SIM is also used for quantitative measurement of the products at low radiation doses.

901,236
PB89-171953 Not available NTIS
National Bureau of Standards (NML), Gaithersburg, MD. Inorganic Analytical Research Div.
Radiochemical and Instrumental Neutron Activation Analysis Procedures for the Determination of Low Level Trace Elements in Human Livers.
Final rept.
R. Zeisler, R. R. Greenberg, and S. F. Stone. 1988, 17p
Sponsored by Environmental Protection Agency, Washington, DC.
Pub. in Jnl. of Radioanalytical and Nuclear Chemistry 124, n1 p47-63 1988.

Keywords: *Trace elements, *Neutron activation analysis, *Liver, Humans, Compton effect, Spectrochemical analysis, Tissues(Biology), Nuclear chemistry, Radiochemistry, Reprints.

A comprehensive approach to the analysis of human livers was developed in a pilot program for the National Environmental Specimen Bank that employed a combination of four analytical techniques. Refinements in this approach were needed for improvement in detection limits, for more effective sample usage, and to reduce the number of analytical steps that were involved. Since neutron activation analysis (NAA) determined most of the elements, expansion of NAA was chosen to achieve these goals. Modifications in the instrumental NAA procedures, including the use of a Compton Suppressor System, gave increased sensitivity for some low level elements, such as arsenic and chromium. Radiochemical procedures that followed the instrumental counts increased the sensitivity for the elements chromium, selenium, arsenic, molybdenum, silver, antimony, and tin. Results are given for two radiochemical procedures that were applied following the modified procedure, either the use of an inorganic ion exchange column or a liquid/liquid extraction, and these are compared to instrumental results.

901,237
PB89-179279 Not available NTIS
National Bureau of Standards (NEL), Gaithersburg, MD. Fire Science and Engineering Div.
Stabilization of Ascorbic Acid in Human Plasma, and Its Liquid-Chromatographic Measurement.
Final rept.
S. A. Margolis, and T. P. Davis. 1988, 7p
Pub. in Clinical Chemistry 34, n11 p2217-2223 1988.

Keywords: *Ascorbic acid, *Stabilization, Chromatographic analysis, Standards, Reprints, *Human plasma, *Clinical analysis, High pressure liquid chromatography, Reference materials.

Two independent HPLC procedures are described for the rapid and accurate analysis of ascorbic acid in human plasma. No sample extraction or phase separation is required. The development of a human plasma reference material for clinical laboratory analysis of ascorbic acid is described. The plasma ascorbic acid content can be determined with as little as 50 uL of sample in 15 min. Analytical recoveries are 100% with direct injection of deproteinted plasma. Extensive stability data under several conditions using dithiothreitol as a preservative (antioxidant) indicate that ascorbic acid remains stable for up to 57 weeks. Round robin analysis of 11 normal human blood samples by two independent methods showed a % CV between 0.1 and 5.3. These clinical samples appear to be stable for no less than 60 days under the described conditions of stabilization and sample treatment. By using these methods, a laboratory can easily automate the analysis for up to 24 hours of injections at room temperature (21 deg C).

901,238
PB89-202550 Not available NTIS
National Bureau of Standards (NML), Gaithersburg, MD. Temperature and Pressure Div.

New International Temperature Scale of 1990 (ITS-90).
Final rept.
B. W. Mangum. 1989, 3p
Pub. in Clinical Chemistry 35, n3 p503-505 1989.

Keywords: *Temperature measurement, Calibrating, Reprints, *Temperature scales.

A new international temperature scale, the ITS-90, will replace the International Practical Temperature Scale of 1968 (amended edition of 1975), IPTS-68(75), on 1 January 1990. Temperatures on the ITS-90 will agree more closely with thermodynamic temperatures; therefore, the ITS-90 represents a substantial improvement over the IPTS-68(75). Fortunately for the clinical laboratory community, the change in the scale will be at most only 0.05 C or less in the range from 0 to 60 C, but corrections in primary calibrations should be made so that the calibrations are based on the ITS-90.

901,239
PB89-234181 Not available NTIS
National Inst. of Standards and Technology (NML), Gaithersburg, MD. Organic Analytical Research Div.
Determination of Serum Cholesterol by a Modification of the Isotope Dilution Mass Spectrometric Definitive Method.
Final rept.
P. Ellerbe, S. Meiselman, L. T. Sniegoski, M. J. Welch, and E. White. 1989, 6p
Pub. in Analytical Chemistry 61, n15 p1718-1723, 1 Aug 89.

Keywords: *Cholesterol, *Blood, *Carbon 13, *Isotopic labeling, *Mass spectroscopy, *Dilution, Measurement, Standards, Methodology, Gas chromatography, Reprints.

An isotope dilution mass spectrometric (ID/MS) method for cholesterol is described that uses capillary gas chromatography with cholesterol-(13)C3 as the labeled internal standard. Labeled and unlabeled cholesterol are converted to the trimethylsilyl ether. Combined capillary column gas chromatography and electron impact mass spectrometry are used to obtain the abundance ratio of the unlabeled and labeled ions from the derivative. Quantitation is achieved by measurement of each sample between measurements of two standards whose unlabeled/labeled ratios bracket that of the sample. Seven pools were analyzed by the method. The method is a modification of the original definitive method for cholesterol.

Clinical Medicine

901,240
PB89-157283 Not available NTIS
National Bureau of Standards (NML), Gaithersburg, MD. Molecular Spectroscopy Div.
Two-Photon Laser-Induced Fluorescence of the Tumor-Localizing Photosensitizer Hematoporphyrin Derivative.
Final rept.
D. King, D. Heller, J. Krasinski, and R. Bodaness. 1986, 4p
See also PB88-175237.
Pub. in AIP (American Institute of Physics) Conference Proceedings, n146 p694-697 1986.

Keywords: *Photosensitivity, *Neoplasms, *Fluorescent dyes, *Drug therapy, Porphyrins, Photons, Excitation, Beams(Radiation), Lasers, Spectra, Emissivity, Tissues(Biology), Penetration, Free radicals, Position(Location).

The tumor localizing photosensitizer hematoporphyrin derivative (HPD) is shown to undergo simultaneous two-photon excitations upon intense laser irradiation at 750 or 1064 nm, a spectral region where there is no significant HPD one-photon absorbance in aqueous solution. Evidence for the two-photon excitation consists in the observation both of the HPD fluorescence spectrum in the region of 615 nm as a result of 750 or 1064 nm excitations and the quadratic dependence of this fluorescence emission intensity upon the excitation laser intensity. Since the penetration depth of ultraviolet and visible light into tissue varies logarithmically with wavelength (red penetrating more deeply than blue), these studies suggest the possibility that two-photon induced localization of tumor-bound HPD

MEDICINE & BIOLOGY

Clinical Medicine

might facilitate the detection of deeper lying tumors than allowed by the current one-photon photolocalization method.

901,241
PB89-171854 Not available NTIS
National Bureau of Standards (NML), Gaithersburg, MD. Ionizing Radiation Physics Div.
Basic Data Necessary for Neutron Dosimetry.
Final rept.
R. S. Caswell, J. J. Coyne, H. M. Gerstenberg, and E. J. Axton. 1988, 7p
Sponsored by Armed Forces Radiobiology Research Inst., Bethesda, MD., and Department of Energy, Washington, DC. Office of Health and Environmental Research.
Pub. in Radiation Protection Dosimetry 23, n1/4 p11-17 1988.

Keywords: *Dosimetry, *Neutron beams, Carbon, Neutron cross sections, Tissues(Biology), Spectra, Reprints, *Radiation therapy, Kerma factor.

Among many developments in basic data for neutron dosimetry, four are highlighted: improvements in kerma factor data for carbon; the use of common data through unification of the European and American protocols for neutron radiation therapy dosimetry; the tabulation of a set of initial spectra of secondary particles produced in tissue and tissue-like materials; and proposed changes in the magnitude and definition of neutron quality factors, which, while not physical quantities, are nevertheless basic to neutron dosimetry.

901,242
PB89-176895 Not available NTIS
National Bureau of Standards (IMSE), Gaithersburg, MD. Metallurgy Div.
Mossbauer Imaging.
Final rept.
S. J. Norton. 1987, 3p
Pub. in Nature 330, n6144 p151-153 1987.

Keywords: *Mossbauer effect, *Diagnosis, *Tissues(Biology), Gamma rays, Images, Doppler effect, Absorption, Rotation, Electromagnetic noise, Spectroscopy analysis, Reprints, *Tomography.

Recoilless gamma-ray resonance, or the Mossbauer effect, is a well-established spectroscopic tool in materials science. Mossbauer spectroscopy shares some of the fundamental characteristics of nuclear magnetic resonance (NMR) spectroscopy, since both rely on nuclear resonance phenomena. A significant recent advance in the latter field is NMR imaging in biomedicine. In Mossbauer spectroscopy, a quantity analogous to the NMR magnetic field is the relative velocity between the gamma-ray source and the absorber. This suggests that an analogous approach to Mossbauer imaging is possible by imposing a velocity gradient on the absorber. This can be achieved simply by rotating the absorbing object relative to the source, generating line integrals of constant Doppler shift, or equivalently, of constant gamma-energy. From such measurements, a spatial map of the gamma-ray absorption coefficient can, in principle, be tomographically reconstructed. Spatial resolution is directly related to the rate of rotation of the absorber, but ultimately is signal-to-noise limited.

901,243
PB89-193858 PC A04/MF A01
National Bureau of Standards (NML), Gaithersburg, MD. Center for Radiation Research.
NBS (National Bureau of Standards) Measurement Services: Calibration of Gamma-Ray-Emitting Brachytherapy Sources.
Final rept.
J. T. Weaver, T. P. Loftus, and R. Loevinger. Dec 88, 63p NBS/SP-250/19
Also available from Supt. of Docs. as SN003-003-02923-8. Library of Congress catalog card no. 88-600609.

Keywords: *Cobalt 60, *Cesium 137, *Iridium 192, *Iodine 125, *Calibrating, *Standards, Radiology, Dosimetry, Radioactive isotopes, Gamma rays, Air, Measurement, Ionization chambers, Accuracy, Exposure, *Brachytherapy.

The calibration of small radioactive sources used for interstitial radiation therapy, short distance therapy (brachytherapy), is performed in terms of the physical quantities exposure or air kerma. (60)Co and (137)Cs sources are calibrated by comparison with NBS working standard sources of the same type, while (192)Ir

and (125)I sources are calibrated by measurement in a reentrant ionization chamber that was calibrated using NBS working standard sources of the same type. The working standard sources were calibrated using the NBS graphite cavity ionization chambers except for (125)I, for which the NBS measurement standard was a free-air chamber. The working standard sources of the two long-lived sources have been measured a number of times over the years; the reliability of the reentrant chamber for the two short-lived sources is assured by use of sealed radium sources as a constancy check. The overall uncertainty (considered to have the approximate significance of a 95% confidence limit) is given as 2% for all the sources except for (125)I seeds, for which it is given as 5%, 6%, and 7%, depending on the type of seed. The stated uncertainty for (125)I seeds does not include possible errors due to low-energy X rays not recognized at the time the standards were established.

Cytology, Genetics, & Molecular Biology

901,244
PB89-157836 Not available NTIS
National Bureau of Standards (NML), Gaithersburg, MD. Organic Analytical Research Div.
Structure of a Hydroxyl Radical Induced Cross-Link of Thymine and Tyrosine.
Final rept.
S. A. Margolis, B. Coxon, E. Gajewski, and M. Dizdaroglu. 1988, 7p
Sponsored by Armed Forces Radiobiology Research Inst., Bethesda, MD.
Pub. in Biochemistry 27, n17 p6353-6359, 23 Aug 88.

Keywords: *Tyrosine, *Proteins, *Crosslinking, *Radiobiology, Microanalysis, Ionizing radiation, Molecular structure, Deoxyribonucleic acids, Alpha amino carboxylic acids, Uracils, Radiation effects, Reprints, *Thymine, *DNA, *Hydroxyl radicals, Hydroxy compounds.

DNA-protein cross-links are formed when living cells or isolated chromatin is exposed to ionizing radiation. Little is known about the actual cross-linked products of DNA and proteins. In the work, a novel hydroxyl radical induced cross-link of thymine and tyrosine has been isolated along with a tyrosine dimer by high-performance liquid chromatography of aqueous mixtures of tyrosine and thymine that had been exposed to hydroxyl radicals generated by ionizing radiation. The isolated compounds have been examined by gas chromatography-mass spectrometry, high-resolution mass spectrometry, and (1)H and (13)C nuclear magnetic resonance spectroscopy. The structure of the thymine-tyrosine cross-link has been identified as the product from the formation of a covalent bond between the methyl group of the thymine and carbon 3 of the tyrosine ring. In addition, the 3,3' tyrosine dimer was isolated and characterized. The mechanism of the formation of these compounds is discussed. The work presents the first complete chemical characterization of a hydroxyl radical-induced DNA base-amino acid cross-link.

901,245
PB89-175269
(Order as PB89-175194, PC A06)
Lister Hill National Center for Biomedical Communications, Bethesda, MD.
New Directions in Bioinformatics.
Bi-monthly rept.
D. R. Masys. 1989, 5p
Included in Jnl. of Research of the National Institute of Standards and Technology, v94 n1 p59-63 Jan-Feb 89.

Keywords: *Information systems, Nucleic acids, *Molecular biology, *Biotechnology, *Data base management systems, Communication networks, Computer networks, Medical information systems.

Development of automated methods to sequence DNA, RNA, proteins, and other macromolecules have yielded oceans of cryptic symbols, for which there is an absolute dependence upon computerized factual databases to acquire, store, retrieve, and analyze data. The Human Genome Project has focused attention on the information science aspects of nucleic acid data, yet for the practicing scientist nucleic acids and

other sequence data are just one piece of an increasingly complex biological puzzle whose solution will be expressed in terms of structure and function. Access to and integration of information across multiple related biological databases is a major challenge facing information system builders, a challenge which holds the promise of creating knowledge synergy from what are today disconnected, stand-alone information sources.

901,246
PB89-175277
(Order as PB89-175194, PC A06)
Allelix Biopharmaceuticals, Mississauga (Ontario).
Use of Structural Templates in Protein Backbone Modeling.
Bi-monthly rept.
L. S. Reid. 1989, 8p
Included in Jnl. of Research of the National Institute of Standards and Technology, v94 n1 p65-72 Jan-Feb 89.

Keywords: *Proteins, Conformation, Data bases, Templates.

Many proteins of interest have low (i.e. less than 50%) sequence similarity to any known structure. In these cases new approaches to prediction of structure are required. The use of sequence profiles which relate sequence to known structure has been proposed as one method to assign local regions of structure. As a first stage, templates or 'icons' of the many relevant substructural motifs found in proteins must be defined. The sequences which gave rise to these structures are then aligned and a weighted profile obtained. Average structures of the 8 and 12 residue helix-turn and turn-helix motifs have been prepared. These coordinate templates were then used to scan through the Brookhaven protein structural database for similar, superimposable fragments. A composite template of 100 similar fragments for each element was found to be internally consistent. All of the sequences, from these structures, were then used to create an overall sequence profile.

901,247
PB89-175285
(Order as PB89-175194, PC A06)
Maryland Univ., Rockville. Center for Advanced Research in Biotechnology.
Comparative Modeling of Protein Structure: Progress and Prospects.
Bi-monthly rept.
J. Moult. 1989, 6p
Included in Jnl. of Research of the National Institute of Standards and Technology, v94 n1 p79-84 Jan-Feb 89.

Keywords: *Proteins, *Molecular structure, Amino acids, Models, Electrostatics, Crystallography, Reliability, *Protein conformation, Databases.

Comparative modeling of protein structure is a process which determines the three-dimensional structure of protein molecules on the basis of amino acid sequence similarity to experimentally known structures. The procedure is facilitated by the growing database of protein structures obtained from crystallography. In the review a series of stages in the modeling process are identified and discussed. These are (i) obtaining a reliable amino acid sequence of the structure of interest, (ii) producing a structurally correct sequence of the structure of interest, (iii) producing a structurally correct sequence alignment, (iii) identifying which structural features are conserved between target and parent structures, (iv) modeling the new pieces of structure, and (v) tests of reliability.

901,248
PB89-202204 Not available NTIS
National Bureau of Standards (NML), Gaithersburg, MD. Chemical Thermodynamics Div.
Comparison of Two Highly Refined Structures of Bovine Pancreatic Trypsin Inhibitor.
Final rept.
A. Wlodawer, J. Deisenhofer, and R. Huber. 1987, 12p
Pub. in Jnl. of Molecular Biology 193, n1 p145-156 1987.

Keywords: *Molecular structure, Enzymes, Crystal structure, Reprints, *Trypsin inhibitors, *Pancreatic juice.

The high resolution structures of bovine pancreatic trypsin inhibitor refined in two distinct crystal forms have been compared. One of the structures was a result of new least squares x-ray refinement of data from crystal form I, while the other was the joint x-ray/neutron structure of crystal form II. After superposition, the molecules show an overall root-mean-squares deviation of 0.40A for the atoms in the main chain, while the deviations for the side chain atoms are 1.53A. The latter number decreases to 0.61A when those side chains which adopted drastically different conformations are excluded from comparison. The discrepancy between atomic temperature factors in the two models was 6.7 sq A, while their general trends are highly correlated. About half of the solvent molecules occupy similar positions in the two models, while the others are different. As expected, solvent molecules with the lowest temperature factors are most likely to be common in the two crystal forms. While the two models are clearly similar, the differences are significantly larger than the errors inherent in the structure determination.

901,249
PB90-123407 Not available NTIS
National Inst. of Standards and Technology (NEL), Gaithersburg, MD. Chemical Process Metrology Div.
Nonlinear Effect of an Oscillating Electric Field on Membrane Proteins.
Final rept.
R. D. Astumian, and B. Robertson. 1989, 11p
Pub. in Jnl. of Chemical Physics 91, n8 p4891-4901, 15 Oct 89.

Keywords: *Electric fields, *Nonlinear systems, Catalysis, Electrochemistry, Relaxation, Kinetics, Reprints, *Membrane proteins, Biological transport.

The nonlinear response of a two-state chemical transition to an oscillating electric field is examined. A reaction for which the analysis is particularly relevant is a conformational transition of a membrane protein exposed to an ac electric field. Even a modest externally applied field leads to a very large local field within the membrane. This gives rise to nonlinear behavior. The applied ac field causes harmonics in the polarization and can cause a dc shift in the state occupancy, both of which can be observed and used to determine kinetic parameters. Fourier coefficients are calculated for the enzyme state probability in the ac field, exactly for infinite frequency, and in powers of the field for finite frequency. Kramers-Kronig relations are proved and response functions are given for the leading terms of the harmonics. The results are extended to the spherical symmetry relevant to suspensions of spherical cells, vesicles, or colloidal particles. If the protein catalyzes a reaction, free energy is transduced from the electric field to the output reaction, even if that reaction is electrically silent. Many transport enzymes are ideal examples. The ac field can cause the enzyme to pump ions or molecules through the membrane against an (electro) chemical potential. The efficiency of the energy transduction can be as high as 25%.

901,250
PB90-136722 Not available NTIS
National Inst. of Standards and Technology (NML), Gaithersburg, MD. Chemical Thermodynamics Div.
Biological Macromolecule Crystallization Database: A Basis for a Crystallization Strategy.
Final rept.
G. L. Gilliland. 1988, 9p
Pub. in Jnl. of Crystal Growth 90, n1-3 p51-59 Jul 89.

Keywords: *Biochemistry, *Molecules, *Crystallization, Molecular weight, Proteins, Viruses, pH, Temperature, Nucleic acids, Reprints, *Databases, *DNA, Osmolar concentration.

A crystallization database, the Biological Macromolecule Crystallization Database, containing crystal data and the crystallization conditions for more than 1000 crystal forms of over 600 biological macromolecules, has been compiled from the scientific literature. Data for proteins, protein:protein complexes, nucleic acids, nucleic-acid:nucleic-acid complexes, protein:nucleic-acid complexes and viruses have been included. The general information cataloged for each macromolecule includes the macromolecular name(s), the molecular weight, the subunit composition, the presence of prosthetic group(s), and the source of the macromolecule. The crystal data include the unit cell parameter, space group, crystal density, crystal habit and size, and diffraction limit and lifetime. The crystallization data consist of the crystallization method, chemical additions to the crystal growth medium, macromolecule concentration, temperature, pH, and growth time. A result of the compilation of the crystallization data

was the development of a general strategy for the crystallization of soluble proteins.

901,251
PB90-136730 Not available NTIS
National Inst. of Standards and Technology (NML), Gaithersburg, MD. Chemical Thermodynamics Div.
Preliminary Crystallographic Study of Recombinant Human Interleukin 1beta.
Final rept.
G. L. Gilliland, E. L. Winborn, Y. Masui, and Y. Hirai. 1987, 2p
Pub. in Jnl. of Biological Chemistry 262, n25 p12323-12324 1987.

Keywords: *Crystallography, Escherichia coli, X ray diffraction, Ammonium sulfate, Reprints, *Interleukin 1, *Recombinant proteins.

Recombinant human interleukin 1 beta (IL-1 beta) which is expressed in Escherichia coli has been crystallized by the method of vapor diffusion using ammonium sulfate as the precipitant. The space group is P41 or P43 with a=b= 55.0 A and c= 77.1 A and one molecule in the asymmetric unit. The crystals diffract to beyond 2.4 A and are suitable for a three-dimensional X-ray structure determination.

Dentistry

901,252
PB89-179238 Not available NTIS
National Bureau of Standards (IMSE), Gaithersburg, MD. Polymers Div.
Comparison of Fluoride Uptake Produced by Tray and Flossing Methods in vitro.
Final rept.
M. K. Guo, L. C. Chow, C. T. Schreiber, and W. E. Brown. 1989, 3p
Sponsored by American Dental Association Health Foundation, Chicago, IL., and National Taiwan Univ., Taipei. Coll. of Medicine.
Pub. in Jnl. of Dental Research 68, n3 p496-498 Mar 89.

Keywords: *Fluoride, Trays, Adsorption, Reprints, *Dental enamel, Dicalcium phosphate dihydrate, Dental floss, Fluoroapatite.

The study compares: (i) the fluoride (F) uptake by enamel in approximal areas of teeth when the F agent was applied in vitro via a tray or a flossing technique; and (ii) the effectiveness of two treatments -- acidulated phosphate fluoride (APF) alone and CaHPO4-2H2O (DCPD)-forming pre-treatment followed by APF. Groups of three teeth (one premolar and two molars) were mounted in impression compounds simulating their oral configuration. In the tray group, teeth received one four-minute treatment by means of custom-formed trays. In the flossing group, the approximal areas of teeth were flossed for 40 sec twice daily for three days with an absorbent floss wetted with the treatment solution. The F uptake was calculated from biopsy data obtained before and after the treatment. The results showed that (i) DCPD-APF produced significantly greater F uptake than APF alone in both the tray and flossing methods, and (ii) that the flossing technique produced significantly greater F uptake in the approximal areas than the tray method for either treatment.

901,253
PB89-186373 Not available NTIS
National Bureau of Standards (IMSE), Gaithersburg, MD. Polymers Div.
Micro-Analysis of Mineral Saturation Within Enamel During Lactic Acid Demineralization.
Final rept.
G. L. Vogel, C. M. Carey, L. C. Chow, T. M. Gregory, and W. E. Brown. 1988, 9p
Sponsored by American Dental Association Health Foundation, Chicago, IL.
Pub. in Jnl. of Dental Research 67, n9 p1172-1180 Sep 88.

Keywords: *Microanalysis, *Lactic acid, Dental caries, Concentration(Composition), Calcium, Phosphate, Reprints, *Dental enamel, Demineralization, Hydroxyapatite, Enamel solubility, Tooth permeability.

In this study, the physicochemical factors responsible for caries-like lesion propagation were investigated by

means of a micro-analytical system used to study the fluid within a lesion during a simulation of the decay process. Four 500-micrometer-thick serial sections prepared from a single human molar were mounted between glass plates with only the natural surface of the tooth exposed. The concentrations of calcium, phosphate, and hydrogen ions of the fluid in the wells were then followed as a function of time as the lesion advanced. The results of this study, in which lactic acid was used to demineralize enamel, were consistent with those previously reported (Vogel et al, 1987a). The solution within the lesion remained saturated during the acid attack. Differences in initial mobilities of the calcium and phosphate and other ions increased the concentrations within the lesion and permanently changed the ratio of these ions in the lesion solution. Based on these results, the authors suggest that the ionic permselectivity of tooth enamel can have a profound effect on the transport of mineral from a caries lesion.

901,254
PB89-201503 Not available NTIS
National Bureau of Standards (IMSE), Gaithersburg, MD. Polymers Div.
Mechanism of Hydrolysis of Octacalcium Phosphate.
Final rept.
B. B. Tomazic, M. S. Tung, T. M. Gregory, and W. E. Brown. 1989, 9p
Sponsored by American Dental Association Health Foundation, Chicago, IL.
Pub. in Scanning Microscopy 3, n1 p119-127 1989.

Keywords: *Hydrolysis, Chemical reactions, Solubility, Transformations, Reprints, *Octacalcium phosphate, Hydroxyapatite.

The chemical and structural properties of hydrolyzed octacalcium phosphate (OCP) appear to be of high relevance to tooth, bone and pathological bioapatites. Hydrolysis of synthetic well-crystallized OCP was studied at constant pH by using the pH stat method over the 6.1 to 8.6 range at 50 C and to a lesser extent at 37 C. Hydrolytic transformation proceeds according to thermodynamic requirements except for some retardation at the highest pH value as a consequence of decreased solubility of OCP which may be rate determining. The product of hydrolysis, OCP-hydrolyzate (OCPH), was characterized by chemical analysis, scanning electron microscopy, x-ray diffraction, electron microprobe (x-ray microanalysis, EDX) and solubility measurement under static and dynamic conditions. Overall findings provide new evidence that OCP may be a precursor phase in the formation of pathologic calcified deposits and normal biomineral, which appear to be complex hydrolyzates of OCP.

901,255
PB89-201511 Not available NTIS
National Bureau of Standards (IMSE), Gaithersburg, MD. Polymers Div.
Formation of Hydroxyapatite in Hydrogels from Tetracalcium Phosphate/Dicalcium Phosphate Mixtures.
Final rept.
A. Sugawara, J. M. Antonucci, S. Takagi, L. C. Chow, and M. Ohashi. 1989, 10p
Sponsored by American Dental Association Health Foundation, Chicago, IL.
Pub. in Jnl. of the Nihon University School of Dentistry 31, n1 p372-381 Mar 89.

Keywords: Dental materials, Reprints, *Hydroxyapatite, *Apatitic calcium phosphate cements, *Dental cements.

Apatitic calcium phosphate cements, formed by the ambient reaction of tetracalcium phosphate (TTCP) with dicalcium phosphates (DCP), have been recently reported. H20 or dilute aq. H3PO4 (0.2%) is used as the liquid vehicle for this reaction. The study ascertained if hydroxyapatite (HAp) can form in self-cured hydrogel composites containing TTCP/DCP mixes. The setting times (ST) and diametral tensile strengths (DTS) of these hydrogel composites were also determined. The hydrogels were of two types: vinyl thermosets derived from the copolymerization of HEMA (2-hydroxyethyl methacrylate) and cross-linking monomers, and polyelectrolyte-based hydrogels formed from aq. poly(alkenoic acids), e.g. poly(acrylic acid). Cylindrical specimens 6 mm D x 3 mm H were prepared and stored in H20 for up to 30 days. The HEMA composites were hardened in 7-15 min by free radical initiation (benzoyl peroxide/tertiary aromatic amine).

MEDICINE & BIOLOGY

Dentistry

After various periods of storage in H20 at 37 C, some of the specimens were examined by X-ray spectroscopy for HAp. HAp formation was not observed in the HEMA composites even after 30 days of H20 storage but was detected in the polyacid cements. Both the H20 content and pH may thus be factors controlling the rate and extent of HAp formation in hydrogel composites containing TTCP/DCP mixtures.

901,256
PB89-201529 Not available NTIS
National Bureau of Standards (IMSE), Gaithersburg, MD. Polymers Div.
Detection of Lead in Human Teeth by Exposure to Aqueous Sulfide Solutions.
Final rept.
A. Sugawara, J. M. Antonucci, G. C. Paffenbarger, and M. Ohashi. 1989, 15p
Sponsored by American Dental Association Health Foundation, Chicago, IL.
Pub. in Jnl. of the Nihon University School of Dentistry 31, n1 p382-396 Mar 89.

Keywords: *Lead(Metal), *Exposure, *Teeth, *Humans, *Detection, Discoloration, Reprints, Dental cements, Sodium sulfide, Aqueous solutions.

A recent study has shown that the presence of lead (Pb) as well as other base metals in esthetic restorative materials, especially dental cements, is detectable by color shifts induced by exposure of hardened specimens to a 0.1% (w/v) aqueous solution of sodium sulfide, Na2S. The present study was initiated to determine the applicability of this simple exposure test to the detection of Pb in human teeth. Extracted whole teeth as well as sectioned, thin specimens were exposed first to either a 0.01% or a 0.001% (w/v) aqueous solution of lead nitrate, Pb(NO3)2 at 37 C for 24 h. After rinsing with distilled H20 and a subsequent 24 h exposure to the 0.1% Na2S solution at 37 C, the tooth specimens were examined visually and by a dental color analyzer for color changes. Neither control specimens exposed to distilled H20 only or to 0.1% Na2S more exhibited any significant change in appearance after 24 h of storage at 37 C. However, specimens exposed first to the Pb(NO3)2 solutions showed discernible delta E values after exposure to the Na2S solution. Delta E was greatest for specimens exposed to the more concentrated Pb(NO3)2 solution. Most of the discoloration in both thin and intact tooth specimens was confined to the outermost layers of the tooth structure. For the intact specimens, the greatest degree of discoloration occurred in the cementum, the most permeable part of the tooth structure.

901,257
PB89-202477 Not available NTIS
National Bureau of Standards (IMSE), Gaithersburg, MD. Polymers Div.
High-Temperature Dental Investments.
Final rept.
J. A. Tesk. 1989, 11p
Pub. in Dental Materials: Properties and Selection, Chapter 18, p351-361 1989.

Keywords: *Dental materials, Polymers, · Castings, High temperature tests, Reprints, *Dental investments.

High-temperature dental investments are materials used primarily for the casting of high-temperature dental alloys, that is, alloys with casting temperatures greater than 1,300 C. While the casting of these alloys into crowns, inlays, partial dentures (fixed or removable), and other restorative devices is certainly the primary use for the investments, they have other dental applications as well. One such use is as fixtures for holding dental prostheses during soldering operations. Another use is for making dies for the fabrication of porcelain facial tooth veneers. Until recently there have been two basic compositional types: phosphate-bonded and ethyl silicate-bonded. However, due to interest in the casting of titanium prostheses, other systems are now being explored and used. This is necessary because the conventional phosphate and ethyl silicate investments react with molten titanium and contaminate the casting. Commercial dental investments used in Japan for titanium are based primarily on magnesium oxide. Investments under development in the United States are based on zirconia. They have been used to make titanium castings for limited clinical evaluations by the Paffenbarger Research Unit at the National Bureau of Standards.

901,258
PB89-202931 Not available NTIS

National Bureau of Standards (IMSE), Gaithersburg, MD. Polymers Div.

Pulpal and Micro-organism Responses to Two Experimental Dental Bonding Systems.
Final rept.
R. L. Blosser, N. W. Rupp, H. R. Stanley, and R. L. Bowen. 1989, 5p
Sponsored by American Dental Association Health Foundation, Chicago, IL.
Pub. in Dental Materials 5, p140-144 Mar 89.

Keywords: *Dental materials, *Dentin, *Bonding, Teeth, Reprints, *Biocompatibility, Dental pulp, Microleakage.

Several new bonding systems have been reported that promote strong adhesion. This in vivo study involves treatment with two experimental bonding systems of Class V cavity preparations in the teeth of three Macaca fascicularis primates and reports the pulpal responses and degree of micro-organism invasion associated with each treatment. The upper and lower left quadrants were treated with clinical materials to establish positive and negative controls. After 4, 25, and 59 days, the teeth were removed and underwent routine histological and bacteriological evaluation. Slight pathological conditions were noted for superficial and deep responses, but all value approached 0.0 by the 59th day. Micro-organisms were seen under only 12% of the restorations. Both experimental systems appear to be safe for human clinical trials.

901,259
PB89-229249 Not available NTIS
National Inst. of Standards and Technology (IMSE), Gaithersburg, MD. Polymers Div.
Quasi-Constant Composition Method for Studying the Formation of Artificial Caries-Like Lesions.
Final rept.
L. C. Chow, and S. Takagi. 1989, 6p
Sponsored by American Dental Association Health Foundation, Chicago, IL.
Pub. in Caries Research 23, p129-134 1989.

Keywords: Demineralizing, Volumetric analysis, Reprints, *Dental caries, *Tooth diseases, Dental enamel solubility.

Caries-like lesions were formed in human tooth enamel in a quasi-constant composition titration system without the use of surface dissolution inhibitors or weak acids. The titration system maintained the composition of the demineralizing solution constant to within 6% on the average. Thus, the rate of lesion formation may be quantitatively assessed from the rate of titration. The system has sufficient sensitivity for measuring the rate of lesion formation in a small area, e.g., 15 sq mm, of a single human enamel specimen.

Microbiology

901,260
PB90-123381 Not available NTIS
National Inst. of Standards and Technology (NML), Gaithersburg, MD. Chemical Thermodynamics Div.
Preliminary Crystal Structure of Acinetobacter glutaminasificans Glutaminase-Asparaginase.
Final rept.
H. L. Ammon, I. T. Weber, A. Wlodawer, R. W. Harrison, G. L. Gilliland, K. C. Murphy, L. Sjolin, and J. Roberts. 1988, 7p
Pub. in Jnl. of Biological Chemistry 263, n1 p150-156 1988.

Keywords: *Crystal structure, *Glutaminase, Reprints, *Asparaginase, *Acinetobacter glutaminasificans, Protein conformation, Amino acid sequence.

The preliminary structure of a glutaminase-asparaginase from Acinetobacter glutaminasificans (AgGA) is reported. The structure was determined at 3.0 A resolution with a combination of phase information from multiple isomorphous replacement at 4 - 5 A resolution, and phase improvement and extension by density modification techniques. Initially polyalanine was fit to the electron density map and was subsequently replaced by a polypeptide with an amino acid sequence in agreement with the sizes and shapes of the side chain electron densities. The crystallographic R-factor is 0.300 following constrained least-squares refinement with data to 2.9 A resolution. The AgGA subunit folds into two domains: the amino-terminal domain

contains a five-stranded beta sheet surrounded by five alpha helices, and the carboxy-terminal domain contains three helices and less regular structure. The connectivity is not fully determined at present, due in part to the lack of a complete amino acid sequence. The AgGA structure has been used successfully to determine the relative orientations of the molecules in crystals of Pseudomonas 7A glutaminase-asparaginase and of Vibrio succinogenes asparaginase, and in a new crystal form of E. coli asparaginase (space group I222, one subunit per asymmetric unit).

901,261
PB90-123712 Not available NTIS
National Inst. of Standards and Technology (IMSE), Gaithersburg, MD. Ceramics Div.
Microbiological Materials Processing.
Final rept.
F. E. Brinckman, and G. J. Olson. 1988, 2p
Pub. in Jnl. of Metals 40, n9 p60-61 1988. ·

Keywords: *Microbiology, *Metals, Materials recovery, Metalliferous minerals, Reprints, *Biotechnology, *Metal recovery.

Microorganisms are increasingly used as agents for processing and recovery of metals. The paper briefly reviews the industrial applications of microbial processing of ores and wastes and describes NBS contributions of the technology including new measurement methodology, standards activities and critical data. Potential future developments and research needs are briefly described.

Nutrition

901,262
PB89-234173 Not available NTIS
National Inst. of Standards and Technology (NML), Gaithersburg, MD. Organic Analytical Research Div.
Determination of Total Cholesterol in Coconut Oil: A New NIST (National Institute of Standards and Technology) Cholesterol Standard Reference Material.
Final rept.
P. Ellerbe, L. T. Sniegoski, M. J. Welch, and E. White V. 1989, 4p
Sponsored by College of American Pathologists, Skokie, IL.
Pub. in Jnl. of Agricultural and Food Chemistry 37, n4. p954-957 1989.

Keywords: *Cholesterol, *Human nutrition, Mass spectroscopy, Plant oils, Blood, Fats, Nutrients, Gas chromatography, Reprints, *Coconut oil, Standard Reference Materials.

A new Standard Reference Material (SRM) consisting of coconut oil with various nutrients added has been developed at the National Institute of Standards and Technology in response to the needs of the food measurement community. SRM 1563 consists of ampules of a coconut oil with added cholesterol and selected fat-soluble vitamins and ampules of the natural coconut oil. Cholesterol has been measured in the material by a modification of the definitive method based on isotope dilution mass spectrometry coupled with gas chromatography, originally developed for the measurement of cholesterol in serum. The cholesterol concentration, as total cholesterol, in the fortified oil was determined to be 64.2 + or - 0.6 mg/100 g of oil. This value, with its precision, complies with the request of the food nutrient measurement community for a standard with an uncertainty within + or - 5% of the certified value at 95% confidence limits. The natural oil was found to contain 0.344 + or - 0.014 mg/100 g of oil.

Pharmacology & Pharmacological Chemistry

901,263
PB89-171870 Not available NTIS
National Bureau of Standards (NML), Gaithersburg, MD. Ionizing Radiation Physics Div.

Intramolecular H Atom Abstraction from the Sugar Moiety by Thymine Radicals in Oligo- and Polydeoxynucleotides.
Final rept.
L. R. Karam, M. Dizdaroglu, and M. G. Simic. 1988, 7p
Sponsored by Armed Forces Radiobiology Research Inst., Bethesda, MD.
Pub. in Radiation Research 116, p210-216 1988.

Keywords: *Uracils, *Free radicals, *Hydrogen, *Sugars, Chemical reactions, Hydroxides, Thymidines, Ribose, Gas chromatography, Mass spectroscopy, Reprints.

Hydroxyl radical addition to uracil (U) has been suggested to lead to strand breaks in polyuridylic acid, an occurrence attributed in part to H atom abstraction by U-OH free radicals from the ribose moiety. The particular reaction is investigated by means of the hydroxyl radical-induced products of thymine (T), pT, TpT, TpTpT, polythymidylic acid (poly-T), (T+dR) poly-dA center dot poly-T, and a mixture of T and 2-deoxyribose (dR). The major monomeric product of T-OH radical in TpT, TpTpT, poly-T, and poly-dA center dot poly-T was found to be 5-hydroxy-6-hydrothymine (H-T-OH), while that in T, pT, and T plus dR was thymine glycol (HO-T-OH). These results indicated that the infra-molecular H atom abstraction from a nearby sugar (in this case, deoxyribose) moiety by base radicals, i.e., T-OH, occurs in oligo- and polydeoxynucleotides of T. In poly-T, the yield of H-T-OH is not much greater than in TpT or TpTpT, indicating that the abstraction of an H atom from the sugar moiety of a nucleotide subunit further than two nucleotides along the chain may not be significant. Additionally, a corresponding decrease in the yield of HO-T-OH with an increase in the yield of H-T-OH suggests that the formations of these two types of thymine products are competitive.

Radiobiology

901,264
PB89-150791   Not available NTIS
National Bureau of Standards (NML), Gaithersburg, MD. Ionizing Radiation Physics Div.
Refinement of Neutron Energy Deposition and Microdosimetry Calculations.
Final rept.
R. S. Caswell, J. J. Coyne, H. M. Gerstenberg, and R. B. Schwartz. 1986, 13p
Sponsored by Department of Energy, Washington, DC. Office of Health and Environmental Research.
Pub. in Proceedings of International Conference on Fast Neutron Physics, Dubrovnik, Yugoslavia, May 26-31, 1986, p122-134.

Keywords: *Monte Carlo method, Energy absorption, Computation, *Microdosimetry, Tissue-equivalent materials.

Calculations describing the deposition of energy by neutrons in tissue-like materials are usually carried out by the 'analytic method' or the 'Monte-Carlo method.' Extensions of the equations of the analytic method to include thin walls as well as thick walls are now available. Furthermore, inclusion of straggling effects in the analytic method is relatively simple and has been programmed for computer calculations. The first step in the analytic method is the calculation of the 'initial spectra of secondary charged particles generated by the neutrons. The authors are preparing tables of initial spectra below ·20 MeV. The calculation of 'lineal energy' or 'y' spectra for neutrons is of interest for microdosimetry. The possibility of carrying out microdosimetric calculations on a nanometer scale using track structure information generated by Wilson and Paretzke is being pursued. A Monte-Carlo code is being generated using the same data base as the authors' analytic method codes. The chief advantage of the Monte-Carlo code is in the correct handling of events where two or three correlated charged particles are emitted. Some results of microdosimetric calculations including straggling are given.

901,265
PB89-171862   Not available NTIS
National Bureau of Standards (NML), Gaithersburg, MD. Ionizing Radiation Physics Div.

Initial Spectra of Neutron-Induced Secondary Charged Particles.
Final rept.
H. M. Gerstenberg, R. S. Caswell, and J. J. Coyne. 1988, 4p
Sponsored by Department of Energy, Washington, DC. Office of Health and Environmental Research.
Pub. in Radiation Protection Dosimetry 23, n1/4 p41-44 1988.

Keywords: *Nuclear cross sections, *Neutron reactions, *Computation, *Carbon, *Tissues(Biology), *Charged particles, Spectra, Hydrogen, Helium, Deuterium, Boron, Nitrogen, Oxygen, Beryllium, Radioactive decay, Ions, Reprints.

Calculations have been made of the initial spectra of secondary charged particles which result from neutron interactions with materials such as tissue. These secondary particles are ions of H, D, He and the heavier recoil ions of Be, B, C, N and O. The spectra of the ions have been determined in 200 keV neutron energy bin sizes between 0 and 20 MeV as well as for 76 almost-logarithmic bins extending from thermal energy to 2 MeV. The primary input for these calculations is the ENDF/B-V nuclear data file from the National Nuclear Data Center at Brookhaven. Additional supplementary information is also needed on the angular distribution of neutron reactions leading to charged particles and on the final excitation of residual nuclei. Recently measured kerma factors in carbon disagree with those calculated using the nuclear cross section data from ENDF/B-V by as much as 25% in the region between 14 MeV and 18 MeV. A recently made evaluation of the carbon cross sections in the energy region between 5 MeV and 32 MeV is used here to calculate kerma and initial spectra at neutron energies of 16.9, 14.9, and 13.9 MeV; the results are then compared with that obtained using the ENDF/B-V data.

Toxicology

901,266
PB90-117888   Not available NTIS
National Inst. of Standards and Technology (NML), Gaithersburg, MD. Ionizing Radiation Physics Div.
Generation of Oxy Radicals in Biosystems.
Final rept.
M. G. Simic, D. S. Bergtold, and L. R. Karam. 1989, 10p
Pub. in Mutation Research 214, p3-12 1989.

Keywords: *Free radicals, *Oxygen, *Toxicity, *Mutations, Reprints, *DNA damage.

Many recent lines of evidence indicate that endogenous free radicals contribute to spontaneous mutagenesis through the direct induction of DNA damage. However, the mechanisms underlying the process are not yet fully understood. A brief overview of the knowledge that is currently available is provided, with emphasis on the generation of oxy radicals in biosystems, the reactions of those radicals with biomolecules, and the induction of oxidative DNA base damage that might lead to mutation.

901,267
PB90-128760   Not available NTIS
National Inst. of Standards and Technology (NML), Gaithersburg, MD. Organic Analytical Research Div.
Anti-T2 Monoclonal Antibody Immobilization on Quartz Fibers: Stability and Recognition of T2 Mycotoxin.
Final rept.
M. L. Williamson, D. H. Atha, D. J. Reeder, and P. V. Sundaram. 1989, 14p
Pub. in Analytical Letters 22, n4 p803-816 1989.

Keywords: Quartz, Bioinstrumentation, Silane, Reprints, *T-2 toxin, *Monoclonal antibodies, *Optical sensors, Binding sites.

Several methods for immobilizing anti-T2 mycotoxin monoclonal antibodies on quartz fibers, for use in optical sensor development, have been evaluated with respect to the surface density and stability of the immobilized proteins. The first method activates matrix hydroxyl groups using p-toluenesulfonyl chloride (TSC). The second method activates these groups using p-nitrophenyl chloroformate (NPCF). The third method requires an initial silanization using 3-aminopropyl-triethoxysilane (APTES) followed by carrier activation

with glutaraldehyde. The activated carrier in all three methods is then reacted with the amino groups of the protein. The first two non-silanizing coupling methods are simple, inexpensive and non-hazardous compared to the third, more complex method in which an initial silanization step is required. The active antibody surface densities and stabilities were monitored at 4 and 50C. Each of these methods produced active antibody surface densities in the range of 181 - 297 ng/sq cm with half lives ranging from 30 to 80 hours at 50C but several months at 4C.

MILITARY SCIENCES

Antimissile Defense Systems

901,268
PB89-173405   Not available NTIS
National Bureau of Standards (NEL), Gaithersburg, MD. Electrosystems Div.
Strategic Defense Initiative Space Power Systems Metrology Assessment.
Final rept.
J. K. Olthoff, and R. E. Hebner. 1989, 4p
Pub. in Transactions of Symposium on Space Nuclear Power Systems (6th), Albuquerque, NM., January 8-12, 1989, p124-127.

Keywords: *Metrology, *Technology assessment, Reliability, Antimissile defense, Measurement, Temperature, Detectors, Calibrating, Reprints, *Strategic Defense Initiative, *Spacecraft power supplies.

A survey of Strategic Defense Initiative (SDI) programs has been performed to determine the measurement requirements of anticipated SDI space power systems. These requirements have been compared to present state-of-the-art metrology capabilities as represented by the calibration capabilities at the National Institute of Standards and Technology. Metrology areas where present state-of-the-art capabilities are inadequate to meet SDI requirements are discussed, and areas of metrology related research which appear promising to meet these needs are examined. Particular attention is paid to the difficulties of long-term, unattended sensor calibrations and measurement reliability.

901,269
PB89-209357   PC A07/MF A01
National Inst. of Standards and Technology (NEL), Gaithersburg, MD. Center for Electronics and Electrical Engineering.
Assessment of Space Power Related Measurement Requirements of the Strategic Defense Initiative.
Technical note (Final).
J. K. Olthoff, and R. E. Hebner. Apr 89, 147p NIST/TN-1259
Also available from Supt. of Docs. as SN003-003-02930-1. Sponsored by Defense Nuclear Agency, Washington, DC., and Strategic Defense Initiative Organization, Washington; DC.

Keywords: Measurement, Sensors, Electromagnetic fields, Lasers, Vibration, Neutron flux, *Strategic Defense Initiative.

A survey has been performed to determine the measurement requirements of space power related parameters for anticipated SDI systems. These requirements have been compared to present state-of-the-art metrology capabilities as represented by the calibration capabilities of the National Institute of Standards and Technology. Metrology areas where present state-of-the-art capabilities are inadequate to meet SDI requirements are discussed, and areas of metrology-related research which appear promising to meet these needs are examined. Particular attention is paid to the difficulties of long-term, unattended sensor calibrations and long-term measurement reliability.

NATURAL RESOURCES & EARTH SCIENCES

Logistics, Military Facilities, & Supplies

Logistics, Military Facilities, & Supplies

901,270
PB89-150965 Not available NTIS
National Bureau of Standards (NML), Gaithersburg, MD. Office of Physical Measurement Services.
Measurement Standards for Defense Technology.
Final rept.
B. C. Belanger, and L. Vestal. 1985, 8p
Pub. in Proceedings of National Conference of Standards Laboratories Workshop and Symposium, Boulder, CO., July 15-18, 1985, p206-213.

Keywords: *Technological intelligence, *National defense, *Standards, *Measurements, Lasers, Infrared surveillance, Millimeter waves, Radar, Electronic warfare, Spacecraft communication, Ammunition, Homing devices, Range finders, Guidance(Motion).

Over the past few years the state of the art of defense technology has advanced rapidly. Measurement requirements to support this technology are particularly demanding in technical areas such as millimeter waves (for radar, electronic warfare, satellite communications, and munitions guidance), lasers (for target designators, rangefinders, and weapons), and IR (for focal plane array space surveillance sensors and tactical missile homing sensors). The paper reviews how the National Bureau of Standards (NBS) and the Department of Defense (DOD) identify measurement and standards requirements and coordinate their planning and describes areas where R&D work is needed to meet future defense needs.

901,271
PB89-177075 Not available NTIS
National Bureau of Standards (NEL), Gaithersburg, MD. Robot Systems Div.
Hierarchically Controlled Autonomous Robot for Heavy Payload Military Field Applications.
Final rept.
H. G. McCain, R. D. Kilmer, S. Szabo, and A, Abrishamian. 1986, 10p
Sponsored by Human Engineering Lab., Aberdeen Proving Ground, MD.
Pub. in Proceedings of International Conference on Intelligent Autonomous Systems, Amsterdam, Netherlands, December 8-11, 1986, p372-381.

Keywords: *Materiel, *Materiels handling, *Robots, Cargo transportation, Research projects, Field Materiel-Handling Robot, Control systems, Real time systems. .

The U.S. Army Human Engineering Laboratory, with assistance from the National Bureau of Standards, Robot Systems Division, is developing a heavy-lift pallet handling robotic system designated as the Field Materiel-Handling Robot (FMR). The initial demonstration of the FMR will be the sensor-driven autonomous acquisition and high speed manipulation of pallets of artillery ammunition. The paper describes the FMR research and development project with emphasis on the robot control architecture and the sensor-driven autonomous operational capabilities.

901,272
PB89-235139 PC A03/MF A01
National Inst. of Standards and Technology (NEL), Boulder, CO. Electromagnetic Fields Div.
Alternative Techniques for Some Typical MIL-STD-461/462 Types of Measurements.
Technical note.
J. E. Cruz, and E. B. Larsen. Mar 89, 43p NIST/TN-1320
Also available from Supt. of Docs. as SN003-003-02946-7. Sponsored by Army Aviation Systems Command, St. Louis, MO.

Keywords: *Antenna radiation patterns, *Standards, *Electromagnetic absorption, *Measurement, Antennas, Electromagnetic radiation, Transmitters, Electromagnetic properties, Electromagnetic fields, *Military equipment.

The report presents antenna factors determined in a screenroom which was partially loaded with radio frequency (rf) absorbing material, using the two-antenna insertion-loss technique. These antenna factors are compared with the antenna factors obtained in an unloaded screenroom, a fully loaded screenroom (anechoic chamber), and at an open field site. In addition, measurements at the eight corners of a cube were made in the partially loaded and fully loaded screen-

room to determine the field deviation at the eight corners of the cube with respect to its center. Also, measurement improvements are quantified for the electric-field strength beneath a single-wire transmission line, in a partially loaded screenroom. Finally, electric-field measurements were made on top of the grounded table in a partially loaded screenroom to determine the field strength variation above the table.

901,273
PB90-128067 Not available NTIS
National Inst. of Standards and Technology (NEL), Gaithersburg, MD. Mathematical Analysis Div.
Analyzing the Economic Impacts of a Military Mobilization.
Final rept.
R. E. Chapman, C. M. Harris, and S. I. Gass. 1989, 34p
Sponsored by Federal Emergency Management Agency, Washington, DC.
Pub. in Proceedings of Institute of Cost Analysis National Conference on Cost Analysis Applications of Economics and Operations Research, Washington, DC., July 5-7, 1989, p353-386.

Keywords: *Military mobilizing, *Economic analysis, Economic mobilization, Readiness, National defense, Command and control, Mathematical models.

A military mobilization is a complex series of events, which if modeled adequately, can specify how a national economy makes the transition from a peacetime to a war-time footing. Problems in modeling such situations have highlighted the importance of evaluating large-scale, policy-oriented models prior to their use by decision makers. The current study outlines a generic procedure for conducting such an evaluation. Specifically, macro-economic modeling and a structured sensitivity analysis can be combined to measure and evaluate the economic impacts of a military mobilization.

Military Operations, Strategy, & Tactics

901,274
PB89-176507 Not available NTIS
National Bureau of Standards (NEL), Boulder, CO. Time and Frequency Div.
Secure Military Communications Can Benefit from Accurate Time.
Final rept.
D. W. Hanson, and J. L. Jespersen. 1986, 12p
Pub. in Proceedings of IEEE (Institute of Electrical and Electronics Engineers) Military Communications Conference, Monterey, CA., October 5-9, 1986, 12p.

Keywords: *Secure communication, *Military communication, Security, Coding, Countermeasures, Detection, Time division multiplexing, Pulse modulation, interception, Deception, National defense.

Some military communications systems have requirements quite different from civilian systems. Among others there are the necessities to protect military communications from detection, interception, exploitation, or disruption by adversaries particularly during times of hostilities. There are a number of techniques available to protect communications against these threats -- some of them can benefit from unambiguous time information. A hypothetical military communications system is used in the paper to discuss these protection techniques and to illustrate how time may assume a useful role in their operation. The paper will also cover sources external to the communications system from which the necessary time information may be obtained.

Nuclear Warfare

901,275
PB89-188809 PC A05/MF A01
National Bureau of Standards (NML), Gaithersburg, MD. Center for Radiation Research.

DCTDOS: Neutron and Gamma Penetration in Composite Duct Systems.
L. V. Spencer. Feb 87, 93p NBSIR-87/3534
Sponsored by Federal Emergency Management Agency, Washington, DC.

Keywords: *Radiation shielding, *Neutron absorption, *Nuclear weapons, Neutron albedo, Ducts, Computer programs, *Computer applications, *Gamma ray absorption.

The paper describes computer methods for estimating neutron and gamma ray fluence rate, dose, and even spectral features due to penetration through a series of duct segments. The procedure links together data on individual segments -- straight sections and bends -- in arbitrary combinations; and the resulting composite can include computations for a room at the end, if there is one. This particular method was developed for rapid estimates in the protection problems against nuclear weapons, but the concepts which are employed are more broadly applicable.

901,276
PB89-200208 PC A10/MF A01
National Inst. of Standards and Technology (NEL), Gaithersburg, MD. Center for Fire Research.
Assessment of Need for and Design Requirements of a Wind Tunnel Facility to Study Fire Effects of Interest to DNA.
W. M. Pitts. May 89, 209p NISTIR-89/4049
Sponsored by Defense Nuclear Agency, Washington, DC.

Keywords: *Fires, *Wind, Spreading, Dispersing, Urban areas, Nuclear explosions, Nuclear weapons, Wind tunnels.

The objective of the study is to recommend whether or not a new wind tunnel facility should be designed and constructed for the investigation of wind-aided fire spread. The focus is on the types of mass fire which can be expected following a nuclear detonation above an urban environment. The final conclusions of the report are (1) the need for an improved understanding of urban fire spread as it relates to nuclear weapon effects is overwhelming, (2) the authors currently have essentially zero predictive capability for fire damage in an urban environment following a nuclear attack, (3) wind tunnel experiments will not provide all of the required information, but offer the opportunity to substantially improve the understanding of the problem, (4) most existing wind tunnels were designed decades ago and are not well-suited for the required experimentation, and (5) some progress can be and is being made in existing facilities, but substantial improvements in understanding require a new facility and a sustained commitment for support.

NATURAL RESOURCES & EARTH SCIENCES

Geology & Geophysics

901,277
PB89-147037 Not available NTIS
National Bureau of Standards (NML), Gaithersburg, MD. Gas and Particulate Science Div.
Application of Synergistic Microanalysis Techniques to the Study of a Possible New Mineral Containing Light Elements.
Final rept.
E. S. Etz, D. E. Newbury, P. J. Dunn, and J. D. Grice. 1985, 5p
Pub. in Microbeam Analysis 1985, p60-64.

Keywords: *Microanalysis, *Minerals, Borate minerals, Carbonate minerals, Electron probes, Chemical elements, Raman spectroscopy, Classifications, Yttrium, Reprints, Light elements, Ion probes.

Complementary microanalysis techniques, including electron probe, ion microprobe and laser-Raman microanalysis, are applied to the compositional and structural characterization of a candidate new mineral containing light elements. The findings from ion probe analysis (SIMS) indicate the presence of boron and carbon as major constituents in addition to yttrium as the chief rare-earth element. The application of Raman microprobe analysis is explored to substantiate the results of optical measurements and x-ray structure determinations concerning the existence of carbonate and borate species. A series of synthetic components and natural minerals are studied by micro-Raman spectroscopy to obtain a spectral data base for structurally complex carbonates, borates, and minerals containing both species. These data are used in the interpretation of the Raman spectrum of the unknown Y/B/C-mineral and its classification as an yttrium carbonborate containing hydrogen-bonded borate groupings.

901,278
PB89-150882 Not available NTIS
National Bureau of Standards (IMSE), Gaithersburg, MD. Metallurgy Div.
Multicritical Phase Relations in Minerals.
Final rept.
B. P. Burton, and P. M. Davidson. 1988, 31p
Pub. in Structural and Magnetic Phase Transitions in Minerals, Chapter 4, p60-90 1988.

Keywords: *Crystallization, *Phase diagrams, *Carbonate minerals, Mathematical models, Hematite, Ilmenite, Heat of mixing, Critical point, Topology, Reprints, Diopside, Jadeite.

Models of multicritical phase relations are reviewed and theoretical results pertaining to the rhombohedral carbonates, hematite-ilmenite, and diopside-jadeite systems are discussed. The microscopic interactions that cause ordering and phase separation in these systems are highly anisotropic, such that ordering is favored in one crystallographic direction but clustering is favored in another. Model calculations which incorporate such interactions predict appropriate phase diagram topologies and appreciate composition and temperature dependence for excess heats of mixing.

901,279
PB89-185953 Not available NTIS
National Bureau of Standards (NML), Boulder, CO. Quantum Physics Div.
Relationships between Fault Zone Deformation and Segment Obliquity on the San Andreas Fault, California.
Final rept.
R. Bilham, and K. Hurst. 1988, 15p
Pub. in Proceedings of China-U.S. Symposium on Crustal Deformation and Earthquakes, Wuhan, People's Republic of China, October 29, 1985, p510-524 1988.

Keywords: *San Andreas Fault, *Geological faults, *Earthquakes, *Creep rate, Earth movements, California, Numerical analysis, *Geologic structures.

Faults of the San Andreas system are considered as a sequence of contiguous straight segments with lengths from 2-30 km. A dominant length of approximately 12 km appears to exist. The segments are not parallel to each other nor to the plate slip vector calculated from global plate motions. Although it is possible to choose a regional slip vector that will approximate the local strike of the fault by invoking slip on adjacent faults, it is not possible to eliminate oblique slip on all segments simultaneously. The detailed consequences of oblique slip on fault segments are described in terms of their impedance, a measure of the resistance of segment to fault motion. Impedance is defined to be the change of area resulting from slip on the fault. Circumstantial evidence suggests that segments with anomalously large or small impedance may play a role in terminating or initiating earthquake rupture. The inferred instantaneous compression rate of creeping sections of the fault is calculated. It is shown that the present creep rate is insufficient to create the observed transpressive features on the southern San Andreas Fault and that much of the observed deformation must occur during earthquakes.

901,280
PB89-185979 Not available NTIS
National Bureau of Standards (NML), Boulder, CO. Quantum Physics Div.
Transducers in Michelson Tiltmeters.
Final rept.
R. Bilham. 1988, 12p
Pub. in Proceedings of China-U.S. Symposium on Crustal Deformation and Earthquakes, Wuhan, Peo-

ple's Republic of China, October 29, 1985, p264-275 1988.

Keywords: *Indicating instruments, *Water distribution, *Earthquakes, *Earth surface, Design criteria, Performance evaluation, Transducers, *Tiltmeters.

A Michelson tiltmeter consists of a horizontal pipe in which a continuous water surface extends from end to end. Tilt of the Earth's surface causes an increase in water depth at one end and a corresponding decrease in water depth at the other. Methods to detect these changes in water level to 0.001 mm accuracy are reviewed and several new methods are discussed. A new transducer is described that senses movements of the image position of a light emitting diode (LED) after it has been reflected from the water surface. The projected image of the LED is followed by a null-seeking servo system based on a silicon photodiode bi-cell whose vertical position is monitored by an LVDT transducer. The absolute depth of the water may be measured directly by the LED follower transducer to approximately 1 micrometers accuracy. In a 1 km tiltmeter this is equivalent to a long term measurement accuracy of 1 nanoradian/year. System accuracy, however, is only as good as the accuracy with which the water level transducer is indexed to the Earth's surface. The method uses an array of vertical extensometers to identify locally generated signals that are typically of non-tectonic origin.

901,281
PB89-227946 Not available NTIS
National Bureau of Standards (NML), Boulder, CO. Quantum Physics Div.
High-Precision Absolute Gravity Observations in the United States.
Final rept.
G. Peter, R. E. Moose, C. W. Wessells, J. Faller, and T. M. Niebauer. 1989, 16p
Pub. in Jnl. of Geophysical Research 94, nB5 p5659-5674 May 89.

Keywords: *Gravity, United States, Precision, Measurement, Reprints, Ground motion.

From May 1987 to June 1988 the National Geodetic Survey (NGS) made approximately 50 observations at 30 sites with one of the six absolute gravimeters built by the Joint Institute for Laboratory Astrophysics between 1983 and 1986. Of the 10 sites where two to three observations were made, the scatter about the mean site values ranged from under plus or minus 1 microGal to plus or minus 4.0 microGal. The data correction methods now employed at NGS allow the establishment of high-precision reference gravity stations in the United States and abroad for monitoring the temporal variations of gravity and studying vertical ground motions.

901,282
PB89-234272 Not available NTIS
National Inst. of Standards and Technology (NML), Boulder, CO. Quantum Physics Div.
Rate of Change of the Quincy-Monument Peak Baseline from a Translocation Analysis of LAGEOS Laser Range Data.
Final rept.
A. Stolz, M. A. Vincent, P. L. Bender, R. J. Eanes, M. M. Watkins, and B. D. Tapley. 1989, 4p
Pub. in Geophysical Research Letters 16, n6 p539-542 Jun 89.

Keywords: *Geological faults, Geodynamics, Earthquakes, California, Motion, Reprints, *Plate tectonics, *San Andreas Fault, Laser range finders, Lageos(Satellite).

Translocation studies of LAGEOS laser range data from Quincy and Monument Peak in California observed during 1984-1987 suggest that plate tectonic motion across the San Andreas fault system in the direction of the baseline between the two stations is uniform at a rate of -30(+ or - 3) mm/a. Changes in the components of the baseline vector were inferred from repeat determinations using the solutions from successive half-year intervals. The changes in the vertical and transverse components of the Quincy-Monument Peak baseline are -0.4(+ or - 5) mm/a and +14(+ or - 5) mm/a respectively. The vertical component determinations attest to the height stability of the laser ranging method. LAGEOS measurements made from Quincy and Monument Peak before 1984 are inaccurate enough to limit their usefulness for plate tectonic studies.

901,283
PB90-136649 Not available NTIS
National Inst. of Standards and Technology (NML), Boulder, CO. Time and Frequency Div.
Tilt Observations Using Borehole Tiltmeters 1. Analysis of Tidal and Secular Tilt.
Final rept.
J. Levine, C. Meertens, and R. Busby. 1989, 13p
Contracts F19628-81-K-0040, F19628-78-C-0065
Sponsored by Air Force Geophysics Lab., Hanscom AFB, MA.
Pub. in Jnl. of Geophysical Research 94, nB1 p574-586, 10 Jan 89.

Keywords: *Boreholes, *Site surveys, Colorado, Wyoming, Yellowstone National Park, Oceans, Topography, Reprints, *Tiltmeters, *Earth tides, *Attitude(Inclination), Erie(Colorado), Secular variations.

The authors have designed a borehole tiltmeter using two horizontal pendulums which have periods of 1 s. They have installed the instruments at seven sites in Colorado and Wyoming to evaluate the secular tilt, the tides, and the coherence between nearby instruments. Using 28 days of data from Boulder, Colorado, the estimates agree with models that include the body tide, the ocean load, and the topographic correction to better than the estimated uncertainty. Tidal measurements at Erie, Colorado, have larger, possibly nonrandom variability that may be caused by a coupling between the tides and long-period tilts. Measurements at Erie, Colorado and in Yellowstone National Park, Wyoming exhibit an annual or biannual periodicity.

Mineral Industries

901,284
PB89-175947 Not available NTIS
National Bureau of Standards (IMSE), Gaithersburg, MD. Ceramics Div.
Microbiological Metal Transformations: Biotechnological Applications and Potential.
Final rept.
G. J. Olson, and R. M. Kelly. 1986, 15p
Pub. in Biotechnol. Prog. 2, n1 p1-15 1986.

Keywords: *Microbiology, *Metals, *Bioengineering, *Mineral deposits, *Copper, *Uranium, Catalysis, Strategic materials, Materials recovery, Hydrometallurgy, Solution mining, Reprints.

The article reviews the recent literature on biological and engineering aspects of biotechnological metals recovery. Microorganisms catalyze many transformations of metals including, solubilization, precipitation and volatilization reactions, often associated with metal reduction, oxidation, alkylation or dealkylation reactions. There is a growing awareness in the microbiological, engineering, and mining fields that such reactions are of importance in metal recovery operations. Currently, copper and uranium are being commercially recovered from ores via biohydrometallurgy, and extension of the technology to other metals, especially precious and strategic, is underway. Efforts to commercialize such activities are hampered by a number of biological and engineering considerations, in part relating to inadequate knowledge of mechanisms of metal solubilization on surfaces by microorganisms and engineering problems related to heterogeneous systems in bioprocessing of solid materials. Various potential process engineering designs are discussed.

901,285
PB89-202113 Not available NTIS
National Bureau of Standards (IMSE), Gaithersburg, MD. Ceramics Div.
Novel Flow Process for Metal and Ore Solubilization by Aqueous Methyl Iodide.
Final rept.
J. S. Thayer, G. J. Olson, and F. E. Brinckman. 1987, 7p
Pub. in Appl. Organomet. Chem. 1, n1 p73-79 1987.

Keywords: *Solution mining, *Strategic materials, Metals, Atomic spectra, Absorption spectra, Bioengineering, Iron, Copper, Leaching, Wastes, Liquid flow, Hydrometallurgy, Metal containing organic compounds, Mineral deposits, Reprints, Iodomethanes.

Mineral Industries

A novel process for bulk metal and metal ore solubilization by aqueous methyl iodide is described. A flow bioreactor system was developed and used in connection with a graphite furnace atomic absorption spectrometer for continuous, on-line quantitation of dissolved metals. Dissolution of binary and ternary metal ores as well as bulk metals was enhanced 5-145x by aqueous methyl iodide in the flow system. Repeated alternating cycles of water and aqueous methyl iodide resulted in increasing enhancement of bulk from dissolution apparently due to a uncovering of fresh new reactive surfaces. Films of copper on circuit boards were also dissolved by methyl iodide. The process has many possible uses in mining and metallurgy especially for recovery of precious and/or strategic metals from difficult to reach locations in ores or wastes.

901,286
PB89-221154 PC A04/MF A01
National Inst. of Standards and Technology (NEL), Gaithersburg, MD. Center for Mfg. Engineering.
Mining Automation Real-Time Control System Architecture Standard Reference Model (MASREM).
Technical note (Final).
J. Albus, R. Quintero, H. M. Huang, and M. Roche.
May 89, 64p NIST/TN-1261-VOL-1
Also available from Supt. of Docs. as SN003-003-02948-3. Sponsored by Bureau of Mines, Pittsburgh, PA.

Keywords: *Real time systems, *Automation, *Mining equipment, *Mining, Models, Control systems, Interfaces, Computer software, Memory devices, Adaptive control systems, Task decomposition, Sensory processing.

The Mining Automation Real-Time Control System Architecture Standards Reference Model (MASREM) defines a logical hierarchical architecture for mining automation. The MASREM architecture defines a set of standard modules and interfaces which facilitates software design, development, validation, and test, and makes possible the integration of software from a wide variety of sources. Standard interfaces also provide the software hooks necessary to incrementally upgrade future mining automation systems as new capabilities develop in computer science, robotics, and autonomous system control.

Natural Resource Surveys

901,287
PB89-201214 Not available NTIS
National Bureau of Standards, Gaithersburg, MD.
Office of the Director.
Environmental Intelligence.
Final rept.
B. D. Kraselsky, and C. C. Gravatt. 1989, 13p
Pub. in Technology in Society 11, p99-111 1989.

Keywords: *Economic analysis, *Environmental surveys, Data acquisition, Remote sensing, Economic surveys, Land surveys, Commerce, Observation, Examination, Regulations, Reprints, *Satellite surveys, Landsat.

Satellite remote sensing is being used to obtain a wide range of information about the land, the oceans, the atmosphere and man-made objects. The information, or environmental intelligence, is used to help solve problems that affect the general population, such as the greenhouse effect, the depletion of the ozone layer and land use planning, as well as for specialized commercial activities such as agriculture and forestry evaluations, mineral and petroleum exploration. The level of commercial involvement in the use of satellite remote-sensing data is increasing, due to technological and economic factors. However, commercial involvement in satellite system operations is plagued with economic impediments, is running into conflict with international obligations that regard this activity as a public good, and is being chilled by ambiguous domestic laws and regulations. The paper identifies the trends in commercial development; the legal, political, technical and economic factors that will affect the rate of the development; and the impact that such development will have on the overall conduct of satellite remote sensing activity.

Soil Sciences

901,288
PB89-186431 Not available NTIS
National Bureau of Standards (NML), Gaithersburg, MD. Gas and Particulate Science Div.
Pahasapaite, a Beryllophosphate Zeolite Related to Synthetic Zeolite Rho, from the Tip Top Pegmatite of South Dakota.
Final rept.
R. C. Rouse, T. J. Campbell, P. J. Dunn, D. Newbury, D. R. Peacor, W. L. Roberts, and F. J. Wicks. 1987, 8p
Pub. in Neues Jahrbuch fur Mineralogie Monatshefte, n10 p433-440 1987.

Keywords: *Minerals, *Ion exchange resins, Calcium, Lithium, Potassium, Sodium, Phosphorus, Beryllium, Crystal structure, X ray diffraction, Refractivity, Density(Mass/volume), Beryl, Reprints, *Pahasapaite.

Pahasapaite, (Ca5.5Li3.6K1.2Na0.2 0 13.5)Li8Be24P24O96(center dot)38H20, is a new zeolite mineral associated with roscherite, tiptopite, and englishite at the Tip Top Mine, Custer, South Dakota. It is cubic, I23, with a = 13.781(4) angstroms and Z=1. The strongest powder X-ray diffraction lines are (d(angstroms), I,hkl): 9.60, 100, 110; 3.684, 90, 321; 3.248, 90, 411, 330; 2.935, 90, 332; 2.702, 60, 510, 431; 2.237, 40, 611, 532; and 4.35, 40, 310. Pahasapaite occurs as light pink, yellow-green, or colorless crystals, which are 1.0 mm in size and show the forms (110) and (111), There is no apparent cleavage, the refractive index is 1.523(2), and the observed and calculated densities are 2.28(4) and 2.241 g/cu m, respectively. A crystal structure determination shows pahasapaite to be a beryllophosphate zeolite with a tetrahedral framework configuration like that in systhetic zeolite rho. The name is from the Lakota Sioux word "Pahasapa," meaning Black Hills, in allusion to the locality.

901,289
PB89-188585 PC A03/MF A01
National Bureau of Standards (NEL), Boulder, CO. Center for Electronics and Electrical Engineering.
Dielectric Mixing Rules for Background Test Soils.
R. G. Geyer. Jun 88, 25p NBSIR-88/3095
Sponsored by Army Belvoir Research and Development Center, Fort Belvoir, VA.

Keywords: *Dielectric properties, *Mixing, *Soil tests, Simulation, Electromagnetic absorption, Microwaves, Remote sensing, Estimates, Soil-water mixtures.

The bulk, or effective dielectric constant of any background test medium (whether naturally occurring or synthetic) determines the electromagnetic visibility of buried objects. Heuristic mixing rules are considered that allow the prediction of complex dielectric behavior in linear, homogeneous, isotropic, and lossy multiphase soil mixtures. Measurement results in bioelectromagnetic and microwave remote sensing suggest a refractive mixing model as that being most suited for dry soils or soil-water mixtures.

901,290
PB89-209274 PC A04/MF A01
National Inst. of Standards and Technology (NEL), Gaithersburg, MD. Center for Building Technology.
Site Characterization for Radon Source Potential.
F. Y. Yokel. Jun 89, 59p NISTIR-89/4106
Sponsored by Department of Housing and Urban Development, Washington, DC. Innovative Projects and Special Technology Div., New Jersey Div. of Housing and Development, Trenton, and Ryland Group, Inc., Columbia, MD.

Keywords: *Site surveys, *Radon, *Soil analysis, *Buildings, Exploration, Concentration(Composition), Soil properties, Moisture content, Permeability, Atmospheric pressure, Wind velocity, Field tests, Gamma ray spectroscopy, Density(Mass/volume), Sources, Numerical analysis, Temperature, Radioactivity, *Soil gases, Environmental transport, Indoor air pollution.

Radon source potential characterization of sites in terms of soil index properties which do not vary with transient conditions such as moisture content, barometric pressure, temperature and wind speed is studied. The invariant index properties which were found to be critical for site characterization are radium activity concentration in the soil, in-place dry density, porosity, and dry gas permeability. These properties can be measured in situ or in the laboratory or estimated on the basis of other soil index properties such as grain-size distribution and Atterberg limits. Various expressions for radon source potential are reviewed and a new expression is formulated on the basis of data from areas of deep glacial terrace deposits. Site exploration methods proposed include use of the Standard Penetration Test together with a laboratory determination of radium activity concentration, and a rapid field measurement procedure using a portable gamma ray spectrometer, a portable nuclear moisture-density meter and retrieval of a soil sample for laboratory determination of particle-size distribution. A plan to develop exploration protocols, test the effectiveness of the source potential prediction, and prepare a draft exploration standard is proposed.

901,291
PB89-211973 Not available NTIS
National Bureau of Standards (NEL), Gaithersburg, MD. Structures Div.
Laboratory Evaluation of an NBS (National Bureau of Standards) Polymer Soil Stress Gage.
Final rept.
R. M. Chung, A. J. Bur, and J. R. Holder. 1985, 6p
Pub. in Proceedings of Symposium on the Interaction of Non-Nuclear Munitions with Structures (2nd), Panama City Beach, FL., April 15-18, 1985, p296-301.

Keywords: *Soil mechanics, *Stress analysis, *Measuring instruments, Laboratory equipment, Determination of stress, Performance evaluation, *Gages.

The NBS polymer gage, which is made of thin sheets of polyvinylidene fluoride (PVDF) sandwiched between polycarbonate sheets, has been tested extensively at the National Bureau of Standards to evaluate its ability to measure soil dynamic stresses due to blast loading.

NAVIGATION, GUIDANCE, & CONTROL

Navigation Systems

901,292
PB89-174080 Not available NTIS
National Bureau of Standards (NML), Boulder, CO. Time and Frequency Div.
Apparent Diurnal Effects in the Global Positioning System.
Final rept.
M. Weiss. 1987, 16p
Pub. in Proceedings of Annual Precise Time and Time Interval (PTTI) Applications and Planning Meeting (19th), Redondo Beach, CA., December 1-3, 1987, p33-48.

Keywords: *Diurnal variations, Bias, Displacement, Errors, Delay time, Resolution, Sensitivity, Damping, Periodic variations, *Global positioning system, *Time transfer.

Since the Global Positioning System (GPS) has been used for common view time and frequency transfer between remote locations various systematic effects have been observed. These effects have been discussed on various occasions appearing as biases between different daily measurements as well as obstructing closure in around-the-world time transfer. GPS satellites are examined from several locations around the world, after linking the ground station clocks GPS. The results are that there are apparent diurnal variations in many of the SV clocks. These systematic effects are studied, the biases in common view time transfer, the lack of closure in around-the-world time transfer, and the diurnal variations in the SV clocks. The diurnal effects are primarily due to errors in the transmitted satellite ephemeris and ionospheric model.

901,293
PB89-185730 Not available NTIS

National Bureau of Standards (NML), Boulder, CO. Time and Frequency Div.
Using Multiple Reference Stations to Separate the Variances of Noise Components in the Global Positioning System.
Final rept.
M. A. Weiss, and D. W. Allan. 1986, 11p
Pub. in Proceedings of Annual Symposium on Frequency Control (40th), Philadelphia, PA., May 28-30, 1986, p394-404.

Keywords: *Periodic variations, *Electromagnetic noise, Space surveillance(Spaceborne), Divergence, Atmospherics, Natural radio frequency interference, Clocks, Correlation, Variability, *Global positioning systems.

The separation of variance technique has been applied to measurements of a clock against received signals from Global Positioning System (GPS) satellites to separate out various noise components in the system. First, the authors show how measurements can be taken from several different locations to obtain estimates of more components of GPS system and to obtain better estimates of the components previously studied. It is shown how to estimate the variances of the GPS system clock, the error in the transmitted correction term between the satellite clock and the GPS system clock, propagation noise in the measurement including ionospheric and tropospheric modelling errors, error in the transmitted ephemeris for the satellite, and the local reference clock. The authors consider the effects of correlations between elements of the data and analyze the confidence one may have in the estimates in light of those correlations. Finally, the multi-station separation of variance technique is applied to recent GPS data. New insights into the GPS system that have been learned using the technique are discussed.

NUCLEAR SCIENCE & TECHNOLOGY

Isotopes

901,294
PB89-146872   Not available NTIS
National Bureau of Standards (NML), Gaithersburg, MD. Inorganic Analytical Research Div.
Analytical Applications of Neutron Depth Profiling.
Final rept.
R. G. Downing, J. T. Maki, and R. F. Fleming. 1987, 14p
Pub. in Jnl. of Radioanalytical and Nuclear Chemistry 112, n1 p33-46 1987.

Keywords: *Neutron irradiation, *Depth detectors, Isotopes, Probes, Nondestructive tests, Metals, Ceramics, Reprints.

Using a low-energy neutron beam as an isotopic probe, neutron depth profiling (NDP) provides quantitative depth profiles in nearly all solid matrix materials. Several of the light elements, such as He, Li, B, and N can be non-destructively analyzed by NDP. The information obtained using NDP is difficult if not impossible to determine by non-nuclear techniques. As a result, NDP is used collaboratively with techniques as SIMS, RBS, FTIR, PGAA, and AES. Profiles measured by NDP are given for semiconductor and optical processing materials, and light weight alloys. Improvements in the technique are discussed with emphasis on the use of intense cold neutron beams.

901,295
PB89-171888   Not available NTIS
National Bureau of Standards (NML), Gaithersburg, MD. Ionizing Radiation Physics Div.
NBS (National Bureau of Standards) Radon-in-Water Standard Generator.
Final rept.
J. M. R. Hutchinson, P. A. Mullen, and R. Colle. 1986, 5p
Pub. in Nuclear Instruments and Methods in Physics Research A247, n2 p385-389, 15 Jun 86.

Keywords: *Radioactivity, *Standards, Water, Reprints, *Radon 222, Radium 226.

NBS has completed the development of a transfer standard for radon-in-water measurements. This standard can be used to generate and accurately dispense radium-free (222)Rn solutions of known concentration. The present finalized version is based on an earlier and previously described prototype. The standard consists of a polyethylene-encapsulated (226)Ra solution source in a small-volume accumulation chamber and an ancillary mixing and dispensing system which is partially automated with motor-driven syringes. The revised source configuration is more stable than the original prototype and the mixing and dispensing system is more compact, rugged, and convenient. The standard generator was calibrated and certified in terms of the (222)Rn concentration or total activity in an aliquot dispensed from the generator when a detailed operating procedure is adequately followed. The overall uncertainty of the calibration was estimated to be approximately $+$ or $- 4\%$.

901,296
PB89-176770   Not available NTIS
National Bureau of Standards (NML), Gaithersburg, MD. Radiometric Physics Div.
Using 'Resonant' Charge Exchange to Detect Traces of Noble Gas Atoms.
Final rept.
J. E. Hardis, W. R. Peifer, C. L. Cromer, A. L. Migdall, and A. C. Parr. 1989, 4p
Pub. in Proceedings of International Symposium on Resonance Ionization Spectroscopy and Its Applications (4th), Gaithersburg, MD., April 10-15, 1988, p237-240 1989.

Keywords: *Resonance, *Krypton, *Rubidium, *Mass spectroscopy, *Ionization, Detection, Trace elements, Rare gases, Interferometers, Charge carriers, Cross sections, Isotopic labeling.

An experiment in progress is described to measure the charge-exchange cross sections of Kr+ incident upon Rb. The column density in the Rb cell will be measured using an optical interferometer. The reaction is expected to generate a significant flux of Kr atoms in the 5s $(3/2)$ (J=2) metastable state, which will be useful as a step in RIMS studies of Kr isotope distributions.

901,297
PB89-201669   Not available NTIS
National Bureau of Standards (NML), Gaithersburg, MD. Gas and Particulate Science Div.
Single Particle Standards for Isotopic Measurements of Uranium by Secondary Ion Mass Spectrometry.
Final rept.
D. S. Simons. 1986, 11p
Pub. in Jnl. of Trace Microprobe Tech. 4, n3 p185-195 1986.

Keywords: *Uranium, *Radioactive isotopes, *Mass spectroscopy, Glass, Standards, Quantitative analysis, Ions, Reprints.

The isotopic abundance ratios of uranium have been determined from individual glass microparticles using secondary ion mass spectrometry (SIMS). Synthetic glasses were prepared using oxides of known isotopic composition as starting materials. Relative precisions and accuracies of better than 1% could be attained for (235)U/(238)U ratios between 0.0020 and 0.0072. These were primarily limited by counting statistics. Even the abundances of the minor isotopes (234)U and (236)U could be determined since molecular ion interferences were not present at measurable levels.

Nuclear Instrumentation

901,298
PB89-147508   Not available NTIS
National Bureau of Standards (NML), Gaithersburg, MD. Ionizing Radiation Physics Div.
Measurement Quality Assurance.
Final rept.
E. H. Eisenhower. 1988, 7p
Pub. in Health Physics 55, n2 p207-213 1988.

Keywords: *Quality assurance, *Measurement standards, *Ionizing radiation, Measurement, Radiation protection, Health physics, Quality control, Laboratories, Reprints.

The quality of a radiation protection program can be no better than the quality of the measurements made to support it. In many cases that quality is unknown, and is merely assumed on the basis of a calibration of a measuring instrument. If that calibration is inappropriate or is performed improperly, the measurement result will be inaccurate and misleading. Assurance of measurement quality can be achieved if appropriate procedures are followed, including periodic quality control actions that demonstrate adequate performance. Several national measurement quality assurance (MQA) programs are operational or under development in specific areas. They employ secondary standards laboratories that provide a high-quality link between the National Bureau of Standards and measurements made at the field use level. The procedures followed by these secondary laboratories to achieve MQA will be described, as well as plans for similar future programs. A growing general national interest in quality assurance, combined with strong specific motivations for MQA in the area of ionizing radiation, will provide continued demand for appropriate national programs.

901,299
PB89-176549   Not available NTIS
National Bureau of Standards (NML), Gaithersburg, MD. Ionizing Radiation Physics Div.
Monte Carlo Calculated Response of the Dual Thin Scintillation Detector in the Sum Coincidence Mode.
Final rept.
K. C. Duvall, and R. G. Johnson. 1988, 3p
Pub. in Proceedings of International Conference on Nuclear Data for Science and Technology, Mito, Japan, May 30-June 3, 1988, p419-421.

Keywords: *Scintillation counters, *Neutron counters, Spectra, Mathematical models, Pulse height analyzers, Monte Carlo method, Neutron flux, Reaction time, Efficiency, Coincidence circuits.

The Dual Thin Scintillator is a unique neutron detector that is being developed for improved fluence and spectrum measurement. Current attention has been directed towards understanding some details of the detector response in the sum coincidence mode of operation where a peaked pulse-height response is exhibited throughout the energy region of interest. As a result of the peaked distribution, the detector efficiency is a weak function of the pulse-height bias, allowing the number of recorded events above the bias to be determined with greater certainty. A Monte Carlo code has been used to calculate the sum coincidence pulse-height response at several energies within the 1 to 15 MeV region. The detector efficiency as a function of neutron energy has also been calculated. The results of the Monte Carlo calculations, which include the effect of multiple scattering on the shape of the response function and efficiency curve are presented.

901,300
PB89-229165   Not available NTIS
National Inst. of Standards and Technology (NML), Gaithersburg, MD. Ionizing Radiation Physics Div.
International Intercomparison of Neutron Survey Instrument Calibrations.
Final rept.
J. B. Hunt, P. Champlong, M. Chemtob, H. Kluge, and R. B. Schwartz. 1989, 8p
Pub. in Radiation Protection Dosimetry 27, n2 p103-110 1989.

Keywords: Reprints, *Neutron monitors, *Survey monitors, Interlaboratory comparisons, Calibration, International separation.

An informal intercomparison of the methods of calibration of neutron area survey meters has been undertaken by NIST, NPL, PTB and ETCA-CEA. The measurement programming was based upon the calibration of two different types of survey instrument in the neutron fields emitted by bare and by heavy water moderated californium spontaneous fission neutron sources. One of the transfer devices was a conventional spherical survey meter and the other was a new type of monitor which attempts to take the neutron spectral distribution into account through the ratio of responses of two different sized moderating spheres. The results are compared and demonstrate that agreement among the various institutions can be obtained within $+$ or $-$ 3% for conventional instruments, but larger differ-

NUCLEAR SCIENCE & TECHNOLOGY

Nuclear Instrumentation

ences may exist for devices which are new and unconventional.

901,301
PB90-117532 Not available NTIS
National Inst. of Standards and Technology (NML), Gaithersburg, MD. Ionizing Radiation Physics Div.
Method for Evaluating Air Kerma and Directional Dose Equivalent for Currently Available Multi-Element Dosemeters in Radiation Protection Dosimetry.
Final rept.
M. Ehrlich. 1989, 7p
Pub. in Radiation Protection Dosimetry 28, n1-2 p89-95 1989.

Keywords: *Radiation protection, Reprints, *Kerma, *Dose equivalents, *Personnel dosimetry, *Dosemeters.

A method is outlined for estimating air kerma and directional dose equivalent from indications on multi-element dosemeters having energy and angle response functions that are different for at least two of the dosemeter's radiation-sensitive elements. The method employs dosemeter calibrations relating response of each element and the corresponding indication ratio(s) to radiation energy and angle of radiation incidence. Indications are measured in the usual way on the elements of each dosemeter irradiated under unknown conditions, and the ratio is formed of the indications of any two elements known to have different response functions. Using the calibration curves, energy-angle pairs are found that correspond to these indication ratios, and are then used to determine the corresponding response values of which there are several per energy-angle pair. Finally, the response values corresponding to any given energy-angle pair are averaged, and air kerma or directional dose equivalent is computed from average response and the measured indications. The method is illustrated for a dosemeter for which photon calibration data were available.

Radioactive Wastes & Radioactivity

901,302
NUREG/CP-0103 PC A10/MF A02
Nuclear Regulatory Commission, Washington, DC. Office of Nuclear Regulatory Research.
Proceedings of the Workshop on Cement Stabilization of Low-Level Radioactive Waste. Held at Gaithersburg, Maryland on May 31-June 2, 1989.
P. R. Reed. Oct 89, 224p NISTIR-89/4178
Also available from Supt. of Docs. Sponsored by National Inst. of Standards and Technology (NEL), Gaithersburg, MD. Building Materials Div.

Keywords: *Meetings, *Cements, *Solidification, *Stabilization, Leaching, Thermal cycling tests, Biodeterioration, Structural forms, Curing, *Low-level radioactive wastes.

The workshop on Cement Stabilization of Low-Level Radioactive Waste was co-sponsored by the U.S. Nuclear Regulatory Commission and National Institute of Standards and Technology and held in Gaithersburg, Maryland on May 31-June 2, 1989. The workshop provided a forum for exchanging information on the solidification and stabilization of low-level radioactive waste in cement among federal and state regulators, nuclear power station operators, cement vendors, national laboratory researchers and consultants. The workshop was structured into a 'Plenary' and four 'Working Group' sessions. Each working group session discussed specific issues: Lessons learned from small- and full-scale waste forms and observations at nuclear power stations; Laboratory test experience and application to problem waste streams; Stabilized waste form testing guidance; and Waste characterization, solidification, and process control programs.

901,303
PB89-215362 PC A07/MF A01
National Inst. of Standards and Technology (NEL), Gaithersburg, MD. Center for Building Technology.
Service Life of Concrete.
J. R. Clifton, and L. I. Knab. Jun 89, 150p NISTIR-89/4086
Sponsored by Nuclear Regulatory Commission, Washington, DC.

154

Keywords: *Concretes, Degradation, Service life, Mathematical models, Corrosion mechanisms, Accelerated tests, Durability, Permeability, *Radioactive waste management, Underground disposal, Low-level radioactive wastes.

The U.S. Nuclear Regulatory Commission (NRC) has the responsibility for developing a strategy for the disposal of low-level radioactive waste (LLW). An approach being considered for their disposal is to place the waste forms in concrete vaults buried in earth. A service life of 500 years is required for the concrete vaults as they may be left unattended for much of their lives. The report examines the basis for making service life predictions based on accelerated testing and mathematical modeling of factors controlling the durability of concrete buried in the ground. Degradation processes are analyzed based on considerations of their occurrence, extent of potential damage, and mechanisms. A recommended research plan for developing methods for predicting the service life of concrete is presented. Concepts of quality and factors affecting quality of concrete are discussed. Permeability is discussed in terms of the water-to-cement ratio, the pore structure of concrete, and the effects of cracks.

Reactor Engineering & Nuclear Power Plants

901,304
PB89-168017 PC A11/MF A01
National Inst. of Standards and Technology (IMSE), Gaithersburg, MD. Reactor Radiation Div.
NBS (National Bureau of Standards) Reactor: Summary of Activities July 1987 through June 1988.
Technical note Jul 87-Jun 88.
C. O'Connor. Jan 89, 230p NIST/TN-1257
Also available from Supt. of Docs. as SN003-003-02920-3. See also PB83-218636.

Keywords: *NBSR reactor, Activation analysis, Crystal structure, Isotopes, Diffraction, Neutron, Radiography, Nondestructive tests.

The report summarizes all those programs which use the NBS Reactor. It covers the period for July 1987 through June 1988. The programs range from the use of neutron beams to study the structure and dynamics of materials through nuclear physics and neutron standards to sample irradiations for activation analysis, isotope production, neutron radiography, and nondestructive evaluation.

Reactor Fuels & Fuel Processing

901,305
PB89-176556 Not available NTIS
National Bureau of Standards (NML), Gaithersburg, MD. Ionizing Radiation Physics Div.
Measurements of the (235)U (n,f) Standard Cross Section at the National Bureau of Standards.
Final rept.
R. G. Johnson, A. D. Carlson, O. A. Wasson, K. C. Duvall, J. W. Behrens, M. M. Meier, B. D. Patrick, and M. S. Dias. 1988, 4p
Pub. in Proceedings of International Conference on Nuclear Data for Science and Technology, Mito, Japan, May 30-June 3, 1988, p1037-1040.

Keywords: *Fissionable materials, *Standards, *Neutron cross sections, *Uranium 235, Nuclear fuels, Linear accelerators, Measurement.

The Neutron Interactions and Dosimetry Group at the National Bureau of Standards (NBS) has had a long-term program for the measurement of standard neutron cross sections. The group has maintained a significant effort on the measurement of one of the most important of these cross sections -- the neutron-induced fission cross section of 235U. Since the ENDF/B-VI evaluation has been recently released, it is appropriate to review the measurements of the (235)U(n,f) cross section which have been made at the NBS using accelerator-based neutron sources. In the 0.1 to 20 MeV region where the cross section is a standard, six separate measurements of the differential-cross section, using a variety of techniques have been made.

Both the NBS 150-MeV Electron Linac and the 3-MV Positive Ion Accelerator have been used as neutron sources. Two of the measurements are relative to the H(n,p) cross section while the remainder are absolute. These measurements will be reviewed and compared to ENDF/B-VI. The current status of the program and possible future improvements will be discussed.

Reactor Physics

901,306
PB89-171946 Not available NTIS
National Bureau of Standards (NML), Gaithersburg, MD. Inorganic Analytical Research Div.
Use of Focusing Supermirror Neutron Guides to Enhance Cold Neutron Fluence Rates.
Final rept.
M. Rossbach, O. Scharpf, W. Kaiser, W. Graf, A. Schirmer, W. Faber, J. Duppich, and R. Zeisler. 1988, 10p
Pub. in Nuclear Instruments and Methods in Physics Research B35, p181-190 1988.

Keywords: *Neutron beams, *Equipment, *Focusing, *Neutron flux, *Mirrors, Gamma spectrometers, Nickel, Titanium, Radioactivation analysis, Neutron cross sections, Transmissivity, Substrates, Reprints.

A simple neutron focusing system was installed and tested at a neutron guide in the external neutron guide laboratory ELLA at KFA in Juelich. The device uses nickel-titanium supermirrors. The production of these supermirrors is described together with results and practical hints for improving their behavior for a certain glass substrate with the measured roughness of 18.5 Angstroms. Results of Monte Carlo calculations of the transmission and focusing properties for different wavelengths are given and compared with measured results. The obelisk shaped 150 cm long supermirror coated tube has an average transmission of 82% and a maximum local gain of 3.

General

901,307
PB90-130279 PC A09/MF A01
National Inst. of Standards and Technology (NML), Gaithersburg, MD. Center for Radiation Research.
Center for Radiation Research (of the National Institute of Standards and Technology) Technical Activities for 1989.
C. E. Kuyatt. Nov 89, 181p NISTIR-89/4183
See also PB89-127294.

Keywords: *Irradiation, *Research projects, *Test facilities, Measurement, Nuclear physics, Radiometry, Ion sources, Ionizing radiation, Instruments, Dosimetry, Calibrating, Standards, Beams(Radiation), Radiochemistry, *Center for Radiation Research.

The report summarizes research projects, measurement method development, calibration and testing, and data evaluation activities that were carried out during Fiscal Year 1989 in the NIST Center for Radiation Research. The activities fall in the areas of radiometric physics, radiation sources and instrumentation, ionizing radiation, and nuclear physics.

OCEAN TECHNOLOGY & ENGINEERING

Biological Oceanography

901,308
PB89-175855 Not available NTIS

National Bureau of Standards (NML), Gaithersburg, MD. Organic Analytical Research Div.
Specimen Banking in the National Status and Trends Program: Development of Protocols and First Year Results.
Final rept.
G. G. Lauenstein, M. M. Schantz, S. A. Wise, and R. Zeisler. 1986, 5p
Pub. in Proceedings of Oceans 86, Conference Record, Washington, DC., September 23-25, 1986, p586-590.

Keywords: *Aquatic animals, *Trace elements, *Chemical analysis, *Organic compounds, Sampling, Performance evaluation, Fishes, Mussels, *Environmental monitoring, *Baseline studies, *Water pollution detection.

The National Oceanic and Atmospheric Administration has initiated a Specimen Bank for estuarine and coastal samples as part of its National Status and Trends Program. During the first year, sample collection protocols were developed for the collection of benthic fish, bivalve molluscs, and associated sediments. Specimens from over 40 sites nationwide have now been submitted for inclusion in the Specimen Bank which is housed at the National Bureau of Standards in Gaithersburg, Maryland. Specimens are preserved at liquid nitrogen temperature with degradation expected to be minimal for decades. Retrospective analysis of specimens will allow the opportunity to derive baseline values for new environmental contaminants and make historical marine samples available for analysis when new and improved analytical procedures become available. Preliminary analyses of selected samples collected during the first year are presently under way for organic and trace element contaminants.

901,309
PB89-177232 Not available NTIS
National Bureau of Standards (IMSE), Gaithersburg, MD. Ceramics Div.
Biodegradation of Tributyltin by Chesapeake Bay Microorganisms.
Final rept.
G. J. Olson, and F. E. Brinckman. 1986, 6p
Sponsored by Office of Naval Research, Arlington, VA.
Pub. in Proceedings of Oceans 86, Conference Record, Washington, DC., September 23-25, 1986, p1196-1201.

Keywords: *Chesapeake Bay, *Biodeterioration, *Marine microorganisms, *Water analysis, Gas chromatography, Chemical analysis, *Tin/tributyl, Flame photometry.

The authors have been studying microbial resistance to butyltin compounds and the degradation of tributyltin spiked into samples of Chesapeake Bay waters. Butyltin species were identified and quantified using a gas chromatograph equipped with a tin-selective flame photometric detector (GC-FPD), providing micrograms/l detection limits. Incubation of these samples under incandescent lamps accelerated biodegradation, suggesting the involvement of photosynthetic microorganisms. At certain times in these degradation experiments, tetrabutyltin was detected by GC-FPD and confirmed by gas chromatography-mass spectrometry.

Dynamic Oceanography

901,310
PB89-171755 Not available NTIS
National Bureau of Standards (NML), Boulder, CO.
Time and Frequency Div.
Gravity Tide Measurements with a Feedback Gravity Meter.
Final rept.
J. Levine, J. C. Harrison, and W. Dewhurst. 1986, 7p
Pub. in Jnl. of Geophysical Research 91, nB12 p12835-12841, 10 Nov 86.

Keywords: *Tides, *Gravimeters, *Measurement, Surface waves, Gravity waves, Tidal currents, Ocean waves, Atmospheric pressure, Amplitude, Reprints.

Gravity-tide data obtained using a calibrated LaCoste and Romberg gravity meter with electrostatic feedback has been analyzed. There is agreement between the measured amplitude and phase of the major semi-diurnal components and the corresponding values to be

expected using current earth models and ocean-load calculations. Both local and global barometric pressure changes make significant contributions to the power in the tidal bands and are included in the fitting function. The admittance estimates at diurnal frequencies can be used to determine the frequency and to set a lower bound on the dissipation of the nearly-diurnal resonance in the tidal response by fitting a resonance function to the observations. These estimates are in reasonable agreement with results obtained by other methods, but are somewhat different from the values to be expected on the basis of more elaborate theoretical estimates.

Physical & Chemical Oceanography

901,311
PB89-177224 Not available NTIS
National Bureau of Standards (IMSE), Gaithersburg, MD. Ceramics Div.
Determination of Ultratrace Concentrations of Butyltin Compounds in Water by Simultaneous Hydridization/Extraction with GC-FPD Detection.
Final rept.
C. L. Matthias, J. M. Bellama, and F. E. Brinckman. 1986, 6p
Sponsored by Office of Naval Research, Arlington, VA., and David W. Taylor Naval Ship Research and Development Center, Annapolis, MD.
Pub. in Proceedings of Oceans 86, Conference Record, Washington, DC., September 23-25, 1986, p1146-1151.

Keywords: *Trace elements, *Water analysis, *Marine atmospheres, Gas chromatography, Extraction, pH, Salinity, Chemical stabilization, *Tin/butyl, Flame photometry, Hydridization.

An improved method for the ultratrace determination of butyltins in water by simultaneous hydridization/extraction followed by gas chromatography coupled with a flame photometric detector is reported. Detection limits for the new system are 5 micrograms/L tributyltin cation (TBT(sup +)). Effects of pH and salinity on the procedure are reported and stability of the derived extracts is discussed.

General

901,312
PB89-187512 Not available NTIS
National Bureau of Standards (NML), Gaithersburg, MD. Office of Standard Reference Data.
Computerized Materials Property Data Systems.
Final rept.
J. Rumble, and J. G. Kaufman. 1986, 3p
Pub. in Proceedings of Oceans '86 Conference Record, Washington, DC., September 23-25, 1986, p370-372.

Keywords: *Marine engineering, *Materials specifications, Offshore structures, Mechanical properties, Structural design, Corrosion, Failure, *Data bases.

Because of the harshness of the marine environment, the selection of materials for structural and other purposes has always been a demanding task, complicated by the enhancement of mechanical failures by corrosion. Not only are data such as corrosion fatigue scarce, locating the available data can be very difficult. The computer offers an opportunity to allow easy and efficient access to technical data. Over the last few years, a comprehensive materials data system has been started to meet the needs of industrial and other materials data users. One result has been to establish the National Materials Property Data Network. Recognizing that databases will be built by the same wide variety of groups that now use only one kind of computer system, the Network plans to offer a gateway service that features a common user interface, a consistent catalog of existing data, and coverage of all types of engineering materials and their properties. Progress towards these goals will be described as well as efforts to make available high-quality data of interest to marine environments.

901,313
PB90-117417 Not available NTIS

National Inst. of Standards and Technology (NEL), Gaithersburg, MD. Structures Div.
Hydrodynamic Forces on Vertical Cylinders and the Lighthill Correction.
Final rept.
G. R. Cook, and E. Simiu. 1989, 18p
Sponsored by Department of the Interior, Washington, DC.
Pub. in Ocean Engineering 16, n4 p355-372 1989.

Keywords: *Hydrodynamics, *Force, *Offshore structures, Damping, Drag, Water waves, Reprints, *Cylinders, *Lighthill correction, Morison equation.

In the paper the expression for the Lighthill correction was derived for finite water depths. Measurements obtained in periodic wave flow at the Naval Civil Engineering Laboratory and in random wave flow at the Delft Hydraulics Laboratory were subjected to an extensive analysis. The results of the analysis showed that for both the periodic and random wave conditions the addition of the Lighthill correction did not improve the Morison equation significantly and had no significant effect on the estimation of the drag force, including the drag force corresponding to very low Keulegan-Carpenter numbers.

ORDNANCE

Ammunition, Explosives, & Pyrotechnics

901,314
PB89-146914 Not available NTIS
National Bureau of Standards (NEL), Boulder, CO.
Electromagnetic Fields Div.
Measurement Procedures for Electromagnetic Compatibility Assessment of Electroexplosive Devices.
Final rept.
J. W. Adams, and D. S. Friday. 1988, 11p
Sponsored by Army Aviation Systems Command, St. Louis, MO., and Naval Surface Weapons Center, Silver Spring, MD.
Pub. in IEEE (Institute of Electrical and Electronics Engineers) Transactions on Electromagnetic Compatibility 30, n4, p484-494 Nov 88.

Keywords: *Initiators(Explosives), *Electromagnetic compatibility, Electromagnetic interference, Measurement, Mathematical models, Statistical theory, Vulnerability, Electromagnetic fields, Electromagnetic pulses, Nuclear explosion effects, Reprints.

Electroexplosive devices (EEDs) are electrically fired explosive initiators used in a wide variety of applications. The nature of most of these applications requires that the devices function with near certainty when required and otherwise remain inactive. Recent concern with pulsed electromagnetic interference (EMI) and the nuclear electromagnetic pulse (EMP) made apparent the lack of methodology for assessing EED vulnerability. A new and rigorous approach for characterizing EED firing levels is developed in the context of statistical linear models and is demonstrated in the paper. The authors combine statistical theory and methodology with thermodynamic modeling to determine the probability.that an EED of a particular type fires when excited by a pulse of a given width and amplitude. The results can be applied to any type of EED for which the hot wire explosive binder does not melt below the firing temperature of the primary explosive.

PHOTOGRAPHY & RECORDING DEVICES

Photographic Techniques & Equipment

901,315
PB89-186340 Not available NTIS
National Bureau of Standards (NEL), Gaithersburg,
MD. Precision Engineering Div.
**Automated Calibration of Optical Photomask
Linewidth Standards at the National Institute of
Standards and Technology.**
Final rept.
J. E. Potzick. 1989, 13p
Pub. in Proceedings of SPIE (Society of Photo-Optical
Instrumentation Engineers) Symposium on Microlitho-
graphy, San Jose, CA., February 26-March 3, 1989,
13p.

Keywords: *Calibrating, *Automation, *Line width,
Standards, Optical equipment, *Photomasks.

An automated system has been developed at the National Institute of Standards and Technology (NIST),
formerly the National Bureau of Standards, for calibrating optical photomask linewidth standards. The
system, controlled by a desktop computer, locates
each feature to be measured in the field of view of the
microscope, centers and focuses the image, scans the
image, and calculates the optical linewidth from the
scan data. The results are checked for errors and the
process repeated until every feature on the photomask
has been calibrated. If statistical tests are passed, a
calibration certificate is printed.

PHYSICS

Acoustics

901,316
PB89-147839 PC A10/MF A01
National Inst. of Standards and Technology (NEL),
Boulder, CO. Electromagnetic Fields Div.
**Measurement of Adapter Loss, Mismatch, and Effi-
ciency Using the Dual Six-Port.**
G. J. Counas, and B. C. Yates. Jul 88, 209p NBSIR-
88/3096
Sponsored by Aerospace Guidance and Metrology
Center, Newark AFS, OH. Metrology Engineering Sec-
tion.

Keywords: *Acoustic measurement, Measuring instruments, Noise, Standards, Efficiency, Accuracy, Adapt-
ers, Temperature, *Dual six-port measurement
system.

A noise measurement system is being developed for
the U.S. Air Force which uses coaxial cryogenic and
ambient noise temperature standards to determine the
noise temperature of the device under test. When the
device under test has a different connector than those
on the noise standards, an adapter has to be used.
Adapter loss and complex reflection coefficient must
be compensated for or noise measurement accuracy
is affected. A technique has been developed which
uses a dual six-port measurement system to determine
the mismatch, loss, and ultimately the efficiency of the
adapter used. This enables correction of measurement results and allows measurements to be made
with an adapter with no degradation of accuracy.

901,317
PB89-179709 Not available NTIS
National Bureau of Standards (NEL), Gaithersburg,
MD. Thermophysics Div.

**Spherical Acoustic Resonators in the Undergradu-
ate Laboratory.**
Final rept.
M. Bretz, M. L. Shapiro, and M. R. Moldover. 1989,
5p
Pub. in American Jnl. of Physics 57, n2 p129-133 Feb
89.

Keywords: *Acoustic resonators, *Acoustic velocity,
*Velocity measurement, Gases, Reprints, Tempera-
ture dependence, Pressure dependence, Acoustic
thermometry.

A spherical acoustic resonator is reported that measures the speed of sound in gases with very high accuracy using comparatively simple instrumentation. The
resonator can be used to illustrate, in a quantitative
fashion, the temperature and pressure dependence of
the speed of sound, the effects of thermal conductivity,
and the splitting of nearly degenerate modes.

901,318
PB89-179808 Not available NTIS
National Bureau of Standards (IMSE), Boulder, CO.
Fracture and Deformation Div.
**Acoustoelastic Determination of Residual
Stresses.**
Final rept.
P. P. Delsanto, R. B. Mignogna, A. V. Clark, D.
Mitrakovic, and J. C. Moulder. 1987, 7p
Sponsored by Office of Naval Research, Arlington,
VA., and Office of Nondestructive Evaluation, Arling-
ton, VA.
Pub. in Residual Stresses in Science and Technology,
v1 p175-181 1987.

Keywords: *Residual stress, *Ultrasonic tests, Acoustic properties, Rayleigh waves, Surface waves, Metal
plates, Crystal lattices, Texture, Nondestructive tests,
Reprints, *Acoustoelasticity.

A general perturbative formalism for the propagation of
Rayleigh waves on the surface of initially deformed anisotropic material plates is applied to the particular
case of an orthotropic distribution of cubic crystallites.
Applied and residual stresses can be determined from
their correlation with the measured Rayleigh waves
propagation velocity. A critical problem, especially in
the case of residual stresses, is the separation of texture and acoustoelastic effects. Two techniques are
proposed for the solution of this problem and their
range of validity is discussed.

901,319
PB89-202220 Not available NTIS
National Bureau of Standards (NML), Gaithersburg,
MD. Temperature and Pressure Div.
**Speed of Sound in a Mercury Ultrasonic Interfer-
ometer Manometer.**
Final rept.
C. R. Tilford. 1987, 11p
Pub. in Metrologia 24, n3 p121-131 1987.

Keywords: *Acoustic velocity, *Sound transmission,
*Interferometers, *Manometers, *Mercury(Metal),
Acoustic measurement, Sound waves, Optical measuring instruments, Pressure measurement, Carbon dioxide lasers, Reprints.

The speed of sound in mercury for frequencies between 9.5 and 10.5 MHz has been measured by comparison with a frequency stabilized infrared laser. The
apparatus and techniques used for both the ultrasonic
and optical measurements are discussed, along with
details of the error analysis.

901,320
PB89-228548 Not available NTIS
National Inst. of Standards and Technology (NEL),
Gaithersburg, MD. Thermophysics Div.
**Acoustic and Microwave Resonances Applied to
Measuring the Gas Constant and the Thermody-
namic Temperature.**
Final rept.
M. R. Moldover. 1989, 8p
Pub. in IEEE (Institute of Electrical and Electronics En-
gineers) Transactions on Instrumentation and Meas-
urement 38, n2 p217-224 Apr 89.

Keywords: *Ideal gas law, *Temperature measurement, *Acoustic resonance, *Microwaves, Thermodynamic properties, Resonant frequency, Resonance,
Acoustic velocity, Reprints.

Techniques are being developed for simultaneously
measuring the frequencies of microwave and acoustic

resonances in a spherical cavity. They will permit the
determination of the thermodynamic temperature with
unprecedented accuracy. Progress to date includes: a
new value for the universal gas constant: R = 8.314
471 + or - 0.000 014 J/(mol times K) (1.7 ppm) with a
five-fold reduction in its standard error, a new value of
the thermodynamic temperature of the triple point of
gallium (302.9169 + or - 0.0005 K), and a microwave
measurement of the volumetric thermal expansion of
an acoustic resonator with an error of about + or - 1.5
ppm. Improved values of the Boltzmann constant,
(1.38 065 13 + or - 0.000 002 5) x 10(-23) J/K, and the
Stefan-Boltzmann constant, (5.670 399 + or - 0.000
038) x 10(-8) W/(m sup 2 times K sup 4), were obtained from R, and further studies of the temperature
scale are in progress.

901,321
PB90-128505 Not available NTIS
National Inst. of Standards and Technology (NEL),
Gaithersburg, MD. Thermophysics Div.
Spherical Acoustic Resonators.
Final rept.
J. B. Mehl, and M. R. Moldover. 1989, 23p
Pub. in Topics of Current Physics: Photoacoustic, Pho-
tothermal and Photochemical Processes in Gases,
Chapter 4, p61-83 1989.

Keywords: *Acoustic resonators, *Acoustic velocity,
*Thermophysical properties, *Gases, Mathematical
models, Acoustic measurement, Reprints, Spherical
configuration.

Gas-filled spherical resonators are excellent tools for
measurements of the speed of sound. The radially
symmetric gas resonances are nondegenerate and
have high quality factors (typically 2,000-10,000).
These resonances can be used with very simple instrumentation and unsophisticated analysis to measure
the speed of sound in a gas with an accuracy on the
order of 0.01%. With data analysis based on a complete theoretical model of the acoustical system, the
accuracy can be increased to better than one part per
million. The model includes the effects of the coupling
between acoustic and thermal waves, thermal and viscous effects at the shell boundary, shell motion, and
imperfect shell geometry. Other boundary effects, including the effects of holes in the resonator wall and
precondensation effects, have also been considered.
The results of the theoretical model are described in
detail and compared with experimental results in the
chapter. There is also a brief review of the thermophysical importance of acoustic measurements, including
a discussion of the determination of ideal-gas specific
heats and information about intermolecular interactions.

Fluid Mechanics

901,322
PB89-150932 Not available NTIS
National Bureau of Standards (NEL), Gaithersburg,
MD. Mathematical Analysis Div.
**Solutal Convection during Directional Solidifica-
tion.**
Final rept.
G. B. McFadden, and S. R. Coriell. 1988, 7p
Sponsored by National Aeronautics and Space Admin-
istration, Washington, DC.
Pub. in Proceedings of National Fluid Dynamics Con-
gress (1st), Cincinnati, OH., July 25-28, 1988, p1572-
1578.

Keywords: *Alloys, *Metals, *Solidification, *Crystallization, *Convection, *Fluid dynamics, Buoyancy, Computation, Gravitation, Mathematical models, Melts, Bifurcation.

During directional solidification of a binary alloy at constant velocity, buoyancy-driven fluid flow may occur
due to the solute gradients generated by the solidification process. Numerical calculations of the solute and
fluid flow fields in the melt have been carried out using
finite differences in a two-dimensional, time-dependent model that assumes a planar crystal-melt interface
and allows time-dependent gravitational accelerations.
The container walls are rigid and perfectly insulating
for solute. For constant vertical gravitational accelerations, as the solutal Rayleigh number is varied, multiple
steady states and time-dependent states may occur.
The bifurcation from the quiescent state may be sub-

critical or transcritical, depending on the aspect ratio of the container. The maximum variation in the solute concentration at the crystal-melt interface was also calculated for various values of the rotation rate of the gravitational acceleration.

901,323
PB89-157259 Not available NTIS
National Bureau of Standards (NEL), Gaithersburg, MD. Thermophysics Div.
Effect of Surface Ionization on Wetting Layers.
Final rept.
R. F. Kayser. 1986, 4p
Pub. in Physical Review Letters 56, n17 p1831-1834 1986.

Keywords: *Ionization, *Liquid phases, *Substrates, *Polarity, *Wetting, *Glass, Langmuir probes, Mathematical models, Mixtures, Carbon disulfide, Dispersants, Reprints.

A generalized surface ionization model of Langmuir's to liquid mixtures of polar and nonpolar components in contact with ionizable substrates is presented. When a predominantly nonpolar mixture is near a miscibility gap, thick wetting layers of the conjugate polar phase form on the substrate. Such charged layers can be much thicker than similar wetting layers stabilized by dispersion forces. The model may explain the 0.4-0.6 micro m thick wetting layers formed by mixtures of nitromethane and carbon disulfide in contact with glass.

901,324
PB89-158117 Not available NTIS
National Bureau of Standards (NEL), Gaithersburg, MD. Chemical Process Metrology Div.
Numerical Computation of Particle Trajectories: A Model Problem.
Final rept.
E. F. Moore, and R. W. Davis. 1986, 6p
Pub. in Gas-Solid Flows, p111-116 1986.

Keywords: *Particle trajectories, *Numerical analysis, Mathematical models, Stokes law(Fluid mechanics), Unsteady flow, Incompressible flow, Computerized simulation, Reprints, *Gas-particle flow.

Computer simulations are useful in gas-solid particle systems where experimental measurements are often difficult. The relative accuracy of various numerical methods becomes important in assessing the significance of results obtained. The note presents a model problem for trajectories of solid particles in a quasi-steady, incompressible axisymmetric flowfield and numerical results obtained from three different methods for tracking particles through this flow. The characteristics of the trajectories, with sudden changes in direction, are reminiscent of particle tracks seen in the wake of a bluff body. Thus, this model problem is highly pertinent to numerical simulations of realistic gas-particle flows. The basic parameter involved here is particle Stokes number. Trajectories of a particle are presented for three values of Stokes number. The relative accuracy of the three particle-tracking methods when the flow is either steady or suddenly reversing direction is discussed.

901,325
PB89-158141 Not available NTIS
National Bureau of Standards (NEL), Gaithersburg, MD. Thermophysics Div.
Shear Induced Anisotropy in Two-Dimensional Liquids.
Final rept.
H. J. M. Hanley, G. P. Morriss, T. R. Welberry, and D. J. Evans. 1988, 26p
Sponsored by Department of Energy, Washington, DC.
Pub. in Physica 149A, p406-431 1988.

Keywords: *Liquids, Shear tests, Photographic techniques, Light scattering, Anisotropy, Reprints, Two dimensional.

The behavior of a dense two-dimensional soft disc liquid under shear is studied via nonequilibrium molecular dynamics. The structure factor for the liquid at a given shear rate is evaluated directly by plotting the particle positions, taken at random from the NEMD simulation at that shear, onto photographic film and using light scattering to obtain a diffraction pattern. The pair correlation function of this system is also extracted directly by histogramming the particle positions with respect to a given central particle as a function of separation and angle. The pair correlation function is compared to that approximated by a Fourier series expansion to rank ten. Results are reported as a function

of shear rate from a shear rate of 0.1 (when the fluid is essentially Newtonian) to 10 (when the fluid can display a string phase). The appearance of the string phase is discussed and shown to be a consequence of the definition of temperature in the simulation algorithm. A modification of the algorithm is proposed. Comparisons between this work and previous work with three-dimensional liquids are given. The two-dimensional structure factor is compared with that obtained from a real colloidal suspension via light scattering.

901,326
PB89-161871 PC A03/MF A01
National Inst. of Standards and Technology (NML), Gaithersburg, MD. Center for Chemical Technology.
Mixing Motions Produced by Pipe Elbows.
T. T. Yeh, and G. E. Mattingly. Jan 89, 41p NISTIR-89/4029

Keywords: *Pipe flow, *Mixing, *Elbows(Pipe), Pipes(Tubes), Velocity measurement, Swirling, Vortex flow, Pressure distribution, Turbulent flow, Laser anemometers.

Experimental measurements have been made, using laser Doppler velocimetry (LDV) of the pipeflows produced by a range of pipe-elbow configurations. The secondary flow characteristics of these pipeflows are described qualitatively and quantitatively together with their decay rates in the downstream piping. The potential these flows have as mixing environments is described on the basis of the profiles of the mean and turbulent velocity components, the change of these with downstream distance, and the pressure losses. Parameters characterizing these flow fields are defined from the measured velocity profiles. It is shown that double elbow 'out-of-plane' combinations where minimal pipelengths separate the elbows can produce very energetic, long-lasting, swirling flows that can serve as effective mixers. Such effectiveness suggests that process designers might consider adding an additional elbow (and the slight increase in pressure loss) to pipe turns so as to take advantage of the enhanced mixedness that can be achieved via close-coupled elbows-out-of-plane.

901,327
PB89-173918 Not available NTIS
National Bureau of Standards (NEL), Boulder, CO. Chemical Engineering Science Div.
Use of Dye Tracers in the Study of Free Convection in Porous Media.
Final rept.
M. C. Jones, and R. A. Perkins. 1988, 7p
Contract DE-AI05-87ER13770
Sponsored by Department of Energy, Washington, DC.
Pub. in Proceedings of Symposium on Energy Engineering Sciences (6th), Argonne, IL., May 4-6, 1988, p152-158.

Keywords: *Porous materials, *Fluorescent dyes, Temperature gradients, Fiber optics, Lasers, Three dimensional flow, Porosity, *Free convection, *Tracer techniques.

A new experimental approach based on age distribution functions is described for the study of free convection in porous media. Fiberoptic probes are being used to track the dispersion of a fluorescent dye by the convective flow. The initial focus is on three-dimensional flows in rectangular boxes with vertical temperature gradients.

901,328
PB89-174023 Not available NTIS
National Bureau of Standards (NEL), Boulder, CO. Thermophysics Div.
Shear Dilatancy and Finite Compressibility in a Dense Non-Newtonian Liquid.
Final rept.
H. J. M. Hanley, J. C. Rainwater, D. J. Evans, and L. Hood. 1988, 3p
Pub. in Proceedings of International Congress on Rheology (10th), Sydney, Australia, August 14-19, 1988, v1 p386-388.

Keywords: *Non-Newtonian fluids, *Compressible flow, *Shear tests, Rheological properties, Dilatancy, Density(Mass/volume), Viscosity, Dynamic pressure, Cylinders.

Nonequilibrium molecular dynamic (NEMD) simulations of model systems with liquids of spherical particles are non-Newtonian. The authors illustrate here that the shear viscosity coefficient, Eta(sub +) of

a liquid of soft spheres is a function of the shear rate, gamma; they show that the liquid is shear dilatant, i.e., that the density, rho, is a decreasing function of gamma at constant pressure, p; they also indicate that the liquid exhibits normal pressure differences. The results are used to analyze fluid behavior in a typical flow problem, namely flow between vertical rotating concentric cylinders.

901,329
PB89-179592 Not available NTIS
National Bureau of Standards (NEL), Gaithersburg, MD. Chemical Process Metrology Div.
Application of Magnetic Resonance Imaging to Visualization of Flow in Porous Media.
Final rept.
A. K. Gaigalas, A. Van Orden, B. Robertson, T. H. Mareci, and L. A. Lewis. 1989, 6p
Sponsored by Nuclear Regulatory Commission, Washington, DC.
Pub. in Nuclear Technology 84, p113-118 Jan 89.

Keywords: *Flow visualization, *Water flow, Porous materials, Aluminum oxide, Bentonite, Kaolin, Absorption, Reprints, *Magnetic resonance imaging.

The flow of water in porous materials has been visualized using nuclear magnetic resonance imaging (MRI). For flow in an initially dry bed, the water gives a large signal that can be detected directly. Flow in a wet bed is visualized indirectly by displacing the pure water with a dilute solution of paramagnetic ions. The solution does not give an MRI signal and so can be contrasted with pure water. Another use of MRI is to observe the absorption of water by a solid. The MRI technique is sensitive and can give accurate and quantitative results for flow with low Peclet number.

901,330
PB89-209262 PC A03/MF A01
National Inst. of Standards and Technology (NEL), Gaithersburg, MD. Center for Computing and Applied Mathematics.
Elimination of Spurious Eigenvalues in the Chebyshev Tau Spectral Method.
G. B. McFadden, B. T. Murray, and R. F. Boisvert.
May 89, 19p NISTIR-89/4090

Keywords: *Chebyshev inequality, *Hydrodynamics, *Eigenvalues, Spectrum analysis, Models, *Chebyshev approximation, *Spectral methods, Computational fluid dynamics, Chebyshev tau method, Chebyshev Galerkin method, Orr-Sommerfeld equations, Vorticity equations.

Spectral methods have been used to great advantage in hydrodynamic stability calculations; the concepts are described in Orszag's seminal application of the Chebyshev tau method to the Orr-Sommerfeld equation for plane Poiseuille flow in 1971. Orszag discusses both the Chebyshev Galerkin and the Chebyshev tau methods, but presents results for the tau method, which is easier to implement than the Galerkin method. The tau method has the disadvantage that two unstable eigenvalues are produced that are artifacts of the discretization. The authors present an extremely simple modification to the Chebyshev tau method which eliminates the spurious eigenvalues. They first study a simplified model of the Orr-Sommerfeld equation discussed by Gottlieb and Orszag. They consider the Chebyshev tau method, which has two spurious eigenvalues, and then describe a modification which eliminates them. Finally, they consider results for the Orr-Sommerfeld equation, where the modified tau method also eliminates the spurious eigenvalues. The simplicity of the modification makes it a convenient alternative to other approaches to the problem.

901,331
PB89-227995 Not available NTIS
National Bureau of Standards (NEL), Boulder, CO. Thermophysics Div.
Development of a Field-Space Corresponding-States Method for Fluids and Fluid Mixtures.
Final rept.
J. R. Fox. 1987, 18p
Sponsored by Department of Energy, Washington, DC.
Pub. in Fluid Phase Equilibria 37, p123-140 1987.

Keywords: *Fluids, *Mixtures, *Phase transformation, Pressure, Temperature, Equations of state, Van der Waals equation, Thermodynamics, Reprints.

PHYSICS

Fluid Mechanics

A field-space corresponding-states transformation is one in which the properties of a fluid, or fluid mixture, are related to the properties of a reference fluid, or fluid mixture, by equations which are analytic in the field variables (i.e., the pressure, temperature and chemical potentials). Such transformations are radically different from traditional corresponding states methods. The most useful of these differences is that the presence of phase coexistence in the target system is a direct reflection of phase coexistence in the reference system and never the result of the transformation itself. An immediate consequence of technical interest is that, if the phase transitions of the reference system are tabulated or known analytically, transforming these states maps all of the phase transitions of the target system directly.

901,332
PB89-228050 Not available NTIS
National Inst. of Standards and Technology (NEL), Gaithersburg, MD. Thermophysics Div.
Simplified Representation for the Thermal Conductivity of Fluids in the Critical Region.
Final rept.
G. A. Olchowy, and J. V. Sengers. 1989, 10p
Contract DE-FG05-88ER13902
Sponsored by Department of Energy, Washington, DC.
Pub. in International Jnl. of Thermophysics 10, n2 p417-426 Mar 89.

Keywords: *Thermal conductivity, *Fluids, *Critical flow, Ethane, Methane, Carbon dioxide, Temperature, sivity, Reprints.

A practical representation for the critical thermal conductivity enhancement is developed by incorporating a finite cutoff into the asymptotic mode-coupling integrals for the diffusivity associated with the critical fluctuations. This procedure yields a simplified approximation to a more complete nonasymptotic solution of the mode-coupling integrals obtained earlier. A comparison is made with thermal conductivity data for carbon dioxide, ethane, and methane.

901,333
PB89-231484 PC A04/MF A01
National Inst. of Standards and Technology (NEL), Boulder, CO. Chemical Engineering Science Div.
Effect of Pipe Roughness on Orifice Flow Measurement.
Technical note.
J. A. Brennan, S. E. McFaddin, C. F. Sindt, and R. R. Wilson. Jul 89, 66p NIST/TN-1329
Contracts GRI-5081-353-0422, GRI-5081-271-1680
Also available from Supt. of Docs. as SN003-003-0. Sponsored by Gas Research Inst., Chicago, Il2951-3.

Keywords: *Orifice flow, *Flow measurement, *Pipes(Tubes), *Roughness, Orifice meters, Surface roughness, Gases, Water, Experimental data.

Flow measurement with orifice flowmeters is simple in concept and can be accurate, but, as demonstrated in flow measurement test facilities, can also result in large errors if any of the significant parameters are not controlled. Many of these parameters are well known and particular care is taken to ensure that they do not cause measurement errors. Others are more subtle and not so easily detected or controlled. One such parameter is the surface finish of the pipe immediately upstream of the orifice plate. Results of an experimental investigation into the effects of this pipe roughness on the orifice discharge coefficient are presented, along with a review of some of the pertinent literature. Measurement errors of approximately 1% can result from using meter tubes that are too rough but still within the specification of the standards.

901,334
PB89-235147 PC A23/MF A01
National Inst. of Standards and Technology (NEL), Gaithersburg, MD.
Measurements of Coefficients of Discharge for Concentric Flange-Tapped Square-Edged Orifice Meters in Water Over the Reynolds Number Range 600 to 2,700,000.
Technical note (Final).
J. R. Whetstone, W. G. Cleveland, G. P. Baumgarten, S. Woo, and M. C. Croarkin. Jun 89, 544p NIST/TN-1264
Also available from Supt. of Docs. as SN003-003-02942-4. Sponsored by American Petroleum Inst., Washington, DC.

Keywords: *Orifice meters, *Flow measurement, *Reynolds number, Fluid flow, Water meters, Standards, Velocity, Pressure, Mathematical models.

Presented is a description of the measurement procedures and standards, data acquisition systems, and data bases developed in the American Petroleum Institute-sponsored orifice discharge coefficient data base project performed at the National Bureau of Standards primary water flow rate measurement facility. Measurements were performed on five orifice meter sizes, 2, 3, 4, 6, and 10 inches, over the beta ratio range of 0.08 to 0.75. The measurement systems and procedures were designed to provide full documentation of the relation between the observations comprising the data base developed and U.S. national measurement standards.

901,335
PB89-235667
(Order as PB89-235634, PC A04)
Great Lakes Fishery Commission, Ann Arbor, MI.
Apparatus for Neutron Scattering Measurements on Sheared Fluids.
Bi-monthly rept.
G. C. Straty. 1989, 3p
Included in Jnl. of Research of the National Institute of Standards and Technology, v94 n4 p259-261 Jul/Aug 89.

Keywords: *Neutron scattering, *Fluids, *Shear properties, *Measuring instruments, Torque, Temperature, Computer applications.

The construction of an apparatus to allow neutron scattering measurements on fluids undergoing shear is reported. The apparatus has been used with the cold neutron small-angle-neutron-scattering (SANS) spectrometer at the NIST research reactor and will be made available to users as a permanent part of the NIST facility.

Optics & Lasers

901,336
PB89-156996 Not available NTIS
National Bureau of Standards (NML), Gaithersburg, MD. Radiometric Physics Div.
Ultrasensitive Laser Spectroscopy and Detection.
Final rept.
R. A. Keller, and J. J. Snyder. 1986, 9p
Pub. in Laser Focus/Electro-Optics 22, n3 p86-94 Mar 86.

Keywords: Mass spectroscopy, Fluorescence, Detection, Reprints, *Laser spectroscopy, Resonance ionization spectroscopy, High resolution.

Recent advances in laser technology have led to the development of several techniques for ultrasensitive high resolution spectroscopy and ultrasensitive detection. A recent symposium and a feature issue of the Journal of the Optical Society of America contain a series of papers on the theory and applications of these new techniques. For simplicity, the authors have chosen most examples in the article from these two references, although they are by no means exhaustive of the subject. They have further restricted their attention to three techniques which they find particularly exciting: quantum noise limited absorption measurements, photon burst detection of fluorescence, and resonance ionization mass spectrometry.

901,337
PB89-157069 Not available NTIS
National Bureau of Standards (NEL), Boulder, CO. Electromagnetic Technology Div.
Fresnel Lenses Display Inherent Vignetting.
Final rept.
M. Young. 1988, 2p
Pub. in Applied Optics 27, n17, p3593-3594, 1 Sep 88.

Keywords: *Optical lenses, Photometry, Radiometry, Reprints, *Fresnel lenses, Optical fibers, Vignetting.

Some of the light refracted by a facet of a Fresnel lens impinges on the axial (or horizontal) portion of the facet and is directed away from the focal point. Loss of this light may be significant in applications where precise radiometric measurements are necessary.

901,338
PB89-157382 Not available NTIS
National Bureau of Standards (NML), Boulder, CO. Time and Frequency Div.
Thermal Shifts of the Spectral Lines in the (4)F3/2 to (4)I11/2 Manifold of an Nd:YAG Laser.
Final rept.
S. Z. Xing, and J. C. Bergquist. 1986, 4p
Sponsored by Air Force Office of Scientific Research, Washington, DC., and Office of Naval Research, Arlington, VA.
Pub. in IEEE (Institute of Electrical and Electronics Engineers) Jnl. of Quantum Electronics 24, n9 p1829-1832 Sep 88.

Keywords: *Spectral lines, *Line spectra, *Frequency shift, Frequency standards, Near infrared radiation, Reprints, *YAG lasers, Temperature dependence.

The authors report the thermal shifts of eleven of the twelve lines from the quartet F (3/2) Stark energy levels to the quartet I (11/2) energy levels in an Nd:YAG laser for a temperature change from 20-200 C. The thermal shift difference between the Stark sublevels R1, R2 in quartet F (3/2) is found to be about -0.6 plus or minus 0.6/cm/100 C. Within the experimental uncertainty, all of the lasting lines either moved to longer wavelength or remained unchanged with increasing temperature.

901,339
PB89-157390 Not available NTIS
National Bureau of Standards (NML), Boulder, CO. Time and Frequency Div.
Precise Test of Quantum Jump Theory.
Final rept.
R. G. Hulet, D. J. Wineland, J. C. Bergquist, and W. M. Itano. 1988, 4p
Sponsored by Air Force Office of Scientific Research, Washington, DC., and Office of Naval Research, Arlington, VA.
Pub. in Physical Review A 37, n11 p4544-4547 Jun 88.

Keywords: Atomic spectroscopy, Optical pumping, Raman spectra, Fluorescence, Tests, Reprints, *Magnesium ions, *Quantum jumps, Quantum optics, Laser spectroscopy, Trapping(Charged particles), Magnesium 24.

Quantum jumps due solely to spontaneous Raman scattering between the Zeeman sublevels of a single (24)Mg(1+) ion have been observed in the fluorescence emitted by the ion. A theory of quantum jumps for the system predicts that coherences between excited levels cause the ratio of the mean duration of the 'fluorescence-on periods' to the mean duration of the 'fluorescence-off periods' to be independent of laser intensity. The measured value agrees with the predicted one to within the measurement precision of 2%. The distribution of the durations of the off periods also agrees with theory.

901,340
PB89-157887 Not available NTIS
National Bureau of Standards (NEL), Gaithersburg, MD. Precision Engineering Div.
Resonance Light Scattering from a Liquid Suspension of Microspheres.
Final rept.
T. R. Lettieri, and E. Marx. 1986, 7p
Pub. in Applied Optics 25, n23 p4325-4331, 1 Dec 86.

Keywords: *Microscopy, Dielectrics, Mie scattering, Electromagnetic scattering, Diameters, Reprints, *Resonance light scattering, *Microspheres.

Resonance light scattering (RLS) spectra have been obtained from a liquid suspension of dielectric microspheres having a narrow size distribution in order to deduce the mean size and distribution width of the spheres. Comparison with single-particle spectra shows that most peaks in the size-distributed spectra are due to several resonances in the a sub n or b sub n Mie scattering coefficients. A mean diameter for the microspheres, obtained by matching experimental resonance wavelengths with those calculated using a vectorized Mie scattering algorithm, is in excellent agreement with optical microscopy and scanning electron microscopy results. However, the measured width of the size distribution is significantly smaller than those from optical microscopy and transmission electron microscopy. Several sources of experimental error which affect the RLS spectra, including angle misalignment and oversized collection apertures, are discussed.

901,341
PB89-158091 Not available NTIS
National Bureau of Standards (NML), Gaithersburg, MD. Atomic and Plasma Radiation Div.
Recent Progress on Spectral Data for X-ray Lasers at the National Bureau of Standards.
Final rept.
J. Reader, J. Sugar, and V. Kaufman. 1988, 4p
Sponsored by Strategic Defense Initiative Organization, Washington, DC., and Department of Energy, Washington, DC. Office of Magnetic Fusion Energy.
Pub. in IEEE (Institute of Electrical and Electronics Engineers) Transactions on Plasma Science 16, n5 p560-563 Oct 88.

Keywords: Energy levels, Spectrum analysis, Reprints, *X ray lasers, Laser-produced plasma, Isoelectronic sequence, Tokamak devices, Multicharged ions.

Recent work on the spectra of laser-produced plasmas and tokamaks has led to the observation of long sequences of isoelectronic ions extending to very high ionic charge states. The measurements of the wavelengths and energy levels provide data that are important for the development of X-ray lasers. In addition to contributing to a knowledge of the energy levels and transitions of possible lasing media, the data provide reference lines for wavelength calibration of X-ray laser experiments and reference data for testing theoretical methods used for predicting the properties of lasing ions.

901,342
PB89-171235 Not available NTIS
National Bureau of Standards (NML), Boulder, CO.
Laser-Noise-Induced Population Fluctuations in Two- and Three-Level Systems.
Final rept.
T. Haslwanter, H. Ritsch, J. Cooper, and P. Zoller. 1988, 8p
Grant NSF-PHY86-04504
Sponsored by National Science Foundation, Washington, DC.
Pub. in Physical Review A 38, n11 p5652-5659, 1 Dec 88.

Keywords: Light scattering, Reprints, *Laser noise, Resonance fluorescence.

Significant fluctuations, above the shot-noise limit, have been observed in the intensity of fluorescent light scattered from atoms excited by an intense noisy laser, where a central role is played by the nonlinearity of the atom-field interaction. By considering the variance of the atomic populations, the authors show that noise spectroscopy is sensitive to the field statistics of the laser. They specifically consider the cases of resonance fluorescence (analogous in the weak-field limit to analysis by a Fabry-Perot interferometer), two-photon excitation, and double optical resonance for variable laser intensity and bandwidth. It is shown that large differences in the noise spectra can occur between lasers characterized by a phase-diffusion or a phase-jump model, although these models would give the same values for the mean atomic populations. The authors believe that observation of population fluctuations can become a useful method for characterizing laser noise.

901,343
PB89-171276 Not available NTIS
National Bureau of Standards (NML), Boulder, CO. Quantum Physics Div.
One-Photon Resonant Two-Photon Excitation of Rydberg Series Close to Threshold.
Final rept.
G. Alber, T. Haslwanter, and P. Zoller. 1986, 7p
Pub. in Jnl. of the Optical Society of America B 5, n12 p2439-2445 Dec 88.

Keywords: Reprints, *Rydberg series, Multi-photon processes, Autoionization, Bound state, Ware packets.

One-photon resonant two-photon excitation of autoionizing Rydberg series close to threshold is studied. Analytical expressions for the time evolution of bound-state amplitudes are derived by adapting ideas from quantum-defect theory. The two limiting cases of excitation of Rydberg wave packets and AC-Stark splitting close to the threshold are discussed in detail.

901,344
PB89-171607 Not available NTIS
National Bureau of Standards (NML), Boulder, CO. Quantum Physics Div.
Scattering of Polarized Light in Spectral Lines with Partial Frequency Redistribution: General Redistribution Matrix.
Final rept.
H. Domke, and I. Hubeny. 1988, 12p
Pub. in Astrophysical Jnl. 334, p527-538, 1 Nov 88.

Keywords: *Spectral lines, Resonance scattering, Legendre functions, Reprints, *Polarized light, Radiative transfer.

The redistribution matrix for resonance scattering of arbitrarily polarized light, described by a vector of Stokes parameters, is derived assuming that the ground state is isotropic (i.e., assuming negligible optical pumping). When specified in the atomic rest frame, the redistribution matrix is found to be composed of several terms with individually separate frequency and angular dependence. The laboratory frame (velocity averaged) redistribution matrix exhibits an analogous structure, but the angular and frequency dependences are intermingled. The authors consider two possibilities for treating the angular dependence in practical applications, namely, an expansion in a series of Legendre polynomials, and an azimuthal expansion. Finally, the concept of azimuthally averaged redistribution matrix is examined, and explicit expressions for resonance lines are given.

901,345
PB89-171672 Not available NTIS
National Bureau of Standards (NEL), Boulder, CO. Electromagnetic Technology Div.
Stability of Birefringent Linear Retarders (Waveplates).
Final rept.
P. D. Hale, and G. W. Day. 1988, 8p
Sponsored by Optical Society of America, Washington, DC.
Pub. in Applied Optics 27, n24 p5146-5153, 15 Dec 88.

Keywords: *Retarders(Devices), Birefringence, Stability, Reprints, *Waveplates.

The effects of changes in temperature, wavelength, and direction of propagation (angle of incidence) on the retardance of zero-order, multiple-order, compound zero-order, and temperature-compensated waveplates are described in detail. A disagreement in the literature regarding the properties of a compound zero-order waveplate is resolved by showing that with respect to temperature and wavelength, it behaves like a true zero-order waveplate, but with respect to angle of incidence it behaves like a multiple-order waveplate. A previously proposed temperature-compensated design is shown to suffer from the same directional limitations. A new design for a retarder consisting of one element of a positive uniaxial crystal and one element of a negative uniaxial crystal is proposed. The retardance of such a waveplate would be much less sensitive to the direction of propagation, but somewhat more sensitive to temperature than a typical compound zero-order waveplate.

901,346
PB89-171680 Not available NTIS
National Bureau of Standards (NEL), Boulder, CO. Electromagnetic Technology Div.
NBS (National Bureau of Standards) Laser Power and Energy Measurements.
Final rept.
T. R. Scott. 1986, 7p
Pub. in Proceedings of the SPIE (Society of Photo-Optical Instrumentation Engineers), Laser Beam Radiometry, Los Angeles, CA., January 14-15, 1988, v888 p48-54.

Keywords: *Calorimeters, *Laser beams, *Power meters, Measuring instruments, Standards, Calibrating, Optical communications, Power measurement, Thermal measurements.

The National Bureau of Standards (NBS) maintains a set of electrically calibrated calorimeters designed and built specifically for laser energy measurements. These calorimeters are used as national reference standards for the calibration of optical power and energy meters. NBS offers laser measurement services based on the standard calorimeter to the public at a variety of laser wavelengths and power ranges. The uncertainties associated with these measurements have recently been re-evaluated.

901,347
PB89-171698 Not available NTIS
National Bureau of Standards (NEL), Boulder, CO. Electromagnetic Technology Div.
Fast-Pulse Generators and Detectors for Characterizing Laser Receivers at 1.06 um.
Final rept.
P. A. Simpson. 1988, 5p
Pub. in Proceedings of the SPIE (Society of Photo-Optical Instrumentation Engineers), Laser Beam Radiometry, Los Angeles, CA., January 14-15, 1988, v888 p43-47.

Keywords: *Detectors, *Generators, *Lasers, Calibrating, Pulsation, Reprints, Optical communication, Pulse amplifiers, Pulse analyzers.

A detector system capable of measuring the waveform of pulses used to calibrate laser receivers at 1.06 micrometers is described. The risetime of the system is 0.8 ns. All p_{arts} of the system are available commercially. Also described is an optical impulse generator at 1.06 micrometers with a risetime of less than 100 ps. This impulse generator can be used to measure the impulse response of the detector system and laser receivers.

901,348
PB89-171714 Not available NTIS
National Bureau of Standards (NEL), Boulder, CO. Electromagnetic Technology Div.
Electrically Calibrated Silicon Bolometer for Low Level Optical Power and Energy Measurements.
Final rept.
R. J. Phelan, and R. M. Craig. 1988, 5p
Pub. in Proceedings of the SPIE (Society of Photo-Optical Instrumentation Engineers), Laser Beam Radiometry, Los Angeles, CA., January 14-15, 1988, v888 p38-42.

Keywords: *Bolometers, Power measurement, Silicon on sapphire, Laser radiation.

A cryogenically cooled, silicon on sapphire, electrically calibrated bolometer has been designed and measured to have a noise equivalent power of 3 x 10 to the -11 power watt per root hertz. The electrical calibration of the bolometer has been agreed with an electrically calibrated pyroelectric to better than 1%.

901,349
PB89-173959 Not available NTIS
National Bureau of Standards (NEL), Boulder, CO. Electromagnetic Technology Div.
Interferometric Dispersion Measurements on Small Guided-Wave Structures.
Final rept.
B. L. Danielson, and C. D. Whittenberg. 1988, 2p
Pub. in Proceedings of Conference on Lasers and Electro-Optics, Anaheim, CA., April 25-29, 1988, p360-361.

Keywords: Fourier analysis, *Optical waveguides, *Optical fibers, *Dispersion, Coherence domain reflectometry, Time domain reflectometry.

A method is described for obtaining dispersion properties of components in microoptic systems. The technique is based on a Fourier analysis of the reflective signatures obtained from a coherence-domain reflectometer.

901,350
PB89-175731 Not available NTIS
National Bureau of Standards (NML), Boulder, CO. Time and Frequency Div.
New FIR Laser Lines and Frequency Measurements for Optically Pumped CD3OH.
Final rept.
R. J. Saykally, K. M. Evenson, D. A. Jennings, L. R. Zink, and A. Scalabrin. 1987, 10p
Pub. in International. Jnl. of Infrared and Millimeter Waves 8, n6 p653-662 1987.

Keywords: *Infrared lasers, Far infrared radiation, Deuterium compounds, Optical pumping, Carbon dioxide lasers, Frequency measurement, Continuous radiation, Reprints, *Methyl alcohol lasers, Methanol.

Twenty new cw FIR laser lines in CD_3OH, optically pumped by a CO_2 laser, are reported. The frequencies of 39 of the stronger laser lines were measured relative to stabilized CO_2 lasers with a fractional uncertain-

PHYSICS

Optics & Lasers

ty, as determined by the reproducibility of the FIR frequency itself, of 2 parts in 10 million.

901,351
PB89-175749   Not available NTIS
National Bureau of Standards (NML), Gaithersburg, MD. Radiation Source and Instrumentation Div.
NBS/NRL (National Bureau of Standards/Naval Research Laboratory) Free Electron Laser Facility.
Final rept.
S. Penner, R. Ayres, R. Cutler, P. Debenham, B. C. Johnson, E. Lindstrom, D. Mohr, J. Rose, M. Wilson, P. Sprangle, and and C. M. Tang. 1988, 8p
See also AD-A194 214.
Pub. in Nuclear Instruments and Methods in Physics Research A272, p73-80 1988.

Keywords: Reprints, *Free electron lasers, *Racetrack microtrons, US NBS, National Institute of Standards and Technology.

A free electron laser is being built at the National Bureau of Standards as a joint project with the Naval Research Laboratory. The electron beam source is the 185 MeV CW racetrack microtron (RTM) presently nearing completion. The accelerator is characterized by extremely good emittance and small energy spread. A new photocathode injector operating on the 32nd subharmonic of the 2380 MHz rf frequency is being developed to increase the peak current to approx = or > 2 A in 3 ps micropulses. The wiggler design has 130 periods of (lambda sub w) = 28 mm with rms wiggler parameter K approx = or < 1. Three-dimensional calculations indicated that power gains of 10-30% per pass can be achieved for optical wavelengths in the range 200 nm to 10.0 micrometers. The design of the RTM and FEL will be described. The FEL is intended for use in a broad program of research applications in biomedicine and materials science.

901,352
PB89-176234   Not available NTIS
National Bureau of Standards (NEL), Gaithersburg, MD. Precision Engineering Div.
Resonance Light Scattering from a Suspension of Microspheres.
Final rept.
T. R. Lettieri, and E. Marx. 1987, 3p
Pub. in AIP (American Institute of Physics) Conference Proceedings, n160 p523-525 1987.

Keywords: *Resonance scattering, *Light(Visible radiation), *Spectra, Dielectrics, Visible spectrum, Spectrum analysis, Diameters, Dimensions, Mie scattering, *Microspheres.

Resonance light scattering spectra have been obtained from a liquid suspension of dielectric microspheres and then used to determine the mean diameter and width of the size distribution of the spheres.

901,353
PB89-176515   Not available NTIS
National Bureau of Standards (NML), Gaithersburg, MD. Radiation Source and Instrumentation Div.
NBS (National Bureau of Standards) Free Electron Laser Facility.
Final rept.
B. C. Johnson, P. H. Debenham, S. Penner, C. M. Tang, and P. Sprangle. 1989, 4p
Pub. in Proceedings of International Symposium on Resonance Ionization Spectroscopy and Its Applications (4th), Gaithersburg, MD., April 10-15, 1989, p247-250.

Keywords: Continuous radiation, Far ultraviolet radiation, Intermediate infrared radiation, *Research facilities, *Free electron lasers, *Tunable lasers, Picosecond pulses, National Institute of Standards and Technology.

A free electron laser (FEL) user facility is being constructed at the National Bureau of Standards in collaboration with the Naval Research Laboratory. The anticipated performance of the FEL is: (1) wavelength variable from approximately 150 nm to 10 micrometers; (2) continuous train of 3 ps-wide pulses at 74.375 MHz; and (3) average power of 10 W to 200 W. One advantage of the NBS-FEL for RIS schemes is the ability to select the wavelength at will. It is also possible to scan the wavelength. The high repetition rate is an additional attractive feature.

901,354
PB89-176929   Not available NTIS

National Bureau of Standards (IMSE), Gaithersburg, MD. Metallurgy Div.

Computing Ray Trajectories between Two Points: A Solution to the Ray-Linking Problem.
Final rept.
S. J. Norton. 1987, 4p
Pub. in Jnl. of the Optical Society of America A-Optics and Image Science 4, n10 p1919-1922 1987.

Keywords: *Ray tracing, *Light transmission, Integral equations, Iteration, Perturbation, Convergence, Reprints, Refractive index, Successive approximations method, Imaging techniques.

The problem of computing the ray trajectory between points a and b, when given the refractive index (or sound velocity) distribution, is complicated by the ignorance of the initial ray direction at the point a, which is needed in defining the path intercepting the end point b when numerically integrating the ray equation. It is shown that this ray-linking problem can be avoided by transforming the ray equation into an implicit integral equation for the true ray path that satisfies the given boundary conditions. The integral equation can be solved for the true path by the method of successive approximations. Simulations suggest that this iterative scheme often converges rapidly to the true path. An explicit expression for a ray path, obeying the boundary conditions, is also derived which is correct to first order in the refractive index perturbation. This path provides an excellent approximation to the true path when the refractive index perturbation is small, and becomes increasingly good as the perturbation goes to zero.

901,355
PB89-177208   Not available NTIS
National Bureau of Standards (IMSE), Gaithersburg, MD. Ceramics Div.
Photoelastic Properties of Optical Materials.
Final rept.
A. Feldman. 1986, 8p
Pub. in Proceedings of SPIE (Society of Photo-Optical Instrumentation Engineers), Laser and Nonlinear Optical Materials, San Diego, CA., August 19-20, 1986, p127-134.

Keywords: *Photoelastic analysis, *Optical properties, *Piezoelectric crystals, Wavelengths, Acoustic measurement, Tensor analysis, Brillouin zones, Interferometers, Birefringence, Polarization(Waves), Refractivity.

The tensorial nature of the photoelastic effect in optical materials is discussed. The commonly used photoelastic constants, the piezo-optic and the elasto-optic constants, are defined, and the general form of these tensors is presented. The most commonly used methods for measuring photoelastic constants, interferometry, polarimetry, the acousto-optic effect, and Brillouin scattering are discussed. The effect of wavelength dispersion is discussed.

901,356
PB89-179139   Not available NTIS
National Bureau of Standards (NML), Boulder, CO. Time and Frequency Div.
CO Laser Stabilization Using the Optogalvanic Lamb-Dip.
Final rept.
M. Schneider, A. Hinz, A. Groh, K. M. Evenson, and W. Urban. 1987, 5p
Pub. in Applied Physics B 44, p241-245 1987.

Keywords: *Frequency stability, Infrared lasers, Intermediate infrared radiation, Stabilization, Reprints, *Carbon monoxide lasers, Optogalvanic spectroscopy.

Frequency stabilization of the CO laser using a CO Lamb-dip is achieved in the range from 5.0-6.3 micrometers. The CO saturation signal is obtained from a low-pressure discharge in absorption and is detected using optogalvanic detection. The frequency stability and reproducibility has been verified to be better than 100 k Hz; this is an improvement of more than one order of magnitude compared with locking techniques using CO laser gain profiles.

901,357
PB89-179212   Not available NTIS
National Bureau of Standards (NML), Gaithersburg, MD. Temperature and Pressure Div.

Measurements of the Nonresonant Third-Order Susceptibility.
Final rept.
G. J. Rosasco, and W. S. Hurst. 1986, 4p
Pub. in AIP (American Institute of Physics) Conference Proceedings 146, p261-264 1986.

Keywords: Kerr electrooptical effect, Roman spectroscopy, Hydrogen, Argon, Optical measurement, *Third order susceptibility, Second harmonic generation, Nonlinearity.

The authors advance evidence for the validity of a long-known dispersion formula for the electronic contribution to the third-order nonlinear susceptibility.

901,358
PB89-179774   Not available NTIS
National Bureau of Standards (NML), Boulder, CO. Quantum Physics Div.
Simple F-Center Laser Spectrometer for Continuous Single Frequency Scans.
Final rept.
D. D. Nelson, A. Schiffman, K. R. Lykke, and D. J. Nesbitt. 1988, 7p
Sponsored by Air Force Office of Scientific Research, Bolling AFB, DC.
Pub. in Chemical Physics Letters 153, n2-3 p105-111, 9 Dec 88.

Keywords: Absorption spectra, Solid state lasers, Spin orbit interactions, Reprints, *Laser spectrometers, *Color center lasers, *F center lasers, Tunable lasers, Bromine atoms.

The authors report a simple and novel scheme for continuous, single frequency scanning of a commercial F-center laser without any computer interfacing. The scheme utilizes galvo tuning of the cavity and intracavity CaF2 Brewster plates with servo loop control of the intracavity etalon. This permits continuous tuning of the F-center frequency over 0.8/cm-1 under complete manual control, as well as arbitrarily long, concatenated scans, and trivial interfacing to a data acquisition system. This scanning spectrometer operation is demonstrated on direct absorption of atomic bromine.

901,359
PB89-186282   Not available NTIS
National Bureau of Standards (NML), Gaithersburg, MD. Radiometric Physics Div.
Apparatus Function of a Prism-Grating Double Monochromator.
Final rept.
R. D. Saunders, and J. B. Shumaker. 1986, 5p
Pub. in Applied Optics 25, n20 p3710-3714 1986.

Keywords: *Monochromators, Spectroradiometers, Spectrometers, Reprints, Slit functions.

The slit function of a double monochromator has been studied using a dye laser and a few ion laser lines. The value of the slit function is roughly nine orders of magnitude below the peak at 300 nm from the line center. The difference between the slit function obtained by monochromator scanning over a fixed spectral line and that obtained by tuning a spectral line through a fixed monochromator setting is illustrated. Also reported is a prominent structure in the slit function which is attributed to the intermediate slit of the instrument.

901,360
PB89-192678   Not available NTIS
National Bureau of Standards (NML), Gaithersburg, MD. Radiation Source and Instrumentation Div.
Research Opportunities Below 300 nm at the NBS (National Bureau of Standards) Free-Electron Laser Facility.
Final rept.
P. H. Debenham, and B. C. Johnson. 1988, 4p
Grant N00014-87-F-0066
See also AD-A193 100. Sponsored by Strategic Defense Initiative Organization, Washington, DC., and Office of Naval Research, Arlington, VA.
Pub. in Free-Electron Laser Applications in the Ultraviolet Topical Meeting, Cloudcroft, NM., March 2-5, 1988, p76-79.

Keywords: Ultraviolet lasers, Near ultraviolet radiation, Light pulses, Reprints, *Free electron lasers, Picosecond pulses, National Institute for Standards and Technology.

Average output power of 25 W in 3 ps pulses at 75 MHz will be available at fundamental wavelengths from 200 to 300 nm beginning in April 1990.

901,361
PB89-193908 PC A03/MF A01
National Bureau of Standards (NML), Boulder, CO.
Time and Frequency Div.
(12)C(16)O Laser Frequency Tables for the 34.2 to 62.3 THz (1139 to 2079 cm(-1)) Region.
Technical note.
M. Schneider, K. M. Evenson, M. D. Vanek, D. A. Jennings, and J. S. Wells. Aug 88, 27p NBS/TN-1321
Also available from Supt. of Docs. as SN003-003-02933-5. Prepared in cooperation with Bonn Univ. (Germany, F.R.).

Keywords: Infrared spectra, Vibrational spectra, Far infrared radiation, Tables(Data), *Carbon monoxide lasers, Laser radiation.

Frequencies for (12)C(16)O laser transitions are tabulated for the spectral range from 34.2 to 62.3 THz (1139 to 2079/cm). The transition frequencies were calculated using molecular constants which were derived by heterodyne frequency measurements on the (12)C(16)O laser.

901,362
PB89-201099 Not available NTIS
National Bureau of Standards (NEL), Gaithersburg, MD. Building Physics Div.
Heuristic Analysis of von Kries Color Constancy.
Final rept.
J. A. Worthey, and M. H. Brill. 1986, 5p
Pub. in Jnl. of the Optical Society of America A 3, n10 p1708-1712 Oct 86.

Keywords: Reflectance, Reprints, *Color constancy, Von Kries adaptation, Robot vision.

The properties of a constancy model based on the proportionality rule of von Kries are examined in a series of simplified examples. It is found that the breadth of receptor sensitivity functions causes metamerism, thwarting color constancy. Overlap of these functions limits the accuracy of von Kries adaptation, for a more subtle reason: it causes non-zero off-diagonal elements in the transformation matrix relating object reflectance to receptor stimuli. Such off-diagonal elements make von Kries adaptation incorrect even when the illuminant is restricted so as to prevent metamerism.

901,363
PB89-212252 Not available NTIS
National Bureau of Standards (IMSE), Gaithersburg, MD. Office of Nondestructive Evaluation.
Collision Induced Spectroscopy: Absorption and Light Scattering.
Final rept.
G. Birnbaum, L. Frommhold, and G. C. Tabisz. 1989, 25p
Sponsored by National Aeronautics and Space Administration, Washington, DC.
Pub. in Proceedings of International Conference on Spectral Line Shapes (9th), Torun, Poland, July 25-29, 1988, p623-647 1989.

Keywords: *Light scattering, *Hydrogen, *Helium, *Deuterium, Molecular spectroscopy, Infrared spectroscopy, Absorption spectra, Planetary atmospheres, *Molecule collisions, Line shape.

The survey deals with collision (or interaction) induced absorption (CIA) and light scattering (CILS) primarily in gases at low densities where bimolecular collisions predominate. The advances in these fields have occurred primarily in the last five years are emphasized. In the area of CIA, these topics include: H2-H2 and H2-He spectra in the far infrared (FIR) and in the fundamental IR bands; dimer features in these spectra; and the role of anisotropic interactions. The effects of the vibrational dependence of the potential function of the H2IR spectra are presented. The interesting line shapes and intensity problems in the FIR and IR spectra of HD due to interference between the collision induced and allowed transitions is discussed in some detail. Collision induced light scattering from simple isotropic and anisotropic molecules, including H2, is discussed. Another section deals with the onset of three-body collisions in both CIA and CILS.

901,364
PB89-221162 PC A99/MF A01

National Inst. of Standards and Technology, Boulder, CO.
Laser Induced Damage in Optical Materials: 1987-Special pub. (Final).
H. E. Bennett, A. H. Guenther, D. Milam, B. E. Newnam, and M. J. Soileau. Oct 88, 653p NIST/SP-756
Also available from Supt. of Docs. as SN003-003-02929-7. See also PB89-129548. Proceedings of a Symposium on 'Optical Materials for High-Power Lasers', Boulder, CO, October 26-28, 1987. Library of Congress catalog card no. 88-600576.Portions of this document are not fully legible. Prepared in cooperation with American Society for Testing and Materials, Philadelphia, i

Keywords: *Optical materials, *Radiation damage, *Meetings, Optical coatings, Optical measurement, Thin films, Surfaces, Mirrors, Substrates, *High power lasers, *Laser damage, Laser radiation.

The Nineteenth Annual Symposium on Optical Materials for High-Power Lasers (Boulder Damage Symposium) was divided into sessions concerning Materials and Measurements, Mirrors and Surfaces, Thin Films, and, finally, Fundamental Mechanisms. As in previous years, the emphasis of the papers presented at the Symposium was directed toward new frontiers and new developments. Particular emphasis was given to materials for high power systems. The wavelength range of prime interest was from 10.6 micrometers to the uv region. Highlights included surface characterization, thin film substrate boundaries, and advances in fundamental laser-matter threshold interactions and mechanisms. The scaling of damage thresholds with pulse duration, focal area, and wavelength was discussed in detail.

901,365
PB89-227938 Not available NTIS
National Bureau of Standards (NML), Boulder, CO. Quantum Physics Div.
Generation of Squeezed Light by Intracavity Frequency Doubling.
Final rept.
S. F. Pereira, M. Xiao, H. J. Kimble, and J. L. Hall. 1988, 4p
Grant N00014-87-K-0156
Sponsored by Office of Naval Research, Arlington, VA.
Pub. in Physical Review A 38, n9 p4931-4934, 1 Nov 88.

Keywords: Frequency multipliers, Reprints, *Squeezed light, Second harmonic generation, Squeezed states(Quantum theory), Quantum optics, Nonlinear optics.

Squeezed states of light are generated by the process of second-harmonic conversion within an optical cavity resonant at both fundamental and harmonic frequencies. Observations of squeezing are made by analyzing the spectral density of photocurrent fluctuations produced by the total field reflected from the nonlinear cavity. Reductions in photocurrent noise of 13% relative to the coherent-state or shot-noise level are achieved for frequency offsets near 4 MHz.

901,366
PB89-228084 Not available NTIS
National Inst. of Standards and Technology (NML), Boulder, CO. Quantum Physics Div.
Optical Novelty Filters.
Final rept.
D. Z. Anderson, and J. Feinberg. 1989, 13p
Pub. in IEEE (Institute of Electrical and Electronics Engineers) Jnl. of Quantum Electronics 25, n3 p635-647 Mar 89.

Keywords: *Optical filters, High-pass filters, Low-pass filters, Bandpass filters, Characteristics, Quantum electronics, Reprints.

A novelty filter detects what is new in a scene and may be likened to a temporal high-pass filter. The report reviews the current status of optical novelty filters and related devices that use four-wave mixing or two-beam coupling in photorefractive media. A detector that shows only what is not new is called a monotony filter and may be likened to a temporal low-pass filter. Demonstrations of high- and low-pass and bandpass temporal image filters are discussed. An analytical treatment of the two-beam coupling devices is given in a Laplace transform framework in the undepleted pump approximation assuming plane wave inputs. This allows a unified treatment of the various filter characteristics.

901,367
PB89-230304 Not available NTIS
National Inst. of Standards and Technology (NML), Gaithersburg, MD. Molecular Spectroscopy Div.
Ultrashort-Pulse Multichannel Infrared Spectroscopy Using Broadband Frequency Conversion in LiIO3.
Final rept.
E. J. Heilweil. 1989, 3p
Sponsored by Air Force Office of Scientific Research, Bolling AFB, DC.
Pub. in Optics Letters 14, n11 p551-553, 1 Jun 89.

Keywords: *Infrared spectroscopy, Frequency converters, Infrared spectra, Broadband, Reprints, Infrared upconversion, Picosecond pulses, Femtosecond pulses, Lithium iodates, YAG lasers, Dye lasers.

A simple probing method for obtaining broadband multichannel infrared spectra with picosecond or higher time resolution is described. Spectrally broad infrared pulses are produced by difference frequency mixing in LiIO3 between the second harmonic of a Nd(+3):YAG laser and the broadband output of a synchronously pumped dye laser. After sample absorption the infrared pulse is upconverted by a second LiIO3 crystal, which yields a visible pulse that is dispersed on a multichannel vidicon detector to obtain transient spectra of 4/cm FWHM resolution.

901,368
PB89-230346 Not available NTIS
National Inst. of Standards and Technology (NEL), Boulder, CO. Electromagnetic Technology Div.
Spatial Filtering Microscope for Linewidth Measurements.
Final rept.
M. Young. 1989, 7p
Pub. in Applied Optics 28, n8 p1467-1473, 15 Apr 89.

Keywords: *Optical microscopes, *Optical measurement, *Line width, High pass filters, Reprints, Integrated optics, Optical waveguides, Spatial filtering.

High mass filtering has been relatively little used in microscopy, yet it may have application to linewidth measurement and visualization of phase objects. The author has designed and built a spatial filtering microscope entirely of conventional microscope objectives. For linewidth measurement, the spatial filter has an optimum width that allows linewidths to be measured within a few percent. Phase lines can also be examined, but phase contrast microscopy may be more suited to weak phase objects such as integrated-optical waveguides.

901,369
PB89-235923 PC A12/MF A01
National Inst. of Standards and Technology (NEL), Gaithersburg, MD. Semiconductor Electronics Div.
Semiconductor Measurement Technology: A Software Program for Aiding the Analysis of Ellipsometric Measurements, Simple Models.
Special pub.
J. F. Marchiando. Jul 89, 256p NIST/SP-400/83
Also available from Supt. of Docs. as SN003-003-02954-8. Library of Congress catalog card no. 89-600743.

Keywords: *Polarimetry, *Ellipsometers, Optical measurement, Least squares method, Iteration, Mathematical models, Computer programs, *Computer calculations, Computer applications.

MAIN1.is a software program for aiding the analysis of ellipsometric measurements. MAIN1 consists of a suite of routines written in FORTRAN that are used to invert the standard reflection ellipsometric equations for simple systems. Here a system is said to be simple if the solid material sample may be adequately characterized by models which assume at least the following: materials are nonmagnetic; samples exhibit depth-dependent optical properties, such as one with layered or related devices that use a substrate that behaves like a semi-infinite half-space; layers are flat and of uniform thickness; and the dielectric function within each layer/substrate is isotropic, homogeneous, local, and linear. Each layer is characterized in part by a thickness, while the optical properties for a given material and wavelength are expressed in terms of a refractive index and extinction coefficient. The ellipsometric equations are formulated as a standard damped nonlinear least-squares problem and then solved by an iterative method when possible. Estimates of the uncer-

PHYSICS

Optics & Lasers

tainties associated with assigning numerical values to the model parameters are calculated as well.

901,370
PB90-118134 Not available NTIS
National Inst. of Standards and Technology (NML), Boulder, CO. Quantum Physics Div.
Precise Laser Frequency Scanning Using Frequency-Synthesized Optical Frequency Sidebands: Application to Isotope Shifts and Hyperfine Structure of Mercury.
Final rept.
M. D. Rayman, C. G. Aminoff, and J. L. Hall. 1989, 11p
Grant NSF-PHY86-04504
Sponsored by Office of Naval Research, Arlington, VA., and National Science Foundation, Washington, DC.
Pub. in Jnl. of the Optical Society of America B 6, n4 p539-549 Apr 89.

Keywords: *Lasers, *Dyes, *Frequency control, *Scanning, Frequency shift, Spectrometers, Modulators, Optical scanners, Automatic control, Frequency synthesizers, Mercury isotopes, Reprints.

Based on an efficient (30%), broadband (approx. 3-5 GHz) electro-optic modulator producing rf optical sidebands locked to a stable cavity, a tunable dye laser can be scanned under computer control with frequency-synthesizer precision. Cavity drift is suppressed in software by using a strong feature in the spectrum for stabilization. Mercury isotope shifts have been measured with a reproducibility of approx. 50 kHz. This accuracy of approx. 1/300 of the linewidth illustrate the power of the technique. Derived hyperfine-structure constants are compared with previous atomic-beam data where available.

901,371
PB90-163932
(Order as PB90-163874, PC **A04**)
National Inst. of Standards and Technology, Gaithersburg, MD.
Search for Optical Molasses in a Vapor Cell: General Analysis and Experimental Attempt.
A. L. Migdall. 1989, 6p
Office of Naval Research, Arlington, VA.
Included in Jnl. of Research of the National Institute of Standards and Technology, v94 n6 p373-378.

Keywords: *Lasers, *Cooling, Viscosity, *Optical molasses, *Vapor cells, Cold atoms.

The authors analyze the application of optical molasses to a thermal vapor cell to make and collect cold atoms. Such an arrangement would simplify the production of cold atoms by eliminating the difficulty of first having to produce and slow an atomic beam. The authors present the results of our calculations, computer models, and experimental work.

Plasma Physics

901,372
PB90-179097 Not available NTIS
National Bureau of Standards (NML), Gaithersburg, MD. American Radiation Div.
Wavelengths and Energy Levels of the K I Isoelectronic Sequence from Copper to Molybdenum.
Final rept.
V. Kaufman, J. Sugar, and W. L. Rowan. 1989, 4p
Pub. in Jnl. of the Optical Society of America B 6, n2 p142-145 Feb 89.

Keywords: *Molecular orbitals, *Rubidium, *Strontium, Copper, Molybdenum, Plasmas(Physics), Spectra, Wavelengths, Energy levels, Arsenic, Bromine, Gallium, Germanium, Krypton, Niobium, Potassium, Selenium, Yttrium, Zinc, Zirconium, Reprints.

Seven lines in the K-like transition array 3p63d-3p53d2 were observed for each of the spectra Cu XI to Mo XXIV (except for Rb and Sr) in radiation from impurity-doped tokamak and laser-generated plasmas. Wavelengths in the range of 70-112 angstroms, measured with an uncertainty of + or - 0.005 angstroms, are given. These are compared with calculations obtained with the relativistic Hartree-Fock code of Cowan. From these comparisons, predicted values for the wavelengths of Rb and Sr were obtained, with an uncertainty of + or - 0.01 angstroms.

901,373
PB89-185904 Not available NTIS
National Bureau of Standards (NML), Boulder, CO. Quantum Physics Div.
Atomic Internal Partition Function.
Final rept.
D. G. Hummer. 1988, 14p
Contract NAGW-766
Sponsored by National Aeronautics and Space Administration, Washington, DC.
Pub. in Proceedings of Conference on Atomic Processes in Plasmas, Santa Fe, NM., October 1987, p1-14 1988.

Keywords: *Atomic energy levels, Equations of state, Thermodynamic equilibrium, *Plasma, Partition functions, Stellar opacity.

Recent work on the evaluation of the atomic internal partition function at densities less than about 0.02 gm/cc is discussed, in which an attempt is made to identify the physical mechanism responsible for determining the level populations of atoms and ions in dense LTE plasmas. Level populations predicted by a phenomenological theory based on the identification are compared w h those obtained by the activity expansion method. It

901,374
PB89-229223 Not available NTIS
National Bureau of Standards (NML), Boulder, CO. Time and Frequency Div.
Heterodyne Frequency Measurements of (12)C(16)O Laser Transitions.
Final rept.
M. Schneider, K. M. Evenson, M. D. Vanek, D. A. Jennings, J. S. Wells, A. Stahns, and W. Urban. 1989, 10p
Pub. in Jnl. of Molecular Spectroscopy 135, p197-206 1989.

Keywords: *Lasers, *Carbon monoxide, *Frequency measurements, Demodulation, Lamb wave tests, Carbon dioxide, Vibrational spectra, Doppler effect, Reprints.

The paper reports the first frequency measurements of a Lamb-dip-stabilized (12)C(16)O laser. The laser was stabilized to the optogalvanic-Lamb dip of the CO molecule excited in a low-pressure dc discharge. The region of study was for transitions with lower state vibrational quantum numbers ranging from nu = 6 to nu = 16. Supplementary Doppler limited measurements are also reported for the range nu = 16 to nu = 34. The frequencies were directly measured in a heterodyne experiment in which two saturation-stabilized CO2 lasers were used as references. By fitting the transition frequencies to the Dunham expression, new coefficients have been determined which fit the new sub-Doppler data with over an order of magnitude more accuracy than previous coefficients.

Radiofrequency Waves

901,375
PB89-186902 Not available NTIS
National Bureau of Standards (NEL), Gaithersburg, MD. Structures Div.
Finite Element Studies of Transient Wave Propagation.
Final rept.
M. Sansalone, N. J. Carino, and N. N. Hsu. 1987, 9p
Pub. in Review of Progress in Quantitative Nondestructive Evaluation, v6A p125-133 1987.

Keywords: *Stress waves, *Elastic shells, Strains, Displacement, Greens function, Differential equations, Impact, Transducers, Reprints, *Finite element method.

The paper shows the versatility and power of the finite element method for solving stress wave propagation problems and provides background information about the finite element program that was used to carry out wave propagation studies at the National Bureau of Standards. The paper illustrates the use of the method to solve the following three problems: stress and displacement fields produced by transient point impact on the surface of an elastic plate; the interaction of transient stress waves produced by point impact with a planar disk-shaped void and a flat-bottom hole within

elastic plates; stress and displacement fields produced by an ultrasonic transducer radiating into an elastic solid. The finite element results are compared to exact Green's function solutions for a point source on an infinite plate, experimentally obtained displacement wave-forms for point impact on a plate containing a planar flaw, and photoelastic pictures of the stress fields produced by an ultrasonic transducer radiating into silica.

901,376
PB90-117466 Not available NTIS
National Inst. of Standards and Technology (NML), Boulder, CO. Time and Frequency Div.
Improved Rotational Constants for HF.
Final rept.
D. A. Jennings, and J. S. Wells. 1988, 2p
Pub. in Jnl. of Molecular Spectroscopy 130, 2p 1988,

Keywords: *High frequencies, *Rotational spectra, *Vibrational spectra, Radio frequencies, Electromagnetic spectra, Electronic spectra, Frequency synthesizers, Radio waves, Radio signals, Reprints.

The J 27 to 26 and J 33 to 32 rotational transition frequencies of the HF ground vibrational state have been measured using frequency synthesis spectroscopy. These new more accurate measurements when combined with recent measurements yield molecular constants with nearly an order of magnitude improvement in accuracy.

901,377
PB90-117698 Not available NTIS
National Inst. of Standards and Technology (NEL), Boulder, CO. Electromagnetic Fields Div.
Implementation of an Automated System for Measuring Radiated Emissions Using a TEM Cell.
Final rept.
G. H. Koepke, M. T. Ma, and W. D. Bensema. 1989, 7p
Pub. in IEEE (Institute of Electrical and Electronics Engineers) Transactions on Instrumentation and Measurement 38, n2 p473-479 Apr 89.

Keywords: *Transverse waves, *Electromagnetic radiation, *Measurement, Electric fields, Automatic control equipment, Electronic test equipment, Electric analyzers, Frequency analyzers, Measuring instruments, Network analyzers, Reprints.

The transverse electromagnetic (TEM) cell is widely used to evaluate the electromagnetic characteristics of electrically small devices. The paper reviews the theoretical basis for a technique to quantify the radiated emissions from any such device in the cell. The technique is well suited to an automated test system provided that the mechanical motions required can be controlled by a computer. The difficulties associated with these mechanical motions are discussed and possible solutions are proposed. The measurement technique is also expanded to include multiple-frequency sources in addition to single-frequency sources.

901,378
PB90-117946 Not available NTIS
National Inst. of Standards and Technology (NEL), Boulder, CO. Electromagnetic Fields Div.
Effect of an Electrically Large Stirrer in a Mode-Stirred Chamber.
Final rept.
D. I. Wu, and D. C. Chang. 1989, 6p
Pub. in IEEE (Institute of Electrical and Electronics Engineers) Transactions on Electromagnetic Compatibility 31, n2 p164-169 May 89.

Keywords: *Electric fields, *Cavity resonators, *Stirrers, *Frequency shift, Apertures, Eigenvectors, Wave dispersion, Oscillations, Pulsation, Frequency stability, Reprints.

In a mode-stirred chamber, the field in the cavity is perturbed with a stirrer or rotating scatterer so that the time-averaged field is constant. The paper investigates the key factor that governs the effectiveness of a stirrer. By examining the fundamental properties associated with a perturbing body in a cavity, the key to effective field perturbation is found in shifting the eigenmode frequencies. The phenomenon is illustrated by examining a 2-D cavity with a 1-D perturbing body. Using the Transmission-Line-Matrix method, the shifting of eigenfrequencies is computed and the variation on the magnitude of the fields is examined for different stirrer sizes. From the analysis, insights one draws include an analogy between the action of a large stirrer and a frequency modulator.

901,379
PB90-128042   Not available NTIS
National Inst. of Standards and Technology (NML), Boulder, CO. Time and Frequency Div.
Frequency Standards Utilizing Penning Traps.
Final rept.
J. J. Bollinger, S. L. Gilbert, W. M. Itano, and D. J. Wineland. 1989, 7p
Sponsored by Air Force Office of Scientific Research; Boiling AFB, DC., and Office of Naval Research, Arlington, VA.
Pub. in Proceedings of Symposium on Frequency Standards and Metrology (4th), Ancona, Italy, September 5-9, 1988, p319-325 1989.

Keywords: *Frequency standards, *Microwaves, *Hyperfine structures, Radiowaves, Atomic spectra, Spectrum analysis, Experimental data, *Penning traps.

ions in a Penning trap provide a promising candidate for a microwave frequency standard. The performance of a frequency standard based on a hyperfine transition in the ground state of a few thousand 9Be+ ions stored in a Penning trap is discussed. An inaccuracy of less than 2x10(-13) and fractional frequency stability of 2x10-11 tou(-1/2) were obtained in a first experimental setup. With the use of sympathetic cooling, a 0.9 mHz linewidth and a second order Doppler shift less than 1x10(-14) were obtained on the Be+ clock transition. This should enable an order of magnitude improvement in the frequency standard.

901,380
PB90-128521   Not available NTIS
National Inst. of Standards and Technology (NEL), Gaithersburg, MD. Electrosystems Div.
DC Electric Field Effects during Measurements of Monopolar Charge Density and Net Space Charge Density Near HVDC Power Lines.
Final rept.
M. Misakian, and R. H. McKnight. 1989, 6p
Sponsored by Department of Energy, Washington, DC.
Pub. in IEEE (Institute of Electrical and Electronics Engineers) Transactions on Power Delivery 4, n4 p2229-2234 Oct 89.

Keywords: *Direct current, *Electric fields, *Charge density, *High voltage, *Power transmission lines, Measurements, Electric current, Electric charge, Power distribution lines, Field strength, Field emission, Polarity, Reprints.

The influence of a dc electric field on the measurement of monopolar charge densities using an aspirator-type ion counter and the measurement of net space charge density using a Faraday cage or filter is examined. Optimum configurations which minimize the effect of the electric field are identified for each type of instrumentation.

901,381
PB90-128612   Not available NTIS
National Inst. of Standards and Technology (NEL), Boulder, CO. Electromagnetic Fields Div.
Proficiency Testing for MIL-STD 462 NVLAP (National Voluntary Laboratory Accreditation Program) Laboratories.
Final rept.
G. R. Reeve. 1988, 3p
Sponsored by Naval Air Systems Command, Washington, DC.
Pub. in Proceedings of EMC EXPO 88 International Conference on Electromagnetic Compatibility, Washington, DC., May 10-12, 1988, pT33.13-T33.15.

Keywords: *Electromagnetic compatibility, *Tests, Standards, Measurement, Requirements, Electronics laboratories, Electric fields, *National Voluntary Laboratory Accreditation Program, *MIL-STD-462.

Some of the difficulties in obtaining accurate results using MIL-STD 462 test procedures are reviewed. Several devices that could be used for verification of test results are presented along with their application to proficiency testing for NVLAP (National Voluntary Laboratory Accreditation Program) certification.

901,382
PB90-128778   Not available NTIS
National Inst. of Standards and Technology (NEL), Boulder, CO. Electromagnetic Fields Div.

Fields Radiated by Electrostatic Discharges.
Final rept.
P. F. Wilson, M. T. Ma, and A. R. Ondrejka. 1988, 5p
Pub. in Proceedings of IEEE (Institute of Electrical and Electronics Engineers) International Symposium on Electromagnetic Compatibility, Seattle, WA., August 2-4, 1988, p179-183.

Keywords: *Electric fields, *Mathematical models, *Dipoles, Electromagnetic theory, Static electricity, Electrostatic charge, Polarity, Electrostatics, Field emission, Field strength, *Electrostatic discharge.

Electrostatic discharge (ESD) can be a serious threat to electronic equipment. To date, metrology efforts have focused primarily on ESD-associated currents in order to develop test simulators. Significantly less work has been done on the ESD radiated fields. The paper examines the fields problem both theoretically and experimentally. Measurements indicate that the electric fields can be quite significant (greater than 150 V/m at a distance of 1.5 m), particularly, for relatively low voltage sparks (less than 6 kV). A theoretical dipole model for the ESD spark is developed to compute the radiated fields. The agreement between theory and experiment is good. The model may be used to predict the fields for a wide range of possible discharge configurations.

Solid State Physics

901,383
PATENT-4 747 684   Not available NTIS
Department of the Army, Washington, DC.
Method of and Apparatus for Real-Time Crystallographic Axis Orientation Determination.
Patent.
S. Weiser. Filed 27 Aug 87, patented 31 May 88, 3p
PB89-230148, PAT-APPL-7-089 893
Supersedes PB88-221460. Prepared in cooperation with National Inst. of Standards and Technology, Gaithersburg, MD.
This Government-owned invention available for U.S. licensing and, possibly, for foreign licensing. Copy of patent available Commissioner of Patents, Washington, DC 20231 $1.50.

Keywords: *Patents, *Crystallography, Orientation, Alignment, Laser beams, PAT-CL-356-31.

A specific small area of a crystal sample is scanned by a laser beam which rotates about an axis substantially perpendicular to the sample surface such that the inter-section of the beam with a plane above and parallel to the surface describes a true spiral or a stepwise spiral pattern. The laser beam is reflected in different amounts for different beam positions to produce a reflectance pattern indicative of crystallographic orientation.

901,384
PB89-146799   Not available NTIS
National Bureau of Standards (IMSE), Gaithersburg, MD. Ceramics Div.
Standard Reference Materials for X-ray Diffraction. Part 1. Overview of Current and Future Standard Reference Materials.
Final rept.
A. L. Dragoo. 1986, 5p
Pub. in Powder Diffr. 1, n4 p294-298 1986.

Keywords: *X-ray diffraction, *Standards, Crystal structure, Lattice parameters, Quartz, Austenite, Silicon, Reprints, *Standard reference materials.

Standard Reference Materials (SRMs) are stable materials which have one or more properties certified by the National Bureau of Standards. A general introduction is given to the types of SRMs and their certification. SRMs for X-ray diffraction are described in detail, including their intended use and their certified and other properties. New SRMs are under consideration as additional quantitative standards, intensity and line shape standards, and materials properties standards.

901,385
PB89-146815   Not available NTIS
National Bureau of Standards (IMSE), Gaithersburg, MD. Ceramics Div.

Magnetic Field Dependence of the Superconductivity in Bi-Sr-Ca-Cu-O Superconductors.
Final rept.
F. J. Adrian, J. Bohandy, B. F. Kim, K. Moorjani, J. S. Wallace, R. D. Shull, L. J. Swartzendruber, and L. H. Bennett. 1988, 5p
Pub. in Physica C 156, p184-188 1988.

Keywords: *Superconductivity, Magnetic fields, Reprints, *High temperature superconductors, *Bismuth calcium strontium cuprates, *Barium yttrium cuprates, *Yttrium barium cuprates, Microwave absorption.

The changes in superconducting behavior of Bi-Sr-Ca-Cu-O and Y-Ba-Cu-O high temperature superconductors upon the application of an applied field of 0.5 T are explored, using field-modulated-microwave absorption. Significant differences in the behavior are noted and discussed.

901,386
PB89-147433   Not available NTIS
National Bureau of Standards (NML), Gaithersburg, MD. Radiation Physics Div.
High Resolution Imaging of Magnetization.
Final rept.
D. T. Pierce, J. Unguris, and R. J. Celotta. 1988, 5p
Sponsored by Office of Naval Research, Arlington, VA.
Pub. in MRS Bulletin XIII, n6 p19-23 Jun 88.

Keywords: *Magnetic domains, Polarization(Spin alignment), Permalloys, Metal films, Iron, Reprints, Scanning electron microscopy, Imaging techniques, Spin orientation, High resolution.

The paper reviews imaging of magnetic domains using scanning electron microscopy with polarization analysis. The measurement of the spin polarization of the secondary electrons provides a vector image of the magnetization with high contrast and spatial resolution independent of surface topography, which is measured simultaneously. Examples of these features are illustrated by data from an iron-silicon single crystal, a ferromagnetic glass, a permalloy tape head, and an ultra-thin Fe film.

901,387
PB89-149181   Not available NTIS
National Bureau of Standards (NML), Boulder, CO. Quantum Physics Div.
Surface Structure and Growth Mechanism of Ga on Si(100).
Final rept.
B. Bourquignon, and S. R. Leone. 1988, 6p
Grants AFOSR-84-0272, NSF-CHE84-08403
Sponsored by Air Force Geophysics Lab., Hanscom AFB, MA., and National Science Foundation, Washington, DC.
Pub. in Proceedings of Symposium on Atomic and Surface Physics, LaPlagne, France, January 17-23, 1988, p228-233.

Keywords: *Semiconductors, *Gallium, *Silicon, Crystallization, Surface properties, Spectroscopy, Bonding, Lasers, Fluorescence, Auger electrons, *Molecular beam epitaxy.

The surface structures and growth mechanism of Ga overlayers on Si(100) are important data for the molecular beam epitaxy (MBE) of GaAs on Si(100). Some experimental results are discussed. It is found that Ga forms a well-ordered first layer with a large binding energy to Si(100). Surface structures are proposed where Ga is dimerized below 0.5 ML, while the Si(100) 2 x 1 reconstruction is removed above 0.5 ML. These structures account for an observed change in desorption energy and preexponential factor at 0.5 ML. (In the paper, 1 ML refers to Ga:Si = 1:1, or 6.8 x 10(sup 14)cm(sup -2).) From these surface structures, covalent bonding as opposed to metallic bonding between the metal atoms and the semiconductor, and bonding to the substrate as opposed to lateral bonding between adatoms, are inferred to be dominant. Islands start to grow at coverages above 1 ML, which depends on the surface temperature T(sub s). These results are contrasted with other recent results on similar systems, namely In and As on Si(100).

901,388
PB89-150833   Not available NTIS
National Bureau of Standards (NEL), Gaithersburg, MD. Semiconductor Electronics Div.

163

PHYSICS

Solid State Physics

Indirect Energy Gap of Si, Doping Dependence.
Final rept.
H. S. Bennett. 1988, 8p
Pub. in Properties of Crystalline Silicon, Electronic Materials Information Services for the Physics and Engineering Communities, p174-181 1988.

Keywords: *Silicon, *Additives, *Energy gap, Optical measurement, Photoluminescence, Electric devices, Superconductivity, Band structure of solids, Reprints.

The doping dependence of the indirect energy gap of silicon is reviewed for the Electronic Materials Information Service of IEE (London). The review is a guide with commentary to assist readers in selecting which values are best for applications. Knowledge in this area is such that intended application for the data on bandgap changes determines in many cases the appropriate values to use. Both data from interpreting electrical and optical measurements are given.

901,389
PB89-150866   Not available NTIS
National Bureau of Standards (NML), Gaithersburg, MD. Surface Science Div.
Angle Resolved XPS (X-ray Photoelectron Spectroscopy) of the Epitaxial Growth of Cu on Ni(100).
Final rept.
W. F. Egelhoff. 1985, 5p
Pub. in Structure of Surfaces, p199-203 1985.

Keywords: *X-ray analysis, *Photoelectrons, *Copper, *Nickel, *Epitaxy, Spectroscopy, Single crystals, Lattice parameters, Reprints.

In angle resolved x-ray photoelectron spectroscopy (XPS) of single crystals the core level peaks exhibit enhanced intensities along major crystal axes. This phenomenon is often referred to as electron channeling (or Kikuchi beams) due to an apparent analogy with effects found in electron microscopy. The present analysis of this phenomenon for epitaxial Cu on Ni(100) demonstrates that the electron channeling (or Kikuchi beams) approach fails completely to describe the data. The actual physical basis for this phenomenon is forward scattering of photoelectrons by overlying atoms in the lattice.

901,390
PB89-153704   PC A04/MF A01
National Bureau of Standards (NEL), Gaithersburg, MD. Precision Engineering Div.
Standard Reference Materials: Description of the SRM 1965 Microsphere Slide.
Final rept.
A. W. Hartman, and R. L. McKenzie. Nov 88, 69p
NIST/SP-260/107
Also available from Supt. of Docs. as SN003-003-02911-4. Library of Congress catalog card no. 88-600908.

Keywords: *Microscopy, Polystyrene, Magnification, Spheres, Arrays, Distortion, Crystal structures, Micrometeorology, *Standard reference materials, *Microspheres, Kubitschek effect, Space manufacturing, Image analysis.

The manual describes a new Standard Reference Material (SRM 1965). The SRM consists of two single-layer groupings of contacting monosize 10-micrometer polystyrene spheres that have been permanently sealed in an air chamber on a microscope slide. One sphere grouping consists of hexagonally ordered arrays, while the other grouping is unordered. The diameter distribution of the sphere material (SRM 1960), which was made under microgravity conditions on a Space Shuttle, is accurately known. SRM 1965 has a dual function: supporting measurements involving microscopy and supporting teachings and experiments in micrometrology. The manual describes how to measure microscope magnification and image distortion at levels below 0.5% and describes micrometrology experiments at the 0.05 micrometer level.

901,391
PB89-157481   Not available NTIS
National Bureau of Standards (IMSE), Gaithersburg, MD. Reactor Radiation Div.
Re-Entrant Spin-Glass Properties of a-(FexCr1-x)75P15C10.
Final rept.
P. Mangin, D. Boumazouza, R. W. Erwin, J. J. Rhyne, and C. Tete. 1987, 3p
Pub. in Jnl. of Applied Physics 61, n6 p3619-3621 1987.

Keywords: Ferromagnetism, Iron, Chromium, Neutron scattering, Reprints, *Spin glass state, Amorphous materials, Spin waves, Order parameters.

The magnetic excitations and instantaneous spatial correlations have been studied in amorphous (Fe(x)Cr(1-x))(75)P(15)C(10) using neutron inelastic scattering and small angle neutron scattering (SANS). The authors report the results for the sample with x = 0.7, which is in the reentrant spin-glass region (RSG) of the magnetic phase diagram. As in other materials displaying RSG properties, the authors found conventional spin-wave behavior for temperatures down to about half the Curie temperature (T(c) = 134 K), but decreasing excitation energies and lifetimes as the temperature is further lowered. They studied the critical scattering near T(c) with SANS, and found that the transverse correlation length 'diverges' provided that the data analysis includes the longitudinal fluctuations.

901,392
PB89-157721   Not available NTIS
National Bureau of Standards (NML), Gaithersburg, MD. Gas and Particulate Science Div.
Refinement of the Substructure and Superstructure of Romanechite.
Final rept.
S. Turner, and J. E. Post. 1988, 7p
Pub. in American Mineralogist 73, p1155-1161 1988.

Keywords: *Crystal structure, *Electron transitions, Barium oxides, Manganese, Minerals, Cells, Anisotropy, Reprints, *Romanechite.

The substructure and superstructure of romanechite, (Ba,H2O)2Mn5O10, were refined using a crystal from Schneeburg, Germany. The subcell is monoclinic, space group C2/m, with a = 13.929(1) Angstroms, b = 2.8459(4) Angstroms, c = 9.678(1) Angstroms, beta = 92.39(1) degrees, Z = 2, and it refined to R = 0.036 for anisotropic temperature factors using 598 reflections. The Mn(2) octahedra of the structure show significant distortion consistent with the concentration of Mn sup 3+ on the Mn(2) site. The supercell consisting of a tripling along b (i.e., the tunnel axis), results from ordering Ba and H2O along the tunnel lengths. Further, long-exposure diffraction patterns show modulated streaks along (201)*, implying only short-range ordering between the tunnels in one direction. A simplest monoclinic unit cell was chosen with space group C2/m and Z = 6. It was refined to R = 0.041 for anisotropic temperature factors using 954 reflections. Ba preferentially occupies the Ba(1) site, and correspondingly the closest framework octahedra (Mn(2) site) are preferentially occupied by Mn sup 3+.

901,393
PB89-158059   Not available NTIS
National Bureau of Standards (NML), Gaithersburg, MD. Radiation Physics Div.
Free-Electron-Like Stoner Excitations in Fe.
Final rept.
D. R. Penn, and P. Apell. 1988, 4p
Pub. in Physical Review B 38, n7 p5051-5054, 1 Sep 88.

Keywords: *Iron, Electron spin, Reprints, Stoner excitations, Spin orientation, Energy losses, Electron energy.

An analysis of spin-polarized electron-energy-loss experiments in Fe is described which identifies the contribution of d-electron Stoner excitations, the usual type of Stoner excitation, suggests that free-electron-like Stoner excitations are more probable than d-electron Stoner excitations, and indicates that exchange events involving large energy losses are as likely as direct scattering.

901,394
PB89-158067   Not available NTIS
National Bureau of Standards (NML), Gaithersburg, MD. Radiation Physics Div.
Domain Images of Ultrathin Fe Films on Ag(100).
Final rept.
J. L. Robins, R. J. Celotta, J. Unguris, D. T. Pierce, B. T. Jonker, and G. A. Prinz. 1988, 3p
Sponsored by Office of Naval Research, Arlington, VA.
Pub. in Applied Physics Letters 52, n22 p1918-1920; 30 May 88.

Keywords: *Iron, *Magnetic domains, Metal films, Thin films, Substrates, Silver, Reprints, *Surface magnetism, Electron spin polarization, Scanning electron microscopy.

Scanning electron microscopy with electron polarization analysis has been used to image domains of ultrathin Fe films grown epitaxially on a Ag(100) substrate. Room-temperature measurements show clearly the existence of large domains of in-plane magnetization for film thickness of 3.4 monolayers or more. No in-plane domains were observed for thinner films.

901,395
PB89-158075   Not available NTIS
National Bureau of Standards (NML), Gaithersburg, MD. Radiation Physics Div.
Spin-Polarized Electron Microscopy.
Final rept.
D. T. Pierce. 1988, 6p
Sponsored by Office of Naval Research, Arlington, VA.
Pub. in Physica Scripta 38, p291-296 1988.

Keywords: Magnetic domains, Microstructure, Reprints, *Electron spin polarization, Scanning electron microscopy, Scanning tunneling microscopy, Secondary electrons, High resolution.

The measurement of the spin polarization of secondary electrons generated by a finely focused (unpolarized) scanning electron microscope (SEM) beam to obtain high-resolution magnetization images is presented. An alternative measurement, using a spin-polarized incident beam in an SEM, has many difficulties which are discussed. To measure spin configurations with higher spatial resolution, the possibility of introducing electron spin polarization in scanning field-emission and tunneling microscopy is considered. The measurement of the spin polarization of secondary electrons generated by a specially prepared single-atom scanning field-emission tip looks promising. The potential advantages and unsolved problems involved in using a ferromagnetic tip or an optically pumped semiconductor tip are described.

901,396
PB89-158158   Not available NTIS
National Bureau of Standards (IMSE), Gaithersburg, MD. Reactor Radiation Div.
Small Angle Neutron Scattering Spectrometer at the National Bureau of Standards.
Final rept.
C. J. Glinka, J. M. Rowe, and J. G. LaRock. 1986, 13p
Pub. in Jnl. of Applied Crystallography 19, pt6 p427-439, 1 Dec 86.

Keywords: *Neutron spectrometers, Neutron scattering, Magnetic domains, Crystal structure, Reprints, Small angle scattering, Position sensitive detectors, Neutron detectors, NBSR reactor.

A small angle neutron scattering spectrometer, suitable for the study of structural and magnetic inhomogeneities in materials in the 10 to 1000 A range, has been constructed at the National Bureau of Standards Research Reactor. The instrument is 8 m long and uses a mechanical velocity selector and pinhole collimation to provide a continuous incident beam whose wavelength is variable from 5 to 10 A. The neutron detector is a 65 x 65 sq cm position-sensitive proportional counter which pivots about the sample position to extend the angular range of the spectrometer. Features unique to this instrument include a multibeam converging collimation system for high resolution measurements and an interactive color graphics terminal with specialized software for the rapid imaging and analysis of data from the two dimensional detector. The design and characteristics of the spectrometer and data acquisition system are described in detail, and examples of data are presented which illustrate its performance.

901,397
PB89-158174   Not available NTIS
National Bureau of Standards (NML), Gaithersburg, MD. Chemical Thermodynamics Div.
Sayre's Equation is a Chernov Bound to Maximum Entropy.
Final rept.
R. W. Harrison. 1987, 3p
Pub. in Acta Crystallographica A43, n3 p428-430 1987.

Keywords: Crystallography, Entropy, Reprints, *Sayre equation.

Sayre's equation is fundamental to a large part of classical direct methods. In the paper, it is shown that this equation can be derived via an integral bound to the entropy integral. While positivity is implicit in this derivation, atomicity is not used.

901,398
PB89-171318 Not available NTIS
National Bureau of Standards (NML), Gaithersburg, MD. Temperature and Pressure Div.
Electronic Structure of the Cd Vacancy in CdTe.
Final rept.
P. H. E. Meijer, P. Pecheur, and G. Toussaint. 1987, 8p
Pub. in Physica Status Solidi B-Basic Research 140, n1 p155-162 1987.

Keywords: *Cadmium tellurides, *Vacancies(Crystal defects), Greens functions, Reprints, *Electronic structure, Tight binding theory, Bound state, Slater method.

By means of a tight binding parametrization of the band structure of Cadmium Telluride, in conjunction with the Green function method, the authors determine the density of state of the perfect crystal. They compare various parametrizations of this compound. The energy levels for the Cd vacancy are calculated, taking into account the change in the wave functions around the missing atom.

901,399
PB89-171342 Not available NTIS
National Bureau of Standards (NML), Boulder, CO. Quantum Physics Div.
Surface Structures and Growth Mechanism of Ga on Si(100) Determined by LEED (Low Energy Electron Diffraction) and Auger Electron Spectroscopy.
Final rept.
B. Bourguignon, K. L. Carleton, and S. R. Leone. 1988, 18p
Grant AFOSR-84-0272
Sponsored by Air Force Office of Scientific Research, Bolling AFB, DC.
Pub. in Surface Science 204, p455-472 1988.

Keywords: *Gallium arsenides, *Gallium, *Silicon, *Surfaces, Substrates, Reprints, *Epitaxial growth, Low energy electron diffraction, Auger electron spectroscopy, Temperature dependence, Binding energy.

The surface structures of gallium overlayers on Si(100) are studied using LEED and Auger electron spectroscopy (AES). At 300 K, Ga grows epitaxially up to at least 5 ML. In the range of low surface temperature T(s) = 300-600 L, the growth obeys a Stranski-Krastanov mechanism, and the coverage at which island growth begins depends on T(s). Structures are proposed for the observed LEED patterns of Ga on Si(100)(3x2, 5x2, 2x2, 8x1 and 2x1). The existence of well-ordered structures suggests that Ga terminated Si(100) surfaces are suitable for epitaxy of GaAs on Si.

901,400
PB89-171359 Not available NTIS
National Bureau of Standards (NML), Gaithersburg, MD. Surface Science Div.
Progress in Understanding Atomic Structure of the Icosahedral Phase.
Final rept.
A. J. Melmed, M. J. Kaufman, and H. A. Fowler. 1986, 8p
Pub. in Jnl. of Physics Colloquium 47, nC-7 p35-40 Nov 86.

Keywords: *Aluminum-manganese alloys, Electron diffraction, Atomic structure, Reprints, *Icosahedral phase, Quasicrystals, Field ion microscopy.

Electron diffraction analysis of the icosahedral phase (i-phase) in AlMn alloys (1-AlMn) is presented which shows clearly that, in addition to the sharp diffraction maxima from the i-phase, there is a considerable amount of diffuse intensity indicating the presence of a certain amount of disorder interspersed within this phase. This result correlates well with previous results from both neutron diffraction and field ion microscopy studies and helps to reduce the number of possible structural models of the i-phase. Analysis of FIM images indicates the presence of a hierarchy of cluster sizes with numerous features which appear to agree qualitatively with recently proposed models. A complete quantitative description of the structure of the i-phase is being pursued by comparing experimentally observed atomic motifs with computer-simulated model surfaces, and preliminary results are presented.

901,401
PB89-171599 Not available NTIS

National Bureau of Standards (NML), Boulder, CO. Quantum Physics Div.
AES and LEED Studies Correlating Desorption Energies with Surface Structures and Coverages for Ga on Si(100).
Final rept.
B. Bourguignon, R. V. Smilgys, and S. R. Leone. 1988, 12p
Grant AFOSR-84-0272
Sponsored by Air Force Office of Scientific Research, Boiling AFB, DC.
Pub. in Surface Science 204, p473-484 1988.

Keywords: *Gallium, *Silicon, *Auger electrons, *Semiconductors(Materials), *Desorption, *Monomolecular films, Chemical bonds, Temperature, Surface properties, Kinetics, Heat of vaporization, Electrodeposition, Spectroscopy, Energy bands, Reprints.

Gallium interactions with silicon (100) are studied with Auger electron spectroscopy and LEED to correlate the desorption energies with surface coverages and structures in isothermal desorption experiments. Some evidence for a temperature-induced change from a Stranski-Karstanov to a Volmer-Weber growth mode between 600 and 700 K is presented. In the temperature range 800-900 K, three different kinetic regimes are observed. Between 0 and 0.5 monolayers (ML), first-order desorption is observed from a well-ordered Ga overlayer (Si:Ga 2x2), with a desorption energy of 2.9 plus or minus 0.2 eV and a pre-exponential factor of 3x10(sup 16 plus or minus 1)s. Between 0.5 and 1 ML, first-order desorption is also observed from a well-ordered Ga layer (Si:Ga 8x1), but the desorption energy decreases to 2.3 plus or minus 0.2 eV with a pre-exponential factor of 8x10(sup 12 plus or minus 1.2)/s. Above 1 ML, zeroth-order desorption from Ga islands on top of an ordered Ga monolayer is observed, and the desorption energy of the combination of surface species is 2.61 plus or minus 0.07 eV with a pre-exponential factor equal to (4 plus or minus 3)x10(sup 13) ML/s. It is suggested that atoms from the islands and the ordered layer are kinetically coupled, and that the islands cover too little of the surface to exhibit the bulk heat of vaporization of liquid Ga, 2.9 eV. The observed kinetic regimes are correlated with the surface structures proposed in the preceding paper.

901,402
PB89-175251
(Order as PB89-175194, PC A06)
A.T. and T. Bell Labs., Murray Hill, NJ.
Applications of the Crystallographic Search and Analysis System CRYSTDAT in Materials Science.
Bi-monthly rept.
T. Siegrist. 1989, 10p
Included in Jnl. of Research of the National Institute of Standards and Technology, v94 n1 p49-58 Jan-Feb 89.

Keywords: *Crystallography, Super conductors, Search structuring, Data retrieval, x ray diffraction, Qualitative analysis, Quantitative analysis, *Crystallinity, *Numerical data bases, CRYSTDAT system, Materials science.

Numerical database systems have recently become available online. Their enhanced search capabilities and fast retrieval of data make them a valuable tool in research. In particular, CRYSTDAT which is a search and analysis system for NBS CRYSTAL DATA has proven to be powerful in the identification of crystalline materials. In conjunction with a single-crystal x-ray diffractometer, a qualitative as well as quantitative phase determination is easily performed. The use of CRYSTDAT is illustrated in several examples.

901,403
PB89-175970 Not available NTIS
National Bureau of Standards (NML), Gaithersburg, MD. Surface Science Div.
Superlattice Magnetoroton Bands.
Final rept.
H. C. A. Oji, S. M. Girvin, and A. H. MacDonald. 1987, 4p
Pub. in Physical Review Letters 58, n8 p824-827 1987.

Keywords: Electron gas, Magnetic fields, Gallium arsenide, Reprints, *Magnetorotons, Fractional quantum Hall effect, Superlattices, Wigner crystallization.

A theory is given for the effect of interlayer coupling in a superlattice on the magnetoroton modes of a two-dimensional electron gas in a strong magnetic field. It was found that for typical fields and layer spacings, the modes broaden into bands. For sufficiently small layer spacings, the gap necessary for the occurrence of the fractional quantum Hall effect vanishes. The authors argue that the vanishing of the gap should be associated with an instability toward the state with a two-dimensional Wigner crystal in each layer and that this state is favored by interlayer coupling.

901,404
PB89-175988 Not available NTIS
National Bureau of Standards (NML), Gaithersburg, MD. Surface Science Div.
Effects of a Gold Shank-Overlayer on the Field Ion Imaging of Silicon.
Final rept.
A. J. Melmed, W. A. Schmidt, J. H. Block, M. Naschitzki, and M. Lovisa. 1986, 4p
Pub. in Jnl. of Physics Colloquium 47, nC-7 p333-336 Nov 86.

Keywords: *Silicon, Thin films, Covering, Gold, Reprints, *Field ion microscopy, Microprobes, Semiconductors.

The importance of the specimen shank in determining the magnitude of photo-illumination and field effects observed for Si specimens in field ion microscopy and atom probe mass analysis is established by measurements with and without an Au shank-overlayer. The field strength Au field ionization of hydrogen and argon over Si is shown to be the same as for metals.

901,405
PB89-176564 Not available NTIS
National Bureau of Standards (NML), Gaithersburg, MD. Radiation Physics Div.
Magnetic Properties of Surfaces Investigated by Spin Polarized Electron Beams.
Final rept.
D. T. Pierce. 1986, 12p
Pub. in Proceedings of International Workshop on Magnetic Properties of Low-Dimensional Systems, Taxco, Mexico, January 6-9, 1986, p58-69.

Keywords: Electron scattering, Chemisorption, Carbon monoxide, Oxygen, Nickel, Magnetic moments, *Surface magnetism, *Magnetic surfaces, Electron spin polarization, Electronic structure.

A spin polarized electron beam incident on a ferromagnetic surface results in elastically and inelastically scattered electrons and in photons via radiative transitions. The spin dependent intensities of each of these provide a sensitive measure of surface magnetization. A comparison between low temperature spin deviations at the surface and in the bulk is given; the variation follows the same power law with temperature but with a larger pre-factor for the surface. The connection between surface electronic structure and surface magnetism and the changes in each induced by chemisorption have been studied by spin polarized inverse photoemission. For oxygen and carbon monoxide on Ni(110), a reduction of the Ni magnetic moment is found, rather than a decrease in exchange coupling and corresponding randomization of the alignment of the moments. Further, in the case of CO, the chemisorption interaction is non-local with one CO molecule eliminating on the average the magnetic moment of two Ni atoms.

901,406
PB89-176978 Not available NTIS
National Bureau of Standards (NEL), Boulder, CO. Electromagnetic Technology Div.
Josephson-Junction Model of Critical Current in Granular Y1Ba2Cu3O(7-delta) Superconductors.
Final rept.
R. L. Peterson, and J. W. Ekin. 1988, 4p
Pub. in Physical Review B 37, n16 p9648-9651, 1 Jun 88.

Keywords: *Superconductors, *Josephson junctions, Magnetic fields, Grain boundaries, Reprints, *High temperature superconductors, *Critical current, *Yttrium barium cuprates, *Barium yttrium cuprates.

The authors calculate the transport critical-current density in a granular superconductor in magnetic fields below about 0.005 T. The field dependence in this region is assumed to be controlled by intragranular or intergranular Josephson junctions. Various model calculations are fitted to transport critical-current data on bulk Y1Ba2Cu3O(7-delta) ceramic superconductors, whose average grain size somewhat exceeds 10 micrometers. The results yield an average junction cross-

PHYSICS

Solid State Physics

sectional area (thickness x length) of 4-6 sq micrometers. If the junctions are at the grain boundaries, a London penetration depth of about 150-300 nm is inferred, consistent with other estimates. The authors conclude that Josephson junctions are limiting the transport critical current in the samples and that they lie at the grain boundaries. The parameters of the fit are not consistent with Josephson junctions at twinning boundaries.

901,407
PB89-176994 Not available NTIS
National Bureau of Standards (NEL), Boulder, CO.
Electromagnetic Technology Div.
Bean Model Extended to Magnetization Jumps.
Final rept.
R. L. Peterson. 1988, 4p
Pub. in Physics Letters A 131, n2 p131-134, 8 Aug 88.

Keywords: *Superconductors, Flux jumping, Magnetization, Reprints, *High temperature superconductors, Bean model.

The author extends the phenomenological Bean model of magnetization in hard superconductors to include the trains of magnetization jumps seen at low temperature in moderate-to-high magnetic fields. As in the original Bean model, no particular mechanisms for flux pinning are invoked. The extended model correctly accounts for the general dependence of the size of the magnetization jumps on sample size and critical current density. The data together with the model show that the shielding fields are approximately equal after each jump.

901,408
PB89-179188 Not available NTIS
National Bureau of Standards (NML), Gaithersburg, MD. Surface Science Div.
Temperature-Dependent Radiation-Enhanced Diffusion in Ion-Bombarded Solids.
Final rept.
D. Marton, J. Fine, and G. P. Chambers. 1988, 4p
Pub. in Physical Review Letters 61, n23 p2697-2700, 5 Dec 88.

Keywords: *Silver, *Nickel, *Diffusion, Crystal defects, Reprints, Ion bombardment, Physical radiation effects, Temperature dependence.

Temperature-dependence radiation-enhanced-diffusion rates for Ag in Ni have been found to decrease at elevated temperatures. The observed narrowing of interface interdiffusion regions with increasing temperature depends on both defect concentration and migration processes which occur in ion-bombarded solids. These findings can be interpreted in terms of a general model of radiation-enhanced diffusion that involves long-lived complex defects which can migrate for large distances and which are themselves subject to annealing.

901,409
PB89-179626 Not available NTIS
National Bureau of Standards (IMSE), Gaithersburg, MD. Reactor Radiation Div.
Significance of Multiple Scattering in the Interpretation of Small-Angle Neutron Scattering Experiments.
Final rept.
J. R. D. Copley. 1988, 6p
Pub. in Jnl. of Appl. Cryst. 21, p639-644 1988.

Keywords: *Neutron scattering, Monte Carlo method, *Computerized simulation, Vacuum, Embedding, Dispersing, Spheres, Wavelengths, Water, Reprints.

The multiple scattering of neutrons in small-angle neutron scattering experiments has been studied using the technique of Monte Carlo simulation. As a test of the approach, investigations have been performed on strongly scattering samples of a system of monodisperse spheres in vacuo: the results are in excellent agreement with semianalytic calculations by Schelten and Schmatz. The scattering by systems of monodisperse spheres embedded in a medium has also been studied: the standard procedure for subtraction of the scattering by the medium slightly oversubtracts the multiple scattering due to the medium alone. The wavelength dependence of the single and multiple scattering in light water has been estimated and compared with experimental measurements by groups at the Institut Laue-Langevin, Grenoble, France.

901,410
PB89-179634 Not available NTIS
National Bureau of Standards (IMSE), Gaithersburg, MD. Reactor Radiation Div.
Occurrence of Long-Range Helical Spin Ordering in Dy-Y Multilayers.
Final rept.
J. J. Rhyne, J. Borchers, R. Du, R. W. Erwin, C. P. Flynn, M. B. Salamon, and S. Sinha. 1987, 6p
See also PB89-179642.
Pub. in Jnl. of Applied Physics 61, n8 p4043-4048 1987.

Keywords: *Dysprosium, *Yttrium, Neutron diffraction, Metal films, Single crystals, Magnetic fields, Magnons, Ferromagnetism, Reprints, *Magnetic ordering, Magnetic films, Molecular beam epitaxy, Multilayers, Temperature dependence, Band theory, Spin waves.

The magnetic ordering of highly perfect single crystal multilayer films of alternate layers of magnetic Dy and non-magnetic Y prepared by molecular beam epitaxy has been studied by neutron diffraction. Results on a series of films with Dy thicknesses of approximately 16 atomic planes (about 45 A) and Y thicknesses ranging from 10 to 22 planes have confirmed the existence of long-range helimagnetic ordering of the Dy 4f spins which is propagated through the intervening Y layers in phase coherence. The propagation vectors in both Dy and Y layers have been calculated from the wave vector of the magnetic satellites and the intensity of the bilayer harmonics. The application of a field along basal plane directions destroys the helical order and produces a ferromagnetic state with all spins aligned along the field direction.

901,411
PB89-179642 Not available NTIS
National Bureau of Standards (IMSE), Gaithersburg, MD. Reactor Radiation Div.
Long-Range Incommensurate Magnetic Order in Dy-Y Multilayers.
Final rept.
J. J. Rhyne, J. Borchers, J. E. Cunningham, R. W. Erwin, C. P. Flynn, M. B. Salamon, and S. Sinha. 1986, 10p
See also PB89-179634.
Pub. in Jnl. of the Less Common Metals 126, p53-62 Dec 86.

Keywords: *Dysprosium, *Yttrium, Ferromagnetism, Neutron diffraction, Metal films, Magnetic fields, Reprints, *Magnetic ordering, Superlattices, Molecular beam epitaxy, Temperature dependence, Multilayers.

Neutron diffraction studies have demonstrated the existence of helical magnetic order in a metallic superlattice of Dy and Y produced by molecular beam epitaxy. Analysis of the magnetic satellite peak widths indicates that phase coherence of the helix exists over at least five bilayer cells, each consisting of 14 atomic planes of Dy and of non-magnetic Y. The multilayer comprises 64 bilayer cells. Results are given for the temperature dependence of the magnetic intensity and for the effect of an applied magnetic field which continuously converts the helical ordering to ferromagnetism.

901,412
PB89-179725 Not available NTIS
National Bureau of Standards (IMSE), Gaithersburg, MD. Ceramics Div.
Synthesis and Magnetic Properties of the Bi-Sr-Ca-Cu Oxide 80- and 110-K Superconductors.
Final rept.
J. S. Wallace, J. J. Ritter, E. Fuller, L. H. Bennett, R. D. Shull, and L. J. Swartzendruber. 1989, 3p
Pub. in Physical Review B 39, n4 p2333-2335, 1 Feb 89.

Keywords: *Superconductors, *High temperature, Synthesis(Chemistry), Bismuth, Strontium, Calcium, Copper oxides, Solids, Magnetic properties, Magnetic hysteresis, Meissner effect, Reprints.

The observation of both the 80- and 110-K superconducting transitions in a nominally BiSrCaCu2OX material produced by a new chemical synthesis procedure as well as by solid-state reaction is reported. Also discussed is a method for the confirmation of the presence of superconductivity when only a small amount is present.

901,413
PB89-179824 Not available NTIS
National Bureau of Standards (NEL), Boulder, CO.
Electromagnetic Technology Div.
High T(sub c) Superconductor/Noble-Metal Contacts with Surface Resistivities in the (10 to the Minus 10th Power) Omega sq cm Range.
Final rept.
J. W. Ekin, T. M. Larson, N. F. Bergren, A. J. Nelson, A. B. Swartzlander, L. L. Kazmerski, A. J. Panson, and B. A. Blankenship. 1988, 3p
Pub. in Applied Physics Letters 52, n21 p1819-1821, 23 May 88.

Keywords: *Superconductors, *Electric contacts, Gold, Silver, Indium, Comparison, Electrical resistivity, Reprints, *High temperature superconductors, *Yttrium barium cuprates, *Barium yttrium cuprates.

Contact surface resistivities (product of contact resistance and area) have been obtained for both silver and gold contacts to high T(sub c) superconductors. This is a reduction by about eight orders of magnitude from the contact resistivity of indium solder connections. The contact resistivity is low enough to be considered for both on-chip and package interconnect applications. The contacts were formed by sputter depositing either silver or gold at low temperatures (< 100 C) on a clean surface of Y1Ba2Cu3O(7-delta) (YBCO) and later annealing the contacts in oxygen. Auger microprobe analysis shows that indium/YBCO contacts contain a significant concentration of oxygen in the indium layer adjacent to the YBCO interface. Silver and gold contacts, on the other hand, contain almost no oxygen and have favorable interfacial chemistry with low oxygen affinity. Silver also acts as a 'switchable' passivation buffer, allowing oxygen to penetrate to the YBCO interface at elevated temperatures, but protecting the YBCO surface at room temperature.

901,414
PB89-179832 Not available NTIS
National Bureau of Standards (NEL), Boulder, CO.
Electromagnetic Technology Div.
Effect of Room-Temperature Stress on the Critical Current of NbTi.
Final rept.
S. L. Bray, and J. W. Ekin. 1989, 4p
Pub. in Jnl. of Applied Physics 65, n2 p684-687, 15 Jan 89.

Keywords: *Stresses, Strains, Comparison, Reprints, *Superconducting wires, *Niobium titanium, *Critical current, Room temperature, Temperature dependence.

The effect of axial tensile stress, applied at room temperature, on the critical current of NbTi superconducting wire was measured and compared with the effect of tensile stress applied at liquid-helium temperature (about 4 K). The results of these measurements indicate that the effect on the critical current is independent of the temperature at which the stress is applied. Thus, the existing 4-K data base can be used to determine I(sub c) degradation from room-temperature fabrication stress, cool-down stress introduced by differential contraction, as well as 4-K stress generated by the Lorentz force when the magnet is energized. To generalize these results for arbitrary matrix-to-superconductor volume ratios, the data are presented in terms of the stress on the NbTi portion of the composite conductor. Methods for determining the stress on the NbTi from the total composite load are presented.

901,415
PB89-180046 Not available NTIS
National Bureau of Standards (IMSE), Gaithersburg, MD.
Critical and Noncritical Roughening of Surfaces (Comment).
Final rept.
C. Rottman, and W. F. Saam. 1987, 1p
Pub. in Physical Review Letters 58, n15 p1588 1987.

Keywords: *Surface roughness, *Crystal lattices, *Lattice properties, *Surfaces, Symmetry, Crystals, Reprints.

The paper comments on 'Critical and Noncritical Roughening of Surfaces,' by Franz S. Rys (Phys. Rev. Lett. 56, 624 (1986)). Rys argues that the special symmetry which must be present for the roughening transition of crystal surfaces to occur is broken with two-body interactions. It is pointed out that only three- and higher-odd-order-body forces break this symmetry.

901,416
PB89-186241 Not available NTIS

National Bureau of Standards (IMSE), Gaithersburg, MD. Reactor Radiation Div.
Alternative Approach to the Hauptman-Karle Determinantal Inequalities.
Final rept.
E. Prince. 1989, 1p
Pub. in Acta Cryst. A45, p144 1989.

Keywords: Crystal structure, Determinants, Inequalities, Reprints, *Structure factors, Electron density, Cholesky factorization, Matrices.

A procedure is described for constructing a series of progressively stronger restrictions on the magnitudes and phases of individual structure factors in terms of sets of other structure factors. The existence of the Cholesky factors of Hauptman-Karle matrices is used to ensure that the electron density is everywhere positive.

901,417
PB89-186258   Not available NTIS
National Bureau of Standards (IMSE), Gaithersburg, MD. Reactor Radiation Div.
Maximum Entropy Distribution Consistent with Observed Structure Amplitudes.
Final rept.
E. Prince. 1989, 4p
Pub. in Acta Cryst. A45, p200-203 1989.

Keywords: Fourier series, Crystal structure, Reprints, *Structure factors, Electron density, Maximum entropy method, Iterative methods.

It is shown that an electron density distribution of the form rho sub k = exp (Sigma (f sub y)(r sub k)(x sub y)) has maximum entropy under the constraint that the expected values of a set of functions, f sub j(r), are constant. For a Fourier map the functions f sub j(r) are the magnitudes of the structure factors for a set of reflections h sub j including F(000). Maximum entropy is an efficient way of expressing the phase implications of a large set of structure amplitudes.

901,418
PB89-186266   Not available NTIS
National Bureau of Standards (IMSE), Gaithersburg, MD. Reactor Radiation Div.
Theoretical Models for High-Temperature Superconductivity.
Final rept.
R. C. Casella. 1988, 10p
Pub. in Il Nuovo Cimento 10, n12 p1439-1448 Dec 88.

Keywords: *Superconductors, Phonons, Bosons, Reprints, *High temperature superconductors, *Band theory, Yttrium barium cuprates, Barium yttrium cuprates.

A semi-phenomenological analysis is given of the effects of certain band structure features on the gap ratios 2 Delta/(k sub B)(T sub c) for high (T sub c) superconductors, including-multigap systems. In addition to phonons, other intermediate bosons (IB) mediating the superconducting interaction are considered. Interesting results emerge when the IB energy exceeds the widths of possible narrow peaks in the density of states associated with subbands presumably belonging to substructures such as stacked Cu-O planes. Comparison with experiment is made. In particular, data obtained by Warren et al. via nuclear-spin relaxation on Ba2YCu3O(7-delta) can be interpreted within the present framework in terms of a model having an IB of energy approx = or > 1 eV, which exceeds the predicted width (approx = or < 0.3 eV) of a peak in the density of states containing the normal-state Fermi level. This suggests that the IB is not a phonon.

901,419
PB89-186274   Not available NTIS
National Bureau of Standards (NML), Gaithersburg, MD. Radiation Physics Div.
Oxygen Partial-Density-of-States Change in the YBa2Cu3Ox Compounds for x(Approx.)6,6.5,7 Measured by Soft X-ray Emission.
Final rept.
C. H. Zhang, T. A. Callcott, K. L. Tsang, D. L. Ederer, J. E. Blendell, C. W. Clark, T. Scimeca, and Y. W. Liu. 1989, 4p
Contract DE-AC05-84OR21400
Sponsored by Department of Energy, Washington, DC., and Air Force Office of Scientific Research, Bolling AFB, DC.
Pub. in Physical Review B 39, n7 p4796-4799, 1 Mar 89.

Keywords: *Superconductors, Emission spectra, Oxygen, Reprints, *High temperature superconduc-

tors, *Yttrium barium cuprates, *Barium yttrium cuprates, Soft x-ray spectra, Density of states, Band theory.

Oxygen K soft x-ray emission spectra are presented for the YBa2Cu3Ox compounds with x nominally equal to 6, 6.5, and 7, and are compared with x-ray emission spectra determined from recent band-structure calculations. The K emission spectrum of O provides a measure of the p-type local partial density of states (p-LPDOS) at the oxygen sites. As x decreases from 7 - delta to about 6.5, a chemical shift of the entire spectrum to lower energy indicates that screening is modified for all oxygen sites. The integrated intensity of the spectra is nearly unchanged by oxygen removal, indicating an increase in p-LPDOS per oxygen site. These results and changes in the spectral shape suggest that itinerant electron density near the O atoms is reduced and bound electron density is increased as oxygen is removed.

901,420
PB89-186860   Not available NTIS
National Bureau of Standards (NML), Gaithersburg, MD. Surface Science Div.
Resonant Excitation of an Oxygen Valence Satellite in Photoemission from High-T(sub c) Superconductors.
Final rept.
R. L. Kurtz, S. W. Robey, R. L. Stockbauer, D. Mueller, A. Shih, and L. Toth. 1989, 4p
Sponsored by Office of Naval Research, Arlington, VA.
Pub. in Physical Review B 39, n7 p4768-4771, 1 Mar 89.

Keywords: *Superconductors, Excitation, Oxygen, Reprints, *High temperature superconductors, *Yttrium barium cuprates, *Barium yttrium cuprates, Lanthanum strontium cuprates, *Photoemission.

A detailed analysis of the intensities of valence-band photoelectron features of superconducting YBa2Cu3O(7-x) and semiconducting La(1.85)(Sr(0.05))CuO4 had revealed a resonance in the peak located at a binding energy of about 9.5 eV for photon energies spanning the onset of O 2s excitations. This demonstrates conclusively that the feature is associated with oxygen excitations. The origin of the satellite is described and its disappearance on superconducting surfaces is explained.

901,421
PB89-186886   Not available NTIS
National Bureau of Standards (NML), Gaithersburg, MD. Surface Science Div.
Dynamical Diffraction of X-rays at Grazing Angle.
Final rept.
T. Jach, P. L. Cowan, Q. Shen, and M. J. Bedzyk. 1989, 9p
Pub. in Physical Review B 39, n9 p5739-5747, 15 Mar 89.

Keywords: *X ray diffraction, Germanium, Standing waves, Reprints, *Grazing incidence, KeV range 01-10.

Details are presented of the theory and experimental observation of dynamical diffraction of X rays at grazing angle from crystal planes normal to a surface. The authors are able to associate different features of the specularly reflected and diffracted-reflected beam fluxes with the contributions from the alpha and beta branches of the dispersion surfaces. The theory predicts surface propagation modes to which internal and external beams can couple only through the diffraction process. An experiment is described in which the specularly reflected and reflected-diffracted beams were simultaneously observed for 8-keV X rays incident on germanium. The agreement with first-order theory is good, but systematic deviations were observed. Calculations are presented that illustrate how eigenstates of the wave fields, which are X ray standing waves with nodal planes normal to the surface of the crystal, can be used to obtain atomic registration at a surface or interface.

901,422
PB89-186928   Not available NTIS
National Bureau of Standards (NML), Boulder, CO. Quantum Physics Div.
Laser Probing of the Dynamics of Ga Interactions on Si(100).
Final rept.
K. L. Carleton, B. Bourguignon, R. V. Smilgys, D. J. Oostra, and S. R. Leone. 1988, 6p
Grant AFOSR-84-0272
Sponsored by Air Force Office of Scientific Research, Bolling AFB, DC.

Pub. in Proceedings of Materials Research Society Symposium, Reno, NV., p45-50 Apr 88.

Keywords: *Gallium, *Silicon, *Kinetics, *Desorption, *Epitaxy, *Semiconducting films, Lasers, Probes, Fluorescence, Auger electrons, Surface properties, Electron diffraction, Monomolecular films, Temperature, Growth, Gallium arsenides.

The kinetics of desorption and scattering of Ga atoms on Si(100) surfaces are probed by laser-induced fluorescence detection of the gas phase species and by Auger analysis of the surface composition. The kinetic parameters are correlated with the structures deduced by Low Energy Electron Diffraction (LEED) and the coverages determined by Auger spectroscopy. The binding energy of Ga on Si(100) is found to be a function of coverage, starting out at 2.9 eV at low coverages and decreasing to 2.3 eV for coverages between 0.5 and 1 monolayer (ML). Ordered growth is always observed for coverages below 1 ML, but above one monolayer the growth of islands occurs on the well-ordered monolayer. The onset of island formation is a strong function of temperature. A model is proposed for the structures and energetics involved in the growth of Ga on Si(100). The results are discussed in terms of the implications for epitaxial growth of GaAs on Si.

901,423
PB89-200448   Not available NTIS
National Bureau of Standards (NEL), Boulder, CO. Electromagnetic Technology Div.
Ag Screen Contacts to Sintered YBa2Cu3Ox Powder for Rapid Superconductor Characterization.
Final rept.
J. Moreland, and L. F. Goodrich. 1989, 4p
Pub. in IEEE (Institute of Electrical and Electronics Engineers) Transactions on Magnetics 25, n2 p2056-2059 Mar 89.

Keywords: *Superconductors, *Electric contacts, Sintering, Silver, Powder(Particles), Electrical resistivity, Electrical measurement, Reprints, *High temperature superconductors, *Yttrium barium cuprates, *Barium yttrium cuprates, Critical current.

A new method was developed for making current contacts and voltage taps to YBa2Cu3Ox sintered pellets for rapid superconductor characterization. Ag wire screens are interleaved between calcined powder sections and then fired at 930 C to form a composite pellet for resistivity and critical current measurements. The Ag diffuses into the powder during the sintering process forming a proximity contact that is permeable to O2. In this configuration, current can be uniformly injected into the ends of the pellet through the bonded Ag screen electrodes. Also, Ag screen voltage contacts, which span a cross section of the pellet, may provide an ideal geometry for detecting voltage drops along the pellet, minimizing current transfer effects.

901,424
PB89-200463   Not available NTIS
National Bureau of Standards (NEL), Boulder, CO. Electromagnetic Technology Div.
Chaos and Catastrophe Near the Plasma Frequency in the RF-Biased Josephson Junction.
Final rept.
R. L. Kautz, and R. Monaco. 1989, 5p
Grant N00014-88-F-0018
Sponsored by Office of Naval Research, Arlington, VA.
Pub. in IEEE (Institute of Electrical and Electronics Engineers) Transactions on Magnetics 25, n2 p1399-1403 Mar 89.

Keywords: *Josephson junctions, *Electric potential, *Plasma frequency, Bessel functions, Cusps(Mathematics), Superconductivity, Bias, Reprints, Chaos, Catastrophe theory.

At bias frequencies much higher than the plasma frequency, the zero-voltage state of the rf-biased Josephson junction is known to span a range of dc bias proportional to the zero-order Bessel function of the rf amplitude. This pattern is modified at frequencies near the plasma frequency by the onset of chaotic instabilities and by the presence of cusp catastrophes.

901,425
PB89-200489   Not available NTIS
National Bureau of Standards (NEL), Boulder, CO. Electromagnetic Technology Div.

PHYSICS

Solid State Physics

Magnetic Evaluation of Cu-Mn Matrix Material for Fine-Filament Nb-Ti Superconductors.
Final rept.
R. B. Goldfarb, D. L. Ried, T. S. Kreilick, and E. Gregory. 1989, 3p
Sponsored by Department of Energy, Washington, DC.
Pub. in IEEE (Institute of Electrical and Electronics Engineers) Transactions on Magnetics 25, n2 p1953-1955 Mar 89.

Keywords: *Superconductivity, *Copper manganese alloys, Magnetic permeability, Niobium intermetallics, Composite materials, Microstructure, Reprints, Superconducting wires, AC losses, Spin glass.

Copper-manganese alloys have been proposed as matrix material for the reduction of coupling losses in fine-filament Nb-Ti superconductor wires. Magnetization and susceptibility measurements show that adverse magnetic effects arising from the spin-glass properties of this matrix are minimal for concentrations of Mn up to at least 4%.

901,426
PB89-200513 Not available NTIS
National Bureau of Standards (NEL), Boulder, CO.
Electromagnetic Technology Div.
Switching Noise in YBa2Cu3Ox 'Macrobridges'.
Final rept.
R. H. Ono, J. A. Beall, M. W. Cromar, P. M. Mankiewich, R. E. Howard, and W. Skocpol. 1989, 6p
Pub. in IEEE (Institute of Electrical and Electronics Engineers) Transactions on Magnetics 25, n2 p976-979 Mar 89.

Keywords: *Superconductors, *Integrated circuits, *Thin films, *Electromagnetic noise, Resistance, Electric bridges, Bias, Switching circuits, Microstructure, Flux density, Reprints, *High temperature superconductors, Magnetic flux, Copper oxide superconductors.

Intermittent switching in the voltage-current characteristics (V/C) of thin film bridges of YBa2Cu3Ox have been observed. At a fixed bias point there are multiple metastable voltage states with lifetimes which depend on the bias current and applied magnetic field. The microbridges are made of thin (<500nm) polycrystalline films of YBa2Cu3Ox which are patterned by liftoff into structures with dimensions ranging from less than 1 micro m to 100 micro m. Details of the fabrication process and the measurements are presented. The results are discussed in the context of fluctuations in the effective resistance of the bridge due to motion of trapped flux.

901,427
PB89-200745 Not available NTIS
National Bureau of Standards (NML), Gaithersburg, MD. Surface Science Div.
Electron and Photon Stimulated Desorption: Benefits and Difficulties.
Final rept.
T. E. Madey, A. L. Johnson, and S. A. Joyce. 1988, 5p
See also PB84-136308. Sponsored by Department of Energy, Washington, DC.
Pub. in Vacuum 38, n8-10 p579-583 1988.

Keywords: *Desorption, *Ionization, *Surface properties, Photons, Electron impact, X ray spectroscopy, Radiation damage, Silicon, Ruthenium, Reprints, Auger electron spectroscopy, Transmission electron microscopy.

Some of the benefits and pitfalls of electron and photon stimulated (ESD/PSD) processes at surfaces are described. The benefits include useful information about the local structure of surface molecules, provided by electron stimulated desorption ion angular distributions (ESDIAD). ESDIAD is an effective ·surface structural tool because the directions of ion desorption are determined by the orientation of the surface bonds ruptured by electron or photon bombardment. Other benefits of electron and photon-stimulated damage processes at surfaces include electron and photon beam lithography in microelectronics. The pitfalls of ESD/PSD include beam damage in surface analysis (by Auger electron spectroscopy, X-ray photoelectron spectroscopy and high resolution transmission electron microscopy), the PSD of gases from vacuum walls in fusion reactors and synchrotron radiation sources, and inaccurate pressure readings due to ESD effects in ionization gauges.

901,428
PB89-201107 Not available NTIS

National Bureau of Standards (IMSE), Gaithersburg, MD.
Physics of Fracture, 1987.
Final rept.
R. Thomson. 1987, 19p
See also PB89-124036.
Pub. in Jnl. of Physics and Chemistry of Solids 48, n11 p965-983 1987.

Keywords: Ductile brittle transition, Crack propagation, Dislocations(Materials), Fracture strength, Brittleness, Reprints, *Fracture mechanics, Crack tips, Atmosphere effects.

The article first presents introductory material which should make it possible for the person unfamiliar with fracture to read the papers of this series. Then material of a basic physical nature regarding cracks in materials is presented. Emphasis is placed on the effects of chemical attack of bonds at a crack tip, and on the basic physical cause for a material to exhibit a tough (desirable) or a brittle (undesirable) overall aspect.

901,429
PB89-201206 Not available NTIS
National Bureau of Standards (IMSE), Gaithersburg, MD. Metallurgy Div.
Mossbauer Hyperfine Fields in RBa2(Cu0.97Fe0.03)3 O(7-x)(R=Y,Pr,Er).
Final rept.
M. Rubinstein, M. Z. Harford, L. J. Swartzendruber, and L. H. Bennett. 1988, 2p
Pub. in Jnl. de Physique 49, n12 pC8-2209-C8-2210 Dec 88.

Keywords: *Mossbauer effect, Hyperfine structure, Ferrates, Substitutes, Iron, Reprints, *High temperature superconductors, Yttrium barium cuprates, Erbium barium cuprates, Praseodymium barium cuprates.

Room temperature (57)Fe Mossbauer spectra of RBa2 (Cu0.97)Fe(0.03)3O(7-x) (R = Y, Pr, Er) were obtained from samples with varying x. A magnetically-split hyperfine field spectrum was observed for the most oxygen-deficient Y sample, for all the Pr samples, and for none of the Er samples.

901,430
PB89-201230 Not available NTIS
National Bureau of Standards (NML), Gaithersburg, MD. Surface Science Div.
Electron-Stimulated-Desorption Ion-Angular Distributions.
Final rept.
A. L. Johnson, S. A. Joyce, and T. E. Madey. 1988, 4p
Sponsored by Department of Energy, Washington, DC.
Pub. in Physical Review Letters 61, n22 p2578-2581, 28 Nov 88.

Keywords: *Ionization, *Chemisorption, *Fluorine, *Ruthenium, Anions, Desorption, Surface chemistry, Reprints.

The first measurements of the electron-stimulated-desorption ion-angular-distributions of negative ions from surface are reported. The angular distribution of F(-) ions for electron-stimulated desorption of PF3, NF3, and (CF3(2)C) on Ru(0001) depend on the molecular geometry and the state of the adsorbed species. The structural information obtained from these negative-ion studies complements that from similar positive-ion studies.

901,431
PB89-201792 Not available NTIS
National Bureau of Standards (IMSE), Gaithersburg, MD. Ceramics Div.
Electron Diffraction Study of the Faceting of Tilt Grain Boundaries in NiO.
Final rept.
J. A. Eastman, M. D. Vaudin, K. L. Merkle, and S. L. Sass. 1989, 13p
Contracts W-31109-ENG-38, DE-FG02-85ER45211
Sponsored by Department of Energy, Washington, DC.
Pub. in Philosophical Magazine A 59, n3 p465-477 1989.

Keywords: *Diffraction, *Grain boundaries, *Crystal dislocations, *Nickel oxides, Crystal lattices, Electron microscopy, Reprints.

The diffraction effects expected from a periodically faceted boundary containing a periodic array of dislocations in the long facet have been analyzed. The characteristic manifestation in reciprocal space of peri-

odic faceting was identified as the occurrence of bundles of reciprocal lattice rods; within each bundle the rods are displaced with respect to each other parallel to their length in a manner related to the facet geometry. Electron diffraction and microscopy were used to study the facet structure of tilt boundaries in NiO, and the boundary structure deduced from the observed diffraction effects was in good agreement with the imaging observations. In those cases where there was only one type of dislocation present in the long facet, it was possible to determine the average boundary plane, the dislocation spacing in the long facet, the facet period and the facet height from electron diffraction observations. The technique is especially useful for large-angle boundaries having high-index rotation axes, where little detailed information can be obtained using imaging techniques.

901,432
PB89-201826 Not available NTIS
National Bureau of Standards (IMSE), Gaithersburg, MD. Reactor Radiation Div.
Statistical Descriptors in Crystallography: Report of the International Union of Crystallography Subcommittee on Statistical Descriptors.
Final rept.
D. Schwarzenbach, S. C. Abrahams, H. D. Flack, W. Gonschorek, T. Hahn, K. Huml, R. E. Marsh, E. Prince, B. E. Robertson, J. S. Rollett, and A. J. C. Wilson. 1989, 13p
Pub. in Acta Crystallographica A45, p63-75 1989.

Keywords: *Crystallography, Definitions, Nomenclature, Statistics, Reprints.

The International Union of Crystallography Subcommittee on Statistical Descriptors has attempted to elucidate the nature of problems encountered in the definition and use of statistical descriptors as applied to crystallography and to propose procedural improvements. The report contains: (a) a dictionary of statistical terms established for use by experimentalists; (b) a description of the statistical basis for refinement procedures; (c) sections dealing with defects in the physical model used for refinement, and with the choice and significance of weighting schemes; and (d) recommendations, some of which may be readily implemented, while others may require a long-term effort to bring them into general use.

901,433
PB89-202030 Not available NTIS
National Bureau of Standards (IMSE), Gaithersburg, MD. Reactor Radiation Div.
Spin-Density-Wave Transition in Dilute YGd Single Crystals.
Final rept.
L. E. Wenger, G. W. Hunter, J. A. Mydosh, J. A. Gotaas, and J. J. Rhyne. 1986, 4p
Pub. in Physical Review Letters 56, n10 p1090-1093 1986.

Keywords: *Yttrium alloys, *Magnons, Gadolinium containing alloys, Neutron diffraction, Specific heat, Single crystals, Phase transformations, Antiferromagnetism; Reprints, *Spin waves, Magnetic susceptibility, Magnetic ordering.

Neutron diffraction, heat capacity, and susceptibility measurements on dilute YGd single crystals with 1.5 to 4.4 at .% Gd show long-range helical magnetic ordering at low temperatures. The neutron data reveal a periodic incommensurate spin structure with the Gd moments in the basal plane and a propagation wave-vector of 0.28c. A semicusp-like behavior is observed in the magnetic specific heats with the maxima occurring at similar temperatures as the susceptibility peaks (H perpendicular to c axis) and the onset of Bragg scattering. These results are quantitatively interpreted within a helical spin-density-wave stabilization of the conduction electrons.

901,434
PB89-202236 Not available NTIS
National Bureau of Standards (IMSE), Gaithersburg, MD. Metallurgy Div.
Roles of Atomic Volume and Disclinations in the Magnetism of the Rare Earth-3D Hard Magnets.
Final rept.
L. H. Bennett, R. E. Watson, and M. Melamud. 1988, 2p
See also DE88015197.
Pub. in Jnl. de Physique 49, n12 pC8-537-C8-538 Dec 88.

Keywords: *Permanent magnets, *Rare earth elements, Atomic properties, Atomic structure, Reprints, Electron orbitals.

Although no clear pattern exists in the experimental data, structural factors are expected to be important to local 3d magnetism in the hard magnets. A measure of two structural factors (local site volume and the presence of disinclinations) are provided so as to disentangle their role in the local magnetism.

901,435
PB89-202444 Not available NTIS
National Bureau of Standards (NEL), Gaithersburg, MD. Thermophysics Div.
Ergodic Behavior in Supercooled Liquids and in Glasses.
Final rept.
D. Thirumalai, R. D. Mountain, and T. R. Kirkpatrick.
1989, 12p
Grants NSF-CHE86-09722, NSF-DMR86-07805
Sponsored by National Science Foundation, Washington, DC.
Pub. in Physical Review A 39, n7 p3563-3574, 1 Apr 89.

Keywords: *Ergodic processes, *Glass, *Liquids, *Phase transformations, Supercooling, Statistical mechanics, Correlation, Mathematical models, Symmetry, Vitreous state, Molecular relaxation, Reprints.

Ergodic behavior in liquids, supercooled liquids, and glasses is examined with a focus on the time scale needed to obtain ergodicity. In addition to broken ergodicity, the possibility that a subtle symmetry is broken as the liquid-to-glass transition takes place is examined. It is suggested that a 'discrete' symmetry, to be referred to as the statistical symmetry, is broken in the glassy phase. This is illustrated by analyzing the distribution of the energy of the particles. Based on this, long-time dynamics and structural relaxation in glasses will be dominated by fluctuations in domains of finite length within which the particles are highly correlated. This is in accord with the ideas of Adams and Gibbs. All of the above arguments are illustrated with the aid of molecular-dynamics simulations of soft-sphere mixtures.

901,436
PB89-202501 Not available NTIS
National Bureau of Standards (NEL), Gaithersburg, MD. Thermophysics Div.
Liquid, Crystalline and Glassy States of Binary Charged Colloidal Suspensions.
Final rept.
R. O. Rosenberg, D. Thirumalai, and R. D. Mountain.
1989, 6p
Grant NSF-CHE86-57396
Sponsored by National Science Foundation, Washington, DC.
Pub. in Jnl. of Physics: Condens. Matter 1, p2109-2114 1989.

Keywords: *Suspending(Mixing), *Colloids, Polystyrene, Spheres, Liquids, Vitreous state, Crystallization, Computerized simulation, Reprints.

The formation of the liquid, crystalline, and glassy states in binary mixtures of aqueous suspensions of charged polystyrene spheres of different sizes is investigated. It is shown that on merely mixing the particles, crystalline (a substitutional body-centered cubic) and glassy states are readily formed for a 1:1 mixture at appropriate values of the density. These results are in agreement with recent experimental studies. The ease of formation of the crystal and the glassy phases is rationalized by an analysis of the local potential energy profiles calculated using the Hessian (dynamical matrix). It is suggested that the eigenvalues of such a Hessian can be used to define a preferred length in any glassy system.

901,437
PB89-202659 Not available NTIS
National Bureau of Standards (IMSE), Gaithersburg, MD. Reactor Radiation Div.
Antiferromagnetic Structure and Crystal Field Splittings in the Cubic Heusler Alloys HoPd2Sn and ErPd2Sn.
Final rept.
W. H. Li, J. W. Lynn, H. B. Stanley, T. J. Udovic, R. N. Shelton, and P. Klavins. 1988, 2p
Sponsored by National Science Foundation, Washington, DC.
Pub. in Jnl. de Physique 49, n12 pC8-373-C8-374 Dec 88.

Keywords: *Antiferromagnetism, *Holmium compounds, *Erbium compounds, Crystal lattices, Electric fields, Neutron diffraction, Low temperature research, Reprints, *Heusler alloys.

Neutron scattering techniques have been employed to investigate the magnetic properties of the cubic Heusler alloys HoPd2Sn and ErPd2Sn. Both materials exhibit an fcc type-II antiferromagnetic order, with T(sub N) =5.0 and 1.0 K for Ho and Er materials, respectively. However, there is an additional modulation of this basic structure. Inelastic neutron scattering measurements have been performed by time-of-flight techniques to determine the crystal field levels of the rare-earth ions. The (cubic) Lea, Leask, and Wolf crystal field parameters were determined to be W = -0.0287(4)meV, x = 0.3248(8) for HoPd2Sn, and W = -0.0450(4)meV, x = 0.3022(6) for ErPd2Sn.

901,438
PB89-202667 Not available NTIS
National Bureau of Standards (IMSE), Gaithersburg, MD. Reactor Radiation Div.
Exchange and Magnetostrictive Effects in Rare Earth Superlattices.
Final rept.
J. J. Rhyne, R. W. Erwin, J. Borchers, M. B. Salamon, R. Du, and C. P. Flynn. 1989, 17p
Pub. in Jnl. of the Less-Common Metals 148, p17-33 1989.

Keywords: *Single crystals, *Rare earth elements, *Thin films, *Yttrium, *Neutron diffraction, *Ferromagnetism, Magnetic properties, Dysprosium, Erbium, Crystal structure, Crystal lattices, Reprints.

Single-crystal multilayer films with alternate heavy rare earth and yttrium layers have been shown by neutron diffraction to exhibit long-range magnetic order. Analysis of the neutron results on Dy/Y and Er/Y super-lattices shows that the phase of the modulated magnetic structures in the dysprosium and erbium is preserved across the intervening yttrium non-magnetic layer and corresponds to a 'pseudo turn-angle' near 51 degrees in the yttrium, which is in accord with theoretical calculations from the band structure. The ferromagnetic transitions occurring in the pure elements are completely suppressed in the multilayers as a result of epitaxial 'clamping' by the yttrium layers, which inhibits the development of sufficient magnetostrictive strain to induce the phase transition. The temperature of the intermediate transitions in Er/Y multilayers is also modified by magnetostriction, and evidence is found for different turn-angles for c axis and basal plane moment components.

901,439
PB89-202675 Not available NTIS
National Bureau of Standards (IMSE), Gaithersburg, MD. Reactor Radiation Div.
Magnetic Structure of Y0.97Er0.03.
Final rept.
J. A. Gotaas, J. J. Rhyne, L. E. Weniger, and J. A. Mydosh. 1988, 2p
Grant NSF-DMR84-00711.
Sponsored by National Science Foundation, Washington, DC.
Pub. in Jnl. de Physique 49, n12 pC8-365-C8-366 Dec 88.

Keywords: *Yttrium alloys, *Erbium containing alloys, Neutron diffraction, Single crystals, Low temperature research, Magnetic moments, Magnetic anisotropy, Reprints, Spin waves, Spin glass.

Neutron diffraction has demonstrated that a single crystal alloy of Y(0.97)ER(0.03) orders below T(sub N) =3.25 K into a sinusoidally modulated slate with a propagation vector along the c-axis. The moments lie predominantly along the c-axis, but a small (< 0.1 mu sub b) basal plane component exhibits the same temperature dependence. Below 2 K; weak third harmonics indicate that the sinusoidal structure is beginning to square up.

901,440
PB89-202972 Not available NTIS
National Bureau of Standards (NML), Gaithersburg, MD. Surface Science Div.
Cross Sections for Inelastic Electron Scattering in Solids.
Final rept.
C. J. Powell. 1989, 8p
Pub. in Ultramicroscopy 28, p24-31 1989.

Keywords: *Electron scattering, *Solids, *Inelastic scattering, *Cross sections, Electron energy, Ionization, X ray spectroscopy, Radiation damage, Beta particle spectroscopy, Mean free path, Reprints, Auger electron spectroscopy, Energy loss spectroscopy, Electronic structure.

An overview is given of available information on cross-sections for inelastic electron scattering in solids with emphasis on the need for cross-section data in electron energy-loss spectroscopy (EELS), X-ray emission spectroscopy (XES), and Auger-electron spectroscopy (AES). After a brief survey of the relevant theory, information is given on inelastic mean free paths of 200-2000 eV electrons in solids (AES), total inner-shell ionization cross sections (AES and XES), partial inner-shell ionization cross section (EELS), and electron-beam-induced damage.

901,441
PB89-211106 PC A04
National Inst. of Standards and Technology, Gaithersburg, MD.
Journal of Research of the National Institute of Standards and Technology, Volume 94, Number 3, May-June 1989.
Bi-monthly rept.
1989, 68p
Also available from Supt. of Docs. as SN703-027-00028-8. See also PB89-211114 through PB89-211130 and PB89-175194.

Keywords: *Superconductivity, *Transformers, *Capacitors, *Research, *Consensus values, High voltage, Calibration.

Contents: A brief review of recent superconductivity research at NIST; Calibration of voltage transformers and high-voltage capacitors at NIST; Consensus values, regressions, and weighting factors.

901,442
PB89-211114
(Order as PB89-211106, PC A04)
National Inst. of Standards and Technology, Gaithersburg, MD.
Brief Review of Recent Superconductivity Research at NIST (National Institute of Standards and Technology).
Bi-monthly rept.
D. R. Lundy, L. J. Swartzendruber, and L. H. Bennett. 1989, 32p
Included in Jnl. of Research of the National Institute of Standards and Technology, v94 n3 p147-178 May-Jun 89.

Keywords: *Superconductivity, Electric contacts, Crystal structures, Phase diagrams, Josephson junctions, Reviews, *High temperature superconductors, Electronic structure, Voltage standards, Yttrium barium cuprates, Barium yttrium cuprates.

A brief overview of recent superconductivity research at NIST is presented. Emphasis is placed on the new high-temperature oxide superconductors, though mention is made of important work on low-temperature superconductors, and a few historical notes are included. For the new high-temperature superconductors, research activities include determination of physical properties such as elastic constants and electronics structure, development of new techniques such as magnetic-field modulated microwave-absorption and determination of phase diagrams and crystal structure. For the low-temperature superconductors, research spans studying the effect of stress on current density to the fabrication of a new Josephson junction voltage standard.

901,443
PB89-212179 Not available NTIS
National Bureau of Standards (NEL), Gaithersburg, MD. Semiconductor Electronics Div.
Electromigration Damage Response Time and Implications for dc and Pulsed Characterization.
Final rept.
J. S. Suehle, and H. A. Schafft. 1989, 3p
Pub. in Proceedings of Annual Reliability Physics Symposium (27th), Phoenix, AZ., April 11-13, 1989, p229-231.

Keywords: *Stress corrosion tests, *Metallizing, Reliability, Reaction time, Vacancies(Crystal defects), Integrated circuits, *Electromigration.

A new measurement interference for highly accelerated electromigration stress tests is identified. Measurements of the median-time-to-failure, t(50) for dc

169

Solid State Physics

for pulsed current stress as a function of pulse repetition frequency, reveal that highly accelerated stress tests may overestimate metallization reliability if t(50) is comparable with the response time of the vacancy concentration. Techniques necessary to make reliable wafer-level t(50) measurements are described.

901,444
PB89-212195 Not available NTIS
National Bureau of Standards (NEL), Gaithersburg, MD. Semiconductor Electronics Div.
Thermal Conductivity Measurements of Thin-Film Silicon Dioxide.
Final rept.
H. A. Schafft, J. S. Suehle, and P. G. A. Mirel. 1989, 6p
Contract DARPA Order-3882
Sponsored by Defense Advanced Research Projects Agency, Arlington, VA.
Pub. in Proceedings of IEEE (Institute of Electrical and Electronics Engineers) International Conference on Microelectronic Test Structures, Edinburgh, Scotland, March 13-14, 1989, v2 n1 p121-126.

Keywords: *Silicon dioxide, *Thin films, *Thermal conductivity, *Measurement, *Integrated circuits, *Semiconducting films, Stresses, Thickness, Ohmic dissipation, Temperature, Microelectronics, Accelerated tests.

Measurements of the thermal conductivity of micrometer-thick films of silicon dioxide are reported for the first time. Results show that the thermal conductivity is much lower than the values reported for bulk specimens and decreases with decreasing film thickness. It means that heating effects may be much larger than expected in accelerated stress tests and in other cases where Joule heating can be a concern.

901,445
PB89-212245 Not available NTIS
National Bureau of Standards (NML), Gaithersburg, MD. Radiometric Physics Div.
Interpolation of Silicon Photodiode Quantum Efficiency as an Absolute Radiometric Standard.
Final rept.
E. F. Zalewski, and W. K. Gladden. 1986, 2p
Pub. in Proceedings of Conference on Precision Electromagnetic Measurements, Gaithersburg, MD., June 23-27, 1986, p134-135.

Keywords: *Photodiodes, *Quantum efficiency, Silicon, Interpolation, Standards, Radiometry, Spectral response, Calibration.

The calibration of the external quantum efficiency of silicon photodiodes over the entire silicon spectral range based on quantum efficiency measurements at a few wavelengths is described.

901,446
PB89-214738 Not available NTIS
National Bureau of Standards (NEL), Boulder, CO. Electromagnetic Technology Div.
MM Wave Quasioptical SIS Mixers.
Final rept.
Q. Hu, C. A. Mears, P. L. Richards, and F. L. Lloyd. 1989, 4p
See also DE89001262. Sponsored by Department of Energy, Washington, DC.
Pub. in IEEE (Institute of Electrical and Electronics Engineers) Transactions on Magnetics 25, n2 p1380-1383 Mar 89.

Keywords: *Mixing circuits, *Integrated circuits, *Log periodic antennas, *Josephson junctions, Electromagnetic noise, Niobium oxides, Niobium, Lead, Indium, Gold, Reprints.

The performance of planar SIS mixers was tested with log-periodic antennas at near millimeter and submillimeter wave frequencies from 90 to 360 GHz. The large (omega)(R sub N)(C) product (about 10 at 90 GHz,) of the Nb/Nb(O sub x)/Pb-In-Au junctions requires an integrated inductive tuning element to resonate the junction capacitance at the operating frequencies. Two types of integrated tuning element were designed with the aid of measurements using a Fourier transform spectrometer. Preliminary results indicate that the tuning elements can give very good mixer performance up to at least 200 GHz. The relatively high mixer noise temperatures compared to those of waveguide SIS mixers in a similar frequency range are attributed mainly to the losses in the optical system, which is being improved.

901,447
PB89-228076 Not available NTIS
National Inst. of Standards and Technology (NEL), Boulder, CO. Thermophysics Div.
Torsional Piezoelectric Crystal Viscometer for Compressed Gases and Liquids.
Final rept.
D. E. Diller, and N. V. Frederick. 1989, 13p
Sponsored by Department of Energy, Washington, DC.
Pub. in International Jnl. of Thermophysics 10, n1 p145-157 Jan 89.

Keywords: *Piezoelectric crystals, *Viscometers, *Gases, *Liquids, Pressure, Temperature, Impedance, Argon, Methane, Resonance, Reprints.

A torsional piezoelectric crystal viscometer for compressed gases and liquids at temperatures to 600 K and at pressures to 70 MPa has been developed. Several torsional crystals were prepared from swept (electrolyzed) quartz to obtain a good performance at high temperatures. Measurements of the bandwidth of the crystal resonance curve were automated using an impedance analyzer. The viscometer was tested on compressed gaseous argon and methane at temperatures to 500 K and at pressures to 50 MPa. The measurements differ from accurate wide-range correlating equations by less than 2%.

901,448
PB89-228431 Not available NTIS
National Bureau of Standards (NEL), Boulder, CO. Electromagnetic Technology Div.
Resistance Measurements at High T(sub c) Superconductors Using a Novel 'Bathysphere' Cryostat.
Final rept.
J. Moreland, Y. Li, R. M. Folsom, and T. E. Capobianco. 1989, 3p
Pub. in IEEE (Institute of Electrical and Electronics Engineers) Transactions on Magnetics 25, n2 p2560-2562 Mar 89.

Keywords: *Superconductors, *Electrical resistance, *Cryostats, Magnetic fields, Critical temperature, Definitions, Thermodynamics, Reprints, *High temperature superconductors, Temperature dependence, Cryogenic equipment, Bathyspheres.

The authors have developed a novel cryostat for variable temperature testing of high temperature superconductors. The cryostat is a bathysphere consisting of an overturned stainless steel Dewar suspended in liquid helium. Results for resistance versus temperature of some high temperature superconductors in a magnetic field are presented. Also, various definitions for thermodynamic and practical T sub c derived from transport resistance measurements are suggested and discussed. These definitions are based on T sub c midpoint, various relative resistance criteria, or absolute resistivity criteria.

901,449
PB89-228449 Not available NTIS
National Inst. of Standards and Technology (NEL), Boulder, CO. Electromagnetic Technology Div.
Evidence for the Superconducting Proximity Effect in Junctions between the Surfaces of YBa2CU3Ox Thin Films.
Final rept.
J. Moreland, R. H. Ono, J. A. Beall, M. Madden, and A. J. Nelson. 1989, 3p
Contract N00014-86-F-0013
Sponsored by Office of Naval Research, Arlington, VA., and Department of Energy, Washington, DC.
Pub. in Applied Physics Letters 54, n15 p1477-1479, 10 Apr 89.

Keywords: Electron tunneling, Thin films, Electrodes, Reprints, *Superconducting junctions, *Superconducting films, *High temperature superconductors, *Yttrium barium cuprates, *Barium yttrium cuprates, Josephson effect, Proximity effect(Electricity).

The authors have used the squeezable electron tunneling junction technique for testing the electrical properties of the surfaces of YBa2Cu3Ox (YBCO) thin-film electrodes. As deposited and annealed, the surfaces of the electrodes were insulating at 4 K. Several methods were used to improve the electrical properties of the electrodes' surfaces including rapid thermal annealing, oxygen sputter etching, and thin Ag coating treatments. The greatest improvement occurred after a deposition of a 5 nm Ag coating and subsequent rapid thermal anneal of one set of YBCO films. Under these conditions it was possible to make a supercon-

ducting Josephson point contact between the surfaces of the electrodes. The authors think that the Ag acts as a normal-metal proximity layer effectively shunting the degraded electrodes' surfaces;

901,450
PB89-228456 Not available NTIS
National Inst. of Standards and Technology (NEL), Boulder, CO. Electromagnetic Technology Div.
Cryogenic Bathysphere for Rapid Variable-Temperature Characterization of High-T(sub c) Superconductors.
Final rept.
J. Moreland, Y. K. Li, R. Folsom, and T. E. Capobianco. 1988, 4p
Pub. in Review of Scientific Instruments 59, n12 p2535-2538 Dec 88.

Keywords: *Superconductors, Electrical resistance, Cryostats, Reprints, *High temperature superconductors, Bathyspheres, Cryogenic equipment, Temperature dependence, Yttrium barium cuprates, Barium yttrium cuprates, Niobium titanium.

A bathysphere consisting of an inverted Dewar flask for submersible operation in cryogenic fluids is used to measure the resistance of superconductors, including high T sub c superconducting copper oxides, as a function of temperature from 4 to 300 K. The authors describe the cryostat incorporating the bathysphere and present data on NbTi (44% Ti) and YBa2Cu3O(7-delta) with respective superconducting transitions temperatures of 9.5 and 91.5 K. There are several advantages of the bathysphere method. The cryostat is of simple, compact design, easily adapted to high-field applications where magnet bore size is a limiting factor. The sample and thermometer are thermolyzed in the dry vapor trapped at the top of the bathysphere. Temperature can be varied rapidly from 300 to 4 K at a rate of 1 K/min with less than a 0.1 K thermal lag between the sample and thermometer.

901,451
PB89-228464 Not available NTIS
National Inst. of Standards and Technology (NEL), Gaithersburg, MD. Precision Engineering Div.
Specimen Biasing to Enhance or Suppress Secondary Electron Emission from Charging Specimens at Low Accelerating Voltages.
Final rept.
M. T. Postek, W. J. Keery, and R. D. Larrabee. 1989, 11p
Pub. in Scanning 11, p111-121 1989.

Keywords: *Transistors, *Electron emission, *Electric potential, Polyethylene, Electron microscopy, Acceleration(Physics), Gallium arsenides, Bias, Silicon, Methylmethacrylates, Reprints.

Biasing of the specimen is shown to produce improved images in the scanning electron microscope at low beam energies (0.8-2.5 keV) when charging effects (induced by the primary electron beam), topographic effects, or detector shadowing effects would otherwise be present. Examples of such improvement are given for gallium arsenide field-effect transistors (positive charging), patterned photoresist layers on silicon wafers (negative charging and shadowing in contact holes), fractured polymethylmethacrylate (negative charging), polyethylene wrapper material (positive charging), and polished diamond tools (positive charging). It is concluded that specimen biasing may be a simpler and more convenient way to achieve some of the advantages of the converted backscattered secondary electron technique for imaging.

901,452
PB89-228472 Not available NTIS
National Inst. of Standards and Technology (NEL), Gaithersburg, MD. Semiconductor Electronics Div.
EXAFS (Extended X-ray Absorption Fine Structure) Study of Buried Germanium Layer in Silicon.
Final rept.
C. E. Bouldin. 1989, 2p
Pub. in Physica B 158, p596-597 1989.

Keywords: *Germanium, *Silicon dioxide, *Single crystals, *Gallium arsenides, Silicon, Thin films, X ray fluorescence, Reprints.

EXAFS measurements are made of a 200 A layer of Ge on a Si substrate. The Ge layer is covered by a 3000 A layer of SiO2. Sensitivity to the buried layer is enhanced through the use of grazing incidence fluorescence detection. A two-channel photodiode defector is used to detect the fluorescence and to discriminate against Bragg peaks from the single-crystal Si substrate. Since the fluorescence signal is isotropic, while the Bragg peaks are directional, one channel of the detector is always free of Bragg peak interference. The average number of Ge-Ge and Ge-Si neighbors in the buried Ge layer, the distances and disorder in the first-shell were determined. Prospects for studying the buried Ge-SiO2 interface are discussed.

901,453
PB89-228522 Not available NTIS
National Inst. of Standards and Technology (NEL), Gaithersburg, MD. Semiconductor Electronics Div.
Effects of Doping-Density Gradients on Band-Gap Narrowing in Silicon and GaAs Devices.
Final rept.
J. R. Lowney, and H. S. Bennett. 1989, 5p
Pub. in Jnl. of Applied Physics 65, n12 p4823-4827, 15 Jun 89.

Keywords: *Semiconductor devices, *Gallium arsenides, *Silicon, *Energy gap, Semiconductor doping, Reprints, Quantum wells.

The limitations of the theory for band-gap narrowing, which is based on uniform material, are considered in devices that have steep doping gradients. Validity criteria are derived that place upper bounds on the dopant and carrier density gradients for the application of the results from uniform theory. The existence of wave-function tailing beyond the potential barriers that occur in devices is studied. At room temperature the effects due to these tails are usually small, but at low temperatures they can become very significant.

901,454
PB89-229082 Not available NTIS
National Bureau of Standards (IMSE), Boulder, CO. Fracture and Deformation Div.
Hysteretic Phase Transition in Y1Ba2Cu3O7-x Superconductors.
Final rept.
H. M. Ledbetter, and S. A. Kim. 1988, 4p
Pub. in Physical Review B 38, n16 p11857-11860, 1 Dec 88.

Keywords: *Superconductors, *Phase transformations, Barium titanates, Hysteresis, Elastic properties, Ultrasonic radiation, Acoustic velocity, Reprints, *High temperature superconductors, *Yttrium barium cuprates, *Barium yttrium cuprates, Holmium barium cuprates, Europium barium cuprates.

The authors studied ultrasonic-wave velocities, both longitudinal and shear, in YBa2Cu3O7-x between 5 and 295 K, during both cooling and warming. Both waves, especially the longitudinal, show thermal hysteresis. The results suggest a hysteretic phase change that occurs between 160 and 170 K during cooling, and between 170 and 260 K during warming. This phase-change hypothesis explains anomalies in several physical properties. The phase change agrees with thermodynamic instability predictions. The authors confirmed the hysteresis in Ho-Ba-Cu-O, where it is smaller than in Y-Ba-Cu-O, and in Eu-Ba-Cu-O, where it is larger. In a comparison perovskite, BaTiO3 they observed zero hysteresis. At T(c), 91 K, sound velocities show no measurable change in either magnitude or slope. This continuity disputes the current popular view that, contrary to thermodynamics, elastic stiffness increases upon cooling through T(c) into the superconducting state. The authors believe that stiffening results from the usual thermal effects after a phase transformation from a stiffer phase.

901,455
PB89-229132 Not available NTIS
National Inst. of Standards and Technology (IMSE), Gaithersburg, MD. Reactor Radiation Div.
Dependence of T(sub c) on the Number of CuO2 Planes per Cluster in Interplaner-Boson-Exchange Models of High-T(sub C) Superconductivity.
Final rept.
R. C. Casella. 1989, 3p
Pub. in Solid State Communications 70, n1 p75-77 1989.

Keywords: *Superconductors, *Critical temperature, *Copper oxides, Bismuth inorganic compounds, Thallium inorganic compounds, Mathematical models, Cuprates, Bosons, Reprints, *High temperature superconductors.

The critical temperature T sub c(n) of systems, such as the Tl and Bi compounds, containing n CuO2 planes

per cluster is discussed theoretically within semi-phenomenological models which are based upon the exchange of (unspecified) high-energy bosons (omega > > omega(phonon)) between different CuO2 planes. Applied to representative data for the Tl2 compounds, the models predict rapid saturation of T sub c(n) at relatively low n, with the possibility of a diminution beyond n = 3. Within the theoretical framework, these qualitative features appear to be quite general for this class of materials.

901,456
PB89-229140 Not available NTIS
National Bureau of Standards (IMSE); Gaithersburg, MD. Reactor Radiation Div.
Hydrogen Sites in Amorphous Pd85Si15HX Probed by Neutron Vibrational Spectroscopy.
Final rept.
J. J. Rush, T. J. Udovic, R. Hempelmann, D. Richter, and G. Driesen. 1989, 10p
Pub. in Jnl. of Physics: Condens. Matter 1, p1061-1070 1989.

Keywords: *Hydrogen, Neutron spectroscopy, Vibrational spectra, Crystal structure, Palladium, Silicides, Hydrides, Reprints, Amorphous materials.

In order to clarify the discrepancy between the Gaussian distribution of H-site energies suggested from different macroscopic measurements and the evidence for two energetically well separated types of H sites obtained from a microscopic H diffusion study of hydrogen in amorphous Pd(85)Si(15)H(x), 0.13 < or = x < or = 8.23. At concentrations below 1% the spectra exhibit distinct features that indicate the occupation of distorted Pd(6) octahedra, along with a range of tetrahedral sites. These observations are consistent with a bimodal distribution of H-site energies in this glassy metal hydride.

901,457
PB89-230353 Not available NTIS
National Inst. of Standards and Technology (NEL), Boulder, CO. Electromagnetic Technology Div.
Flux Creep and Activation Energies at the Grain Boundaries of Y-Ba-Cu-O Superconductors.
Final rept.
M. Nikolo, and R. B. Goldfarb. 1989, 4p
Pub. in Physical Review B 39, n10 p6615-6618, 1 Apr 89.

Keywords: *Superconductors, *Flux, *Creep properties, *Activation energy, *Grain boundaries, Magnetic hysteresis, Temperature, Reprints.

The ac susceptibility of sintered YBa2CuO7-delta pellets as a function of temperature and ac magnetic-field amplitude and frequency was measured. The imaginary part of the susceptibility chi double prime exhibits two peaks. A narrow peak is located at the critical temperature of the grains. A broad peak at lower temperature is attributed to hysteresis losses at the grain boundaries. There is a small shift in this coupling peak to higher temperature as the frequency increases from 10 to 1000 Hz. The shift is explained in terms of Anderson flux creep on a time scale of milliseconds. The shift depends on the amplitude of the measuring field. The activation energy for flux creep ranges from 11.9 + or - 1.0 eV in zero-field limit (0.8 Am(sup -1) (0.01 Oe)) to 1.2 + or - 0.3 eV at 800 Am(sup -1) (10 Oe). The data is extrapolated to find the value for an intergrain decoupling field of 1-2 kAm(sup -1) (13-25 Oe), above which flux creep presumably becomes flux flow at the grain boundaries.

901,458
PB89-231088 Not available NTIS
National Inst. of Standards and Technology (IMSE), Gaithersburg, MD. Ceramics Div.
Thermomechanical Detwinning of Superconducting YBa2Cu3O7-x Single Crystals.
Final rept.
D. L. Kaiser, F. W. Gayle, R. S. Roth, and L. J. Swartzendruber. 1989, 3p
Pub. in Jnl. of Materials Research 4, no. 4, p745-747, Jul/Aug 89.

Keywords: *Single crystals, *Superconductors, *Twinning, *Crystal defects, Yttrium oxides, Barium oxides, Copper oxides, Perovskites, Thermomagnetic effects, Anisotropy, Reprints.

A method for the complete removal of twins from single crystals of superconducting YBa2Cu3O(7-x) is described. The process depends on ferroelastic behavior found to exist in the phase, and should be gen-

erally applicable to the layered perovskite-type phases containing accommodation twins resulting from a tetragonal-to-orthorhombic transformation on cooling. The twin-free, superconducting single crystals will enable investigations of a-b anisotropy of properties as well as crystal structure determination without complication by the presence of microtwins.

901,459
PB89-231328 Not available NTIS
National Inst. of Standards and Technology (NML), Gaithersburg, MD. Surface Science Div.
Photon-Stimulated Desorption as a Measure of Surface Electronic Structure.
Final rept.
J. A. Yarmoff, and S. A. Joyce. 1989, 4p
Pub. in Jnl. of Vacuum Science and Technology A7, n3 p2445-2448 May/Jun 89.

Keywords: *Surfaces, Fluorine, Silicon, Adsorption, Reprints, *Photon stimulated desorption, *Electronic structure.

The yield of ions desorbed via excitation of the Si 2p and F 1s levels was measured for fluorine adsorbed on Si(111). These photon-stimulated desorption (PSD) spectra were compared with the absorption measured via secondary or Auger electrons. It was seen that the PSD is dominated by direct excitations, and that the PSD spectra are sensitive to the final-state density at the local atomic site associated with the initial excitation. At the Si 2p edge, the PSD is sensitive to the oxidation state of the bonding silicon atom. At the F 1s edge, it is sensitive to the correspondence between the polarization vector of the incident light and the bond direction. Measurements of the kinetic energies of the desorbed ions were used to ascertain details of the desorption mechanism.

901,460
PB89-234314 Not available NTIS
National Inst. of Standards and Technology (NML), Boulder, CO. Quantum Physics Div.
Universal Resputtering Curve.
Final rept.
W. L. Morgan. 1989, 3p
Pub. in Applied Physics Letters 55, n2 p106-108, 10 Jul 89.

Keywords: *Sputtering, Monte Carlo method, Deposition, Simulation, Substrates, Reprints, *Resputtering, Universal curves.

The process of resputtering material being sputter deposited onto a substrate is investigated via Monte Carlo simulations and simple analytical models. This resputtering comprises contributions from self-sputtering and from neutralized ions reflected from the target being sputtered. The results of these models are in reasonable agreement with recent measurements over a wide variety of gases and metal targets. When plotted versus a dimensionless mass parameter, the intrinsic resputtered fraction lies on a seemingly universal curve. The reason for this becomes clear through the development of simple analytical models.

901,461
PB90-112400 PC A04/MF A01
National Inst. of Standards and Technology, Gaithersburg, MD.
Directional Solidification of a Planar Interface in the Presence of a Time-Dependent Electric Current.
L. N. Brush, S. R. Coriell, and G. B. McFadden. Sep 89, 53p NISTIR-89/4161

Keywords: *Crystal growth, *Electric current, Peltier effect, Thermoelectricity, Ohmic dissipation, Resistance heating, Bismuth, Mathematical models, Indium antimonides, Germanium containing alloys, Tin alloys, *Directional solidification(Crystals), Numerical solution, Electromigration.

The paper develops a numerical method to study the motion of a planar crystal-melt interface during the directional solidification of a binary alloy in the presence of a time-dependent electric current. The model includes the Thomson effect, the Peltier effect, Joule heating and electromigration of solute in the coupled set of equations governing heat flow in the crystal and melt, and solute diffusion in the melt. For a variety of time-dependent currents, the temperature fields and interface velocity are calculated as functions of time for indium antimonide and bismuth, and for the binary alloys, germanium-gallium and tin-bismuth. For the

PHYSICS

Solid State Physics

alloys, it also calculates the solid composition of a function of position, and thus makes quantitative predictions of the effect of an electrical pulse on the solute distribution in the solidified material. In addition, for a sinusoidal current of small amplitude, it compares the numerical solutions with approximate analytical solutions valid to first order in the current amplitude. By using the numerical approach the specific mechanisms which play dominant roles in interface demarcation by current pulsing can be identified.

901,462
PB90-117334   Not available NTIS
National Inst. of Standards and Technology (IMSE), Gaithersburg, MD. Reactor Radiation Div.
Comparison of Interplaner-Boson-Exchange Models of High-Temperature Superconductivity - Possible Experimental Tests.
Final rept.
R. C. Casella. 1989, 3p
Pub. in Applied Physics Letters 55, n9 p908-910, 28 Aug 89.

Keywords: *Superconductivity, Bismuth inorganic compounds, Thallium inorganic compounds, Copper oxides, Comparison, Reprints, *High temperature superconductors, *Boson-exchange models, Cuprates.

Semiquantitative tests are considered of models in which high T(c) superconductivity follows from the exchange of unspecified intermediate bosons (IB) with omega >> omega(phonon) between fermionic pairs in different CuO(2) layers. Earlier predictions of the dependence of T(c)(n) on the number n of layers per cluster and extended to include the possibility that T(c)(n = 1) < about 10K in the Tl1201 and 2201 compounds. In common with other authors rapid saturation of T(c)(n) with increasing n is found here. Nonetheless, the experimental value of T(c)(n = 1) in the Tl and Bi systems can play an important role in discriminating between the various models empirically.

901,463
PB90-117490   Not available NTIS
National Inst. of Standards and Technology (NML), Gaithersburg, MD. Radiation Physics Div.
Structure of Cs on GaAs(110) as Determined by Scanning Tunneling Microscopy.
Final rept.
P. N. First, R. A. Dragoset, J. A. Stroscio, R. J. Celotta, and R. M. Feenstra. 1989, 5p
Sponsored by Office of Naval Research, Arlington, VA.
Pub. in Jnl. of Vacuum Science and Technology A 7, n4 p2868-2872 Jul/Aug 89.

Keywords: *Cesium, Monomolecular films, Gallium arsenides, Crystal structure, Absorption, Substrates, Reprints, *Surface structure, Scanning tunneling microscopy, Semiconductors, Room temperature.

Submonolayer coverages of Cs adsorbed at room temperature on the GaAs(110) surface are examined with scanning tunneling microscopy. Linear chains, formed by two adjoining rows of Cs atoms, are observed along the (1,-1,0) direction for coverages as low as 0.03 monolayer. The one-dimensional Cs chains are observed to be several hundred A long, and are seen in images of both the occupied and unoccupied electronic states. At higher coverages, approaching 0.15 monolayer, stable linear structures consisting of three neighboring Cs rows have been found.

901,464
PB90-117540   Not available NTIS
National Inst. of Standards and Technology (IMSE), Boulder, CO. Fracture and Deformation Div.
Reentrant Softening in Perovskite Superconductors.
Final rept.
T. Datta, H. M. Ledbetter, C. E. Violet, C. Almasan, and J. Estrada. 1988, 4p
Pub. in Physical Review B 37, n13 p7502-7505, 1 May 88.

Keywords: *Superconductors, Elastic properties, Shear modulus, Acoustic velocity, Reprints, *High temperature superconductors, Yttrium barium cuprates, Barium yttrium cuprates, Lanthanum strontium cuprates.

A model of reentrant elastic softening is suggested that achieves three useful results. First, and principally, it reconciles existing sound-velocity-elastic-constant measurements with thermodynamics. Second, it leads to Debye characteristic temperatures that agree with those from specific-heat and phonon density-of-states

determinations. Third, it links elastic-constant-temperature behavior in Y-Ba-Cu-O and La-Sr-Cu-O. The model predicts a superconducting-state elastic stiffness lower than the normal state.

901,465
PB90-117615   Not available NTIS
National Inst. of Standards and Technology (IMSE), Boulder, CO. Fracture and Deformation Div.
Gruneisen Parameter of Y1Ba2Cu3O7.
Final rept.
H. Ledbetter. 1989, 3p
Pub. in Physica C 159, p486-490 1989.

Keywords: *Superconductors, Bulk modulus, Elastic properties, Reprints, *High temperature superconductors, *Yttrium barium cuprates, *Barium yttrium cuprates, *Gruneisen constant.

Contrary to reports that the Gruneisen parameter of Y1Ba2Cu3O7 is approximately 3.0, the author argues that the parameter is approximately 1.5., a value consistent with metal oxides. The author offers three arguments. One depends on a lower bulk modulus (B), near 101 GPa, than found in high-pressure X-ray diffraction studies, which yield bulk-modulus values up to 200 GPa. The second depends on an ionic-crystal-model calculation of the Gruneisen parameter. The third depends on the Anderson-Gruneisen parameter determined by measuring dB/dT.

901,466
PB90-117789   Not available NTIS
National Inst. of Standards and Technology (IMSE), Gaithersburg, MD. Reactor Radiation Div.
Magnetic Structure of Cubic Tb0.3Y0.7Ag.
Final rept.
J. A. Gotaas, M. R. Said, J. S. Kouvel, and T. O. Brun. 1988, 2p
See also DE89-005783.
Pub. in Jnl. de Physique 49, n12 pC8-1103-C8-1104 Dec 88.

Keywords: *Phase transformations, Terbium alloys, Yttrium alloys, Silver alloys, Cryogenics, Reprints, *Antiferromagnetic materials, Spin glass, Cubic lattices.

Tb(0.3)Y(0.7)Ag undergoes a magnetic phase transition at about 36 K to an antiferromagnetic structure which neutron diffraction has shown to have two components. A commensurate antiferromagnetic component is similar to the (pi pi 0) structure found in TbAg, but with a correlation range of only 42 A at 4 K while an incommensurate modulated component is like that found in HoAg, but with a finite correlation range of 290 A.

901,467
PB90-118019   Not available NTIS
National Inst. of Standards and Technology (NML), Gaithersburg, MD. Radiation Physics Div.
Influence of the Surface on Magnetic Domain-Wall Microstructure.
Final rept.
M. R. Scheinfein, J. Unguris, R. J. Celotta, and D. T. Pierce. 1989, 4p
Sponsored by Office of Naval Research, Arlington, VA.
Pub. in Physical Review Letters 63, n6 p668-671, 7 Aug 89.

Keywords: *Magnetic domains, Microstructure, Surfaces, Permalloys, Iron, Reprints, *Domain walls, Magnetic films, Scanning electron microscopy.

The magnetization orientations in domain walls at the surfaces of an Fe crystal, a ferromagnetic glass, and a Permalloy film, measured by scanning electron microscopy with polarization analysis, exhibit asymmetric surface Neel wall profiles which are at least twice as wide as interior Bloch walls in bulk and are described quantitatively by the authors' micromagnetic calculations without assuming any special surface parameters. Misinterpretation of domain-wall widths, Bitter patterns, and magnetic-force-microscopy images can result from overlooking the extreme effect of the surface on magnetic microstructure.

901,468
PB90-123480   Not available NTIS
National Inst. of Standards and Technology (IMSE), Gaithersburg, MD. Ceramics Div.

Neutron Study of the Crystal Structure and Vacancy Distribution in the Superconductor Ba2Y Cu3 O(sub g-delta).
Final rept.
F. Beech, S. Miraglia, A. Santoro, and R. S. Roth. 1987, 4p
Pub. in Physical Review B 35, n16 p8778-8781, 1 Jun 87.

Keywords: *Superconductors, *Crystal structure, *Vacancies(Crystal defects), Orthorhombic lattices, Neutron diffraction, Oxygen, Cryogenics, Reprints, *High temperature superconductors, *Barium yttrium cuprates, *Yttrium barium cuprates, Room temperature.

Two samples of the high temperature superconductor Ba2YCu3O(g-delta) with delta=2.0 and 2.2, have been studied at room temperature and at 10K, with the neutron powder diffraction method and profile analysis. The structure of the compound is orthorhombic. In the compound with delta=2.0 all oxygen sites are fully occupied. When delta=2.2 there are oxygen vacancies, but these are confined to one set of positions only, specifically to the oxygen atoms of the chains, located on the b-axis. No detectable change of the structure has been observed between room and low temperature.

901,469
PB90-123613   Not available NTIS
National Inst. of Standards and Technology (IMSE), Gaithersburg, MD. Ceramics Div.
Bulk Modulus and Young's Modulus of the Superconductor Ba2Cu3YO7.
Final rept.
S. Block, G. J. Piermarini, R. G. Munro, and W. Wong-Ng. 1987, 5p
Pub. in Advanced Ceramic Materials 2, n3B p601-605 1987.

Keywords: *Bulk modulus, *Modulus of elasticity, *Superconductors, *Barium oxides, *Copper oxides, *Yttrium oxides, *Ceramics, X ray diffraction, Isotherms, Equations of state, Compressibility, Reprints.

The isothermal equation of state of the high temperature superconducting ceramic material Ba2Cu3YO7 has been measured in a diamond anvil pressure cell using an energy dispersive x-ray diffraction method. The unit cell lattice parameters (a,b,c) were found to have compressions of (2.0%, 2.3%, 1.1%), respectively over the pressure range from one atmosphere to 10.6 GPa at room temperature. Subsequent equation of state analysis of the approximately linear compression of the volume determined that the isothermal bulk modulus was 196 + or - 17 GPa. Young's modulus was estimated to be 235 + or - 20 GPa assuming that the Poisson's ratio for Ba2Cu3YO7 was 0.3, which is typical of many ceramics.

901,470
PB90-123662   Not available NTIS
National Inst. of Standards and Technology (IMSE), Gaithersburg, MD. Reactor Radiation Div.
Characterization of Structural and Magnetic Order of Er/Y Superlattices.
Final rept.
J. Borchers, M. B. Salamon, R. Du, C. P. Flynn, R. W. Erwin, and J. J. Rhyne. 1988, 4p
Pub. in Superlattices and Microstructures 4, n4 pt5 p439-442 1988.

Keywords: *Erbium, *Ytterbium, Phase transformations, Neutron diffraction, Reprints, *Superlattices, Molecular beam epitaxy, Magnetic ordering, Multilayers.

Coherent, crystalline superlattices of erbium and yttrium have been prepared by molecular beam epitaxy techniques. Magnetometer measurements indicate that the transition temperatures for the superlattices are significantly lower than those for pure erbium. The ferromagnetic phase observed in erbium is completely supressed. Neutron diffraction studies show that the periodic magnetic order of Er propagates through the Y layers, but has a temperature independent magnetic wavelength. These results suggest that the erbium magneto-elastic energy has been altered in the superlattice samples.

901,471
PB90-123803   Not available NTIS
National Inst. of Standards and Technology (IMSE), Gaithersburg, MD. Reactor Radiation Div.

Magnetic Order of Pr in PrBa2Cu3O7.
Final rept.
W. H. Li, J. W. Lynn, S. Skanthakumar, T. W. Clinton, A. Kebede, C. S. Jee, J. E. Crow, and T. Mihalisin.
1989, 4p
Pub. in Physical Review B 40, n7 p5300-5303, 1 Sep 89.

Keywords: Specific heat, Neutron diffraction, Magnetic moments, Reprints, *Praseodymium barium cuprates, *Barium praseodymium cuprates, *Magnetic ordering, *Praseodymium ions, Magnetic susceptibility, Antiferromagnetic materials.

The magnetic order of Pr in nonsuperconducting PrBa2Cu3O7 has been studied by specific-heat, susceptibility, and neutron-diffraction measurements. The basic ordering consists of a simple antiferromagnetic arrangement, with a saturated moment of 0.74 (mu sub s) and a Neel temperature T(N) of about 17 K, which is two orders of magnitude higher than expected from either dipolar or Rudeman-Kittel-Kasuya-Yosida interactions alone. The small moment, along with the large value of the low-temperature electronic specific-heat coefficient gamma of 196 mJ/mole(K squared), suggests that there is substantial f-electron character at the Fermi level.

901,472
PB90-123829   Not available NTIS
National Inst. of Standards and Technology (IMSE), Gaithersburg, MD. Reactor Radiation Div.
Pressure Dependence of the Cu Magnetic Order in RBa2Cu3O6+x.
Final rept.
J. W. Lynn, W. H. Li, S. F. Trevino, and Z. Fisk.
1989, 4p
Pub. in Physical Review B 40, n7 p5172-5175, 1 Sep 89.

Keywords: Neutron diffraction, Neel temperature, Reprints, *Neodymium barium cuprates, *Barium neodymium cuprates, *Magnetic ordering, *Copper ions, Pressure dependence, Antiferromagnetic materials, High temperature superconductors.

Neutron-diffraction measurements have been carried out as a function of hydrostatic pressure to study the magnetic order of the Cu spins in NdBa2Cu3O(6.35) and NdBa2Cu3O6.1). In the high-temperature phase, where the Cu planes order antiferromagnetically, it is found that the Neel temperature T(N1) is very strongly dependent on pressure, increasing at the rate of about 23 K/kbar. The authors attribute this phenomenal sensitivity to the two-dimensional-like behavior of this magnetic system. In the low-temperature phase, which is associated with magnetic ordering of the chains, only a small change in the ordering temperature T(N2) is observed.

901,473
PB90-123878   Not available NTIS
National Inst. of Standards and Technology (NML), Boulder, CO. Quantum Physics Div.
Initial Stages of Heteroepitaxial Growth of InAs on Si(100).
Final rept.
D. J. Oostra, R. V. Smilgys, and S. R. Leone. 1989, 3p
Sponsored by Air Force Office of Scientific Research, Bolling AFB, DC.
Pub. in Applied Physics Letters 55, n13 p1333-1335, 25 Sep 89.

Keywords: *Indium arsenides, Silicon, Substrates, Adsorption, Desorption, Reprints, *Epitaxial growth, Auger electron spectroscopy, Laser induced fluorescence, Semiconductor materials.

Adsorption and desorption of In on partially and fully As-terminated Si(100) is investigated by laser-induced fluorescence detection and Auger electron spectroscopy using the methods of temperature programmed desorption and isothermal desorption. Desorption measurements show that As is bound to the surface more strongly than In. For In, a 2/3 order kinetic desorption mechanism is observed. This and Si auger intensity attenuation measurements indicate a strong tendency for In to form three-dimensional islands on the As-terminated surface. The activation energy for In diffusion from the islands ranges from 1.5 to 1.8 eV, depending on the As coverage. The results have important implications for growth of InAs on Si(100).

901,474
PB90-128133   Not available NTIS

National Inst. of Standards and Technology (NEL), Boulder, CO. Electromagnetic Technology Div.
Offset Criterion for Determining Superconductor Critical Current.
Final rept.
J. W. Ekin. 1989, 3p
Pub. in Applied Physics Letters 55, n9 p905-907, 28 Aug 89.

Keywords: *Superconductors, Standards, Criteria, Reprints, *Critical current, High temperature superconductors.

Critical-current criteria based on electric field or resistivity can present a number of problems in defining critical current, especially for high T(c) superconductors in the vicinity of the critical temperature or upper critical field. The resulting critical-current density J(c) can be quite arbitrary, since it depends strongly on criterion level at high fields and temperatures. These J(c) definitions also create problems in distinguishing between superconductors and high-conductivity normal metals such as copper. They can also bias J(c) data when superconductors are compared that have different values of normal-state resistivity. To minimize these problems, an intrinsic J(c) criterion is proposed, which effectively separates superconducting and normal-state properties. Based on the long-standing concept of a flux-flow resistivity, J(c) is defined as the current where the tangent to the E-J curve at a given electric field level extrapolates to zero electric field. This determines an offset J(c) that minimizes the above problems. The criterion is particularly useful near T(c) or near the effective upper critical field where the E-J characteristic starts to approach ohmic behavior.

901,475
PB90-128216   Not available NTIS
National Inst. of Standards and Technology (NML), Gaithersburg, MD. Surface Science Div.
Direct Observation of Surface-Trapped Diffracted Waves.
Final rept.
T. Jach, D. B. Novotny, M. J. Bedzyk, and Q. Shen. 1989, 4p
Pub. in Physical Review B 40, n8 p5557-5560, 15 Sep 89.

Keywords: *x ray diffraction, Surface roughness, Germanium, Interfaces, Reprints, Grazing incidence.

The authors have made the first direct observation of a diffracted x-ray beam that occurs only at an interface in the grazing-angle diffraction geometry. The beam, which is unable to propagate into the bulk of either component of the interface, was detected at the surface of a Ge crystal. A roughened area etched into the surface permitted phase matching to a beam in the vacuum. The diffracted beam that is observed to escape in this configuration shows a wave-vector dependence that cannot be qualitatively explained by purely kinematic models of scattering from rough surfaces.

901,476
PB90-128240   Not available NTIS
National Inst. of Standards and Technology (NML), Gaithersburg, MD. Radiation Physics Div.
Vector Imaging of Magnetic Microstructure.
Final rept.
M. H. Kelley, J. Unguris, M. R. Scheinfein, D. T. Pierce, and N. J. Celotta. 1989, 6p
Pub. in Microbeam Analysis - 1989, p391-396 1989.

Keywords: *Magnetic domains, Whiskers(Single crystals), Microstructure, Polarization(Spin alignment), Cobalt, Iron, Reprints, *Imaging techniques, Electron spin polarization.

An ability to study the properties of microscopic magnetic structures and to investigate magnetic properties with submicron spatial resolution is important both for its fundamental scientific value and its usefulness in applied magnetic technology. Many current techniques for the investigation of magnetic structures suffer either from poor spatial resolution or from the inability clearly to separate-contrast due to magnetic structures from that due to topographic or other physical features. The authors describe a method of magnetic imaging that overcomes many of the difficulties of other current techniques and that allows quantitative analysis at high spatial resolution of the vectorial properties of sample magnetization.

901,477
PB90-130261   PC A03/MF A01

National Inst. of Standards and Technology (NEL), Gaithersburg, MD. Applied and Computational Mathematics Div.
Effect of a Crystal-Melt Interface on Taylor-Vortex Flow.
G. B. McFadden, S. R. Coriell, B. T. Murray, M. E. Glicksman, and M. E. Selleck. Oct 89, 22p NISTIR-89/4192
Prepared in cooperation with Rensselaer Polytechnic Inst., Troy, NY. Dept. of Materials Engineering.

Keywords: *Crystal growth, *Hydrodynamics, Solidification, Morphology, Stability, Couette flow, Taylor instability.

The linear stability of circular Couette flow between concentric infinite cylinders is considered for the case that the stationary outer cylinder is a crystal-melt interface rather than a rigid surface. A radial temperature difference is maintained across the liquid gap, and equations for heat transport in the crystal and melt phases are included to extend the ordinary formulation of the problem. The stability of the two-phase system depends on the Prandtl number. For small Prandtl number the linear stability of the two-phase system is given by the classical results for a rigid-walled system. For increasing values of the Prandtl number, convective heat transport becomes significant and the system becomes increasingly less stable. Previous results in a narrow-gap approximation are extended to the case of a finite gap, and both axisymmetric and non-axisymmetric disturbance modes are considered. The two-phase system becomes less stable as the finite gap tends to the narrow-gap limit. The two-phase system is more stable to non-axisymmetric modes with azimuthal wavenumber n = 1; the stability of these n = 1 modes is sensitive to the latent heat of fusion.

901,478
PB90-136706   Not available NTIS
National Inst. of Standards and Technology (IMSE), Gaithersburg, MD. Reactor Radiation Div.
Mn-Mn Exchange Constants in Zinc-Manganese Chalcogenides.
Final rept.
T. M. Giebultowicz, J. J. Rhyne, and J. K. Furdyna. 1987, 3p
Pub. in Jnl. of Applied Physics 61, n8 pt2A p3537-3539, 15 Apr 87.

Keywords: Neutron scattering, Inelastic scattering, Single crystals, Polycrystalline, Reprints, *Magnetic semiconductors, *Antiferromagnetic materials, *Manganese zinc sulfides, *Manganese zinc selenides, *Manganese zinc tellurides, Exchange interactions.

Excited levels of isolated nearest-neighbor Mn-Mn pairs in Zn(1-x)Mn(x)S, Zn(1-x)Mn(x)Se, and Zn(1-x)Mn(x)Te have been studied by inelastic neutron scattering. The measurements have been carried out on several single crystal and polycrystalline samples with x = or <0.05 at various temperatures. Values obtained for the exchange constants 2J(NN) are -2.78, -2.12, and -1.64 meV for the three studied systems, respectively.

901,479
PB90-136714   Not available NTIS
National Inst. of Standards and Technology (IMSE), Gaithersburg, MD. Reactor Radiation Div.
Neutron Diffraction Study of the Wurtzite-Structure Dilute Magnetic Semiconductor Zn0.45Mn0.55Se.
Final rept.
T. M. Giebultowicz, J. J. Rhyne, J. K. Furdyna, and U. Debska. 1987, 3p
Pub. in Jnl. of Applied Physics 61, n8 pt2A p3540-3542, 15 Apr 87.

Keywords: Neutron diffraction, Reprints, *Magnetic semiconductors, *Manganese zinc selenides, *Antiferromagnetic materials, Magnetic ordering, Exchange interactions.

First results are reported of neutron diffraction studies of magnetic ordering phenomena in a wurtzite-structured diluted magnetic (semimagnetic) semiconductor Zn(0.45)Mn(0.55)Te. At low temperatures the system exhibits a short-range antiferromagnetic ordering, closely related to the antiferromagnetic structure seen in the wurtzite form of beta-MnS. This type of ordering indicates short-range and antiferromagnetic only interactions between the Mn spins, in agreement with present theories of exchange mechanism in DMS materials. Similarly as in cubic DMS systems, the magnet-

PHYSICS

Solid State Physics

ic correlation range in Zn(0.45)Mn(0.55)Te exhibits a pronounced anisotropy.

901,480
PB90-136748 Not available NTIS
National Inst. of Standards and Technology (NEL), Boulder, CO. Electromagnetic Technology Div.
Critical Current Measurements of Nb3Sn Superconductors: NBS (National Bureau of Standards) Contribution to the VAMAS (Versailles Agreement on Advanced Materials and Standards) Interlaboratory Comparison.
Final rept.
L. F. Goodrich, and S. L. Bray. 1989; 11p
Sponsored by Department of Energy, Washington, DC.
Pub. in Cryogenics 29, p699-709 Jul 89.

Keywords: *Superconductors, Electrical measurement, Reprints, *Niobium stannides, *Critical current, Interlaboratory comparisons, Comparative evaluations, VAMAS.

Critical current measurements on several Nb3Sn superconductors were made as part of an interlaboratory comparison (round robin). These measurements were made in conjunction with twenty-four laboratories from the European Economic Community, Japan and the USA as part of the Versailles Agreement on Advanced Materials and Standards (VAMAS). The results of the NBS measurements, including the effect of sample mounting techniques on the measured critical current, are given. A systematic study of the effect of measurement mandrel (tubular sample-holder made from G10 fiberglass-epoxy composite) geometry revealed that a seemingly small change in that geometry can result in a 40% change in the measured critical current at a magnetic field of 12 T. Specifically, the radial thermal contraction of the measurement mandrel depends on its wall thickness and, thus, so does the conductor prestrain (at 4 K) and, ultimately, the measured critical current. Techniques for reducing variation in the measured critical current are suggested.

Structural Mechanics

901,481
PB89-157788 Not available NTIS
National Bureau of Standards (IMSE), Boulder, CO. Fracture and Deformation Div.
Conventional and Quarter-Point Mixed Elements in Linear Elastic Fracture Mechanics.
Final rept.
P. R. Heyliger. 1988, 15p
Pub. in Engineering Fracture Mechanics 31, n1 p157-171 1988.

Keywords: *Finite element analysis, Displacement, Stress analysis, Numerical analysis, Stiffness methods, Elastic analysis, Reprints, *Elastic fracture mechanics, Quarter-point element, Stress intensity factor, J-integral technique.

The numerical performance of conventionally configured and quarter-point mixed elements is examined for planar problems in linear elastic fracture mechanics. Because of the independent approximations of the displacement and stress components characteristic of the mixed formulation, the shifting of a mid-side node of an element to the quarter-point results in a singular strain but a finite stress at the corner node. For the example problems considered, the conventional and quarter-point mixed elements provide lower and upper bounds, respectively, for the strain energy of a cracked body. Mode I stress intensity factors are computed for several representative geometries using the crack extension, stiffness derivative and J-integral techniques. The quarter-point elements yield superior results only for the J-integral technique, with each of the three techniques giving excellent results for very coarse mesh configurations.

901,482
PB89-157903 Not available NTIS
National Bureau of Standards (IMSE), Gaithersburg, MD. Metallurgy Div.
Elastic Interaction and Stability of Misfitting Cuboidal Inhomogeneities.
Final rept.
W. C. Johnson, and P. W. Voorhees. 1987, 10p
Pub. in Jnl. of Applied Physics 61, n4 p1610-1619 1987.

Keywords: *Elastic analysis, *Stability, Energy methods, Matrix methods, Elastic properties, Plates(Structural members), Precipitates, Heterogeneity, Reprints.

The elastic interaction energy between several rectangular parallel-epipeds and the elastic self energy of a cuboid are calculated to first-order in the difference in elastic constants between the precipitate and matrix. The system is assumed to be isotropic and the precipitates to possess a dilatational misfit. The functionality and magnitude of the interaction energy per precipitate is extremely sensitive to the precipitate morphology and number of precipitates considered. The interaction energy between three or more precipitates cannot be accurately estimated by summing the pairwise interactions. The influence of the interaction energy on the stability of precipitate arrays is discussed.

901,483
PB89-166110 PC A03/MF A01
National Inst. of Standards and Technology (NEL), Gaithersburg, MD. Center for Building Technology.
Method to Measure the Tensile Bond Strength between Two Weakly-Cemented Sand Grains.
L. I. Knab, and N. E. Waters. Nov 88, 41p NISTIR-88/3983
Sponsored by Air Force Weapons Lab., Kirtland AFB, NM.

Keywords: *Bonding strength, *Sands, *Tension tests, Adhesive strength, Bonding, Cements, Grain structure, Tensile strength, Breaking load, Tensile stress.

A method to measure the tensile bond strength between two weakly-cemented sand grains was developed. Special microloading testing equipment was developed to measure the force required to pull apart two sand grains. To illustrate the method, bond strengths were measured in tension for six pairs of cemented sand grains. A wide range in the bond failure stress occurred and was attributed primarily to (a) difficulties in identifying and measuring the actual bond failure surface area and, (b) eccentricity in the specimens during loading. The method developed is seen as a starting point and can be used as a basis for further development. Improved techniques need to be developed to identify and measure the actual bond failure surface area and to reduce, or at least measure and account for, the eccentricity introduced.

901,484
PB89-229124 Not available NTIS
National Bureau of Standards (IMSE), Boulder, CO. Fracture and Deformation Div.
Higher Order Beam Finite Element for Bending and Vibration Problems.
Final rept.
P. R. Heyliger, and J. N. Reddy. 1988, 18p
Pub. in Jnl. of Sound and Vibration 126, n2 p309-326 1988.

Keywords: *Beams (Supports), *Finite element analysis, Vibration, Bending, Shear properties, Deformation, Reprints, Timoshenko beams, Rectangular configuration, Numerical solution.

The finite element equations for a variationally consistent higher order beam theory are presented for the static and dynamic behavior of rectangular beams. The higher order theory correctly accounts for the stress-free conditions on the upper and lower surfaces of the beam while retaining the parabolic shear strain distribution. The need for a shear correction coefficient is therefore eliminated. Full integration of the shear stiffness terms is shown to result in the recovery of the Kirchoff constraint for thin beams without introducing spurious locking constraints. The accuracy of this formulation is demonstrated by using several numerical examples for the cases of small and large displacements. For a hinged-hinged beam, the linear thickness-shear mode frequency can be matched with the Timoshenko frequency to yield a shear coefficient of 0.824. Matching the bending frequencies between the two theories indicates a shear coefficient for the Timoshenko theory that changes with mode number and slenderness ratio. The influence of in-plane inertia and slenderness ratio on the non-linear frequency is examined for beams with a number of different support conditions.

General

901,485
PB89-147466 Not available NTIS
National Bureau of Standards (IMSE), Gaithersburg, MD. Reactor Radiation Div.
Use of Multiple-Slot Multiple Disk Chopper Assemblies to Pulse Thermal Neutron Beams.
Final rept.
J. R. D. Copley. 1988, 10p
Pub. in Nuclear Instruments and Methods in Physics Research A273, p67-76 1988.

Keywords: *Neutron beams, *Thermal neutrons, Neutron scattering, Reprints, *Neutron choppers, Time-of-flight method, High resolution.

Single-slot disk choppers are commonly used to pulse thermal neutron beams, but their use in high resolution applications is limited because the maximum transmitted beam intensity, given that the chopper is rotating at its maximum possible speed, is proportional to the square of the burst time of the chopper, which is itself proportional to the width of the incident beam. The transmitted intensity can be doubled, with no associated increase in the burst time, by doubling the beam width and using counter-rotating choppers to double the effective chopping speed. In order to increase the intensity still further, without degrading the resolution, the author proposes the use of multiply-slotted choppers in combination with a multiply-slotted beam mask. With two slots in each of two choppers and a mask, the intensity is four times that of the single-slot single chopper arrangement. A further doubling is achieved if a system of three choppers and a mask, each fitted with four slots, is employed. The effects of relative phasing errors in multiple chopper systems are examined in detail, and the implications of nonzero chopper separation and nonzero radial slot extent are briefly discussed.

901,486
PB89-149124 Not available NTIS
National Bureau of Standards (NML), Gaithersburg, MD. Center for Radiation Research.
QCD Vacuum.
Final rept.
M. Danos. 1987, 15p
Pub. in Proceedings of NATO (North Atlantic Treaty Organization) Advanced Study Institute, Maratea, Italy, June 1-4, 1986, p817-831 1987.

Keywords: Quantum electrodynamics, *Quantum chromodynamics, Bose-Einstein condensation, Quasi particles, Vacuum.

QCD is inherently a strong-field system in the infrared. Hence the physical vacuum itself without external charges is already non-perturbatively polarized; it has an energy density lower than that of the QED-type so-called perturbative vacuum by the amount of the 'bag energy density.' It will be shown that the physical vacuum can be described by an analogue of the BCS-state of superconductivity. This vacuum exhibits color confinement.

901,487
PB89-149157 Not available NTIS
National Bureau of Standards (NEL), Gaithersburg, MD. Fire Science and Engineering Div.
Evaporation of a Water Droplet Deposited on a Hot High Thermal Conductivity Solid Surface.
Final rept.
M. di Marzo, and D. D. Evans. 1987, 8p
See also PB86-247871.
Pub. in Proceedings of National Heat Transfer Conference and Exhibition, Heat and Mass Transfer in Fire, (24th), Pittsburgh, PA., August 9-12, 1987, p11-18.

Keywords: *Evaporative cooling, *Drops(Liquids), Thermal conductivity, Thermal diffusivity, Heat transfer, Heat flux, Spray queaching, Models, Interfacial temperature, Semi-infinite body, Molar fraction.

A model is presented that predicts major features of the evaporation of water droplets deposited on a hot non-porous solid surface. In the temperature range of interest, nucleate boiling heat transfer is fully suppressed, hence the model is only concerned with the evaporative process. In the model, the solid material is assumed to have high thermal conductivity and diffusivity, so that the surface temperature under the water droplet can be considered uniform. The temperature of this portion of a larger solid surface covered by the liquid is calculated from the classic solution for contact temperature between two semi-infinite bodies. The liquid-vapor interfacial temperature and the water-vapor molar fraction in the air at the exposed surface of the water droplet, are deduced from the coupled heat and mass transfer, energy balance at the interface. Spatial and temporal integration of the overall droplet energy equation is used to predict the droplet evaporation time and the instantaneous evaporation rate. Model predictions for the total evaporation time and temporal variation of the droplet volume agree well with experiments performed using a heated aluminum block.

901,488
PB89-149249 Not available NTIS
National Bureau of Standards (NEL), Gaithersburg, MD. Fire Measurement and Research Div.
Cooling Effect Induced by a Single Evaporating Droplet on a Semi-Infinite Body.
Final rept.
M. di Marzo, and D. D. Evans. 1987, 4p
Pub. in Proceedings of Fall Technical Meeting on Chemical and Physical Processes in Combustion, San Juan, PR., December 15-17, 1986, p23.1-23.4 1987.

Keywords: *Heat transfer, *Drops(Liquids), Evaporation, Water, Research projects, *Cooling effect, Solid fuels, Hot metal surfaces, Semi-infinite body.

The publication is an extended abstract of work performed to model the evaporation process of a water droplet deposited on a hot metal surface. This work is the first phase of an extensive research program aimed at developing accurate droplet cooling models of burning solid fuel surfaces.

901,489
PB89-153878 PC A09/MF A01
National Inst. of Standards and Technology (NML), Boulder, CO. Time and Frequency Div.
Trapped Ions and Laser Cooling 2: Selected Publications of the Ion Storage Group, Time and Frequency Division, NIST, Boulder, CO.
Technical note.
D. J. Wineland, W. M. Itano, J. C. Bergquist, and J. J. Bollinger. Oct 88, 198p NIST/TN-1324
Also available from Supt. of Docs. as SN003-003-02918-1. See also PB86-110855. Sponsored by Office of Naval Research, Arlington, VA., and Air Force Office of Scientific Research, Bolling AFB, DC.

Keywords: *Frequency standards, *Atomic spectroscopy, Time standards, functions, *Ion traps, *Laser cooling, Laser spectroscopy, Trapping(Charged particles), Quantum jumps, Reports.

The technical note is a collection of selected reprints of the Ion Storage Group for the period July 1985 to September 1986. Major topics include the following: Spectroscopy and frequency standards; Quantum jumps; Nonneutral plasma studies; General articles; Apparatus.

901,490
PB89-156988 Not available NTIS
National Bureau of Standards (NEL), Gaithersburg, MD. Electrosystems Div.
Effect of an Oil-Paper Interface Parallel to an Electric Field on the Breakdown Voltage at Elevated Temperatures.
Final rept.
E. F. Kelley, R. E. Hebner, W. E. Anderson, J. A. Lechner, and J. L. Blue. 1988, 11p
Sponsored by Department of Energy, Washington, DC.
Pub. in IEEE (Institute of Electrical and Electronics Engineers) Transactions on Electrical Insulation 23, n2 p249-259 Apr 88.

Keywords: *Electrical faults, *Interfaces, Insulating oil, Webull density functions, Reprints, *Breakdown, Paper(Material), Finite element method.

The paper reports the measurement of the electrical breakdown location in the vicinity of an oil-paper interface over the temperature range from room temperature to 150 deg C. The data indicated that the electrical breakdown occurred at the interface from 15% to 43% of the time, depending on the details of the particular set of measurements. A theoretical analysis shows that this experimental result is consistent with the electric field enhancement, the area over which the enhancement occurs, and the spread in the breakdown voltages for nominally identical tests.

901,491
PB89-157002 Not available NTIS
National Bureau of Standards (NML), Gaithersburg, MD. Electricity Div.
U.S. Perspective on Possible Changes in the Electrical Units.
Final rept.
K. Jaeger, and B. N. Taylor. 1987, 4p
Pub. in IEEE (Institute of Electrical and Electronics Engineers) Transactions on Instrumentation and Measurement IM-36, n2 p672-675 Jun 87.

Keywords: *Units of measurement, *Electrical potential, *Electrical resistance, Standards, Electrical measurement, Reprints.

The paper summarizes the U.S. view regarding possible changes in the U.S. legal units of voltage and resistance. Such changes, about 9 and 1.5 ppm respectively, would result if the Consultative Committee on Electricity adopted a new value for the Josephson frequency-voltage ratio and a value for the quantized Hall resistance consistent with the SI and these values were used internationally for defining and maintaining national units of voltage and resistance.

901,492
PB89-157077 Not available NTIS
National Bureau of Standards (NEL), Boulder, CO. Electromagnetic Technology Div.
Current Ripple Effect on Superconductive D.C. Critical Current Measurements.
Final rept.
L. F. Goodrich, and S. L. Bray. 1988, 7p
Sponsored by Department of Energy, Washington, DC. Office of Fusion Energy.
Pub. in Cryogenics 28, p737-743 Nov 88.

Keywords: *Superconductors, Titanium alloys, Direct current, Alternating current, Power supplies, Electromagnetic noise, Reprints, *Critical current, Niobium alloys, Ripples.

The effect of current ripple or noise on d.c. critical current measurements was systematically studied. Measurements were made on multifilamentary Nb-Ti superconductor. A low-noise, battery-powered current supply was required in the study in order to make the pure d.c. critical current measurements. Also, an electronic circuit that stimulates a superconductor's general current-voltage characteristic was developed and used as an analysis tool. In order to make critical current measurements in which current ripple was present, the battery supply was modified to allow the introduction of controlled amounts of a.c. ripple. The results of this work are general and quantitatively applicable to the evaluation of critical current data and measurement systems. A theoretical model was developed to further support and explain the ripple effect. An unexpected benefit of this work was a more precise method for general critical current data acquisition. Problems common to all large conductor critical current measurements are discussed.

901,493
PB89-157408 Not available NTIS
National Bureau of Standards (NML), Boulder, CO. Time and Frequency Div.
Perpendicular Laser Cooling of a Rotating Ion Plasma in a Penning Trap.
Final rept.
W. M. Itano, L. R. Brewer, D. J. Larson, and D. J. Wineland. 1988, 9p
Sponsored by Office of Naval Research, Arlington, VA., Air Force Office of Scientific Research, Washington, DC., and National Science Foundation, Washington, DC.
Pub. in Physical Review A 38, n11 p5698-5706, 1 Dec 88.

Keywords: Temperature measurement, Reprints, *Beryllium ions, Beryllium 9, Laser cooling, Penning traps, Ion storage.

The steady-state temperature of an ion plasma in a Penning trap, cooled by a laser beam perpendicular to the trap axis, has been calculated and measured. The rotation of the plasma, due to crossed E and B fields, strongly affects the velocity distribution of the ions, as seen by a laser beam intersecting the plasma at some distance from the axis of rotation, is skewed, and this leads to a change in the velocity distribution (and hence temperature) at which a steady state is attained. The calculated temperature is a function of the intensi-

ty, frequency, and position of the laser beam, and of the rotation frequency of the plasma. Temperatures of (9)Be(1+) plasmas were measured for a wide range of experimental parameters. The lowest and highest temperatures were approximately 40 mK and 2 K. The measured and calculated temperatures are in agreement.

901,494
PB89-157424 Not available NTIS
National Bureau of Standards (NML), Boulder, CO. Time and Frequency Div.
Atomic-Ion Coulomb Clusters in an Ion Trap.
Final rept.
D. J. Wineland, J. C. Bergquist, W. M. Itano, J. J. Bollinger, and C. H. Manney. 1987, 4p
Sponsored by Office of Naval Research, Arlington, VA., and Air Force Office of Scientific Research, Washington, DC.
Pub. in Physical Review Letters 59, n26 p2935-2938, 28 Dec 87.

Keywords: Atomic spectroscopy, Reprints, *Mercury ions, Ion traps, Laser cooling, Ion storage, Wigner crystallization.

Small numbers of laser-cooled Hg(1+) ions, which are confined in a Pual radio-frequency ion trap, to crystallize into regular arrays or clusters were observed. The structure of these clusters was investigated by direct imaging, optical spectroscopy, and numerical calculations. The spectroscopy of such 'pseudomolecules' is unique in that individual atoms of the molecule can be probed separately. The ratio of Coulomb potential energy per ion to k(B)T in these clusters is observed to be as high as 120.

901,495
PB89-157895 Not available NTIS
National Bureau of Standards (NEL), Gaithersburg, MD. Precision Engineering Div.
Electromagnetic Pulse Scattered by a Sphere.
Final rept.
E. Marx. 1987, 6p
Pub. in IEEE (Institute of Electrical and Electronics Engineers) Transactions on Antennas and Propagation 35, n4 p412-417 1987.

Keywords: *Electromagnetic pulses, *Electromagnetic scattering, Integral equations, Magnetic fields, Spheres, Reprints, Transient radiation effects.

The magnetic field integral equation for transient electromagnetic scattering by a perfectly conducting sphere is solved by the stepping-in-time procedure. The contribution of the self-patch, where the integrand is singular, and of the neighboring patches to the surface current density are computed separately from the contributions of other patches to correct for inaccuracies. The term of the integrand with the time-derivative of the current density is shown to make a contribution to these corrections that cannot be neglected in calculating the initial (small) values of the surface current density. The improvement in the values of the scattered fields due to these corrections is significant but not dramatic.

901,496
PB89-158109 Not available NTIS
National Bureau of Standards (NML), Gaithersburg, MD. Atomic and Plasma Radiation Div.
Calculation of Tighter Error Bounds for Theoretical Atomic-Oscillator Strengths.
Final rept.
D. V. I. Roginsky, and A. W. Weiss. 1988, 7p
Pub. in Physical Review A 38, n4 p1760-1766, 15 Aug 88.

Keywords: Hartree-Fock approximation, Electron transitions, Transition probabilities, Lithium, Sodium, Reprints, *Oscillator strengths, Beryllium ions, Magnesium ions.

The authors report a series of calculations of error bounds to the Hartree-Fock approximation for the 2s-2p and 3s-3p transitions in lithium and sodium, and the singly ionized ions Be(1+) and Mg(1+). The purpose is to test the efficacy of various modifications of Weinhold's effective-bounds formula for several simple but realistic examples of multielectron systems. The authors therefore assume the overlap error epsilon to be known, adopting overlap values from extensive variational calculations. The authors find angular momentum projection of the transition operator to be effective in tightening the bounds, while the variational optimiza-

175

PHYSICS
General

tion of a mixture of length and velocity operators is not. The authors also found a wave-function projection based on Brillouin's theorem to be especially effective. When used in conjunction with angular momentum projection, this last procedure has yielded bounds for lithium very close to the known Hartree-Fock error.

901,497
PB89-161541   Not available NTIS
National Bureau of Standards (NEL), Gaithersburg, MD. Thermophysics Div.
Thermodynamic Values Near the Critical Point of Water.
Final rept.
L. Haar, and J. S. Gallagher. 1986, 3p
Pub. in Proceedings of International Conference on Properties of Steam (10th), Moscow, USSR, September 3-7, 1984, p167-169 1986.

Keywords: *Thermodynamic properties, *Water, *Critical point, Steam, Equations of state, Specific heat, Temperature, Pressure, Volume, Measurement.

In the immediate neighborhood of the critical point the thermodynamic consistency between very accurate measurements for the heat capacities, the speed of sound, and pressure-volume-temperature values is examined. For this purpose several thermodynamic formulations are employed, including: the Haar, Gallagher, and Kell (HGK) formulation and the Pollak formulation, which everywhere are analytic, and the scaled equation for the critical region reported by Levelt Sengers, and co-workers, which is an expansion about a non-analytic critical point. It is shown that HGK and the scaled equation are in accord with the measurements for the different thermodynamic properties; also, the recent PVT measurements by Hanafusa and co-workers are the most accurate yet made close to the critical point. A description is provided of the HGK formulation and specification of the liquid-vapor co-existence curve.

901,498
PB89-161558   Not available NTIS
National Bureau of Standards (NML), Gaithersburg, MD. Ionizing Radiation Physics Div.
NBS (National Bureau of Standards) Decay-Scheme Investigations of (82)Sr-(82)Rb.
Final rept.
D. D. Hoppes, B. M. Coursey, F. J. Schima, and D. Yang. 1987, 9p
Pub. in Applied Radiation and Isotopes 38, n3 p195-203 1987.

Keywords: *Decay schemes, Gamma rays, Positrons, Half life, Gamma ray spectroscopy, Impurities, Standards, Reprints, *Strontium 82, *Rubidium 82, Liquid scintillation detectors, Nuclear medicine, Annihilation, Calibration.

Measurements of photon- and positron-emission rates for equilibrium mixtures of (82)Sr-(82)Rb indicate a gamma-ray probability per decay for the 776-keV gamma ray of 0.152 plus or minus 0.003 if the positron rate is derived from annihilation-radiation measurements and 0.145 plus or minus 0.002 if positrons are measured in a liquid scintillator. The half life of (82)Sr was measured as 25.36 plus or minus 0.03 days. Methods for measuring a (82)Sr impurity by gamma-ray spectrometry and by liquid-scintillation counting are described.

901,499
PB89-161566   Not available NTIS
National Bureau of Standards (NML), Boulder, CO. Time and Frequency Div.
Frequency Measurement of the J = 1 <- 0 Rotational Transition of HD (Hydrogen Deuteride).
Final rept.
K. M. Evenson, D. A. Jennings, J. M. Brown, L. R. Zink, K. R. Leopold, M. D. Vanek, and I. G. Nolt. 1988, 2p
Pub. in Astrophysical Jnl. 330, pL135-L136, 15 Jul 88.

Keywords: *Rotational spectra, Far infrared radiation, Planetary atmospheres, Interstellar matter, Reprints, Hydrogen deuteride.

The frequency of the astronomically important J = 1 <- 0 rotational transition of hydrogen deuteride (HD) at 2.7 THz (90/cm) has been measured with tunable far-infrared radiation with an accuracy of 150 kHz. This frequency is now known to sufficient accuracy for use in future astrophysical heterodyne observations of HD in planetary atmospheres reported by Bezard et al. in 1986 and the interstellar medium reported by Bussoletti et al. in 1975.

901,500
PB89-171185   Not available NTIS
National Bureau of Standards (NML), Boulder, CO. Quantum Physics Div.
Systems Driven by Colored Squeezed Noise: The Atomic Absorption Spectrum.
Final rept.
H. Ritsch, and P. Zoller. 1988, 12p
Pub. in Physical Review A 38, n9 p4657-4668, 1 Nov 88.

Keywords: *Atomic spectra, *Absorption spectra, Stochastic processes, Reprints, *Squeezed light, Squeezed states (Quantum theory), Bloch equations, Matrices.

Stochastic density-matrix equations are derived for an atom strongly driven by finite-bandwidth squeezed light. The quantum properties of the light are accounted for by a doubling of dimensions of the stochastic process for c-number electric field amplitudes. Saturation properties and the weak-field absorption spectrum of a two-level atom embedded in finite-bandwidth squeezed light and driven by a coherent field are calculated. The effect of finite bandwidth of the squeezed light in obtaining subnatural linewidth in the atomic absorption spectra is discussed, based on nonperturbative solutions of the stochastic optical Bloch equations.

901,501
PB89-171540   Not available NTIS
National Bureau of Standards (NML), Boulder, CO. Quantum Physics Div.
Electron-Transport, Ionization, Attachment, and Dissociation Coefficients in SF6 and Its Mixtures.
Final rept.
A. V. Phelps, and R. J. Van Brunt. 1988, 9p
Sponsored by Department of Energy, Washington, DC.
Pub. in Jnl. of Applied Physics 64, n9 p4269-4277, 1 Nov 88.

Keywords: *Sulfur hexafluoride, Collision cross sections, Mixtures, Nitrogen, Oxygen, Neon, Gas ionization, Dissociation, Reprints, Electron-molecule collisions.

An improved set of electron-collision cross sections is derived for SF6 and used to calculate transport, ionization, attachment, and dissociation coefficients for electron kinetic energy distributions computed from numerical solutions of the electron-transport (Boltzmann) equation using the two-term, spherical harmonic expansion approximation were used to obtain electron-transport and reaction coefficients as functions of E/N and the fractional concentration of SF6. Here E is the electric field strength and N is the gas number density. Attachment rate data for low concentrations of SF6 in N2 are used to test the attachment cross sections. Particular attention is given to the calculation of transport and reaction coefficients at the critical E/N = (E/N) sub c at which the ionization and attachment rates are equal.

901,502
PB89-171557   Not available NTIS
National Bureau of Standards (NML), Boulder, CO. Quantum Physics Div.
Electron-Impact Excitation of the Resonance Transition in CA(1+).
Final rept.
J. Mitroy, D. C. Griffin, D. W. Norcross, and M. S. Pindzola. 1988, 12p
Contract DOE-EA-77-A-01-6010
Sponsored by Department of Energy, Washington, DC.
Pub. in Physical Review A 38, n7 p3339-3350.

Keywords: Electron irradiation, Perturbation theory, Excitation, Reprints, *Calcium ions, *Electron-ion collisions.

Detailed calculations of the electron-impact excitation of Ca(1+) are performed using both perturbation theory and the close-coupling approach. Particular attention is focused on the resonance (4s-4p) excitation since experimental emission-cross-section data are available for this transition. The results of the most sophisticated model, a six-state (4s, 3d, 4p, 5s, 4d, 5p) close-coupling calculation with semiempirical Hartree-Fock target wave functions and including one- and two-body core-polarization potentials are in better agreement with the experimental cross section and resonance-fluorescence polarization data than any other calculation. At incident electron energies below

the 5s, 4d, and 5p thresholds, the six-state calculations are essentially in agreement with the experimental data, although rich resonance structures predicted by theory are not seen experimentally due to the finite energy resolution. At energies above the 5s, 4d, and 5p thresholds the six-state emission cross sections exceed the experimental cross sections by about 20%, once allowance is made for cascades from the 5s, 4d, and 5p levels.

901,503
PB89-171565   Not available NTIS
National Bureau of Standards (NML), Boulder, CO. Quantum Physics Div.
Electron-Impact Excitation of Al(2+).
Final rept.
J. Mitroy, and D. W. Norcross. 1989, 8p
Contract DOE-EA-77-A-01-6010
Sponsored by Department of Energy, Washington, DC.
Pub. in Physical Review A 39, n2 p537-544, 15 Jan 89.

Keywords: Electron irradiation, Approximation, Excitation, Reprints, *Aluminum ions, *Electron-ion collisions, Bound state, Oscillator strengths.

Detailed calculations of the electron-impact excitation of the sodiumlike ion Al(2+) were performed using both the unitarized Coulomb-Born approximation and the close-coupling approach. Calculations were undertaken at both the five-state (3s, 3p, 3d, 4s, and 4p) and nine-state levels of approximation. Calculations using Hamiltonians both with and without (semiempirical) core-polarization potentials were completed. Particular attention was paid to the resonance (3s-3p) excitation. The inclusion of core-polarization potentials resulted in cross sections for the resonance excitation that are 10% smaller in magnitude. As an additional check on the calculations, binding energies of Al(2+) bound states and oscillator strengths for Al(2+) transitions, were also computed and compared with the results of measurements and other calculations.

901,504
PB89-171581   Not available NTIS
National Bureau of Standards (NML), Boulder, CO. Quantum Physics Div.
Comment on 'Possible Resolution of the Brookhaven and Washington Eotvos Experiments'.
Final rept.
T. M. Niebauer, and J. E. Faller, and P. L. Bender. 1988, 1p
Pub. in Physical Review Letters 61, n19 p2272, 7 Nov 88.

Keywords: Gravitation, Reprints, *Eotvos experiment, Fifth force.

The comment points out several things that the authors of a recent article (Physical Review Letters 60, 1225 (1988)) overlooked or misinterpreted. Particular emphasis is on the significance of the authors' earlier paper (Physical Review Letters 59, 609 (1987)) to this discussion.

901,505
PB89-171631   Not available NTIS
National Bureau of Standards (NML), Boulder, CO. Time and Frequency Div.
Recoilless Optical Absorption and Doppler Sidebands of a Single Trapped Ion.
Final rept.
J. C. Bergquist, W. M. Itano, and D. J. Wineland. 1987, 3p
Sponsored by Air Force Office of Scientific Research, Arlington, VA., and Office of Naval Research, Arlington, VA.
Pub. in Physical Review A 36, n1 p428-430, 1 Jul 87.

Keywords: Near ultraviolet radiation, Reprints, *Mercury ions, Trapping(Charged particles), Laser cooling, Laser spectroscopy, High resolution, Quadrupoles, Doppler sidebands.

Spectroscopic measurements of the electric-quadrupole-allowed 5d(10)6s doublet s(1/2) to 5d(9)6s(2) doublet D(5/2) transition near 282 nm on a single, laser-cooled Hg(1+) ion give a recoil-free absorption line (carrier) and well-resolved motional sidebands. From the intensity ratio of the sidebands to the carrier, the effective temperature of the Hg(1+) ion was determined to be near the theoretical minimum of 1.7 mK. A fraction resolution of better than 3 x 10 to the -11th power for this ultraviolet transition is achieved.

901,506
PB89-172361 Not available NTIS
National Bureau of Standards (NEL), Boulder, CO.
Mathematical Analysis Div.
Classical Chaos, the Geometry of Phase Space, and Semiclassical Quantization.
Final rept.
W. P. Reinhardt. 1985, 77p
Grants NSF-CHE80-11442, NSF-PHY82-00805
Sponsored by National Science Foundation, Washington, DC.
Pub. in Mathematical Analysis of Physical Systems, Chapter 8, p169-245 1985.

Keywords: Hamiltonian functions, Schrodinger equation, Reprints, *Chaos, *Phase space, *Quantization, Two degrees of freedom.

Basic concepts relating to the onset of chaos in deterministic dynamics are illustrated using conservative Hamiltonian dynamics in two degrees of freedom--that is, dynamics in a four-dimensional phase space. The correspondence between regular and chaotic dynamics and concepts of integrability and nonintegrability are established, and semiclassical quantization (i.e., determination of estimates of energy eigenvalues of the corresponding Schrodinger operator using only classical dynamical information as input) is carried out utilizing the invariant tori of integrable (regular) classical dynamics. The origins of classical chaos are briefly discussed in terms of orbit bifurcation and in terms of the Painleve analysis of the complex time singularities of the orbits themselves. Expectations of quantum correspondence principle ramifications of classical chaos are discussed in terms of Percival's 'irregular spectrum' and in terms of the statistics of nearest-neighbor level spacings.

901,507
PB89-172407 Not available NTIS
National Bureau of Standards (IMSE), Gaithersburg, MD.
Grain Boundaries with Impurities in a Two-Dimensional Lattice-Gas Model.
Final rept.
R. Kikuchi, and J. W. Cahn. 1987, 11p
Pub. in Physical Review B 36, n1 p418-428, 1 Jul 87.

Keywords: *Grain boundaries, *Lattice parameters, *Gases, *Impurities, Models, Thermodynamic properties, Statistical mechanics, Temperature, Adsorption.

Grain boundaries are examined between two 2-dimensional two-component square grains. Although the grains are assumed ideal solid solutions, the model grain boundary shows complicated thermodynamic behavior. There are high adsorption regions, both a low temperature and again a high temperature where the grain boundary resembles a molten zone. The sign of the adsorption can be different in these two regions. In between adsorption is low.

901,508
PB89-175210
(Order as PB89-175194, PC A06)
National Inst. of Standards and Technology, Gaithersburg, MD.
NIST (National Institute of Standards and Technology)/Sandia/ICDD Electron Diffraction Database: A Database for Phase Identification by Electron Diffraction.
Bi-monthly rept.
M. J. Carr, W. F. Chambers, D. Melgaard, V. L. Himes, J. K. Stalick, and A. D. Mighell. 1989, 7p
Sponsored by Sandia National Labs., Albuquerque, NM., and I and M Systems Ltd., Albuquerque, NM.
Included in Jnl. of Research of the National Institute of Standards and Technology, v94 n1 p15-20 Jan-Feb 89.

Keywords: *Electron diffraction, Phase, Electron microscopes, Crystallography, Inorganic compounds, Searching, Matching, Data retrieval, X ray spectroscopy, *Numerical data bases, Search profiles.

A new database containing crystallographic and chemical information designed especially for application to electron diffraction search/match and related problems has been developed. The database described in the report contains what the authors believe to be the only complete collection of inorganic compounds data structured for phase identification by electron diffraction available. Nevertheless, the database is small enough to reside on a personal computer or laboratory microcomputer dedicated to electron dif-

fraction and energy dispersive x-ray spectroscopy analysis in an electron microscope laboratory. It is anticipated that many different search/match schemes will be able to use this database.

901,509
PB89-176002 Not available NTIS
National Bureau of Standards (NML), Gaithersburg, MD. Atomic and Plasma Radiation Div.
Spectra and Energy Levels of Br XXV, Br XXIX, Br XXX, and Br XXXI.
Final rept.
U. Feldman, J. F. Seely, C. M. Brown, J. O. Ekberg, M. C. Richardson, W. E. Behring, and J. Reader. 1986, 4p
Pub. in Jnl. of the Optical Society of America B 3, n11 p1605-1608 Nov 86.

Keywords: *Atomic energy levels, *Emission spectra, Reprints, *Bromine ions, Soft x rays, Multicharged ions, M1-transitions.

Emission lines of highly-ionized bromine in the wavelength region 17 A to 93 A have been identified in spectra recorded at the University of Rochester's OMEGA laser facility. The wavelengths of 2s-2p transitions in nitrogenlike Br XXIX, carbonlike Br XXX, and boronlike Br XXXI are presented. The wavelengths of the magnetic dipole transitions within the 2s(2)2p(3) ground configuration of Br XXIX are predicted from the experimental energy levels. Transitions from the n = 4 and 5 levels of sodiumlike Br XXV were also identified, and the ionization energy of Br XXV was determined.

901,510
PB89-176010 Not available NTIS
National Bureau of Standards (NML), Gaithersburg, MD. Atomic and Plasma Radiation Div.
Laser-Produced Spectra and QED (Quantum Electrodynamic) Effects for Fe-, Co-, Cu-, and Zn-Like Ions of Au, Pb, Bi, Th, and U.
Final rept.
J. F. Seely, J. O. Ekberg, C. M. Brown, U. Feldman, W. E. Behring, J. Reader, and M. C. Richardson. 1986, 3p
Pub. in Physical Review Letters 57, n23 p2924-2926, 8 Dec 86.

Keywords: Far ultraviolet radiation, Reprints, *Gold ions, *Lead ions, *Bismuth ions, *Thorium ions, *Uranium ions, *Laser-produced plasma, Isoelectronic sequence, Multicharged ions, Extreme ultraviolet radiation, Quantum electrodynamics.

Transitions in the Fe I, Co I, Cu I, and Zn I isoelectronic sequences of the elements Au, Pb, Bi, Th, and U have been identified in the extreme ultraviolet spectra from laser-produced plasmas. The measured wavelengths are compared with calculated values, and the QED contributions to the CU I transition energies are determined.

901,511
PB89-176440 Not available NTIS
National Bureau of Standards (NML), Gaithersburg, MD. Electricity Div.
Laser-Cooling and Electromagnetic Trapping of Neutral Atoms.
Final rept.
W. D. Phillips, A. L. Migdall, and H. J. Metcalf. 1986, 4p
See also PB88-175096.
Pub. in AIP (American Institute of Physics) Conference Proceedings, n146 p362-365 1986.

Keywords: Atomic beams, Reviews, *Atom traps, Laser cooling, Magnetic traps, Sodium atoms.

Until recently it has been impossible to confine and trap neutral atoms using electromagnetic fields. While many proposals for such traps exist, the small potential energy depth of the traps and the high kinetic energy of available atoms prevented trapping. The authors review various schemes for atom trapping, the advances in laser cooling of atomic beams which have now made trapping possible, and the successful magnetic trapping of cold sodium atoms.

901,512
PB89-176531 Not available NTIS
National Bureau of Standards (NML), Gaithersburg, MD. Ionizing Radiation Physics Div.

2.5 MeV Neutron Source for Fission Cross Section Measurement.
Final rept.
K. C. Duvall, O. A. Wasson, and M. Hongchang. 1986, 4p
Pub. in Proceedings of International Conference on Nuclear Data for Science and Technology, Mito, Japan, May 30-June 3, 1988, p355-358.

Keywords: *Neutron sources, Fission cross sections, Neutron beams, Deuterium target, Deuteron reactions, Uranium 235 target, MeV range 01-10.

A 2.5-MeV neutron source has been established on the beamline of a 100-kV, 0.5-ma ion accelerator. The neutron source is produced by the D(d,n)(3)He reaction with a yield of about 10 million n.sec. A fission chamber containing six uranium tetrafluoride deposits has been designed for use in the (235)U(n,f) cross section measurement at 2.5 MeV. A description of the 2.5-MeV neutron source facility is presented along with details of the associated particle detection and neutron beam characteristics. Preparation for the fission cross section measurement are discussed.

901,513
PB89-176572 Not available NTIS
National Bureau of Standards (NML), Gaithersburg, MD. Radiation Physics Div.
State Selection in Electron-Atom Scattering: Spin-Polarized Electron Scattering from Optically Pumped Sodium.
Final rept.
J. J. McClelland, M. H. Kelley, and R. J. Celotta. 1986, 4p
Pub. in Proceedings of Symposium on Physics of Ionized Gases (13th), Sibenik, Yugoslavia, September 1-5, 1986, p15-18.

Keywords: *Electron scattering, *Sodium, Polarization(Spin alignment), Optical pumping, *Electron-atom collisions.

When an electron collides with an atom, there are generally many quantum channels through which the scattering can take place. These can involve several energetically accessible channels, several angular momentum states degenerate in energy, and several spin states also degenerate in energy. A theoretical calculation of the scattering must examine each of the channels individually, and then perform an average over those not separated in the experiment with which comparison is to be made.

901,514
PB89-176903 Not available NTIS
National Bureau of Standards (IMSE), Gaithersburg, MD. Metallurgy Div.
Fast Magnetic Resonance Imaging with Simultaneously Oscillating and Rotating Field Gradients.
Final rept.
S. J. Norton. 1987, 11p
Pub. in IEEE (Institute of Electrical and Electronics Engineers) Transactions on Medical Imaging 6, n1 p21-31 1987.

Keywords: *Nuclear magnetic resonance, Gradients, Magnetic fields, Oscillations, Reprints, *Image reconstruction, Tomography.

The report presents an approach to magnetic resonance imaging employing a magnetic field gradient that rotates 180 degrees in the image plane while the gradient magnitude oscillates rapidly during the rotation. A single free induction decay recorded during this rotation contains all information needed to reconstruct a two-dimensional image. In effect, each sinusoidal oscillation of the gradient provides information corresponding to one projection in more conventional Fourier-projection approaches. Since the data acquisition can be achieved in a period less than T2, the method offers the potential of great speed, which is limited only by the gradient modulation frequency. An explicit image reconstruction formula is derived that gives, when evaluated, a reconstruction of the magnetization equal to the true magnetization convolved with a space-invariant point spread function. This point spread function is derived and characterizes the resolving power and sidelobe response of the technique. Moreover, it derives a similar reconstruction formula which is valid when known inhomogeneities in the static field H0 and T2 are present. Finally, it shows how the general approach can be extended to three dimensions.

PHYSICS
General

901,515
PB89-176937 Not available NTIS
National Bureau of Standards (NML), Gaithersburg, MD. Electricity Div.
Cooling and Trapping Atoms.
Final rept.
W. D. Phillips, and H. J. Metcalf. 1987, 7p
Sponsored by Office of Naval Research, Arlington, VA.
Pub. in Scientific American 256, n3 p50-58 1987.

Keywords: Reprints, *Atom traps, *Laser cooling, Magnetic traps.

Thermal motion of atoms interferes with many measurements of atomic properties. The authors describe techniques whereby lasers are used to cool the atomic motion. Laser-cooled atoms can then be confined in electromagnetic traps for long periods of time.

901,516
PB89-176986 Not available NTIS
National Bureau of Standards (NEL), Boulder, CO.
Electromagnetic Technology Div.
Cool It.
Final rept.
J. Lehman. 1988, 4p
Pub. in Science Teacher, p29-32 Mar 88.

Keywords: *Cryogenics, *Education, Instructors, Training devices, Reprints, Instructional materials, Teaching methods, Science instruction.

Sometimes a well-rounded curriculum for the physical sciences is burdened with bringing into the classroom topics that do not necessarily correspond to daily observations and intuition. Heat, energy, resistance, kinetics, and countless other topics challenge not only the student's imagination, but also an instructor's ability to present such topics in realistic and interesting ways. Cryogenics, known informally as the science of cold, offers many avenues for learning beyond the obvious effects of cold temperature. The paper includes some justification, pedagogy, and motivation for use of cryogenics in science curricula.

901,517
PB89-177042 Not available NTIS
National Bureau of Standards (NEL), Gaithersburg, MD. Robot Systems Div.
Algebraic Representation for the Topology of Multicomponent Phase Diagrams.
Final rept.
D. J. Orser. 1986, 30p
Pub. in Proceedings of Computer Modeling of Phase Diagrams Symposium, Toronto, Ontario, Canada, October 13-17, 1985, p301-330 1986.

Keywords: *Phase diagrams, *Topology, Models, Thermodynamics, Incidence matrices, Algebra, Equilibrium, Data bases.

The paper describes a methodology based on treating phase diagrams as topological structures and develops a representation along the following lines: for each topologically distinct phase diagram there exists a finite incidence algebra whose elements correspond to the invariant (vertices), monovariant (edges), bivariant (surfaces), etc., transition equilibria of the diagram. The elements of the incidence algebra are sets for which a relation and two binary operations are defined. This defines a calculus of phase diagram equilibria and provides an efficient method for a computer to retrieve the topological relationships between an equilibrium and the rest of the diagram or between any two equilibria. Its application to a multicomponent data base and its potential for qualitative thermodynamic modeling are discussed.

901,518
PB89-179147 Not available NTIS
National Bureau of Standards (NML), Boulder, CO.
Time and Frequency Div.
Laser Cooling.
Final rept.
D. J. Wineland, and W. M. Itano. 1987, 8p
Sponsored by Office of Naval Research, Arlington, VA., and Air Force Office of Scientific Research, Bolling AFB, DC.
Pub. in Physics Today, p1-8 Jun 87.

Keywords: Radiation pressure, Reprints, *Ion traps, *Atom traps, *Laser cooling, *Trapping(Charged particles), Ion storage.

Theory and experiments on laser cooling of neutral atoms and atomic ions is presented. A brief history of the mechanical forces of light is given first. Then, a simple theory of radiation pressure cooling (laser cooling) of free and bound atoms is presented along with the temperature limits imposed by recoil. Experiments on trapped ions and neutral atoms are then described. Finally, the future of laser cooling is briefly discussed.

901,519
PB89-179204 Not available NTIS
National Bureau of Standards (NML), Gaithersburg, MD. Temperature and Pressure Div.
Magnetic Resonance of (160)Tb Oriented in a Terbium Single Crystal at Low Temperatures.
Final rept.
P. Roman, W. D. Brewer, E. Klein, H. Marshak, K. Freitag, and P. Herzog. 1986, 4p
Pub. in Physical Review Letters 56, n18 p1976-1979, 5 May 86.

Keywords: *Nuclear magnetic resonance, Quadrupole moment, Single crystals, Gamma rays, Reprints, *Terbium 160, *Oriented nuclei, Ion implantation, Thermometers, Terbium 159, Rare earth nuclei.

The first observation of magnetic resonance of oriented rare earth nuclei in a rare earth host is reported. Radioactive 160Tb implanted in a single crystal of ferromagnetic terbium was subjected to magnetic resonance detected by perturbation of the gamma-ray anisotropy. The open Tb 4f shell gives rise to a strong electric quadrupole interaction in addition to the magnetic interaction; the resulting resonance signal has 8 components, of which the first was detected at 480.0(4) MHz. The derived quadrupole interaction frequency is 167.7(2.6) MHz, giving $Q = 3.56(10)$ b.

901,520
PB89-184113 (Order as PB89-184089, PC A04)
National Inst. of Standards and Technology, Boulder, CO.
Relation between Wire Resistance and Fluid Pressure in the Transient Hot-Wire Method.
Bi-monthly rept.
H. M. Roder, and R. A. Perkins. 1989, 4p
Included in Jnl. of Research of the National Institute of Standards and Technology, v94 n2 p113-116 Mar-Apr 89.

Keywords: *Thermal conductivity, *Fluids, Plantinum, Electrical resistance, Pressure, Calibrating, Measurement, *Hot wire flowmeters.

The resistance of metals is a function of applied pressure, and the dependence is large enough to be significant in the calibration of transient hot-wire thermal conductivity instruments. For the highest possible accuracy, the instrument's hot wires should be calibrated in situ. If this is not possible, the author recommended that a value of gamma, the relative resistance change with pressure, of -2x10 sup (-5) MPa(-1) be used to account for the pressure dependence of the platinum wire's resistance.

901,521
PB89-185615 PC A13/MF A01
National Bureau of Standards (NML), Gaithersburg, MD. Center for Basic Standards.
Technical Activities 1987, Center for Basic Standards.
P. L. M. Heydemann. Oct 87, 290p NBSIR-87/3587
See also PB86-140043.

Keywords: *Standards, *Physics, *Research management, Electric current, Temperature, Mass, Length, Time, Frequencies, Quantum theory, X rays, *Center for Basic Standards.

The report summarizes the research and technical activities of the Center for Basic Standards during the Fiscal Year 1987. These activities include work in the areas of electricity, temperature and pressure, mass and length, time and frequency, quantum metrology, and quantum physics.

901,522
PB89-185912 Not available NTIS
National Bureau of Standards (NML), Boulder, CO.
Quantum Physics Div.
Current Research Efforts at JILA (Joint Institute for Laboratory Astrophysics) to Test the Equivalence Principle at Short Ranges.
Final rept.
J. E. Faller, T. M. Niebauer, M. P. McHugh, and D. A. Van Baak. 1988, 14p
Sponsored by Air Force Geophysics Lab., Hanscom AFB, MA.
Pub. in Proceedings of Moriond Conference 5th Force Neutrino Physics (23rd), Les Arcs, Savoie, France, January 23-30, 1988, p457-470.

Keywords: *Gravitation, Gravity, *Equivalence principle, Fifth force, Free fall.

The authors are presently engaged in three different experiments to search for a possible breakdown of the equivalence principle at short ranges. The first of these experiments, which has been completed, is the so-called Galilean test in which the differential free-fall of two objects of differing composition was measured using laser interferometry. The authors observed that the differential acceleration of two test bodies was less than 5 parts in 10 billion. The experiment set new limits on a suggested baryon dependent Fifth Force at ranges longer than 1 km. With a second experiment, the authors are investigating substance dependent interactions primarily for ranges up to 10 meters using a fluid supported torsion balance; the apparatus has been built and is now undergoing laboratory tests. A proposal has been made to measure the gravitational signal associated with the changing water level at a large pumped storage facility in Ludington, Michigan. Measuring the gravitational signal above and below the pond will yield the value of the gravitational constant, G, at ranges from 10-100 m.

901,523
PB89-185920 Not available NTIS
National Bureau of Standards (NML), Boulder, CO.
Quantum Physics Div.
Fundamental Tests of Special Relativity and the Isotropy of Space.
Final rept.
S. A. Lee, L. U. Andersen, N. Bjerre, O. Poulsen, E. Riis, and J. L. Hall. 1987, 4p
Grant NSF-PHY86-04504
Sponsored by National Science Foundation, Washington, DC.
Pub. in Proceedings of International Conference on Laser Spectroscopy (8th), Are, Sweden, June 22-26, 1987, p52-55.

Keywords: *Special relativity, Atomic beams, Frequency shift, Anisotropy, Tests, *Two photon absorption, Light speed, Neon atoms, Spectral shift, Isotropy.

A two-photon absorption experiment was performed using a fast Ne+ atom beam merged to be coaxial with a standing wave laser field. Intermediate state resonance was achieved by controlling the accelerator-determined Doppler shift. A diurnal optical frequency shift would be expected if the speed of light were anisotropic. Preliminary measurements relative to an I2 reference line yield a diurnal frequency shift < 2 kHz, corresponding to a '1-way' speed of light anisotropy epsilon $< 3 \times 10$ to the -9 power.

901,524
PB89-186738 Not available NTIS
National Bureau of Standards (NML), Gaithersburg, MD. Center for Radiation Research.
Quasifree Electron Scattering on Nucleons in a Momentum-Dependent Potential.
Final rept.
J. S. O'Connell, and B. Schroder. 1988, 3p
Pub. in Physical Review C 38, n5 p2447-2449 Nov 88.

Keywords: *Electron scattering, Momentum transfer, Carbon, Reprints, *Electron-nucleon interactions, Energy losses, Effective mass.

Systematics on the location of the quasifree peak observed in electron-nucleus scattering as a function of momentum transfer are related to the momentum-dependent mean field in which the struck nucleon moves. Data on carbon are used as an example.

901,525
PB89-200455 Not available NTIS
National Bureau of Standards (NEL), Boulder, CO.
Electromagnetic Technology Div.
Battery-Powered Current Supply for Superconductor Measurements.
Final rept.
S. L. Bray, L. F. Goodrich, and W. P. Dube. 1989, 4p
Sponsored by Department of Energy, Washington, DC.
Pub. in Review of Scientific Instruments 60, n2 p261-264 Feb 89.

Keywords: *Superconductors, *Power supplies, Electrical measurement, Reprints, *Critical current, Circuit diagrams, Battery operated.

To measure the critical current of superconductors, a high output current supply is required. In addition to high current capability, the supply should be designed to reduce ground loop problems, respond linearly to an input control signal, and minimize output noise. A current supply with these qualifications has been constructed and tested. Although the supply was originally designed for testing conventional superconductors at high current levels, it has also been successfully used in measurements on the high-critical-temperature ceramic superconductors where the maximum current output was less than 1 A. The supply can produce 1000 A output current with a noise level of approximately 0.05 A peak-to-peak. The specifics of the current supply's design and performance are given.

901,526
PB89-200471   Not available NTIS
National Bureau of Standards (NEL), Boulder, CO.
Electromagnetic Technology Div.
Current Capacity Degradation in Superconducting Cable Strands.
Final rept.
L. F. Goodrich, and S. L. Bray. 1989, 4p
Sponsored by Department of Energy, Washington, DC.
Pub. in IEEE (Institute of Electrical and Electronics Engineers) Transactions on Magnetics 25, n2 p1949-1952 Mar 89.

Keywords: Superconducting magnets, Deformation, Degradation, Reprints, *Superconducting cables, Critical current, Superconducting super collider, Niobium titanium, Aspect ratio.

The electromagnetic properties of NbTi strands extracted from Rutherford cables were studied to clarify the effect of mechanical deformation, caused by the cabling process, on the current capacity of the strands. Three different cables were studied, all of which are prototypes for the Superconducting Super Collider's dipole magnets. The extracted cable strands were instrumented to allow measurement of the voltage across several key regions of mechanical deformation as a function of current and the orientation of the applied magnetic field. The resulting data are presented in terms of the strand's voltage profile as well as its critical current in order to more thoroughly characterize the conductor's electromagnetic properties.

901,527
PB89-200497   Not available NTIS
National Bureau of Standards (NEL), Boulder, CO.
Electromagnetic Technology Div.
Nb3Sn Critical-Current Measurements Using Tubular Fiberglass-Epoxy Mandrels.
Final rept.
L. F. Goodrich, S. L. Bray, and T. C. Stauffer. 1989, 4p
Sponsored by Department of Energy, Washington, DC.
Pub. in IEEE (Institute of Electrical and Electronics Engineers) Transactions on Magnetics 25, n2 p2375-2378 Mar 89.

Keywords: Superconductors, Strains, Thermal expansion, Contraction, Electrical measurement, Reprints, *Superconducting wires, *Niobium stannides, *Critical current, Fiberglass epoxy composites, Niobium tin.

A systematic study of the effect of sample mounting techniques on the superconducting critical-current measurement was made in conjunction with the VAMAS (Versailles Agreement on Advanced Materials and Standards) interlaboratory comparison (round robin) measurements. A seemingly small change in mandrel geometry can result in a 40% change in the measured critical current of a Nb3Sn sample at 12 T. This is a result of a change in the conductor pre-strain at 4 K caused by variation in thermal contraction between thick- and thin-walled fiberglass-epoxy composite (G-10) tubes.

901,528
PB89-201065   Not available NTIS
National Bureau of Standards (NML), Gaithersburg, MD. Atomic and Plasma Radiation Div.
4s(2) 4p(2)-4s4p(3) Transition Array and Energy Levels of the Germanium-Like Ions Rb VI - Mo XI.
Final rept.
U. Litzen, and J. Reader. 1989, 6p
See also PB87-109666. Sponsored by Department of Energy, Washington, DC.
Pub. in Physica Scripta 39, n468-473 1989.

Keywords: *Atomic energy levels, *Ultraviolet spectra, Far ultraviolet radiation, Reprints, *Rubidium ions,

*Strontium ions, *Zirconium ions, *Niobium ions, *Molybdenum ions, *Yttrium ions, Multicharged ions.

Spectra of the germanium-like ions Rb VI, Sr VII, Y VIII, Zr IX, Nb X, and Mo XI have been investigated in the region 280-790 A. Identification of lines of the transition array 4s(2) 4p(2)-4s4p(3) has yielded all levels of the two configurations except 4s 4p(3) quintet S. The level structure has been studied by means of ab initio and parametric calculations.

901,529
PB89-201073   Not available NTIS
National Bureau of Standards (NML), Gaithersburg, MD. Atomic and Plasma Radiation Div.
Resonance-Enhanced Multiphoton Ionization of Atomic Hydrogen.
Final rept.
L. R. Brewer, F. Buchinger, M. Ligare, and D. E. Kelleher. 1989, 12p
Sponsored by U.S. Air Force Office of Scientific Research, Bolling AFB, DC.
Pub. in Physical Review A 39, n8 p3912-3923, 15 Apr 89.

Keywords: Resonance, Reprints, *Hydrogen atoms, *Multi-photon processes, *Photoionization, Laser radiation.

The resonance-enhanced four-photon ionization of atomic hydrogen was measured. The degenerate four-photon process occurs via a three-photon resonance between the 1s and the 2p levels, with subsequent one-photon ionization near threshold. The highly asymmetric resonance-enhanced profile was studied, i.e., the photo ion yield as a function of laser detuning from three-photon resonance between the 1s and 2p levels. In particular, the authors have determined the width, shift, peak, and asymmetry of the profile as a function of laser intensity. The experimental results are compared to theoretical models. These models involve both the properties of the atom in an intense near-resonant radiative field, and a detailed model for the multimode laser field, particularly the field fluctuations due to mode beating. The asymmetric resonance-enhanced photoionization profile is well reproduced by both a random-phase and a chaotic-light model.

901,530
PB89-201578   Not available NTIS
National Bureau of Standards (NML), Gaithersburg, MD. Molecular Spectroscopy Div.
Influence of the ac Stark Effect on Multiphoton Transitions in Molecules.
Final rept.
W. L. Meerts, I. Ozier, and J. T. Hougen. 1989, 8p
Pub. in Jnl. of Chemical Physics 90, n9 p4681-4688, 1 May 89.

Keywords: *Molecular beams, *Stark effect, Electronic spectra, Photons, Linear differential equations, Computerized simulation, Resonance absorption, Boundary value problems, Mathematical models, Reprints.

A multiphoton mechanism for molecular beam transitions is presented which relies on a large first-order ac Stark effect to modulate the energy separation of the initial and final states of the multiphoton transition, but which does not require the presence of any intermediate level(s). The algebraic formalism is checked by computer solution of an initial value problem involving four real coupled linear differential equations. It is then used to explain the multiphoton transitions previously observed in molecular beam electric resonance studies on the two symmetric top molecules OPF3 and CH3CF3, where the number of photons involved in a given transition varies from 1-40. Application of the analysis to other experiments is briefly discussed.

901,531
PB89-201644   Not available NTIS
National Bureau of Standards (NML), Gaithersburg, MD. Gas and Particulate Science Div.
Modeling of the Bremsstrahlung Radiation Produced in Pure Element Targets by 10-40 keV Electrons.
Final rept.
J. A. Small, S. D. Leigh, D. E. Newbury, and R. L. Myklebust. 1987, 11p
Pub. in Jnl. of Applied Physics 61, n2 p459-469 1987.

Keywords: *x ray analysis, *Bremsstrahlung, Mathematical models, Electron irradiation, Microanalysis, Reprints, KeV range 10-100.

A new global relation has been developed for predicting electron-excited bremsstrahlung intensities over a wide range of accelerating voltages, atomic numbers, and x-ray energies. The new relation was determined from the mathematical modeling of extensive data and is designed for calculating bremsstrahlung intensities in analytical procedures, such as those requiring peak-to-background measurements, where the direct measurement of the bremsstrahlung intensities is impracticable. The distribution of errors between the data and the model is symmetrical, centered around zero error with 63% of the values falling between plus or minus 10% relative error.

901,532
PB89-201685   Not available NTIS
National Bureau of Standards (NML), Gaithersburg, MD. Radiometric Physics Div.
NBS (National Bureau of Standards) Scale of Spectral Radiance.
Final rept.
J. H. Walker, A. T. Hattenburg, and R. D. Saunders. 1987, 10p
Pub. in Metrologia 24, n2 p79-88 1987.

Keywords: *Radiance, *Spectroradiometers, Precision, Blackbody radiation, Measuring instruments, Radiometry, Electromagnetic scattering, Linearity, Reprints, *Calibration standards.

The paper describes the measurement methods and instrumentation used in the realization and transfer of the NBS scale of spectral radiance. The application of the basic measurement equation to both blackbody and tungsten strip lamp sources is discussed. The polarizance, response linearity, spectral responsivity function, and size-of-source effect of the spectroradiometer are described. The analysis of sources of error and estimates of uncertainty are presented. The assigned uncertainties in spectral radiance range from about 1.75% at 225 nm to 0.25% at 2400 nm.

901,533
PB89-202048   Not available NTIS
National Bureau of Standards (NML), Gaithersburg, MD. Office of Standard Reference Data.
Properties of Steam.
Final rept.
H. J. White. 1986, 2p
Pub. in Mechanical Engineering 108, n8 p36-37 1986.

Keywords: *Steam, *Thermodynamic properties, *Standards, Water vapor, Steam tables(Thermodynamics), Enthalpy, Entropy, Density, Reprints, *International Association for the Properties of Steam.

The paper briefly outlines the activities of the International Association for the Properties of Steam (IAPS) and the ASME Research Committee on the Properties of Steam, which serves as the U.S. National Committee for the Properties of Steam to IAPS. Emphasis is based on the internationally agreed upon reference data published by IAPS in the form of releases. A list of current releases is provided as well as archival papers which back up the releases.

901,534
PB89-202147   Not available NTIS
National Bureau of Standards (NEL), Boulder, CO.
Electromagnetic Technology Div.
VAMAS (Versailles Project on Advanced Materials and Standards) Intercomparison of Critical Current Measurement in Nb3Sn Wires.
Final rept.
K. Tachikawa, K. Itoh, H. Wada, D. Gould, H. Jones, C. R. Walters, L. F. Goodrich, J. W. Ekin, and S. L. Bray. 1989, 7p
Pub. in IEEE (Institute of Electrical and Electronics Engineers) Transactions on Magnetics 25, n2 p2368-2374 Mar 89.

Keywords: Electrical measurement, Reprints, *Superconducting wires, *Niobium stannides, *Critical current, Niobium tin, Interlaboratory comparisons.

The VAMAS (Versailles Agreement on Advanced Materials and Standards) technical working party in the area of superconducting and cryogenic structural materials has recently carried out the first world-wide intercomparison of critical current, I sub c, measurement on multifilamentary Nb3Sn wires. Three sample wires were supplied from each of the European Communities, Japan and USA. The total number of participant labs were 24 (EC 11, Japan 8 and USA 5). There were

179

few restrictions for the I sub c measurement at participant labs. The standard deviations of the I sub c values reported from these labs varied among test samples and were 6-21% of averaged I sub c values at 12 Tesla.

901,535
PB89-202154 Not available NTIS
National Bureau of Standards (NML), Gaithersburg, MD. Electricity Div.
History of the Present Value of 2e/h Commonly Used for Defining National Units of Voltage and Possible Changes in National Units of Voltage and Resistance.
Final rept.
B. N. Taylor. 1987, 6p
Pub. in IEEE (Institute of Electrical and Electronics Engineers) Transactions on Instrumentation and Measurement 36, n2 p659-664 1987.

Keywords: *Electrical measurement, *Electric potential, *Electrical resistance, Metrology, Hall effect, Units of measurement, Reprints.

The national standards laboratories of most major industrialized countries employ the Josephson effect to define and maintain their national or laboratory unit of voltage V(LAB). The value of the Josephson frequency-voltage ratio commonly used for this purpose, 2e/h = 483594 GHz/V(LAB), is now known to be about eight parts-per-million less than the absolute or SI value. Consequently, the different national units of voltage are smaller than the SI unit by the same amount. One of the purposes of the paper is to review how this value of 2e/h was selected and hence the origin of the present inconsistency between national voltage units and the SI unit. The motivation for such an historical study is the hope that it can benefit the selection of a new, more accurate value of 2e/h planned for the near future. Also discussed is the status of national units of resistance and the effect of defining and maintaining such units using a value of the quantized Hall resistance consistent with the SI as may also be suggested in the near future.

901,536
PB89-202170 Not available NTIS
National Bureau of Standards (NEL), Boulder, CO. Thermophysics Div.
Determination of Binary Mixture Vapor-Liquid Critical Densities from Coexisting Density Data.
Final rept.
L. J. Van Poolen, and J. C. Rainwater. 1987, 21p
Pub. in International Jnl. of Thermophysics 8, n6 p695-715 1987.

Keywords: *Mixtures, *Critical density, Nitrogen, Methane, Thermodynamic equilibrium, Thermodynamic properties, Experimental data, Liquids, Two phase flow, Vapors, Reprints, *isochoric processes.

Two-phase vapor-liquid equilibrium ((VLE) isochores for binary mixtures are defined as the thermodynamic paths along which the overall density and composition are fixed. Data along such isochores are generated from a modified Leung-Griffiths model fit to experimental data for the binary system nitrogen-methane. The behavior of the liquid volume fraction along these isochores is found to be similar to that for pure fluids. Rectilinear diameters for varying overall densities (fixed composition) are seen to be nearly coincident. Straight-line diameters and the critical liquid volume fraction method are utilized to predict critical densities using data near and removed from the critical point. Both methods give acceptable results but the critical liquid volume fraction method is more accurate. A critical literature review of the need for binary mixture critical densities is presented and a proposed experimental procedure is given for the determination of mixture critical densities.

901,537
PB89-202469 Not available NTIS
National Bureau of Standards (IMSE), Gaithersburg, MD. Polymers Div.
Computer Simulation Studies of the Soliton Model. 3. Noncontinuum Regimes and Soliton Interactions.
Final rept.
K. J. Wahlstrand. 1988, 7p
Pub. in Polymer 29, n2 p256-262 1988.

Keywords: *Elementary excitations, *Klein-Gordon equation, Computerized simulation, Correlation, interactions, Temperature, Field theory(Physics), Energy

levels, Scaling, Phonons, Mathematical models, Reprints, *Solitons, Coupling constants.

The Klein-Gordon soliton model is studied by performing stochastic molecular dynamics computer simulations for a wide range of both temperatures and coupling constants. Three general regimes of behavior are found: the continuum limit or non-interacting regime; the pinned or transition state theory limit, where soliton-phonon interactions are important; and the general noncontinuum regime, where soliton-soliton interactions or 'multiple soliton effects' are important. In the noncontinuum regime the correlation function changes as a function of temperature and coupling constant. This will lead to deviations from the continuum limit temperature scaling and soliton energy scaling observed in the dynamics of Klein-Gordon systems.

901,538
PB89-202535 Not available NTIS
National Bureau of Standards (NML), Gaithersburg, MD. Radiation Physics Div.
Computation of the ac Stark Effect in the Ground State of Atomic Hydrogen.
Final rept.
L. Pan, K. T. Taylor, and C. W. Clark. 1988, 4p
Pub. n Physical Review Letters 61, n23 p2673-2676, 5 Dec 88.

Keywords: *Hydrogen, *Stark effect, Perturbation theory, Ground state, Greens function, Reprints, Coulomb potential.

Using a Sturmian function expansion the authors have computed the nth-order coefficients E sub n(omega), for a 2 less than or = n less than or = 22, in the perturbation expansion of the ac Stark effect in the hydrogen 1s state. An effective convergence similar to that in the dc case is observed. A parametrization of these coefficients, based upon the analytic structure of the Coulomb Green's function, separates the rapid oscillatory resonant behavior and the smooth background variation with respect to omega.

901,539
PB89-202600 Not available NTIS
National Bureau of Standards (NML), Boulder, CO. Quantum Physics Div.
Quantum-Defect Parametrization of Perturbative Two-Photon Ionization Cross Sections.
Final rept.
M. G. J. Fink, and P. Zoller. 1989, 15p
Sponsored by National Science Foundation, Washington, DC.
Pub. in Physical Review A 39, n6 p2933-2947, 15 Mar 89.

Keywords: *Hydrogen, Photons, Perturbation theory, Energy levels, Quantum theory, Reprints, *Ionization cross sections, Rydberg states, Coulomb potential.

A multichannel quantum-defect theory (MQDT) parametrization of two-photon ionization (2PI) of atoms, suitable for use with ab initio calculations of perturbative 2PI amplitudes in the presence of intermediate-and/or final-state Rydberg resonances is developed. The rapid energy variation of such amplitudes is extracted in terms of elementary functions of energy and a set of MQDT parameters, which are smooth functions of energy. As an example, results of numerical computations of MQDT parameters for 2PI of atomic hydrogen (1s, 2s, and 3s) by circularly polarized light are presented. MQDT parameters for the many-channel case are derived, and an analysis of their phase-shifted Coulomb functions, and simple analytic formulas are given for the energy dependence of the amplitude as a function of a small set of physical parameters.

901,540
PB89-211981 Not available NTIS
National Inst. of Standards and Technology (IMSE), Gaithersburg, MD. Reactor Radiation Div.
Applications of Mirrors, Supermirrors and Multilayers at the National Bureau of Standards Cold Neutron Research Facility.
Final rept.
C. F. Majkrzak, C. J. Glinka, and S. K. Satija. 1989, 15p
Pub. in Proceedings of SPIE (Society of Photo-Optical Instrumentation Engineers) Thin-Film Neutron Optical Devices: Mirrors, Supermirrors, Multilayer Monochromators, Polarizers, and Beam Guides, v983, p129-143 1989.

Keywords: Neutron spectrometers, Neutron scattering, Polarization(Spin alignment), *Neutron reflectors,

Neutron reflectometers, Supermirrors, Multilayers, Polarized beams.

It is expected that thin-film mirrors, multilayers and supermirrors will play an important role as optical elements to transport, focus, monochromate, and polarize neutrons at the cold neutron research facility which is presently under construction at the National Bureau of Standards. In the paper, specific applications of these reflecting devices to three instruments, a spin-polarized, inelastic scattering spectrometer with novel, adjustable resolution properties, a focusing small angle scattering machine, and a neutron reflectometer are described.

901,541
PB89-211999 Not available NTIS
National Bureau of Standards (IMSE), Gaithersburg, MD. Reactor Radiation Div.
Calculations and Measurement of the Performance of Converging Neutron Guides.
Final rept.
J. R. D. Copley, and C. F. Majkrzak. 1989, 12p
Pub. in Proceedings of SPIE (Society of Photo-Optical Instrumentation Engineers), Thin-Film Neutron Optical Devices: Mirrors, Supermirrors, Multilayer Monochromators, Polarizers, and Beam Guides, v983 p93-104 1989.

Keywords: Neutron beams, Convergence, Mirrors, *Neutron guides, Neutron reflectors, Supermirrors.

If properly designed, converging guides may be used to increase the current density of neutrons in a beam from a parallel neutron guide. The disadvantage of this type of device is that there is an increase in beam divergence. The spatial and angular distributions of the beam emerging from a converging guide are generally nonuniform, and the performance of the guide is dependent on the neutron's wavelength. Analytic and numerical methods of calculation are briefly discussed, and the results of selected Monte Carlo numerical calculations are presented. Measurements on a scaled down version of a converging guide system are reported and found to compare well with calculation.

901,542
PB89-212062 Not available NTIS
National Bureau of Standards (NML), Boulder, CO. Quantum Physics Div.
Liquid-Supported Torsion Balance: An Updated Status Report on Its Potential for Tunnel Detection.
Final rept.
J. E. Faller, P. T. Keyser, and M. P. McHugh. 1988, 22p
Sponsored by Air Force Weapons Lab., Kirtland AFB, NM.
Pub. in Proceedings of Technical Symposium on Tunnel Detection (3rd), Golden, CO., January 12-15, 1988, p412-433.

Keywords: *Torsion balances, *Tunnel detection, Variometers, Precision, Measurement, *Gradiometers.

At the Joint Institute for Laboratory Astrophysics, the authors have been developing the liquid-supported torsion balance (LSTB), also known as the fluid-fiber torsion balance, for a variety of precision measurement applications. Recent work has concentrated on curvature variometers and 'split-disc' vertical gradiometers as an auxiliary to 'Fifth-Force' tests. The authors review the history and design of the LSTB and describe the theoretical and experimental progress since the last Symposium on Tunnel Detection. Significant increases in experimental sensitivity to changes in the curvature of the level surface have been made together with advances in understanding and reducing various noise terms. The current limiting noise terms and sensitivity to tunnel evolution are discussed.

901,543
PB89-212138 Not available NTIS
National Bureau of Standards (NEL), Gaithersburg, MD. Thermophysics Div.
Simulation Study of Light Scattering from Soot Agglomerates.
Final rept.
R. D. Mountain. 1988, 6p
Pub. in Springer Proceedings in Physics, v33 p49-54 1988.

Keywords: *Light scattering, *Agglomeration, Soot, Computerized simulation, Distribution functions, Dynamics.

The article presents an example of current research utilizing Langevin dynamics simulation methods. This technique is used to generate agglomerates with soot-like structures. The light scattering properties of the agglomerates are examined in order to determine the information available in light scattering measurements and how this information relates to the structure of the clusters.

901,544
PB89-218374 PC A04/MF A01
National Inst. of Standards and Technology (NIST), Boulder, CO. Thermophysics Div.
Vapor-Liquid Equilibrium of Binary Mixtures in the Extended Critical Region. I. Thermodynamic Model.
Technical note.
J. C. Rainwater. Apr 89, 73p NIST/TN-1328
Also available from Supt. of Docs. as SN003-003-02945-9. Sponsored by Department of Energy, Washington, DC. Office of Basic Energy Sciences.

Keywords: *Thermodynamic equilibrium, *Models, Thermodynamics, Density, Vapor phases, Liquids, *Binary mixtures.

The thermodynamic model of Leung and Griffiths for binary mixture vapor-liquid equilibrium near the critical locus, as modified by Moldover, Rainwater and co-workers, is extended to accommodate fluid pairs of greater dissimilarity.

901,545
PB89-227987 Not available NTIS
National Inst. of Standards and Technology (NEL), Boulder, CO. Thermophysics Div.
Asymptotic Expansions for Constant-Composition Dew-Bubble Curves Near the Critical Locus.
Final rept.
J. C. Rainwater. 1989, 12p
Sponsored by Department of Energy, Washington, DC. Pub. in International Jnl. of Thermophysics 10, n2 p357-368 1989.

Keywords: *Critical point, *Bubbles, *Dew, Critical temperature, Critical pressure, Curves(Geometry), Asymptotic series, Approximation, Thermodynamic properties, Reprints, Binary mixtures, Liquid vapor equilibrium.

Explicit function representations are developed for constant-composition dew and bubble curves near critical according to the modified Leung-Griffiths theory. The critical point in temperature-density space is shown to be a point of maximum concave upward curvature, rather than an inflection point as previously conjectured.

901,546
PB89-228019 Not available NTIS
National Bureau of Standards (NEL), Boulder, CO. Thermophysics Div.
Mean Density Approximation and Hard Sphere Expansion Theory: A Review.
Final rept.
L. J. Chen, J. F. Ely, and G. A. Mansoori. 1987, 27p
Sponsored by Gas Research Inst., Chicago, IL.
Pub. in Fluid Phase Equilibria 37, p1-27 1987.

Keywords: *Liquids, *Mixtures, Density, Computer simulation, Approximation, Van der Waals equation, Thermodynamic equilibrium, Thermodynamic properties, Reprints, *Mean density approximation, *Hard sphere expansion theory, Liquid vapor equilibrium.

The review surveys research dealing with the Mean Density Approximation (MDA) and the Hard Sphere Expansion (HSE) theory developed by Leland and co-workers. MDA and its modifications provide a simple way to predict radial distribution functions of mixtures from the pure fluid information. Comparisons with computer simulation data show the MDA to be superior to the van der Waals approximation for mixture radial distribution functions. Derivations of the HSE theory and the HSE Conformal Solution Theory (HSE-CST) are also described. For Lennard-Jones mixtures, the HSE theory is proven to be superior to the van der Waals theory by using a proper method to determine the hard-sphere diameter. The major problem associated with the development of a consistent method to determine the hard-sphere diameter of the HSE-CST and the requirements regarding the extension of the HSE-CST to polar mixtures are discussed.

901,547
PB89-226027 Not available NTIS

National Bureau of Standards (NEL), Boulder, CO. Thermophysics Div.
Method for Improving Equations of State Near the Critical Point.
Final rept.
D. D. Erickson, T. W. Leland, and J. F. Ely. 1987, 21p
Sponsored by Gas Research Inst., Chicago, IL.
Pub. in Fluid Phase Equilibria 37, p185-205 1987.

Keywords: *Equations of state, *Carbon dioxide, *Pentanes, *Critical point, Thermodynamic properties, Ideal gases, Reprints.

Accurate nonanalytic equations of state have been developed for pentane and carbon dioxide for use as reference equations in the critical region. This was done using a method which transforms any analytic equation of state into a nonanalytic equation of state. For this study, a 32 constant analytic BWR equation of state was transformed into a nonanalytic equation of state. The nonanalytic equation of state is more accurate in the critical region for the prediction of the PVT properties of carbon dioxide than is a Schmidt-Wagner analytic equation of state developed at the National Bureau of Standards. The nonanalytic equation of state is slightly less accurate over the whole PVT surface than the Schmidt-Wagner equation of state, but the predicted nonanalytic surface makes a smooth transition into the classical region. This is an improvement over the scaled fundamental equation of state which breaks down at a small distance away from the critical point.

901,548
PB89-228035 Not available NTIS
National Inst. of Standards and Technology (NEL), Boulder, CO. Thermophysics Div.
Prediction of Shear Viscosity and Non-Newtonian Behavior in the Soft-Sphere Liquid.
Final rept.
H. J. M. Hanley, J. C. Rainwater, and M. Huber. 1988, 10p
Sponsored by Department of Energy, Washington, DC. Pub. in International Jnl. of Thermophysics 9, n6 p1041-1050 Nov 88.

Keywords: *Non-Newtonian fluids, *Shear properties, *Viscosity, Rheological properties, Relaxation time, Molecular relaxation, Liquids, Correlation, Reprints.

It is shown that a shear rate-dependent viscosity coefficient, normal pressure differences, and shear dilatancy can be predicted in a soft-sphere liquid given only the equilibrium radial distribution function and a relaxation time. Calculations are made using the relaxation-time theory of Hess and Hanley, and the results are compared with simulation data from nonequilibrium molecular dynamics.

901,549
PB89-226100 Not available NTIS
National Bureau of Standards (NML), Boulder, CO. Quantum Physics Div.
Spectroscopic Detection Methods.
Final rept.
U. Hefter, and K. Bergmann. 1988, 61p
Pub. in Atomic and Molecular Beam Methods, Chapter 9, p193-253 1988.

Keywords: *Atomic energy levels, *Molecular energy levels, Molecular beams, Alignment, Reprints, *Laser induced fluorescence, *Vibrational states, *Rotational states, Multi-photon processes, Raman effect, Optical fibers, Magic angle detectors.

The authors discuss the basic formulae relevant to laser induced fluorescence detection of atomic and molecular level population and alignment as well as velocities. Discussion of other techniques such as two-photon ionization and Raman scattering is also included. Experimental hardware such as high-speed optical devices and optical fibers are described.

901,550
PB89-228118 Not available NTIS
National Bureau of Standards (NML), Boulder, CO. Quantum Physics Div.
State Selection via Optical Methods.
Final rept.
K. Bergmann. 1988, 52p
Pub. in Atomic and Molecular Beam Methods, Chapter 12, p293-344 1988.

Keywords: *Atomic energy levels, *Molecular energy levels, Experimental design, Optical pumping, Selec-

tion, Alignment, Atomic beams, Molecular beams, Reprints, *Rotational states, *Vibrational states, Laser applications.

The author describes the experimental techniques of laser state selection in atomic and molecular beams that have been developed since 1975. The bibliography contains about 200 references concerning selection of rotational-vibrational states as well as alignment in atoms, simple molecules, and molecular ions.

901,551
PB89-228365 Not available NTIS
National Bureau of Standards (NML), Boulder, CO. Quantum Physics Div.
Precision Experiments to Search for the Fifth Force.
Final rept.
J. E. Faller, E. Fischbach, Y. Fujii, K. Kuroda, H. J. Paik, and C. C. Speake. 1989, 9p
Sponsored by Department of Energy, Washington, DC. Pub. in IEEE (Institute of Electrical and Electronics Engineers) Transactions on Instrumentation and Measurement 38, n2 p180-188 Apr 89.

Keywords: *Gravitation, Experimental design, Precision, Reprints, *Fifth force, *Basic interactions.

The suggestion of a possible new fifth force of Nature has prompted a large number of high precision experiments to search for its presence. After reviewing the motivation for this suggestion, the authors describe some of the experiments that are presently underway and the results that have been obtained to date.

901,552
PB89-228361 Not available NTIS
National Inst. of Standards and Technology (NML), Boulder, CO. Quantum Physics Div.
Ionization and Current Growth in N2 at Very High Electric Field to Gas Density Ratios.
Final rept.
V. T. Gylys, B. M. Jelenkovic, and A. V. Phelps. 1989, 12p
Pub. in Jnl. of Applied Physics 65, n9 p3369-3380, 1 May 89.

Keywords: *Nitrogens, *Gas discharges, *Gas ionization, Electric current, Electric fields, Electric measuremen, Reprints, Electron impact, Low pressure, High field.

Measurements and analyses have been made of electron impact ionization and of current growth in pulsed, low-current, prebreakdown discharges in parallel-plane geometry in N2 at very high electric field to gas density ratios E/n and low products of the gas density n and electrode separation d. Measurements were made of the transported charge on the tie scales of electron transit, ion transit, and metastable decay. Measurements were also made of the growth of steady-state discharge currents as a function of discharge voltage. The contributions of avalanches resulting from ion and metastable-induced secondary electrons were determined from the ratio of electron-excited N2(1+) 391.4-nm emission integrated over all avalanches to the integrated emission during the laser-initiated electron pulse.

901,553
PB89-229090 Not available NTIS
National Bureau of Standards (NEL), Boulder, CO. Chemical Engineering Science Div.
Performance of He II of a Centrifugal Pump with a Jet Pump Inducer.
Final rept.
D. E. Daney, P. R. Ludtke, and A. Kashani. 1989, 6p
Sponsored by National Aeronautics and Space Administration, Moffett Field, CA. Ames Research Center. Pub. in Cryogenics 29, p563-568 May 89.

Keywords: *Centrifugal pumps, *Jet pumps, Liquid helium, Weightlessness, Superfluidity, Cavitation, Performance, Reprints, Helium II.

The tendency of turbopumps operating in He II to cavitate makes their use in zero gravity questionable because of the zero net positive suction head (NPSH) available at the pump inlet. The authors investigated a jet pump, positioned at the inlet of a centrifugal pump with a screw inducer, as a means of operating a centrifugal pump at zero or lower NPSH. Pump performance in He II was measured as a function of NPSH for six different combinations of primary and secondary nozzles. Suction heads down to -91 mm were meas-

PHYSICS

General

ured for a 3% reduction in developed head. These are referenced to the leading edge of the screw inducer, which is 100 mm above the jet pump inlet. Because cavitation at the primary jet always precedes cavitation in the jet pump secondary nozzle, they also tested reverse (pressure driven) flow through a porous plug as a means of obtaining a subcooled primary jet. These brief tests were inconclusive.

901,554
PB89-230395   Not available NTIS
National Bureau of Standards (NML), Gaithersburg, MD. Electricity Div.
Improved Transportable DC Voltage Standard.
Final rept.
B. F. Field, and M. R. McCaleb. 1989, 6p
Sponsored by Department of Defense Calibration Coordination Group, Redstone Arsenal, AL.
Pub. in IEEE (Institute of Electrical and Electronics Engineers) Transactions on Instrumentation and Measurement 38, n2, p324-329, Apr 89.

Keywords: *Electrical measurement, *Standards, *Avalanche diodes, Calibrating, Prototypes, Reprints.

Zener-diode-based dc voltage standards can be excellent transport standards for the unit of dc voltage because of their resistance to physical shock and temperature changes. The problems of transporting a unit of voltage and the properties of available Zener standards were studied to develop a set of characteristics considered essential for an optimum transport standard. The report lists some of the results of the requirements study, explains the design of the improved transport standard, discusses the efforts to select Zener diodes for the standards, and presents data obtained from prototype Zener reference modules to be used in the standard.

901,555
PB89-230411   Not available NTIS
National Bureau of Standards (NML), Gaithersburg, MD. Electricity Div.
Low Field Determination of the Proton Gyromagnetic Ratio in Water.
Final rept.
E. R. Williams, G. R. Jones, S. Ye, R. Liu, H. Sasaki, P. T. Olsen, W. D. Phillips, and H. P. Layer. 1989, 5p
Pub. in IEEE (Institute of Electrical and Electronics Engineers) Transactions on Instrumentation and Measurement 38, n2, p233-237, Apr 89.

Keywords: *Fundamental constants, Nuclear magnetic resonance, Electrical resistance, Solenoids, Precision, Reprints, *Gyromagnetic ratio, Quantum Hall effect, Fine structure constant.

The authors measured the proton gyromagnetic ratio in H2O by the low field method, gamma'(p)(low). The result gamma'(p)(low) = 2.67 513 376 10 to the 8th power/s(T(NBS)) (0.11 ppm), leads to a value of the fine structure constant of 1/alpha = 137.0 359 840 (0.037 ppm) and a value for the quantized Hall resistance in SI units of R(H) = 25 812.80460 ohm (0.037 ppm). To achieve this result, they measured the dimensions of a 2.1-m solenoid with an accuracy of 0.04 micrometer, and then measured the NMR frequency of a water sample in the field of the solenoid.

901,556
PB89-230437   Not available NTIS
National Bureau of Standards (NML), Gaithersburg, MD. Electricity Div.
NBS (National Bureau of Standards) Determination of the Fine-Structure Constant, and of the Quantized Hall Resistance and Josephson Frequency-to-Voltage Quotient in SI Units.
Final rept.
M. E. Cage, R. F. Dziuba, R. E. Elmquist, B. F. Field, G. R. Jones, P. T. Olsen, W. D. Phillips, J. Q. Shields, R. L. Steiner, B. N. Taylor, and E. R. Williams. 1989, 6p
Pub. in IEEE (Institute of Electrical and Electronics Engineers) Transactions on Instrumentation and Measurement 38, n2, p284-289, Apr 89.

Keywords: *Fundamental constants, Reprints, *Fine structure constant, *Quantum Hall effect, *Josephson effect, *Plancks constant, *Electron charge.

Results from NBS experiments to realize the ohm and the watt, to determine the proton gyromagnetic ratio by the low field method, to determine the time dependence of the NBS representation of the ohm using the quantum Hall effect, and to maintain the NBS representation of the volt using the Josephson effect, are appropriately combined to obtain an accurate value of the fine-structure constant and of the quantized Hall resistance in SI units, and values in SI units of the Josephson frequency-to-voltage quotient, Planck constant, and elementary charge.

901,557
PB89-230445   Not available NTIS
National Bureau of Standards (NML), Gaithersburg, MD. Electricity Div.
New Realization of the Ohm and Farad Using the NBS (National Bureau of Standards) Calculable Capacitor.
Final rept.
J. Q. Shields, R. F. Dziuba, and H. P. Layer. 1989, 3p
Pub. in IEEE (Institute of Electrical and Electronics Engineers) Transactions on Instrumentation and Measurement 38, n2, p249-251, Apr 89.

Keywords: *Electric measuring instruments, Ohmmeters, Electrical measurement, Electrical resistance, Capacitors, Reprints, *National Institute of Standards and Technology, *Absolute ohm, *Absolute farad.

Results of a new realization of the ohm and farad using the NBS calculable capacitor and associated apparatus are reported. The results show that both the NBS representation of the ohm and the NBS representation of the farad are changing with time: the ohm at the rate of-0.054 ppm/year and the Farad at the rate of 0.010 ppm/year. The realization of the ohm is of particular significance at this time because of its role in assigning an SI value to the quantized Hall resistance. The estimated uncertainty of the ohm realization is 0.022 ppm while the estimated uncertainty of the farad realization is 0.014 ppm.

901,558
PB89-234165   Not available NTIS
National Inst. of Standards and Technology (NML), Gaithersburg, MD. Atomic and Plasma Radiation Div.
Line Identifications and Radiative-Branching Ratios of Magnetic Dipole Lines in Si-like Ni, Cu, Zn, Ge, and Se.
Final rept.
R. U. Datla, J. R. Roberts, N. Woodward, S. Lippman, and W. L. Rowan. 1989, 4p
Sponsored by Department of Energy, Washington, DC.
Pub. in Physical Review A 40, n3 p1484-1487, 1 Aug 89.

Keywords: Plasmas(Physics), Line spectra, Reprints, *M1-transitions, Multicharged ions, Branching ratio, Nickel ions, Copper ions, Zinc ions, Germanium ions, Selenium ions, Tokamak devices, Calibration.

Magnetic dipole transitions within the ground-state configuration of Si-like Ni, Cu, Zn, Ge, and Se have been observed. Seven lines have been newly identified. Observations are made on the Texas Experimental Tokamak by laser-ablation injection of each of these elements into the plasma. The radiative-branching-ratio technique, utilizing lines originating from the same upper level, 3p(2) (1)O2, is used for the radiometric calibration of a 1-m normal-incidence spectrometer. This calibration is in good agreement with the absolute radiometric calibration obtained by using an argon miniarc.

901,559
PB89-234306   Not available NTIS
National Inst. of Standards and Technology (NML), Boulder, CO. Quantum Physics Div.
Sodium Doppler-Free Collisional Line Shapes.
Final rept.
M. J. O'Callaghan, and A. Gallagher. 1989, 16p
Grant NSF-PHY86-04504
Sponsored by National Science Foundation, Washington, DC.
Pub. in Physical Review A 39, n12 p6190-6205, 15 Jun 89.

Keywords: *Sodium, Excitation, Reprints, Atom collisions, Line broadening.

Measurements of resonant, two-step excitation of sodium in the presence of very-low-pressure buffer gases (He, Ar, Xe, and N2) are reported. These data are analyzed to isolate the effects of line broadening and velocity-changing (VC) collisions in the single-collision limit. Velocity dependence of the several line broadenings involved is obtained from the data analysis. The inelastic VC kernels are also obtained as a function of initial velocity.

901,560
PB90-117292   Not available NTIS
National Inst. of Standards and Technology (NML), Gaithersburg, MD. Center for Radiation Research.
Cross Section and Linear Polarization of Tagged Photons.
Final rept.
J. Asai, H. S. Caplan, D. M. Skopik, W. Del Bianco, and L. C. Maximon. 1988, 9p
Pub. in Canadian Jnl. of Physics 66, p1079-1087 1988.

Keywords: *Bremsstrahlung, *Photon cross sections, *Polarization(Waves), Monochromators, Marking, Reprints, *Tagged photon method, Coordinate transformations, Linear polarization.

Formulae for bremsstrahlung cross sections and polarizations are usually presented in coordinate systems unsuitable for application by experimental physicists to devices such as photon-tagging monochromators. The paper presented the transformations between the different coordinate systems, along with examples of the calculated cross sections and polarizations in a form convenient from the experimental standpoint. These examples also give the predicted characteristics of the photon tagger currently under construction at the Saskatchewan Accelerator Laboratory.

901,561
PB90-117565   Not available NTIS
National Inst. of Standards and Technology (NML), Gaithersburg, MD. Center for Radiation Research.
Intrinsic Sticking in dt Muon-Catalyzed Fusion: Interplay of Atomic, Molecular and Nuclear Phenomena.
Final rept.
M. Danos, A. A. Stahlhofen, and L. C. Biedenharn. 1989, 46p
Pub. in Annals of Physics 192, n1 p158-203, 15 May 89.

Keywords: *Nuclear fusion, Molecules, Atoms, Deuterium, Tritium, Reprints, *Muon-catalyzed fusion, Branching ratio.

A comprehensive reaction theory for the resonant muon catalyzed fusion of deuterium and tritium is formulated. Emphasis is put on non-perturbative, many body treatment of the long range Coulomb force and its interference with the nuclear forces, with the aim of providing the theoretical framework for an accurate calculation of the branching ratio dt(mu) -> ((alpha + mu) + n)/(alpha + mu + n) essential for muon catalyzed fusion.

901,562
PB90-117581   Not available NTIS
National Inst. of Standards and Technology (NML), Gaithersburg, MD. Quantum Metrology Div.
Marked Differences in the 3p Photoabsorption between the Cr and Mn(1+) Isoelectronic Pair: Reasons for the Unique Structure Observed in Cr.
Final rept.
J. W. Cooper, C. W. Clark, C. L. Cromer, T. B. Lucatorto, B. F. Sonntag, E. T. Kennedy, and J. T. Costello. 1989, 4p
Contract AFOSR-ISSA-87-0050
Sponsored by Air Force Office of Scientific Research, Bolling AFB, DC.
Pub. in Physical Review A 39, n11 p6074-6077, 1 Jun 89.

Keywords: *Chromium, *Atomic structure, *Absorption spectra, Atomic energy levels, Reprints, *Manganese ions, *Photoabsorption, Inner-shell excitation, Inner-shell ionization, Giant resonance.

Chromium is the only member of the transition-group elements with a well-developed Rydberg structure appearing in its 3p-absorption spectrum. New high-resolution measurements of Mn, Mn(1+) and Cr have revealed weak Rydberg structure in Mn(1+) and an analysis of the data shows that the anomalous Cr 3p absorption spectra is due to a unique relationship between its energy levels, and not to the previously supposed fact that Cr has an unpaired 4s electron.

901,563
PB90-117730   Not available NTIS
National Inst. of Standards and Technology (NML), Gaithersburg, MD. Center for Radiation Research.

Comment on 'Feasibility of Measurement of the Electromagnetic Polarizability of the Bound Nucleon'.
Final rept.
E. Hayward. 1989, 3p
Pub. in Physical Review C 40, n1 p467-469 1989.

Keywords: *Nucleons, Measurement, Feasibility, Polarization(Spin alignment), Reprints, *Polarizability, Bound state, Total cross sections, Photon absorption, Photon scattering.

It is shown that the polarizability of the bound nucleon is, within the errors in the total cross section measurements, the same as that of a free nucleon.

901,564
PB90-123449 Not available NTIS
National Inst. of Standards and Technology (NML), Gaithersburg, MD. Ionizing Radiation Physics Div.
(109)Pd and (109)Cd Activity Standardization and Decay Data.
Final rept.
C. Ballaux, B. M. Coursey, and D. D. Hoppes. 1988, 9p
Pub. in Applied Radiation and Isotopes 39, n11 p1131-1139 1988.

Keywords: Radioactive decay, Internal conversion, Beta particles, Radiotherapy, Standardization, Reprints, *Cadmium 109, *Palladium 109, Gamma radiation, Activity levels.

Sources of (109)Cd and (109)Pd were measured with liquid-scintillation counters, a NaI (Tl) well detector, a 4 pi(Si(Li)) assembly, a Ge(Li) detector, and ionization chambers. For the 88.034-keV transition, a total internal-conversion coefficient of 26.21 plus or minus 0.14 and a gamma-ray emission probability of 3.675 plus or minus 0.018% were measured. (109)Pd, because of its short half life, 13,404 plus or minus 0.008 h, and high beta-ray energy, has some promise of applications in radiotherapy. The uncertainty in the present activity standardization was only 0.41%.

901,565
PB90-123506 Not available NTIS
National Inst. of Standards and Technology (NML), Gaithersburg, MD. Ionizing Radiation Physics Div.
Advances in the Use of (3)He in a Gas Scintillation Counter.
Final rept.
J. W. Behrens, H. Ma, and O. A. Wasson. 1986, 1p
Pub. in Transactions of the American Nuclear Society 53, p163 1986.

Keywords: *Scintillation counters, *Helium 3, Reprints, *Gas scintillation detectors, *Neutron detectors.

The development of a (3)He gas scintillator for use as a neutron detector is described.

901,566
PB90-123605 Not available NTIS
National Inst. of Standards and Technology (NML), Gaithersburg, MD. Ionizing Radiation Physics Div.
Electron Stopping Powers for Transport Calculations.
Final rept.
M. J. Berger. 1988, 24p
Sponsored by Department of Energy, Washington, DC.
Pub. in Monte Carlo Transport of Electrons and Photons, Chapter 3, p57-80 1988.

Keywords: Excitation, Positrons, Reviews, Reprints, *Electron transport, *Electron collisions, *Stopping power.

The paper reviews the calculation and tabulation of collision stopping powers for electrons, mainly at energies above 10 keV. The discussion includes mean excitation energies and the density-effect correction.

901,567
PB90-123670 Not available NTIS
National Inst. of Standards and Technology (NML), Gaithersburg, MD. Quantum Metrology Div.
Determination of Short Lifetimes with Ultra High Resolution (n,gamma) Spectroscopy.
Final rept.
H. G. Borner, J. Jolie, F. Hoyler, S. Robinson, M. S. Dewey, G. Greene, E. Kessler, and R. D. Deslattes. 1988, 5p
Pub. in Physics Letters B 215, n1 p45-49 1988.

Keywords: *Nuclear energy levels, Neutron reactions, Gamma rays, Reprints, *Excited states, *Lifetime,

Gamma spectroscopy, Doppler broadening, High resolution.

It is shown how high resolution (n,gamma) spectroscopy allows the determination of lifetimes of nuclear excited levels, through the observation of Doppler broadening. The Doppler broadening of these gamma-rays is due to the recoil from the feeding of gamma-ray transition.

901,568
PB90-123720 Not available NTIS
National Inst. of Standards and Technology (NML), Boulder, CO. Quantum Physics Div.
Approximate Formulation of Redistribution in the Ly(alpha), Ly(beta), H(alpha) System.
Final rept.
J. Cooper, R. J. Ballagh, and I. Hubeny. 1989, 17p
Grant NSF-PHY86-04504
Sponsored by National Science Foundation, Washington, DC.
Pub. in Astrophysical Jnl. 344, n2 p949-965, 15 Sep 89.

Keywords: *Hydrogen, Absorption spectra, Emission spectra, Approximation, Distribution functions, Reprints, Lyman lines, Line broadening, Redistribution.

Simple approximate formulae are given for the coupled redistribution of Ly(alpha), Ly(beta), and H(alpha), by using well-defined approximations to an essentially exact formulation. These formulae incorporate all the essential physics including Raman scattering, lower state radiative decay, and correlated terms representing emission during a collision which must be retained in order that the emission coefficients are properly behaved in the line wings. Approximate expressions for the appropriate line broadening parameters are collected. Finally, practical expressions for the source functions are given. These are formulated through newly introduced non-impact redistribution functions, which are shown to be reasonably approximated by existing (ordinary and generalized) redistribution functions.

901,569
PB90-123738 Not available NTIS
National Inst. of Standards and Technology (IMSE), Gaithersburg, MD. Reactor Radiation Div.
Neutron Scattering and Its Effect on Reaction Rates in Neutron Absorption Experiments.
Final rept.
J. R. D. Copley, and C. A. Stone. 1989, 12p
Pub. in Nuclear Instruments and Methods in Physics Research A281, p593-604 1989.

Keywords: *Neutron absorption, *Neutron scattering, Monte Carlo method, Neutron radiography, Reprints, Activation analysis, Self-shielding.

In general there is a systematic error in the results of any in-beam neutron absorption experiment because of neutron scattering in the sample. The error is largest when scattering predominates over absorption, when the transmission of the sample is small, and when the lateral dimensions of the sample are large compared with its thickness. The ratio of the reaction rate in the sample to the rate calculated ignoring both scattering and self-shielding due to absorption may be significantly less than or greater than unity, depending on the scattering properties of the sample and its size and shape. The conclusions are derived from Monte Carlo calculations based on a very general expression for the rate of a neutron absorption reaction in a sample which scatters and absorbs neutrons. The expression for the reaction rate, written as a sum over orders of scattering within the sample, was obtained using the technique adopted by V.F. Sears in his study of multiple scattering effects in neutron scattering experiments. The authors' calculations were performed for slab samples placed in a narrow, monoenergetic, monodirectional neutron beam. The assumed scattering cross section was isotropic and also static, meaning that no change in neutron energy was permitted to occur within the sample. The work has potentially important implications in various areas of neutron research including activation analysis and radiography.

901,570
PB90-123746 Not available NTIS
National Inst. of Standards and Technology (NML), Gaithersburg, MD. Radiometric Physics Div.

Neonlike Ar and Cl 3p-3s Emission from a theta-pinch Plasma.
Final rept.
R. C. Elton, R. U. Datla, J. R. Roberts, and A. K. Bhatia. 1989, 3p
Pub. in Physical Review A 40, n7 p4142-4144, 1 Oct 89.

Keywords: Spectral lines, Emission spectra, Reprints, *Argon ions, *Chlorine ions, Multicharged ions, Extreme ultraviolet radiation, X ray lasers, Theta pinc

Time-resolved extreme-ultraviolet emission from sixteen 3p-3s transitions, some of the type in which lasing has been demonstrated in heavier elements, is measured for neonlike Ar(8+) and Cl(7+). These observations are made on a hydrogen theta-pinch plasma with a 5% admixture of argon or freon (for Cl). Fourteen 3d-3p spectral lines are also detected. The measured intensities are compared to theoretical predictions. At major lines agree within plus or minus 30%. Hence, there is no evidence of anomalously intense lines originating on 2p(5)3p J = 2 upper levels compared to J = 0, as observed in gain experiments.

901,571
PB90-123761 Not available NTIS
National Inst. of Standards and Technology (NML), Boulder, CO. Quantum Physics Div.
Exoergic Collisions of Cold Na*-Na.
Final rept.
A. Gallagher, and D. Pritchard. 1989, 4p
Grant N00014-83-K-0695
Sponsored by Office of Naval Research, Arlington, VA.
Pub. in Physical Review Letters 63, n9 p957-960, 28 Aug 89.

Keywords: *Sodium, Reprints, *Atomic traps.

Rates were calculated for two exothermic excited-state collision processes involving sodium at ultracold temperatures. The authors predict that the rate for fine-structure changing collisions exceeds that of radiative redistribution with release of enough energy to cause loss from current optical traps. A semiclassical treatment is used which accounts for the frequency dependence of absorption and spontaneous emission in midcollision.

901,572
PB90-123837 Not available NTIS
National Inst. of Standards and Technology (NML), Boulder, CO. Quantum Physics Div.
Spectroscopy of Autoionizing States Contributing to Electron-Impact Ionization of Ions.
Final rept.
A. Muller, G. Hofmann, K. Tinschert, E. Salzborn, G. H. Dunn, and R. Becker. 1989, 3p
Pub. in Nuclear Instruments and Methods in Physics Research B40/41, p232-234 1989.

Keywords: Excitation, Reprints, *Electron ion collisions, *Autoionization, Ionization cross sections, Electron impact, Colliding beams.

In electron-ion crossed-beam experiments the authors have used a fast electron-energy scanning technique to detect fine details in ionization cross sections. They obtained data with a relative point to point uncertainty of less than 0.1%. The electron energy spread at 100 eV (15 mA beam current) is 0.4 eV. Thus they were able to measure state-resolved excitation-autoionization contributions and to demonstrate new ionization mechanisms involving dielectronic capture of the projectile electron with subsequent emission of several electrons.

901,573
PB90-123845 Not available NTIS
National Inst. of Standards and Technology (NML), Boulder, CO. Quantum Physics Div.
Electron-Impact Ionization of La(q+) Ions (q=1,2,3).
Final rept.
A. Mueller, K. Tinschert, G. Hofmann, E. Salzborn, G. H. Dunn, S. M. Younger, and M. S. Pindzola. 1989, 15p
Contract DE-AI05-86ER53237
Sponsored by Department of Energy, Washington, DC. Office of Energy Research.
Pub. in Physical Review A 40, n7 p3584-3598, 1 Oct 89.

Keywords: Excitation, Reprints, *Electron ion collisions, *Lanthanum ions, Ionization cross sections,

PHYSICS

General

Multicharged ions, Colliding beams, Electron impact, Autoionization.

Experimental cross sections for single ionization of La(1+), La(2+), and La(3+) ions as well as for double ionization of La(1+) and La(2+) ions and for triple ionization of La(2+) ions are presented in an electron-impact energy range from threshold up to 1000 eV. By using a fast energy-scanning technique with step widths of about 40 meV and an energy resolution of 0.4 eV, fine details in the cross sections and especially resonant contributions could be observed. Resonant excitation of La(2+) at 95 eV with subsequent emission of three electrons appears in the authors' interpretation to account for 30% of the total measured double-ionization cross section. The data are interpreted on the basis of calculations of energy levels and distorted-wave calculations of direct ionization from various electron sub-shells and of excitation-autoionization processes. A detailed analysis of the La(3+) single ionization both for ground-state and 5p(5)4f metastable ions is presented.

901,574
PB90-123902   Not available NTIS
National Inst. of Standards and Technology (NML), Gaithersburg, MD. Ionizing Radiation Physics Div.
Applications of ETRAN Monte Carlo Codes.
Final rept.
S. M. Seltzer. 1988, 26p
Sponsored by Department of Energy, Washington, DC.
Pub. in Monte Carlo Transport of Electrons and Photons, Chapter 9, p221-246 1988.

Keywords: Monte Carlo method, Bremsstrahlung, Electrons, Radiation shielding, Aerospace environment, Reprints, *ETRAN computer code, *Radiation transport, Electron transport, Gamma detection, Liquid scintillation detectors, Energy losses, Uses.

Some applications of ETRAN Monte Carlo calculations to radiation transport problems are described. These include calculations of the response of gamma-ray detectors, the shielding from the dose due to electrons and bremsstrahlung in space-radiation environments, the characteristics of bremsstrahlung beams for use in radiation processing, and the distortion of energy-loss spectra due to energy lost in the walls of containers used in liquid-scintillation counting of beta emitters.

901,575
PB90-123910   Not available NTIS
National Inst. of Standards and Technology (NML), Gaithersburg, MD. Ionizing Radiation Physics Div.
Cross Sections for Bremsstrahlung Production and Electron-Impact Ionization.
Final rept.
S. M. Seltzer. 1988, 34p
Sponsored by Department of Energy, Washington, DC.
Pub. in Monte Carlo Transport of Electrons and Photons, Chapter 4, p81-114 1988.

Keywords: *Bremsstrahlung, *Ionization, Positrons, Reprints, *Electron impact, *Electron transport, Ionization cross sections, Electron-atom collisions, Electron-molecule collisions.

The paper discusses the cross section for two processes that are important in electron transport calculations: bremsstrahlung production by electrons and positrons in the fields of atomic nuclei and orbital electrons, and electron-impact ionization of atoms and molecules. In the discussion of bremsstrahlung, the author outlines the synthesis of various theoretical results that were involved in the development of a comprehensive set of cross sections, differential in emitted photon energy, for the production of bremsstrahlung by electrons. Illustrates the generally good agreement between these cross sections and the results of measurement, and indicates the differences in the corresponding positron cross sections. In the case of electron-impact ionization, the author describes exploratory calculations of the shell-by-shell cross section, differential in ejected electron energy, based on a Weizsacker-Williams treatment.

901,576
PB90-123928   Not available NTIS
National Inst. of Standards and Technology (NML), Gaithersburg, MD. Ionizing Radiation Physics Div.
Overview of ETRAN Monte Carlo Methods.
Final rept.
S. M. Seltzer. 1988, 29p
Sponsored by Department of Energy, Washington, DC.
Pub. in Monte Carlo Transport of Electrons and Photons, Chapter 7, p153-181 1988.

Keywords: Monte Carlo method, Bremsstrahlung, Cross sections, Sampling, Reprints, *ETRAN computer code, *Radiation transport, Electron transport.

The paper outlines the sampling methods and the underlying cross sections used in the ETRAN Monte Carlo codes for coupled electron/photon transport calculations. The structure, capabilities, and limitations of current versions are briefly discussed, and some future improvements are indicated.

901,577
PB90-123936   Not available NTIS
National Inst. of Standards and Technology (NML), Boulder, CO. Quantum Physics Div.
Collisional Losses from a Light-Force Atom Trap.
Final rept.
D. Sesko, T. Walker, C. Monroe, A. Gallagher, and C. Weiman. 1989, 4p
Grant NSF-PHY86-04504
Sponsored by National Science Foundation, Washington, DC., and Office of Naval Research, Arlington, VA.
Pub. in Physical Review Letters 63, n9 p961-964, 28 Aug 89.

Keywords: Energy transfer, Reprints, *Atom collisions, *Atom traps, Cesium atoms, Energy losses.

The authors have studied the collisional loss rates for very cold cesium atoms held in a spontaneous-force optical trap. In contrast with previous work, it was found that collisions involving excitation by the trapping light fields are the dominant loss mechanism. It was also found that hyperfine-changing collisions between atoms in the ground state can be significant under some circumstances.

901,578
PB90-128034   Not available NTIS
National Inst. of Standards and Technology (NML), Boulder, CO. Time and Frequency Div.
Ion Trapping Techniques: Laser Cooling and Sympathetic Cooling.
Final rept.
J. J. Bollinger, L. R. Brewer, J. C. Bergquist, W. M. Itano, D. J. Larson, S. L. Gilbert, and D. J. Wineland. 1988, 11p
Sponsored by Air Force Office of Scientific Research, Boiling AFB, DC., and Office of Naval Research, Arlington, VA.
Pub. in Proceedings of Workshop on Intense Positron Beams, Idahofalls, Idaho, June 18-19, 1987, p63-73 1988.

Keywords: Radiation pressure, Positrons, *Ion trapping, *Laser cooling, *Beryllium ions, *Mercury ions, *Sympathetic cooling, Penning traps.

Radiation pressure from lasers has been used to cool and compress (9)Be(1+) ions stored in a combination of static electric and magnetic fields (Penning trap) to temperatures less than 10 mK and densities greater than 10 to the 7th power/cm(-3) in a magnetic field of 1.4 T. A technique called sympathetic cooling can be used to transfer this cooling and compression to other ion species. An example of (198)Hg(1+) ions sympathetically cooled by laser cooled (9)Be(1+) ions is given. The possibility of making an ultracold positron source via sympathetic cooling is also discussed.

901,579
PB90-128075   Not available NTIS
National Inst. of Standards and Technology (NML), Boulder, CO. Quantum Physics Div.
Redistribution in Astrophysically Important Hydrogen Lines.
Final rept.
J. Cooper, R. J. Ballagh, and I. Hubeny. 1989, 30p
Grant NGL-06-003-057
Sponsored by National Aeronautics and Space Administration, Washington, DC.
Pub. in Proceedings of International Conference on Spectral Line Shapes (9th), Torun, Poland, July 25-29, 1988, v5 p275-304 1989.

Keywords: *Hydrogen, Line spectra, Emission spectra, Absorption spectra, *Lyman lines.

Theory is specifically developed for the coupled Ly-alpha, Ly-beta, H-alpha system, and equations of statistical equilibrium and absorption and emission coefficients are given. All correlated events are examined and emission during a collision is found to be important in the line wings. Stimulated emission and absorption is also included within a broadband approximation. The major approximation is also included within a

broadband approximation. The major approximation, adopted for convenience, is to ignore lower state interaction. (For H-alpha estimated errors in the redistribution formulae from this approximation are the order of 20% in the line center and the order of 40% in the wings.)

901,580
PB90-128083   Not available NTIS
National Inst. of Standards and Technology (NML), Gaithersburg, MD. Quantum Metrology Div.
Performance of a High-Energy-Resolution, Tender X-ray Synchrotron Radiation Beamline.
Final rept.
P. L. Cowan, S. Brennan, T. Jach, D. W. Lindle, and B. A. Karlin. 1989, 5p
Pub. in Review of Scientific Instruments 60, n7 p1603-1607 Jul 89.

Keywords: Reprints, *NSLS, *Beamlines, Synchrotron radiation sources, X-ray sources, High resolution.

Beamline X-24A at the National Synchrotron Light Source was designed for optimal performance in the x-ray spectral region 500-5000eV. This choice of energy range placed a number of constraints on the beamline design, requiring a crystal monochromator in a windowless UHV environment. Although this increased the complexity of the design, there were compelling scientific reasons for the desire to work in this range. In addition to tunability over the selected energy range, a primary goal was to obtain the highest possible energy resolution in the primary beam. The authors have achieved incident energy resolution significantly better than the typical core-level lifetime broadening for this energy range. This has permitted studies of processes that are not broadened by lifetimes, such as resonant scattering and back-reflection x-ray standing-wave effects. In addition to high resolution, it was designed to collect and focus as much flux as possible from the bending magnet source.

901,581
PB90-128091   Not available NTIS
National Inst. of Standards and Technology (NML), Boulder, CO. Time and Frequency Div.
Laser Cooling to the Zero-Point Energy of Motion.
Final rept.
F. Diedrich, J. C. Bergquist, W. M. Itano, and D. J. Wineland. 1989, 4p
Pub. in Physical Review Letters 62, n4 p403-406, 23 Jan 89.

Keywords: Atomic spectroscopy, Atomic structure, Reprints, *Laser cooling, *Mercury ions, Trapping(Charged particles), Laser spectroscopy, High resolution.

A single trapped (198)Hg(1+) ion was cooled by scattering laser radiation that was tuned to the resolved lower motional sideband of the narrow doublet S(1/2) - doublet D(5/2) transition. The different absorption strengths on the upper and lower sidebands after cooling indicated that the ion was in the ground state of its confining well approximately 95% of the time.

901,582
PB90-128257   Not available NTIS
National Inst. of Standards and Technology (NML), Gaithersburg, MD. Atomic and Plasma Radiation Div.
Branching Ratio Technique for Vacuum UV Radiance Calibrations: Extensions and a Comprehensive Data Set.
Final rept.
J. Z. Klose, and W. L. Wiese. 1989, 17p
Pub. in Jnl. of Quantitative Spectroscopy and Radiative Transfer 42, n5 p337-353 1989.

Keywords: Spectral lines, Transition probabilities, Reprints, *Vacuum ultraviolet radiation, *Branching ratio, *Multicharged ions, *Calibration, High temperature, Plasma.

The branching-ratio technique for calibrations in the VUV is reviewed in detail. The basic method is described, followed by extensions and applications. Lists of transitions suitable for the technique are given for H-, He-, Li-, and Be-like ions, along with pertinent data for their application.

901,583
PB90-128299   Not available NTIS
National Inst. of Standards and Technology (NML), Gaithersburg, MD. Radiation Physics Div.

184

Progress on Spin Detectors and Spin-Polarized Electron Scattering from Na at NIST.
Final rept.
J. J. McClelland. 1989, 13p
Sponsored by Department of Energy, Washington, DC.
Pub. in Proceedings of Symposium on Polarization and Correlation in Electronic and Atomic Collisions, Hoboken, NJ., p1-13 Aug 89.

Keywords: *Electron scattering, *Thorium, Electron spin, Detectors, *Sodium atoms, *Electron spin polarization, Sherman functions, S matrix.

Recent progress in the Electron Physics Group at NIST is discussed. Improvements have been made on the low-energy diffuse-scattering spin analyzer, reducing instrumental asymmetries and boosting the effective Sherman function. A figure of merit of 0.00023 has been achieved. Thorium has been used as a target in a 100 ke V retarding Mott spin analyzer, resulting in an effective Sherman function as high as 0.49. This increased Sherman function, together with an increased scattering intensity, results in a factor of 2 increase in the figure of merit. Good agreement is seen between experiment and theoretical predictions of the Sherman function for thorium. A hierarchical description of the T-matrix is discussed as a context for interpreting recent results on spin-polarized electron scattering from optically pumped sodium. Results are presented for elastic and superelastic scattering at 20 eV incident energy.

901,584
PB90-128307   Not available NTIS
National Inst. of Standards and Technology (NML), Gaithersburg, MD. Radiation Physics Div.
Superelastic Scattering of Spin-Polarized Electrons from Sodium.
Final rept.
J. J. McClelland, M. H. Kelley, and R. J. Celotta.
1989, 9p
Sponsored by Department of Energy, Washington, DC.
Pub. in Physical Review A 40, n5 p2321-2329, 1 Sep 89.

Keywords: *Electron scattering, Polarization(Spin alignment), Reprints, *Sodium atoms, *Electron spin polarization, EV range 01-10, EV range 10-100.

Superelastic scattering of spin-polarized electrons from laser-excited sodium atoms has been measured at incident energies of 2.0, 17.9, and 52.3 eV over the angular range 10 deg - 120 deg. Circularly polarized excitation of the sodium atoms was used to produce pure 3 doublet P(3/2) (F=3,Mf=plus or minus 3) states, which are deexcited by collisions with spin-polarized electrons. The spin polarization of both the target electron and the incident electron allows the resolution of triplet and singlet contributions to L(1), the angular momentum transferred in the collision perpendicular to the scattering plane, and the measurement of r, the ratio of triplet to singlet cross sections. At low energy, agreement with theory is good over the entire angular range for r, but only at small angles for L(1). At high energy, agreement is excellent over the full angular range.

901,585
PB90-128513   Not available NTIS
National Inst. of Standards and Technology (NML), Gaithersburg, MD. Electricity Div.
Drift Tubes for Characterizing Atmospheric Ion Mobility Spectra.
Final rept.
M. Misakian, W. Anderson, and O. Laug. 1989, 4p
Sponsored by Department of Energy, Washington, DC.
Pub. in Proceedings of International Symposium on High Voltage Engineering (6th), New Orleans, LA., August 28-September 1, 1989, p1-4.

Keywords: *Ionic mobility, *Atmospheric pressure, *Spectra, Measurement, Alternating current, Direct current, Transport properties, Atmospheric tides, Electric fields, Spectral emittance, *Drift tubes.

Two drift tubes constructed of insulating cylinders with conductive guard rings on the inside walls are examined to determine their suitability for measuring ion mobility spectra at atmospheric pressure. One drift tube is of the pulse time-of-flight (TOF) type with adjustable drift distance and the other is an ac-TOF drift tube similar in principle to a device reported by Van de Graaff. The latter drift tube is evaluated using sinusoidal and alternating-polarity pulse-voltage waveforms for gating the shutters.

901,586
PB90-133158   PC A16/MF A02

National Inst. of Standards and Technology (NML), Gaithersburg, MD. Center for Atomic, Molecular and Optical Physics.
Center for Atomic, Molecular, and Optical Physics Technical Activities, 1989.
K. B. Gebbie. Dec 89, 364p NISTIR-89/4184
See also PB89-132302.

Keywords: *Atomic physics, *Physical chemistry, Fundamental constants, Synchrotron radiation, Plasmas(Physics), Atomic structure, Surface chemistry, Molecular spectroscopy, Quantum theory, Gamma rays, Metrology, Frequency standards, Time standards, Astrophysics, Lasers, Gravity, *Molecular physics, Laser cooling, Atom traps, Soft x-rays, Surface reactions, Calibration.

The report summarizes the research and technical activities of the Center for Atomic, Molecular and Optical Physics (CAMOP) during the Fiscal Year 1989. The activities include work in the areas of fundamental constants, radiation physics, surface science, molecular spectroscopy, electron and optical physics, atomic and plasma spectroscopy, time and frequency, quantum metrology, and quantum physics.

901,587
PB90-136797   Not available NTIS
National Inst. of Standards and Technology (NML), Gaithersburg, MD. Chemical Thermodynamics Div.
Monte Carlo Simulation of Domain Growth in the Kinetic Ising Model on the Connection Machine.
Final rept.
J. G. Amar, and F. Sullivan. 1989, 9p
Pub. in Computer Physics Communications 55, p287-295 1989.

Keywords: *Ferromagnetism, *Computerized simulation, Mathematical models, Monte Carlo method, Algorithms, Reprints, Computer applications, Parallel processing.

A fast multispin algorithm for the Monte Carlo simulation of the two-dimensional spin-exchange kinetic Ising model has been adapted for use on the Connection Machine and applied as a first test in a calculation of domain growth. Features of the code include: the use of demon bits, the simulation of several runs simultaneously to improve the efficiency of the code, the use of virtual processors to simulate easily and efficiently a larger system size, the use of the (NEWS) grid for fast communication between neighboring processors and updating of boundary layers, the implementation of an efficient random number generator much faster than that provided by Thinking Machines Corp., and the use of the LISP function 'funcall' to select which processors to update. Overall speed of the code when run on a (128 X 128) processor machine is about 130 million attempted spin-exchanges per second, about 9 times faster than the comparable code, using hardware vectorized-logic operations and 64-bit multispin coding on the Cyber 205. The same code can be used on a larger machine (65536 processors) and should produce speeds in excess of 500 million attempted spin-exchanges per second.

901,588
PB90-163924
                        (Order as PB90-163874, PC A04)
National Inst. of Standards and Technology, Gaithersburg, MD.
Measuring the Root-Mean-Square Value of a Finite Record Length Periodic Waveform.
E. C. Teague. 1989, 5p
Included in Jnl. of Research of the National Institute of Standards and Technology, v94 n6 p367-371 1989.

Keywords: *Waveforms, *Surface roughness, Measurement, Profiles, Random error, Root mean square value, Uncertainty.

The paper presents a discussion of the uncertainty in measuring the root-mean-square, rms, value of a periodic waveform which results from the use of a finite record length. The analysis was motivated by seeking to understand the source of a random uncertainty component which was present in some measurements of the absolute arithmetic average, R sub a deviation from a mean line of profiles of precision roughness specimens. The profiles of these specimens had an approximately triangular waveform with two wavelengths and amplitudes. For the longer wavelength specimens the random phasing of the waveform with respect to the recording interval proved to be a major source of uncertainty in the measurements.

SPACE TECHNOLOGY

Manned Spacecraft

901,589
PB89-193940   PC A05/MF A01
National Inst. of Standards and Technology (NEL), Gaithersburg, MD. Robot Systems Div.
NASA/NBS (National Aeronautics and Space Administration/National Bureau of Standards) Standard Reference Model for Telerobot Control System Architecture (NASREM).
Technical note (Final).
J. S. Albus, H. G. McCain, and R. Lumia. Apr 89, 85p NIST/TN-1235-89
Also available from Supt. of Docs. as SN003-003-02928-9. See also PB88-123773. Sponsored by National Aeronautics and Space Administration, Greenbelt, MD. Goddard Space Flight Center.

Keywords: *Robots, Control systems, Space stations, System analysis, Functional analysis, Computer communications, Interfaces, Memory devices, *Computerized control systems, Hierarchical control, NASA standard reference model.

The document describes the NASA Standard Reference Model (NASREM) Architecture for the Space Station Telerobot Control System. It defines the functional requirements and highlevel specifications of the control system for the NASA Space Station document for the functional specification, and a guideline for the development of the control system architecture, of the IOC Flight Telerobot Servicer. The NASREM telerobot control system architecture defines a set of standard modules and interfaces which facilitate software design, development, validation, and test, and make possible the integration of telerobotics software from a wide variety of sources. Standard interfaces also provide the software hooks necessary to incrementally upgrade future Flight Telerobot Systems as new capabilities develop in computer science, robotics, and autonomous system control.

901,590
PB89-231013   Not available NTIS
National Inst. of Standards and Technology (NEL), Gaithersburg, MD. Center for Fire Research.
Expert Systems Applied to Spacecraft Fire Safety.
Final rept.
R. L. Smith, and T. Kashiwagi. 1989, 12p
Contract NASA-C-32000-M
Sponsored by National Aeronautics and Space Administration, Cleveland, OH. Lewis Research Center.
Pub. in NASA Contractor Report 182266, p1-12, Jun 89.

Keywords: *Fire safety, Fire detection systems, Fire protection, Space stations, Fire extinguishers, Flammability, Reduced gravity, Ventilation, Air Flow, Decisions, Reprints, *Spacecraft electronic equipment, *Expert systems, Knowledge bases(Artificial intelligence).

Expert systems are problem-solving programs that combine a knowledge base and a reasoning mechanism to simulate a human 'expert.' The development of an expert system to manage fire safety in spacecraft, in particular the NASA Space Station Freedom, is difficult but clearly advantageous in the long-term. The report discusses some needs in low-gravity flammability characteristics, ventilating-flow effects, fire detection, fire extinguishment, and decision models, all necessary to establish the knowledge base for an expert system.

901,591
PB90-123811   Not available NTIS
National Inst. of Standards and Technology (NEL), Gaithersburg, MD. Robot Systems Div.
Teleoperation and Autonomy for Space Robotics.
Final rept.
R. Lumia, and J. S. Albus. 1988, 7p
Pub. in Robotics 4, n1 p27-33 Mar 88.

SPACE TECHNOLOGY

Manned Spacecraft

Keywords: *Manned space flight, *Robots, Technology transfer, Standards, Reprints, Teleoperators, Control systems, Autonomy.

A logical enhancement to manned space flight includes the use of robots in space. To achieve this goal, there must be a phased program where the capabilities of the robot can evolve as technology advances. The present paper will review some of the ways in which robots can be used in space. Then, a system architecture standard will be suggested which supports the evolution of robot control from teleoperation to autonomy. Finally, some areas of technology transfer will be discussed which are relevant to land-based robot operation.

Spacecraft Trajectories & Flight Mechanics

901,592
PB89-156962   Not available NTIS
National Bureau of Standards (NML), Gaithersburg, MD. Inorganic Analytical Research Div.
Detection of Uranium from Cosmos-1402 in the Stratosphere.
Final rept.
R. Leifer, Z. R. Juzdan, W. R. Kelly, J. D. Fassett, and K. R. Eberhardt. 1987, 3p
Pub. in Science 238, n4826 p512-514, 23 Oct 87.

Keywords: *Atmospheric entry, Stratosphere, Uranium, Mass spectroscopy, Ablation, Reprints, *Space power reactors, *Cosmos 1402 satellite, Spacecraft power supplies.

A series of balloon flights were launched in 1984 to intercept the debris from the ablation of Cosmos 1402 which unexpectedly re-entered the earth's atmosphere February 7, 1983. Based on isotopic uranium analyses of filters collected between 26 and 36 km in February and March 1984, the authors are able to show unequivocally that the reactor from the Russian satellite cosmos 1402 did in fact burn in the high stratosphere.

Unmanned Spacecraft

901,593
PB89-156152   PC A05/MF A01
National Bureau of Standards (NEL), Boulder, CO. Center for Electronics and Electrical Engineering.
Development of Near-Field Test Procedures for Communication Satellite Antennas, Phase 1, Part 2.
A. C. Newell. Aug 88, 93p NBSIR-87/3061
See also Part 1, PB86-164357.

Keywords: *Spacecraft antennas, Electromagnetic fields, Antennas, Tests, Measurement, Communication satellites, Near field.

The purpose of the program is to define and further develop the capabilities of near-field antenna test techniques, specifically for the requirements associated with the development and verification testing of reconfigurable, multibeam, frequency reuse, commercial satellite antennas. The report, Phase I, Part 2, focuses on the planar near-field measurement method and covers the determination of sampling criteria and scan limits, development of diagnostic and design assist methods, development of beam alignment techniques, development of swept-frequency equivalent tests, and specification of hardware requirements for the measurement system. The basis for the choice of the best measurement technique was established with the planar near-field measurement method receiving the best score for the directive antennas considered.

901,594
PB89-234231   Not available NTIS
National Inst. of Standards and Technology (NML), Boulder, CO. Quantum Physics Div.

Antenna for Laser Gravitational-Wave Observations in Space.
Final rept.
J. E. Faller, P. L. Bender, J. L. Hall, D. Hils, R. T. Stebbins, and M. A. Vincent. 1989, 5p
Contract NAGW-822
Sponsored by National Aeronautics and Space Administration, Washington, DC.
Pub. in Advances in Space Research 9, n9 p(9)107-(9)111 1989.

Keywords: *Astronomical observatories, Continuous radiation, Sensitivity, Reprints, *Gravitational waves, *Gravitational wave antennas, Infrasonic frequencies.

Progress during the past two years on a proposed Laser Gravitational-Wave Observatory in Space (LAGOS) is discussed. Calculated performance for a 1 million km sized antenna over the frequency range of 0.00001 to 1 Hz is given. The sensitivity from 0.001 to 0.1 Hz is expected to be 1 x 10 to the -21st power/(Hz to the 0.5 power). Noise sources such as accelerations of the drag-free test masses by random molecular impacts and by fluctuations in the net thermal radiation pressure will limit the sensitivity at lower frequencies. The scientific objectives are the observation of CW gravitational waves from large numbers of binary systems and the detection of pulses which may have been emitted during the period of galaxy formation.

901,595
PB89-234249   Not available NTIS
National Inst. of Standards and Technology (NML), Boulder, CO. Quantum Physics Div.
Conceptual Design for a Mercury Relativity Satellite.
Final rept.
P. L. Bender, N. Ashby, M. A. Vincent, and J. M. Wahr. 1989, 4p
Contract NAGW-822
Sponsored by National Aeronautics and Space Administration, Washington, DC.
Pub. in Advances in Space Research 9, n9 p(9)113-(9)116 1989.

Keywords: *Gravitation, *Relativity, Polar orbits, Spacecraft tracking, Celestial mechanics, Mercury(Planet), Doppler effect, Tests, Reprints, *Mercury spacecraft, Orbit calculation, Satellite design.

It was shown earlier that 1 x 10 to the -14th power Doppler data and 3 cm accuracy range measurements to a small Mercury Relativity Satellite in a polar orbit with 4-hour period can give high-accuracy tests of gravitational theory. A particular conceptual design has been developed for such a satellite, which would take less than 10% of the approach mass for a possible future Mercury Orbiter Mission. The spacecraft is similar to the Pioneer Venus Orbiter, but scaled down by about a factor of 4 in linear dimensions. The orbit parameters for individual eight-hour arcs and the gravity field of Mercury through degree and order 10 are determined mainly from the Doppler data. A 50 MHz K-band sidetone system provides the basic ranging accuracy. The spacecraft mass is 50 kg or less.

TRANSPORTATION

Railroad Transportation

901,596
PB89-189229   PC A05/MF A01
National Inst. of Standards and Technology (IMSE), Boulder, CO. Fracture and Deformation Div.
Ultrasonic Railroad Wheel Inspection Using EMATs (Electromagnetic-Acoustic Transducers), Report No. 18.
R. E. Schramm, and A. V. Clark. Dec 88, 86p NISTIR-88/3906
See also PB88-194519. Sponsored by Federal Railroad Administration, Washington, DC.

Keywords: *Ultrasonic testing, *Railroad cars, *Wear, Cracks, Nondestructive tests, Transducers, Residual stress, Birefringence, Wheels, EMAT, Roll-by inspection.

The report is number 18 in a series covering the research performed by the National Institute of Standards and Technology (formerly National Bureau of Standards) for the Federal Railroad Administration. The issue collects seven reprints and preprints of papers written by the Fracture and Deformation Division over the last two years on the ultrasonic nondestructive evaluation of railroad wheels for the presence of residual stress and cracks. All the work concentrated on the use of electromagnetic-acoustic transducers (EMATs). Tensile residual stresses and tread cracks are major factors in wheel failure. Two ultrasonic techniques are applicable to these wear defects: (1) Birefringence: A stress field effects the velocity of a shear horizontal wave depending on its polarization. Precise velocity measurements in a wheel rim may allow calculation of the amount and direction of stresses; (2) Pulse-echo: A Rayleigh (surface) wave transducer mounted inside the rail can introduce a signal to interrogate the circumference of a wheel as it rolls by. An echo indicates a flaw's presence and size.

901,597
PB90-123894   Not available NTIS
National Inst. of Standards and Technology (IMSE), Boulder, CO. Fracture and Deformation Div.
EMATs (Electromagnetic Acoustic Transducers) for Roll-By Crack Inspection of Railroad Wheels.
Final rept.
R. E. Schramm, P. J. Shull, A. V. Clark, and D. V. Mitrakovic. 1989, 7p
Sponsored by Federal Railroad Administration, Washington, DC.
Pub. in Review of Progress in Quantitative Nondestructive Evaluation, v8A p1063-1089 1989.

Keywords: *Cracks, *Inspection, *Rolling stock, *Wheels, Nondestructive tests, Ultrasonic tests, Railroad cars, Rayleigh waves, Safety, Reprints.

Railroad safety depends on many factors. The integrity of the wheels on rolling stock is one that is subject to nondestructive evaluation. For some years, ultrasonic testing has been applied to the detection of cracks in wheel treads, with particular attention to automatic, in-rail, roll-by methods. A system using relatively low frequency Rayleigh waves generated by electromagnetic-acoustic transducers (EMATs) is being developed. The current design uses a permanent magnet to maintain a compact structure and minimize the size of the pocket machined into the rail. Measurements thus far indicate a responsiveness, even to small flaws. With the development of a signal processing and analysis system, field tests should soon be possible.

URBAN & REGIONAL TECHNOLOGY & DEVELOPMENT

Emergency Services & Planning

901,598
PB89-189203   PC A10/MF A01
National Inst. of Standards and Technology (NEL), Gaithersburg, MD. Center for Computing and Applied Mathematics.
Evaluating Emergency Management Models and Data Bases: A Suggested Approach.
R. E. Chapman, S. I. Gass, J. J. Filliben, and C. M. Harris. Mar 89, 215p NISTIR-88/3826
Prepared in cooperation with George Mason Univ., Fairfax, VA. Dept. of Operations Research and Applied Statistics. Sponsored by Federal Emergency Management Agency, Washington, DC.

Keywords: *Mathematical models, Evaluation, Assessments, Management, Tables(Data), Graphs(Charts), *Emergency planning, Emergency preparedness, Crises, Data bases.

Large-scale models and data bases are key informational resources for the Federal Emergency Management Agency (FEMA). In order to carry out its emergency missions, it is necessary for FEMA to determine which models, modeling techniques and data bases are appropriate for what purposes and which ones need modification, updating and maintenance. The development of evaluation guidelines is therefore of direct benefit to FEMA in discharging its emergency management duties. The purpose of the report is two-fold. First, it provides the reader with a generic set of guidelines which can be used to evaluate large-scale, computer-based models and data bases. Second, the guidelines are illustrated through a critical evaluation of the Dynamic General Equilibrium Model (DGEM). DGEM is currently being used by FEMA to analyze a variety of emergency management problems. The evaluation of DGEM serves both, as a step-by-step procedure for conducting an indepth model evaluation and as an introduction to a non-proprietary model which has broad applicability to the analysis of macroeconomic issues.

Fire Services, Law Enforcement, & Criminal Justice

901,599

PB89-176283   Not available NTIS
National Bureau of Standards (NEL), Gaithersburg, MD. Law Enforcement Standards Lab.
ACSB (Amplitude Companded Sideband): What Is Adequate Performance.
Final rept.
W. A. Kissick, and M. J. Treado. 1986, 9p
Sponsored by National Inst. of Justice, Washington, DC.
Pub. in Proceedings of International Carnahan Conference on Security Technology: Electronic Crime Countermeasures, Gothenburg, Sweden, August 12-14, 1986, p219-227.

Keywords: *Compandor transmission, *Law enforcement, *Voice communication, Single sideband transmission, Modulation, Speech recognition, Intelligibility,

Performance evaluation, Amplitude companded sideband.

Amplitude companded sideband (ACSB) is a new modulation technique that uses a much smaller channel width than does conventional frequency modulation (FM). ACSB has been proposed for the land mobile communications needs of law enforcement agencies. Among the requirements of such a communications system is adequate speech intelligibility under a variety of conditions. The paper explores this aspect of 'adequate performance.' First, the basic principles of ACSB are described, with emphasis on those features that affect speech quality. Second, the results of ACSB equipment testing are given. Next, the appropriate performance measures for ACSB are reviewed. Last, a subjective voice quality scoring method is used to determine the values of the performance measures that equate to the minimum level of intelligibility. It is assumed that the intelligibility of an FM system operating at 12 dB SINAD represents that minimum.

PERSONAL AUTHOR INDEX

SAMPLE ENTRY

Wack, J. P., and Carnahan, L. J.
Computer Viruses and Related Threats: A Management
Guide
PB90-111683   900,654

Author name(s)
Title

NTIS order number

Abstract number

AARON, D. A.
Experimental Investigation and Modeling of the Flow Rate
of Refrigerant 22 Through the Short Tube Restrictor.
PB89-229041   901,118

ABBOTT, D. C.
Photospheres of Hot Stars. 3. Luminosity Effects at Spectral Type 09.5.
PB89-202592   900,020

ABOU-GAMRA, Z.
Redox Chemistry of Water-Soluble Vanadyl Porphyrins.
PB89-150999   900,300

ABRAHAMS, S. C.
Statistical Descriptors in Crystallography: Report of the
International Union of Crystallography Subcommittee on
Statistical Descriptors.
PB89-201826   901,432

ABRISHAMIAN, A.
Hierarchically Controlled Autonomous Robot for Heavy Payload Military Field Applications.
PB89-177075   901,271

ABRISHAMIAN, H.
Robot Crane Technology.
PB90-111667   900,146

ABU-ZAID, M.
Effect of Water on Piloted Ignition of Cellulosic Materials.
PB89-189187   900,127

ADAIR, R. T.
ANA (Automatic Network Analyzer) Measurement Results
on the ARFTG (Automatic RF Techniques Group) Traveling
Experiment.
PB89-173777   900,715

Performance Evaluation of Radiofrequency, Microwave, and
Millimeter Wave Power Meters.
PB89-193916   900,814

ADAM, G.
On-Line Arc Welding: Data Acquisition and Analysis Using a
High Level Scientific Language.
PB90-117391   900,972

ADAMS, F. C.
Dependence of Interface Widths on Ion Bombardment Conditions in SIMS (Secondary Ion Mass Spectrometry) Analysis of a Ni/Cr Multilayer Structure.

PB89-172506   900,364

ADAMS, J. W.
Electromagnetic Fields in Loaded Shielded Rooms.
PB89-180426   900,780

Measurement Procedures for Electromagnetic Compatibility
Assessment of Electroexplosive Devices.
PB89-146914   901,314

Techniques for Measuring the Electromagnetic Shielding
Effectiveness of Materials. Part 1. Far-Field Source Simulation.
PB89-161525   900,680

ADRIAN, F. J.
Magnetic Field Dependence of the Superconductivity in Bi-
Sr-Ca-Cu-O Superconductors.
PB89-146815   901,385

AGRON, E.
Mathematical Software: PLOD. Plotted Solutions of Differential Equations.
PB89-147425   901,194

AHLUWALIA, J. C.
Thermodynamics of the Hydrolysis of Sucrose.
PB89-227904   901,227

AL-SHEIKHLY, M.
Dichromate Dosimetry: The Effect of Acetic Acid on the Radiolytic Reduction Yield.
PB89-147490   900,248

ALBER, G.
One-Photon Resonant Two-Photon Excitation of Rydberg
Series Close to Threshold.
PB89-171276   901,343

ALBERS, J. H.
Experimental Verification of the Relation between Two-
Probe and Four-Probe Resistances.
PB89-231211   900,794

ALBERTY, R. A.
Standard Chemical Thermodynamic Properties of Polycyclic
Aromatic Hydrocarbons and Their Isomer Groups 1. Benzene Series.
PB89-186480   900,412

Standard Chemical Thermodynamic Properties of Polycyclic
Aromatic Hydrocarbons and Their Isomer Groups. 2.

Pyrene Series, Naphthopyrene Series, and Coronene
Series.
PB89-226591   900,459

ALBUS, J.
Mining Automation Real-Time Control System Architecture
Standard Reference Model (MASREM).
PB89-221154   901,286

ALBUS, J. S.
NASA/NBS (National Aeronautics and Space Administration/National Bureau of Standards) Standard Reference
Model for Telerobot Control System Architecture
(NASREM).
PB89-193940   901,589

Robot Crane Technology.
PB90-111667   900,146

Teleoperation and Autonomy for Space Robotics.
PB90-123611   901,591

ALFASSI, Z. B.
Absolute Rate Constants for Hydrogen Abstraction from
Hydrocarbons by the Trichloromethylperoxyl Radical.
PB89-171532   900,357

ALLAN, D. W.
Comparison of Time Scales Generated with the NBS (National Bureau of Standards) Ensembling Algorithm.
PB89-174072   900,628

Dual Frequency P-Code Time Transfer Experiment.
PB89-174064   900,614

In Search of the Best Clock.
PB90-117367   900,632

Millisecond Pulsar Rivals Best Atomic Clock Stability.
PB89-185722   900,629

NBS (National Bureau of Standards) Calibration Service
Providing Time and Frequency at a Remote Site by Weighting and Smoothing of GPS (Global Positioning System)
Common View Data.
PB89-212211   900,631

NIST Automated Computer Time Service.
PB90-117311   900,676

Study of Long-Term Stability of Atomic Clocks.
PB89-174098   900,367

PA-6

PB89-202246 *900,214*

EVANS, J. M.
Assessment of Robotics for Improved Building Operations and Maintenance.
PB89-189146 *900,092*

EVENSON, K. M.
(12)C(16)O Laser Frequency Tables for the 34.2 to 62.3 THz (1139 to 2079 cm(-1)) Region.
PB89-193908 *901,361*

CO Laser Stabilization Using the Optogalvanic Lamb-Dip.
PB89-179139 *901,356*

Detection of the Free Radicals FeH, CoH, and NiH by Far Infrared Laser Magnetic Resonance.
PB90-117342 *900,495*

Far-Infrared Laser Magnetic Resonance Spectrum of the CD Radical and Determination of Ground State Parameters.
PB90-117359 *900,496*

Far-Infrared Laser Magnetic Resonance Spectrum of Vibrationally Excited C2H(1).
PB89-147474 *900,292*

Frequency Measurement of the J = 1 < - 0 Rotational Transition of HD (Hydrogen Deuteride).
PB89-161565 *901,499*

Heterodyne Frequency Measurements of (12)C(16)O Laser Transitions.
PB89-229223 *901,374*

New FIR Laser Lines and Frequency Measurements for Optically Pumped CD3OH.
PB89-175731 *901,350*

EWING, M. B.
Microwave Measurements of the Thermal Expansion of a Spherical Cavity.
PB89-147458 *900,291*

FABER, W.
Use of Focusing Supermirror Neutron Guides to Enhance Cold Neutron Fluence Rates.
PB89-171946 *901,306*

FAETH, G. M.
Structure and Radiation Properties of Large-Scale Natural Gas/Air Diffusion Flames.
PB89-157572 *900,589*

FAHR, A.
Reactions of Phenyl Radicals with Ethene, Ethyne, and Benzene.
PB89-150908 *900,297*

FALLER, A.
High-Precision Absolute Gravity Observations in the United States.
PB89-227946 *901,281*

FALLER, J. E.
Antenna for Laser Gravitational-Wave Observations, in Space.
PB89-234231 *901,594*

Comment on 'Possible Resolution of the Brookhaven and Washington Eotvos Experiments'.
PB89-171581 *901,504*

Current Research Efforts at JILA (Joint Institute for Laboratory Astrophysics) to Test the Equivalence Principle at Short Ranges.
PB89-185912 *901,522*

Liquid-Supported Torsion Balance: An Updated Status Report on Its Potential for Tunnel Detection.
PB89-212062 *901,542*

Precision Experiments to Search for the Fifth Force.
PB89-228365 *901,551*

FALTYNEK, R. A.
Effect of pH on the Emission Properties of Aqueous tris (2,6-dipicolinato) Terbium (III) Complexes.
PB89-151155 *900,250*

Element-Specific Epifluorescence Microscopy in vivo Monitoring of Metal Biotransformations in Environmental Matrices.
PB89-177216 *901,220*

FANCONI, B. M.
Institute for Materials Science and Engineering, Polymers: Technical Activities 1987.
PB89-188601 *900,566*

Institute for Materials Science and Engineering, Polymers: Technical Activities 1986.
PB89-166094 *900,003*

FANG, Q. T.
Preparation of Multistage Zone-Refined Materials for Thermochemical Standards.
PB89-166795 *900,203*

FARABAUGH, E. N.
Cathodoluminescence of Defects in Diamond Films and Particles Grown by Hot-Filament Chemical-Vapor Deposition.
PB90-117961 *901,069*

FARAHANI, M.
Hydroxyl Radical Induced Cross-Linking between Phenylalanine and 2-Deoxyribose.
PB89-147029 *900,547*

Radiation-Induced Crosslinks between Thymine and 2-D-Deoxyerythropentose.

PB89-146682 *900,247*

FASSETT, J. D.
Analytical Applications of Resonance Ionization Mass Spectrometry (RIMS).
PB89-161590 *900,189*

Detection of Uranium from Cosmos-1402 in the Stratosphere.
PB89-156962 *901,592*

Determining Picogram Quantities of U in Human Urine by Thermal Ionization Mass Spectrometry.
PB89-146906 *900,175*

Development of the NBS (National Bureau of Standards) Beryllium Isotopic Standard Reference Material.
PB89-231070 *900,221*

Isotope Dilution Mass Spectrometry for Accurate Elemental Analysis.
PB89-230336 *900,220*

Resonance Ionization Mass Spectrometry of Mg: The 3pnd Autoionizing Series.
PB89-150817 *900,296*

FATIADI, A. J.
Facile Synthesis of 1-Nitropyrene-d9 of High Isotopic Purity.
PB90-123753 *900,240*

FAVREAU, J. P.
Application of Formal Description Techniques to Conformance Evaluation.
PB89-211906 *900,652*

Automatic Generation of Test Scenario (Skeletons) from Protocol-Specifications Written in Estelle.
PB89-177125 *900,615*

User Guide for the NBS (National Bureau of Standards) Prototype Compiler for Estelle (Revised).
PB89-196158 *900,619*

FAWZY, W. M.
Rotational Energy Levels and Line Intensities for (2S+ 1)Lambda-(2S+ 1) Lambda and (2S+ 1)(Lambda + or -)-(2S+ 1)Lambda Transitions in a Diatomic Molecule van der Waals Bonded to a Closed Shell Partner.
PB90-117441 *900,498*

FEENSTRA, R. M.
Structure of Cs on GaAs(110) as Determined by Scanning Tunneling Microscopy.
PB90-117490 *901,463*

FEIGERLE, C. S.
Vibrationally Resolved Photoelectron Angular Distributions for H2 in the Range 17 eV< or= h(nu)< or= 39 eV.
PB89-176952 *900,365*

FEINBERG, J.
Optical Novelty Filters.
PB89-228084 *901,366*

FELDMAN, A.
Cathodoluminescence of Defects in Diamond Films and Particles Grown by Hot-Filament Chemical-Vapor Deposition.
PB90-117961 *901,069*

Photoelastic Properties of Optical Materials.
PB89-177208 *901,355*

FELDMAN, P. A.
Rotational Modulation and Flares on RS Canum Venaticorum and BY Draconis Stars X: The 1981 October 3 Flare on V711 Tauri (= HR 1099).
PB89-208316 *900,021*

FELDMAN, U.
Laser-Produced Spectra and QED (Quantum Electrodynamic) Effects for Fe-, Co-, Cu-, and Zn-Like Ions of Au, Pb, Bi, Th, and U.
PB89-176010 *901,510*

Scheme for a 60-nm Laser Based on Photopumping of a High Level of Mo(6+) by a Spectral Line of Mo(11+).
PB89-186415 *900,407*

Spectra and Energy Levels of Br XXV, Br XXIX, Br XXX, and Br XXXI.
PB89-176002 *901,509*

FELTON, C. M.
Low Noise Frequency Synthesis.
PB89-174056 *900,716*

FENIMORE, C.
Method for Fitting and Smoothing Digital Data.
PB90-128794 *900,830*

Thermal-Expansive Growth of Prebreakdown Streamers in Liquids.
PB89-149074 *900,803*

FERNANDEZ, M. T.
Fundamental Configurations in Mo IV Spectrum.
PB89-147011 *900,284*

FERNANDEZ-PELLO, A. C.
Fire Propagation in Concurrent Flows.
PB89-151781 *900,867*

Fire Propagation in Concurrent Flows.
PB89-188577 *900,597*

FERRETT, T. A.
Autoionization Dynamics in the Valence-Shell Photoionization Spectrum of CO.
PB89-176960 *900,386*

Vibrationally Resolved Photoelectron Angular Distributions for H2 in the Range 17 eV< or= h(nu)< or= 39 eV.
PB89-176952 *900,385*

Vibrationally Resolved Photoelectron Studies of the 7(sigma) (-1) Channel in N2O.
PB89-176945 *900,257*

FIALA, J. C.
Interfaces to Teleoperation Devices.
PB89-181739 *900,993*

FICKETT, F. R.
Effects of Grain Size and Cold Rolling on Cryogenic Properties of Copper.
PB90-128604 *901,176*

FIELD, B. F.
Guidelines for Implementing the New Representations of the Volt and Ohm Effective January 1, 1990.
PB89-214761 *900,817*

Improved Transportable DC Voltage Standard.
PB89-230395 *901,554*

Josephson Array Voltage Calibration System: Operational Use and Verification.
PB89-230403 *900,820*

National Institute of Standards and Technology (NIST) Information Poster on Power Quality.
PB89-237986 *900,754*

NBS (National Bureau of Standards) Determination of the Fine-Structure Constant, and of the Quantized Hall Resistance and Josephson Frequency-to-Voltage Quotient in SI Units.
PB89-230437 *901,556*

Possible Quantum Hall Effect Resistance Standard.
PB89-149058 *900,801*

FIELDS, B. A.
Elevated Temperature Deformation of Structural Steel.
PB89-172621 *901,098*

FIELDS, R. J.
Elevated Temperature Deformation of Structural Steel.
PB89-172621 *901,098*

Texture Monitoring in Aluminum Alloys: A Comparison of Ultrasonic and Neutron Diffraction Measurements.
PB89-171409 *901,159*

FILIPPELLI, M.
Biotransformation of Mercury by Bacteria Isolated from a River Collecting Cinnabar Mine Waters.
PB89-229280 *900,864*

FILLA, B. J.
Enthalpies of Desorption of Water from Coal Surfaces.
PB89-173868 *900,838*

FILLIBEN, J. J.
Evaluating Emergency Management Models and Data Bases: A Suggested Approach.
PB89-189203 *901,598*

FINE, J.
Analytical Expression for Describing Auger Sputter Depth Profile Shapes of Interfaces.
PB89-157176 *900,309*

Status of Reference Data, Reference Materials and Reference Procedures in Surface Analysis.
PB89-157705 *900,332*

Temperature-Dependent Radiation-Enhanced Diffusion in Ion-Bombarded Solids.
PB89-179188 *901,408*

FINK, M. G. J.
Quantum-Defect Parametrization of Perturbative Two-Photon Ionization Cross Sections.
PB89-202600 *901,539*

FINNEY, J. L.
Repulsive Regularities of Water Structure in Ices and Crystalline Hydrates.
PB89-196753 *900,414*

FINZEL, B. C.
Use of an Imaging Proportional Counter in Macromolecular Crystallography.
PB90-136599 *900,538*

FIRST, P. N.
Structure of Cs on GaAs(110) as Determined by Scanning Tunneling Microscopy.
PB90-117490 *901,463*

FISCHBACH, E.
Precision Experiments to Search for the Fifth Force.
PB89-228365 *901,551*

FISHER, W. S.
Evaluating Office Lighting Environments: Second Level Analysis.
PB89-189153 *900,073*

FISK, Z.
Pressure Dependence of the Cu Magnetic Order in RBa2Cu3O6+ x.
PB90-123829 *901,472*

FITTING, D. W.
Measuring In-Plane Elastic Moduli of Composites with Arrays of Phase-Insensitive Ultrasound Receivers.
PB90-136672 *900,970*

PERSONAL AUTHOR INDEX

Refinement of Neutron Energy Deposition and Microdosimetry Calculations.
PB89-150791 *901,264*

SCHWARZ, F.
Biological Standard Reference Materials for the Calibration of Differential Scanning Calorimeters: Di-alkylphosphatidylcholine in Water Suspensions.
PB89-166779 *900,415*

SCHWARZ, F. P.
Biological Thermodynamic Data for the Calibration of Differential Scanning Calorimeters: Heat Capacity Data on the Unfolding Transition of Lysozyme in Solution.
PB90-117920 *900,513*

Differential Scanning Calorimetric Study of Brain Clathrin.
PB90-117912 *900,225*

SCHWARZENBACH, D.
Statistical Descriptors in Crystallography: Report of the International Union of Crystallography Subcommittee on Statistical Descriptors.
PB89-201826 *901,432*

SCIMECA, T.
Oxygen Partial-Density-of-States Change in the YBa2Cu3Ox Compounds for x(Approx.)6,6.5,7 Measured by Soft X-ray Emission.
PB89-166274 *901,419*

SCIRE, F.
Optical Roughness Measurements for Industrial Surfaces.
PB89-176655 *900,979*

SCOTT, T. R.
NBS (National Bureau of Standards) Laser Power and Energy Measurements.
PB89-171680 *901,346*

NBS (National Bureau of Standards) Standards for Optical Power Meter Calibration.
PB89-176200 *900,726*

Optical Power Measurements at the National Institute of Standards and Technology.
PB89-187579 *900,918*

SCRIBNER, C.
Brick Masonry: U.S. Office Building in Moscow.
PB89-187504 *900,160*

SCROGER, M. G.
NIST (National Institute of Standards and Technology) Measurement Services: The Calibration of Thermocouples and Thermocouple Materials.
PB89-209340 *900,897*

SCULL, L. L.
Low Temperature Mechanical Property Measurements of Silica Aerogel Foam.
PB90-128638 *901,061*

Tensile and Fatigue-Creep Properties of a Copper-Stainless Steel Laminate.
PB90-128646 *901,063*

SEELY, J. F.
Laser-Produced Spectra and QED (Quantum Electrodynamic) Effects for Fe-, Co-, Cu-, and Zn-Like Ions of Au, Pb, Bi, Th, and U.
PB89-176010 *901,510*

Spectra and Energy Levels of Br XXV, Br XXIX, Br XXX, and Br XXXI.
PB89-176002 *901,509*

SEILER, J. F.
Interim Criteria for Polymer-Modified Bituminous Roofing Membrane Materials.
PB89-166025 *900,114*

Report of Roof Inspection: Characterization of Newly-Fabricated Adhesive-Bonded Seams at an Army Facility.
PB90-113276 *900,107*

Results of a Survey of the Performance of EPDM (Ethylene Propylene Diene Terpolymer) Roofing at Army Facilities.
PB89-209316 *900,136*

SEKERKA, R. F.
Effect of Anisotropic Thermal Conductivity on the Morphological Stability of a Binary Alloy.
PB89-228985 *901,155*

SELLECK, M. E.
Effect of a Crystal-Melt Interface on Taylor-Vortex Flow.
PB90-130261 *901,477*

SELTZER, S. M.
Applications of ETRAN Monte Carlo Codes.
PB90-123902 *901,574*

Cross Sections for Bremsstrahlung Production and Electron-Impact Ionization.
PB90-123910 *901,575*

Overview of ETRAN Monte Carlo Methods.
PB90-123928 *901,576*

Pattern Recognition Approach in X-ray Fluorescence Analysis.
PB90-128786 *900,234*

SEMANCIK, S.
Coadsorption of Water and Lithium on the Ru(001) Surface.
PB89-202956 *900,440*

Economical Ultrahigh Vacuum Four-Point Resistivity Probe.
PB89-147086 *900,870*

Fundamental Characterization of Clean and Gas-Dosed Tin Oxide.

PB89-202964 *900,785*
Influence of Electronic and Geometric Structure on Desorption Kinetics of Isoelectronic Polar Molecules: NH3 and H2O.
PB89-176473 *900,381*

Surface Properties of Clean and Gas-Dosed SnO2 (110).
PB89-179576 *900,393*

SEMERJIAN, H. G.
Dynamic Light Scattering and Angular Dissymmetry for the In situ Measurement of Silicon Dioxide Particle Synthesis in Flames.
PB89-179584 *900,246*

FT-IR (Fourier Transform-Infrared) Emission/Transmission Spectroscopy for In situ Combustion Diagnostics.
PB89-211866 *900,600*

Laser Excited Fluorescence Studies of Black Liquor.
PB89-176416 *900,243*

Laser Induced Fluorescence for Measurement of Lignin Concentrations in Pulping Liquors.
PB89-122530 *901,184*

Remote Sensing Technique for Combustion Gas Temperature Measurement in Black Liquor Recovery Boilers.
PB89-179568 *900,392*

SENGERS, J. M. H. L.
Semi-Automated PVT Facility for Fluids and Fluid Mixtures.
PB89-157184 *900,875*

SENGERS, J. V.
Capillary Waves of a Vapor-Liquid Interface Near the Critical Temperature.
PB89-228555 *900,476*

Simplified Representation for the Thermal Conductivity of Fluids in the Critical Region.
PB89-228050 *901,332*

Van der Waals Fund, Van der Waals Laboratory and Dutch High-Pressure Science.
PB89-185755 *900,401*

SERIES, R. W.
Interlaboratory Determination of this Calibration Factor for the Measurement of the Interstitial Oxygen Content of Silicon by Infrared Absorption.
PB90-117300 *900,224*

SESKO, D.
Collisional Losses from a Light-Force Atom Trap.
PB90-123936 *901,577*

SETTLE, F. A.
Expert-Database System for Sample Preparation by Microwave Dissolution. 1. Selection of Analytical Descriptors.
PB89-229108 *900,216*

SHAPIRO, I. I.
Microarcsecond Optical Astrometry: An Instrument and Its Astrophysical Applications.
PB89-171268 *900,013*

SHAPIRO, M. L.
Spherical Acoustic Resonators in the Undergraduate Laboratory.
PB89-179709 *901,317*

SHELTON, R. N.
Antiferromagnetic Structure and Crystal Field Splittings in the Cubic Heusler Alloys HoPd2Sn and ErPd2Sn.
PB89-202659 *901,437*

SHEN, Q.
Direct Observation of Surface-Trapped Diffracted Waves.
PB90-128216 *901,475*

Dynamical Diffraction of X-rays at Grazing Angle.
PB89-186886 *901,421*

SHERIDAN, P.
Performance Standards for Microanalysis.
PB89-201651 *900,211*

SHERIDAN, P. J.
Determination of Experimental and Theoretical k (sub ASi) Factors for a 200-kV Analytical Electron Microscope.
PB90-128653 *900,232*

Methods for the Production of Particle Standards.
PB89-201636 *901,047*

SHERMAN, G. J.
PVT Relationships in a Carbon Dioxide-Rich Mixture with Ethane.
PB89-229181 *900,478*

SHIELDS, J. Q.
NBS (National Bureau of Standards) Determination of the Fine-Structure Constant, and of the Quantized Hall Resistance and Josephson Frequency-to-Voltage Quotient in SI Units.
PB89-230437 *901,556*

New Realization of the Ohm and Farad Using the NBS (National Bureau of Standards) Calculable Capacitor.
PB89-230445 *901,557*

SHIELDS, J. R.
Flammability Characteristics of Electrical Cables Using the Cone Calorimeter.
PB89-162572 *900,741*

SHIH, A.
Resonant Excitation of an Oxygen Valence Satellite in Photoemission from High-T(sub c) Superconductors.

PB89-186860 *901,420*
Synchrotron Radiation Study of BaO Films on W(001) and Their Interaction with H2O, CO2, and O2.
PB89-157697 *900,252*

SHINN, N. D.
Cr(110) Oxidation Probed by Carbon Monoxide Chemisorption.
PB89-228423 *900,239*

Oxygen Chemisorption on Cr(110): 1. Dissociative Adsorption.
PB89-202980 *900,441*

Oxygen Chemisorption on Cr(110): 2. Evidence for Molecular O2(ads).
PB89-202998 *900,442*

Stimulated Desorption from CO Chemisorbed on Cr(110).
PB89-203004 *900,443*

Synchrotron Photoemission Study of CO Chemisorption on Cr(110).
PB89-231336 *900,262*

SHIRLEY, J.
New Cavity Configuration for Cesium Beam Primary Frequency Standards.
PB89-171649 *900,714*

SHIVAPRASAD, S. M.
Adsorption Properties of Pt Films on W(110).
PB89-146864 *900,281*

SHIVES, T. R.
Tensile Tests of Type 305 Stainless Steel Mine Sweeping Wire Rope.
PB90-130297 *901,112*

SHNEIER, M. O.
Building Representations from Fusions of Multiple Views.
PB89-177059 *900,991*

SHOEMAKER, C.
Turning Workstation in the AMRF (Automated Manufacturing Research Facility).
PB89-185607 *900,954*

SHULL, P. J.
EMATs (Electromagnetic Acoustic Transducers) for Roll-By Crack Inspection of Railroad Wheels.
PB90-123694 *901,597*

SHULL, R. D.
Magnetic Field Dependence of the Superconductivity in Bi-Sr-Ca-Cu-O Superconductors.
PB89-146815 *901,385*

Synthesis and Magnetic Properties of the Bi-Sr-Ca-Cu Oxide 80- and 110-K Superconductors.
PB89-179725 *901,412*

SHUMAKER, J. B.
Apparatus Function of a Prism-Grating Double Monochromator.
PB89-186282 *901,359*

SIDDAGANGAPPA, M. C.
Production and Stability of S2F10 in SF6 Corona Discharges.
PB89-231039 *900,822*

SIECK, L. W.
Ionic Hydrogen Bond and Ion Solvation. 5. OH-...(1-)O Bonds. Gas Phase Solvation and Clustering of Alkoxide and Carboxylate Anions.
PB89-157531 *900,328*

Relative Acidities of Water and Methanol and the Stabilities of the Dimer Anions.
PB89-150981 *900,299*

Thermochemistry of Solvation of SF6(1-) by Simple Polar Organic Molecules in the Vapor Phase.
PB89-202527 *900,437*

SIEGRIST, T.
Applications of the Crystallographic Search and Analysis System CRYSTDAT in Materials Science.
PB89-175251 *901,402*

SIEGWARTH, J. D.
Vortex Shedding Flowmeter for Fluids at High Flow Velocities.
PB90-128661 *900,606*

SIEWERT, T.
Standards for Real-Time Radioscopy.
PB90-128687 *900,924*

SIEWERT, T. A.
Ferrite Number Prediction to 100 FN in Stainless Steel Weld Metal.
PB89-201586 *901,106*

Improved Standards for Real-Time Radioscopy.
PB90-128679 *900,923*

Influence of Molybdenum on the Strength and Toughness of Stainless Steel Welds for Cryogenic Service.
PB89-173512 *901,100*

On-Line Arc Welding: Data Acquisition and Analysis Using a High Level Scientific Language.
PB90-117391 *900,972*

Role of Inclusions in the Fracture of Austenitic Stainless Steel Welds at 4 K.
PB89-173504 *901,099*

KEYWORD INDEX

KEYWORD INDEX

KEYWORD INDEX

KEYWORD INDEX

KEYWORD INDEX

KEYWORD INDEX

Analysis of Ultrapure Reagents from a Large Sub-Boiling Still Made of Teflon PFA.
PB89-186357 *900,202*

Laser Microprobe Mass Spectrometry: Description and Selected Applications.
PB89-201628 *900,210*

Single Particle Standards for Isotopic Measurements of Uranium by Secondary Ion Mass Spectrometry.
PB89-201669 *901,297*

Isotope Dilution Mass Spectrometry for Accurate Elemental Analysis.
PB89-230338 *900,220*

Development of the NBS (National Bureau of Standards) Beryllium Isotopic Standard Reference Material.
PB89-231070 *900,221*

Determination of Serum Cholesterol by a Modification of the Isotope Dilution Mass Spectrometric . Definitive Method.
PB89-234181 *901,239*

Determination of Selenium and Tellurium in Copper Standard Reference Materials Using Stable Isotope Dilution Spark Source Mass Spectrometry.
PB90-123472 *900,230*

High-Accuracy Gas Analysis via Isotope Dilution Mass Spectrometry: Carbon Dioxide in Air.
PB90-123951 *900,032*

Identification of Carbonaceous Aerosols via C-14 Accelerator Mass Spectrometry,. and Laser Microprobe Mass Spectrometry.
PB90-136540 *900,236*

Technical Examination, Lead Isotope Determination, and Elemental Analysis of Some Shang and Zhou Dynasty Bronze Vessels.
PB90-136862 *901,178*

Laser Induced Vaporization Time Resolved Mass Spectrometry of Refractories.
PB90-136904 *900,540*

MATERIALS
Intelligent Processing of Materials: Report of an Industrial Workshop Conducted by the National Institute of Standards and Technology.
PB89-151823 *900,942*

Integrated Knowledge Systems .for Concrete and Other Materials.
PB89-176119 *900,582*

MATERIALS HANDLING
Material Handling Workstation Implementation.
PB89-159644 *900,988*

Material Handling Workstation: Operator Manual.
PB89-159651 *900,989*

AMRF (Automated Manufacturing Research Facility) Material Handling System Architecture.
PB89-177091 *900,952*

MATERIALS HANDLING EQUIPMENT
Material Handling Workstation, Recommended Technical Specifications for Procurement of Commercially Available Equipment.
PB89-162564 *900,998*

MATERIALS SCIENCE
Journal of Research of the National Institute of Standards and Technology, Volume 94, Number 1, January-February 1989. Special Issue: Numeric Databases in Materials and Biological Sciences.
PB89-175194 *901,186*

Importance of Numeric Databases to Materials Science.
PB89-175202 *901,187*

International Cooperation and Competition in Materials Science and Engineering.
PB89-208332 *901,191*

MATERIALS SPECIFICATIONS
Preliminary Performance Criteria for Building Materials, Equipment and Systems Used in Detention and Correctional Facilities.
PB89-148514 *900,109*

Computerized Materials Property Data Systems.
PB89-187512 *901,312*

ASTM (American Society for Testing and Materials) Committee Completes Work on EPDM Specification.
PB89-212260 *900,140*

MATERIALS TESTS
Techniques for Measuring the Electromagnetic Shielding Effectiveness of Materials. Part 1. Far-Field Source Simulation.
PB89-161525 *900,680*

Techniques for Measuring the Electromagnetic Shielding Effectiveness of Materials. Part 2. Near-Field Source Simulation.
PB89-161533 *900,681*

Measurements of Tribological Behavior of Advanced Materials: Summary of U.S. Results on VAMAS (Versailles Advanced Materials and Standards) Round-Robin No. 2.
PB90-130295 *901,003*

MATERIEL
Hierarchically Controlled Autonomous Robot for Heavy Payload Military Field Applications.
PB89-177075 *901,271*

MATERIELS HANDLING
Hierarchically Controlled Autonomous Robot for Heavy Payload Military Field Applications.

PB89-177075 *901,271*

MATHEMATICAL MODELS
Calculating Flows through Vertical Vents in Zone .Fire Models under Conditions of Arbitrary Cross-Vent Pressure Difference.
PB89-148126 *900,108*

Experimental Validation of a Mathematical Model for Predicting Moisture Transfer in Attics.
PB89-150783 *900,057*

Generalized Mathematical Model for Machine Tool Errors.
PB89-150874 *900,977*

S2F10 Formation in Computer Simulation Studies of the Breakdown of SF6.
PB89-157523 *900,327*

Hand Calculations for Enclosure Fires.
PB89-173983 *900,164*

Computer Fire Models.
PB89-173991 *900,165*

Vapor-Liquid Equilibrium of Nitrogen-Oxygen Mixtures and Air at High Pressure.
PB89-174932 *900,368*

Analytical Modeling of Device-Circuit Interactions for the Power Insulated Gate Bipolar Transistor (IGBT).
PB89-176259 *900,777*

Influence of Reaction Reversibility on Continuous-Flow Extraction by Emulsion Liquid Membranes.
PB89-176461 *900,244*

Evaluating Emergency Management Models and Data Bases: A Suggested Approach.
PB89-189203 *901,598*

Technical Reference Guide for FAST (Fire and Smoke Transport) Version 18.
PB89-216366 *900,602*

Modeling Dynamic Surfaces with Octrees.
PB90-112335 *901,206*

High Accuracy Modeling of Photodiode Quantum Efficiency.
PB90-117599 *900,733*

Fields Radiated by Electrostatic Discharges.
PB90-122778 *901,382*

Model for Particle Size and Phase Distributions in Ground Cement Clinker.
PB90-136847 *901,062*

MATHEMATICAL & STATISTICAL METHODS
Error Bounds for Linear Recurrence Relations.
AD-A201 256/5 *901,182*

Finite Unions of Closed Subgroups of the n-Dimensional Torus.
PB89-143283 *901,193*

Analytical Model for the Steady-State and Transient Characteristics of the Power Insulated-Gate Bipolar Transistor.
PB89-146680 *900,767*

Measurement Procedures for Electromagnetic Compatibility Assessment of Electroexplosive Devices.
PB89-148914 *901,314*

Generalized Mathematical Model for Machine Tool Errors.
PB89-150874 *900,977*

Iterative Technique to Correct Probe Position Errors in Planar Near-Field to Far-Field Transformations.
PB89-153686 *900,695*

Problems with Interval Estimation When Data Are Adjusted via Calibration.
PB89-157812 *901,209*

Estimation of the Error Probability Density from Replicate Measurements on Several Items.
PB89-157820 *901,210*

Numerical Evaluation of Certain Multivariate Normal Integrals.
PB89-158166 *901,195*

Internal Revenue Service Post-of-Duty Location Modeling System: Programmer's Manual for PASCAL Solver.
PB89-161905 *900,001*

Internal Revenue Service Post-of-Duty Location Modeling System: Programmer's Manual for FORTRAN Driver Version 5.0.
PB89-161913 *900,002*

Real Time Generation of Smooth Curves Using Local Cubic Segments.
PB89-171623 *901,196*

Minimax Approach to Combining Means, with Practical Examples.
PB89-171847 *901,211*

How to Estimate Capacity Dimension.
PB89-172522 *901,197*

Laser Induced Fluorescence for Measurement of Lignin Concentrations in Pulping Liquors.
PB89-172530 *901,184*

Designs for Assessment of Measurement Uncertainty: Experience in the Eastern Lake Survey.
PB89-173827 *900,862*

Variances Based on Data with Dead Time between the Measurements.
PB89-174049 *901,204*

Comparison of Time Scales Generated with the NBS (National Bureau of Standards) Ensembling Algorithm.
PB89-174072 *900,628*

Efficient Algorithms for Computing Fractal Dimensions.
PB89-175871 *901,198*

Tests of the Recalibration Period of a Drifting Instrument.
PB89-176275 *900,199*

Electron Mean Free Path Calculations Using a Model Dielectric Function.
PB89-177026 *901,141*

Note on the Capacitance Matrix Algorithm, Substructuring, and Mixed or Neumann Boundary Conditions.
PB89-177034 *901,199*

Algebraic Representation for the Topology of Multicomponent Phase Diagrams.
PB89-177042 *901,517*

Millisecond Pulsar Rivals Best Atomic Clock Stability.
PB89-185722 *900,629*

Using Multiple Reference Stations to Separate the Variances of Noise Components in the Global Positioning System.
PB89-185730 *901,293*

Calibration with Randomly Changing Standard Curves.
PB89-186381 *900,888*

Probabilistic Models for Ground Snow Accumulation.
PB89-188894 *900,100*

Evaluating Emergency Management Models and Data Bases: A Suggested Approach.
PB89-189203 *901,598*

Estimation of an Asymmetrical Density from Several Small Samples.
PB89-201131 *901,212*

Statistical Analysis of Experiments to Measure Ignition of Cigarettes.
PB89-201149 *900,135*

Graphical Analyses Related to the Linewidth Calibration Problem.
PB89-201669 *900,783*

Statistical Descriptors in Crystallography: Report of the International Union of Crystallography Subcommittee on Statistical Descriptors.
PB89-201826 *901,432*

Elimination of Spurious Eigenvalues in the Chebyshev Tau Spectral Method.
PB89-209262 *901,330*

Expected Complexity of the 3-Dimensional Voronoi Diagram.
PB89-209332 *901,200*

Consensus Values, Regressions, and Weighting Factors.
PB89-211130 *901,213*

Electronic Mail and the 'Locator's' Dilemma.
PB89-211957 *901,205*

Simulation Study of Light Scattering from Soot Agglomerates.
PB89-212138 *901,543*

NBS (National Bureau of Standards) Calibration Service Providing Time and Frequency at a Remote Site by Weighting and Smoothing of GPS (Global Positioning System) Common View Data.
PB89-212211 *900,631*

Computation and Use of the Asymptotic Covariance Matrix for Measurement Error Models.
PB89-215321 *901,214*

AutoMan: Decision Support Software for Automated Manufacturing Investments. User's Manual.
PB89-221873 *900,963*

User's Reference Guide for ODRPACK: Software for Weighted Orthogonal Distance Regression Version 1.7.
PB89-229066 *901,215*

Analysis of Ridge-to-Ridge Distance on Fingerprints.
PB89-230478 *901,216*

Comparison of Far-Field Methods for Determining Mode Field Diameter of Single-Mode Fibers Using Both Gaussian and Petermann Definitions.
PB90-117474 *900,756*

Guide to Available Mathematical Software Advisory System.
PB90-123654 *901,201*

Shortest Paths in Simply Connected Regions in R2.
PB90-123688 *901,202*

Merit Functions and Nonlinear Programming.
PB90-123944 *901,208*

Bootstrap Inference for Replicated Experiments.
PB90-128273 *900,914*

Method for Fitting and Smoothing Digital Data.
PB90-128794 *900,830*

Allocating Staff to Tax Facilities: A Graphics-Based Microcomputer Allocation Model.
PB90-128891 *900,645*

Polytope Volume Computation.
PB90-128982 *901,203*

Monte Carlo Simulation of Domain Growth in the Kinetic Ising Model on the Connection Machine.
PB90-136797 *901,587*

KW-46

KEYWORD INDEX

KEYWORD INDEX

KEYWORD INDEX

TITLE INDEX

SAMPLE ENTRY

Computer Viruses and Related Threats: A Management Guide			Title			
PB90-111683	900,654	PC A03/MF A01	NTIS order number	Abstract number	Availability Price Code

---

2.5 MeV Neutron Source for Fission Cross Section Measurement.
PB89-176531   *901,512*   Not available NTIS

4s(2) 4p(2)-4s4p(3) Transition Array and Energy Levels of the Germanium-Like Idns Rb VI - Mo XI.
PB89-201065   *901,528*   Not available NTIS

(12)C(16)O Laser Frequency Tables for the 34.2 to 62.3 THz (1139 to 2079 cm(-1)) Region.
PB89-193908   *901,361*   PC A03/MF A01

13C NMR Method for Determining the Partitioning of End Groups and Side Branches between the Crystalline and Non-Crystalline Regions in Polyethylene.
PB89-202451   *900,569*   Not available NTIS

(109)Pd and (109)Cd Activity Standardization and Decay Data.
PB90-123449   *901,564*   Not available NTIS

Absolute Cross Sections for Molecular Photoabsorption, Partial Photoionization, and Ionic Photofragmentation Process.
PB89-186464   *900,410*   Not available NTIS

Absolute Infrared Transition Moments for Open Shell Diatomics from J Dependence of Transition Intensities: Application to OH.
PB89-227912   *900,463*   Not available NTIS

Absolute Isotopic Abundance Ratios and Atomic Weight of a Reference Sample of Nickel.
PB90-163890   *900,543*
                    (Order as PB90-163874, PC A04)

Absolute Isotopic Composition and Atomic Weight of Terrestrial Nickel.
PB90-163908   *900,544*
                    (Order as PB90-163874, PC A04)

Absolute Rate Constants for Hydrogen Abstraction from Hydrocarbons by the Trichloromethylperoxyl Radical.
PB89-171532   *900,357*   Not available NTIS

AC-DC Difference Calibrations at NBS (National Bureau of Standards).

PB89-201560   *900,816*   Not available NTIS

AC Electric and Magnetic Field Meter Fundamentals.
PB89-173470   *900,746*   Not available NTIS

AC Impedance Method for High-Resistivity Measurements of Silicon.
PB89-231203   *900,793*   Not available NTIS

Accurate Determination of Planar Near-Field Correction Parameters for Linearly Polarized Probes.
PB89-156871   *900,704*   Not available NTIS

Accurate Energies of nS, nP, nD, nF and nG Levels of Neutral Cesium.
PB89-202121   *900,431*   Not available NTIS

Accurate RF Voltage Measurements Using a Sampling Voltage Tracker.
PB89-201552   *900,815*   Not available NTIS

Acoustic and Microwave Resonances Applied to Measuring the Gas Constant and the Thermodynamic Temperature.
PB89-228548   *901,320*   Not available NTIS

Acoustic Emission: A Quantitative NDE Technique for the Study of Fracture.
PB89-211924   *900,921*   Not available NTIS

Acoustical Technique for Evaluation of Thermal Insulation.
PB89-193866   *900,919*   PC A03/MF A01

Acoustoelastic Determination of Residual Stresses.
PB89-179808   *901,318*   Not available NTIS

ACSB (Amplitude Companded Sideband): What Is Adequate Performance.
PB89-176283   *901,599*   Not available NTIS

Activation Analysis Opportunities Using Cold Neutron Beams.
PB89-156970   *900,183*   Not available NTIS

Activities of the International Association for the Properties of Steam between 1979 and 1984.
PB89-212153   *900,446*   Not available NTIS

Adhesion to Dentin by Means of Gluma Resin.

PB89-157168   *900,039*   Not available NTIS

Adhesive Bonding of Composites.
PB90-123696   *900,050*   Not available NTIS

Adsorption of High-Range Water-Reducing Agents on Selected Portland Cement Phases and Related Materials.
PB90-124306   *900,583*   PC A03/MF A01

Adsorption of Water on Clean and Oxygen-Predosed Nickel(110).
PB90-123555   *900,522*   Not available NTIS

Adsorption of 4-Methacryloxyethyl Trimellitate Anhydride (4-META) on Hydroxyapatite and Its Role in Composite Bonding.
PB89-179220   *900,041*   Not available NTIS

Adsorption Properties of Pt Films on W(110).
PB89-146864   *900,281*   Not available NTIS

Advanced Ceramics: A Critical Assessment of Wear and Lubrication.
PB89-188589   *901,045*   PC A06/MF A01

Advanced Heat Pumps for the 1990's Economic Perspectives for Consumers and Electric Utilities.
PB90-118043   *900,089*   Not available NTIS

Advances in NIST (National Institute of Standards and Technology) Dielectric Measurement Capability Using a Mode-Filtered Cylindrical Cavity.
PB89-231146   *900,907*   Not available NTIS

Advances in the Use of (3)He in a Gas Scintillation Counter.
PB90-123506   *901,565*   Not available NTIS

Aerodynamics of Agglomerated Soot Particles.
PB89-147482   *900,586*   Not available NTIS

AES and LEED Studies Correlating Desorption Energies with Surface Structures and Coverages for Ga on Si(100).
PB89-171599   *901,401*   Not available NTIS

Ag Screen Contacts to Sintered YBa2Cu3Ox Powder for Rapid Superconductor Characterization.
PB89-200448   *901,423*   Not available NTIS

TITLE INDEX

TITLE INDEX

TITLE INDEX

TITLE INDEX

Window U-Values: Revisions for the 1989 ASHRAE (American Society of Heating, Refrigerating and Air-Conditioning Engineers) Handbook - Fundamentals.
PB89-229215   *900,145*   Not available NTIS

Working Implementation Agreements for Open Systems Interconnection Protocols.
PB89-221196   *900,624*   PC A10/MF A01

PB89-235931   *900,642*   PC A16/MF A01

Workstation Controller of the Cleaning and Deburring Workstation.
PB89-189286   *900,955*   PC A04/MF A01

X-Band Atmospheric Attenuation for an Earth Terminal Measurement System.
PB90-100736   *900,626*   PC A03/MF A01

ZIP: The ZIP-Code Insulation Program (Version 1.0) Economic Insulation Levels for New and Existing Houses by Three-Digit ZIP Code. Users Guide and Reference Manual.
PB89-151765   *900,058*   PC A03/MF A01

ZIP: ZIP-Code Insulation Program (for Microcomputers).
PB89-159446   *900,060*   CP D01

SAMPLE ENTRY

NIST/SP-500/166
Computer Viruses and Related Threats: A Management Guide
PB90-111683   *900,654*   **PC A03/MF A01**

Report or series number
Title

NTIS order number   *Abstract number*   Availability
                                              Price Code

PB90-111683
Compuer Viruses and Related Threats: A Management Guide
PB90-111683   *900,654*   **PC A03/MF A01**

Report or series number
Title

NTIS order number   *Abstract number*   Availability
                                              Price code

AD-A201 256/5
Error Bounds for Linear Recurrence Relations.
AD-A201 256/5   *901,192*   PC A03/MF A01

AD-A202 820/7
Alignment Effects in Electronic Energy Transfer and Reactive Events.
AD-A202 820/7   *900,267*   PC A03/MF A01

AFOSR-TR-88-1316
Alignment Effects in Electronic Energy Transfer and Reactive Events.
AD-A202 820/7   *900,267*   PC A03/MF A01

ARO-20608.9-MA
Error Bounds for Linear Recurrence Relations.
AD-A201 256/5   *901,192*   PC A03/MF A01

CAM-8901
Advanced Ceramics: A Critical Assessment of Wear and Lubrication.
PB89-188569   *901,045*   PC A06/MF A01

EPRI-RP-2033-26
Proposed Methodology for Rating Air-Source Heat Pumps That Heat, Cool, and Provide Domestic Water Heating.
PB90-112368   *900,267*   PC A06/MF A01

ESA-SP-281-V-2
Proceedings of the Celebratory Symposium on a Decade of UV (Ultraviolet) Astronomy with the IUE Satellite, Volume 2.
N89-16535/1   *900,014*   PC A19/MF A01

FIPS PUB 134-1
Coding and Modulation Requirements for 4,800 Bit/Second Modems, Category: Telecommunications Standard.
FIPS PUB 134-1   *900,660*   PC A03/MF A01

FIPS PUB 149
General Aspects of Group 4 Facsimile Apparatus, Category: Telecommunications Standard.
FIPS PUB 149   *900,661*   PC E08

FIPS PUB 150
Facsimile Coding Schemes and Coding Control Functions for Group 4 Facsimile Apparatus, Category: Telecommunications Standard.
FIPS PUB 150   *900,662*   PC E08

FIPS PUB 152
Standard Generalized Markup Language (SGML).
FIPS PUB 152   *900,627*   PC E19

FIPS PUB 154
High Speed 25-Position Interface for Data Terminal Equipment and Data Circuit-Terminating Equipment, Category: Telecommunications Standard.
FIPS PUB 154   *900,663*   PC E09

FIPS PUB 155
Data Communication Systems and Services User-Oriented Performance Measurement Methods, Category: Telecommunications Standard.
FIPS PUB 155   *900,664*   PC E11

GRI-88/0290
Advanced Ceramics: A Critical Assessment of Wear and Lubrication.
PB89-188569   *901,045*   PC A06/MF A01

ICST/SNA-87/3
User Guide for the NBS (National Bureau of Standards) Prototype Compiler for Estelle (Revised).
PB89-196158   *900,619*   PC A05/MF A01

ISBN-0-86318-585-7
Journal of Physical and Chemical Reference Data, Volume 17, 1988, Supplement No. 3. Atomic Transition Probabilities Scandium through Manganese.
PB89-145197   *900,276*   Not available NTIS

ISBN-0-86318-587-3
Journal of Physical and Chemical Reference Data, Volume 17, 1988, Supplement No. 2. Thermodynamic and Transport Properties of Molten Salts: Correlation Equations for Critically Evaluated Density, Surface Tension, Electrical Conductance, and Viscosity Data.
PB89-145205   *900,277*   Not available NTIS

ISBN-1-55977-069-0
Promoting Technological Excellence: The Role of State and Federal Extension Activities.
PB90-120742   *900,171*   PC A05/MF A01

N89-16535/1
Proceedings of the Celebratory Symposium on a Decade of UV (Ultraviolet) Astronomy with the IUE Satellite, Volume 2.

APPENDIX A
List of Depository Libraries in the United States

ALABAMA

Alexander City

Alexander City State Junior College Thomas D. Russell Library (1967)*

Auburn

Auburn University Ralph Brown Draughon Library (1907)

Birmingham

Birmingham Public Library (1895)
Birmingham–Southern College Library (1932)
Jefferson State Junior College James B. Allen Library (1970)
Miles College C. A. Kirkendoll Learning Resource Center (1980)
Samford University Library (1884)

Enterprise

Enterprise State Junior College Learning Resources Center (1967)

Fayette

Brewer State Junior College Learning Resources Center Library (1979)

Florence

University of North Alabama Collier Library (1932)

Gadsden

Gadsden Public Library (1963)

Huntsville

University of Alabama in Huntsville Library (1964)

Jacksonville

Jacksonville State University Houston Cole Library (1929)

Mobile

Mobile Public Library (1963)
Spring Hill College Thomas Byrne Memorial Library (1937)
University of South Alabama Library (1968)

Montgomery

Alabama Public Library Service (1984)

*Year designated.

Alabama Supreme Court and State Law Library (1884)
Auburn University at Montgomery Library (1971) REGIONAL
Air University Library Maxwell Air Force Base (1963)

Normal

Alabama Agricultural and Mechanical University J. F. Drake Memorial Learning Resources Center (1963)

Troy

Troy State University Library (1963)

Tuscaloosa

University of Alabama Library (1860) REGIONAL
University of Alabama School of Law Library (1967)

Tuskegee

Tuskegee University Hollis Burke Frissell Library (1907)

ALASKA

Anchorage

Anchorage Law Library (1973)
Anchorage Municipal Libraries Z. J. Loussac Public Library (1978)
University of Alaska at Anchorage Library (1961)
U.S. Alaska Resources Library (1981)
U.S. District Court Library (1983)

Fairbanks

University of Alaska Elmer E. Rasmuson Library (1922)

Juneau

Alaska State Library (1900)
University of Alaska–Juneau Library (1981)

Ketchikan

Ketchikan Community College Library (1970)

AMERICAN SAMOA

Pago Pago

Community College of American Samoa Library (1985)

ARIZONA

Coolidge

Central Arizona College Instruction Materials Center (1973)

Flagstaff

Northern Arizona University Cline Library (1937)

Glendale

Glendale Public Library (1986)

Holbrook

Northland Pioneer College Learning Resources Center (1985)

Mesa

Mesa Public Library (1983)

Phoenix

Department of Library Archives, and Public Records (unknown) REGIONAL
Grand Canyon College Fleming Library (1978)
Phoenix Public Library (1917)
U.S. Court of Appeals 9th Circuit Library (1984)

Prescott

Yavapai College Library (1976)

Tempe

Arizona State University College of Law Library (1977)
Arizona State University Library (1944)

Tucson

Tucson Public Library (1970)
University of Arizona Library (1907) REGIONAL

Yuma

Yuma City-County Library (1963)

ARKANSAS

Arkadelphia

Ouachita Baptist University Riley Library (1963)

Batesville

Arkansas College Library (1963)

Clarksville

University of the Ozarks Dobson Memorial Library (1925)

Conway

Hendrix College Olin C. Bailey Library (1903)

Fayetteville

University of Arkansas Mullins Library (1907)
University of Arkansas School of Law Library (1978)

Little Rock

Arkansas State Library (1978) REGIONAL
Arkansas Supreme Court Library (1962)
Central Arkansas Library System Main Library (1953)
University of Arkansas at Little Rock Library Ottenheimer Library (1973)
University of Arkansas at Little Rock, School of Law Library (1979)

Magnolia

Southern Arkansas University Magale Library (1956)

Monticello

University of Arkansas at Monticello Library (1956)

Pine Bluff

University of Arkansas at Pine Bluff Watson Memorial Library (1976)

Russellville

Arkansas Tech University Tomlinson Library (1925)

Searcy

Harding University Beaumont Memorial Library (1963)

State University

Arkansas State University Dean B. Ellis Library (1913)

Walnut Ridge

Southern Baptist College Felix Goodson Library (1967)

CALIFORNIA

Anaheim

Anaheim Public Library (1963)

Arcadia

Arcadia Public Library (1975)

Arcata

Humboldt State University Library (1963)

Bakersfield

California State College Bakersfield Library (1974)
Kern County, Beale Memorial Library (1943)

Berkeley

University of California General Library (1907)
University of California Law Library (1963)

Carson

California State University Dominguez Hills Educational Resources
  Center (1973)
Carson Regional Library (1973)

Chico

California State University, Merriam Library (1962)

Claremont

Claremont Colleges' Libraries Honnold Library (1913)

Compton

Compton Public Library (1972)

Culver City

Culver City Library (1966)

Davis

University of California Shields Library (1953)
University of California at Davis Law Library (1972)

Downey

Downey City Library (1963)

Fresno

California State University, Fresno, Henry Madden Library (1962)
Fresno County Free Library (1920)

Fullerton

California State University at Fullerton Library (1963)
Western State University College of Law Library (1984)

Garden Grove

Garden Grove Regional Library (1963)

Gardena

Gardena Public Library (1966)

Hayward

California State University, Hayward Library (1963)

Huntington Park

Huntington Park Library (1970)

Inglewood

Inglewood Public Library (1963)

Irvine

University of California at Irvine Main Library (1963)

La Jolla

University of California at San Diego Central University Library (1963)

Lakewood

Angelo Iacoboni Public Library (1970)

Lancaster

Lancaster Library (1967)

La Verne

University of La Verne College of Law Library (1979)

Long Beach

California State University at Long Beach Library (1962)
Long Beach Public Library (1933)

Los Angeles

California State University at Los Angeles John F. Kennedy Memorial
  Library (1956)
Los Angeles County Law Library (1963)
Los Angeles Public Library (1891)
Loyola Marymount University Charles Von der Ahe Library (1933)
Loyola Law School Law Library (1979)
Occidental College Library (1941)
Southwestern University School of Law Library (1975)
University of California, University Research Library (1932)
University of California, Los Angeles Law Library (1958)
University of Southern California Doheny Memorial Library (1933)
University of Southern California Law Library (1978)
U.S. Court of Appeals Ninth Circuit Library (1981)
Whittier College School of Law Library (1978)

Malibu

Pepperdine University Payson Library (1963)

Menlo Park

Department of Interior Geological Survey Library (1962)

Montebello

Montebello Regional Library (1966)

Monterey

U.S. Naval Postgraduate School Dudley Knox Library (1963)

Monterey Park

Bruggemeyer Memorial Library (1964)

Northridge

California State University at Northridge Oviatt Library (1958)

Norwalk

Norwalk Regional Library (1973)

Oakland

Mills College Library (1966)
Oakland Public Library (1923)

Ontario

Ontario City Library (1974)

Palm Springs

Palm Springs Public Library (1980)

Pasadena

California Institute of Technology Millikan Memorial Library (1933)
Pasadena Public Library (1963)

Pleasant Hill

Contra Costa County Library (1964)

Redding

Shasta County Library (1956)

Redlands

University of Redlands Armacost Library (1933)

Redwood City

Redwood City Public Library (1966)

Reseda

West Valley Regional Branch Library (1966)

Richmond

Richmond Public Library (1943)

Riverside

Riverside City and County Public Library (1947)
University of California at Riverside Library (1963)

Sacramento

California State Library (1895) REGIONAL
California State University at Sacramento Library (1963)
Sacramento County Law Library (1963)
Sacramento Public Library (1860)
University of the Pacific McGeorge School of Law Library (1978)

San Bernardino

Don A. Turner County Law Library (1984)
San Bernardino County Library (1964)

San Diego

San Diego County Law Library (1973)

San Diego County Library (1973)
San Diego Public Library (1895)
San Diego State University Library (1962)
University of San Diego Kratter Law Library (1967)

San Francisco

Golden Gate University School of Law Library (1979)
Hastings College of Law Library (1972)
San Francisco Public Library (1889)
San Francisco State University J. Paul Leonard Library (1955)
Supreme Court of California Library (1979)
U.S. Court of Appeals Ninth Circuit Library (1971)
University of San Francisco Richard A. Gleeson Library (1963)

San Jose

San Jose State University Library (1962)

San Leandro

San Leandro Community Library Center (1961)

San Luis Obispo

California Polytechnic State University Robert E. Kennedy Library (1969)

San Mateo

College of San Mateo Library (1967)

San Rafael

Marin County Free Library (1975)

Santa Ana

Orange County Law Library (1975)
Santa Ana Public Library (1959)

Santa Barbara

University of California at Santa Barbara Library (1960)

Santa Clara

University of Santa Clara Orradre Library (1963)

Santa Cruz

University of California at Santa Cruz McHenry Library (1963)

Santa Rosa

Sonoma County Library (1896)

Stanford

Stanford University Libraries (1895)
Stanford University Robert Crown Law Library (1978)

Stockton

Public Library of Stockton and San Joaquin County (1884)

Thousand Oaks

California Lutheran University Library (1964)

Torrance

Torrance Public Library (1969)

Turlock

California State University, Stanislaus Library (1964)

Vallejo

Solano County Library John F. Kennedy Library (1982)

Valencia

Valencia Regional Library (1972)

Ventura

Ventura County Library Services Agency (1975)

Visalia

Tulare County Free Library (1967)

Walnut

Mount San Antonio College Educational Resources Library Center (1966)

West Covina

West Covina Regional Library (1966)

Whittier

Whittier College Wardman Library (1963)

COLORADO

Alamosa

Adams State College Library (1963)

Aurora

Aurora Public Library (1984)

Boulder

University of Colorado at Boulder Norlin Library (1879) REGIONAL

Colorado Springs

Colorado College Tutt Library (1880)
University of Colorado at Colorado Springs Library (1974)
U.S. Air Force Academy Library (1956)

Denver

Auraria Library (1978)
Colorado Supreme Court Library (1978)
Denver Public Library (1884) REGIONAL
Department of the Interior Library (1962)
Regis College Dayton Memorial Library (1915)
U.S. Court of Appeals Tenth Circuit Library (1973)
University of Denver Penrose Library (1909)
University of Denver College of Law Westminster Law Library (1978)

Fort Collins

Colorado State University Libraries (1907)

Golden

Colorado School of Mines Arthur Lakes Library (1939)

Grand Junction

Mesa College Lowell Heiny Library (1978)
Mesa County Public Library (1975)

Greeley

University of Northern Colorado James A. Michener Library (1966)

Gunnison

Western State College Leslie J. Savage Library (1932)

La Junta

Otero Junior College Wheeler Library (1963)

Lakewood

Jefferson County Public Library Lakewood Library (1968)

Pueblo

Pueblo Library District (1893)
University of Southern Colorado Library (1965)

CONNECTICUT

Bridgeport

Bridgeport Public Library (1884)
University of Bridgeport School of Law Library Wahlstrom Library (1979)

Danbury

Western Connecticut State University Ruth A. Haas Library (1967)

Danielson

Quinebaug Valley Community College Audrey P. Beck Library (1968)

Enfield

Enfield Central Library (1967)

Hartford

Connecticut State Library (unknown) REGIONAL
Hartford Public Library (1945)
Trinity College Library (1895)
University of Connecticut School of Law Library (1978)

Middletown

Wesleyan University Olin Library (1906)

Mystic

Mystic Seaport Museum, Inc., G. W. Blunt White Library (1964)

New Britain

Central Connecticut State University Elihu Burritt Library (1973)

New Haven

Southern Connecticut State University Hilton C. Buley Library (1968)
Yale Law Library (1981)
Yale University Seeley G. Mudd Library (1859)

New London

Connecticut College C. E. Shain Library (1926)
U.S. Coast Guard Academy Library (1939)

Stamford

Ferguson Library (1973)

Storrs

University of Connecticut Homer Babbidge Library (1907)

Waterbury

Post College Traurig Library and Learning Resources Center (1977)
Silas Bronson Public Library (1869)

West Haven

University of New Haven Peterson Library (1971)

DELAWARE

Dover

Delaware State College William C. Jason Library-Learning Center (1962)
State Law Library in Kent County (unknown)

Georgetown

Delaware Technical and Community College Library (1968)

Newark

University of Delaware Library (1907)

Wilmington

Wider University School of Law Library (1976)

DISTRICT OF COLUMBIA

Washington

Administrative Conference of the United States Library (1972)
Advisory Commission on Intergovernmental Relations Library (1977)
American University Washington College of Law Library (1983)
Catholic University of America Robert J. White Law Library (1979)
Comptroller of the Currency Library (1986)
Department of the Army Pentagon Library ANRAL(1969)
Department of Commerce Library (1955)
Department of Education (1988)
Department of Health and Human Services Library (1954)
Department of Housing and Urban Development Library (1969)
Department of the Interior Library Natural Resources Library (1895)
Department of Justice Main Library (1895)
Department of Labor Library (1976)
Department of the Navy Library (1895)
Department of State Library (1895)
Department of State Law Library (1966)
Department of Transportation Main Library (1982)
Department of Transportation, U.S. Coast Guard Law Library (1982)
Department of the Treasury Library (1895)
District of Columbia Court of Appeals Library (1981)
District of Columbia Public Library (1943)
Equal Employment Opportunity Commission Library (1984)
Executive Office of the President, Office of Administration, Library & Information Service Division (1965)
Federal Deposit Insurance Corporation Library (1972)
Federal Election Commission Law Library (1975)
Federal Energy Regulatory Commission Library (1983)
Federal Labor Relations Authority Law Library (1982)
Federal Mine Safety & Health Review Commission Library (1979)
Federal Reserve System Board of Governors Research Library (1978)
Federal Reserve System Law Library (1976)
General Accounting Office Technical Library (1974)
General Services Administration Library (1975)
Georgetown University Library (1969)
Georgetown University Law Center Fred O. Dennis Law Library (1978)
George Washington University Melvin Gelman Library (1983)
George Washington University National Law Center Jacob Burns Law Library (1978)
Library of Congress Congressional Research Service (1978)
Library of Congress Serial and Government Publications (1977)
Merit Systems Protection Board Library (1979)
National Defense University Library (1895)
Pension Benefit Guaranty Corporation Legal Dept. Library (1984)
U.S. Court of Appeals Judges' Library (1975)
U.S. Court of Appeals for the Federal Circuit Library (1986)
U.S. Information Agency Library (1984)
U.S. Office of Personnel Management Library (1963)
U.S. Postal Service Library (1895)
U.S. Senate Library (1979)
U.S. Supreme Court Library (1978)
University of the District of Columbia Library Learning Resources Division (1970)
Veterans' Administration Central Office Library (1967)

FLORIDA

Boca Raton

Florida Atlantic University S. E. Wimberly Library (1963)

Clearwater

Clearwater Public Library (1972)

Coral Gables

University of Miami Otto G. Richter Library (1939)

Daytona Beach

Volusia County Library Center (1963)

De Land

Stetson University duPont-Ball Library (1887)

Fort Lauderdale

Broward County Library (1967)
Nova University Law Library (1967)

Fort Pierce

Indian River Community College Library (1975)

Gainesville

University of Florida College of Law Library (1978)
University of Florida Libraries (1907) REGIONAL

Jacksonville

Haydon Burns Public Library (1914)
Jacksonville University Swisher Library (1962)
University of North Florida Thomas G. Carpenter Library (1972)

Lakeland

Lakeland Public Library (1928)

Leesburg

Lake-Sumter Community College Library (1963)

Melbourne

Florida Institute of Technology Library (1963)

Miami

Florida International University Library Tamiami Trail (1970)
Miami-Dade Public Library (1952)

North Miami

Florida International University Bay Vista Campus Library (1977)

Opa Locka

St. Thomas University Library (1977)

Orlando

University of Central Florida Library (1966)

Palatka

Saint Johns River Community College Library (1963)

Panama City

Bay County Public Library (1983)

Pensacola

University of West Florida John C. Pace Library (1966)

Port Charlotte

Charlotte-Glades Library System (1973)

Saint Petersburg

Saint Petersburg Public Library (1965)
Stetson University College of Law Charles A. Dana Law Library (1975)

Sarasota

Selby Public Library (1970)

Tallahassee

Florida Agricultural and Mechanical University Coleman Memorial Library (1936)
Florida State University College of Law Library (1978)
Florida State University Strozier Library (1941)
Florida Supreme Court Library (1974)
State Library of Florida (1929)

Tampa

Tampa-Hillsborough County Public Library (1965)
University of South Florida Library (1962)
University of Tampa Merl Kelce Library (1953)

Winter Park

Rollins College Olin Library (1909)

GEORGIA

Albany

Dougherty County Public Library (1964)

Americus

Georgia Southwestern College James Earl Carter Library (1966)

Athens

University of Georgia Libraries (1907) REGIONAL
University of Georgia School of Law Library (1979)

Atlanta

Atlanta-Fulton Public Library (1880)
Atlanta University Center Robert W. Woodruff Library (1962)
Emory University School of Law Library (1968)
Emory University Woodruff Library (1928)
Georgia Institute of Technology Price Gilbert Memorial Library (1963)
Georgia State Library (unknown)
Georgia State University William Russell Pullen Library (1970)
Georgia State University College of Law Library (1983)
U.S. Court of Appeals 11th Circuit Library (1980)

Augusta

Augusta College Reese Library (1962)
Medical College of Georgia Library (1986)

Brunswick

Brunswick-Glynn County Regional Library (1965)

Carrollton

West Georgia College Irvine Sullivan Ingram Library (1962)

Columbus

Columbus College Simon Schwob Memorial Library (1975)

Dahlonega

North Georgia College Stewart Library (1939)

Dalton

Dalton College Library (1978)

Macon

Mercer University Stetson Memorial Library (1964)
Mercer University Walter F. George School of Law Library (1978)

Marietta

Kennesaw College Library (1968)

Milledgeville

Georgia College Ina Dillard Russell Library (1950)

Rome

Berry College Memorial Library (1970)

Savannah

Chatham-Effingham Liberty Regional Library (1857)

Statesboro

Georgia Southern College Zoah S. Henderson Library (1939)

Valdosta

Valdosta State College Library (1956)

GUAM

Agana

Nieves M. Flores Memorial Library (1962)

Mangilao

University of Guam Robert F. Kennedy Memorial Library (1978)

HAWAII

Hilo

University of Hawaii at Hilo Edwin H. Mookini Library (1962)

Honolulu

Hawaii Medical Library Incorporated (1968)
Hawaii State Library (1929)
Municipal Reference & Records Center (1965)
Supreme Court Law Library (1973)
University of Hawaii Hamilton Library (1907) REGIONAL
University of Hawaii William S. Richardson School of Law Library (1978)

Laie

Brigham Young University Hawaii Campus, Joseph F. Smith Library (1964)

Lihue

Lihue Public Library (1967)

Pearl City

Leeward Community College Library (1967)

Wailuku

Maui Public Library (1962)

IDAHO

Boise

Boise Public Library and Information Center (1929)
Boise State University Library (1966)
Idaho State Law Library (unknown)
Idaho State Library (unknown)

Caldwell

College of Idaho Terteling Library (1930)

Moscow

University of Idaho College of Law Library (1978)
University of Idaho Library (1907) REGIONAL

Nampa

Northwest Nazarene College John E. Riley Library (1984)

Pocatello

Idaho State University Eli Oboler Library (1908)

Rexburg

Ricks College Davis O. McKay Library (1946)

Twin Falls

College of Southern Idaho Library (1970)

ILLINOIS

Bloomington

Illinois Wesleyan University, Sheean Library (1964)

Bourbonnais

Olivet Nazarene University Benner Library & Learning Resource Center (1946)

Carbondale

Southern Illinois University at Carbondale Morris Library (1932)
Southern Illinois University School of Law Library (1978)

Carlinville

Blackburn College Lumpkin Library (1954)

Carterville

Shawnee Library System (1971)

Champaign

University of Illinois Law Library (1965)

Charleston

Eastern Illinois University Booth Library (1962)

Chicago

Chicago Public Library (1876)
Chicago State University Paul and Emily Douglas Library (1954)
DePaul University Law Library (1979)
Field Museum of Natural History Library (1963)
Illinois Institute of Technology Chicago-Kent College of Law Library (1978)
Illinois Institute of Technology Paul V. Galvin Library (1982)
John Marshall Law School Library (1981)
Loyola University of Chicago E. M. Cudahy Memorial Library (1966)
Loyola University School of Law Library (1979)
Northeastern Illinois University Ronald Williams Library (1961)
Northwestern University School of Law Library (1978)
University of Chicago Law Library (1964)
University of Chicago Library (1897)
University of Illinois at Chicago Library (1957)
William J. Campbell Library of the U.S. Courts (1979)

Decatur

Decatur Public Library (1954)

De Kalb

Northern Illinois University Founders' Memorial Library (1960)
Northern Illinois University College of Law Library (1978)

Des Plaines

Oakton Community College Library (1976)

Edwardsville

Southern Illinois University at Edwardsville Lovejoy Memorial Library (1959)

Elsah

Principia College Marshall Brooks Library (1957)

Evanston

Northwestern University Library (1876)

Freeport

Freeport Public Library (1905)

Galesburg

Galesburg Public Library (1896)

Jacksonville

MacMurray College Henry Pfeiffer Library (1929)

Lake Forest

Lake Forest College Donnelley Library (1962)

Lebanon

McKendree College Holman Library (1968)

Lisle

Illinois Benedictine College Theodore F. Lownik Library (1911)

Macomb

Western Illinois University Government Publications & Legal Reference Library (1962)

Moline

Black Hawk College Learning Resources Center (1970)

Monmouth

Monmouth College Hewes Library (1860)

Mount Carmel

Wabash Valley College Bauer Media Center (1975)

Mount Prospect

Mount Prospect Public Library (1977)

Normal

Illinois State University Milner Library (1877)

Oak Park

Oak Park Public Library (1963)

Oglesby

Illinois Valley Community College Jacobs Memorial Library (1976)

Palos Hills

Moraine Valley Community College Learning Resources Center (1972)

Peoria

Bradley University Cullom-Davis Library (1963)
Peoria Public Library (1883)

River Forest

Rosary College Library Rebecca Crown Library (1966)

Rockford

Rockford Public Library (1895)

Romeoville

Lewis University Library (1952)

Springfield

Illinois State Library (unknown) REGIONAL

Streamwood

Poplar Creek Public Library (1980)

University Park

Governors' State University Library (1974)

Urbana

University of Illinois Documents Library (1907)

Wheaton

Wheaton College Buswell Memorial Library (1964)

Woodstock

Woodstock Public Library (1963)

INDIANA

Anderson

Anderson College Charles E. Wilson Library (1959)
Anderson Public Library (1963)

Bloomington

Indiana University Library (1881)
Indiana University Law Library (1978)

Crawfordsville

Wabash College Lilly Library (1906)

Evansville

Evansville and Vanderburgh County Public Library (1928)
University of Southern Indiana Library (1969)

Fort Wayne

Allen County Public Library (1896)
Indiana University-Purdue University at Fort Wayne (1965)

Franklin

Franklin College Library (1976)

Gary

Gary Public Library (1943)
Indiana University Northwest Library (1966)

Greencastle

De Pauw University Roy O. West Library (1879)

Hammond

Hammond Public Library (1964)

Hanover

Hanover College Duggan Library (1892)

Huntington

Huntington College Richlyn Library (1964)

Indianapolis

Butler University Irwin Library (1965)
Indianapolis-Marion County Public Library (1906)
Indiana State Library (unknown) REGIONAL
Indiana Supreme Court Law Library (1975)
Indiana University School of Law Library (1967)
Indiana University-Purdue University Library (1979)

Kokomo

Indiana University at Kokomo Learning Resource Center (1969)

Muncie

Ball State University Alexander M. Bracken Library (1959)
Muncie Public Library (1906)

New Albany

Indiana University Southeast Library (1965)

Notre Dame

Notre Dame Law School Kresge Law Library (1965)
University of Notre Dame Memorial Library (1883)

Rensselaer

Saint Joseph's College Library (1964)

Richmond

Earlham College Lilly Library (1964)
Morrison-Reeves Library (1906)

South Bend

Indiana University at South Bend Library (1965)

Terre Haute

Indiana State University Cunningham Memorial Library (1906)

Valparaiso

Valparaiso University Moellering Memorial Library (1930)
Valparaiso University Law Library (1978)

West Lafayette

Purdue University Libraries (1907)

IOWA

Ames

Iowa State University Library (1907)

Cedar Falls

University of Northern Iowa Library (1946)

Cedar Rapids

Cedar Rapids Public Library (1986)

Council Bluffs

Free Public Library (1885)
Iowa Western Community College Herbert Hoover Library (1972)

Davenport

Davenport Public Library (1973)

Des Moines

Drake University Cowles Library (1966)
Drake University Law Library (1972)
Public Library of Des Moines (1888)
State Library of Iowa (unknown)

Dubuque

Carnegie-Stout Public Library (unknown)
Loras College Wahlert Memorial Library (1967)

Fayette

Upper Iowa University Henderson-Wilder Library (1974)

Grinnell

Grinnell College Burling Library (1874)

Iowa City

University of Iowa College of Law Library (1968)
University of Iowa Libraries (1884) REGIONAL

Lamoni

Graceland College Frederick Madison Smith Library (1927)

Mason City

North Iowa Area Community College Library (1976)

Mount Vernon

Cornell College Russell D. Cole Library (1896)

Orange City

Northwestern College Ramaker Library (1970)

Sioux City

Sioux City Public Library (1894)

KANSAS

Atchison

Benedictine College North Campus Library (1965)

Baldwin City

Baker University Collins Library (1908)

Colby

Colby Community College H. F. Davis Memorial Library (1968)

Emporia

Emporia State University William Allen White Library (1909)

Hays

Fort Hays State University Forsyth Library (1926)

Hutchinson

Hutchinson Public Library (1963)

Lawrence

University of Kansas Law Library (1971)
University of Kansas Spencer Research Library (1869) REGIONAL

Manhattan

Kansas State University Farrell Library (1907)

Pittsburg

Pittsburg State University Leonard H. Axe Library (1952)

Salina

Kansas Wesleyan University Memorial Library (1930)

Shawnee Mission

Johnson County Library (1979)

Topeka

Kansas State Historical Society Library (1877)
Kansas State Library (unknown)
Kansas Supreme Court Law Library (1975)
Washburn University of Topeka Law Library (1971)

Wichita

Wichita State University Ablah Library (1901)

KENTUCKY

Ashland

Boyd County Public Library (1946)

Barbourville

Union College Abigail E. Weeks Memorial Library (1958)

Bowling Green

Western Kentucky University Helm-Cravens Library (1934)

Columbia

Lindsey Wilson College Katie Murrell Library (1987)

Crestview Hills

Thomas More College Library (1970)

Danville

Centre College Grace Doherty Library (1884)

Frankfort

Kentucky Department of Libraries and Archives (1967)
Kentucky State Law Library (unknown)
Kentucky State University Blazer Library (1972)

Hazard

Hazard Community College Library (1968)

Highland Heights

Northern Kentucky University W. Frank Steely Library (1973)

Lexington

University of Kentucky Law Library (1968)
University of Kentucky Libraries (1907) REGIONAL

Louisville

Louisville Free Public Library (1904)
University of Louisville Ekstrom Library (1925)
University of Louisville Law Library (1975)

Morehead

Morehead State University Camden-Carroll Library (1955)

Murray

Murray State University Waterfield Library (1924)

Owensboro

Kentucky-Wesleyan College Library Learning Center (1966)

Richmond

Eastern Kentucky University John Grant Crabbe Library (1966)

Williamsburg

Cumberland College Norma Perkins Hagan (1988)

LOUISIANA

Baton Rouge

Louisiana State Library (1976)
Louisiana State University Middleton Library (1907) REGIONAL
Louisiana State University Paul M. Hebert Law Center Library (1929)
Southern University Law School Library (1979)
Southern University Library (1952)

Eunice

Louisiana State University at Eunice LeDoux Library (1969)

Hammond

Southeastern Louisiana University Sims Memorial Library (1966)

Lafayette

University of Southwestern Louisiana Library (1938)

Lake Charles

McNeese State University Lether E. Frazar Memorial Library (1941)

Monroe

Northeast Louisiana University Sandel Library (1963)

Natchitoches

Northwestern State University of Louisiana Watson Memorial Library (1887)

New Orleans

Law Library of Louisiana (unknown)
Loyola University Government Documents Library (1942)
Loyola University Law Library (1978)
New Orleans Public Library (1883)
Our Lady of Holy Cross College Library (1968)
Southern University in New Orleans Leonard S. Washington Memorial Library (1962)
Tulane University Law Library (1976)
Tulane University Howard-Tilton Memorial Library (1942)
U.S. Court of Appeals 5th Circuit Library (1973)
University of New Orleans Earl K. Long Library (1963)

Pineville

Louisiana College Richard W. Norton Memorial Library (1969)

Ruston

Louisiana Technical University Prescott Memorial Library (1896) REGIONAL

Shreveport

Louisiana State University at Shreveport Library (1967)
Shreve Memorial Library (1923)

Thibodaux

Nicholls State University Ellender Memorial Library (1962)

MAINE

Augusta

Maine Law and Legislative Reference Library (1973)
Maine State Library (unknown)

Bangor

Bangor Public Library (1884)

Brunswick

Bowdoin College Library (1884)

Castine

Maine Maritime Academy Nutting Memorial Library (1969)

Lewiston

Bates College George and Helen Ladd Library (1883)

Orono

University of Maine Raymond H. Fogler Library (1907) REGIONAL

Portland

Portland Public Library (1884)
University of Maine School of Law Garbrecht Law Library (1964)

Presque Isle

University of Maine at Presque Isle Library/Learning Resources Center (1979)

Sanford

Louis B. Goodall Memorial Library (1984)

Waterville

Colby College Miller Library (1884)

MARYLAND

Annapolis

Maryland State Law Library (unknown)
U.S. Naval Academy Nimitz Library (1895)

Baltimore

Enoch Pratt Free Library (1887)
Johns Hopkins University Milton S. Eisenhower Library (1882)
Morgan State University Soper Library (1940)
University of Baltimore Langsdale Library (1973)
University of Baltimore Law Library (1980)
University of Maryland School of Law Marshall Law Library (1969)
U.S. Court of Appeals 4th Circuit Library (1982)

Bel Air

Harford Community College Library (1967)

Beltsville

Department of Agriculture National Agricultural Library (1895)

Bethesda

Department of Health and Human Services National Library of Medicine (1978)
Uniformed Services University of Health Sciences Learning Resource Center (1983)

Catonsville

University of Maryland, Baltimore County Albin O. Kuhn Library & Gallery (1971)

Chestertown

Washington College Clifton M. Miller Library (1891)

College Park

University of Maryland McKeldin Library (1925) REGIONAL

Cumberland

Allegany Community College Library (1974)

Frostburg

Frostburg State University Library (1967)

Patuxent River

Patuxent River Central Library (1968)

Rockville

Montgomery County Department of Public Libraries (1951)

Salisbury

Salisbury State College Blackwell Library (1965)

Towson

Goucher College Julia Rogers Library (1966)
Towson State University Cook Library (1979)

Westminster

Western Maryland College Hoover Library (1886)

MASSACHUSETTS

Amherst

Amherst College Library (1884)
University of Massachusetts University Library (1907)

Boston

Boston Athenaeum Library (unknown)
Boston Public Library (1859) REGIONAL
Boston University School of Law Pappas Law Library (1979)
Northeastern University Dodge Library (1962)
State Library of Massachusetts (unknown)
Suffolk University Law Library (1979)
Supreme Judicial Court Social Law Library (1979)
U.S. Court of Appeals First Circuit Library (1978)

Brookline

Public Library of Brookline (1925)

Cambridge

Harvard College Library (1860)
Harvard Law School Library (1981)
Massachusetts Institute of Technology Library (1946)

Chestnut Hill

Boston College Thomas P. O'Neill Jr., Library (1963)

Chicopee

College of Our Lady of the Elms Alumnae Library (1969)

Lowell

University of Lowell Lydon Library (1952)

Medford

Tufts University Wessel Library (1899)

Milton

Curry College Levin Library. (1972)

New Bedford

New Bedford Free Public Library (1858)

Newton Centre

Boston College Law School Library (1979)

North Dartmouth

Southeastern Massachusetts University Library (1965)

North Easton

Stonehill College Cushing-Martin Library (1962)

Springfield

Springfield City Library (1966)
Western New England College Law Library (1978)

Waltham

Brandeis University Library (1965)
Waltham Public Library (1982)

Wellesley

Wellesley College Library (1943)

Wenham

Gordon College Jenks Learning Resource Center (1963)

Williamstown

William College Sawyer Library (unknown)

Worcester

American Antiquarian Society Library (1814)
University of Massachusetts Medical Center Library (1972)
Worcester Public Library (1859)

MICHIGAN

Albion

Albion College Stockwell-Mudd Library (1966)

Allendale

Grand Valley State College Zumberge Library (1963)

Lansing

Library of Michigan (unknown) REGIONAL
Thomas M. Cooley Law School Library (1978)

Livonia

Livonia Public Library (1967)
Schoolcraft College Library (1962)

Madison Heights

Madison Heights Public Library (1982)

Marquette

Northern Michigan University Lydia M. Olson Library (1963)

Monroe

Monroe County Library System (1974)

Mount Clemens

Macomb County Library (1968)

Mount Pleasant

Central Michigan University Library (1958)

Muskegon

Hackley Public Library (1894)

Petoskey

North Central Michigan College Library (1962)

Port Huron

Saint Clair County Library (1876)

Center

Rochester

Oakland University Kresge Library (1964)

Royal Oak

Royal Oak Public Library (1984)

Saginaw

Hoyt Public Library (1890)

Sault Ste. Marie

Lake Superior State College Kenneth Shouldice Library (1982)

Traverse City

Northwestern Michigan College Mark Osterlin Library (1964)

University Center

Delta College Library (1963)

A-15

Warren

Warren Public Library Arthur J. Miller Branch (1973) ,

Ypsilanti

Eastern Michigan University Library (1965)

MICRONESIA

East Caroline Islands

Community College of Micronesia Library (1982)

MINNESOTA

Bemidji

Bemidji State University A.C. Clark Library (1963)

Blaine

Anoka County Library (1971)

Collegeville

Saint John's University Alcuin Library (1954)

Cottage Grove

Washington County Library-Park Grove Branch (1983)

Duluth

Duluth Public Library (1909)
University of Minnesota Duluth Library (1984)

Eagan

Dakota County Library—Westcott Branch (1983)

Edina

Southdale-Hennepin Area Library (1971)

Mankato

Mankato State University Memorial Library (1962)

Marshall

Southwest State University Library (1986)

Minneapolis

Minneapolis Public Library (1893)
University of Minnesota Law School Library (1976)

University of Minnesota Wilson Library (1907) REGIONAL

Moorhead

Moorhead State University Livingston Lord Library (1956)

Morris

University of Minnesota, Morris, Rodney A. Briggs Library (1963)

Northfield

Carleton College Library (1930).
Saint Olaf College Rolvaag Memorial Library (1930)

Saint Cloud

Saint Cloud State University, Learning Rescources Center (1962)

Saint Paul

Hamline University School of Law Library (1978)
Minnesota Historical Society Library (1867)
Minnesota State Law Library (unknown)
Saint Paul Public Library (1914)
University of Minnesota Saint Paul Campus Library (1974)
William Mitchell College of Law Library (1979)

Saint Peter

Gustavus Adolphus College Library (1941)

Winona

Winona State University Maxwell Library (1969)

MISSISSIPPI

Cleveland

Delta State University W. B. Roberts Library (1975)

Columbus

Mississippi University for Women John Clayton Fant Memorial Library (1929)

Hattiesburg

University of Southern Mississippi Joseph A. Cook Memorial Library (1935)

Jackson

Jackson State University Henry Thomas Sampson Library (1968)
Millsaps College Millsaps-Wilson Library (1963)
Mississippi College School of Law Library (1977)
Mississippi Library Commission (1947)
Mississippi State Law Library (unknown)

Lorman

Alcorn State University J. D. Boyd Library (1970)

Mississippi State

Mississippi State University Mitchell Memorial Library (1907)

University

University of Mississippi Library (1883) REGIONAL
University of Mississippi James O. Eastland Law Library (1967)

MISSOURI

Cape Girardeau

Southeast Missouri State University Kent Library (1916)

Columbia

University of Missouri at Columbia Library (1862) REGIONAL
University of Missouri-Columbia Law Library (1978)

Fulton

Westminster College Reeves Library (1875)

Hillsboro

Jefferson College Library (1984)

Jefferson City

Lincoln University Inman E. Page Library (1944)
Missouri State Library (1963)
Missouri Supreme Court Library (unknown)

Joplin

Missouri Southern State College Library (1966)

Kansas City

Kansas City Missouri Public Library (1881)
Rockhurst College Greenlease Library (1917)
University of Missouri at Kansas City General Library (1938)
University of Missouri Kansas City Leon E. Bloch Law Library (1978)

Kirksville

Northeast Missouri State University Pickler Memorial Library (1966)

Liberty

William Jewell College Charles F. Curry Library (1900)

Maryville

Northwest Missouri State University B. D. Owens Library (1982)

Rolla

University of Missouri-Rolla Curtis Laws Wilson Library (1907)

Saint Charles

Lindenwood College Margaret Leggat Butler Library (1973)

Saint Joseph

Saint Joseph Public Library (1891)

Saint Louis

Maryville College Library (1976)
Saint Louis County Library (1970)
Saint Louis Public Library (1866)
Saint Louis University Law Library (1967)
Saint Louis University Pius XII Memorial Library (1866)
U.S. Court of Appeals Eighth Circuit Library (1972)
University of Missouri at Saint Louis Thomas Jefferson Library (1966)
Washington University John M. Olin Library (1906)
Washington University Law Library (1978)

Springfield

Drury College, Walker Library (1874)
Southwest Missouri State University Duane G. Meyer Library (1963)

Warrensburg

Central Missouri State University Ward Edwards Library (1914)

MONTANA

Billings

Eastern Montana College Library (1958)

Bozeman

Montana State University Renne Library (1907)

Butte

Montana College of Mineral Science and Technology Library (1901)

Havre

Northern Montana College Vande Bogart Library (1980)

Helena

Carroll College Library (1974)
Montana State Library (1966)
State Law Library of Montana (1977)

Missoula

University of Montana Maurene & Mike Mansfield Library (1909)
REGIONAL

NEBRASKA

Blair

Dana College Dana-LIFE Library (1924)

Crete

Doane College Perkins Library (1944)

Fremont

Midland Lutheran College Luther Library (1924)

Kearney

Kearney State College Calvin T. Ryan Library (1962)

Lincoln

Nebraska Library Commission (1972)
Nebraska State Library (unknown)
University of Nebraska-Lincoln College of Law Library (1981)
University of Nebraska-Lincoln D. L. Love Memorial Library (1907)
 REGIONAL

Omaha

Creighton University Reinert/Alumni Library (1964)
Creighton University School of Law Library (1979)
Omaha Public Library W. Dale Clark Library (1880)
University of Nebraska at Omaha University Library (1939)

Scottsbluff

Scottsbluff Public Library (1925)

Wayne

Wayne State College U.S. Conn Library (1970)

NEVADA

Carson City

Nevada State Library (unknown)
Nevada Supreme Court Library (1973)

Las Vegas

Clark County Law Library (1988)
Las Vegas-Clark County Library (1974)
University of Nevada at Las Vegas James Dickinson Library (1959)

Reno

National Judicial College Law Library (1979)
Nevada Historical Society Library (1974)

University of Nevada-Reno Library (1907) REGIONAL
Washoe County Library (1980)

NEW HAMPSHIRE

Concord

Franklin Pierce Law Center Library (1973)
New Hampshire State Library (unknown)

Durham

University of New Hampshire Library (1907)

Hanover

Dartmouth College Library (1884)

Henniker

New England College Danforth Library (1966)

Manchester

Manchester City Library (1884)
New Hampshire College H. A. B. Shapiro Memorial Library (1976)
Saint Anselm College Geisel Library (1963)

Nashua

Nashua Public Library (1971)

NEW JERSEY

Bayonne

Bayonne Free Public Library (1909)

Bloomfield

Bloomfield Public Library (1965)

Bridgeton

Cumberland County Library (1966)

Camden

Rutgers University Camden Library (1966)
Rutgers University School of Law Library (1979)

Convent Station

College of Saint Elizabeth Mahoney Library (1938)

East Brunswick

East Brunswick Public Library (1977)

East Orange

East Orange Public Library (1966)

Elizabeth

Free Public Library of Elizabeth (1895)

Glassboro

Glassboro State College Savitz Library (1963)

Hackensack

Johnson Free Public Library (1966)

Irvington

Irvington Public Library (1966)

Jersey City

Jersey City Public Library (1879)
Jersey City State College Forrest A. Irwin Library (1963)

Lawrenceville

Rider College Franklin F. Moore Library (1975)

Madison

Drew University Library (1939)

Mahwah

Ramapo College Library (1971)

Mount Holly

Burlington County Library (1966)

New Brunswick

New Brunswick Free Public Library (1908)
Rutgers University Alexander Library (1907)

Newark

Newark Public Library (1906) REGIONAL
Rutgers-The State University of New Jersey John Cotton Dana Library (1966)
Rutgers University Law School Ackerson Law Library (1979)
Seton Hall University Law Library (1979)

Newton

Sussex County Library (1986)

Passaic

Passaic Public Library (1964)

Phillipsburg

Phillipsburg Free Public Library (1976)

Plainfield

Plainfield Public Library (1971)

Pomona

Stockton State College Library (1972)

Princeton

Princeton University Library (1884)

Randolph

County College of Morris Sherman H. Masten Learning Resource Center (1975)

Rutherford

Fairleigh Dickinson University Messler Library (1953)

Shrewsbury

Monmouth County Library (1968)

South Orange

Seton Hall University McLaughlin Library (1947)

Teaneck

Fairleigh Dickinson University Weiner Library (1963)

Toms River

Ocean County College Learning Resources Center (1966)

Trenton

New Jersey State Library (unknown)
Trenton Free Public Library (1902)

Union

Kean College of New Jersey Nancy Thompson Library (1971)

Upper Montclair

Montclair State College Harry A. Sprague Library (1967)

Wayne

Wayne Public Library (1972)

West Long Branch

Monmouth College Guggenheim Memorial Library (1963)

Woodbridge

Woodbridge Public Library (1965)

NEW MEXICO

Albuquerque

University of New Mexico Medical Center Library (1973)
University of New Mexico School of Law Library (1973)
University of New Mexico General Library (1896) REGIONAL

Hobbs

New Mexico Junior College Pannell Library (1969)

Las Cruces

New Mexico State University Library (1907)

Las Vegas

New Mexico Highlands University Donnelly Library (1913)

Portales

Eastern New Mexico University Golden Library (1962)

Santa Fe

New Mexico State Library (1960) REGIONAL
New Mexico Supreme Court Law Library (unknown)

Silver City

Western New Mexico University Miller Library (1972)

Socorro

New Mexico Institute of Mining & Technology Martin Speare Memorial Library (1984)

NEW YORK

Albany

Albany Law School Schaffer Law Library (1979)
New York State Library (unknown) REGIONAL
State University of New York at Albany University Library (1964)

Auburn

Seymour Library (1972)

Binghamton

State University of New York at Binghamton Glenn G. Bartle Library (1962)

Brockport

State University of New York at Brockport Drake Memorial Library (1967)

Bronx

Fordham University Library (1937)
Herbert H. Lehman College Library (1967)
New York Public Library (1973)
State University of New York Maritime College Stephen B. Luce Library (1947)

Bronxville

Sarah Lawrence College Esther Raushenbush Library (1969)

Brooklyn

Brooklyn College Library (1936)
Brooklyn Law School Library (1974)
Brooklyn Public Library Business Library (1984)
Brooklyn Public Library (1908)
Pratt Institute Library (1891)
State University of New York Health Center at Brooklyn Library (1958)

Buffalo

Buffalo and Erie County Public Library (1895)
State University of New York at Buffalo Charles B. Sears Law Library (1978)
State University of New York at Buffalo Lockwood Memorial Library (1963)

Canton

Saint Lawrence University Owen D. Young Library (1920)

Corning

Corning Community College Arthur A. Houghton Jr. Library (1963)

Cortland

State University of New York College at Cortland Memorial Library (1964)

Delhi

State University Agricultural and Technical College Library (1970)

East Islip

East Islip Public Library (1973)

Elmira

Elmira College Gannett Tripp Learning Center (1956)

Farmingdale

State University of New York at Farmingdale Greenley Library (1917)

Flushing

CUNY Law School at Queens College CUNY Law Library (1983)
Queens College Paul Klapper Library (1939)

Garden City

Adelphi University Swirbul Library (1966)

Geneseo

State University of New York at Geneseo Milne Library (1967)

Greenvale

Long Island University B. Davis Schwartz Memorial Library (1964)

Hamilton

Colgate University, Everett Needham Case Library (1902)

Hempstead

Hofstra University Library (1964)
Hofstra University School of Law Library (1979)

Huntington

Touro College Jacob D. Fuchsberg Law Center Library (1985)

Ithaca

Cornell University Library (1907)
Cornell Law Library (1978)
New York State College of Agriculture and Human Ecology Albert R. Mann Library (1943)

Jamaica

Queens Borough Public Library (1926)
Saint John's University Library (1956)
Saint John's University School of Law Library (1978)

Kings Point

U.S. Merchant Marine Academy Schuyler Otis Bland Library (1962)

Long Island City

Fiorello H. LaGuardia Community College Library (1981)

Middletown

Thrall Library (1986)

Mount Vernon

Mount Vernon Public Library (1962)

New Paltz

State University College at New Paltz Sojourner Truth Library (1965)

New York City

City College of City University of New York Cohen Library (1884)
College of Insurance Library (1965)
Columbia University Libraries (1882)
Columbia University School of Law Library (1981)
Cooper Union for the Advancement of Science and Arts Library (1930)
Fordham University School of Law Leo T. Kissam Memorial Library (1987)
Medical Library Center of New York (1976)
New York Law Institute Library (1909)
New York Law School Library (1979)
New York Public Library Astor Branch (1907)
New York Public Library Lenox Branch (1884)
New York University Law Library (1974)
New York University Elmer Holmes Bobst Library (1967)
U.S. Court of Appeals Second Circuit Library (1976)
Yeshiva University Chutick Law Library Cardozo School of Law (1979)
Yeshiva University Pollack Library (1979)

Newburgh

Newburgh Free Library (1909)

Niagara Falls

Niagara Falls Public Library (1976)

Oakdale

Dowling College Library (1965)

Oneonta

State University College at Onenonta James M. Milne Library (1966)

Oswego

State University of New York at Oswego Penfield Library (1966)

Plattsburgh

State University College at Plattsburgh Benjamin F. Feinberg Library (1967)

Potsdam

Clarkson University Harriet Call Burnap Memorial Library (1938)
State University College at Potsdam Frederick W. Crumb Memorial Library (1964)

Poughkeepsie

Vassar College Library (1943)

Purchase

State University of New York at Purchase Library (1969)

Rochester

Rochester Public Library (1963)
University of Rochester Rush Rhees Library (1880)

Saint Bonaventure

Saint Bonaventure University Friedsam Memorial Library (1938)

Saratoga Springs

Skidmore College Library (1964)

Schenectady

Union College Schaffer Library (1901)

Southampton

Long Island University Southhampton Campus Library (1973)

Sparkill

St. Thomas Aquinas College Lougheed Library (1984)

Staten Island

Wagner College Horrmann Library (1953)

Stony Brook

State University of New York at Stony Brook Main Library (1963)

Syracuse

Onondaga County Public Library (1978)
Syracuse University Bird Library (1878)
Syracuse University College of Law H. Douglas Barclay Law Library (1978)

Troy

Troy Public Library (1869)

Uniondale

Nassau Library System (1965)

Utica

Utica Public Library (1885)
SUNY College of Technology Library (1977)

West Point

U.S. Military Academy Library (unknown)

White Plains

Pace University Law School Library (1976)

Yonkers

Yonkers Public Library Getty Square Branch (1910)

Yorktown Heights

Mercy College Library (1976)

NORTH CAROLINA

Asheville

University of North Carolina at Asheville D. Hiden Ramsey Library (1965)

Bolling Springs

Gardner-Webb College Dover Memorial Library (1974)

Boone

Appalachian State University Carol Grotnes Belk Library (1963)

Bules Creek

Campbell University Carrie Rich Memorial Library (1965)

Chapel Hill

University of North Carolina at Chapel Hill Davis Library (1884) REGIONAL
University of North Carolina Law Library (1976)

Charlotte

Public Library of Charlotte and Mecklenburg County (1964)
Queens College Everett Library (1927)
University of North Carolina at Charlotte Atkins Library (1964)

Cullowhee

Western Carolina University Hunter Library (1953)

Davidson

Davidson College Library (1893)

Durham

Duke University School of Law Library (1978)
Duke University William R. Perkins Library (1890)
North Carolina Central University Law School Library (1979)
North Carolina Central University James E. Shepard Memorial Library (1973)

Elon College

Elon College Iris Holt McEwen Library (1971)

Fayetteville

Fayetteville State University Charles W. Chesnutt Library (1971)

Greensboro

North Carolina Agricultural and Technical State University F. D. Bluford Library (1937)
University of North Carolina at Greensboro Walter Clinton Jackson Library (1963)

Greenville

East Carolina University J. Y. Joyner Library (1951)

Laurinburg

Saint Andrews Presbyterian College DeTamble Library (1969)

Lexington

Davidson County Public Library (1971)

Mount Olive

Mount Olive College Moye Library (1971)

Pembroke

Pembroke State University Mary H. Livermore Library (1956)

Raleigh

Department of Cultural Resources Division of State Library (unknown)
North Carolina State University D. H. Hill Library (1923)
North Carolina Supreme Court Library (1972)

Rocky Mount

North Carolina Wesleyan College Library (1969)

Salisbury

Catawba College Library (1925)

Wilmington

University of North Carolina at Wilmington William M. Randall Library
(1965)

Wilson

Atlantic Christian College Hackney Library (1930)

Winston-Salem

Forsyth County Public Library (1954)
Wake Forest University Z. Smith Reynolds Library (1902)

NORTH DAKOTA

Bismarck

North Dakota State Library (1971)
North Dakota Supreme Court Law Library (unknown)
State Historical Society of North Dakota State Archives & Historical
  Research Library (1907)
Veterans' Memorial Public Library (1967)

Dickinson

Dickinson State College Stoxen Library (1968)

Fargo

Fargo Public Library (1964)
North Dakota State University Library (1907) REGIONAL

Grand Forks

University of North Dakota Chester Fritz Library (1890)

Minot

Minot State University Memorial Library (1925)

Valley City

Valley City State University Allen Memorial Library (1913)

NORTHERN MARIANA ISLANDS

Saipan

Northern Marianas College Olympio T. Borja Memorial Library (1988)

OHIO

Ada

Ohio Northern University J. P. Taggart Law Library (1965)

Akron

Akron-Summit County Public Library (1952)
University of Akron Bierce Library (1963)
University of Akron School of Law Library (1978)

Alliance

Mount Union College Library (1888)

Ashland

Ashland College Library (1938)

Athens

Ohio University Alden Library (1886)

Batavia

University of Cincinnati Clermont College Library (1973)

Bluffton

Bluffton College Musselman Library (1951)

Bowling Green

Bowling Green State University Jerome Library (1933)

Canton

Malone College Everett L. Cattel Library (1970)

Chardon

Chardon Public Library (1971)

Cincinnati

Public Library of Cincinnati and Hamilton County (1884)
University of Cincinnati Central Library (1929)
University of Cincinnati College of Law Marx Law Library (1978)
U.S. Court of Appeals 6th Circuit Library (1986)

Cleveland

Case Western Reserve University Freiberger Library (1913)
Case Western Reserve University School of Law Library (1979)
Cleveland Public Library (1886)
Cleveland State University Cleveland-Marshall College of Law,
  Joseph W. Bartunek III Law Library (1978)
Cleveland State University Library (1966)
Municipal Reference Library (1970).

Cleveland Heights

Cleveland Heights-University Heights Public Library (1970)

Columbus

Capital University Law School Library (1980)
Capital University Library (1968)
Ohio State University College of Law Library (1984)
Ohio State University Libraries (1907)
Ohio Supreme Court Law Library (1973)
Public Library of Columbus and Franklin County (1885)
State Library of Ohio (unknown) REGIONAL

Dayton

Dayton and Montgomery County Public Library (1909)
University of Dayton Roesch Library (1969)
Wright State University Library (1965)

Delaware

Ohio Wesleyan University L. A. Beeghly Library (1845)

Elyria

Elyria Public Library (1966)

Findlay

Findlay College Shafer Library (1969)

Gambier

Kenyon College Library (1873)

Granville

Denison University Libraries William H. Doane Library (1884)

Hiram

Hiram College Teachout-Price Memorial Library (1874)

Kent

Kent State University Libraries (1962)

Marietta

Marietta College Dawes Memorial Library (1884)

Marion

Marion Public Library (1979)

Middletown

Miami University Middletown Gardner-Harvey Library (1970)

New Concord

Muskingum College Library (1966)

Oberlin

Oberlin College Library (1858)

Oxford

Miami University Libraries King Library (1909)

Portsmouth

Shawnee State University Library (1987)

Rio Grande

Rio Grande College and Community College Jeanette Albiez Davis
  Library (1966)

Springfield

Warder Public Library (1884)

Steubenville

Franciscan University of Steubenville John Paul II Library (1971)
Public Library of Steubenville and Jefferson County (1950)

Tiffin

Heidelberg College Beeghly Library (1964)

Toledo

Toledo-Lucas County Public Library (1884)
University of Toledo College of Law Library (1981)
University of Toledo Library (1963)

University Heights

John Carroll University Grasselli Library (1963)

Westerville

Otterbein College Courtright Memorial Library (1967)

Wilmington

Wilmington College S. Arthur Watson Library (1986)

Wooster

College of Wooster Andrews Library (1966)

Worthington

Worthington Public Library (1984)

Youngstown

Public Library of Youngstown and Mahoning County (1923)
Youngstown State University William F. Maag Library (1971)

OKLAHOMA

Ada

East Central Oklahoma State University Linscheid Library (1914)

Alva

Northwestern Oklahoma State University J. W. Martin Library (1907)

Bethany

Southern Nazarene University R. T. Williams Learning Resources
Center (1971)

Durant

Southeastern Oklahoma State University Henry G. Bennett Memorial
Library (1929)

Edmond

Central State University Library (1934)

Enid

Public Library of Enid and Garfield County (1908)

Langston

Langston University G. Lamar Harrison Library (1941)

Lawton

Lawton Public Library (1987)

Norman

University of Oklahoma Libraries Bizzell Memorial Library (1893)
University of Oklahoma Law Library (1978)

Oklahoma City

Metropolitan Library System Main Library (1974)
Oklahoma City University Dulaney Browne Library (1963)
Oklahoma Department of Libraries (1893) REGIONAL

Shawnee

Oklahoma Baptist University Library (1933)

Stillwater

Oklahoma State University Library (1907) REGIONAL

Tahlequah

Northeastern Oklahoma State University John Vaughan Library (1923)

Tulsa

Tulsa City-County Library System (1963)
University of Tulsa College of Law Library (1979)
University of Tulsa McFarlin Library (1929)

Weatherford

Southwestern Oklahoma State University Al Harris Library (1958)

OREGON

Ashland

Southern Oregon State College Library (1953)

Bend

Central Oregon Community College Library/Media Service
(1985)

Corvallis

Oregon State University Library (1907)

Eugene

University of Oregon Law Library (1979)
University of Oregon Library (1883)

Forest Grove

Pacific University Harvey W. Scott Memorial Library (1897)

Klamath Falls

Oregon Institute of Technology Learning Resources Center-Library
(1962)

La Grande

Eastern Oregon State College Walter M. Pierce Library (1954)

McMinnville

Linfield College Northup Library (1965)

Monmouth

Western Oregon State College Library (1967)

Pendleton

Blue Mountain Community College Library (1983)

Portland

Lewis and Clark College Aubrey R. Watzek Library (1967)
Library Association of Portland (1884)
Northwestern School of Law Lewis and Clark College Paul L. Boley
Law Library (1979)
Portland State University Millar Library (1963) REGIONAL
Reed College Library (1912)
U.S Department of Energy Bonneville Power Administration Library
(1962)

Salem

Oregon State Library (unknown)
Oregon Supreme Court Law Library (1974)
Willamette University College of Law Library (1979)
Willamette University Main Library (1969)

PENNSYLVANIA

Allentown

Muhlenberg College Haas Library (1939)

Altoona

Altoona Area Public Library (1969)

Bethel Park

Bethel Park Public Library (1980)

Bethlehem

Lehigh University Libraries Linderman Library (1876)

Blue Bell

Montgomery County Community College Learning Resources Center (1975)

Bradford

University of Pittsburgh at Bradford Bradford Campus Library (1979)

California

California University of Pennsylvania Louis L. Manderino Library (1986)

Carlisle

Dickinson College Boyd Lee Spahr Library (1947)
Dickinson School of Law Sheeley-Lee Law Library (1978)

Cheyney

Cheyney University Leslie Pinckney Hill Library (1967)

Collegeville

Ursinus College Myrin Library (1963)

Coraopolis

Robert Morris College Library (1978)

Doylestown

Bucks County Free Library (1970)

East Stroudsburg

East Stroudsburg University Kemp Library (1966)

Erie

Erie County Library System (1897)

Greenville

Thiel College Langenheim Memorial Library (1963)

Harrisburg

State Library of Pennsylvania (unknown) REGIONAL

Haverford

Haverford College Magill Library (1897)

Hazleton

Hazleton Area Public Library (1964)

Indiana

Indiana University of Pennsylvania Stapleton Library (1962)

Johnstown

Cambria County Library System Glosser Memorial Library Building (1965)

Lancaster

Franklin and Marshall College Shadek-Fackenthal Library (1895)

Lewisburg

Bucknell University Ellen Clarke Bertrand Library (1963)

Mansfield

Mansfield University Library (1968)

Meadville

Allegheny College Lawrence Lee Pelletier Library (1907)

Millersville

Millersville University Helen A. Ganser Library (1966)

Monessen

Monessen Public Library (1969)

New Castle

New Castle Public Library (1963)

Newtown

Bucks County Community College Library (1968)

Norristown

Montgomery County-Norristown Public Library (1969)

Philadelphia

Drexel University W. W. Hagerty Library (1963)

Free Library of Philadelphia (1897)
Saint Joseph's University Drexel Library (1974)
Temple University Paley Library (1947)
Temple University Law Library (1979)
Thomas Jefferson University Scott Memorial Library (1978)
U.S. Court of Appeals Third Circuit Library (1973)
University of Pennsylvania Biddle Law Library (1974)
University of Pennsylvania Library (1886)

Pittsburgh

Allegheny County Law Library (1977)
Carnegie Library of Pittsburgh (1895)
Carnegie Library of Pittsburgh Allegheny Regional Branch (1924)
Duquesne University Law Library (1978)
La Roche College John J. Wright Library (1974)
U.S. Bureau of Mines Library (1962)
University of Pittsburgh Hillman Library (1910)
University of Pittsburgh Law Library (1979)

Pottsville

Pottsville Free Public Library (1967)

Reading

Reading Public Library (1901)

Scranton

Scranton Public Library (1895)

Shippensburg

Shippensburg University Ezra Lehman Memorial Library (1973)

Slippery Rock

Slippery Rock University Bailey Library (1965)

Swarthmore

Swarthmore College McCabe Library (1923)

University Park

Pennsylvania State University Libraries Pattee Library (1907)

Villanova

Villanova University Law School Pulling Law Library (1964)

Warren

Warren Library Association Warren Public Library (1885)

Waynesburg

Waynesburg College Library (1964)

West Chester

West Chester University Francis Harvey Green Library (1967)

Wilkes-Barre

King's College D. Leonard Corgan Library (1949)

Williamsport

Lycoming College Library (1970)

York

York College of Pennsylvania Schmidt Library (1963)

Youngwood

Westmoreland County Community College Learning Resources Center (1972)

PUERTO RICO

Mayaguez

University of Puerto Rico Mayaguez Campus Library (1928)

Ponce

Catholic University of Puerto Rico Encarnacion Valdes Library (1966)
Catholic University of Puerto Rico School of Law Library (1978)

Rio Piedras

University of Puerto Rico J. M. Lazaro Library (1928)

REPUBLIC OF PANAMA

Balboa Heights

Panama Canal Commission Technical Resources Center (1963)

RHODE ISLAND

Barrington

Barrington Public Library (1986)

Kingston

University of Rhode Island Library (1907)

Newport

U.S. Naval War College Library (1963)

Providence

Brown University John D. Rockefeller Jr. Library (unknown)
Providence College Phillips Memorial Library (1969)
Providence Public Library (1884)
Rhode Island College James P. Adams Library (1965)
Rhode Island State Law Library (1979)
Rhode Island State Library (1895)

Warwick

Warwick Public Library (1966)

Westerly

Westerly Public Library (1909)

Woonsocket

Woonsocket Harris Public Library (1977)

SOUTH CAROLINA

Charleston

Baptist College at Charleston L. Mendel Rivers Library (1967)
The Citadel Military College Daniel Library (1962)
College of Charleston Robert Scott Small Library (1869)

Clemson

Clemson University Cooper Library (1893)

Columbia

Benedict College Library Payton Learning Resources Center (1969)
South Carolina State Library (1895)
University of South Carolina Coleman Karesh Law Library (1983)
University of South Carolina Thomas Cooper Library (1884)

Conway

University of South Carolina Coastal Carolina College Kimbel Library (1974)

Due West

Erskine College McCain Library (1968)

Florence

Florence County Library (1967)
Francis Marion College James A. Rogers Library (1970)

Greenville

Furman University Library (1962)
Greenville County Library (1966)

Greenwood

Lander College Larry A. Jackson Library (1967)

Orangeburg

South Carolina State College Miller F. Whittaker Library (1953)

Rock Hill

Winthrop College Dacus Library (1896)

Spartanburg

Spartanburg County Public Library (1967)

SOUTH DAKOTA

Aberdeen

Northern State College Beulah Williams Library (1963)

Brookings

South Dakota State University H. M. Briggs Library (1889)

Pierre

South Dakota State Library (1973)
South Dakota Supreme Court Library (1978)

Rapid City

Rapid City Public Library (1963)
South Dakota School of Mines and Technology Devereaux Library (1963)

Sioux Falls

Augustana College Mikkelsen Library (1969)
Sioux Falls Public Library (1903)

Spearfish

Black Hills State College Library Learning Center (1942)

Vermillion

University of South Dakota I. D. Weeks Library (1889)

TENNESSEE

Bristol

King College E. W. King Library (1970)

Chattanooga

Chattanooga-Hamilton County Bicentennial Library (1908)
U.S. Tennessee Valley Authority Technical Library (1976)

Clarksville

Austin Peay State University Felix G. Woodward Library (1945)

Cleveland

Cleveland State Community College Library (1973)

Columbia

Columbia State Community College John W. Finney Memorial Library (1973)

Cookeville

Tennessee Technological University Jere Whitson Memorial Library (1969)

Jackson

Lambuth College Luther L. Gobbel Library (1967)

Jefferson City

Carson-Newman College Library (1964)

Johnson City

East Tennessee State University Sherrod Library (1942)

Knoxville

Knoxville County Public Library System Lawson McGhee Library (1973)
University of Tennessee at Knoxville John C. Hodges Library (1907)
University of Tennessee Law Library (1971)

Martin

University of Tennessee at Martin Paul Meek Library (1957)

Memphis

Memphis-Shelby County Public Library and Information Center (1896)
Memphis State University Cecil C. Humphreys School of Law Library (1979)
Memphis State University Libraries (1966)

Murfreesboro

Middle Tennessee State University Todd Library (1912)

Nashville

Fisk University Library (1965)
Public Library of Nashville and Davidson County (1884)
Tennessee State Library and Archives (unknown)
Tennessee State University Brown-Daniel Library (1972)
Vanderbilt University Alyne Queener Massey Law Library (1976)
Vanderbilt University Library (1884)

Sewanee

University of the South Jessie Ball duPont Library (1873)

TEXAS

Abilene

Abilene Christian University Margaret and Herman Brown Library (1978)

Hardin-Simmons University Rupert and Pauline Richardson Library (1940)

Arlington

Arlington Public Library (1970)
University of Texas at Arlington Library (1963)

Austin

Texas State Law Library (1972)
Texas State Library (unknown) REGIONAL
University of Texas at Austin Perry-Castaneda Library (1884)
University of Texas at Austin Edie and Lew Wasserman Public Affairs Library (1966)
University of Texas at Austin Tarlton Law Library (1965)

Baytown

Lee College Library (1970)

Beaumont

Lamar University Mary and John Gray Library (1957)

Brownwood

Howard Payne University Walker Memorial Library (1964)

Canyon

West Texas State University Cornette Library (1928)

College Station

Texas Agricultural and Mechanical University David G. Evans Library (1907)

Commerce

East Texas State University James Gilliam Gee Library (1937)

Corpus Christi

Corpus Christi State University Library (1976)

Corsicana

Navarro College Gaston T. Gooch Library (1965)

Dallas

Dallas Baptist University Vance Memorial Library (1967)
Dallas Public Library (1900)
Southern Methodist University Fondren Library (1925)
University of Texas Health Science Center-Dallas Library (1975)

Denton

North Texas State University Library (1948)

Edinburg

Pan American University Library (1959)

San Antonio

Saint Mary's University Academic Library (1964)
Saint Mary's University Sarita Kennedy East Law Library (1982)
San Antonio College Library (1972)
San Antonio Public Library (1899)
Trinity University Elizabeth Coates Maddux Library (1964)
University of Texas at San Antonio Library (1973)

San Marcos

Southwest Texas State University Library (1955)

Seguin

Texas Lutheran College Blumberg Memorial Library (1970)

Sherman

Austin College Abell Library (1963)

Texarkana

Texarkana College Palmer Memorial Library (1963)

Victoria

Victoria College/University of Houston Victoria Campus Library (1973)

Waco

Baylor University Law Library (1982)
Baylor University Moody Memorial Library (1905)

Wichita Falls

Midwestern State University Moffett Library (1963)

UTAH

Cedar City

Southern Utah State College Library (1964)

Ephraim

Snow College Lucy A. Phillips Library (1963)

Logan

Utah State University Merrill Library and Learning Resources Center (1907) REGIONAL

Ogden

Weber State College Stewart Library (1962)

Provo

Brigham Young University Harold B. Lee Library (1908)
Brigham Young University Law Library (1972)

Salt Lake City

University of Utah Eccles Health Sciences Library (1970)
University of Utah Law Library (1966)
University of Utah Marriott Library (1893)
Utah State Library (unknown)
Utah State Supreme Court Law Library (1975)

VERMONT

Burlington

University of Vermont Bailey/Howe Library (1907)

Castleton

Castleton State College Calvin Coolidge Library (1969)

Johnson

Johnson State College John Dewey Library (1955)

Lyndonville

Lyndon State College Samuel Reed Hall Library (1969)

Middlebury

Middlebury College Egbert Starr Library (1884)

Montpelier

Vermont Department of Libraries (1845)

Northfield

Norwich University Library (1908)

South Royalton

Vermont Law School Library (1978)

VIRGIN ISLANDS

Saint Croix

Florence Williams Public Library (1968)

Saint Thomas

College of the Virgin Islands Ralph M. Paiewonsky Library (1973)
Enid M. Baa Library and Archives (1968)

VIRGINA

Alexandria

Dept. of the Navy Office of Judge Advocate General Law Library (1963)

Arlington

George Mason University School of Law Library (1981)
U.S. Patent & Trademark Office Science Library (1986)

Blacksburg

Virginia Polytechnic Institute and State University Carol M. Newman Library (1907)

Bridgewater

Bridgewater College Alexander Mack Memorial Library (1902)

Charlottesville

University of Virginia Alderman Library (1910) REGIONAL
University of Virginia Arthur J. Morris Law Library (1964)

Chesapeake

Chesapeake Public Library (1970)

Danville

Danville Community College Learning Resources Center (1969)

Emory

Emory and Henry College Kelly Library (1884)

Fairfax

George Mason University Fenwick Library (1960)

Fredericksburg

Mary Washington College E. Lee Trinkle Library (1940)

Hampden-Sydney

Hampden-Sydney College Eggleston Library (1891)

Hampton

Hampton University Huntington Memorial Library (1977)

Harrisonburg

James Madison University Carrier Library (1973)

Hollins College

Hollins College Fishburn Library (1967)

Lexington

Virginia Military Institute Preston Library (1874)
Washington and Lee University University Library (1910)
Washington and Lee University Wilbur C. Hall Law Library (1978)

Martinsville

Patrick Henry Community College Library (1971)

Norfolk

Norfolk Public Library (1895)
Old Dominion University Library (1963)
U.S. Armed Forces Staff College Library (1963)

Petersburg

Virginia State University Johnston Memorial Library (1907)

Quantico

Federal Bureau of Investigation Academy Library (1970)
Marine Corps Education Center MCDEC James Carson Breckinridge
  Library (1967)

Reston

Department of the Interior Geological Survey Library (1963)

Richmond

U.S. Court of Appeals Fourth Circuit Library (1973)
University of Richmond Boatwright Memorial Library (1900)
University of Richmond Law School Library (1982)
Virginia Commonwealth University James Branch Cabell Library (1971)
Virginia State Law Library (1973)
Virginia State Library & Archives (unknown)

Salem

Roanoke College Library (1886)

Williamsburg

College of William and Mary Marshall-Wythe Law Library (1978)
College of William and Mary Swem Library (1936)

Wise

Clinch Valley College John Cook Wyllie Library (1971)

WASHINGTON

Bellingham

Western Washington University Mable Zoe Wilson Library (1963)

Cheney

Eastern Washington University JFK Library (1966)

Des Moines

Highline Community College Library (1983)

Ellensburg

Central Washington University Library (1962)

Everett

Everett Public Library (1914)

Olympia

Evergreen State College Daniel J. Evans Library (1972)
Washington State Law Library (1979)
Washington State Library (unknown) REGIONAL

Port Angeles

North Olympic Library System (1965)

Pullman

Washington State University Holland Library (unknown)

Seattle

Seattle Public Library (1908)
University of Washington Suzallo Library (1890)
University of Washington Mariah Gould Gallagher Law Library (1969)
U.S. Court of Appeals 9th Circuit Library (1981)

Spokane

Gonzaga University School of Law Library (1979)
Spokane Public Library (1910)

Tacoma

Tacoma Public Library (1894)
University of Puget Sound Collins Memorial Library (1938)
University of Puget Sound School of Law Library (1978)

Vancouver

Fort Vancouver Regional Library (1962)

Walla Walla

Whitman College Penrose Memorial Library (1890)

WEST VIRGINIA

Athens

Concord College J. Frank Marsh Library (1924)

Bluefield

Bluefield State College Hardway Library (1972)

Charleston

Kanawha County Public Library (1952)
West Virginia Library Commission (1975)
West Virginia Supreme Court Law Library (1977)

Elkins

Davis and Elkins College Library (1913)

Fairmont

Fairmont State College Library (1884)

Madison

Madison Public Library (1965)
State Historical Society of Wisconsin Library (1870) REGIONAL
University of Wisconsin-Madison Memorial Library (1939)
University of Wisconsin-Madison Law Library (1981)
Wisconsin State Law Library (unknown)

Milwaukee

Alverno College Library/Media Center (1971)
Marquette University Law Library (1987)
Medical College of Wisconsin, Inc. Todd Wehr Library (1980)
Milwaukee County Law and Reference Library (1934)
Milwaukee Public Library (1861) REGIONAL
Mount Mary College Haggerty Library (1964)
University of Wisconsin-Milwaukee Library (1960)

Oshkosh

University of Wisconsin-Oshkosh Forrest R. Polk Library (1956)

Platteville

University of Wisconsin-Platteville Karrman Library (1964)

Racine

Racine Public Library (1898)

Ripon

Ripon College Library (1982)

River Falls

University of Wisconsin-River Falls Chalmer Davee Library (1962)

Sheboygan

Mead Public Library (1983)

Stevens Point

University of Wisconsin-Stevens Point Library (1951)

Superior

Superior Public Library (1908)
University of Wisconsin-Superior Jim Dan Hill Library (1935)

Waukesha

Waukesha Public Library (1966)

Wausau

Marathon County Public Library (1971)

Whitewater

University of Wisconsin-Whitewater Harold Anderson Library (1963)

WYOMING

Casper

Natrona County Public Library (1929)

Cheyenne

Wyoming State Law Library (1977)
Wyoming State Library (unknown) REGIONAL

Gillette

Campbell County Public Library (1980)

Laramie

University of Wyoming, Coe Library (1907)
University of Wyoming Law Library (1978)

Powell

Northwest Community College John Taggart Hinckley Library (1967)

Riverton

Central Wyoming College Learning Resources Center (1969)

Rock Springs

Western Wyoming Community College Library (1969)

Sheridan

Sheridan College Griffith Memorial Library (1963)

APPENDIX B
List of District Offices of the U.S. Department of Commerce

ALABAMA

Birmingham–2015 2nd Avenue North, 3rd Floor, Berry Building, 35203, Area Code 205 Tel 254–1331, FTS 8 229–1331

ALASKA

Anchorage–222 W. 7th Avenue, P.O. Box 32, 99513, Area Code 907 Tel 271–5041, FTS 8 271–5041

ARIZONA

Phoenix–Federal Building & U.S. Courthouse, 230 N. 1st Avenue, Room 3412, 85025, Area Code 602 Tel 261–3285, FTS 8 261–3285

ARKANSAS

Little Rock–Suite 811, Savers Federal Building, 320 W. Capitol Avenue, 72201, Area Code 501 Tel 378–5794, FTS 8 740–5794

CALIFORNIA

Los Angeles–Room 810, 11777 San Vicente Blvd., 90049–5076, Area Code 213 Tel 209–6616, FTS 8 793–6616

•San Diego–6363 Greenwich Drive, Suite 145, 92122, Area Code 619 Tel 557–5395, FTS 8 895–5395

San Francisco–Federal Building, Box 6013, 450 Golden Gate Avenue, 94102, Area Code 415 Tel 556–5860, FTS 8 556–5868

Santa Ana–116A W. 4th Street, Suite #1, 92701, Area Code 714 Tel 836–2461, FTS 8 799–2461

COLORADO

Denver–Room 119, U.S. Customhouse, 721 19th Street, 80202, Area Code 303 Tel 844–3246, FTS 8 564–3246

CONNECTICUT

Hartford–Room 610–B, Federal Office Building, 450 Main Street, 06103, Area Code 203 Tel 240–3530, FTS 8 244–3530

DISTRICT OF COLUMBIA

Room 1066, HCHB, Department of Commerce, 14th Street & Constitution Avenue, N.W., 20230, Area Code 202 Tel 377–3181, FTS 8 377–3181

FLORIDA

•Clearwater–128 N. Osceola Avenue, 33515, Area Code 813 Tel 461–0011, FTS 8 826–3738

•Jacksonville–Independence Square, Suite 200A, 32216, Area Code 904 Tel 791–2796, FTS 8 946–2796

Miami–Suite 224, Federal Building, 51 S.W. First Avenue, 33130, Area Code 305 Tel 536–5267, FTS 8 350–5267

•Orlando–Room 346, University of Central Florida, 32802, Area Code 305 Tel 648–1608, FTS 8 820–6235

•Tallahassee–Collins Building, Room 401, 107 W. Gaines Street, 32304, Area Code 904 Tel 488–6469, FTS 8 965–7194

GEORGIA

Atlanta–Room 625, 1365 Peachtree Street, N.E., 30309, Area Code 404 Tel 347–2271, FTS 8 257–2271

Savannah–120 Barnard Street, A–107, 31402, Area Code 912 Tel 944–4204, FTS 8 248–4204

HAWAII

Honolulu–4106 Federal Building, P.O. Box 50026, 300 Ala Moana Blvd., 96850, Area Code 808 Tel 541–1782, FTS 8 551–1785

IDAHO

Boise (Denver, Colorado District)–700 W. State Street, 83720, Area Code 208 Tel 334–2470, FTS 8 554–9254

ILLINOIS

Chicago–1406 Mid Continental Plaza Building, 55 E. Monroe Street, 60603, Area Code 312 Tel 353–4450, FTS 8 353–4450

•Palatine–W. R. Harper College, Algonquin & Rodelle Road, 60067, Area Code 312 Tel 397–3000, x532

•Rockford–515 N. Court Street, P.O. Box 1747, 61110–0247, Area Code 815 Tel 987–8123, FTS 8 363–4347

INDIANA

Indianapolis–One N. Capitol, Suite 520, 46204, Area Code 317 Tel 269–6214, FTS 8 331–6214

IOWA

Des Moines–817 Federal Building, 210 Walnut Street, 50309, Area Code 515 Tel 284–4222, FTS 8 862–4222

KANSAS

•Wichita (Kansas City, Missouri District)–River Park Place, Suite 565, 727 North Waco, 67203, Area Code 316 Tel 269–6160, FTS 8 752–6160

KENTUCKY

Louisville–U.S. Post Office and Courthouse Building, 601 W. Broadway, 40202, Area Code 502 Tel 582–5066, FTS 8 352–5066

LOUISIANA

New Orleans–432 World Trade Center, No. 2 Canal Street, 70130, Area Code 504 Tel 589–6546, FTS 8 682–6546

•Denotes trade specialist at post or duty station

MAINE

•Augusta (Boston, Massachusetts District)–77 Sewell Street, 04330, Area Code 207 Tel 622–8249, FTS 8 833–6249

MARYLAND

Baltimore–415 U.S. Customhouse, Gay and Lombard Streets, 21202, Area Code 301 Tel 962–3560, FTS 8 922–3560

MASSACHUSETTS

Boston–World Trade Center, Suite 307, Commonwealth Pier Area, 02210, Area Code 617 Tel 565–8563, FTS 8 835–8563

MICHIGAN

Detroit–1140 McNamara Building, 477 Michigan Avenue, 48226, Area Code 313 Tel 226–3650, FTS 8 226–3650

•Grand Rapids–300 Monroe N.W., Room 409, 49503, Area Code 616 Tel 456–2411, FTS 8 372–2411

MINNESOTA

Minneapolis–108 Federal Building, 110 S. 4th St., 55401, Area Code 612 Tel 349–1638, FTS 8 777–1638

MISSISSIPPI

Jackson–328 Jackson Mall Office Center, 300 Woodrow Wilson Blvd., 39213, Area Code 601 Tel 965–4388, FTS 8 490–4388

MISSOURI

Kansas City–Room 635, 601 E. 12th Street, 64106, Area Code 816 Tel 426–3141, FTS 8 867–3141

St. Louis–7911 Forsyth Blvd., Suite 610, 63105, Area Code 314 Tel 425–3302–4, FTS 8 279–3302

NEBRASKA

Omaha–1133 O Street, 68137, Area Code 402 Tel 221–3664, FTS 8 864–3664

NEVADA

Reno–1755 E. Plumb Lane, Room 152, 89502, Area Code 702 Tel 784–5203, FTS 8 470–5203

NEW JERSEY

Trenton–3131 Princeton Pike Building, Suite 100, 08648, Area Code 609 Tel 989–2100, FTS 8 423–2100

NEW MEXICO

Albuquerque–5000 Marble Avenue, N.E., Suite 320, 87110, Area Code 505 Tel 262–6024, FTS 8 474–2386

NEW YORK

Buffalo–1312 Federal Building, 111 W. Huron Street, 14202, Area Code 716 Tel 846–4191, FTS 8 437–4191

New York–Federal Office Building, 26 Federal Plaza, Foley Square, 10278, Area Code 212 Tel 264–0634, FTS 8 264–0634

•Rochester–111 E. Avenue, Suite 220, 14604, Area Code 716 Tel 263–6480, FTS 8 963–6480

NORTH CAROLINA

Greensboro–203 Federal Building, 324 W. Market Street, P.O. Box 1950, 27402, Area Code 919 Tel 333–5345, FTS 8 699–5345

OHIO

Cincinnati–9504 Federal Office Building, 550 Main Street, 45202, Area Code 513 Tel 684–2944, FTS 8 684–2944

Cleveland–Room 600, 668 Euclid Avenue, 44114, Area Code 216 Tel 522–4750, FTS 8 942–4750

OKLAHOMA

Oklahoma City–5 Broadway Executive Park, Suite 200, 6601 Broadway Extension, 73116, Area Code 405 Tel 231–5302, FTS 8 736–5302

•Tulsa–440 S. Houston Steet, 74127, Area Code 918 Tel 581–7650, FTS 8 745–7650

OREGON

Portland–Room 618, 1220 S.W. 3rd Avenue, 97204, Area Code 503 Tel 221–3001, FTS 8 423–3001.

PENNSYLVANIA

Philadelphia–9448 Federal Building, 600 Arch Street, 19106, Area Code 215 Tel 597–2850, FTS 8 597–2850

Pittsburgh–2002 Federal Building, 1000 Liberty Avenue, 15222, Area Code 412 Tel 644–2850, FTS 8 722–2850

PUERTO RICO

San Juan (Hato Rey)–Room 659, Federal Building, Chardon Avenue, 00918, Area Code 809 Tel 753–4555, Ext. 555, FTS 8 753–4555

RHODE ISLAND

•Providence (Boston, Massachusetts District)–7 Jackson Walkway, 02903, Area Code 401 Tel 528–5104, Ext. 22, FTS 8 838–5104

SOUTH CAROLINA

•Charleston–17 Lockwood Drive, 29401, Area Code 803 Tel 724–4361, FTS 8 677–4361

Columbia–Strom Thurmond Federal Building, Suite 172, 1835 Assembly Street, 29201, Area Code 803 Tel 765–5345, FTS 8 677–5345

TENNESSEE

•Memphis–22 N. Front Street, Suite 200, 38101, Area Code 901 Tel 521–4137, FTS 8 222–4137

Nashville–Suite 1114, Parkway Towers, 404 James Robertson Parkway, 37219-1505, Area Code 615 Tel 736–5161, FTS 8 852–5161

TEXAS

Austin–400 First City Center, 816 Congress Avenue, 78701, Area Code 512 Tel 472–5059, FTS 8 770–5939

Dallas–Room 7A5, 1100 Commerce Street, 75242, Area Code 214 Tel 767–0542, FTS 8 729–0542

Houston–2625 Federal Building, Courthouse, 515 Rusk Street, 77002, Area Code 713 Tel 229–2578, FTS 8 526–4578

UTAH

Salt Lake City–U.S. Customhouse, 350 S. Main Street, 84101, Area Code 801 Tel 524–5116, FTS 8 588–5116

VIRGINIA

Richmond–8010 Federal Building, 400 North 8th Street, 23240, Area Code 804 Tel 771–2246, FTS 8 925–2246

WASHINGTON

Seattle–3131 Elliott Avenue, Suite 290, 98121, Area Code 206 Tel 442–5616, FTS 8 399–5615

Spokane–West 808 Spokane Falls Boulevard, Room 650, 99201, Area Code 509 Tel 353–2922, FTS 8 439–2922

WEST VIRGINIA

Charleston–42 New Federal Building, 500 Quarrier Street, 26301, Area Code 304 Tel 347–5123, FTS 8 930–5123

WISCONSIN

Milwaukee–605 Federal Building, 517 E. Wisconsin Avenue, 53202, Area Code 414 Tel 291–3473, FTS 8 362–3473

NIST-114A
(REV. 3-89)

U.S. DEPARTMENT OF COMMERCE
NATIONAL INSTITUTE OF STANDARDS AND TECHNOLOGY

BIBLIOGRAPHIC DATA SHEET

11. ABSTRACT (A 200-WORD OR LESS FACTUAL SUMMARY OF MOST SIGNIFICANT INFORMATION.   IF DOCUMENT INCLUDES A SIGNIFICANT BIBLIOGRAPHY OR LITERATURE SURVEY, MENTION IT HERE.)

The 21st Supplement to Special Publication 305 contains full bibliographic citations
including keywords and abstracts for National Institute of Standards and Technology (NIST)
(formerly National Bureau of Standards (NBS)) 1989 papers published and entered into the
National Technical Information Service (NTIS) collection. (Also included are NBS/NIST
papers published prior to 1989 but not reported in previous supplements of this annual
catalog.) Four indexes are included to allow the user to identify NBS/NIST papers by
personal author, keywords, title, and NTIS order/report number.

AILABILITY

UNLIMITED

FOR OFFICIAL DISTRIBUTION. DO NOT RELEASE TO NATIONAL TECHNICAL INFORMATION SERVICE (NTIS).

ORDER FROM SUPERINTENDENT OF DOCUMENTS, U.S. GOVERNMENT PRINTING OFFICE,
WASHINGTON, DC 20402.

ORDER FROM NATIONAL TECHNICAL INFORMATION SERVICE   S , SPRINGFIELD, VA 22161.

ANNOUNCEMENT OF NEW PUBLICATIONS OF THE NATIONAL INSTITUTE OF STANDARDS AND TECHNOLOGY

Superintendent of Documents
Government Printing Office
Washington, DC 20402

Dear Sir:

Please add my name to the announcement list of new publications as issued by the National Institute of Standards and Technology.

Name _____

Company _____

Address _____

City _____ State _____ Zip Code _____

(Notification key N519)

Superintendent of Documents **Publications and Subscriptions** Order Form

Order Processing Code:
*6870

Charge your order.
It's easy!

PS

To fax your orders and inquiries—(202) 275-0019

PUBLICATIONS   **Please Type or Print** (Form is aligned for typewriter use.)

Qty.	Stock Number	Title	Price Each	Total Price
1	021-602-00001-9	Catalog—Bestselling Government Books	FREE	FREE
		Total for Publications		

SUBSCRIPTIONS

Qty.	List ID	Title	Price Each	Total Price
		Total for Subscriptions		
		Total Cost of Order		

NOTE: Prices include regular domestic postage and handling. Publication prices are good through 1/91. After that date, please call Order and Information Desk at 202-783-3238 to verify prices. Subscription prices are subject to change at any time. International customers please add 25%.

Please Choose Method of Payment:

(Company or personal name)   (Please type or print)

☐ Check payable to the Superintendent of Documents

(Additional address/attention line)

☐ GPO Deposit Account

☐ VISA or MasterCard Account

(Street address)

(City, State, ZIP Code)   (Credit card expiration date)   *Thank you for your order!*

(   )
(Daytime phone including area code)   (Signature)   7/89

Mail To: Superintendent of Documents, Government Printing Office, Washington, DC 20402-9325

NTIS® ORDER FORM

☎ **TELEPHONE ORDERS** TELEX 89-9405 Telecopier (703) 321-8547 Subscriptions: (703) 487-4630
   Call (703) 487-4650 (See reverse side for RUSH and EXPRESS ordering options)

- HANDLING FEE: A handling fee is required for each order except for Express, Rush, Subscription, QuikORDER, or Pickup orders.
- SHIPPING: **U.S.:** Printed reports and microfiche copies are shipped First Class Mail or equivalent.
  FOREIGN: Regular service: Printed reports and microfiche copies are shipped surface mail.
  Air Mail service to Canada and Mexico: add $3 per printed report; 75¢ per microfiche copy.
  Air Mail service to all other addresses: add $6 per printed report; 75¢ per microfiche copy.
  SUBSCRIPTIONS and standing orders are sent surface mail; contact NTIS for air mail rates.

1 Address Information

DTIC Users Code:_____ Contract No. __ ___Last six digits___ __

PURCHASER: DATE:_____

SHIP TO (Enter ONLY if different from purchaser):

Last Name	First Initial
Title	
Company/Organization	
Address	
City/State/ZIP	
Attention	
Telephone number	

Last Name	First Initial
Title	
Company/Organization	
Address	
City/State/ZIP	
Attention	
Telephone number	

2 Method of Payment

☐ Charge my NTIS Deposit Account __ __ __ __ __ __ __   ☐ Check/Money order enclosed for $_____

Charge my   ☐ Amer. Express ☐ VISA ☐ MasterCard   ☐ Please bill **ADD $7.50 per order** (See below for restrictions)†

Account No._____   Exp._____

Purchase Order No._____

Signature:_____
(Required to validate all orders)

3 Order Selection (For computer products, see reverse)

Enter NTIS order number(s) (Ordering by title only will delay your order)	Customer†† Routing (up to 8 digits)	QUANTITY		UNIT PRICE	Foreign Air Mail	TOTAL PRICE
		Printed Copy	Micro-fiche			
1.						
2.						
3.						
4.						
5.						
6.						
7.						

☐ **OVER** - Order continued on reverse

† Billing Service: This service is restricted to customers in the United States, Canada, and Mexico for an additional $7.50 per order. A late payment charge will be applied to all billings more than 30 days overdue.

†† Customer Routing Code: NTIS can label each item for routing within your organization. If you want this service, put your routing code in this box.

SUBTOTAL From Other Side	
Regular Service Handling Fee per order ($3 U.S., Canada, and Mexico; $4 others)	
Billing Fee if required ($7.50)	
GRAND TOTAL	

Rev. 5/20/88

3 Order Selection (Cont.)

Enter the NTIS order number(s) (Ordering by title only will delay your order)	Customer Routing	QUANTITY Printed Copy	QUANTITY Micro-fiche	UNIT PRICE	Foreign Air Mail	TOTAL PRICE
8.						
9.						
10.						
11.						
12.						
13.						
14.						
15.						
16.						
17.						
18.						
19.						
					Subtotal	

ENTER this amount on the other side of this form. ⇨

4 Computer Products

If you have questions about a particular computer product, please call our Federal Computer Products Center at (703) 487-4763.

Enter the NTIS order number(s) (Ordering by title only will delay your order)	Customer Routing	TAPE DENSITY (9 track) 1600 bpi	TAPE DENSITY (9 track) 6250 bpi	TOTAL PRICE
20.				
21.				
22.				
23.				
			Subtotal	

All magnetic tapes are sent air mail or equivalent service to both U.S. and foreign addresses.

ENTER this amount on the other side of this form. ⇨

SPECIAL RUSH and EXPRESS ORDERING OPTIONS

Telephone: (800) 336-4700
In Virginia call
(703) 487-4700

RUSH SERVICE–Add $10 per item: Orders are processed within 24 hours and sent First Class or equivalent. Available to U.S. addresses.

EXPRESS SERVICE–Add $20 per item: Orders are processed within 24 hours AND delivered by overnight courier. Available to U.S. addresses only.

Rev. 5/20/88

NIST Technical Publications

Periodical

Journal of Research of the National Institute of Standards and Technology — Reports NIST research and development in those disciplines of the physical and engineering sciences in which the Institute is active. These include physics, chemistry, engineering, mathematics, and computer sciences. Papers cover a broad range of subjects, with major emphasis on measurement methodology and the basic technology underlying standardization. Also included from time to time are survey articles on topics closely related to the Institute's technical and scientific programs. Issued six times a year.

Nonperiodicals

Monographs — Major contributions to the technical literature on various subjects related to the Institute's scientific and technical activities.

Handbooks — Recommended codes of engineering and industrial practice (including safety codes) developed in cooperation with interested industries, professional organizations, and regulatory bodies.

Special Publications — Include proceedings of conferences sponsored by NIST, NIST annual reports, and other special publications appropriate to this grouping such as wall charts, pocket cards, and bibliographies.

Applied Mathematics Series — Mathematical tables, manuals, and studies of special interest to physicists, engineers, chemists, biologists, mathematicians, computer programmers, and others engaged in scientific and technical work.

National Standard Reference Data Series — Provides quantitative data on the physical and chemical properties of materials, compiled from the world's literature and critically evaluated. Developed under a worldwide program coordinated by NIST under the authority of the National Standard Data Act (Public Law 90-396). NOTE: The Journal of Physical and Chemical Reference Data (JPCRD) is published quarterly for NIST by the American Chemical Society (ACS) and the American Institute of Physics (AIP). Subscriptions, reprints, and supplements are available from ACS, 1155 Sixteenth St., NW., Washington, DC 20056.

Building Science Series — Disseminates technical information developed at the Institute on building materials, components, systems, and whole structures. The series presents research results, test methods, and performance criteria related to the structural and environmental functions and the durability and safety characteristics of building elements and systems.

Technical Notes — Studies or reports which are complete in themselves but restrictive in their treatment of a subject. Analogous to monographs but not so comprehensive in scope or definitive in treatment of the subject area. Often serve as a vehicle for final reports of work performed at NIST under the sponsorship of other government agencies.

Voluntary Product Standards — Developed under procedures published by the Department of Commerce in Part 10, Title 15, of the Code of Federal Regulations. The standards establish nationally recognized requirements for products, and provide all concerned interests with a basis for common understanding of the characteristics of the products. NIST administers this program as a supplement to the activities of the private sector standardizing organizations.

Consumer Information Series — Practical information, based on NIST research and experience, covering areas of interest to the consumer. Easily understandable language and illustrations provide useful background knowledge for shopping in today's technological marketplace.
Order the above NIST publications from: Superintendent of Documents, Government Printing Office, Washington, DC 20402.
Order the following NIST publications — FIPS and NISTIRs — from the National Technical Information Service, Springfield, VA 22161.

Federal Information Processing Standards Publications (FIPS PUB) — Publications in this series collectively constitute the Federal Information Processing Standards Register. The Register serves as the official source of information in the Federal Government regarding standards issued by NIST pursuant to the Federal Property and Administrative Services Act of 1949 as amended, Public Law 89-306 (79 Stat. 1127), and as implemented by Executive Order 11717 (38 FR 12315, dated May 11, 1973) and Part 6 of Title 15 CFR (Code of Federal Regulations).

NIST Interagency Reports (NISTIR) — A special series of interim or final reports on work performed by NIST for outside sponsors (both government and non-government). In general, initial distribution is handled by the sponsor; public distribution is by the National Technical Information Service, Springfield, VA 22161, in paper copy or microfiche form.

ADMINISTRATION & MANAGEMENT

AERONAUTICS & AERODYNAMICS

AGRICULTURE & FOOD

ASTRONOMY & ASTROPHYSICS

ATMOSPHERIC SCIENCES

BEHAVIOR & SOCIETY

BIOMEDICAL TECHNOLOGY & HUMAN FACTORS ENGINEERING

BUILDING INDUSTRY TECHNOLOGY

BUSINESS & ECONOMICS

CHEMISTRY

CIVIL ENGINEERING

COMBUSTION, ENGINES, & PROPELLANTS

COMMUNICATION

COMPUTERS, CONTROL & INFORMATION THEORY

DETECTION & COUNTERMEASURES

ELECTROTECHNOLOGY

ENERGY

ENVIRONMENTAL POLLUTION & CONTROL

HEALTH CARE

INDUSTRIAL & MECHANICAL ENGINEERING

LIBRARY & INFORMATION SCIENCES

MANUFACTURING TECHNOLOGY

MATERIALS SCIENCES

MATHEMATICAL SCIENCES

MEDICINE & BIOLOGY

MILITARY SCIENCES

MISSILE TECHNOLOGY

NATURAL RESOURCES & EARTH SCIENCES

NAVIGATION, GUIDANCE, & CONTROL

NUCLEAR SCIENCE & TECHNOLOGY

OCEAN TECHNOLOGY & ENGINEERING

ORDNANCE

PHOTOGRAPHY & RECORDING DEVICES

PHYSICS

PROBLEM-SOLVING INFORMATION FOR STATE & LOCAL GOVERNMENTS

SPACE TECHNOLOGY

TRANSPORTATION

URBAN & REGIONAL TECHNOLOGY & DEVELOPMENT

Lightning Source UK Ltd.
Milton Keynes UK
UKHW021256171218
334146UK00012B/757/P